# SELECTED PAPERS

# SELECTED PAPERS

BY

## S. LEFSCHETZ

INCLUDING THE BOOK

L'ANALYSIS SITUS

CHELSEA PUBLISHING COMPANY
BRONX, NEW YORK, N. Y.

INTERNATIONAL STANDARD BOOK NUMBER 0-8284-0234-5
LIBRARY OF CONGRESS CATALOG CARD NUMBER 73-113137
LIBRARY OF CONGRESS CLASSIFICATION NUMBER QA611
DEWEY DECIMAL CLASSIFICATION NUMBER 514'.08

PRINTED ON LONG-LIFE ALKALINE PAPER



PRINTED IN THE UNITED STATES OF AMERICA

# PREFACE

THE present volume contains a selection of the major mathematical papers of S. Lefschetz, as well as—in its entirety—his important book *L'Analysis Situs et la Géométrie Algébrique*, published originally in the Collection Borel. Not included in this volume are the chapters which Professor Lefschetz contributed to the Report of the Committee on Rational Transformations of the National Research Council, since the report has just been re-issued [*Selected Topics in Algebraic Geometry*, by V. Snyder, S. Lefschetz, L. A. Dye, et al. (Chelsea Publishing Company, New York, 1970)].

A discussion of Professor Lefschetz' work, if one were to be offered here, would require a great many pages written by experts in more than one field of mathematics; indeed, to quote Professor Norman E. Steenrod, "the full story of his influence [in algebraic topology] would require a review of the entire field." Fortunately, there is a survey, some forty-three pages in length, by Professors W. V. D. Hodge and Norman E. Steenrod, to which the reader can be referred. This survey, "An Appreciation of the Work and Influence of S. Lefschetz," constitutes Part I of the book *Algebraic Geometry and Topology: A Symposium in honor of S. Lefschetz,* eds. R. H. Fox, D. C. Spencer and A. W. Tucker (Princeton, Princeton University Press, 1957). Professor Hodge there discusses Lefschetz' contribution to algebraic geometry and Professor Steenrod, his contributions to algebraic topology. (Lefschetz' work in differential equations and applied mathematics—to which he is continuing to make contributions—lies outside the scope of the present volume and that of the survey referred to.) We confine ourselves to a brief quotation from Professor Hodge:

> A number of the discoveries which he has made are of more vital interest to mathematicians at the present day than they have been since they aroused widespread excitement on their first appearance some thirty or thirty-five years ago. . . . Lefschetz by his work on the topology and transcendental theory of algebraic varieties has been a major influence in turning the minds of geometers in new and fruitful directions, and in so doing he has achieved what is given to few to do.

The reader's attention should be called to two papers that are of special interest over and above their purely mathematical content. The first is "A Page of Mathematical Autobiography" (Chapter I, [130]), which is one of the rare first-hand accounts of an important mathematical discovery. The second is the Bordin Prize memoir (Chapter II, [24]) which, says Professor Hodge, "is well worth attention from those who derive inspiration from the study of the way in which a master works."

No special significance, incidentally, should be attached to the order in which the papers are presented in this volume. The numbers in brackets, here and elsewhere, refer to the bibliography of Professor Lefschetz' publications that appears on pages 629 ff., where full bibliographical details concerning the papers will be found.

The bibliography is in two parts, the first of which is a reproduction, made with the kind permission of Princeton University Press, of "Bibliography of the Publications of S. Lefschetz to June 1955," prepared by Miss Mary L. Carll, which constitute pages 44-49 of the book already cited, *Algebraic Geometry and Topology.*

Thanks are due to the American Mathematical Society for their kind permission to reproduce the papers [130] and [53] from the Bulletin of the American Mathematical Society and the papers [24], [33], and [36] from the Transactions of the American Mathematical Society. Thanks are also due to the Annals of Mathematics for their kind permission to reproduce the papers [42], [48], [49], [51], [57], and [71]; to the Duke Mathematical Journal for the paper [60]; and to the Johns Hopkins Press for the paper [55] from the American Journal of Mathematics. Thanks are also due to the publishing firm Gauthier-Villars for their gracious permission to reproduce the book *L'Analysis Situs et la Géométrie Algébrique* [29].

Finally, very special thanks are due to Professor Lefschetz himself for his help and generous cooperation.

T. C.

# TABLE OF CONTENTS

# I

# I

## A PAGE OF MATHEMATICAL AUTOBIOGRAPHY

### INTRODUCTION

As my natural taste has always been to look forward rather than backward this is a task which I did not care to undertake. Now, however, I feel most grateful to my friend Mauricio Peixoto for having coaxed me into accepting it. For it has provided me with my first opportunity to cast an objective glance at my early mathematical work, my algebro-geometric phase. As I see it at last it was my lot to plant the harpoon of algebraic topology into the body of the whale of algebraic geometry. But I must not push the metaphor too far.

The time which I mean to cover runs from 1911 to 1924, from my doctorate to my research on fixed points. At the time I was on the faculties of the Universities of Nebraska (two years) and Kansas (eleven years). As was the case for almost all our scientists of that day my mathematical isolation was complete. This circumstance was most valuable in that it enabled me to develop my ideas in complete mathematical calm. Thus I made use most uncritically of early topology à la Poincaré, and even of my own later developments. Fortunately someone at the Académie des Sciences (I always suspected Émile Picard) seems to have discerned "the harpoon for the whale" with pleasant enough consequences for me.

To close personal recollections, let me tell you what made me turn with all possible vigor to topology. From the $\rho_0$ formula of Picard, applied to a hyperelliptic surface $\Phi$ (topologically the product of 4 circles) I had come to believe that the second Betti number $R_2(\Phi)=5$, whereas clearly $R_2(\Phi)=6$. What was wrong? After considerable time it dawned upon me that Picard only dealt with *finite* 2-cycles, the only useful cycles for calculating periods of certain double integrals. Missing link? The cycle at infinity, that is the plane section of the surface at infinity. This drew my attention to cycles carried by an algebraic curve, that is to *algebraic* cycles, and · · · the harpoon was in!

My general plan is to present the first concepts of algebraic geometry, then follow up with the early algebraic topology of Poincaré plus some of my own results on intersections of cycles. I will then discuss the topology of an algebraic surface. The next step will be a

---

An address delivered at Brown University on April 14, 1967. Submitted by invitation of the editors; received by the editors September 7, 1967.

summary presentation of the analytical contributions of Picard, Severi and Poincaré leading to my work, application of topology to complex algebraic geometry concluding with a rapid consideration of the effect on the theory of abelian varieties.

This is not however a cold recital of results achieved duly modernized. To do this would be to lose the "autobiographical flavor" of my tale. I have therefore endeavored to place myself back in time to the period described and to describe everything as if I were telling it a half century ago. From the point of view of rigor there is no real loss. Analytically the story is fairly satisfactory and to make it so in the topology all that is needed is to accept the results amply described in my Colloquium Lectures [**10**].

To place the story into focus I must say something about what we knew and accepted in days gone by. That is I must describe our early background.

In its early phase (Abel, Riemann, Weierstrass), algebraic geometry was just a chapter in analytic function theory. The later development in this direction will be fully described in the following chapters. A new current appeared however (1870) under the powerful influence of Max Noether who really put "geometry" and more "birational geometry" into algebraic geometry. In the classical mémoire of Brill-Noether (Math. Ann., 1874), the foundations of "geometry on an algebraic curve" were laid down centered upon the study of linear series cut out by linear systems of curves upon a fixed curve $f(x, y) = 0$. This produced birational invariance (for example of the genus $p$) by essentially algebraic methods.

The next step in the same direction was taken by Castelnuovo (1892) and Enriques (1893). They applied analogous methods to the creation of an entirely new theory of algebraic surfaces. Their basic instrument was the study of linear systems of curves on a surface. Many new birationally invariant properties were discovered and an entirely new and beautiful chapter of geometry was opened. In 1902 the Castelnuovo-Enriques team was enriched by the brilliant personality of Severi. More than his associates he was interested in the contacts with the analytic theory developed since 1882 by Émile Picard. The most important contribution of Severi, his theory of the base (see §12) was in fact obtained by utilizing the Picard number $\rho$ (see §11).

The theory of the great Italian geometers was essentially, like Noether's, of algebraic nature. Curiously enough this holds in good part regarding the work of Picard. This was natural since in his time Poincaré's creation of algebraic topology was in its infancy.

Indeed when I arrived on the scene (1915) it was hardly further along.

About 1923 I turned my attention to "fixed points" which took me away from algebraic geometry and into the more rarefied air of topology. I cannot therefore refer even remotely to more recent doings in algebraic geometry. I cannot refrain, however, from mention of the following noteworthy activities:

I. The very significant work of W. V. D. Hodge. I refer more particularly to his remarkable proof that an $n$-form of $V^q$ which is of the first kind cannot have all periods zero (see Hodge [**13**]).

II. The systematic algebraic attack on algebraic geometry by Oscar Zariski and his school, and beyond that of André Weil and Grothendieck. I do feel however that while we wrote algebraic GEOMETRY they make it ALGEBRAIC geometry with all that it implies.

*References.* For a considerable time my major reference was the Picard-Simart treatise [2]. In general however except for the writings of Poincaré on topology my Borel series monograph [**9**] is a central reference. The best all around reference not only to the topics of this report but to closely related material is the excellent Ergebnisse monograph of Zariski [**11**]. Its bibliography is so comprehensive that I have found it unnecessary to provide an extensive one of my own.

## TABLE OF CONTENTS

## I. General Remarks on Algebraic Varieties

**1. Definition. Function field.** It was the general implicit or explicit understanding among algebraic geometers of my day that an algebraic $n$-variety $V^n$ ($n$ dimensional variety) is the partial or complete irreducible intersection of several complex polynomials or "hypersurfaces" of a projective space $S^{n+k}$, in which $V^n$ had no singularities (it was homogeneous). Thus $V^n$ was a compact real $2n$-manifold $M^{2n}$ (complex dimension $n$). It could therefore be considered as its own Riemann manifold as I shall do throughout.

For convenience in analytical operations one customarily represents $V^n$ by a general projection in cartesian $S^{n+1}$

$$F(x_1, x_2, \cdots, x_n, y) = 0, \tag{1.1}$$

where $F$ is an irreducible complex polynomial of degree $m$. In this representation, the variety, now called $F$, occupies no special position relative to the axes.[1] As a consequence (1.1) possesses the simplest singularities. For a curve they consist of double points with distinct tangents, for a surface: double curve with generally distinct tangent planes along this curve.

Incidentally, the recent brilliant reduction of singularities by Hironaka [12] has shown that the varieties as just described are really entirely general.

Returning to our $V^n$ the study of its topology will lean heavily upon the properties of the pencil of hypersurfaces $\{H_y\}$ cut out by the hyperplanes $y = \text{const}$. The particular element of the pencil cut out by $y = c$ is written $H_c$. As my discourse will be mostly on surfaces I will only describe (later) certain pecularities for varieties.

*Function field.* Let the complex rational functions $R(x_1, \cdots, x_n, y)$ be identified mod $F$. As a consequence they constitute an algebraic extension of the complex field $K$ written $K(F)$, called the function field of $F$.

Let $F^*$ be the nonsingular predecessor of $F$ in $S^{n+k}$ and let $(u_1, \cdots, u_{n+k})$ be cartesian coordinates for $S^{n+k}$. On $F^*$ they determine elements $\xi_h$, $h \leqq n+k$ of $K(F)$. The system

$$u_h = \xi_h, \qquad h \leqq n + k$$

is a parametric representation of $F^*$. $F^*$ is a *model* of $K(F)$.

Any two models $F_1^*$, $F_2^*$ are birationally equivalent: birationally transformable into one another. The properties that will mainly

[1] That is, $F$ has only those singularities which arise from a general projection on $S^{n+1}$ of a nonsingular $V^n \subset S^{n+k}$.

interest us are those possessing a certain degree of birational invariance (details in §17).

*Terminology.* Since only algebraic curves, surfaces, varieties will be dealt with, I drop the mention "algebraic" and merely say curve, . . . .

The symbol $\mathcal{V}^n$ represents a (usually complex) $n$ dimensional vector space.

2. **Differential forms.** Let $\alpha$, $\beta$, $\cdots$, denote elements of the function field $K(F)$. I shall refer to various differentials: zero, one, two, $\cdots$ forms $\omega^0$, $\omega^1$, $\omega^2$, $\cdots$, in the sense of Élie Cartan of type

$$\omega^k = \sum \alpha_{i_1,\cdots,i_k} d\alpha_{i_1} \cdots d\alpha_{i_k},$$

every $\alpha$ in $K(F)$, as zero, one, two, $\cdots$, forms. They are calculated by the rules of calculus, remembering that the $d\alpha_j$ are skew-symmetric, that is $d\beta d\alpha = -d\alpha d\beta$.

Note that $d\omega^k$ is an $\omega^{k+1}$ called *exact* and that if $d\omega^k = 0$ one says that $\omega^k$ is *closed.*

Special terms are: $\omega^k$ is of the first kind when it is holomorphic everywhere on $F$; of the second kind when it is holomorphic at any point of $F$ mod some $d\alpha$; of the third kind if neither of the first nor of the second kind.

The evaluation of the number of kinds one or two constitutes one of the main problems to be discussed.

3. **Differential forms on a curve.** Let the curve be

$$f(x, y) = 0 \tag{3.1}$$

and let $m$ be its degree. We refer to it as "the curve $f$." Under our convention, $f$ has no other singularities than double points with distinct tangents and is identified in a well-known sense with its Riemann surface. Its one-forms are said to be *abelian.* An *adjoint* to $f$ is a polynomial $\phi_n(x, y)$ ($n$ is its degree) vanishing at all double points.

The following are classical properties:

*One-forms of the first kind.* They are all reducible to the type

$$\frac{\phi_{m-3}dx}{f'_y}. \tag{3.2}$$

They form a $\mathcal{V}^p$, where $2p = R_1$, the first Betti number of the Riemann surface $f$. Of course the collection $\{\phi_{m-3}\}$ forms likewise a $\mathcal{V}^p$.

*One-forms of the second kind.* Same type of reduction to (3.2) mod a $d\omega^0$, save that $\phi_{m-3}$ is replaced by some $\phi_s$. Their vector space mod $dK(f)$ is a $\mathcal{V}^{2p}$.

*One-forms of the third kind.* They have a finite number of logarithmic points with residues whose sum is zero.

*Some special properties of one-forms of the first kind.* Let

$$\psi = \sum_{h=0}^{r} \alpha_k \psi_h(x, y) = 0 \tag{3.3}$$

be a linear system of polynomials linearly independent mod $f$ and of common degree. Let the general $\psi$ intersect $f$ in a set of points $P_1, \cdots, P_s$ which includes all the variable points and perhaps some fixed points. The collection of all such sets is a *linear series of degree* $n$ and *dimension* $r$. The series is *complete* when its sets do not belong to an amplified series of the same degree: designation $g_n^r$ (concepts and terminology of Brill and Noether).

(3.4) THEOREM OF ABEL. *Let $du$ be any one-form of the first kind; let $\{P_h\}$ be any element of a $g_n^r$ and let $A$ be a fixed point of $f$. Then with integration along paths on $f$:*

$$\sum \int_A^{P_h} du = v$$

*is a constant independent of the element $\{P_h\}$ of $g_n^r$.*

Still another classic, a sort of inverse of Abel's theorem is this:

(3.5) THEOREM OF JACOBI. *Let $\{du_h\}$ be a base for the one-forms of the first kind. Then for general values of the constants $v_h$ (exceptions noted) the system*

$$\sum_{k=1}^{p} \int_A^{P_k} du_h = v_h$$

*in the $p$ unknowns $P_k$, $k \leqq p$, has a unique solution.*

*Periodic properties.* Let $\{du_h\}$ be as just stated and let $\{\gamma_\mu^1\}$, $\mu \leqq 2p$ be an integral homology base (see (5.4)) for the module of one-cycles of $f$. The expression

$$\pi_{h\mu} = \int_{\gamma_\mu^1} du_h$$

is the *period* of $\int du_h$ as to the cycle $\gamma_\mu^1$. Let the matrix

$$\Pi = [\pi_{h\mu}]; \qquad h, \mu \leqq 2p; \qquad \pi_{h+p,\mu} = \bar{\pi}_{h\mu},\ h \leqq p.$$

By means of integration on the Riemann surface $f$, Riemann has obtained the following comprehensive result (formulation of Scorza):

(3.6) THEOREM OF RIEMANN. *There exists an integral skew-symmetric $2p\times 2p$ matrix $M$ with invariant factors unity such that*

$$i\Pi M\Pi' = \begin{bmatrix} 0 & A \\ A^* & 0 \end{bmatrix}, \qquad (A^* = \bar{A}') \tag{3.7}$$

*is a positive definite hermitian matrix.*

*Riemann matrices.* This is the name given by Scorza to a matrix like $\Pi$ satisfying a relation (3.7) except that $M$ is merely rational skew-symmetric. The theory of such matrices has been extensively developed by Scorza [6]. He called $M$: *principal matrix* of $\Pi$.

It may very well happen that there is more than one rational skew-symmetric matrix $M$ satisfying a relation (3.7) but without necessarily the positive definite property. These matrices are called singularity matrices. They form a rational vector space whose dimension $k$ is the *singularity index* of the Riemann matrix (Scorza).

## II. TOPOLOGY

4. **Results of Poincaré.** Let $M^n$ be a compact orientable $n$-manifold which admits a cellular subdivision with $\alpha_k$ $k$-cells (well-known property for varieties). The characteristic is the expression

$$\chi(M^n) = \sum (-1)^k \alpha_k. \tag{4.1}$$

The following two relations were proved by Poincaré:

$$\chi(M^n) = \sum (-1)^k R_k \tag{4.2}$$

$$R_k = R_{n-k} \tag{4.3}$$

where $R_k$ is the $k$th integral Betti number of $M^n$: maximum number of linearly independent $k$-cycles with respect to homology (=with respect to bounding).

5. **Intersections.** In my work on algebraic geometry I freely used the intersection properties described below; they were actually justified and proved topologically invariant a couple of years later in my paper in the 1926 Transactions and much more fully in [10].

Let $M^n$ be as before and let $\gamma^p$ and $\gamma^q$ be integral $p$- and $q$-cycles of $M^n$. One may define the intersection $\gamma^p\cdot\gamma^q$ and it is a $(p+q-n)$-cycle.

(5.1) *If $\gamma^p$ or $\gamma^q\sim 0$ (bounds), then also $\gamma^p\cdot\gamma^q\sim 0$.*

The more important situation arises when $p+q=n$. The intersection (geometric approximation) is then a zero-cycle

$$C^0 = \sum s_j A_j$$

where the $s_j$ are integers. The *intersection number*

$$(\gamma^p, \gamma^{n-p}) = \sum s_j$$

is independent of the approximation. One proves readily

(5.2) $$(\gamma^p, \gamma^{n-p}) = (-1)^{(n-p)p}(\gamma^{n-p}, \gamma^p).$$

A basic result is:

(5.3) THEOREM. *A n.a.s.c. in order that* $\lambda\gamma^p \sim 0$, $\lambda \neq 0$ *is that*

$$(\gamma^p, \gamma^{n-p}) = 0$$

*for every* $\gamma^{n-p}$ [9, p. 15], [10, p. 78].

(5.4) HOMOLOGY BASE. The collection $\{\gamma_h^p\}$, $h \leqq R_p$ is a homology base for the $p$-cycles when the $\gamma_h^p$ are independent and every $\gamma^p$ satisfies a relation

$$\lambda\gamma^p \sim \sum s_h \gamma_h^p, \qquad \lambda \neq 0.$$

(5.5) *A n.a.s.c. in order that the* $\{\gamma_h^p\}$, $h \leqq R_p$ *be a homology base for* $p$*-cycles is the existence of a set of* $R_p$ *cycles* $\{\gamma_k^{n-p}\}$ *such that the determinant*

$$|(\gamma_h^p, \gamma_k^{n-p})| \neq 0.$$

*Then* $\{\gamma_k^{n-p}\}$ *is likewise a homology base for* $(n-p)$*-cycles.*

6. **The surface** $F$. **Orientation.** Let $P$ be a point of $F$ and let $u = u' + iu''$, $v = v' + iv''$ be local coordinates for $P$. Orient $F$ by naming the real coordinates in the order $u'$, $u''$, $v'$, $v''$. There results a unique and consistent orientation throughout the surface $F$. Hence $F$ is an orientable $M^4$.

Similarly if $C$ is a curve of $F$ and $u$ is a local coordinate at a nonsingular point $Q$ of $C$. The resulting orientation turns $C$ into a definite two-cycle, still written $C$.

Let $D$ be a second curve through $Q$, for which $Q$ is nonsingular and not a point of contact of the two curves. Then $Q$ contributes $+1$ to both the intersection number $(C, D)$ and to the number $[CD]$ of geometric intersections of $C$ and $D$. This holds also, through certain approximations when $Q$ is a multiple intersection. Hence always

(6.1) $$(C, D) = [CD].$$

I will return to these questions later.

**7. Certain properties of the surface $F$. Its characteristic.** To be a little precise let for a moment $F^*$ denote the nonsingular predecessor of $F$ in projective $S^{k+2}$. One may always choose a model $F^*$ of the function field $K(F)$ whose hyperplane sections are in general of a fixed genus $p>0$. We pass now to a cartesian representation of degree $m$:

$$F(x, y, z) = 0 \tag{7.1}$$

which is a general projection of $F^*$ and in particular in general position relative to the axes. The general scheme that follows is due to Picard. Let $\{H_y\}$ be the pencil cut out by the planes $y=\text{const.}$, and let $a_h$, $h \leqq N$, be the values for which the planes $y=a_h$ are tangent to $F$. Then the following properties hold:

I. Every $H_y$, $y$ not an $a_k$, is of fixed genus $p$.

II. Every $H_y$ is irreducible.

III. The plane $y=a_k$ has a unique point of contact $A_k$ with $F$ and $A_k$ is a double point of $H_{a_k}$ with distinct tangents. Hence the genus of $H_{a_k}$ is $p-1$.

IV. Among the branch points of the function $z(x)$ taken on $H_y$ exactly two $\rightarrow A_k$ as $y \rightarrow a_k$.

V. The fixed points $P_1, \cdots, P_m$ of $H_y$ are all distinct.

I denote by $S_y$ the sphere of the complex variable $y$.

*Characteristic.* Cover $H_y$ with a cellular decomposition among whose vertices are the fixed points $P_h$ of the curve.

Then if $H_y^* = H_y - \sum P_h$, $\chi(H_y^*) = 2-2p-m$. Decompose also $S_y$ into cells with the $a_k$ as vertices. Were it not for these points, and since a sphere has characteristic two, $H_y^*$ promenading over $S_y$ would generate a set $E = S_y \times H_y^*$ of characteristic

$$\chi(E^*) = 2(2 - 2p - m).$$

Now in comparison with $H_y^*$, $H_{a_k}$ has lost two one-cycles, and has two points replaced by one. Hence

$$\chi(\overset{*}{H}_{a_k}) = \chi(\overset{*}{H}_y) + 1.$$

Upon remembering to add the missing points $P_h$ we have then

$$\chi(F) = \chi(E^*) + N + m = (N - m - 4p) + 4 = I + 4 \tag{7.2}$$

a formula due to J. W. Alexander (different proof). The number $I = N - m - 4p$ is the well-known *invariant of Zeuthen-Segre.*

8. **One-cycles of $F$.** The first step was taken by Picard who proved this noteworthy result:

(8.1) THEOREM. *Every one-cycle $\gamma^1$ of $F$ is $\sim$ a cycle $\gamma^1$ contained in an $H_y$.*

The next important observation made by Picard was that $H_y$ contained a certain number $r$ of one-cycles which are invariant as $y$ varies. That is such a cycle $\gamma^1$ situated say in $H_a$ ($a$ not an $a_k$) has the property that as $y$ describes any closed path from $a$ to $a$ on the sphere $S_y$ the cycle $\gamma^1$ returns to a position $\gamma^1 \sim \gamma^1$ in $H_y$. This draws attention to the nature of the variation $\mathcal{V}\gamma^1$ of any cycle $\gamma^1$ under the same conditions.

Draw lacets $aa_k$ on $S_y$. Owing to (7, IV) as $y$ describes $aa_k$ a certain cycle $\delta_k^1$ of $H_y$ tends to the point of contact $A_k$ of the plane $y = a_k$ and hence is $\sim 0$ on $H_{a_k}$. This is the *vanishing cycle* as $y \to a_k$. A simple lacet consideration shows that as $y$ turns once positively around $a_k$ the variation $\mathcal{V}\gamma^1$ of the cycle $\gamma^1$ is given by

$$\mathcal{V}\overset{1}{\gamma} = (\overset{1}{\gamma}, \overset{1}{\delta_k})\overset{1}{\delta_k}. \tag{8.2}$$

Hence

(8.3) THEOREM. *N.a.s.c. for invariance of the cycle $\gamma^1$ is that every*

$$(\overset{1}{\gamma}, \overset{1}{\delta_k}) = 0.$$

A noteworthy generalization is obtained when $\gamma^1$ is replaced by a one-chain $L$ uniquely determined in term of $y$ provided that $y$ crosses no lacet.[2] As $y$ turns as above around $a_k$ the variation of $L$ is

$$\mathcal{V}(L) = (L, \overset{1}{\delta_k})\overset{1}{\delta_k}. \tag{8.4}$$

Noteworthy special cases are

I. $L$ is an oriented arc joining in $H_y$ two fixed points of $H_y$.

II. Let $C$ be an algebraic curve of $F$ and let $M_1, \cdots, M_n$ be its intersections with $H_y$. Then $L$ is a set of paths from a $P_j$ to every point $M_k$ in $H_y$.

(8.5) THEOREM. *The number of invariant cycles of $H_y$ is equal to the Betti number $R_1(F)$ and both are even*: $r = R_1 = 2q$.

[2] In modern terminology, $L$ will be a relative cycle.

This property was first proved in [7], although it was often admitted before. I give here an outline of the proof (not too different from the proof of [7]).[3]

To make the proof clearer I will use the following special notations:

$\Gamma$ a 3-cycle of $F$; $\{\Gamma_h\}$ base for the $\Gamma$'s; $\gamma = \Gamma H_y$: (one-cycle of $H_y$);

$\{\alpha_h\}$, $h \leqq 2p$, base for one-cycles of $H_y$;

$\{\beta_j\}$, $j \leqq 2p - r$, base for the one cycles of $H_y$, none invariant; $\beta$ any linear combination of the $\beta_j$;

Matrices such as $[(\beta_h, \Gamma_k)]_F$ will be written $[\beta\Gamma]_F$.

*Proof that* $r = R_1$. $\gamma = \Gamma H_y$ is invariant; conversely $\gamma$ invariant is a $\Gamma H_y$. Moreover $\gamma \sim 0$ in $H_y$ and $\Gamma \sim 0$ in $F$ are equivalent. Hence $\{\gamma_h\}$, $h \leqq R_1$, is a base for invariant cycles and therefore $r = R_1$.

*Proof that r is even.* Since no $\beta$ is invariant, $[\beta\delta]$ is of rank $2p - r$. Hence there exist $2p - r$ cycles $\delta$ which are independent in $H_y$. Denote them by $\bar{\delta}_h$, $h \leqq 2p - r$. Since $(\gamma_k \delta_h) = 0$ for every $k$, the $\delta_h$ depend on the $\beta_j$ in $H_y$. Hence one may take $\{\gamma_h; \bar{\delta}_k\}$ as base for the one-cycles of $H_y$. Hence

$$[\gamma\delta] = \begin{bmatrix} [\gamma\gamma] & 0 \\ 0 & [\delta\delta] \end{bmatrix} \tag{8.6}$$

is nonsingular. It follows that $[\gamma\gamma]$ is likewise nonsingular. Since it is skew-symmetric, a well-known theorem of algebra states that $r$ is even.

9. **The two-cycles of** $F$. From the expression (7.2) of the characteristic we have

$$\chi(F) = I + 4 = R_2 - 2R_1 + 2.$$

Hence

$$R_2 = I + 2R_1 + 2. \tag{9.1}$$

Besides this formula it is of interest to give an analysis of the 2-cycles.

Given a $\gamma^2$ one may assume it such that it meets every $H_y$ in at most a finite set of points. Let $Q$ be one of these and let $P$, $Q$ be a directed path from the fixed point $P$ to the point $Q$ in $H_y$. Call $L$ the sum of these paths. As $y$ describes $S_y - \sum$ lacets $aa_k$, $L$ generates a 3-chain $C^3$ whose boundary $\partial C^3$ consists of these chains:

(a) As $y$ describes $aa_k$ the vanishing one-cycle $\delta_k^1$ of $H_y$ generates a 2-chain $\Delta_k$ whose boundary

[3] The point here is to prove that an invariant cycle, which is also a vanishing cycle, is necessarily zero.

$$\partial\Delta_k = (\overset{1}{\delta}_k)_{H_a}.$$

The corresponding contribution to $\partial C^3$ is $\mu_k\Delta_k$, where (Zariski)

(9.2) $\mu_1 = (L, \overset{1}{\delta}_1), \qquad \mu_k = (L + \mu_1\overset{1}{\delta}_2 + \cdots + \mu_{k-1}\overset{1}{\delta}_{k-1}, \overset{1}{\delta}_k).$

(b) A part $(H_a)$ of $H_a$.

(c) $-\gamma^2$ itself.

Hence

$$\partial C^3 = -\gamma^2 + \sum \mu_k\Delta_k + (H_a) \sim 0$$

and so

(9.3) $$\gamma^2 \sim \sum \mu_k\Delta_k + (H_a).$$

Since the right side is a cycle, and $y=a$ is arbitrary we have

(9.4) $$\sum \mu_k\overset{1}{\delta}_k \sim 0 \qquad \text{in } H_y.$$

Conversely when (9.4) holds, (9.3) is a 2-cycle. Thus to obtain $R_2$ it is merely necessary to compute the number of linearly independent relations (9.4) and add to them one unit for all $\mu_k$ zero, that is for the cycle $H_a$ itself. This yields again (9.1).

For purposes of counting certain double integrals Picard required the number of *finite* 2-cycles independent relative to homologies in $F-H_\infty$. This is the number $R_2(F-H)$ and he found effectively

(9.5) $$R_2(F - H) = R_2 - 1.$$

10. **Topology of algebraic varieties.** I have dealt with it at length in both [**8**] and [**9**]. Questions of orientation and intersection are easily apprehended from the case of surfaces. I shall only recall here a few properties that are not immediate derivatives from the case of a surface.

The designations $V^n$, $H_y$ are the same as in Chapter I. The following properties are taken from [**9**, Chapter V]. The symbol $\gamma^k$ will represent a $k$-cycle of $V^n$.

I. *Every $\gamma^k$, $k<n-1$, of $H_y$ is invariant.*

II. *Every $\gamma^k$, $k<n$, of $V^n$ is $\sim\gamma^{k'}$ in $H_y$.*

III. *When $k\leqq n-2$, $\gamma^k\sim 0$ in $V^n$ and $\gamma^{k'}\sim 0$ in $H_y$ are equivalent relations.*

IV. *Under the same conditions $R_k(V^n)=R_k(H_y)$.*

## III. Analysis with Little Topology

This is a rapid résumé of the extensive contributions of Picard, Severi and Poincaré upon which I applied topology (see IV). I will continue to consider the same surface $F$ and all notations of II.

11. **Émile Picard and differentials on a surface.** During the period 1882–1906 Picard developed almost single-handedly the foundations of this theory. His evident purpose was to extend the Abel-Riemann theory and this he accomplished in large measure. Reference: Picard-Simart [**2**].

Picard studied particularly closed $\omega^1$, that is

$$\omega^1 = \alpha dx + \beta dy, \qquad \partial\alpha/\partial y = \partial\beta/\partial x$$

and $\omega^2$. The choice of closed $\omega^1$ is very appropriate since then $\int\omega^1$ is an element of $K(F)$, and analytic function theory plus topology are fairly readily available.[4]

For closed one-forms the same three kinds as for abelian differentials are distinguished, save that for the third kind logarithmic curves replace logarithmic points.

Significant results are

I. Closed one-forms of the first kind make up a $\mathfrak{V}^q$ (Castelnuovo) ($q = \frac{1}{2}R_1$ as I have shown).

II. For the second kind same property save that they form a $\mathfrak{V}^{2q}$ mod $dK(F)$. (Picard)

III. Regarding the third kind Picard obtained this noteworthy result: There exists a least number $\rho \geqq 1$ such that any set of $\rho + 1$ curves are logarithmic for some closed $\omega^1$ having no other poles.

The 2-forms admit again three kinds: (a) first kind: holomorphic everywhere; (b) second kind: holomorphic to within a $d\omega^1$ about each point; (c) the rest. The third kind is characterized by the possession of periods: *residues* over some 2-cycle $\gamma^2$ bounding an arbitrarily small neighborhood of a one-cycle on a curve.

The 2-forms of the first kind were already found by Max Noether. They are of the type

$$\omega^2 = \frac{Q(x, y, z)dxdy}{F'_z}$$

where $Q$ is an adjoint polynomial of degree $m-4$. These $\omega^2$ (or the associated $Q$) make up a $\mathfrak{V}^{p_g}$, where $p_g$ is the *geometric* genus of $F$, studied at length by Italian geometers.

[4] Strictly speaking, $\int\omega^1$ is in $K(F)$ only if $\omega^1$ has no residues or periods, but since $d\omega^1 = 0$, $\int\omega^1$ is invariant under a continuous variation in the path of integration.

Let $\mathcal{V}^{\rho_0}$ be the vector space of the $\omega^2$ of the second kind mod $d\omega^1$. Picard utilized his topological description of *finite* 2-cycles to arrive at the following formula:

$$\rho_0 = I + 4q - \rho + 2. \tag{11.1}$$

12. **Severi and the theory of the base.** The central idea here is a notion of *algebraic dependence* between curves on the surface $F$. I must first describe this concept.

Let the nonsingular surface $F$ be in an $S^{k+2}$. A linear system of hypersurfaces of the space cuts out on $F$ a *linear* system of curves $|C|$. This system is *complete* if its curves are not curves of an amplified linear system.

We owe to the Italian School the following property: Every sufficiently ample complete system $|C|$ is part of a collection $\{C\}$ of $\infty^q$ such systems. The elements $|C|$ of the collection are in an algebraic one-one correspondence with the points of an abelian variety $V^q$, unique for $F$ and called sometimes the *Picard variety* of $F$ (see § 18).

A system $\{C\}$, $\infty^2$ at least, without fixed points and with irreducible generic curve is said to be *effective*. Its curves are also called *effective*.

Note the following properties:

(a) An effective system is fully individualized by any one of its curves.

(b) The generic curves of an effective system have the same genus, written $[C]$.

(c) The curves $C$, $D$ of two effective systems intersect in a set of distinct points whose number is denoted by $[CD]$. In particular we write $[C^2]$ for $[CC]$ and $[C^2]$ is the *degree* of $C$.[5]

(d) With $C$, $D$ as before let two curves $C$, $D$ taken together be individuals of an effective system $\{A\}$. This system is unique and we write

$$A = B + C. \tag{12.1}$$

(e) Any two curves $A_1$, $A_2$ of an effective system $\{A\}$ may be joined in $\{A\}$ by a continuous system $\infty^1$ of curves of $\{A\}$, whose genus, except for those of $A_1$ and $A_2$, is fixed and equal to $[A]$.

As an application of (e) let $A$, $B$, $C$ be effective and $A = B + C$. Following Enriques, connect $A$ to $B + C$ as indicated in (e). There follows a relation

[5] This degree should not be confused with the degree of $C$ as an algebraic curve in projective space.

$$\chi(A) + [BC] = \chi(B) + \chi(C) - [BC].$$

Hence if we define

$$\phi(A) = \chi(A) + [A^2] = 2 - 2[A] + [A^2],$$

we verify at once that

$$\phi(A) = \phi(B + C) = \phi(B) + \phi(C).$$

That is $\phi(A)$ is *an additive function on effective systems.*

When (12.1) holds between effective systems we set

$$C = A - B$$

and we have

$$\phi(C) = \phi(A - B) = \phi(A) - \phi(B).$$

Note also that as regards the symbols $[BC]$ we may operate as with numbers, that is

$$[(B \pm C)D] = [BD] \pm [CD].$$

*Virtual systems*: Let $\{A\}$, $\{B\}$ be effective systems. Without imposing any further condition define a virtual system $\{C\} = \{A-B\}$ as the pair of symbols $\{\phi(A)-\phi(B)\}$, $[(A-B)^2]$. This defines automatically $[C]$ and $[C^2]$. It is also clear that they are the same for $A-B$ and $A+D-(B+D)$ whatever $D$ effective. In other words $\{C\}$ depends only upon the difference $A-B$. The symbol $\{C\}$ is called a virtual algebraic system of curves and $[C]$, $[C^2]$ are the related virtual genus and degree.

It may very well happen that while $A$, $B$ are effective there exist curves $C$, not necessarily effective such that $B+C$ ($B$ together with $C$) is a member of $\{A\}$. If so $C$ is considered as a curve of the virtual system $\{C\}$ and has virtual characters $[C]$ and $[C^2]$, not necessarily its actual characters.

If we define $\{0\} = \{A-A\}$, as a virtual curve 0 is unique. One readily finds that $[0]=1$, $[0^2]=0$.

To sum up, the totality of effective and virtual curves form a module $M_S$ over the integers: the Severi module. Within $M_s$ a relation

$$\lambda_1 C_1 + \cdots + \lambda_s C_s = 0 \tag{12.2}$$

has a definite meaning. It is a relation of *algebraic dependence* between curves of $F$ in the sense of Severi.

The following remarkable result was proved by Severi:

(12.3) THEOREM OF SEVERI. *The module of curves of $F$ has a base consisting of $\rho$ effective curves $C_1, C_2, \cdots, C_\rho$, where $\rho$ is the Picard number relative to closed $\omega^1$ of the third kind.*

That is any curve $C$ satisfies a relation

$$\lambda C = \lambda_1 C_1 + \cdots + \lambda_\rho C_\rho$$

where $\lambda$ and the $\lambda_h$ are integers and $\lambda \neq 0$.

Severi also proved

(12.4) *The base may be chosen minimal, that is such that*

$$\lambda C = \lambda \sum \lambda_h C_h.$$

*Moreover there exist effective curves $D_1, D_2, \cdots, D_{\sigma-1}$ such that actually*

$$C = \sum \lambda_h C_h + \sum \mu_j D_j.$$

*One assumes, as one may that $\sigma$ is the least possible.*

Severi also proved the following criteria:

(12.5) *A n.a.s.c. in order that the curves $C_1, C_2, \cdots, C_s$ be algebraically independent is that, with $H$ a plane section, the matrix*

$$\begin{bmatrix} [C_h & C_k] \\ [C_h & H] \end{bmatrix}$$

*be of rank s.*

(12.6) *N.a.s.c. in order that $\{C_h\}$, $h \leqq \rho$, be a base is that the determinant $|[C_h C_k]| \neq 0$ and that its order $\rho$ be the highest order for which this holds. Moreover, $\rho$ is the Picard number.*

13. **Poincaré and normal functions.** Through an ingenious application of the theorems of Abel and Jacobi Poincaré arrived at a rapid derivation of some of the major results of Picard and the Italian geometers. I shall mainly deal with the part referring to Severi's theory of the base.

Let me first put in a most convenient form due to Picard and Castelnuovo the $\omega^1$ of the first kind of the curve $H_y$. A base for them may be chosen of type

$$(13.1) \qquad du_s = \frac{Q_s(x, y, z)dx}{F'_z}, \qquad s \leqq p$$

where $Q_s$ is an adjoint polynomial of degree $m-3$ in $x$ and $z$. For the first $p-q$ the polynomial is of degree $m-3$ in $x$, $y$ and $z$. For $s=p-q+1, \cdots, p$, it is of degree $m-2$ in $x, y, z$. Actually within this last range one may choose the $Q_h$ so that the $du_h$ only have constant periods relative to the invariant cycles and zero relative to the rest. As for the first $p-q$ they will have zero periods relative to the invariant cycles.

Let $C$ be a curve on $F$ and $M_1, \cdots, M_n$ its intersections with $H_y$. The sums from a fixed point $P_1$ of $H_y$ to the $M_k$

$$\sum_k \int_{P_1}^{M_k} du_s = v_s, \qquad s \leqq p,$$

(integration in $H_y$) are Poincaré's normal functions.

Let $L$ be the set of integration paths and with $\delta_k^1$ as in (8) let

$$\Omega_{ks} = \int_{\delta_k^1} du_s. \tag{13.2}$$

Then with the $\mu_k$ as in (9.2) we find

$$\begin{aligned} v_h &= \sum \frac{\mu_k}{2\pi i} \int_a^{a_k} \frac{\Omega_{kh}(Y)dY}{Y-y} \qquad h = 1, 2, \cdots, p-q \\ v_{p-q+j} &= \alpha_j(\text{constant}) \qquad j = 1, 2, \cdots, q. \end{aligned} \tag{13.3}$$

REMARK. The only condition imposed upon the points $M_k$ is that they be rationally defined together on $H_y$. They may represent for example the following special cases: (a) any sum of multiples of the fixed points $P_h$ of $H_y$, in particular they may represent just $\mu P_h$; (b) if $C$ is reducible say $C = C_1 + C_2$ with $M_{1h}$ and $M_{2h}$ as respective intersections one might have any set $t_1 \sum M_{1h} + t_2 \sum M_{2h}$, and similarly for several reducible curves; (c) any combination of the preceding two cases. In what follows, "curve" must be understood to include all these special cases.

As usual when dealing with abelian sums the $v_s$ are only determined mod periods of the related $u_s$.

(13.4) THEOREM OF POINCARÉ. *N.a.s.c. to have a set of $v_s$ given by (13.3) represent a curve by means of Jacobi's inversion theorem are*

$$\sum_h \mu_h \Omega_{hs}(a) = 0, \qquad s = 1, 2, \cdots, p. \tag{13.4a}$$

(13.4b) *Let $P(x, y, z)$ be any linear combination of the $P_s(x, y, z)$ divisible by $y-a$ and let*

$$du = \frac{P(x, y, z)dx}{F_z'}, \qquad \Omega_k(y) = \int_{\delta_k^1} du.$$

*Then one must have*

$$\sum \mu_k \int_a^{a_k} \Omega_k(y)dy = 0. \tag{13.4c}$$

(13.5) *Comparison with Severi's results.* Let the collection $\{\mu_s\}$ of the $\mu_s$ occurring in any set of normal functions be designated by $\mu$. The collection $\{\mu\}$ is a module $U$. Let $U_0$ be the submodule of all the elements $\mu^0$ corresponding to the $\sum t_h P_h$, $t_h$ an integer. The quotient $U_1 = U/U_0$ is the factor-module corresponding to all the curves which are not a plane section or more generally a $\sum t_j P_j$. The $U_1$ module has a base made up of $\rho - 1$ algebraically independent curves and a minimum base consisting of $\rho + \sigma - 2$ curves. By adding the $\mu(H)$ one has respectively $\rho$ and $\rho + \sigma - 1$ for base and minimum base.

The quotient module $U_1 = U/U_0$ is the module of all $\mu$ of curves none a plane section $H$. The module $U_1 + H = M_p$ is the *Poincaré* module and it is isomorphic with the Severi module $M_s$.

(13.6) REMARK. In order to get rapidly to the "heart of the matter" I have assumed at the outset that in (13.1) the polynomials $Q_{p-q+j}$ were of degree $m-2$ in $x$, $y$, $z$. This was based upon rather deep results of Picard and Castelnuovo. Poincaré however merely assumed that the degree of $Q_{p-q+j}$ was $m-2+\nu_j$. As a consequence in (13.3) the constants $\alpha_j$ must be replaced by polynomials $\alpha_j(y)$ of degree $\nu_j$. Then Poincaré shows that on the strength of the theorems of Abel and Jacobi every $\nu_j = 0$ hence the $\alpha_j(y)$ must be constants and one has in fact the form (13.3).

Notice also that from the form of the $Q_{m-3+j}$ one may find another adjoint polynomial $R_{m-3+j}$ of degree $m-3$ in $y$, $z$ and $m-2$ in $x$, $y$, $z$ such that

$$dw_j = \frac{Q_{m-3+j}dx + R_{m-3+j}dy}{F_z'}$$

is a closed $\omega^1$ of the first kind. The set $\{dw_j\}$ is then shown to be a base for such differentials. This proves rapidly that their "independent number" is $q$. Finally since the $\alpha_j$ are arbitrary constants the form of (13.3) shows implicitly that a complete (maximal) algebraic system of curves consists of $\infty^q$ linear systems in one-one correspondence with the points of an abelian $q$ dimensional variety (see IV, §17).

In outline this shows how normal functions enabled Poincaré to obtain with ease a number of the major results of Picard and the Italian geometers.

## IV. Analysis with Topology

14. **On the Betti number $R_1$.** In II I recalled my proof that $R_1$ is even and $R_1 = 2q$, the number of invariant cycles of the curve $H_y$. This gave incidentally a direct topological proof that the number of independent one-cycles in any curve of a sufficiently general system was fixed and equal to $R_1$. It showed also that the irregularity $q$ of a surface, in the sense of Castelnuovo and Enriques was actually a topological character. As I will show in §17, a topological proof that $q$ is an "absolute invariant" is immediate. Notice also that the distribution of complete algebraic systems in $\infty^q$ linear systems, referred to in (13.6) is also shown to have topological character.

(14.1) Let $\{du_k\}$, $k \leqq q$, be a base for the closed $\omega^1$ of the first kind of $F$. On $H_y$ they coincide with the $u_{p-q+h}$ of §13. Let $\pi_{k\mu}$, $\mu \leqq 2q$, be the periods of $u_k$ relative to a homology base $\{\gamma_\mu^1\}$, $\mu \leqq 2q$, for the one-cycles of $F$. From the fact that the periods of the differentials of the first kind of $H_y$ form a Riemann matrix, we infer:

(14.2) Theorem. *The matrix $\pi$ of the periods of the $u_k$ and their conjugates $\bar{u}_k$ as to the $\gamma_\mu^2$ is a Riemann $2q \times 2q$ matrix.*

15. **On algebraic two-cycles.** A collection of mutually homologous 2-cycles is a *homology class*. In this manner algebraic cycles yield algebraic homology classes. Through addition they generate a module $M_L$. Thus in relation to the collection of curves on a surface $F$ there are three definite modules: $M_S$ (Severi module), $M_P$ (Poincaré module) and $M_L$ (Lefschetz module).

(15.1) Theorem. *The three modules $M_S$, $M_P$ and $M_L$ are identical.*

This property will be a final consequence of an extensive argument.

Returning to Poincaré's normal functions (III, §13) a glance at his two conditions for a set of normal functions to represent an algebraic curve reveals immediately that Poincaré's first condition simply means that

$$(15.2) \qquad \begin{cases} \gamma^2 = \sum \mu_k \Delta_k + (H_a) \\ \sum \mu_k \overset{1}{\delta_k} \sim 0 \qquad \text{in } H_y \end{cases}$$

is a cycle. As to the second condition it says merely that if

$$\omega^2 = \frac{Q(x, y, z)dxdy}{F_z'}$$

is of the first kind, that is if $Q$ is adjoint of order $m-4$, then

$$\int_{\gamma 2} \omega^2 = 0.$$

Hence Poincaré's conditions are equivalent to the following result:

(15.3) THEOREM. *A n.a.s.c. for a cycle $\gamma^2$ to be algebraic is that the period of every 2-form of the first kind relative to $\gamma^2$ be zero.*

(15.4) REMARK. Among all the "algebraic" curves there were included all the sums $\sum m_j P_j$, where the $P_j$ are the fixed points of $H_y$. It is evident that for these special "2-cycles" $\int \omega^2$ is zero.

(15.5) COROLLARY. *Severi's number $\sigma$ is merely the order of the torsion group of the two-cycles (or equally of the torsion group of the one-cycles).*

For if $\gamma^2$ is a torsion 2-cycle we have $\lambda\gamma^2 \sim 0$, $\lambda \neq 0$, and hence

$$\int_{\gamma 2} \omega^2 = 0$$

for every $\omega^2$ of the first kind.

(15.6) THEOREM. *The number $\rho$ is the Betti number of algebraic cycles.*

This is a consequence of the following property:

(15.7) *Let $C_1, \cdots, C_s$ be a set of curves and let $\overline{C}_h$ be the cycle of $C_h$. Then*

$P_a$: *algebraic independence of the $C_h$*

$P_h$: *homology independence of the $\overline{C}_h$*

*are equivalent properties.*

From obvious considerations $P_a$ implies $P_h$. Conversely let $P_h$ hold. We must show that $\overline{C} \sim 0$ implies $\lambda C = 0$, $\lambda \neq 0$. Here I follow Albanese's rapid argument. Let $C = A - B$, $A$ and $B$ effective. Since $\overline{A} \sim \overline{B}$ and $[CD] = (\overline{C}, \overline{D})$ we have

$$[A^2] = [AB] = [B^2]; \qquad [AH] = [BH]$$

where $H$ is a plane section. Hence Severi's independence criterion is violated between $A$ and $B$. Consequently $\lambda A = \mu B$, $\lambda\mu \neq 0$. From $[AH] = [BH]$ follows $\lambda = \mu$ and therefore $\lambda(A - B) = 0 = \lambda C$, $\lambda \neq 0$. This proves (15.7).

It follows that $M_S = M_L$ and as $M_P = M_S$, (15.1) is proved.

Notice that we may now give the following *very simple* definition of virtual curve: it is merely an algebraic 2-cycle. Simplicity is even augmentable by replacing everywhere the symbol = of algebraic dependence (=) by the homology symbol $\sim$.

16. **On 2-forms of the second kind.** The basic result is the proof of the formula

$$\rho_0 = R_2 - \rho. \tag{16.1}$$

I shall just indicate an outline of my proof. I shall also show that the process outlined obtains incidentally Picard's fundamental result for $\rho$ concerning logarithmic curves of a closed $\omega^1$ of third kind. The steps follow closely an analogous outline in my monograph [**9**].

For convenience I call $\omega^1$ and $\omega^2$ *regular* when

$$\omega^1 = \frac{Pdy - Qdx}{\phi(y)F_z'}, \qquad \omega^2 = \frac{Pdxdy}{\phi(y)F_z'},$$

where $P$ and $Q$ are adjoint polynomials and $\phi(y)$ is a polynomial.

If $\omega^2 = d\omega^1$, $\omega_2$ is said to be *improper*. Thus $\rho_0$ is the dimension of the vector space of the $\omega^2$ of the second kind mod those which are improper.

By *reduction* of $\omega^2$ I understand the subtraction of an improper $\omega^2$.

I. *The periods and residues of a normal 2-form are arbitrary.*

II. *One may reduce any $\omega^2$ of the second kind to the regular type.*

III. *A regular $\omega^2$ such that $\int\omega^2$ has neither residues nor periods is reducible to a regular $d\omega^1$.*

Except for the presence of the polynomial $\phi(y)$ the proofs of the preceding propositions are very close to those of Picard. It is true that allowing $\phi(y)$ in regular $\omega^1$ and $\omega^2$ considerably simplifies every step (see [**9**, Note I]).

IV. *Let $C$ be a curve of order $s$. We may choose coordinates such that $C$ does not pass through any of the fixed points $P_j$ of $H_y$, nor through the points of contact of the planes $y = a_k$. One may form an $\omega^1 = Rdx$, $R \in K(F)$ possessing on $H_y$ the s-logarithmic points of $CH_y$ with logarithmic period $2\pi i$ and say $P_1$ with logarithmic period $-2\pi is$. One may even select $R$ so that $(\partial R/\partial y)dx$ has no periods. From this follows that there is an $S(x, y, z) \in K(F)$ such that*

$$\omega^2 = d(Rdx + Sdy)$$

is regular.

Take now $C_1, C_2, \cdots, C_t$ and the axes so chosen that they all behave like $C$. Let $\omega_h^2$ be analagous to $\omega^2$ for $C_h$.

*Owing to* III *it is now readily shown that n.a.s.c. in order that some linear combination*

$$\omega^2 = \sum \alpha_h \omega_h^1$$

*be without periods is that the* $C_s$ *and* $H$ *be logarithmic curves of a closed* $\omega^1$.

Since $R_2$ is finite there is a least $\rho-1$ such that for $s=\rho$ the curve $C_h$, $H$ are logarithmic curves of a closed $\omega^1$ of the third kind. Hence

(16.2) *Picard's fundamental property for* $\rho$ *is a consequence of the finiteness of the Betti number* $R_2$.

V. *To proceed one may form* $\rho-1$ *linearly independent* $\omega^2$ *which are improper. Since the total number of distinct periods is equal to* $R_2(F-H)$ $=R_2-1$ *we have then* $\rho_0=R_2-\rho$, *as asserted.*

(16.3) *On Picard's treatment of* $\rho_0$ *and* $\rho$. Owing to lack of topological technique Picard proved directly that $\rho_0$ was finite by showing through strong algebraic operations that if

$$\omega^2 = \frac{Q(x, y, z)dxdy}{F_z'}$$

where $Q$ is adjoint, is of the second kind, the degree of $Q$ was bounded.

Although Picard did not observe it, his later treatment of $\omega^2$ of the second kind contained implicitly (argument of 16.2) the proof that $\rho$ had the property by which he defined it relative to closed $\omega^1$ of the third kind.

**17. Absolute and relative birational invariance.** Take again a general $n$-variety

$$F(x_1, \cdots, x_n, y) = 0. \tag{17.1}$$

Let $\{\xi_0, \cdots, \xi_r\}$ be a homogeneous base for the function field $K(F)$. Then the system

$$\tau y_h = \xi_h, \qquad h = 0, 1, 2, \cdots, r > n$$

represents a model $F_1$ of $F$ in the projective space $S^r$, with homogeneous coordinates $y_h$. If $\{\eta_0, \cdots, \eta_s\}$, $s>n$, is a second homogeneous base for $K(F)$, the system

$$\sigma z_k = \eta_k, \qquad k = 0, 1, \cdots, s$$

represents a second model $F_2$ of $F$ in a projective space $S^s$. Since $\{\xi_h\}$ and $\{\eta_k\}$ are homogeneous bases for $K(F)$ $F_1$ and $F_2$ are birationally transformable into one another. The simple example of two elliptic curves of degrees 3 and 4 show however that the corresponding structures need not be homeomorphic. The difficulty is caused by the presence of singularities. A standard device for curves enables one to "forget" singularities and restore homeomorphism. No such device is known for a $V^n$, $n>1$.

For simplicity let me limit the argument to surfaces. I have really considered a surface as a nonsingular model in some projective space. Let $F_1$, $F_2$ be two such distinct models and suppose that the *field* $K(F)$ *is not that of a ruled surface.* Then according to Castelnuovo and Enriques a birational transformation $T$: $F_1 \rightarrow F_2$ may take a finite number $\delta_{12}$ of *exceptional* points of $F$ into *disjoint* nonsingular rational curves. There exists an analogous $\delta_{21}$ for $T^{-1}$. Let a point $P$ of $F_1$ be sent by $T$ into a curve $C$ of $F_2$. Since $C$ is rational and nonsingular it is topologically a sphere. Hence its characteristic $\chi(C)=2$. Hence the gain in $\chi(F_1)$ through $\delta_{12}$ exceptional points is $\delta_{12}$. Therefore

$$\chi(F_1) + \delta_{12} = \chi(F_2) + \delta_{21}. \tag{17.2}$$

Now a character, numerical or other of $F$ is said to be an *absolute* invariant if it is unchanged under all transformations such as $T$. A *relative invariant* is one that may change under certain transformations $T$.

Let me examine some of the characters that have been introduced. It is readily shown that under $T$ both $R_2$ and $\rho$ are increased by the same amount $\delta_{12}-\delta_{21}$. Hence both are relative invariants and $\rho_0 = R_2 - \rho$ is an absolute invariant.

Since

$$\chi(F) = R_2 - 2R_1 + 2$$

and both $\chi$ and $R_2$ vary in the same way, $\chi$ is a relative invariant and $R_1$ is an absolute invariant.

Therefore:

(17.3) *The dimensions of the spaces of closed* $\omega^1$ *of the first and second kinds and of* $\omega^2$ *of the second kind are absolute invariants.*

18. **Application to abelian varieties.** Let $\Pi$ and $M$ be a Riemann matrix and its principal matrix (see §3).

Introduce the following vectors:

$$u = (u_1, \cdots, u_{2p}), \qquad u_{p+j} = \bar{u}_j$$

$$\pi_\mu = (\pi_{1\mu}, \cdots, \pi_{2p,\mu}), \qquad \mu = 1, 2, \cdots, 2p; \qquad \pi_{p+j,\mu} = \bar{\pi}_{j\mu}.$$

Through the hyperplanes

$$u = \sum s_\mu \pi_\mu,$$

$s_\mu$ an integer, real $2p$-space is partitioned in a familiar way into paralellotopes. A suitable fundamental domain $D$ is

$$u = \sum t_\mu \pi_\mu, \qquad 0 \leqq t_\mu < 1.$$

The identification of congruent boundary points turns this domain into a $2p$-ring $R^{2p}$ (product of $2p$ circles):

Corresponding to $\Pi$ and $M$ there may be defined a whole family of functions $\theta$ of various orders. Each such function $\phi$ is a holomorphic function in the domain $D$. Those of a given order, say $n$, are characterized by the property that $\phi_n(u+\pi_\mu)=\phi_n(u)$ times a fixed linear exponential function of $u$. I have shown that one may find an $n$ such that if $\{\theta_n^j(u+\alpha)\}$, $\alpha$ a fixed $p$-vector, $j=0, 1, \cdots, r$, is a finite linear base for the $\theta_n(u+\alpha)$ then the system

$$kx_j = \theta_n^j(u + \alpha)$$

represents a nonsingular $p$-variety $V^p$ in projective $S^r$, and this $V^p$ is in analytic homeomorphism with the ring $R^{2p}$. This is an abelian $p$-variety (see [8])

The topological relation $V^p \leftrightarrow R^{2p}$ assigns an exceptionally simple topology to $V^p$. Let the edges of $D$ oriented from the origin out be designated by $1, 2, \cdots, 2p$. Any $i_h$ defines a one-cycle represented by $(i_h)$; any two edges $i_h$, $i_k$ define a 2-cycle represented by $(i_h, i_k)$, etc. The $(i_h, i_k)$, $i_h < i_k$ form a base for the 2-cycles of $V^p$, etc.

I am mainly concerned with the 2-cycles. In view of $(\nu, \mu) = -(\mu, \nu)$ a general 2-cycle is represented by a homology

$$\gamma^2 \sim \sum m_{\mu\nu}(\mu, \nu), \qquad m_{\mu\nu} = -m_{\nu\mu}.$$

On the other hand

$$\overset{2}{\omega}_{jk} = du_j du_k; \qquad j, k \leqq p; \qquad j < k$$

is a closed 2-form of the first kind of $V^p$ and $\{du_h du_k\}$ is a base for all such forms.

(18.2) REMARKS. *On a general n-variety* $V^n$, $n \geqq 2$. Considerations of the same type as in §12 may be extended automatically to algebraic dependence of hypersurfaces of $V^n$ (its $V^{n-1}$), and also to their $(2n-2)$-cycles. Algebraic and homology dependence give rise to a number $\rho(V^n)$. I single out especially the following proposition from [9, p. 104] (Corollary):

(18.3) THEOREM. *Let* $\Phi$ *be a fixed surface of* $V^n$ *(general intersection of hyperplane sections of* $V^n$*) and let* $C_1, \cdots, C_s$ *be hypersurfaces which cut* $\Phi$ *in curves* $C_h^*$, $h \leqq s$. *Then the following relations are all equivalent: relations of algebraic dependence between the* $C_h$ *in* $V^n$, *the same between the* $C_h^*$ *in* $\Phi$; *relations of homology between the standard oriented cycles* $\overline{C}_h$, $C_h$ *in* $V^n$; *the same for the* $C_h^*$ *in* $\Phi$.

Returning now to the abelian variety $V^p$ let $\Phi$, $C$, $C^*$ be this time the same as above but for $V^p$. Now the $\omega_{hk}^2$ taken on $\Phi$ become $\omega^2$ of the first kind for $\Phi$. If $\{C_s\}$, $s \leqq \rho(V^p)$, is a base as to $=$, or equivalently as to $\sim$ and algebraic $(2p-2)$-cycles of $V^p$, then the same holds for the curves $C_s^*$ in $\Phi$. Hence by theorem (15.2):

$$\int_{C_s^*} du_j du_l = 0; \qquad j, l \leqq p. \tag{18.4}$$

On the other hand since the $(\mu, \nu)$ are cycles in $\Phi$ we have in $\Phi$

$$C_s^* \sim \sum \overset{s}{m}_{\mu\nu}(\mu, \nu), \qquad m_{\mu\nu} = -m_{\nu\mu}.$$

Hence (18.4) yields

$$\sum \overset{s}{m}_{\mu\nu}\pi_{j\mu}\pi_{k\nu} = 0; \qquad j, k \leqq p. \tag{18.5}$$

This really means that the $\rho$ matrices $[\overset{s}{m}_{\mu\nu}]$ are linearly independent singularity matrices for the Riemann matrix $\Pi$. If the singularity index of $\Pi$ is $k$, then one must have

$$\rho \leqq k. \tag{18.6}$$

The possible inequality is due to the fact that an algebraic 2-cycle of $\Phi$ must satisfy a relation such as (18.4) not merely with respect to the closed $\omega^2$ of the first kind of $V^p$ in $\Phi$, but also with respect to all $\omega^2$ of the first kind of $\Phi$, and one cannot exclude the possible existence of such $\omega^2$ other than the closed taken on $\Phi$. However, the following two properties hold:

(a) There is a base for the forms $M$ made up of principal forms.

(b) Each principal $M$ gives rise to a particular system of functions $\phi$ à la $\theta$. These functions are said to be *intermediary*.

(c) If $\{M_j\}$, $j \leqq k$, is a base for the matrices $M$, and $\phi_j$ is an intermediary function relative to $M_j$ then $\phi_j = 0$ represents a hypersurface of $V^p$ and these hypersurfaces are algebraically independent.

It follows that $\rho \geqq k$ and therefore

$$\rho = k. \tag{18.7}$$

This is the result that I was looking for.

Actually the relations between the hypersurfaces *as* cycles and their Severi independence are the same as for their sections with the surface $\Phi$. That is,

(18.8) THEOREM. *For hypersurfaces of $V^p$ algebraic dependence and homology in $V^p$ are equivalent relations.*

BIBLIOGRAPHY


1. Émile Picard, *Traité d'analyse*. Vol. 2, Gauthier-Villars, Paris.

2. Émile Picard and George Simart, *Théorie des fonctions algébriques de deux variables indépendantes*. Vols. I, II, Gauthier-Villars, Paris, 1895, 1906.

3. Francisco Severi, *Sulla totalità delle curve algebriche traceiate sopra una superficie algebrica*, Math. Ann. **62** (1906), 194–226.

4. ———, *La base minima pour la totalité des courbes tracées sur une surface algébrique*, Ann. Sci. École Norm. Sup. **25** (1908), 449–468.

5. Henri Poincaré, *Sur les courbes tracées sur une surface algébrique*, Ann. Sci. École Norm. Sup. **27** (1910), 55–108.

6. Gaetano Scorza, *Intorno alla teoria generale delle matrici di Riemann e ad alcune sus applicaciones*, Rend. Circ. Mat. Palermo **41** (1916), 263–380.

7. Solomon Lefschetz, *Algebraic surfaces, their cycles and integrals*, Ann. of Math. **21** (1920), 225–258.

8. ———, *On certain numerical invariants of algebraic varieties with application to abelian varieties*, Trans. Amer. Math. Soc. **22** (1921), 327–482.

9. ———, *L'analysis situs et la géométrie algébrique*, Borel Series, 1924.

10. ———, *Topology*, Amer. Math. Soc. Colloq. Publ., Vol. 12, Amer. Math. Soc., Providence, R. I., 1930; reprint Chelsea, New York, 2nd ed., 1950.

11. Oscar Zariski, *Algebraic surfaces*, Ergebnisse der Math., Springer-Verlag, Berlin, 1935; reprint Chelsea, New York, 1948.

12. Heisuki Hironaka, *Resolution of singularities of an algebraic variety over a field of characteristic zero*, Ann. of Math. (2) **79** (1964), 109–329.

13. W. V. D. Hodge, *Theory and application of harmonic integrals*, Cambridge Univ. Press, New York, 1941.

# II

# II

# ON CERTAIN NUMERICAL INVARIANTS OF ALGEBRAIC VARIETIES WITH APPLICATION TO ABELIAN VARIETIES*

CONTENTS



* Presented to the Society November 27, 1920. This is a translation with minor modifications of the memoir awarded the Bordin Prize by the Paris Academy of Sciences for the year 1919. The publication has been assisted by grants from the American Association for the Advancement of Science and the University of Kansas.

## Introduction

In the development of the theory of algebraic functions of one variable the introduction by Riemann of the surfaces that bear his name has played a well-known part. Owing to the partial failure of space intuition with the increase in dimensionality, the introduction of similar ideas into the field of algebraic functions of several variables has been of necessity slow. It was first done by Emile Picard, whose work along this line will remain a classic. A little later came the capital writings of Poincaré in which he laid down the foundations of Analysis Situs, thus providing the needed tools to obviate the failure of space intuition.

Meanwhile other phases of the theory were investigated in Italy and to some extent also in France, receiving an especially powerful impetus at the hands of Castelnuovo, Enriques and Severi. It is however a rather remarkable fact that in their work topological considerations are all but absent, practically never going beyond the study of linear cycles, and this on the whole by very indirect methods. The reason is to be found of course in the fact that the investigations of the Italian School all center around linear systems and the problems that they naturally suggest.

It seems certain that a further use of topological considerations will bring many new results.* In such a development the notion of *algebraic cycles*, or cycles homologous to those formed by algebraic varieties, is destined to play an important part. It is in fact the central concept dominating the first half of this memoir.

After some general theorems on the analysis situs of algebraic varieties and on their multiple integrals, we concentrate our attention on surfaces, where, as could be expected, the results are most expressive. It is found first that a double integral of the first kind has no periods relatively to algebraic cycles, that is, to cycles homologous to those formed by algebraic curves. This result is then inverted by means of certain functions introduced by Poincaré. It follows that Picard's number $\rho$ is equal to the number of two-cycles without periods for the integrals of the first kind, while Severi's number $\sigma$ is the number of cycles whose multiple is homologous to zero. For higher varieties the result obtained is less precise, $\rho$ being merely at most equal to the number of

*In a monograph entitled *L'Analysis Situs et la Géométrie Algébrique* in course of publication in the Borel Series (Paris, Gauthiers-Villars), I have undertaken a thorough and systematic study of the topology of algebraic surfaces and varieties. Many new and interesting results, discovered since this memoir was written, will be found there. (Added in 1922).

cycles without periods for the double integrals of the first kind. These general theorems are followed by various applications to the determination of $\rho$ and $\sigma$ in many interesting cases. We may mention in particular what seems to be the first complete proof of a theorem first considered by Noether, to the effect that an arbitrary algebraic surface of order greater than three in ordinary space, possesses only curves that are complete intersections.

The second and main part of this work is entirely devoted to Abelian varieties. Let $V_p$ be such a variety, of rank one and genus $p$,

$$\Omega \equiv \| \omega_{j\mu} \| \qquad (j = 1, 2, \cdots, p;\ \mu = 1, 2, \cdots, 2p)$$

its period matrix. There exist in general $1 + k$ linearly independent alternate forms with rational coefficients,

$$\sum c_{\mu\nu} x_\mu y_\nu \qquad (c_{\mu\nu} = -c_{\nu\mu}), \tag{1}$$

which vanish when the $(x)$'s and the $(y)$'s are replaced by the elements of any two rows of $\Omega$. In particular if a certain inequality due to Riemann is also satisfied, (1) is said to be a principal form of $\Omega$. Instead of the forms (1) we may consider others analogous to them, but no longer constrained to being alternate. The number of linearly independent forms of this more general type will be denoted by $1 + h$ and we always have $h \geqq k \geqq 0$. The properties of matrices such as $\Omega$ have recently been the object of searching investigations by Scorza and Rosati.

Now it is found at once that $V_p$ possesses exactly $1 + k$ two-cycles without periods with respect to the integrals of the first kind, and hence for $V_p$, $\rho \leqq 1 + k$. On the other hand let $\phi(u_1, u_2, \cdots, u_p)$ be an intermediary function

$$\phi(u + \omega_\mu) = e^{-2\pi i(\sum_{(j)} \alpha_{j\mu} u_j + \beta_j)} \cdot \phi(u)$$

the relations which express its behavior relatively to $\Omega$. By considering the effect of adding the $(\omega_\mu)$'s followed by the $(\omega_\nu)$'s and then reversing the order we obtain at once $\sum_{(j)} (\alpha_{j\mu}\, \omega_{j\nu} - \alpha_{j\nu}\, \omega_{j\mu}) = m_{\mu\nu} = -m_{\nu\mu}$ the $(m)$'s being integers. We thus have corresponding to $\phi$ a certain alternate form with integral coefficients

$$\sum m_{\mu\nu} x_\mu y_\nu . \tag{2}$$

On the other hand $\phi = 0$ determines a certain algebraic hypersurface of $V_p$, and the intersection of $p - 1$ hypersurfaces of the same system is a certain algebraic curve. By evaluating the cycle formed by this curve in terms of a certain simple fundamental system of cycles and then applying Riemann's inequality to the curve, we succeed in showing that the inverse of (2) is a certain principal form of $\Omega$ whence follows readily enough $\rho \geqq 1 + k$, and therefore finally $\rho = 1 + k$. Thus the result obtained in the first part for algebraic surfaces holds for Abelian varieties. All this is to be compared with certain investigations of Appell, Humbert, Bagnera and de Franchis, leading in a wholly different manner to the same conclusion for $p = 2$. There are also some related memoirs by Poincaré.

The consideration of Abelian varieties possessing certain complex multiplications leads us to the determination of $h, k, \rho$ in a wide range of cases. We touch here in many points investigations of Scorza, Rosati, and Frobenius, our methods being more nearly related to those of the last-named author. In the simplest type examined

$$\Omega \equiv \| 1, \alpha_j, \alpha_j^2, \cdots, \alpha_j^{2p-1} \| \qquad (j = 1, 2, \cdots, p),$$

where the $(\alpha)$'s together with their conjugates form the roots of an irreducible equation of degree $2p$ with integer coefficients. The determination of $h, k, \rho$, is shown to be closely related to certain properties of the group of the equation. In particular if this group is Abelian it is shown that in general

$$1 + h = 2(1 + k) = 2\rho = 2p,$$

and $V_p$ does not contain any Abelian subvariety of genus $< p$.

After considering some new properties of Abelian varieties of rank $> 1$, we pass to the study of the curves

$$y^q = \prod_{(i)} (x - a_i)^{\,i} \qquad (q \text{ odd prime})$$

and of their Jacobi varieties with which the memoir terminates. We have succeeded in determining the total group of birational transformations when the $(a)$'s are arbitrary. In particular we have showed that in some well-defined cases there are no systems of reducible integrals, that the group is of order $q$ and that the invariants of the Jacobi variety are given by

$$1 + h = 2(1 + k) = 2\rho = 2(q - 1).$$

A remarkable class of curves met with occurs when $q - 1$ is divisible by three. They are birationally transformable to the form

$$x^m y^n + y^m z^n + z^m x^n = 0 \qquad (m^2 - mn + n^2 = q).$$

Their group is a collineation group of order $3q$ if $q > 7$. For $q = 7$, we deal with the famous Klein quartic $x^3 y + y^3 z + z^3 x = 0$, whose group as shown by Klein is the classical simple collineation group of order 168. In all cases we have for the Jacobi varieties $1 + h = 2(1 + k) = 2\rho = 6p = 3(q - 1)$, and there are systems of reducible integrals of genus $(1/3)p$. Special cases have been considered by various authors (Klein, Ciani, Snyder, Scorza).

# PART I

## PRELIMINARY THEORY

## Chapter 1. Cycles and integrals of algebraic varieties

### § 1. Some theorems of analysis situs

1. We collect here some notations to be used frequently in the sequence.

$V_d$, fundamental $d$-dimensional variety;

$S_d$, $d$-space;

$\{C_u\}$, linear pencil depending upon the parameter $u$ and generic in a linear system $|C|$ of hypersurfaces of $V_d$;

$R_i$, $i$th connectivity index of $V_d$;

$M N \cdots P$, intersection of the varieties $M, N, \cdots, P$.

$[M]$, arithmetic genus of $M$, the number of points of the group when $M$ is a group of points.

In general we shall denote indifferently by $M$ an algebraic variety or its Riemann-image. At times we shall find it convenient to distinguish between them; and then we shall denote the image by $(M)$. Finally by hypersurface of $V_d$ is to be understood a $(d-1)$-dimensional algebraic variety contained in it.

2. Let $V_d$ be an algebraic variety with ordinary singularities, $|C|$ an irreducible linear system, $\infty^d$ at least, without base points, contained in $V_d$, and such that no point of $V_d$ is multiple for any $C$ through it. Taking then a pencil $\{C_u\}$ we mark in the complex $u$ plane the *critical* points $a_1, a_2, \cdots, a_n$, or points where $C_u$ acquires a conical node and trace cuts $a_0 a_i$. For the sake of simplification we shall denote the hypersurfaces $C_{a_0}$, $C_{a_i}$ by $C_0$, $C_i$. We propose to prove the following three propositions:

(*a*) Any $i$-cycle of $V_d (i < d)$ is homologous to a cycle wholly within a $C$, and when $i < d - 1$, the two cycles bound at the same time in their respective manifolds.

(*b*) Any $d$-cycle is the sum of two others of which one is wholly within a $C$ and the other is composed of a $d$-dimensional manifold contained in $C_0$ plus the loci of certain $(d-1)$-cycles of $C_u$ when $u$ describes the lines $a_0 a_i$.

(*c*) The $i$th connectivity indices, $i \leqq d - 2$, of $V_d$ and $C$ are equal.

For linear cycles (*a*) and (*c*) have already been proved by Picard, Castelnuovo, and Enriques, for two-cycles of algebraic surfaces by Poincaré, whose method adapted to the general case shall be used here.* We shall also apply his notations $\equiv$, $\sim$, and say with him: "$A$ is homologous to $B$ (or $A \sim B$), modulo $M$", to indicate that $A - B$ bounds in $M$.

*Journal de mathématiques pures et appliquées, ser. 6, vol. 2 (1906), p. 157. See also the writer's paper in the Annals of Mathematics, vol. 21 (1920) as well as the monograph already referred to.

3. Since the propositions to be proved are true for $d = 2$ we can assume that they hold for a $V_{d-1}$ and prove them for a $V_d$. Any $i$-cycle of $C_u$ $(i \leqq d - 2)$ is homologous to a cycle contained in the base manifold $B$ of $\{C_u\}$. The deformation of the cycle in $B$ must be continuous when $u$ varies, hence it remains homologous to itself, which shows that *any cycle of less than* $(d - 1)$ *dimensions of* $C_u$ *is invariant when* $u$ *varies.* Even if the cycle has more than $(d - 2)$ dimensions this still applies when it is homologous to a cycle in $B$.

Let then $K_u$ be the $(2d - 1)$-dimensional locus of $C_u$ when $u$ describes the semi-straight line: *argument* $(u - a_0) =$ *Const.* The manifold $K_u$ cuts an $i$-cycle $\Gamma_i$ $(i \leqq d)$ in a certain $(i - 1)$-dimensional manifold. By considering the homeomorphism which exists between $C_u$ and $C_0$ when the above semi-straight line does not contain any critical point, we may show that the $(i - 1)$-cycle just considered is homologous to a cycle in $C_0$. Hence by a slight change in Poincaré's reasoning (loc. cit.) we can show that the cycle $\Gamma_i$ is homologous to a sum of cycles situated in the hypersurfaces $C_h$ plus a cycle $\Gamma'_i$ obtained as follows: a certain cycle $\gamma^h_{i-1}$ of $C_u$, ($u$ is assumed on $a_0\, a_h$), reduced to a point for $u = a_h$, has for locus when $u$ describes the line $a_0\, a_h$ a manifold $\Delta^h_i$ whose boundary is the position of $\gamma^h_{i-1}$ in $C_0$. The cycle $\Gamma'_i$ is the sum of $\sum \lambda_h \Delta^h_i$ and a manifold $\Delta_i$ situated wholly in $C_0$. According to a remark made above no cycle of less than $(d - 1)$ dimensions of $C_u$ disappears at the critical points, hence if $i < d$, $\Gamma_i$ is homologous to a sum of cycles situated in several $(C_u)$'s. But when $C_u$ is displaced it can neither gain nor lose cycles of $V_d$. Hence the cycles situated in several $(C_u)$'s are all homologous to cycles within one of them, for example $C_0$. This completes the proof for the first part of (*a*).

In the case of a $d$-cycle $\Gamma_d$, besides the cycle in $C_0$ there is the cycle $\sum \lambda_h \Delta^h_d + \Delta_d$ which proves (*b*). Let now $\Gamma_i \sim 0$ $(i \leqq d - 2)$ and denote by $M_{i+1}$ a manifold bounded by $\Gamma_i$. By considering the intersections of $M_{i+1}$ and $\Gamma_i$ with $K_u$ we can show that there is a manifold $M'_{i+1}$, bounded by the cycle $\Gamma'_{i+1}$ of $C_0$, to which $\Gamma'_i$ has been reduced. The manifold $M'_{i+1}$ will be the sum of a manifold in $C_0$ and of another locus of certain $i$-cycles of $C_u$ when $u$ describes the cuts. Just as before, since no cycles of less than $(d - 1)$ dimensions disappear at the critical points, the part of $M'_{i+1}$ exterior to $C_0$ can be suppressed. The cycle $\Gamma'_{i+1}$ will therefore form the boundary for a manifold wholly in $C_0$. This completes the proof of (*a*).

4. Let $\Gamma_{i-2}$ $(i \leqq d + 1)$ be an invariant $(i - 2)$-cycle of $C_u$. It will be an arbitrary cycle of $C_u$ for $i < d + 1$. Associated with the complex $u$ plane it will generate an $i$-dimensional manifold $\Gamma_i$ bounded by the locus of $\Gamma_{i-2}$ when $u$ describes the cuts. But the parts of this boundary generated by $\Gamma_{i-2}$ when $u$ describes the opposite borders of the same cut destroy each other, hence $\Gamma_i$ is an $i$-cycle. It is besides evident that if $\Gamma_i \sim 0$, also $\Gamma_{i-2} \sim 0$ mod.

$C_u$. Conversely, if $\Gamma_{i-2}$ bounds an $(i-1)$-dimensional manifold on $C_u$, $\Gamma_i$ bounds its $(i+1)$-dimensional locus when $u$ describes the complex $u$ plane. There might be doubt only if $i-2=d-2$, for then parts of the locus corresponding to the cuts might also have to be considered as forming part of the boundary of the $(i+1)$-dimensional manifold. But in that case we replace $\Gamma_{d-2}$ by a homologous cycle in $CC_u$, ($C$ fixed). This amounts to replacing $V_d$ by $C$ for which the difficulty does not present itself any more. A certain cycle $\sim \Gamma_i$ will then bound in $C$ and therefore also in $V_d$. In this manner $R_{d-1}$ independent cycles of $V_d$ in $C_u$ generate $R_{d-1}$ independent $(d+1)$ cycles of $V_d$, and as $R_{d-1}=R_{d+1}$, (Poincaré, loc. cit.), every $(d+1)$-cycle depends upon those so obtained. Let us start then with a $\Gamma_{d-i}$ of $V_d$ in a $C^i$. It yields first a $\Gamma_{d-i+2}$ in a $C^{i-1}$, which in its turn yields a $\Gamma_{d-i+4}$ in a $C^{i-2}$ etc., so that ultimately we arrive at a $\Gamma_{d+i}$ of $V_d$ of which $\Gamma_{d-i}$ is the trace on $C^i$, and if a multiple of one of the cycles is $\sim 0$, then also a multiple of the other is $\sim 0$. Owing to Poincaré's relation $R_{d-i}=R_{d+i}$, every $(d+i)$-cycle depends upon those thus obtained or conversely all $(d-i)$-cycles of $V_d$ in $C^i$ are dependent upon the traces of its $(d+i)$-cycles. Now a $(d-i)$-cycle in $C^i$ can just as well be considered as the trace of a $(d+i-2)$-cycle in a $C$ passing through $C^i$, and all $(d-i)$-cycles of $C^i$ depend upon the traces of these $(d+i-2)$-cycles of $C$. The reverse is true, namely, all the $(d+i-2)$-cycles of $C$ depend upon the cycles generated by the $(d-i)$-cycles in the $(C^i)$'s. It follows that the index $R_{d+i-2}$ of $C$ is equal to $R_{d+i}$, or, owing to Poincaré's relations, as applied to $C$, the indices $R_{d-i}$ of both manifolds are equal. This proves (*c*).

## § 2. Effective and algebraic cycles

5. It is important to distinguish between the cycles which do and those which do not intersect a given hypersurface $C$. If we merely limit our considerations to an assigned cycle or hypersurface we can not go very far, for the one can be deformed and the other displaced in its continuous system, their intersection thereby being changed. To arrive at a valuable concept we must compare a continuous system of hypersurfaces to a set of homologous cycles. We shall say by definition that an $i$-cycle $\Gamma_i$ is *effective* relatively to the hypersurface $C$ if there is a homologous cycle having no point of intersection with a hypersurface of the same continuous system as $C$.

Let then $|C|$ be a linear system such as we have already considered, and assume that a cycle $\Gamma_i (i \leqq d)$ meets a $C$ along a $\Gamma_{i-2}$.

In the intersection $CC_1$ of $C$ with another hypersurface $C_1$ of the system there exists, (No. 2), a $\Gamma'_{i-2}$ trace of a cycle $\Gamma''_i$ of $C$, $\sim$ mod. $C$ to a multiple of $\Gamma_{i-2}$. Replacing if necessary $\Gamma_i$ by an adequate multiple, we may say that $\Gamma'_{i-2} \sim \Gamma_{i-2}$ mod. $C$. Now make $C_1$ tend towards $C$ in $|C|$, while maintaining $|CC_1|$ and $\Gamma'_{i-2}$ fixed. $\Gamma''_i$ will tend towards a cycle $\Gamma'_i$ of $C_1$, whose trace on $CC_1$ will still be $\Gamma'_{i-2}$. We thus have a cycle $\Gamma'_i$ of $C_1$, whose trace on $CC_1$ bounds with $\Gamma_{i-2}$ a certain $M_{i-1}$ of $C$.

Let now $E_2$ be a two-dimensional elementary sensed manifold or "element" passing through a point $A$ of $C$ and transverse to $C$. Together with a sensed $(2d-2)$-fold element of $C$ through $A$ it determines the sensing of $V_d$, and since $C$ and $V_d$ are both two sided, when $A$ describes any closed path in $C$, $E_2$ may be displaced at the same time without ever acquiring a line segment in common with $C$, so as to return to its former position with sense unchanged. The following analogous occurrence in ordinary space will make this clearer: When the origin $A$ of a directed segment $AB$ describes a closed path on a two-sided surface, $AB$ may be displaced without ever becoming tangent to the surface, so as to return exactly to its original position.

To come back to our question, let $A$ describe $M_{i-1}$ and displace $E_2$ in the manner just indicated so that moreover when $A$ is in $\Gamma_{i-2}$ or $\Gamma'_{i-2}$, $E_2$ will be in $\Gamma_i$ or $\Gamma'_i$ respectively. In these conditions a circuit surrounding $A$ in $E_2$ will generate a manifold $M_i$ meeting nowhere $C$, and if we suppress in $\Gamma_i - \Gamma'_i + M_i$ the part generated by the area of $E_2$ which is bounded by $\zeta$ when $A$ describes $\Gamma_{i-2}$ or $\Gamma'_{i-2}$, we obtain a cycle $\sim (\Gamma_i - \Gamma'_i)$, since we can make it tend towards this cycle as a limit, and this new $i$-cycle is *effective* relatively to $C$. But $\Gamma'_i$, which is in $C_1$, is $\sim$ to a cycle wholly in $C$. Hence, *there exists a combination of any i-cycle* ($i \leqq d$) *and of cycles wholly in* $C$, *effective with respect to* $C$.

Algebraic cycles. These cycles, which occur when $i$ is even, are simply the cycles homologous to a combination of those formed by the algebraic manifolds of $V_d$. The cycles formed by algebraic curves have been considered by Poincaré although he made no application of them.

6. A non-bounding $d$-cycle of $C$ is also one of $V_d$, for when $C$ varies in its system it neither gains nor loses $(d-2)$-cycles and therefore $d$-cycles, so that they are invariant. Hence the number of $d$-cycles non-effective relatively to $C$ is equal to the $d$th index of connectivity of the hypersurface, hence also to its $(d-2)$th index and therefore, according to what we have already shown, to $R_{d-2}$. If we denote by $R'_i$ the number of $i$-cycles effective with respect to $C$, we shall have $R'_d = R_d - R_{d-2}$.

Let now $i \leqq d-1$. Any $i$-cycle of $V_d$ is homologous to a cycle in $C^{d-i}$. Replacing then $V_d$ by $C^{d-i}$, which has the same number of effective $i$-cycles with respect to a $C^{d-i+1}$ as $V_d$ with respect to $C$, we obtain

$$R_i - R'_i = R_{i-2}, \qquad R_2 - R'_2 = 1 \qquad (2 < i \leqq d).$$

The second formula follows from the fact that we have to take into account the non-effective algebraic cycle $C^{d-1}$.

In the next chapter it will be shown that on an algebraic surface the algebraic curves are non-effective cycles with respect to a curve in a properly chosen system. Moreover, from the above it follows that if $C$ is a plane section, $A$ any other curve, $\Gamma$ any two-dimensional cycle, a certain cycle $mA + n\Gamma\,(n \neq 0)$ is effective with respect to $C$. The proof holds as well if $C$ is generic in a certain linear system $|C|$ analogous to the system of hypersurfaces of similar name. This establishes the second formula for $d = 2$. When $d > 2$, a multiple of any $\Gamma_2$ is homologous to a cycle in $C$, which shows that we may replace $V_d$ by $V_{d-1}$; hence this second formula holds for all cases.

We thus see that $R'_i$ does not depend upon the special system $|C|$ considered, provided it has certain general properties specified in No. 2.

7. From the formulas obtained follow these:

$$R_{2i} = \sum_{h=0}^{i} R'_{2h}, \qquad 2i \leqq d$$

$$R_{2i+1} = \sum_{h=0}^{i} R'_{2h+1}, \qquad 2i + 1 \leqq d;$$

$$(R'_1 = R_1, \qquad R_0 = 1).$$

Hence if $R_{2i} = 1$, its minimum value, $R'_{2h} = 0, R_{2h} = 1, h \leqq i$, and if $R_{2i+1} = 0$, $R'_{2h+1} = R_{2h+1} = 0$, $h \leqq i$. In particular if a single $R_{2i+1} = 0$, *$R_1 = 0$ and $V_d$ is regular.*

Finally let us notice explicitly that a *multiple of an $i$-cycle non-effective with respect to $C$ is the sum of an effective cycle and of a cycle in $C^{d-i+1}$.* For $i = 2$, the multiple of any non-effective cycle is the sum of an effective cycle and the algebraic cycle $C^{d-1}$ taken a certain number of times.

Let us mention here a formula derived first by Alexander* and giving $R_d$ in terms of the $(R_i)$'s and of the Zeuthen-Segre invariant $I_d$ of $V_d$. It can be written

$$R_d = I_d + 2\sum_{i=1}^{d-1}(-1)^{d-i+1} R_i + 2(-1)^d (d-1).$$

8. Let $A_1, \cdots, A_s$ be hypersurfaces of $V_d$ and assume $|C|$ so chosen that it contains their sum, leaving a residue $|K|$, linear system of the same type as $|C|$. Let now $C_0 = A_1 + A_2 + \cdots + A_s + K$, and, instead of considering a pencil $\{C_u\}$ generic in $|C|$, let us take one containing $C_0$. Reasoning in regard to this pencil as previously we find that the multiple of any $d$-cycle is the sum of a cycle effective with respect to the hypersurfaces $A_r$, $K$, and of a cycle contained within one or more of these hypersurfaces, and non-effective with respect to their totality. The number of effective cycles will now be $R'_d - r_d$, $r_d \geqq 0$. Thus there will be $r_d$ cycles effective with respect to $K$,

* Rendiconti dei Lincei, ser. 5 vol. 23 (1914), pp. 55–62.

but not with respect to the $(A)$'s. Since $r_d \leqq R'_d$, when any set of hypersurfaces whatever is considered, this integer has a maximum $\rho_d$ which is obviously a numerical invariant of $V_d$. It is the maximum number of cycles contained in the hypersurfaces of $V_d$ and yet effective with respect to an irreducible linear system, $\infty^d$ at least, without base points.

A similar invariant $\rho_i$ may be defined for $i$-cycles. The invariants $\rho_i$ of $V_d$ and of $C$ are the same when $i \leqq d-2$. Hence $\rho_i$ is the same for $V_d$ and $C^k$, $k \leqq d-i-1$. In particular $\rho_2$ is the same for $V_d$ and $C$. We shall see that for any $V_d$, $1+\rho_2$ is equal to Picard's number $\rho$.

9. *Torsion indices.* It may happen that two $i$-cycles $\Gamma_i$, $\Gamma'_i$, without being homologous to each other, are nevertheless such that $\lambda\Gamma_i \sim \lambda\Gamma'_i$, $\lambda > 1$. The cycle $\Delta_i = \Gamma_i - \Gamma'_i$ is called a *zero-divisor.* Poincaré has shown that when this occurs, $V_d$ undergoes a sort of internal torsion and the $(\lambda)$'s are related to his so-called *torsion-coefficients.* He further established that the torsion coefficients for $i$-cycles are equal to those for the $(2d-i-1)$-cycles.

The operations $+\Delta_i$ applied to the $i$-cycles give rise to an abelian group whose order $\sigma_i$ is the product of the torsion coefficients. We shall call it the $i$th *torsion index.* These indices satisfy the relation $\sigma_i = \sigma_{2d-i-1}$, $\sigma_{2d-1} = 1$. In particular, $\sigma_1 = \sigma_{2d-2} = \sigma$ is the invariant of Severi, as we shall see later. That is in an algebraic surface it turns out that the zero divisors for the cycles or algebraic curves lead all to the same numerical invariant. Algebraic surfaces with $\sigma > 1$, of which examples have been given by Severi and Godeaux, have then the property of possessing linear cycles that do not bound, but of which a certain multiple bounds.

Since any $i$-cycle, $i \leqq d-2$, is homologous to a cycle in $C$ and also since such cycles when in $C$ form cycles of $V_d$, $V_d$ and $C$ have the same zero divisors for these cycles and therefore *equal invariants* $\sigma_i$, $i \leqq d-2$.

10. The notions of *base, intermediary base, minimum base* are commonly useful. Let $A_1, A_2, \cdots$ be a set of entities forming a modulus, i.e., having these properties: (*a*) The sum or difference of any two belong to the set, (*b*) the zero of the set is well defined, (*c*) $A+B$ does not differ from $B+A$. The entities $A_1, \cdots, A_n$ are said to form a *base* if for every $A$ there is a relation $\lambda A = \lambda_1 A_1 + \lambda_2 A_2 + \cdots + \lambda_n A_n$, where the $(\lambda)$'s are integers. If $\lambda$ always divides all the $(\lambda_i)$'s the base is called *intermediary,* while if for every $A$ we can take $\lambda = 1$, we have to deal with a *minimum base.* Examples of such systems are given by the $i$-cycles of a closed manifold, the hypersurfaces of a $V_d$, etc.

Repeating for the $i$-cycles a discussion of Severi's* for algebraic curves it can be shown that the $i$-cycles of $V_d$ possess ordinary and minimum bases composed of $R_i$ and $R_i + \sigma_i - 1$ cycles respectively.

---

*Annales de l'Ecole Normale supérieure, ser. 2, vol. 25 (1908), pp. 449–468. The nomenclature is his.

11. If $V_d$ possesses arbitrary singularities but is birationally transformable into $V'_d$, variety with ordinary singularities, the neighborhood of the extraordinary singularities of $V_d$ may be considered as forming a certain number of infinitesimal algebraic varieties. We shall thus be led to consider infinitesimal $i$-cycles contained in these varieties and corresponding to finite cycles on $V_d$.

In general an invariant $m$ may take a different value according as infinitesimal cycles are or are not taken into consideration. Reserving the ordinary notation for the first case we shall denote the invariant in the second by $[m]$. This notation has already been used by Bagnera and de Franchis for the number $\rho$. We shall then have $R_i - [R_i] = \rho_i - [\rho_i]$.

*Remark.* If $V_d$ has ordinary singularities, in considering the $i$-cycles contained in a given hypersurface, the infinitesimal cycles must be left out, that is, no attention must be paid to accidental singularities of the hypersurface.

## § 3. Integrals of the second kind

12. Much interest is added to the discussion by introducing integrals of the second kind. The author has recently published a series of results (most of them without proofs) relating to triple integrals of a $V_3$* and a memoir containing the proofs is ready for publication. In another memoir he has given the extension to double integrals of a $V_3$†, of the Picard theory of double integrals of the second kind of algebraic surfaces. The extension from $d = 2$ to $d = 3$ presents many difficulties which seem to disappear in the extension from $d = 3$ to any other value of $d$. We shall then indicate mostly without proofs the statements for a $V_d$:

Let $F_d(x_1, x_2, \cdots, x_d, t) = 0$ be the equation of $V_d$ still assumed irreducible and with ordinary singularities, $|H|$ the system of hyperplane sections, $H_{x_i}$ the section by $x_i =$ const. The integral

$$(1) \qquad \int\int \cdots \int U(x_1, x_2, \cdots, x_d, t)\, dx_1\, dx_2 \cdots dx_d \quad (U \text{ a rational function}),$$

is of the *second kind* if to every hypersurface of infinity $A$ of $U$ corresponds an integral of type called *improper* of the second kind,

$$(2) \qquad \int\int \cdots \int \sum \frac{\partial U_i}{\partial x_i}\, dx_1\, dx_2 \cdots dx_d \quad (U_i \text{ a rational function}),$$

such that their difference is finite in the neighborhood of an arbitrary point of $A$. If (1) is not of the type (2) it is said to be *proper* of the second kind.

Proper and improper integrals preserve their character under birational

*Comptes Rendus, vol. 165 (1917), pp. 850–854.

†Annali di matematica, ser. 3, vol. 26 (1917), pp. 227–261.

transformations. The number $\rho_0^d$ of proper integrals of which no linear combination is improper is an absolute invariant. If the system of hyperplane sections is adequately selected every integral of the second kind can be reduced by subtraction of an improper integral to the *normal* form

$$(3) \qquad \int\int\cdots\int\frac{P(x_1, x_2, \cdots x_d, t)}{F'_t}\, dx_1\, dx_2 \cdots dx_d,$$

where $P$ is an adjoint polynomial.

Whether of the second kind or not, (3) has *periods* and *residues,* but only relatively to finite cycles or cycles effective relatively to the hypersurface at infinity. Indeed, as Picard has shown, if the cycle is infinite the corresponding value of (3) is indefinite or if definite may vary when the cycle is deformed.

In connection with (3) we may also consider

$$(4) \qquad \int\int\cdots\int\frac{P(x_1, x_2, \cdots, x_d, t)}{F'_t}\, dx_1\, dx_2 \cdots dx_{d-1},$$

attached to $H_{x_d}$. Its periods satisfy a linear homogeneous differential equation—the equation of Picard-Fuchs. By expressing the periods of (3) in terms of those of (4) we may show that when $P$ is properly chosen they assume an arbitrary value.

13. Let then $r'_i$ be the number corresponding to $R'_i$ for $H$. On the strength of what has been stated we may derive this formula, in which $N$ denotes the class of $V_d$, $R'_d = N - 2(r'_{d-1} - R'_{d-1}) - (r'_{d-2} - R'_{d-2})$, as being the exact number of integrals without residues, but with some non-zero periods. These integrals are of the second kind and every proper integral is reducible to them.

By means of this relation we may derive Alexander's formula for $R_d$. Indeed, if $I_s$ is the Zeuthen-Segre invariant of $H^{d-s}$ we have for $V_d$:

$$I_d = N - 2I_{d-1} - I_{d-2}$$

$$\therefore\ R'_d - I_d = 2R'_{d-1} + R'_{d-2} - 2(r'_{d-1} + I_{d-1}) - (r'_{d-2} - I_{d-2})$$

and similarly for $H^{d-s}$ $(s = 3, 4, \cdots, d-1)$,

$$r'_s - I_s = 2R'_{s-1} + R'_{s-2} - 2(r'_{s-1} - I_{s-1}) - (r'_{s-2} - I_{s-2}),$$

$$r'_2 - I_2 = 2R_1 + 1,$$

whence

$$R'_d = I'_d + (-1)^d \cdot (d-1) + \sum_{i=0}^{d-2} (-1)^{d-i} \cdot (d-i) \cdot R'_{i+1},$$

which reduces to Alexander's formula for $R_d$ if we remember that

$$R'_i = R_i - R_{i-2}$$

and set, as we should, $R_0 = 1$, $R_{-1} = 0$.

Let us return to our integral of the second kind under normal form. If its periods are all zero, it is improper of the second kind. We shall show below that there are exactly $\rho_d$ improper integrals with non-zero periods, hence for the number of proper integrals

$$\rho_0^d = R'_d - \rho_d = I_d + 2(-1)^d(d-1) - \rho_d + \sum_{i=0}^{d-2}(-1)^{d-i}(R_{i+1} - R_{i-1})$$

$$= I_d + 2(-1)^d(d-1) - \rho_d - R_{d-2} + 2\sum_{i=1}^{d-1}(-1)^{d-i+1}R_i.$$

For $d = 2$ this becomes a well-known formula given by Picard and for $d = 3$ it becomes the formula of the author's Comptes Rendus note, the $\lambda$ of the note being the same as $\rho_d$.

14. We will now outline the proof that *there exist $\rho_d$ improper integrals in normal form and with non-zero periods.*

Let then $A_1, A_2, \cdots, A_n$ be hypersurfaces such that there are $\rho_d$ finite cycles $\Gamma_1, \Gamma_2, \cdots, \Gamma_{\rho_d}$ non-effective relatively to some of the $(A)$'s. We shall assume (as we may) that the simple hypersurface of intersection of $F'_t = 0$ with $V_d$ is among the $(A)$'s. There are $R'_d - \rho_d$ cycles, $\Gamma'_1, \Gamma'_2, \cdots, \Gamma'_{R'_d - \rho_d}$, effective relatively to $H$, or as a special case finite, which do not meet the $(A)$'s.

It is sufficient to show that from any integral of the second kind in the normal form we may subtract an improper one so that the difference will be without periods relatively to the $(\Gamma)$'s. Indeed, by means of a change of variables it can be shown that (4), extended to a closed continuum, where it is finite and where $F'_t \neq 0$, yields a zero period.* It follows that an improper integral under the normal form has no period relatively to the $(\Gamma')$'s and therefore the $R'_d - \rho_d$ integrals that may be formed having no periods with respect to the $(\Gamma)$'s, but having some with respect to the $(\Gamma')$'s, are certainly proper. As by assumption all proper integrals are reducible to them, we must have $\rho_0^d = R'_d - \rho_d$. Therefore there must be $\rho_d = R'_d - \rho_0^d$ improper integrals in the normal form with non-zero periods.

Let $\Gamma_{ik}$ be the $(d-2)$-cycle intersection of $\Gamma_i$ and $A_k$. A multiple of $\Gamma_{ik}$ is homologous mod. $A_k$ to a $\overline{\Gamma}_{ik}$ situated in $H_{x_1}A_k$, wherein it is invariant when $x_1$ varies. Replacing if need be $\Gamma_i$ by a certain multiple of itself we may assume that actually $\Gamma_{ik} \sim \overline{\Gamma}_{ik}$ mod. $A_k$. The possible exceptional singularities of $A_k$ need not be taken into account for the theorems of No. 2 are readily seen to hold even then since $\Gamma_{ik}$ is of dimensionality $d-2$ and $\Gamma_i$ may be assumed not to go through the exceptional singularities.

By reasoning as in No. 5, we may reduce $\Gamma_i$ to a cycle cutting $A_k$ in $\overline{\Gamma}_{ik}$ and nowhere else. Henceforth assume that $\Gamma_i$ is that very cycle, $\Gamma_{ik}$ thus coinciding with $\overline{\Gamma}_{ik}$. We can then arrange matters so that the vicinity of $\Gamma_{ik}$

* The proof for $d = 2$ is given in the author's Annali di matematica paper, no. 8.

in $\Gamma_i$ be entirely in $H_{x_1}$. It constitutes a small region bound by a $(d-1)$-cycle $\Delta_{ik}$. Suppress these regions from $\Gamma_i$ and let $M_i$ be what is left.

We observe that if $U_1$ is a rational function infinite on the $(A)$'s only

$$\int\int\cdots\int_{M_i}\frac{\partial U_1}{\partial x_1}dx_1\,dx_2\,\cdots\,dx_d=\sum\int\int\cdots\int_{\Delta_{ik}}U_1\,dx_2\,dx_3\,\cdots\,dx_d$$

$$\int\int\cdots\int_{M_i}\frac{\partial U}{\partial x_{h+1}}\,dx_1\,dx_2\,\cdots\,dx_d=0\,.$$

Thus the first $d$-uple integral is a sum of residues of

(5) $$\int\int\cdots\int U_1\,dx_2\,dx_3\,\cdots\,dx_d\,,$$

attached to $H_{x_1}$, relatively to the cycles $\Gamma_{ik}$.

It may be shown that the residues at finite distance of (5)* are as arbitrary as if $U_1$ were no more constrained to be rational in $x_1$. This follows in principle from the fact that an integral of a rational function belonging to a $V_d$ and with arbitrary residues can be formed by operations rational with respect to the coefficients of its equation.

A corollary is the existence of an integral (5) with constant residues with respect to the cycles such as $\Gamma_{ik}$. In that case

(6) $$\frac{\partial}{\partial x_1}\int\int\cdots\int U_1\,dx_2\,dx_3\,\cdots\,dx_d$$

behaves like an integral of the second kind at finite distance. From this can be deduced the existence of rational functions $U_2, U_3, \cdots, U_d$, such that

$$\sum_{i=1}^{d}\frac{\partial U_i}{\partial x_i}=\frac{P(x_1, x_2, \cdots, x_d, t)}{\phi(x_1)F'_t}$$

where $P$ is an adjoint polynomial and $\phi$ an ordinary polynomial. From the relations established above, and since the integral at the left in (7), taken over $\Gamma_i - M_i$, is very small, follows then

(7) $$\int\int\cdots\int_{\Gamma_i}\frac{P\cdot dx_1\,dx_2\,\cdots\,dx_d}{\phi\cdot F'_t}=\sum\int\int\cdots\int_{\Delta_{ik}}U_1\,dx_2\,dx_3\,\cdots\,dx_d\,.$$

Hence the period of (7) as to $\Gamma_i$, like the residues of (5) as to $\Gamma_{ik}$, is completely arbitrary. Thus we may obtain an improper integral of the second kind (7) with arbitrary periods relatively to the $(\Gamma)$'s. As (7) may be shown to be reducible to the normal form without gain or loss of periods, we see finally that its periods relatively to the $(\Gamma')$'s are zero since these last cycles are effective relatively to the $(A)$'s. As the periods of (7) relatively

*Quarterly journal of mathematics, vol. 49 (1917), pp. 333–343.

to the $(\Gamma)$'s are perfectly arbitrary we may obtain an improper integral in the normal form which when subtracted from a given integral of the second kind in that form will yield one without periods with respect to the $(\Gamma)$'s, and this proves our theorem.

All that precedes may be extended to integrals of "total differentials," that is, to integrals of the type $\int\int\cdots\int\sum P_{i_1 i_2 \ldots i_s}\, dx_{i_1}\, dx_{i_2} \cdots dx_{i_s}$, where the $(P)$'s are rational and satisfy Poincaré's conditions of integrability. For example, when $s = 2$, these conditions are

$$\frac{\partial P_{ik}}{\partial x_l} + \frac{\partial P_{kl}}{\partial x_i} + \frac{\partial P_{li}}{\partial x_k} = 0.$$

The integral will be proper or improper of the second kind according as it determines an integral of one or the other type on an arbitrary $H$. This makes it possible to define these integrals by recurrence. The number of proper integrals of the second kind is $\rho_0^s = R_s' - \rho_s = R_s - R_{s-2} - \rho_s$.

In particular for $s = 2$ we have $\rho_0^2 = R_2' - \rho_2 = R_2' - (\rho - 1) = R_2 - \rho$. This special case $s = 2$ has already been considered* by the author of the present paper.

## § 4. Periods of integrals of the first kind

15. Integrals of the first kind are defined as usual as those which are everywhere finite. While restrictions were necessary before as to cycles relatively to which integrals of the second kind could properly be assumed to have periods, they cease to be necessary when we deal with integrals of the first kind. However, the question may be asked, *how many of the $R_s$ periods which an s-uple integral of the first kind may have, are linearly independent in the field of rational numbers,* in the sense that they do not have to satisfy a linear homogeneous equation with integral coefficients. Let us assume at first $s = d$ and consider the integral of the first kind

$$\int\int\cdots\int \frac{P(x_1, x_2, \cdots, x_d, t)}{F_t'}\, dx_1\, dx_2 \cdots dx_d. \tag{8}$$

What can be said as to its period relatively to a $d$-cycle $\Gamma_d$ non-effective relatively to $H$? By combining $\Gamma_d$ with effective cycles we may derive a cycle contained in $H_{x_1}$. Grant that this has already been done for $\Gamma_d$. It will intersect the $H^2$ at infinity of $H_{x_1}$ in a $(d-2)$-cycle $\Gamma_{d-2}$. We may isolate on $\Gamma_d$ the vicinity of $\Gamma_{d-2}$, the boundary of the isolated part $\Gamma_d'$ being a manifold $M_{d-1}$ — a tube for $d = 3$. The integral (8) extended to $\Gamma_d - \Gamma_d'$ gives zero since $dx_1 = 0$ at all points of this continuum and it is at finite distance. Let us make a projective transformation reducing the $H^2$ at in-

*Annali di matematica, loc. cit. In the formula there given $R_2$ should be replaced by $R_2'$, as only finite cycles are considered.

finity of $H_{x_1}$ to $H_{x_1} H_{x_2}$. Keeping notations unchanged we see that the period of (8) relatively to $\Gamma_d$ is equal to the value of (8) extended over $\Gamma'_d$, a portion of the cycle near $\Gamma_{d-2}$, and this will certainly give zero if

$$(9) \qquad \int\int \cdots \int \frac{P(x_1, x_2, \cdots, x_d, t)}{F'_t} dx_3\, dx_4 \cdots dx_d$$

has no period relatively to $\Gamma_{d-2}$. Now if $\Gamma_{d-2}$ is finite, that is, if it is effective relatively to $H$, the period in question is certainly zero. For, maintaining $x_1$ fixed, we see that the periods of (9) vanish when $x_2 = \infty$ because $P$ is an adjoint canonical polynomial (that is of degree $m - d - 2$, where $m$ is the degree of $V_d$). But as $x_2$ varies the period relatively to $\Gamma_{d-2}$, cycle of $V_d$ in $H_{x_1} H_{x_2}$, is rational in $x_2$, finite everywhere since the cycle is finite, hence constant and zero at infinity, and therefore identically zero.

The same result can be obtained differently. The doubtful part of $\Gamma_d$ is composed of $\infty^2$ $(d - 2)$-cycles differing very little from $\Gamma_{d-2}$, and the value of (8) extended over $\Gamma'_d$ is equal to the constant period of (9) relatively to $\Gamma_{d-2}$ integrated over a finite two-dimensional continuum of the complex $(x_1, x_2)$ space, and therefore again zero.

We have thus obtained $R'_{d-2}$ $d$-cycles with respect to which (8) has no periods. The same reasoning may be pursued by replacing (8), $V_d$ and $\Gamma_d$ by (9), $H^2$ and $\Gamma_{d-2}$, and so on. *We shall have finally*

$$R'_{d-2} + R'_{d-6} + \cdots = R_{d-2} - R_{d-4} + R_{d-6} \cdots$$

*non-effective d-cycles relatively to which integrals of the first kind have no periods.* Similarly there are $R_{s-2} - R_{s-4} + \cdots$ non-effective $s$-cycles with respect to which $s$-uple integrals of the first kind have no periods. In these expressions the last term is either $\pm R_1$ or $\pm R_0 = \pm 1$.

16. It is easy to show that $s$-tuple integrals of the first kind have no periods relatively to the $\rho_s$ cycles of No. 8. To show this, transform $V_d$ birationally into a variety $V'_d$ with ordinary singularities in such a way that the transformed $A'_k$ of $A_k$ be part of the hyperplane section $x_1 = 0$. The finite cycles contained in the $(A'_k)$'s will now yield zero for period since at every point of them $dx_1 = 0$. The same is true for finite cycles of $V_d$ that have become infinitesimal on $V'_d$,—a circumstance that may well happen—, for these infinitesimal cycles may be assumed at finite distance also. Thus *there will be in all $\rho_s + R_{s-2} - R_{s-4} + R_{s-6} \cdots$ distinct s-cycles relatively to which integrals of the first kind have no periods.*

Are there any others with the same property? Probably not, though we have succeeded in proving it only for algebraic surfaces and for abelian varieties with a period matrix as general as possible.

## CHAPTER 2. INVARIANTS $\rho$, $\sigma$ OF ALGEBRAIC VARIETIES

### § 1. Poincaré's normal functions for algebraic surfaces

17. Let $F(x, y, z) = 0$ represent an algebraic surface $V_2$, of order $m$, irregularity $\frac{1}{2}R_1 = q$, and denote by $|H|$ and $p$ the system of plane sections and the genus of a generic $H$. As is well known, a set of $p$ independent integrals of the first kind of $H_y$, arbitrary curve of the pencil $\{H_y\}$, can be obtained thus: For the first $p - q$ integrals we take

$$u_h = \int \frac{P_h(x, y, z)\,dx}{F'_z} \qquad (h = 1, 2, \cdots, p - q),$$

where $P_h$ is an adjoint polynomial of order $m - 3$, and for the remaining we take the $q$ integrals of the first kind $u_{p-q+1}, u_{p-q+2}, \cdots, u_p$ of $V_2$. Let $A_1, A_2, \cdots, A_m$ be the base points of $\{H_y\}$, $C$ any algebraic curve of $V_2$, $M_1, M_2, \cdots, M_n$ the variable points of the group $CH_y$ or these points associated with some of the $(A)$'s. Consider the abelian sums

$$v_h(y) = \sum_{j=1}^{n} \int_{A_1}^{M_j} du_h \qquad (h = 1, 2, \cdots, p),$$

first introduced by Poincaré,* who gave for them expressions which the present choice of integrals of the first kind reduces to

$$(1) \qquad \begin{aligned} v_h(y) &= \sum_{k=1}^{N} \frac{\lambda_k}{2\pi i} \int_b^{b_k} \frac{\Omega_{hk}(Y)}{Y - y}\,dY \qquad (h = 1, 2, \cdots, p - q), \\ v_{p-q+s} &= \beta_s \qquad (s = 1, 2, \cdots, q). \end{aligned}$$

In these expressions the $(\lambda)$'s are integers, called *characteristic integers*, $b$ arbitrary, $b_1, b_2, \cdots, b_N$ the critical values of $y$ for $\{H_y\}$, the $(\beta)$'s constants, and finally $\Omega_{hk}$ the period of $u_h$ relatively to the cycle of $H_y$ that vanishes for $y = b_k$. The corresponding periods of $u_{p-q+s}$ are obviously zero. The paths of integration form a system of non-intersecting cuts in the $y$ plane.

In the expressions as given by Poincaré there appear certain rational functions at the right. These expressions can only be polynomials, for their infinities are the same as those of the $(u)$'s. Moreover, by means of the birational space transformation $x = x'/y'$, $y = 1/y'$, $z = z'/y'$, we may easily verify that $v_h(y)$, $h \leqq p - q$, vanishes at infinity, while the other $(v)$'s are finite there, hence the form of the expressions (1) follows. Besides it can be shown directly that $v_{p-q+s}$ is constant for it is uniform in $y$ and finite everywhere (Severi).

*Annales de l'Ecole Normale supérieure, ser. 2, vol. 27 (1910), pp. 55–108. Sitzungsberichte der Berliner mathematischen Gesellschaft, vol. 10 (1911), pp. 28–55.

18. We obtain a first set of relations to be satisfied by the quantities entering in the $(v)$'s when we express the fact that the $(v)$'s are independent of $b$ or that $\partial v_k/\partial b = 0$.

If in the relation obtained we replace $b$, which is arbitrary, by $y$, it becomes

$$\sum_{k=1}^{N} \lambda_k \Omega_{hk}(y) = 0 \qquad (h = 1, 2, \cdots, p-q). \tag{2}$$

This condition once satisfied, the $(v)$'s as given by (1) are regular everywhere, infinity included, except at the critical points $b_i$, and this is as it should be.

Another condition is obtained thus: Let there be a value $y = \alpha$ for which

$$\sum_{\lambda=1}^{p-q} \frac{\gamma_h P_h(x, y, z)}{(y-\alpha)^s}$$

remains finite, the $(\gamma)$'s being constants. Then $\sum \gamma_h v_h(y) \cdot (y-\alpha)^{-s}$ must also remain finite. We can always select our integrals so that $s = 1$ for all the $(\alpha)$'s (Poincaré), but this is immaterial.

A set of functions satisfying these two conditions and behaving like the $(v)$'s at the critical points forms what Poincaré has called a set of *normal functions*. He has proved the following fundamental theorem: *To a set of normal functions always corresponds an algebraic curve for which they are Abelian sums.* Exceptionally, the curve is reduced to one or more of the points $A_i$.

Given two algebraic curves $B$, $C$, of $V_2$, we will write with Severi $B = C$, if there exists an algebraic curve $E$ such that $B + E$, $C + E$ are both total curves of the same continuous system.* When $B$ varies in its continuous system the characteristic integers corresponding to the total group $BH_y$ are fixed. They are only determined however up to a system of integers corresponding to a period, as the periods obviously form sets of normal functions. The set of characteristic integers of $B + E$ is obtained by adding the corresponding integers of $B$ and $E$. Hence if $B = C$, the differences between their characteristic integers form a set corresponding to a period, it being understood that characteristic integers are taken for the total groups $BH_y$, $CH_y$.

## § 2. **Algebraic cycles of a surface. Fundamental theorem**

19. We propose first to interpret the conditions that the $(v)$'s must satisfy from a somewhat different viewpoint.

We begin with the second condition. When $\sum \gamma_h P_h(x, y, z) = (y - \alpha) \cdot Q(x, y, z)$, $Q$ is an adjoint polynomial of order $m - 4$ and the double integral,

$$\iint \frac{Q(x, y, z)}{F'_z}\, dxdy \tag{3}$$

* These systems may be reducible but must be connected. A thorough treatment of this point has been given by Albanese, Annali di matematica, ser. 3, vol. 24 (1915), pp. 159-233. Our "equivalence" is the same as his "virtual equivalence".

is of the first kind. Let $\omega_k(y)$ be the period of

$$(4) \qquad \int \frac{Q(x, y, z)}{F_2'} dx$$

corresponding to $\Omega_{hk}$ for $u_h$. We must have

$$(5) \qquad \sum_{h,k} \gamma_h \lambda_k \int_b^{b_k} \frac{\Omega_{hk}(Y)\,dY}{Y-\alpha} = \sum_k \lambda_k \int_b^{b_k} \omega_k(y)\,dy = 0.$$

Conversely if (3) is of the first kind,

$$\int \frac{(y-\alpha)\,Q(x, y, z)}{F_2'} dx$$

is a linear combination of $u_1, u_2, \cdots, u_{p-q}$, identically zero for $y = \alpha$, and the relation (5) corresponding to it must be satisfied. Hence (5) is true whatever the integral of the first kind (3) considered.

20. Let $\Gamma$ be any two-cycle of $V_2$. We may assume without loss of generality that any $H_y$ meets it in a finite number of points since this can always be obtained by slightly deforming the cycle if necessary. Since $\Gamma$ is two sided we may sense it by assigning a positive direction of rotation for its small circuits. Let now $y$ describe a small positive circuit in its plane. Of the points of intersection of $\Gamma$ and $H_y$, a certain number, say $n'$, will describe positive circuits on $\Gamma$, while the $n''$ remaining points will do just the opposite. The number $n' - n''$ does not depend upon the position of the circuit in the $y$ plane, for when a point of one type disappears one of the other disappears at the same time. The number $n' - n''$ is therefore a definite simultaneous character of the two manifolds $\Gamma$, $H_y$. Poincaré introduced it for manifolds $M_s$, $M_{k-s}$ in an $M_k$ (the indices indicate the dimensionality), and he denoted it by $N(M_s\, M_{k-s})$. The $N$ really is superfluous, and we shall simply write $(M_s\, M_{k-s})$, or $(\Gamma\, H_y) = n' - n''$. All we need to know concerning it here is that if $\Gamma \sim 0$, also $(\Gamma\, H_y) = 0$.

We observe at once that if $\Gamma$ is an algebraic cycle $(C)$, of the two numbers $n'$, $n''$, one is necessarily zero, the other being equal to $\pm [CH_y]$, or in absolute value equal to the order of $C$. This follows from the well-known property of a multiply sheeted Riemann surface, that when the independent variable describes a small circuit in its plane, the corresponding points in the various sheets describe circuits sensed alike relatively to the surface. We shall make once for all the convention that $(C)$ is to be so sensed that $(C\, H_y) = [CH_y]$.

Returning to the cycle $\Gamma$, let us join $A_1$ to all the points of intersection of $\Gamma$ and $H_y$ by lines in $H_y$. Their locus as $y$ describes its plane without crossing the cuts $bb_k$ is a manifold $M_3$ whose boundary is composed of $\Gamma$, of part or all of $H_y$, and of a manifold $M_2$ that may be described thus: When $y$ goes from

$y_0$ on one side of $bb_k$ to the point $y'_0$ right opposite on the other, by turning around $b_k$ without crossing the cut, the aggregate of lines whose locus is $M_3$ returns to a new position in $H_{y_0} = H_{y'_0}$, which differs from the old position by a multiple of the linear cycle $\delta_k$ of $H_y$ that reduces to a point when $y = b_k$. The locus of $\delta_k$ when $y$ describes $bb_k$ is a two-dimensional manifold $\Delta_k$, somewhat similar to a cone with its vertex at the point of contact of the plane $y = b_k$, and its base in the position of $\delta_k$ in $H_b$, and we have

$$M_2 = -\sum \lambda_k \Delta_k .$$

$$\therefore \quad (6) \qquad \Gamma \sim \sum \lambda_k \Delta_k + \text{part of } (H_b) .$$

A question may be raised as to this discussion: The difference $n' - n''$ is constant, but $n'$, $n''$, may vary. Let $\zeta$ be a line separating regions in the $y$ plane with different values of these integers, so that along $\zeta$ some of the points $B'_i$ of one type, coincide with as many, $B''_i$, of the other. Is the $M_2$ locus of the lines $A_1B'_i$, $A_1B''_i$, when $y$ describes $\zeta$, part of the boundary of $M_3$? We may deform $\zeta$, without making it cross the critical points, so as to reduce it to a sum of the loops $bb_k$. This amounts to deforming the given cycle into one for which the $M_2$ here considered is reduced to a sum of the manifolds $\Delta_k$, which still leads to a relation such as (6).

To return to our problem, in the particular case where $\Gamma = (C)$, the mode of derivation of the $(v)$'s shows at once that the coefficients $\lambda_k$ in (6) are the characteristic integers of the curve $C$. Now from the theorem of Poincaré-Cauchy for functions of several variables it follows at once that the periods of an integral of the first kind relatively to homologous cycles are equal. But (3) extended over part of $(H_b)$ gives zero, hence

$$\int\int_{(C)} \frac{Q(x, y, z)}{F'_z} dxdy = \sum \lambda_k \int\int_{\Delta_k} \frac{Q(x, y, z)}{F'_z} dxdy = \sum \lambda_k \int_b^{b_k} \omega_k(y)\, dy .$$

The relation (5) means then that the period of (3) relatively to the algebraic cycle $(C)$ is zero, in accordance with Ch. I. Thus the conditions satisfied by the $(\lambda)$'s express merely that *integrals of the first kind have no periods relatively to the algebraic cycles.*

21. The converse of the property just obtained constitutes the

FUNDAMENTAL THEOREM: *A two-cycle* $\Gamma$ *without periods of integrals of the first kind is algebraic.*

For if $\Gamma$ satisfies (6), the corresponding period of (3) is

$$\sum \lambda_k \int_b^{b_k} \omega_k(y)\, dy ,$$

and since it vanishes there is an algebraic curve $C$ such that to the variable part of $CH_y$ or to this part associated with some of the points $A_i$ correspond the characteristic integers $\lambda_k$. Let $\mu_k^i$ be the $k$th characteristic integer of $A_i$. There is a certain two-cycle $\Gamma^i = \sum \mu_k^i \Delta_k +$ part of $(H_b) \sim 0$ forming the boundary of the locus of $A_1 A_i$ as $y$ describes its plane under the same conditions as above, and the $(\mu^i)$'s are the characteristic integers of the normal functions corresponding to $A_i$. We may find integers $t_i$ such that $\Gamma + \sum t_i \Gamma^i \sim \Gamma \sim \sum_k (\lambda_k + \sum_i t_i \mu_k^i) \Delta_k +$ part of $(H_b)$, the coefficients of the $(\Delta)$'s at the right being now the characteristic integers of the total group $CH_y$.

We may assume $C$ to be variable in a continuous system with no fixed base points at the $(A)$'s for, if this is not so, we may replace $\Gamma$ and $C$ by $\Gamma + t(H)$ and $C + tH$, and with $t$ sufficiently great the assumption will certainly be satisfied, especially since the $(A)$'s are after all arbitrary points of the surface. Let $C'$ be a curve of the same system as $C$ and not going through the $(A)$'s. $C$ and $C'$ have the same characteristic integers, hence

$$(C') \sim \sum_k (\lambda_k + \sum_i t_i \mu_k^i) \Delta_k + \text{part of } (H_b).$$

$(C') - \Gamma$ is then $\sim$ to a part of $(H_b)$, and if, as it is proper to assume, $H_b$ is irreducible, $(C') - \Gamma \sim (tH)$, which proves our theorem.

22. Given two algebraic curves $C, D$ of $V_2$, if $C = D$, then also $(C) \sim (D)$. For if $E$ is such that $C + E$ and $D + E$ are total curves of the same continuous system the cycles $(C + E)$ and $(D + E)$, reducible to each other by continuous deformation, are homologous, whence at once $(C) \sim (D)$.

I say that *conversely if* $(C) \sim (D)$, *then also* $C = D$. For we have then

$$((C - D)\, H_y) = (C\, H_y) - (D\, H_y) = [CH_y] - [DH_y] = 0,$$

and therefore $C, D$ *have the same order.*

The axes being arbitrary, we may assume that the curves do not go through the $(A)$'s. Then since $(C) \sim (D)$ their characteristic integers are equal, and so are their normal functions $v_h$, $h \leqq p - q$. We can add to $C$ any curve of the same continuous system as $tH$, and take $t$ so great that the complete continuous systems determined by $C + tH$ and $D + tH$ contain $\infty$ linear systems. It may as well then be assumed that $C$, $D$ already satisfy this condition. We may therefore choose a curve $C' = C$ not passing through the $(A)$'s and for which the $(v_{p-q+k})$'s are the same as for $D$. As $C'$ and $D$ have already the same normal functions of index $\leqq p - q$, these functions will all be the same for both. Since the orders are equal, $C'$ and $D$ are total curves of the same linear system (Poincaré), and therefore $C' = D = C$, as was to be proved.

More generally the two relations $\sum t_i (C_i) \sim 0$, $\sum t_i C_i = 0$ are equivalent. For by writing $t_i = t_i' - t_i''$, where $t_i'$ and $t_i''$ are non-negative integers, they become $\sum t_i' (C_i) \sim \sum t_i'' (C_i)$, $\sum t_i' C_i = \sum t_i'' C_i$, and these relations owing to the positive sign of all coefficients are obviously equivalent.

*Remark.* Since $(C\ H) \neq 0$, $(C)$ is non-effective relatively to $H$ if $C$ is not a virtual curve. More generally, $(C)$ is non-effective relatively to a curve belonging to a sufficiently general linear system.

23. A zero divisor $\Gamma$ for two cycles is algebraic (No. 21). Since it can be considered as the difference between two algebraic cycles $(C) + \Gamma$ and $(C)$, the corresponding algebraic curve can be considered as the difference between two curves $D$ and $C$. If $\lambda\Gamma \sim 0$ for some integer $\lambda > 1$, but not for $\lambda = 1$, then also $\lambda(D - C) = 0$, this being untrue for $\lambda = 1$, so that $D - C$ is a zero divisor for the curves. Conversely if $D - C$ is a zero divisor for the curves, $(D) - (C)$ is a zero divisor for the two-cycles. If then we introduce with Severi *virtual curves,* there exists a holoedric isomorphism between the operations $+\ C$ applied to the curves and $+\ (C)$ applied to the cycles. We recall that, as shown by Picard and Severi, there is a maximum $\rho$ of the number of algebraically independent curves of $V_2$, and that Severi has denoted by $\sigma$ the order of the Abelian group formed by the zero divisors for algebraic curves, so that there are exactly $\sigma - 1$ such divisors. According to what precedes, *the number $\rho$ is therefore equal to the number of distinct two-cycles without periods of integrals of the first kind, and $\sigma$ is equal to the torsion indices $\sigma_1$, $\sigma_2$ (equal according to Poincaré).*

24. The case when the geometrical genus $p_g = 0$ leads at once to an interesting result. For then any cycle is algebraic, hence $R_2 = \rho$,

$$\rho_0 = \rho_0^2 = R_2 - \rho = 0.$$

Thus if $p_g = 0$, $\rho_0 = 0$, that is, if an *algebraic surface is without double integrals of the first kind it is also without double integrals of the second kind.* This has already been proved by Bagnera and de Franchis for irregular surfaces and for some regular surfaces,* and there is a remark by Poincaré at the end of his mémoire in the *Annales de l'Ecole Normale* which easily leads to the same result.

25. Assume $p_g > 0$ and consider $p_g$ independent integrals of the first kind

$$\iint \frac{Q_h(x, y, z)}{F'_z}\, dxdy \qquad (h = 1, 2, \cdots, p_g).$$

We must have

$$\sum_{k=1}^{N} \lambda_k \int_b^{b_k} \omega_{hk}(y)\, dy = 0 \qquad (h = 1, 2, \cdots, p_g),$$

where $\omega_{hk}$ corresponds to $\omega_k$ for (3). Are these relations distinct? If they were not there would have to exist constants $c_h$ not all zero such that

$$\sum_{h=1}^{p_g} c_h \int_b^{b_k} \omega_{hk}(y)\, dy = 0 \qquad (k = 1, 2, \cdots, N),$$

** Rendiconti del Circolo Matematico di Palermo*, vol. 30 (1910), pp. 185–238.

and therefore an integral of the first kind

$$\sum c_h \int\int \frac{Q_h(x, y, z)}{F'_z} dxdy$$

without periods. *If we assume that this is impossible* the $p_g$ relations obtained are all distinct.

26. We are thus led to inquire: *Can a double integral of the first kind be without periods?* It is very likely that this question must be answered in the negative though we have succeeded in proving it only in some very special cases. For the present we merely wish to establish two relations which must exist in case the answer were affirmative.

If (3) is without periods we must have (Picard),

$$\frac{Q(x, y, z)}{F'_z} = \frac{\partial}{\partial x}\left(\frac{A(x, y, z)}{\phi(u)F'_z}\right) + \frac{\partial}{\partial y}\left(\frac{B(x, y, z)}{\phi(u)F'_z}\right),$$

where $A$, $B$ are polynomials adjoint to $V_2$ and $\phi$ is a polynomial in

$$u = ax + by + cz,$$

an arbitrary linear form in $x, y, z$. In fact this result has been established by Picard solely for $u = y$, but the extension is at once obtained by a transformation of coördinates.

Let $U$, $V$ denote the functions in parenthesis at the right and let us introduce with Picard the integral of total differentials

$$\int(-U + \epsilon)dy + Vdx; \qquad \epsilon = \int^{x,z} \frac{Q(x, y, z)}{F'_z} dx.$$

It has most of the properties of those with algebraic integrands since $\epsilon$ is holomorphic everywhere except in the vicinity of the planes $y = b_k$, and there only fails to be so for certain determinations. We may reproduce almost verbatim a discussion of Severi's relating to integrals of total differentials of the second kind.* In particular we may subtract an integral $\int - Rdy + Sdx$ of total differentials with rational integrands $R$, $S$, so as to suppress the curves of infinity other than $y = b_i$ and also the periods with respect to the linear cycles of $V_2$. We will then have

$$\frac{Q}{F'_z} = \frac{\partial}{\partial x}(U - R) + \frac{\partial}{\partial y}(V - S) = \frac{\partial}{\partial x}\frac{A(x, y, z)}{F'_z} + \frac{\partial}{\partial y}\frac{B(x, y, z)}{F'_z},$$

where $A$, $B$ are adjoint polynomials, and now the Abelian integral

$$\int\left(\frac{1}{F'_z}\right)(Ady - Bdx),$$

* Mathematische Annalen, vol. 61 (1905), pp. 20–49.

when attached to any algebraic curve $C$ of $V_2$, becomes one of the first kind for that curve. Its periods with respect to an invariant cycle in a plane section are zero. It follows that it possesses all the properties of an integral of total differentials which do not require the use of the conditions of integrability in their proof. In particular we have $A = xE + A_1$, $B = yE + B_1$, where $A_1, B_1, E$ are polynomials of order $m - 3$. However, since the invariant periods are now zero we may take $E$, and therefore $A_1$ and $B_1$, to be adjoint polynomials, which is certainly not the case for an integral of total differentials. Finally it may be shown that there exists an adjoint polynomial $C(x, y, z)$ such that

$$(7) \qquad AF'_x + BF'_y + CF'_z \equiv DF, \qquad A'_x + B'_y + C'_z + D \equiv Q.$$

These are the relations which we wished to derive. We observe that

$$\iint \frac{Q(x, y, z)}{F'_z} dxdy = \iint \frac{Q(x, y, z)}{F'_y} dzdx = \iint \left[\frac{\partial}{\partial x}\frac{A}{F'_y} + \frac{\partial}{\partial y}\frac{B}{F'_y}\right] dzdx,$$

and hence $C = zE + C_1$ where $C_1$ is adjoint of order $m - 3$.

By introducing homogeneous variables we may replace the necessary conditions for the existence of a double integral of the first kind without periods (7) by

$$(8) \qquad AF'_x + BF'_y + CF'_z + DF'_t \equiv 0, \qquad A'_x + B'_y + C'_z + D'_t \equiv Q,$$

where $A, B, C, D$ are polynomials of order $m - 3$. We verify at once by means of a projective transformation that $\alpha A + \beta B + \gamma C + \epsilon D = 0$ cuts out on the plane $\alpha x + \beta y + \gamma z + \delta t = 0$ an adjoint of order $m - 3$ to the section of $V_2$ by the plane. Let us assume then that $(0, 0, 0, 1)$ is a point on the double curve. The surfaces $A = B = C = 0$ go through this point, hence it is a double point for $AF'_x + BF'_y + CF'_z = 0$, and therefore for $DF'_t = 0$, which shows that $D = 0$ also goes through the point. Hence $A, B, C, D$ are adjoint polynomials of order $m - 3$.

27. In the whole discussion of this chapter it is possible to replace $|H|$ by any simple linear system $|E|$, $\infty^2$ at least, irreducible and such that when an $E$ acquires a new singularity this consists in general in an ordinary double point. We simply take a generic pencil $\{E_u\}$ and replace everywhere $y$ by $u$, the modifications being insignificant.

By considering an adequate $|E|$ on a $V_2$ with *arbitrary* singularities our results can be extended to it. Since $\rho_0 = R_2 - \rho$ is an absolute invariant the cycles which are gained or lost by a birational transformation are all algebraic as we already had occasion to state.

If $|E|$ is not general enough much, if not all, of the discussion is still valid. However, one may be led to neglect certain infinitesimal cycles and thus be led to the numbers $[\rho]$, $[\sigma]$ instead of $\rho$, $\sigma$.

In fact all this holds whenever we deal with a rational or irrational pencil such that a certain point $A_1$ may be uniquely determined on the generic curve of it. We can thus derive Severi's results as to the base for a surface which represents the pairs of points of two algebraic curves. The Abelian sums such as in § 1 lead at once to Hurwitz's equations for correspondences between the two curves.

28. We have studied more especially the relations between the numbers $\rho$, $\sigma$ and certain cycles. The one to one correspondence between algebraic cycles and curves shows that to a minimum base for the cycles corresponds a minimum base for the curves and conversely. But the curves of a minimum base may well be *virtual*. It may be shown that from a base containing such curves we may always derive one composed exclusively of effective curves. There will be an effective minimum base composed of $\rho + \sigma - 1$ curves if $\rho > 1$, and of $\sigma$ curves for $\rho = 1$, and in both cases an ordinary effective base composed of $\rho$ curves (Severi).

## § 3. **Extension to any $V_d$**

29. For the sake of simplicity let us take $d = 3$. Let then $F(x, y, z, t) = 0$ be the equation of an irreducible $V_3$, with ordinary singularities, of order $m$, irregularity $\frac{1}{2}R_1 = q$ and genus of the plane sections equal to $p$. We designate again by $|H|$ the system of hyperplane sections, and will call $A_1, A_2, \cdots, A_m$ the fixed points of the curves $H_y H_z$. As before we shall have $p$ integrals of the first kind attached to these curves of which $p - q$ are

$$u_h = \int \frac{P_h(x, y, z, t)}{F'_t} dx \qquad (h = 1, 2, \cdots, p - q),$$

where $P_h$ is an adjoint polynomial of order $m - 3$ in $x$, $y$, $t$, and the $q$ remaining integrals $u_{p-q+1}, u_{p-q+2}, \cdots, u_p$ are those of total differentials of the first kind of $V_3$ (Castelnuovo-Enriques).

Let $C$ be an algebraic surface of $V_3$, $M_1$, $M_2$, $\cdots$, $M_n$ the variable points of the group $CH_y H_z$ or these points associated with some of the $(A)$'s. We have

$$v_h(y, z) = \sum_{s=1}^{n} \int_A^{M_s} du_h = \sum_{k=1}^{N} \frac{\lambda_k}{2\pi i} \int_b^{b_k} \frac{\Omega_{hk}(Y, z)\, dY}{Y - y},$$

$$v_{p-q+j} = \beta_j \qquad (h = 1, 2, \cdots, p - q;\ j = 1, 2, \cdots, q).$$

In these expressions the $(b_k)$'s are the critical values of $y$ for the pencil $\{H_y H_z\}$ of $H_z$ and the $(\beta)$'s are constants. Let us consider an arbitrary double integral of the first kind of $H_z$ of the type $\int\int (1/F'_t) Q(x, y, z, t)\, dx dy$, where $Q$ is an adjoint polynomial of order $m - 4$ in $x$, $y$, $t$. It may be easily shown that any integral of the first kind of $H_z$ is linearly dependent upon those of this type. We must then have

$$\sum \lambda_k \int_b^{b_k} \omega_k(y, z)\,dy = 0, \qquad \sum \lambda_k\, \omega_k(y, z) \equiv 0,$$

the $(\omega)$'s having always the same meaning. But besides these conditions the following must also be fulfilled: When $z$ describes a closed path the set of $(\lambda)$'s is in general returned to a different set composed of $(\lambda')$'s, the differences $\lambda_k - \lambda'_k$ forming the set of characteristic integers corresponding to the variations of the cycle $(CH_z)$ in the manner of No. 20. As this variation is $\sim 0$, or, if we please, as $CH_z$ is a uniquely determined curve on each $H_z$, the differences $\lambda_k - \lambda'_k$ must form a set corresponding to a system of periods, and this must be true whatever the closed path described by $z$.

Conversely, if all these conditions are fulfilled there exists an algebraic surface $C$ corresponding to the functions $v_h(y, z)$. Indeed, for every $z$ these functions determine in a unique way, that is, rationally in terms of $z$, a certain linear system of curves on $H_z$. By imposing an adequate behavior at the $(A)$'s we may always determine in a unique manner in this linear system, a curve the locus of which will be a surface $C$. More explicitly, the curve in $H_y$ will be the intersection of the surface $F(x, y, z, t) = 0$ of the space $(x, y, t)$ with other surfaces $\Phi(x, y, t) = 0$, where these various surfaces have no other curves in common. The $(\Phi)$'s however may be taken to be polynomials with coefficients rational in $z$, whence the existence of the surface $C$ follows immediately.

30. Let $C_1$, $C_2$, be two algebraic surfaces of $V_3$. If $(C_1) \sim (C_2)$, mod. $V_3$, $(C_1 H_z) \sim (C_2 H_z)$ mod. $H_z$, and therefore also mod. $V_3$. It follows that

$$C_1 H_z = C_2 H_z$$

in $H_z$ and consequently $C_1$ and $C_2$ are of the same order and have the same set of characteristic integers.

We may find two surfaces $D'$, $D''$, belonging to a continuous system with $\infty^q$ linear systems and such that $C_1 + D'$ and $C_2 + D''$ belong totally to the same linear system, whence at once $C_1 = C_2$, this relation having a sense similar to that given to it for a $V_2$. Thus if $(C_1) \sim (C_2)$, then also $C_1 = C_2$. The converse is obviously true. We can see then that the relations

$$\sum t_i (C_i) \sim 0 \text{ mod. } V_3, \ \sum t_i (C_i H_z) \sim 0 \text{ mod. } H_z, \ \sum t_i C_i H_z = 0, \ \sum t_i C_i = 0,$$

are all equivalent.* From the existence of minimum and ordinary bases

* See a different proof of the equivalence of the relations $C_1 H_z = C_2 H_z$ and $C_1 = C_2$ given by Severi in the *Atti del Reale Instituto Veneto*, vol. 75 (1916), p. 1138. He had already stated this proposition without proof at the end of his memoir of the *Annales de l'Ecole Normale supérieure*, loc. cit.—The question may be raised as to the sense to be assigned to the algebraic four-cycle $(C)$. We know (No. 20) the sense to be assigned to the cycle $(CH_z)$. The four-cycle in question is then to be considered as the locus of $(CH_z)$ properly sensed associated to the $z$-plane sensed in a definite manner.

for the cycles follows the existence of similar bases for the curves. The variety will have numbers $\rho$, $\sigma$ and if $\rho'$, $\sigma'$ are those of $H$, then $\rho \leqq \rho'$, $\sigma \leqq \sigma'$, whatever $H$ may be.

31. The periods of the double integral attached to $H_z$

$$\iint \frac{P(x, y, z, t)}{F'_t} dxdy, \tag{9}$$

where $P$ is the adjoint polynomial, are of the form

$$\sum_{k=1}^{N} \lambda_k \int_b^{b_k(z)} \Omega_k(y, z)\, dy. \tag{10}$$

Let $z = c$ be a value $b$ for which two of the upper limits, say $b_1$ and $b_2$, coincide. The author has shown (Paris Comptes Rendus, loc. cit.) that when $z$ turns around $c$, (10) is increased by a multiple of the period

$$\int_b^{b_1} \Omega_1(y, z)\, dy - \int_b^{b_2} \Omega_2(y, z)\, dy; \qquad \Omega_1 = \Omega_2.$$

This will be true whether (10) is a period or not. If we develop in series of powers of $Y^{-1}$ the differential coefficient of

$$\iint \frac{P(x, y, z, t)}{(y - Y) F'_t} dxdy$$

and apply this to each term of the series we find that the period

$$\sum \lambda_k \int_b^{b_k} \frac{\Omega_k(y, z)}{y - Y} dy$$

increases by a multiple of

$$\int_b^{b_1} \frac{\Omega_1(y, z)}{y - Y} dy - \int_b^{b_2} \frac{\Omega_2(y, z)}{y - Y} dy.$$

In particular, for $|y|$ great enough, $v_h(y)$ $(h \leqq p - q)$ is increased by a multiple of

$$\frac{1}{2\pi i} \int_b^{b_1} \frac{\Omega_{h1}(Y, z)}{Y - y} dY - \frac{1}{2\pi i} \int_b^{b_2} \frac{\Omega_{h2}(Y, z)}{Y - y} dY.$$

I say that this and the similar expressions are not submultiples of the periods of $u_h$. For it vanishes for $z = c$ and if it were a submultiple of the periods the $(u)$'s would lose a period and $H_y H_z$ would lose a linear cycle for $z = c$, $y$ arbitrary, which is certainly not the case if $\{H_z\}$ is generic in $|H|$.

It follows at once that the zero divisors for the curves of $H_z$ are invariant. For otherwise a multiple of the increments of some set of $(v)$'s when $z$ turns around $c$ would form a period, hence this would be true for all sets of $(v)$'s in direct contradiction to what has just been found.

Thus the zero divisors for the curves of $H_z$ are also zero divisors of $V_3$, whence $\sigma \geqq \sigma'$, and therefore finally $\sigma = \sigma'$, that is, $V_3$ *and its hyperplane sections have equal invariants* $\sigma$. As they have equal invariants $\sigma_1$, it follows that $\sigma = \sigma_1$. Here we cannot affirm any more that $\sigma = \sigma_2$. This is due to the possible existence in $H_z$ of two-cycles not zero divisors for it but zero divisors for $V_3$.

32. We pass to the proof of a theorem specially important in the determination of $\rho$, $\sigma$, for the surfaces in $V_3$.

Let $c_1, c_2, \cdots, c_\nu$ be the critical values of $z$ for $\{H_z\}$. For each of these two of the $(b)$'s and no more coincide. Let us make the pencil vary continuously in a net $\Sigma$ containing it. The $(C)$'s will be displaced on an irreducible curve $D$ if as we shall assume $\Sigma$ is arbitrary in $|H|$, for when $V_3$ assumed in an $S_4$ is transformed by reciprocal polars, $D$ is changed into a plane section of the transformed variety.

To every point $c_j$ corresponds a well-defined two-cycle $\Gamma_j$ of $H_z$ which becomes zero for $z = c_j$ and such that when $z$ turns around $c_j$ any other cycle is increased by a multiple of it. I say that *if one of these cycles is algebraic so are the others.* For $\{H_z\}$ may be made to vary within $\Sigma$ so as to permute $c_j$ with $c_k$, thus bringing $\Gamma_j$ assumed within an $H_z$ very near $H_{c_j}$ into $\Gamma_k$ within an $H_z$ very near $H_{c_k}$. Noticing that the two extreme surfaces have the same geometric genus, we infer from the theorem of No. 21 that if $\Gamma_j$ is algebraic at the beginning of the deformation it will also be so at the end, that is $\Gamma_k$ is then also algebraic, which proves our statement.

Suppose first that *no* $\Gamma_k$ is *algebraic.* Then any algebraic cycle $\Delta$ of $H_z$ must be invariant. For if, when $z$ turns around $c_i$, $\Delta$ is returned to a cycle $\Delta'$ of $H_z$, $\Delta - \Delta'$ must be algebraic and as $\Delta - \Delta' \sim \mu\Gamma_k$, $\mu\Gamma_k$ must also be algebraic. Now if integrals of the first kind have no period relatively to $\mu\Gamma_k$, they have none relatively to $\Gamma_k$, and $\Gamma_k$ is also algebraic. We have then a contradiction unless $\mu = 0$, $\Delta \sim \Delta'$, which shows that $\Delta$ is invariant. It follows that under our assumptions the complete systems of curves on $H_z$ are invariant. If we take then $\rho'$ independent continuous systems we may, as shown in No. 30, determine $\rho'$ surfaces of $V_3$ whose intersections with $H_z$ are independent curves of the surface. These $\rho'$ surfaces will be themselves independent, hence $\rho \geqq \rho'$ and therefore since $\rho \leqq \rho'$, finally $\rho = \rho'$.

Suppose now that *the* $(\Gamma)$*'s are all algebraic.* What happens then is an immediate corollary of the following proposition, whose proof we shall do no more than outline here*: *Any two-cycle of* $H_z$ *is dependent upon the* $(\Gamma)$*'s and the invariant cycles.*—The $R_2$ invariant cycles are distinct mod. $V_3$, hence whatever $\Delta$, two-cycle of $H_z$, there is an integer $\lambda$ such that $\lambda\Delta$ is homologous to an invariant cycle mod. $V_3$, that is $\Delta' \sim \lambda\Delta - \text{invariant cycle} \sim 0$, mod. $V_3$.

* A topological proof will be found in the Borel Series monograph.

Now generalizing a discussion of Picard's (see Picard et Simart, Traité, vol. 2, p. 388), pertaining to the periods of certain integrals, and with a ready passage from periods to cycles, which shall be omitted here, it is found that the number of $(\Gamma)$'s distinct mod. $H_z$, (they all bound on $V_3$), is precisely equal to the maximum number of cycles distinct mod. $H_z$, but bounding on $V_3$. Hence $\Delta'$ depends upon the $(\Gamma)$*'s*, from which our assertion follows.

The application to our problem is immediate: *If the* $(\Gamma)$*'s are all algebraic, any two-cycle of* $H_z$ *depends upon the invariant and the algebraic cycles.*

A proposition of my Annali di matematica paper taken together with a result established in a note of the Rendiconti dei Lincei of 1917 leads to this proposition: The number $\rho_0 = \rho_0^2$ of $V_3$ is equal to the number of non-algebraic two-cycles, or which is the same, to the number of invariant non-algebraic cycles, of $H_z$. We also know from Picard's work that the same number for $H_z$, which we shall denote by $\rho_0'$, is equal to its total number of non-algebraic cycles. Hence if $\rho_0' > \rho_0$, $H_z$ possesses some non-invariant non-algebraic cycles, none of the $(\Gamma_k)$'s are algebraic and $\rho = \rho'$. We may therefore state:

THEOREM. *If the number of proper integrals of the second kind of a generic hyperplane section exceeds that of* $V_3$, *they have equal numbers* $\rho$.

33. Thus for $H$ we have $\sigma' = \sigma$, and if $\rho_0' > \rho_0$ also $\rho' = \rho$. In these conditions the trace on an arbitrary $H$ of an ordinary base for the surfaces of $V_3$ forms a base for the curves of $H$. Let us show that if the first is a *minimum base* so is the second, this being true even if only effective surfaces or curves are considered.

Indeed let $C_1, C_2, \cdots, C_r$ be the surfaces of a minimum base and $\gamma$ an arbitrary curve of a generic $H_z$. As we have seen, the complete continuous systems of curves of $H_z$ are all invariant. Hence the complete continuous system $\{\gamma + kH \cdot H_z\}$ ($k$ a sufficiently large integer) is invariant and contains $\infty^q$ linear systems. One of these systems will be determined by the $(\beta)$'s, $(\beta_k = v_{p-q+k})$, and a curve $\gamma'$ of it may be defined rationally on $H_z$ by means of an adequate behavior at the points $A_i$, for example. The locus of $\gamma'$ is an algebraic surface $C$ going, say, $l$ times through the curve at infinity of $H_z$, and we have relations such as these: $C = \Sigma \lambda_i C_i$, $H = \Sigma \mu_i C_i$, (on $V_3$); $CH_z = \Sigma \lambda_i C_i H_z$, $HH_z = \Sigma \mu_i C_i H_z$, (on $H_z$). Hence on $H_z$, $\gamma = CH_z - (l + k) HH_z = \Sigma [\lambda_i - (l + k)\mu_i] C_i H_z$, as was to be proved.

34. The preceding propositions remain true if we replace $|H|$ by $|E|$, $\infty^3$ at least, irreducible, simple, without base points, with irreducible characteristic curve $E^2$, such that if the surface $E_0$ of the system acquires singularities other than those of the generic $E$, they consist in general in one isolated double point, and that finally the system of the surfaces $E_0$ is irreducible. If $|E|$ has a base group but still irreducible variable intersection, we have the following:

Let $D_1, D_2, \cdots, D_k$ be the finite or infinitesimal curves of the base group forming base for all the surfaces of the group which are not traces of the surfaces of $V_3$, $D'_1, D'_2, \cdots, D'_k$, those forming minimum base in the same conditions. Then if $\rho'_0 > \rho_0$ we have $\rho' = \rho + k$, $\sigma' = \sigma + k' - k$. Moreover, the trace of an ordinary (minimum) base of $V_3$ added to the $(D)$'s (to the $(D')$'s) forms an ordinary (minimum) base for $E$.

35. The extension to a $V_d$, $d > 3$, is immediate. Since $V_d$ and its hyperplane sections have the same two-cycles the restriction $\rho'_0 > \rho_0$ in the theorems just proved is now superfluous. We still have $\sigma = \sigma_1$.

The case where the two-cycles of $V_d$ are all algebraic is especially noteworthy. $H$ need then not be generic in its system but only without new singularities. For, let $H_0$ be a special $H$, but with the same singularities as the generic $H$, $C_1, C_2, \cdots, C_\gamma$ a minimum base for $V_d$. The two-cycles $(C_i H^{d-2})$ form a minimum base for the two-cycles of $H$, and as $H_0$ and $H$ are homeomorphic the limiting positions $(C_i H^{d-3} H_0)$ of these cycles form a minimum base for the two-cycles of $H_0$.

Let now $\gamma$ be any hypersurface of $H_0$. We have mod. $H_0$

$$(\gamma H^{d-3}) \sim \sum \lambda_i (C_i H^{d-3} H_0) \qquad (\lambda_i \text{ integer}).$$

But the equivalence between the relations $C = D$, $(CH^{d-2}) \sim (DH^{d-2})$, proved for $d = 2, 3$, can easily be shown to hold for any $d$. Applying it to $H_0$ we obtain $\gamma = \Sigma \lambda_i C_i H_0$, and therefore the trace of a minimum base of $V_d$ on $H_0$ is a minimum base for $H_0$. Clearly the minimum base could everywhere be replaced by an ordinary one. It follows that $H_0$ has the same numbers $\rho$, $\sigma$ as $V_d$.

If $H_0$ had special singularities of less than $d - 2$ dimensions we could only affirm that its numbers $[\rho]$, $[\sigma]$ are equal to $\rho$, $\sigma$ respectively. Similar considerations lead to this result: Let $|E|$ be a linear system analogous to that of the same name described in No. 34 for a $V_3$ and without base group. If the invariant $\rho_0^{d-1}$ of the generic $E$ is greater than that of $V_d$, their invariants $\rho_{d-1}$ are equal. Their numbers $\sigma_{d-n}$, $\rho_{d-i}$, $i \geqq 2$ are equal without this restriction.

36. Returning to a $V_3$, we may remark that there is a case where the condition $\rho'_0 > \rho_0$ is certainly verified,—it is when $V_3$ possesses a triple integral of the first kind

$$\iiint \frac{Q(x, y, z, t)}{F'_t}\, dxdydz \tag{11}$$

with periods not all zero. Indeed its periods relatively to cycles within $H_z$ are all zero, as follows from the discussion at the end of the first chapter. Hence there must be periods relatively to finite cycles or special cycles such as described in proposition ($b$) of No. 2. This in turn requires that the double integral of the first kind attached to $H_z$,

$$(12) \qquad \int\int \frac{Q(x, y, z, t)}{F'_t} dxdy$$

be not without periods. But the periods of (12) relatively to algebraic cycles are zero, and those relatively to invariant cycles are rational in $z$, finite everywhere, infinitesimal for $z$ infinite, and therefore also zero. Hence the cycles $\Gamma_k$ of No. 32 are not algebraic and $\rho' = \rho$ for $|H|$ or for any linear system without base group such as $|E|$, already described. This may also be shown by remarking that (12) is not improper of the second kind and since it is without invariant periods it cannot be derived from an integral of the second kind of $V_3$, consequently $\rho'_0 > \rho_0$.

Let us now take a $V_d$ and call $\Gamma$ a $d$-cycle relatively to which one of its $d$-uple integrals of the first kind has a non-zero period. $\Gamma$ may be the sum of two cycles, the first $\Gamma'$ being contained in $H$ and the second $\Gamma''$ of the special type described in proposition ($b$), No. 2. If the period relatively to $\Gamma''$ is not zero the reasoning just used for $V_3$ applies here and we shall find that $H$ possesses a $(d - 1)$-uple integral of the first kind with at least one non-zero period. On the other hand if the period relatively to $\Gamma'$ is not zero we may by reasoning as in No. 15, Chapter 1, show that there is a $(d - 2)$-uple integral of the first kind of $H^2$ having at least one non-zero period. Continuing this we shall come either to an $H^{d-3}$ or to an $H^{d-2}$ with integrals of the first kind of multiplicity $d - 3$ or $d - 2$ having at least one non-zero period. In the first case it follows at once that $H^{d-2}$ has the same number $\rho$ as $H^{d-3}$ and therefore as $V_d$, its bases being the traces of those of $V_d$. In the second case the reasoning applied for $V_3$ holds as between $H^{d-3}$ and $H^{d-2}$, leading to the identical conclusion.

As an application, a generic $V_2$, complete intersection of $d - 2$ hypersurfaces in a $V_d$, each generic in an adequate linear system, has for bases the traces of those of $V_d$, provided there is a $d$-uple integral of the first kind of $V_d$ with periods not all zero.

The most interesting case is that of an Abelian variety of *genus* $p$ and rank *one*, for if $u_1, u_2, \cdots, u_p$ are the variables occurring in its parametric representation $\int\int \cdots \int du_1\, du_2 \cdots du_p$ is an integral of the first kind with at least one non-zero period.

37. For a $V_d$, $d > 2$, contrary to what happens for an algebraic surface, the number of two-cycles without periods for the integrals of the first kind is only a maximum of $\rho$. At all events we have succeeded in proving that these two numbers are equal only for Abelian varieties, as we shall see in the second part, and for a certain type of $V_3$, which may be defined thus: let $P_a$, $P_g$ be

respectively the arithmetic and geometric genus of a $V_3$. *If* $P_g - P_a + q = 0$, *all the two-cycles are algebraic, and therefore* $\rho$ *reaches its maximum.* Indeed as shown by Severi, in this case the system adjoint to $|H|$ cuts out the complete canonical system on $H$. Hence every double integral of the first kind of $H_z$ is a linear combination of those of type $\int\int (1/F'_t)Q(x, y, z, t)\,dxdy$, where $Q$ is adjoint of order $m - 4$, whose periods relatively to invariant cycles all vanish. It follows then that all these cycles are algebraic cycles of $H_z$ and, being invariant, they are also algebraic cycles of $V_3$. In this case $\rho_0 = R_2 - \rho = 0$, hence $\rho = R_2$.

## § 4. Some applications

38. *Complete intersections.* Let $V_2$ be a non-singular algebraic surface, complete intersection of $(r - 2)$ varieties in an $S_r$. I say that $V_2$ has no double integrals of the first kind without periods. For let $F(x, y, z) = 0$ be the equation of its projection $V'_2$ in an $S_3$. It is sufficient to show that there can be no relation $AF'_x + BF'_y + CF'_z \equiv -DF$, where $A$, $B$, $C$ are adjoint of order $m - 2$. For according to this relation the $N$ contacts of the planes $y = Ct$, tangent to $V'_2$, must be on $B = 0$. Now if $|K|$ designates the canonical system and $|H|$ the complete system of the plane sections, the adjoint surfaces of order $m - 2$ cut out on $V'_2$ the system $|K + 2H|$. They intersect the first polar of an arbitrary point of $S_3$ in a number of points exterior to the double curve, at most equal to*

$$[(K + 2H)(K + 3H)] = [K^2] + 5[KH] + 6[H^2],$$

and hence it is only necessary to prove that this number is $< N$. Let $m_1, m_2, \cdots, m_{r-2}$, be the orders of the varieties of $S_r$ of which $V_2$ is the intersection. Then $m = m_1 m_2 \cdots m_{r-2}$, and the adjoint system is cut out on $V_2$ by the varieties of order $n = \Sigma m_i - r - 1$; therefore $[K^2] = mn^2$, $[KH] = mn$, $[H^2] = m$. Hence the number considered above is equal to $mn^2 + 5mn + 6m$. To find the value of $N$ we observe that in $S_r$ the hyperplanes of a linear net which are tangent to $V_2$ touch the surface at the points where it is intersected by a variety of order $n + 3$, hence $N = m(n + 3)^2 > mn^2 + 5mn + 6m$, which proves our assertion.

This still applies when all the $(m)$'s are equal to unity except for one which is $> 3$, that is, when we deal with a non-singular algebraic surface of order $> 3$ in ordinary space.

39. Now a non-singular $V_{r-1}$ in $S_r$, $r > 3$, has for minimum base the trace of a minimum base of $S_r$, that is, a hyperplane section. Its index $R_2 = 1$ as it is for $S_r$ and its two-cycles are all algebraic. Its complete systems of hyper-

* It is in general less than this number by a certain multiple of the number of pinch points owing to the peculiar behavior of the first polars at these points.

surfaces are all linear. Here then, equivalence à la Severi between two hypersurfaces implies that they are contained in the same linear system, and as they are all multiples of a hyperplane section, $V_{r-1}$ *contains only hypersurfaces which are complete intersections.* The same reasoning holds for its intersection with another non-singular $V_{r-1}$ of $S_r$, provided that this intersection is non-singular and also that $r > 4$, and so on. We thus see that *a non-singular* $V_d$, $d > 2$, *complete intersection in an* $S_r$, *contains only hypersurfaces that are themselves complete intersections.*

Let now $d = 2$. Any $V_3$, complete intersection and non-singular, passing through $V_2$, possesses no proper double integrals of the second kind since its $R_2 = \rho$, whereas $V_2$ certainly possesses such an integral if its order is $> 1$ and no space of less than four dimensions can be passed through it, or, as this is usually stated, if it is *normal* in an $S_r$, $r > 3$. Hence *a non-singular* $V_2$, *complete intersection and normal in an* $S_r$, $r > 3$, *contains in general only curves which are complete intersections.* This is true also for a non-singular surface of order $> 3$ in ordinary space.* For ordinary space this theorem has been stated and proved many years ago by Noether, but his proof based on enumeration of constants has long been considered unsatisfactory. The above is believed to be the first complete proof ever given of this important proposition.

As shown previously, even limiting ourselves to ordinary space, we can go farther. Let $|E|$ be a linear system, simple, irreducible, $\infty^3$ at least, with a base group composed of $k$ curves independent on $E$ and such that no point is multiple for any $E$ through it. The generic $E$ contains no other curves than those cut out by surfaces passing through curves of the base group, and for it $\rho = k + 1$, $\sigma = 1$.

40. The part of our results relating to a $V_{r-1}$ in $S_r$, $r > 3$, has already been obtained by Severi. The cubic variety in $S_4$, (Segre variety), has been treated at length by G. Fano. It furnishes a very interesting illustration. Indeed it may be verified that when there are less than six double points the variety belongs to a linear system, $\infty^4$ at least, and with the properties described more than once, hence it contains only complete intersections. For the hyperplane sections whose geometric genus is zero, $\rho = 7$. Intersections with hypersurfaces of order $\geqq 2$ contain in general only complete intersections.

41. *Double plane.* Let $z^2 = f(x, y)$ be the equation of a double plane. One may always transform it birationally so that the branch curve be of even order $2m$, without multiple components and with multiple points all of even order and with distinct tangents. Any double integral of the first kind is of

* During the final preparation of this memoir the author's attention was attracted by an incomplete proof of these theorems due to G. Fano, (Torino Atti, vol. 44 (1909), pp. 415–430) which has some points in common with the one given here. Fano admits that a double integral of the first kind cannot be without periods, and this can scarcely be considered as axiomatic.

the form $\int\int [f(x, y)]^{-\frac{1}{2}} Q(x, y, t)\, dxdy$, where $Q$ is a polynomial of order $m - 3$ and has an $(i-1)$-uple point at a $2i$-uple point of $f(x, y)$, (Enriques). If the periods are zero one may show that equations (8) now become $A'_x + B'_y + C'_t \equiv Q(x, y, t)$, $Af'_x + Bf'_y + Cf'_t \equiv 0$, where $A$, $B$, $C$ are of order $m - 1$ and behave like $Q$ at the multiple points of $f$. Here again the points of intersection, other than the multiple points of $f = 0$, of the curves $f'_x = 0, f'_y = 0$, must be on $C = 0$. Hence $C = 0$ and $f'_x = 0$ have in common at least

$$(2m-1)^2 - \sum (2i-1)^2 + \sum (2i-1)(i-1) \\ = (2m-1) - \sum i(2i-1)$$

points, the summation being extended to all the multiple points. To show that our double integral cannot be deprived of periods it is sufficient to show that this number exceeds $(m-1)(2m-1)$, which, since $f'_x = 0$ may be assumed irreducible, will lead to an impossibility. But the inequality to be proved reduces at once to $m(2m-1) > \Sigma\, i(2i-1)$ which is actually verified, for it may be obtained by expressing that the number of isolated double points equivalent to the multiple points of $f(x, y)$ does not exceed its value $m(2m-1)$ when $f$ is a product of $2m$ linear factors. Thus *a double plane has no double integral of the first kind without periods.*

42. When does a double plane belong to a linear system with the properties described in No. 34? I say that this will certainly occur if $f(x, y)$ is divisible by a factor $\phi(x, y)$ such that the curve $\phi(x, y) = 0$ be generic in a linear system $|\gamma|$, $\infty^2$ at least, not composed with the curves of a pencil, and such that there is no point not a base point multiple for any $\gamma$ through it. For let $f \equiv \phi \cdot \psi$, and $\phi + k\phi_1 = 0$ the equation of a generic pencil of $|\gamma|$. First we verify that the variable curve of intersection of the two surfaces $z^2 = \phi\psi$, $z^2 = (\phi + k\phi_1)\psi$, is irreducible. Indeed if it were not the two surfaces would have the same tangent planes at the points $z = \phi = \phi_1 = 0$ which are not base points for $|\gamma|$, and the curves $\phi = \phi_1 = 0$ would be tangent at those points, which is not the case. Next the multiple points of the double plane are the multiple points of $f = 0$, and a fixed $(2m-2)$-uple base point at infinity, hence there is no point other than a base point multiple for all the double planes having for branch curve $\psi = 0$ associated with a curve $\gamma$, and we can apply our general theorems to them.

43. To find the minimum base we must first determine whether the components of the branch curve are independent curves of the double plane. Let $f \equiv \phi_1 \cdot \phi_2 \cdots \phi_h$, where the $(\phi)$'s are irreducible polynomials prime to each other, and let $C_i$ be the curve $\phi_i = 0$ of the double plane. Denoting by $|H|$, $\{H_y\}$ the same elements as usual, we consider $m - 1$ independent integrals of the first kind $u_1, u_2, \cdots, u_{m-1}$ of $H_y$ and observe that since in

the case here considered the double plane is *regular* (Castelnuovo-Enriques), a relation $\Sigma\, t_i\, C_i = 0$ would necessarily lead to another

$$\sum_{i,j} t_i \int_{(x,z)}^{M_{ij}} du_s = 0 \qquad (s = 1, 2, \cdots, m-1),$$

where the points $M_{ij}$ are the points of the group $C_i\, H_y$, and the lower limit of integration is arbitrary on $H_y$. Now $H_y$ is hyperelliptic of genus $m-1$, and, as far as the integrals at the left are concerned, one may be made zero, while their sum vanishes. We also know that by combining them linearly with suitable integral coefficients we obtain $2m-2$ independent periods of $u_s$. Hence the only relation of the type considered corresponds to

$$t_1 = t_2 = \cdots = t_h = t,$$

and as we have $t\, \Sigma\, C_i = tH \neq 0$, the $(C)$'s are actually independent.

We conclude from this that if $p_g > 0$ and one of the $(C)$'s is variable in a suitable linear system, the minimum base is composed of: (*a*) The vicinity of the singular point at infinity on the $z$ axis which counts for $h$ curves. (*b*) The vicinities of the multiple points of the branch curve which will count for, say, $k$ curves. The value of $k$ will depend upon the nature of the singular points, etc. In the simplest case where the curve has only ordinary double points $k$ will be equal to their number.

(*c*) The $h$ components of the branch curve. We have therefore $\rho = 2h + k$, $\sigma = 1$, $[\rho] = h$. Every curve of the double plane is cut out by a surface which goes through several of the $(C)$'s and through no other curve of the double plane.

As to $\rho_0$, it may be computed thus: If $N$ is the class of the branch curve, $d$ the equivalence in double points of its multiple points we find by means of $\{H_y\}$, $[R'_2] = N - 6m + 5 + d$. Therefore

$$\rho_0 = [R'_2] - [\rho] + 1 = N - 6m + 6 + d - h.$$

This discussion can be applied to derive some of Picard's results for double planes as well as some due to Enriques for surfaces of linear genus $p_1 = 1$. Let us take for example the surface $z^2 = a(y)x^3 + b(y)x^2 + c(y)x + d(y)$, where $a, b, c, d$, are polynomials, a surface considered by Picard.* If these polynomials are arbitrary and prime to each other the double plane contains only complete intersections and this relation cannot be satisfied by substituting for $z$ and $x$ rational functions of $y$.

Enriques showed that the surfaces named after him are reducible to double planes, the determination of their minimum base as he gave it is then immediate.† These double planes contain also only complete intersections.

* Picard-Simart, *Traité des fonctions algébriques de deux variables*, vol. 2, p. 268.

† Rendiconti dei Lincei, ser. 2, vol. 23 (1914), pp. 291–297.

## § 5. Varieties representing the groups of points of other varieties

44. We insert here a short discussion of these varieties that will be found useful in some applications. They present also the added interest that we can form very readily varieties of this type having quite arbitrary torsion indices.

Let then $V^1, V^2, \cdots, V^n$ be algebraic varieties with ordinary singularities, $d_i$ the dimensionality of $V^i$, $V_d$ a variety representing without exception the groups of $n$ points of which each is on a $V^i$, so that $d = \Sigma\, d_i$.

By generalizing somewhat the reasoning at the beginning of this paper, it may be shown that all the $i$-cycles are sums of those obtained by associating an $i_1$-cycle of $V^1$, an $i_2$-cycle of $V^2$, $\cdots$ $(i_1 + i_2 + \cdots = i)$ in the same manner as $V^1, V^2, \cdots$. Let then $R_i^h$, $R_i$ and $\sigma_i^h$, $\sigma_i$ be the $i$th indices of connectivity and torsion of $V^h$ and $V_d$ respectively. We shall have $R_i = \Sigma\, R_{i_1}^1 R_{i_2}^2 \cdots R_{i_n}^n$, the summation being extended to all the partitions of the $(i)$'s such that $i = i_1 + i_2 + \cdots + i_n$, $i_h \leqq 2d_h$, and with the convention that $R_{2d_h}^h = R_0^h = 1$.

As to the $(\sigma)$'s suffice to state that $\sigma_1 = \sigma_1^1\, \sigma_1^2 \cdots \sigma_1^h$. Now Godeaux has shown that there are algebraic surfaces whose invariant $\sigma = \sigma_1$ is any odd prime, and an example of $\sigma = 2$ has been given by Severi. This shows that there are algebraic varieties for which $\sigma$ has any value whatever. We shall have occasion to consider in the second part Abelian varieties of rank $> 1$ whose invariant $\sigma$ is arbitrarily assigned and in fact equal to the rank. Suitable surfaces contained in these varieties will have the same value of $\sigma$, i.e., equal to an arbitrary integer.

*Remark.* All the varieties so far known, including those to be constructed in the second part, for which $\sigma > 1$, represent involutions on varieties whose $\sigma = 1$. The question presents itself—are there any others? For the present it must remain unanswered.

45. Let us now pass to the determination of $\rho$. Assume first $n = 2$. $V_d$ contains a continuous system $\infty^{d_1}$, $\Sigma_1$, of varieties identical with $V^2$ and another $\infty^{d_2}$, $\Sigma_2$, of varieties identical with $V^1$. A hypersurface of $V_d$ representing the pairs of points of $V^2$ and a hypersurface of $V^1$ will be called *fundamental* for $V^1$. Similarly there will be hypersurfaces fundamental for $V^2$. The hypersurfaces fundamental for $V_i$ possess an ordinary (minimum) base composed of $\rho_i$, $(\rho_i + \sigma_i - 1)$, hypersurfaces. Assume that $E_i$ is fundamental for $V^i$. Then certainly there can be no relation $E_1 = E_2$ since the homology $(E_1) \sim (E_2)$ is not verified. Hence the hypersurfaces of the two fundamental sets are not related.

To find the number of hypersurfaces independent of those of the fundamental sets, I say that if $d_1 \geqq 3$, $V^1$ may be replaced by a $V_{d_1 - 1}$ having the same Picard variety $W_1$ as $V^1$. For let $\{C_\mu\}$ be an arbitrary linear pencil of hypersurfaces fundamental for $V_1$. If $A$, $B$ are two hypersurfaces such that $AC_\mu = BC_\mu$ on any $C_\mu$, then $A - B$ is fundamental for $V^1$. This can be shown in a manner very similar to that used to prove the analogous theorem for any $V_3$ and the pencil $\{H_z\}$. If $\{C_\mu\}$ were generic in a suitable linear system we would have $A = B$, but here we may only conclude that $A - B$ is a multiple of $C_\mu$. It follows then that as far as we are concerned $V^1$ may be replaced by $C_\mu$ whose Picard variety will be $W_1$, if $d_1 \geqq 3$.

We propose to show next that just as in the case $d_1 = d_2 = 1$ already investigated by Severi,* the number of hypersurfaces independent of those of the two fundamental sets is equal to Scorza's simultaneous index $\lambda_{12}$ of the Picard varieties $W_1$, $W_2$ of $V_1$ and $V_2$. This index will be defined in the second part (No. 47). We may then assume $d_1 \leqq 2$, $d_2 \leqq 2$, $d_1 + d_2 > 2$. One of the $(d)$'s, say $d_1$, will have the value *two*.

Let $A$ be an arbitrary hypersurface of $V_d$, $\{C_1\}$ a complete continuous system of curves of $V^1$ containing $\infty^{q_1}$ linear systems, ($q_i$ irregularity of $V^i$), $C_1^1$ the fundamental hypersurface determined by $C_1$. To the points $C_1^1 A$ may correspond either all the points of $V^2$ or only a $V_{d_2-1}$ of $V^2$. The first case may be rejected at once for then there corresponds to the points in question only a finite number of points on $C_1$, and to $A$ only a curve on $V^1$; hence $A$ is fundamental. In the second case, to which we may limit ourselves then, there corresponds to $C_1^1 A$ a certain algebraic curve $C_2$ on $V^2$ if $d_2 = 2$, or a group of points if $d_2 = 1$. But $C_1$ determines a point of $W_1$ and $C_2$ one of $W_2$, for by adding if necessary a suitably chosen fundamental hypersurface to $A$, we may arrange matters so that $C_2$ belongs to a continuous system containing $\infty^{q_2}$ linear systems. We have here an algebraic correspondence between the points of $W_1$ and $W_2$ and the theory of Hurwitz-Severi may be extended at once, leading to the announced result. The extension to $n > 2$ is immediate. It is sufficient to consider $V_d$ as corresponding to the group of two points one of which is on $V^1$ and the other on the $V_{d-d_1}$, obtained by associating $V^2, \cdots, V^n$. We shall have finally $\rho = \Sigma\, \rho^{(i)} + \Sigma\, \lambda_{ik}$, where $\rho^{(i)}$ is the number $\rho$ of $V^i$ and $\lambda_{ik}$ the simultaneous index of the Picard varieties of $V^i$ and $V^k$.

*Torino Memorie, ser. 2, vol. 54 (1903), pp. 1–49.

# PART II

## ABELIAN VARIETIES

## CHAPTER I. GENERAL PROPERTIES

### § 1. A summary of certain fundamental theorems and definitions

46. An *Abelian* variety of genus $p$, $V_p$, is a variety whose non-homogeneous point coördinates are equal to $2p$-ply periodic meromorphic functions of $p$ arguments $u_1, u_2, \cdots, u_p$, or whose homogeneous point coördinates are proportional to theta's of the same order and continuous characteristic. The variety is algebraic (Weierstrass) and of dimensionality $p$. When the periods are those of a set of independent integrals of the first kind of a curve of genus $p$, $V_p$ is called a *Jacobi* variety.

An array with $p$ rows and $2p$ columns

$$\Omega \equiv \| \omega_{j\mu} \| \quad (j = 1, 2, \cdots, p;\ \mu = 1, 2, \cdots, 2p)$$

is the period matrix of a $V_p$ provided that:

(*a*) There exists an alternate bilinear form with rational coefficients

$$\sum_{\mu,\, \nu=1}^{2p} c_{\mu\nu}\, x_\mu y_\nu, \qquad c_{\mu\nu} = -\, c_{\nu\mu}, \tag{1}$$

vanishing identically when the $(x)$'s and the $(y)$'s are replaced by the elements of any two rows of $\Omega$.

(*b*) If we set $\xi_\mu + i\eta_\mu = \Sigma_{(j)} \lambda_j\, \omega_{j\mu}$, then, for all non-zero values of the $(\lambda)$'s, $\Sigma\, c_{\mu\nu}\, \xi_\mu\, \eta_\nu > 0$. According to Scorza (*b*) is equivalent to this: Denoting by $\bar{x}$ the conjugate of any number $x$, the Hermitian form in the $(\lambda)$'s,

$$\sum A_{jk}\, \lambda_j\, \bar{\lambda}_k, \qquad A_{jk} = -\frac{1}{2i} \sum c_{\mu\nu}\, \omega_{j\mu}\, \bar{\omega}_{k\nu}, \tag{2}$$

must be definite positive. When these conditions are satisfied $\Omega$ is called by the same author a *Riemann matrix*.

47. We propose now to recall some concepts and definitions incipient in the works of various authors, but formally introduced and only fully developed in recent writings of Scorza and also Rosati.* The nomenclature which we shall use is Scorza's.

The alternate form (1) is called a *Riemann form* of $\Omega$. If the condition (*b*) is satisfied for that form, it is said to be a *principal* form of the matrix. There

* See Scorza's memoir in the Rendiconti di Palermo, vol. 41 (1916), for numerous bibliographical indications. The very interesting method which dominates his and Rosati's investigations had been used previously without their being apparently aware of it by Cotty in his Paris thesis (1912) for the case $p = 2$. Cotty attributes the idea to Humbert. We shall have opportunity to apply it on several occasions.

may be several independent forms alternate of $\Omega$, and if $1 + k$ is their number, $k$ is called the *index of singularity*. If $k \neq 0$ the matrix is said to be *singular*.

There may be $1 + h$ bilinear forms with rational coefficients

$$\sum_{\mu,\,\nu=1}^{2p} c_{\mu\nu}\, x_\mu\, y_\nu \tag{3}$$

vanishing identically when the $(x)$'s and the $(y)$'s are replaced by the elements of any two rows of $\Omega$. The integer $h$ is called the *index of multiplication*, and we have $h \geqq k$.

Finally, given two Riemann matrices $\Omega$, $\Omega'$, if there exist $\lambda$ bilinear forms with rational coefficients

$$\sum c_{\mu\nu}\, x_\mu\, y_\nu \quad (\mu = 1, 2, \cdots, 2p;\ \nu = 1, 2, \cdots, 2p';\ p,\ p' \text{ genera of } \Omega, \Omega') \tag{4}$$

which vanish identically when the $(x)$'s are replaced by the elements of a row of $\Omega$ and the $(y)$'s by those of a row of $\Omega'$, $\lambda$ is called the *index of simultaneity* of the two matrices.

The forms of types (1), (3), (4), generate moduli and admit bases. The corresponding numbers $\sigma$ are all equal to *one*.

Two matrices $\Omega$, $\Omega'$ are *isomorphic* if we can pass from one to the other by a linear transformation with arbitrary coefficients applied to the rows and by a linear transformation with rational coefficients applied to the columns. If this last transformation is of determinant one and integral coefficients (unimodular), $\Omega$, $\Omega'$ are *equivalent*. In all cases the determinants of the two transformations must not be zero. When two matrices are isomorphic, the most general Abelian varieties which correspond to them in the sense defined below, are in algebraic correspondence with each other. When the matrices are equivalent, their Abelian varieties can be transformed birationally into each other.

A Riemann matrix $\Omega$ is said to be *impure* if it is isomorphic to one of type

$$\begin{Vmatrix} \Omega_1, & 0 \\ 0, & \Omega_2 \end{Vmatrix},$$

where $\Omega_1$, $\Omega_2$, are Riemann matrices of genus $p$ and the ciphers represent matrices with elements all zero. If $\Omega$ is not impure, it is said to be *pure*. When $\Omega$ is impure the field of Abelian functions which belongs to it contains functions of genus $< p$, (Poincaré). Finally, in the same case, an Abelian variety belonging to $\Omega$ represents the pairs of points, one of which is on an Abelian variety belonging to $\Omega_1$ and the other on one belonging to $\Omega_2$.

The results at the end of the preceding part allow us to replace the determination of the base invariants of an impure variety by that of the invariants of varieties of genus $< p$. We shall return to this point later.

*Remark.* All the properties which we have recalled here are invariant in regard to isomorphism.

48. The matrix $\Omega$ is equivalent to a matrix in the so-called *canonical* form

$$A \equiv \left\| \begin{matrix} \frac{1}{e_1}, & 0, & \cdots & 0 & a_{11}, & a_{12}, & \cdots & a_{1p} \\ 0, & \frac{1}{e_2}, & \cdots & 0 & a_{21}, & a_{22}, & \cdots & a_{2p} \\ \cdot & \cdot & \cdot & \cdot & \cdot & \cdot & \cdot & \cdot \\ 0, & \cdot & \cdots & \frac{1}{e_p} & a_{p1}, & \cdot & \cdots & a_{pp} \end{matrix} \right\| \quad (a_{jk} = a_{kj}),$$

where the integers $e_\mu$ are the elementary divisors of (1), assumed now with integral coefficients. The matrix has a principal form $\Sigma_{\mu=1}^{p} e_\mu (x_\mu y_{p+\mu} - x_{p+\mu} y_\mu)$ derived from (1) up to an integral factor, and condition (*b*) gives now this: If $a_{jk} = a'_{jk} + ia''_{jk}$, the quadratic form $\Sigma\, a''_{jk} x_j x_k$ must be definite positive, or, which is the same, $\Sigma\, a''_{jk} x_j x_k = 1$ must be a real generalized ellipsoid. The arbitrariness of the $(a'')$'s shows that in general the above alternate form is unique—that is, that for an arbitrary Riemann matrix $k = 0$.

Let us indicate a certain property of the Hermitian form (2), interesting especially in some applications. Designate by $C$ an arbitrary alternate Riemann form of $\Omega$, for example (1), then apply a transformation $B$ of non-zero determinant defined by

$$\omega'_{j\mu} = \sum_{\nu=1}^{2p} b_{\mu\nu}\, \omega_{j\nu} \qquad (\mu = 1, 2, \cdots, 2p;\ j = 1, 2, \cdots, p).$$

To the matrix of the $(\omega')$'s will correspond an alternate form $C'$ defined in the notation of Frobenius by $C' = \bar{B}^{-1} C B^{-1}$, where $\bar{B}$ is the transposed of $B$.

Consider now the transformation $D$ defined by the matrix

$$\left\| \begin{matrix} \omega_{11}, & \omega_{12}, & \cdots & \omega_{1,2p} \\ \cdot & \cdot & \cdot & \cdot \\ \omega_{p1}, & \cdot & \cdots & \omega_{p,2p} \\ \bar{\omega}_{11}, & \cdot & \cdots & \bar{\omega}_{1,2p} \\ \cdot & \cdot & \cdot & \cdot \\ \bar{\omega}_{p1}, & \cdot & \cdots & \bar{\omega}_{p,2p} \end{matrix} \right\|$$

whose determinant as we know is not zero. Applying finally to the $(\omega)$'s the transformation $\bar{D}^{-1}$, the form $C$ will be replaced by a form $C''$ of same genus $q \leqq p$ as $C$, that is, depending essentially upon the same number of linearly distinct variables, ($2q$ for the $(x)$'s, $2q$ for the $(y)$'s), as $C$. But owing to the relations $\Sigma\, c_{\mu\nu}\, \omega_{j\mu}\, \omega_{k\nu} = \Sigma\, c_{\mu\nu}\, \bar{\omega}_{j\mu}\, \bar{\omega}_{k\nu} = 0$, we find that the matrix of the coefficients of $C''$ is equal to the product of $2^{2p}$ by

$$\left\| \begin{matrix} 0, & A \\ \bar{A}, & 0 \end{matrix} \right\|,$$

where $A$ is the matrix $||A_{jk}||(j, k = 1, 2, \cdots, p)$. If we call genus of a Hermitian form the number of essentially distinct variables upon which it depends, we see *that the genus of an alternate Riemann form of $\Omega$ is equal to that of the corresponding Hermitian form.*

50. *Rank of Abelian Varieties.* Let us transform $\Omega$ into a canonical matrix and, using the same notations as in Krazer's *Lehrbuch der Thetafunctionen,* consider the equations

$$\tau x_j = \theta_k^j[{}^g_h](u) \qquad (j = 1, 2, \cdots, s > p), \tag{5}$$

where the $(\theta)$'s are linearly independent theta-functions of sufficiently high order $k$ belonging to $A$. The $(x)$'s are point coördinates of an Abelian variety belonging to $A$, or if we wish to $\Omega$. Let $v_1, v_2, \cdots, v_p$ be the variables corresponding to $\Omega$. To every system of values of the $(v)$'s corresponds one and only one of the $(u)$'s modulo the periods of $A$, and conversely to every system of values of the $(u)$'s corresponds one and only one of the $(v)$'s modulo the periods of $\Omega$. The relations $v_h = \Sigma_{\mu=1}^{2p} t_\mu \, \omega_{h\mu} \, (h = 1, 2, \cdots, p)$, where the $(t)$'s are real variables such that $0 \leqq t_\mu < 1$, define for each system of values of the $(u)$'s, or the $(v)$'s, a unique point of an $S_{2p}$ of which the $(t)$'s are considered as non-homogeneous point coördinates. The point-set thus obtained fills up a generalized cube $U_{2p}$, at least if we assume, as we may, that the axes of the $(t)$'s in $S_{2p}$ are rectangular. We shall make frequent use of this generalized cube. We could have defined it in terms of the $(u)$'s but for certain applications it is more advantageous to relate directly $\Omega$, and $U_{2p}$.

Now, to every point of $U_{2p}$ corresponds one and only one of $V_p$. However, it may be that to any point of $V_p$ there correspond more than one of $U_{2p}$. Let $r$ be the number of these points. It has been called by Enriques and Severi the *rank* of $V_p$.

Let us show that *the most general Abelian variety belonging to $V_p$ is of rank one.* If $V_p$ is such a variety it will be possible to express every periodic function belonging to $\Omega$ as a rational function of the $(\theta_k)$'s. Assume this condition fulfilled and let $(u)$, $(u')$ be two points of $U_{2p}$ which correspond to the same point of $V_p$. We shall have

$$\frac{\theta_k^1(u)}{\theta_k^1(u')} = \frac{\theta_k^2(u)}{\theta_k^2(u')} = \cdots = \frac{\theta_k^s(u)}{\theta_k^s(u')}.$$

But the periodic functions are in fact rational in the ratios of $(\theta_k)$'s, hence they take the same values at $(u)$ and $(u')$. This will be the case in particular for the ratios of the functions $\theta_k\,(u - e)$, since they are periodic. It follows that the above system will still be satisfied when the $(u)$'s and the $(u')$'s are replaced by the quantities $u_i - e_i$, $u' - e_i$ respectively, the $(e)$'s being any constants. Hence, if we replace $u_i$ by $u_i + du_i$ we must, in virtue of the functional relationship thus established, replace $u'_i$ by $u'_i + du_i$. Otherwise stated

$$du'_i = \frac{\partial u'_i}{\partial u_1} du_1 + \frac{\partial u'_i}{\partial u_2} du_2 + \cdots + \frac{\partial u'_i}{\partial u_p} du_p = du_i$$

$$\therefore \frac{\partial u'_i}{\partial u_k} = 0, \quad i \neq k; \qquad \frac{\partial u'_i}{\partial u_i} = 1; \qquad u'_i = u_i + \omega_i.$$

The $(\omega)$'s form a system of constants which may be added to the $(u)$'s without changing the values of the ratios $\theta^j_k[{}^g_h](u)/\{\theta^l_k[{}^g_h](u)\}$. They are therefore simultaneous periods of these periodic functions, and the points $(u)$, $(u')$ coincide, which means that $r = 1$.

Thus to every Riemann matrix, corresponds a class of Abelian varieties of rank one and they are obviously birationally equivalent.

51. Let us show that *amongst the varieties of rank one, there exists always one without singularities situated in a suitable space and in point to point correspondence without exception with* $U_{2p}$.

Let us start again from the representation (5) where we shall assume the $(\theta_k)$'s such that any other $\theta$ of the same order and characteristic is a linear combination of them. Two circumstances may prevent that $V_p$ as represented by equations (5) satisfy the desired conditions.

(*a*) All the $(\theta_k)$'s vanish for at least one point, $(u)$.

(*b*) There exists at least one point, $(u)$, for which the hyperplanes of $S_{s-1}$ represented by

$$\left\| \begin{matrix} x_1 & x_2 & \cdots & x_s \\ \theta^1_k & \theta^2_k & \cdots & \theta^s_k \\ \dfrac{\partial \theta^1_k}{\partial u_1} & \cdot & \cdots & \cdot \\ \cdot & \cdot & \cdot & \cdot \\ \dfrac{\partial \theta^1_k}{\partial u_p} & \cdot & \cdots & \cdot \end{matrix} \right\| = 0$$

have in common a space of more than $p$ dimensions or, if we prefer, there exists at least one point, $(u)$, for which the array

$$\left\| \begin{matrix} \theta^1_k & \theta^2_k & \cdots & \theta^s_k \\ \dfrac{\partial \theta^1_k}{\partial u_1} & \cdot & \cdots & \cdot \\ \cdot & \cdot & \cdot & \cdot \\ \dfrac{\partial \theta^1_k}{\partial u_p} & \cdot & \cdots & \cdot \end{matrix} \right\|$$

is of rank $< p + 1$.

Let $W$ be the manifold of $U_{2p}$, or, if we please, the algebraic sub-variety of $V_p$ for which either circumstance presents itself. The question is to show that for a suitable choice of the representation (5) there is no manifold of this nature. Let us remark first that among the functions $\theta_{nk}[{g \atop h}]$ are found the $(\theta_k[{(g+a)/n \atop (h+b)/n}])^n$ $(a, b$, integers). Hence if, whatever the representation (5), $W$ contained points of type (*a*), the functions $\theta_k[{(g+a)/n \atop (h+b)/n}](u)$, would all have a point in common. Since for $n$ great enough there is a characteristic $[{(g+a)/n \atop (h+b)/n}]$ differing as little as we please from any given one, by passing to the limit, we could conclude that the $(\theta)$'s of a given order all vanish at a point $(u_0)$ of $V_p$. This is equivalent to stating that for any $\theta_k$,

$$\theta_k(u_1^0 - e_1, u_2^0 - e_2, \cdots, u_p^0 - e_p) = 0,$$

whatever the $(e)$'s, or else $\theta_k \equiv 0$.

Thus, when the characteristic is arbitrary, case (*a*) will certainly not present itself.

Let us show now that case (*b*) may as well be avoided. Consider the representation

$$\tau x_i = \theta_{nk}^j[{g \atop h}](u) \qquad (j = 1, 2, \cdots, s'),$$

where $s'$ is the dimension of the linear system of $(\theta)$'s of order $nk$ and given characteristic and let $W'$ be the singular subvariety corresponding to it. For $k$ great enough, the functions $\theta_k[{g/n \atop h/n}](u)$ will not all vanish at the same point. But the functions $(\theta_k^i[{(g+a)/n \atop (h+b)/n}](u))^n$ are linear combinations of the $(\theta_{nk}^i)$'s of our representation. Let $s$ be the dimension of the system of the $(\theta_k)$'s of given characteristic. We can choose $s$ functions $\theta_k[{(g+a)/n \atop (h+b)/n}](u)$ not vanishing at a given point $A$ of $W'$. Let us form the matrix which defines $W'$ by taking for functions relatively to its $s$ first columns, the $n$th powers of these $s$ functions. One can then verify at once that $A$ must belong to the $W$ which corresponds to the representation by the $\theta_k[{(g+a)/n \atop (h+b)/n}](u)$. But these subvarieties $W$ are all the transformed of one and the same one by ordinary transformations of the first kind, transformations in as great a number as desired provided $n$ is great enough.* When $n$ exceeds a certain limit, they certainly do not have any common points, hence for the representation then obtained by means of the functions $\theta_{nk}[{g \atop h}](u)$ with arbitrary characteristic, neither case (*a*) nor case (*b*) will present itself, and the theorem is proved.

The importance of this theorem is due to the fact that a number of propositions on varieties of more than two dimensions are not applicable when the singularities are not ordinary. We may mention in particular many of Severi's results as well as the results of our first part. In the sequence, unless otherwise stated, in speaking of an Abelian variety, it will always be under-

* A birational transformation of the first kind of $V_p$ is defined by equations such as

$$u'_i = u_i + \text{const.} \ (i = 1, 2, \cdots, p).$$

stood that we mean the variety of rank one without singularities in point to point correspondence without exception with $U_{2p}$, and it is this variety which we shall designate by $V_p$.

## § 2. Connectivity of varieties of rank one

52. Let now $u_1, u_2, \cdots, u_p$ be the variables corresponding to $\Omega$ and consider the generalized cube $U_{2p}$ as defined by the relations

$$u_i = \sum_{\mu=1}^{2p} t_\mu \, \omega_{i\mu} \qquad (i = 1, 2, \cdots, p),$$

where the $(t)$'s describe the interval $0 \cdots 1$, end points included. $U_{2p}$ will have for $s$-dimensional elements, the aggregates of points obtained when $2p - s$ of the $(t)$'s are equal to zero or one and the others vary. Two elements composed of congruent points modulo the periods are homologous. Hence, a system of $\binom{2p}{s}$ $s$-dimensional elements not congruent to each other form a minimum base for all elements of the same dimensionality. Besides, every one of them represents a closed $s$-dimensional manifold, that is, an $s$-cycle, and as these cycles are independent, we shall have a minimum base for $s$-cycles composed of independent cycles. Hence, for $U_{2p}$, and therefore for $V_p$, $R_s = \binom{2p}{s}$, $\sigma_s = 1$. In particular, $R_2 = p(2p - 1)$, $\sigma_1 = \sigma = 1$. As a verification for hyperelliptic surfaces, the values $\rho_0 = 5, \rho = 1$, were given by Picard, and indeed, $R_2 = 6 = \rho_0 + \rho$.

53. We shall attack the same problem by a slightly different method, which will be found very useful below.

Let us designate by $1, 2, \cdots, 2p$, the edges of $U_{2p}$, abutting on the origin, and sense each of them in such manner that starting from the origin the direction of advancement be positive. The face $(\mu, \nu)$ will be sensed by sensing its periphery so that the edge $(\mu)$ be a positive segment of it. We then have $(\mu, \nu) \sim -(\nu, \mu)$. From there we can pass to the sensing of a 3-dimensional face $(\mu, \nu, \pi)$, etc. We are after all merely dealing here with Heegaard's "sensed corners".*

To each combination of indices, $i_1, i_2, \cdots, i_s$ correspond two opposite $s$-dimensional elements. Let $M_s$ be an analytical $s$-cycle. Its projection on $(i_1, i_2, \cdots, i_s)$ is composed of that element counted several times. *I say that if we project it each time on only one of the elements corresponding to the indices* $i_1, i_2, \cdots, i_s$, $M_s$ *is homologous to the sum of its projections.* Let us cut up $M_s$, as we may, into a sum of elements homologous to a hypersphere, then project one of these, $M'_s$, as well as its boundary $M_{s-1}$ upon $(i_1, i_2, \cdots, i_s)$. This will be done by passing a space $S_{2p-s}$ through an arbitrary point $A$ of $M'_s$

*Bulletin de la Société Mathématique de France, vol. 44 (1916), pp. 161–242.

and taking its intersection $B$ with $(i_1, i_2, \cdots, i_s)$. An arbitrary line (straight or otherwise) joining $A$ to $B$, will be called a projecting line of $A$. It is in no wise necessary that this projecting line be in the $S_{2p-s}$, but only that in some manner it should be uniquely and continuously defined for all points $A$.

By continuous deformation we may reduce $M'_s$ to its projection $M''_s$ provided we replace it in $M_s$ by this projection increased by the $s$-dimensional manifold $M'''_s$ generated by the projecting lines of $M_{s-1}$. The manifold $M'_s + M'''_s - M''_s$ will form an $s$-cycle $\overline{M}_s$, reducible by deformation to a point, hence the conclusion: If $P$ is an $S_{2p-s}$ of the $S_{2p}$ which contains $U_{2p}$, Poincaré's character belonging to $\overline{M}_s$ and $P$ will be $(\overline{M}_s\, P) = 0$. By taking in particular for $P$ a space normal to $(i_1, i_2, \cdots, i_s)$ it is seen that the projection of $\overline{M}_s$ on $(i_1, i_2, \cdots, i_s)$ is composed of mutually opposite elements. Hence, finally, when $M'_s$ is replaced in $M_s$ by $M''_s - M'''_s$, the projections on the $s$-dimensional elements of $U_{2p}$ are not changed. Proceeding similarly with all the $s$-dimensional elements such as $M'_s$ of which $M_s$ is composed, we shall succeed in replacing finally this cycle by $(i_1, i_2, \cdots, i_s)$ counted a certain number of times and an $s$-cycle whose projection on $(i_1, i_2, \cdots, i_s)$ has less than $s$ dimensions while on any other $s$-dimensional element, the projection is the same as for $M_s$. We may reason similarly with the part of this new cycle exterior to $(i_1, i_2, \cdots, i_s)$ relatively to an element upon which it has an $s$-dimensional projection and so on, so that finally $M_s$ will have been replaced by the sum of its projections. *The theorem is therefore proved.*

*Remark.* The proof holds also when we consider the projections on elements derived from those such as $(i_1, i_2, \cdots, i_s)$ by translation.

The $s$-cycles are equal to sums of multiples of the cycles $(i_1, i_2, \cdots, i_s)$ which therefore form a minimum base. To establish the independence of the cycles of this base, it is sufficient to remark with Picard, that the integral $\int\int\cdots\int dt_{i_1}\, dt_{i_2} \cdots dt_{i_s}$ has a period $+1$ relatively to $(i_1, i_2, \cdots, i_s)$ and 0 relatively to any other cycle. We have thus a system of $\binom{2p}{s}$ integrals with a period matrix relatively to the cycles in question of rank equal to the number of these periods. As these integrals have no periods relatively to the cycles homologous to zero, the $\binom{2p}{s}$ cycles thus obtained are independent and form a minimum base, hence again $R_s = \binom{2p}{s}$, $\sigma_s = 1$.

54. The $s$-uple integrals of the first kind of $V_p$, as is well known for $p = 2$ and easily proved for any $p$, are all linear combinations of the integrals

$$\int\int \cdots \int du_{j_1}\, du_{j_2} \cdots du_{j_s}.$$

Their number is therefore $i_s = \binom{p}{s}$. To say that there is a certain number of $s$-cycles with periods of integrals of the first kind all zero is therefore the same as to say that there are as many linear relations with integral coefficients between the determinants of order $s$ derived from the array

$$\begin{matrix} \omega_{j_1,1} & \omega_{j_1,2} & \cdots & \omega_{j_1,2p} \\ \omega_{j_2,1} & \cdot & \cdots & \cdot \\ \cdot & \cdot & \cdot & \cdot \\ \omega_{j_s,1} & \cdot & \cdots & \omega_{j_s,2p}, \end{matrix}$$

the coefficients being independent of the indices $j_1, j_2, \cdots, j_{2p}$.

In the case $s = 2$, the number of these independent relations is precisely the number $1 + k$ of alternate Riemann forms belonging to $\Omega$. We have therefore for $V_p$ according to Part I, $\rho \leqq 1 + k$. The consideration of intermediary functions will allow us to show that $\rho \geqq 1 + k$. Assuming this result, for the present, we have then $\rho = 1 + k$. In other words, *every two-cycle of $V_p$ without periods of integrals of the first kind, is algebraic.* For $p = 2$, this has already been established by Bagnera and de Franchis. Thus, the theory of Part I leads us naturally to the consideration of the alternate Riemann forms of $\Omega$.

55. Let us reduce $\Omega$ to the canonical matrix $A$. In general, there will be no other relations between the periods of the unique $p$-uple integral of the first kind of $V_p$ than the following: Let

$$|a_{i_\alpha, j_\beta}|, \qquad |a_{j_\beta, i_\alpha}| \qquad (\alpha, \beta = 1, 2, \cdots, s)$$

be two minors of the determinant $|a_{jh}|$ $(j, h = 1, 2, \cdots, p)$, symmetrical relatively to the principal diagonal, then denote for the present by $\omega_{i\mu}$ the term of the matrix $A$ at the intersection of the $i$th row and $\mu$th column. To the first of our minors corresponds the following determinant of order $p$ derived from $A$:

$$D = \begin{vmatrix} \omega_{1h_1}, & \omega_{1h_2}, & \cdots, & \omega_{1h_{p-s}}, & a_{1j_1}, & a_{1j_2}, & \cdots, & a_{1j_s} \\ \omega_{2h_1}, & \omega_{2h_2}, & \cdots, & \omega_{2h_{p-s}}, & a_{2j_1}, & a_{2j_2}, & \cdots, & a_{2j_s} \\ \cdot & \cdot & \cdot & \cdot & \cdot & \cdot & \cdot & \cdot \\ \cdot & \cdot & \cdot & \cdot & \cdot & \cdot & \cdot & \cdot \\ \omega_{ph_1}, & \cdots, & \cdots, & \omega_{ph_{p-s}}, & a_{pj_1}, & \cdots, & \cdots, & a_{pj_s} \end{vmatrix},$$

where the $(h)$'s are integers in increasing order such that $h_1, h_2, \cdots, h_{p-s}, j_1, j_2, \cdots, j_s$, form in some order the sequence, $1, 2, \cdots, p$. Similarly there corresponds to the other minor a determinant of order $p$, $D'$, derived from $A$ by adjunction of certain columns of indices $k_1, k_2, \cdots, k_{p-s}$, at most equal to $p$. . One verifies at once that $e_{h_1} \cdot e_{h_2} \cdots e_{h_{p-s}} \cdot D = e_{k_1} \cdot e_{k_2} \cdots e_{k_{p-s}} \cdot D'$. When $A$ is as general as possible, every linear relation with integral coefficients existing between the determinants of order $p$ derived from $A$ reduces to a combination of those of that type which result immediately from the fact that $|a_{jh}|$ $(j, h = 1, 2, \cdots, p)$ is a symmetrical determinant. Their number is equal to half the number of non-symmetrical minors derived from the preceding determinant, that is to

$$\frac{1}{2}\sum_{j=1}^{p}\binom{p}{j}\left[\binom{p}{j}-1\right]=\frac{1}{2}\binom{2p}{p}-2^{p-1}.$$

This is the number of $p$-cycles with zero periods for the $p$-uple integral of the first kind of $V_p$ when the variety is as general as possible.

Let us start from the relation

$$(1+i)^{2p}=2^p\, i^p=\sum_{j=0}^{p}\binom{2p}{j}i^j.$$

By considering separately the two cases where $p$ is even or odd, we find that

$$\frac{1}{2}\binom{2p}{p}-2^{p-1}=R_{p-2}-R_{p-4}+R_{p-6}-\cdots,$$

and therefore according to Part I, $\rho_p=0$.

A similar though somewhat more complicated discussion which we have not carried out in detail leads no doubt to $\rho_s=0$. Thus *the numbers* $\rho_s$ *of the most general Abelian variety of rank one are all zero.*

For a special $V_p$ the number of $s$-cycles with zero periods for the integrals of the first kind may exceed the above. Let generally,

$$k_s+R_{s-2}-R_{s-4}+\cdots$$

be this number, $k_s \geqq 0$. The integer $k_s$ may be called the $s$-dimensional index of singularity of $V_p$ or $\Omega$. We always have

$$\rho_s \leqq k_s, \qquad \rho_s^0 \geqq \binom{2p}{s}-\binom{2p}{s-2}-k_s.$$

We may of course introduce an $s$-dimensional index of multiplication. All these indices would no doubt be useful for the classification of Abelian varieties and many of the propositions given by Scorza for $s=2$ could be extended to them but we shall not discuss this any further.

## § 3. **Intermediary functions**

57. An entire function $\phi(u_1, u_2, \cdots, u_p)$ belonging to a matrix $\Omega$ with $p$ rows and $2p$ columns is said to be an intermediary function if

$$\phi(u+\omega_\mu)=e^{-2\pi i(\sum_{j=1}^{p}\alpha_{j\mu}u_j+\beta_\mu)}\cdot\phi(u) \qquad (\mu=1,2,\cdots,2p).$$

By comparing the two possible values for $\phi(u+\omega_\mu+\omega_\nu)$ the following equations of condition are obtained:

$$\sum_{j=1}^{p}(\alpha_{j\mu}\,\omega_{j\nu}-\alpha_{j\nu}\,\omega_{j\mu})=m_{\mu\nu} \qquad (\mu,\nu=1,2,\cdots,2p).$$

where the $(m)$'s are integers. Since $m_{\mu\nu} = -m_{\nu\mu}$ the form

$$F = \sum_{\mu,\nu=1}^{2p} m_{\mu\nu}\, x_\mu\, y_\nu$$

is an alternate form with integral coefficients. We shall call it the *fundamental form* belonging to $\phi$. It will play a very important part in the sequence.

58. The second derivatives of $\log \phi$ are meromorphic periodic functions belonging to $\Omega$. We shall assume that they are effectively $2p$-ply periodic, which is permissible, for otherwise we could replace $\Omega$ by the matrix of genus $< p$, formed by the primitive periods. $\Omega$ will then be a Riemann matrix of genus $p$.

If $\phi$ is a function which vanishes nowhere, the second derivatives of $\log \phi$ are entire periodic functions and therefore they are constants, hence $\phi$ is of the form $e^{G(u)}$, where $G$ is a quadratic polynomial in the $(u)$'s. Let us assume that $\phi$ is not of this type. The second derivatives of $\log \phi$ will then be of type $\theta_m/\theta'_m$ and there will be a hypersurface $E$ of $V_p$ determined by $\phi = 0$, or, as we shall say, cut out by $\phi$. By a $\theta_m$ we mean a $\theta$-function belonging to a canonical matrix $A$ equivalent to $\Omega$. In terms of the $(u)$'s it is a function

$$\theta_m(\lambda_1 u_1 + \lambda_2 u_2 + \cdots + \lambda_p u_p);$$

in reality an intermediary function derived from a $\theta$.

The hypersurface cut out by $\theta'_m$ is composed of $E$ and of another hypersurface $E'$ upon which $\theta_m$ vanishes also. As $\theta_m$, $\theta'_m$, cut out algebraic hypersurfaces, their common part $E'$ is also algebraic. Hence, $E$, the residue of an algebraic hypersurface with respect to another, is itself algebraic and therefore *any intermediary function cuts out in $V_p$ an algebraic hypersurface.*

59. For the sequence, it is very important to show that an arbitrary intermediary function is always reducible to a $\theta$ by a linear change of variables. This reduction has been very simply effected for $p = 2$ by Humbert, Bagnera and de Franchis. Humbert's method may be extended to any $p$, but as it consists in taking $\Omega$ always in the canonical form, we do not obtain thus in any simple manner the relationship between fundamental and principal forms. As to the method of Bagnera and de Franchis, we have not succeeded in extending it. The difficulty consists in showing that a certain form is principal. The method to be followed here, completely different from that of these authors, seems to present considerable interest in itself and probably will be useful in other applications as well.

This method consists in the following: We shall obtain the expression of the algebraic cycle $(E^{p-1})$ in terms of the cycles $(\mu, \nu)$, and from the expression in question, we shall deduce that the *inverse* of the fundamental form is a principal form of Riemann.

60. In the first place, the hypersurface $(E)$ itself is a $(2p-2)$-cycle. We shall first seek the expression of this cycle $(E)$ in terms of the $p(2p-1)$ fundamental cycles. Let $(E) \sim \Sigma m_{i_1 i_2 \cdots i_{2p-2}} (i_1, i_2, \cdots, i_{2p-2})$. In the sum at the right, only one of the cycles corresponding to a given combination of indices must be taken. We shall agree to take always the cycle for which the $(i)$'s are in increasing order. The question is to evaluate the coefficients in the homology. We shall see that if $E$ is irreducible, a hypothesis which will be made until further notice, $(E)$ considered as a manifold entirely in $U_{2p}$ is sensed everywhere in the same manner with respect to the $(2p-2)$ dimensional elements of $U_{2p}$, which translates itself into the fact that its projections on $(i_1, i_2, \cdots, i_{2p})$ do not include any $(2p-2)$ dimensional elements which destroy each other. It will be necessary then, first to determine that sense, then the number of times that the projection on $(i_1, i_2, \cdots, i_{2p-2})$ covers it up.

61. We must first examine a little closer the notion of sense. Let there be given in an $S_q$ a system of $q$ directed rectangular axes and corresponding variables $t_1, t_2, \cdots, t_q$ and in that space an analytical manifold of $k < q$ dimensions, $M_k$, without any singular points. We wish to define what is meant by the sensing of that element with respect to the $S_k$ defined by the equations $t_{i_1} = t_{i_2} = \cdots = t_{i_{q-k}} = 0$, which we will designate by $(i_{q-k+1}, i_{q-k+2}, \cdots, i_q)$, it being understood that $t_{i_{q-k+1}}, \cdots, t_{i_q}$ are variable in that space.

Let $M'_{q-k}$ be an $S_{q-k}$ transverse to $M_k$ and passing through one of its points $A$ and take a system of $q$ axes, of which $q-k$, $As_1, As_2, \cdots, As_{q-k}$ are in $M'_{q-k}$ and $k$ others $As_{q-k+1}, As_{q-k+2}, \cdots, As_q$ are in $M_k$. More precisely the coordinates of any point of $M_k$ are, in the vicinity of $A$, analytical functions of certain variables $s_{q-k+i}$ $(i = 1, 2, \cdots, k)$. We may define the "axes in $M_k$" as the tangents at $A$ to the lines $ds_{q-k+i} \neq 0$, $ds_{q-k+j} = 0$ $(j \neq i)$, sensed positively in the direction of increreasing $(s)$'s.—Assume first that the axes $t_{i_h}$ are congruent to those just defined, i.e. if $t'_h$ are the variables for this second set, that there exists a continuous series of infinitesimal transformations of the group with parameters $a_{ik}, b_i$,

$$t'_i = \sum_{k=1}^{q} a_{ik} t_{i_k} + b_i, \qquad |a_{ik}| = +1 \quad (i, k = 1, 2, \cdots, q)$$

reducing the first axes to the second. This continuous series of transformations defines a displacement in $S_q$. The Jacobian is

$$\frac{D(s_1, s_2, \cdots, s_q)}{D(t_{i_1}, t_{i_2}, \cdots, t_{i_q})} = +1$$

If there exists such a displacement for which the Jacobian

$$\frac{D(t'_1, t'_2, \cdots, t'_{q-k})}{D(t_{i_1}, t_{i_2}, \cdots, t_{i_{q-k}})}$$

changes its sign an even number of times, we shall say that $M_k$ is sensed positively with respect to the initial space, $(i_{q-k+1}, \cdots, i_q)$. In such a displacement, a sufficiently small $k$-dimensional element within the displaced space $(i_{q-k+1}, \cdots, i_q)$ is never projected within its initial position into an element of $< k$ dimensions or is so an even number of times. Finally, the product

$$\frac{D(s_1, s_2, \cdots, s_{q-k})}{D(t_{i_1}, t_{i_2}, \cdots, t_{i_{q-k}})} \cdot \frac{D(s_1, s_2, \cdots, s_q)}{D(t_{i_1}, t_{i_2}, \cdots, t_{i_q})} > 0.$$

If on the contrary the sign considered above changes an odd number of times—and it is easy to show that its parity is perfectly defined—we shall say that the sense in question is negative. The product of the two Jacobians is then negative.

If the axes $(s)$ and $(t)$ are not congruent, we compare $M_k$ to the $S_k$ corresponding to the indices $i_2, i_1, i_3, \cdots, i_q$ which is opposed to $(i_{q-k+1}, \cdots, i_q)$ and we define the sense as being opposite to that then obtained. We see then that the sense is always given by the sign of the Jacobian product, which may therefore be used as definition of this sense. The axes $(s)$ will be called intrinsic axes of $M_k$.

62. Let us return to our problem. To show first that $(E)$ has an invariant sense relatively to $(i_3, i_4, \cdots, i_{2p})$ it is sufficient to establish that if $s_1, s_2, \cdots, s_{2p}$ is a system of intrinsic axes and $s_1, s_2$, the variables belonging to the transverse plane the Jacobians $D(s_1, s_2)/D(t_{i_1}, t_{i_2})$ vanish only on a manifold of $(2p - 4)$ dimensions at most, of $(E)$. Let

$$\phi = \phi' + i\phi'', \qquad u_j = u_j' + iu_j'',$$

where $\phi', \phi'', u_j', u_j''$ are real, and take as coördinates $s_1, s_2$, two distinct linear combinations of $\phi', \phi''$. The question is reduced to showing then that the Jacobian $D(\phi', \phi'')/D(t_1, t_2)$, for example, vanishes only on a $(2p - 4)$-dimensional manifold of $(E)$. But the left hand side is a linear combination with constant coefficients of products of terms such as $\partial\phi'/\partial u_j'$, $\partial\phi'/\partial u_j''$, $\partial\phi''/\partial u_j'$, $\partial\phi''/\partial u_j''$, that is, by virtue of classical relations, of terms $\partial\phi/\partial u_j$. It is therefore an entire function of $u_1, u_2, \cdots, u_p$ and the dimension of the manifold in question is at most $2p - 4$.

It follows from this that $m_{i_3 i_4 \cdots i_{2p}}$ has the sign of the product

$$\frac{D(t_{i_1}, t_{i_2})}{D(s_1, s_2)} \cdot \frac{D(t_{i_1}, t_{i_2}, \cdots, t_{i_{2p}})}{D(s_1, s_2, \cdots, s_{2p})}.$$

But the second factor is equal to $(-1)^n D(t_1, t_2, \cdots, t_{2p})/D(s_1, s_2, \cdots, s_{2p})$ where $n$ is the number of transpositions of the permutation $\binom{i_1, i_2, \ldots, i_{2p}}{1, 2, \ldots, 2p}$. Finally, we can always choose the $(s)$'s in such a way that the coefficient of

$(-1)^n$ will be positive. In these conditions $m_{i_3 i_4 \cdots i_{2p}}$ will have the sign of $(-1)^n D(t_{i_1}, t_{i_2})/D(s_1, s_2)$.

To determine the absolute value of $m_{i_3 i_4 \cdots i_{2p}}$, we remark that the number of times that the projection of an $(E)$ on an $(i_3, i_4, \cdots, i_{2p})$ covers it up, is equal to the absolute value of

$$\frac{1}{2\pi i}\int d \log \phi$$

taken over the positive contour of $(i_1, i_2)$. The value as found by an elementary integration is* $\sum_{j=1}^{p} (\alpha_{j i_1} \omega_{j i_2} - \alpha_{j i_2} \omega_{j i_1}) = m_{i_1 i_2}$. Now let us take in the transverse plane, a circuit $\zeta$ surrounding its point of incidence and such that

$$\frac{1}{2\pi i}\int_{\zeta} d \log \phi = +1 .$$

We shall assume the axes $s_1, s_2$, so chosen that $\zeta$ is sensed positively in the direction $s_1, s_2$.

Let $\zeta'$ be the projection of $\zeta$ on $(i_1, i_2)$ when the point of incidence of the transverse plane comes in one of the points where $(E)$ cuts $(i_1, i_2)$. We have

$$\frac{1}{2\pi i}\int_{\zeta} d \log \phi = \frac{1}{2\pi i}\int_{\zeta_1} d \log \phi = +1 .$$

Hence the integral taken over the contour of $(i_1, i_2)$ in the sense from $i_1$ towards $i_2$, is positive if $\zeta'$ and the contour have the same sense, negative in the contrary case. Therefore

$$m_{i_1 i_2} \frac{D(t_{i_1}, t_{i_2})}{D(s_1, s_2)} > 0 ,$$

and finally $m_{i_3 i_4 \cdots i_{2p}} = (-1)^n m_{i_1 i_2}$. If $n'$ corresponds to the indices $i_2, i_1, \cdots, i_{2p}$ we have $(-1)^n = -(-1)^{n'}$, $(-1)^n m_{i_1 i_2} = (-1)^{n'} m_{i_2 i_1}$, and we can decide without inconvenience to take $i_1 < i_2$ since we must choose only one of these combinations. In assuming, as we already had occasion to state, $i_3 < i_4 < \cdots < i_{2p}$, $n$ will be the number of transpositions which bring $i_1, i_2$, into the position $1, 2$. Finally then

$$(E) \sim \sum (-1)^n m_{i_1 i_2} (i_3, i_4, \cdots, i_{2p}) .$$

63. Let us show that with a suitable convention nothing is changed, even if $E$ is reducible.

I say first, that two arbitrary hypersurfaces, $(E')$, $(E'')$ have two sides sensed alike with respect to all the $(i_3, i_4, \cdots, i_{2p})$. For, taking always a

* Poincaré, as I have found since writing the above, has already used analagous considerations in a memoir of the Acta mathematica, vol. 26 (1902), pp. 43–98.

representation analogous to (5) of No. 50 for $V_p$, there exists a polynomial $F(x_1, x_2, \cdots, x_s)$ vanishing on $E'$ and $E''$. Replacing the $(x)$'s by the $(\theta)$'s, we see that $E'$, $E''$, are part of an irreducible hypersurface cut out by a certain $\theta$. Hence $(E')$, $(E'')$, have an invariant sense at all their points. Moreover, there exists an algebraic variety of the $S_{s-1}$, containing $V_p$, cutting out on $V_p$ an irreducible hypersurface, $G$, tangent to $E'$, $E''$ in two ordinary points, $A$, $B$. If we prefer, we may say that there is a suitable $\theta$ cutting out $G$. Let us now take on $(E')$, $(E'')$, $(G)$, intrinsic coördinates with the corresponding axes congruent to each other, and let $s_j'$, $s_j''$, $s_j$ be these coördinates. The senses with respect to $(i_3, i_4, \cdots, i_{2p})$ of $(E')$, $(G)$ in $A$ and of $(E'')$, $(G)$ in $B$ will be the same or opposite according as the Jacobians $D(s_1', s_2')/D(s_1, s_2)$, $D(s_1'', s_2'')/D(s_1, s_2)$ are positive or negative, the first in $A$ and the second in $B$. As these signs do not depend upon the $(i)$'s, our affirmation follows.

The preceding discussion makes it possible to define a typical sensing for algebraic hypersurfaces of $V_p$ to be taken as positive, and this we shall do in the future.

Let us assume that $(E')$, $(E'')$, are two irreducible parts of the hypersurface $(E)$ cut out by $\phi$ and that they intersect in a point, $A$, ordinary for both. Through $A$ we draw a plane parallel to $(i_1, i_2)$ wherein we trace a small circuit $\zeta$ surrounding $A$. By attributing to $\zeta'$ a suitable direction, we shall have $\int_\zeta d \log \phi = + 2 \cdot 2\pi i$. Through a point near $A$, let us draw a plane transverse to $(E')$, then surround its point of incidence with a small circuit $\zeta'$, avoiding $(E'')$ and sensed in such a way that

$$\int_{\zeta'} d \log \phi = 2\pi i.$$

Let $\zeta''$ be an analogous circuit for $(E'')$. As $\int_\zeta d \log \phi = \int_{\zeta'} + \int_{\zeta''}$, $\zeta'$, $\zeta''$ are sensed similarly with respect to all the $(i_1, i_2)$. If, therefore, we desire once for all to take for every hypersurface only the cycle formed by its positive side, the Jacobians $D(t_{i_1}, t_{i_2})/D(s_1, s_2)$ will still have the same sign at all points of $(E)$ provided the intrinsic axes are congruent at all points. These signs will always be those of the coefficients $m_{i_1 i_2}$, and our reasoning still holds, leading again to the same homology as previously.

Let us assume, finally, that $(E')$, $(E'')$, have no simple points in common, and let $\psi$ be an intermediary function cutting out a hypersurface $G$, which meets $E'$, $E''$ in two ordinary points, $A'$, $A''$. By reasoning as we have just done, with $\phi\psi$ taking now the place of $\phi$, and comparing $E'$ and $E''$ with $G$, we show at once that we can always arrange matters so that the Jacobians $D(t_{i_1}, t_{i_2})/D(s_1, s_2)$ have the same sign on $(E')$ and $(E'')$, the corresponding intrinsic axes being always assumed congruent. The rest can be completed as previously.

*Remark.* An immediate consequence of the similar sensing of the hypersurfaces of $V_p$ is that the signs of the coefficients $m_{i_1 i_2}$ depend only upon the indices *but not upon the fundamental form considered.* The sign of $(-1)^n m_{i_1 i_2}$ indicates the sensing of the algebraic hypersurfaces of $V_p$ relatively to $(i_3, i_4, \cdots, i_{2p})$.

64. It is clear that, in the same manner as above for a $2p - 2$ dimensional element, the sensing of a $2p - 4$ dimensional element with respect to $(i_5, i_6, \cdots, i_{2p})$ can be considered as determined by the sign of the product

$$\frac{D(t_{i_1}, t_{i_2}, t_{i_3}, t_{i_4})}{D(s_1, s_2, s_3, s_4)} \cdot \frac{D(t_{i_1}, t_{i_2}, \cdots, t_{i_{2p}})}{D(s_1, s_2, \cdots, s_{2p})},$$

where the coördinates $s_1, s_2, s_3, s_4$ belong to the transverse $S_4$ and the others belong to the element itself.

We can show next that the image in $U_{2p}$ of an irreducible $p - 2$ dimensional algebraic subvariety of $V_p$ is a $2p - 4$ dimensional manifold sensed relatively to the $(i_5, i_6, \cdots, i_{2p})$, in a manner invariant in all its points and independent of the subvariety chosen.

Let us consider finally several such subvarieties belonging to the intersection of two algebraic hypersurfaces $E_1, E_2$. At all their points we can take intrinsic coördinates, $s_1, s_2, \cdots, s_{2p}$, such that $s_1, s_2$ belong to the plane transverse to $(E_1)$, $s_3, s_4$ to the plane transverse to $(E_2)$, and such that, moreover, the Jacobians

$$\frac{D(t_1, t_2, \cdots, t_{2p})}{D(s_1, s_2, \cdots, s_{2p})}, \frac{D(t_{i_1}, t_{i_2})}{D(s_1, s_2)}, \frac{D(t_{i_1}, t_{i_2})}{D(s_3, s_4)}$$

should be, the first equal to one, and the other two of invariant sign at all points of the intersection of $(E_1)$ with $(E_2)$. This is equivalent to associating in a definite way, $(E_1)$, $(E_2)$, in their common points. The known identity

$$\frac{D(t_{i_1}, t_{i_2}, t_{i_3}, t_{i_4})}{D(s_1, s_2, s_3, s_4)} = \sum \pm \frac{D(t_{i_1}, t_{i_2})}{D(s_\alpha, s_\beta)} \cdot \frac{D(t_{i_3}, t_{i_4})}{D(s_\gamma, s_\delta)}$$

shows that we change the sense of the intersection whenever we change the sense of one of the hypersurfaces but not when we change that of both. We shall agree always to associate the sides of $(E_1)$, $(E_2)$ which have the same sense. The side thus obtained for their intersection will be said to be its positive side. The sum of the irreducible parts similarly sensed will be what we shall call the intersection $(E_1 E_2)$. The following are two of its properties: (*a*) The cycles $(E_1 E_2)$ and $(E_2 E_1)$ are identical; (*b*) let there be two cycles $\Delta_1, \Delta_2$, nowhere sensed in opposition to $(E_1)$, $(E_2)$, and respectively reducible to these two cycles by a deformation, during which they never acquire a common $2p - 3$ dimensional part. We may define their common $(2p - 4)$-cycle $\Delta_1 \Delta_2$ as we have done for $(E_1)$, $(E_2)$, and this cycle is homologous to $(E_1 E_2)$.

65. Let us designate for the present by $(i_1, i_2, \cdots, i_s)$ not merely an element forming part of the boundary of $U_{2p}$ but more generally any element which may be derived from it by a translation parallel to an edge—such an element forming of course a homologous cycle.

The hypersurfaces $E_1$, $E_2$, are cut out by intermediary functions belonging to two fundamental forms

$$F_1 = \sum m^1_{\mu\nu} x_\mu y_\nu, \qquad F_2 = \sum m^2_{\mu\nu} x_\mu y_\nu.$$

We can take for cycles $\Delta_1$, $\Delta_2$, the following:

$$\Delta_1 = \sum (-1)^n m^1_{i_1 i_2} (i_3, i_4, \cdots, i_{2p}), \Delta_2 = \sum (-1)^n m^2_{i_1 i_2} (i'_3, i'_4, \cdots, i'_{2p}),$$

where at the right the cycles are so chosen that $\Delta_1$, $\Delta_2$ do not have any $(2p-3)$-dimensional manifold in common.

Owing to the large degree of arbitrariness in the definition of the projecting lines in No. 53, we may first reduce $(E_1)$ to $\Delta_1$, then $(E_2)$ to $\Delta_2$ in such a manner that during the deformation the two cycles never acquire a common $(2p-3)$-dimensional manifold. The property (*b*) is then applicable here and we shall have $(\Delta_1 \Delta_2) \sim (E_1 E_2)$.

The question is therefore reduced to the determination of the $(2p-4)$-dimensional cycles common to the $(i_3, i_4, \cdots, i_{2p})$. Now consider the elements $(i_3, i_4, \cdots, i_{2p})$, $(i'_3, i'_4, \cdots, i'_{2p})$. They are respectively represented by two systems of equations

$$x_{i_1} = \text{const.}, \qquad x_{i_2} = \text{const.}; \qquad x_{i'_1} = \text{const.}, \qquad x_{i'_2} = \text{const.},$$

where the two sequences $i_1, i_2, \cdots, i_{2p}$ and $i'_1, i'_2, \cdots, i'_{2p}$ form each in a certain order the sequence of integers, $1, 2, \cdots, 2p$. These equations will be compatible if and only if the four indices, $i_1, i_2, i'_1, i'_2$ are all different. We will then have intersection of the cycles $(i_3, i_4, \cdots, i_{2p})$, $(i'_3, i'_4, \cdots, i'_{2p})$ only if all their indices except two are the same. Let $i_\alpha$, $i_\beta$ be those of the first which do not belong to the second, $i'_\gamma$, $i'_\delta$ those of the second which do not belong to the first, and $n$, $n'$ the numbers of transpositions bringing these pairs of indices into the position $1, 2$, from their right place in the sequence $1, 2, \cdots, 2p$. We can replace $(E_1)$, $(E_2)$, by the sums of homologous cycles. In the one belonging to $(E_1)$ we have the term

$$(-1)^{n'} m_{i'_\gamma i'_\delta} (i_3, i_4, \cdots, i_{2p})$$

and in that belonging to $(E_2)$, the term $(-1)^n m_{i_\alpha i_\beta} (i'_3, i'_4, \cdots, i'_{2p})$, and in no other terms of the sum do the same cycles appear. Let $j_5, j_6, \cdots, j_{2p}$ be the common indices. These two terms can be replaced by

$$(-1)^{n+n'} m_{i'_\gamma i'_\delta} (i_\alpha, i_\beta, j_5, \cdots, j_{2p})$$

and $(-1)^{n+n'} m_{i_\alpha i_\beta}(i'_\gamma, i'_\delta, j_5, \cdots, j_{2p})$ respectively. Their intersection is composed of the cycle

$$m^2_{i_\alpha i_\beta} m^1_{i'_\gamma i'_\delta} \frac{D(t_{i_\alpha}, t_{i_\beta}, t_{i'_\gamma}, t_{i'_\delta}, t_{j_5}, \cdots, t_{j_{2p}})}{D(t_1, t_2, \cdots, t_{2p})} (j_5, j_6, \cdots, j_{2p}),$$

the Jacobian being introduced as above for reasons of sensing. This term is of the type

$$(-1)^n m^1_{i_1 i_2} m^2_{i_3 i_4} (i_5, i_6, \cdots, i_{2p}),$$

where $n$ is always the same number and can be taken to be equal to the sum of the transpositions bringing the pairs $i_1, i_2$ and $i_3, i_4$ each from their proper places in the sequence $1, 2, \cdots, 2p$ into the position $1, 2$. If we assume each of the three systems of indices $i_1, i_2; i_3, i_4; i_5, i_6, \cdots, i_{2p}$, in increasing order, we shall have

$$(E_1 E_2) \sim \sum (-1)^n m^1_{i_1 i_2} m^2_{i_3 i_4} (i_5, i_6, \cdots, i_{2p})$$

and in particular if we decide always to take $i_1 < i_3$

$$(E^2) \sim 2\sum (-1)^n m_{i_1 i_2} m_{i_3 i_4} (i_5, i_6, \cdots, i_{2p}).$$

This can be generalized at once and we shall have with conventions similar to those already made

$$(E_1 E_2 \cdots E_s) \sim \sum (-1)^n m^1_{i_1 i_2} m^2_{i_3 i_4} \cdots m^s_{i_{2s-1} i_{2s}} (i_{2s+1}, i_{2s+2}, \cdots, i_{2p})$$

$$(E^s) \sim s! \sum (-1)^n m_{i_1 i_2} m_{i_3 i_4} \cdots m^s_{i_{2s-1} i_{2s}} (i_{2s+1}, i_{2s+2}, \cdots, i_{2p}).$$

It is of course understood that the various irreducible parts of $(E_1 E_2 \cdots E_s)$ are everywhere sensed alike.*

66. The two most interesting cases are $s = p$ and $s = p - 1$. In the first case we obtain, except for the sign, $[E_1 E_2 \cdots E_p]$ which is the number of zeros common to $p$ intermediary functions. In particular†

$$\pm [E^p] = p! \sum (-1)^n m_{i_1 i_2} m_{i_3 i_4} \cdots m_{i_{2p-1} i_{2p}} = p!\,\pi,$$

where the quantity $\pi$ is the *pfaffian* of the skew symmetric determinant $|m_{jk}|$ $(j, k = 1, 2, \cdots, 2p)$ of the fundamental form $F$.

For $s = p - 1$ we obtain *the expression for the algebraic two-cycle*:

$$(E^{p-1}) \sim (p-1)! \sum (-1)^n m_{i_1 i_2} m_{i_3 i_4} \cdots m_{i_{2p-3} i_{2p-2}} (i_{2p-1}, i_{2p}).$$

The coefficient of $(i_{2p-1}, i_{2p})$ in the sum multiplied by $\pi$ is equal to the coeffi-

* $(E^s)$ is actually a $(2p - 2s)$-cycle, for the functions $\phi(u - e)$ cut out hypersurfaces of $\{E\}$, and $s$ of them with the $(e)$'s arbitrary will intersect in a $2p - 2s$ dimensional manifold, as may be shown for example by recurrence.

† For ordinary $\theta$'s this reduces to a classical formula due to Poincaré.

cient $M_{i_{2p-1}i_{2p}}$ of $m_{i_{2p-1}i_{2p}}$ in the expansion of the determinant of $F$. Hence, finally

$$\pi(E^{p-1}) \sim (p-1)!\sum M_{\mu\nu}(\mu,\nu) \qquad (\mu<\nu).$$

67. Since $(E^{p-1})$ is algebraic, the corresponding periods of the integrals of the first kind of $V_p$ are zero, hence $\Phi = \sum M_{\mu\nu}x_\mu y_\nu$, which up to a numerical factor is the *inverse* of $F$, is an alternate Riemann form of $\Omega$. We shall see that it is a principal form of $\Omega$ or else is such a form changed in sign.*

Assume, indeed, first that $(E^{p-1})$ is irreducible. We mean thereby, explicitly, that there exist $p-2$ hypersurfaces of the same algebraic system as $E$, whose intersection with $E$ is an irreducible curve. Let $\zeta$ be the contour making $(E^{p-1})$ simply connected, and let $X = x+iy$ be one of the point coördinates on $(E^{p-1})$. Finally, set

$$u = u' + iu'' = \sum_{j=1}^{p}\lambda_j u_j; \qquad \xi_\mu + i\eta_\mu = \sum_{j=1}^{p}\lambda_j\omega_{j\mu}.$$

We have, according to Riemann,

$$\iint_{\pm(E^{p-1})}\left[\left(\frac{\partial u'}{\partial x}\right)^2+\left(\frac{\partial u'}{\partial y}\right)^2\right]dxdy = \int_\zeta u'\,du'' > 0.$$

But by means of a classical transformation of the first integral

$$\int_\zeta u'\,du'' = \iint_{\pm(E^{p-1})}\frac{D(u',u'')}{D(x,y)}\,dxdy = \iint_{\pm(E^{p-1})}du'\,du''$$

$$= \pm\frac{(p-1)!}{\pi}\sum M_{\mu\nu}\iint_{(\mu,\nu)}du'\,du''$$

$$= \pm\frac{(p-1)!}{\pi}\sum M_{\mu\nu}(\xi_\mu\eta_\nu - \xi_\nu\eta_\mu) > 0,$$

which proves our affirmation in the case considered.

The signs $\pm$ come from the fact that it is only for one of the sides of $(E^{p-1})$ that the double integral will be positive.

Assume now that $E^{p-1}$ is composed of several irreducible curves $C_1, C_2, \cdots, C_n$. The cycle $(E^{p-1})$ is composed of the sum of the cycles $(C_j)$ all assumed sensed in the same manner. We know that the following condition will then be fulfilled: Let $C$ be an arbitrary algebraic curve tangent to $C_j$ at a simple point $A$ and to $C_h$ at a simple point $B$, then trace in a tangent plane at $A$ and $B$, two small circuits $\zeta'$, $\zeta''$ around these points. If we impose upon $\zeta'$, $\zeta''$ directions corresponding to the positive circuits of $C_j$ and $C_h$, they will both be sensed alike with respect to $C$. More precisely speaking,

* Since this was written the same result has been derived in a more direct and elegant manner by Castelnuovo. See Rendiconti dei Lincei, ser. 5, vol. 30 (1921), pp. 50–55.

we may consider a small positive circuit $\zeta$ of $(C)$ starting from the interior of a small tube of axis $\zeta'$ (a tube of which the transverse dimensions are very small with respect to $\zeta'$), then bring it into a similar position relatively to $\zeta''$ without changing its sense. The integral $\int_\zeta u'\,du'' - \int_{\zeta''} u'\,du''$ will be at first very small with respect to each term of the difference, and similarly when we consider $\int_\zeta u'\,du'' - \int_{\zeta''} u'\,du''$ and the final position of $\zeta$. Since, besides, the first integral does not change its sign, the two integrals $\int_{\zeta'}$ $\int_{\zeta''}$ have the same sign. This sign will be that of every positive circuit bounding on $(C_j)$ or $(C_h)$ and we may assume that it is the sign $+$.

Let us draw in particular the contours $\zeta_j$ making the $(C)$'s simply connected. They bound on these manifolds, hence we can sense them. If we do this in the manner just defined, we shall then have $\int_{\zeta_i} du'\,du'' > 0$;

$$\therefore\ \sum \int_{\zeta_j} u'\,du'' = \sum \int\int_{(Cj)} du'\,du'' = \int\int_{\pm(E^{p-1})} du'\,du'' > 0,$$

and the rest is ended as previously. Hence finally:

THEOREM. *The inverse of a non-singular fundamental form is up to the sign a principal form of* $\Omega$.

*Remark.* If $\phi$ is of genus $p_1 < p$, we may, as we have already stated, reduce $\Omega$ to the form

$$\begin{Vmatrix} \Omega_1, & 0 \\ 0, & \Omega_2 \end{Vmatrix},$$

where $\Omega_1$ is of genus $p_1$, $\phi$ being a function of the sole variables belonging to $\Omega_1$. $F$ will then be a fundamental form belonging to $\Omega_1$, hence degenerate for $\Omega$, and the inverse $\Phi$ of $F$ considered as non-degenerate form of genus $p_1$, will be a principal form for $\Omega_1$, and therefore will be a degenerate alternate form of $\Omega$.

67. Before continuing with the reduction of intermediary forms, let us establish a generalization of Riemann's classical inequality. Its interest lies in the fact that it is the only one known to date for multiple integrals of the first kind.

Let $v_1, v_2, \cdots, v_s$ be distinct linear combinations of the variables $u_1, u_2, \cdots, u_p$, and $x_1, x_2, \cdots, x_s$, $s$ arbitrary point coördinates on $E_1, E_2, \cdots, E_{p-s}$, then set $v_j = v_j' + iv_j''$, $x_j = x_j' + ix_j''$. By starting with the relations

$$\frac{\partial v_j}{\partial x_h} = \frac{\partial v_j'}{\partial x_h'} + i\frac{\partial v_j''}{\partial x_h'}, \qquad i\frac{\partial v_j}{\partial x_h} = \frac{\partial v_j'}{\partial x_h''} + i\frac{\partial v_j''}{\partial x_h''},$$

$$\frac{\partial \bar v_j}{\partial \bar x_h} = \frac{\partial v_j'}{\partial x_h'} - i\frac{\partial v_j''}{\partial x_h'}, \qquad -i\frac{\partial \bar v_j}{\partial \bar x_h} = \frac{\partial v_j'}{\partial x_h''} - i\frac{\partial v_j''}{\partial x_h''}$$

we find that

$$\frac{D(v_1', v_2', \cdots, v_s', v_1'', \cdots, v_s'')}{D(x_1', x_2', \cdots, x_s', x_1'', \cdots, x_s'')} = \left(-\frac{1}{2}\right)^s \left|\frac{D(v_1, v_2, \cdots, v_s)}{D(x_1, x_2, \cdots, x_s)}\right|^2$$

and therefore

$$\iint \cdots \int_{(E_1 E_2 \cdots E_{p-s})} dv_1' \, dv_2' \cdots dv_s' \, dv_1'' \cdots dv_s'' \neq 0 .$$

The sign of this multiple integral depends only upon the sensing of the $(E)$'s but not upon these hypersurfaces nor even upon $V_p$. Suppose in particular that $V_p$ possesses $p$ independent elliptic integrals which we shall still denote by $u_1, u_2, \cdots, u_p$ and take for hypersurfaces $E_j$, an hypersurface which is here algebraic, defined by

$$u_{s+j} = \text{const.} \qquad (j = 1, 2, \cdots, p - s).$$

The multiple integral is then reduced to the product by $(-1)^{\frac{s(s-1)}{2}}$ of $s$ integrals of type

$$\iint_{(E_1 E_2 \cdots E_{p-1})} du_1' \, du_1''$$

which have all the same sign. Hence

$$(-1)^{\frac{s(s-1)}{2}} \iint \cdots \int_{(E_1 E_2 \cdots E_{p-s})} dv_1' \, dv_2' \cdots dv_s' \, dv_1'' \cdots dv_s''$$

is positive if $s$ is even and has the sign of the double integral if $s$ is odd. Let $\tau_{i_{2(p-s)+1} i_{2(p-s)+2} \cdots i_{2p}}$ be the period of the multiple integral with which we are dealing, with respect to $(i_{2(p-s)+1}, \cdots, i_{2p})$. The quantity

$$(-1)^{\frac{s(s-1)}{2}} \sum (-1)^n m^1_{i_1 i_2} m^2_{i_3 i_4} \cdots m^{p-s}_{i_{2(p-s)-1} i_{2(p-s)}} \tau_{i_{2(p-s)+1} \cdots i_{2p}}$$

is positive if $s$ is even and has the same sign as for $s = 1$ if $s$ is odd. This is the generalization which we had in view.

68. It is easy to show now that we can always transform $\Omega$ into a canonical matrix $A$ in such a way that $\phi$ becomes, up to a quadratic exponential factor, a $\theta$ belonging to $A$. We can obviously assume that $\phi$ is of genus $p$.

Reduce $\Omega$ to a canonical matrix $A$ in such a way that $\Phi$ becomes the form

$$\Phi' = \sum_{\mu=1}^{p} e_\mu (x_\mu y_{p+\mu} - x_{p+\mu} y_\mu).$$

The transformed of $F$ is then

$$F' = m e_p \sum_{\mu=1}^{p} \frac{(x_\mu y_{p+\mu} - x_{p+\mu} y_\mu)}{e_\mu},$$

where $m$ is up to the sign the greatest common divisor of the coefficients of $F$. Let $u_1', u_2', \cdots, u_p'$ be the new variables and $\theta_m(u')$ a $\theta$ of order $m$ belonging to $A$ if $m$ is positive or the inverse of a $\theta$ of order $-m$ if $m$ is negative. Let finally $\psi(u')$ be the transformed of $\phi$. The meromorphic function

$$\theta_m^{-1}(u')\cdot\phi(u) = \theta_m^{-1}(u')\psi(u') = \chi(u')$$

behaves like an intermediary function belonging to $A$ with a fundamental form identically zero. By considering the relations for the corresponding coefficients $\alpha_{j\mu}$, it is found that $\alpha_{j\mu} = \alpha_{\mu j}$ $(\mu, j \leqq p)$. Hence the product of $\chi$ by a suitable quadratic exponential behaves in a similar manner with $\alpha_{j\mu} = 0$ $(j, \mu \leqq p)$. The relations satisfied by the other $(\alpha)$'s show that they are also zero. Hence, if $\alpha'_{j\mu}$, $\alpha''_{j\mu}$ correspond to $\theta_m(u')$, and to $e^{G(u')}\cdot\psi$, where the exponential is the one already mentioned, we shall have $\alpha'_{\mu\nu} = \alpha''_{\mu\nu}$ whence follows that the entire function $e^{G(u')}\cdot\psi$ behaves relatively to $A$ like a $\theta$-function. *It is therefore a theta of order* $m$, which proves our affirmation.

The integer $m$, which is the greatest common divisor of the coefficients of $F$, is also called the *order* of $\phi$.

The dimension of the linear system to which $\theta_m$ belongs, or which is the same, of the one to which $\phi$ belongs, is the pfaffian of $F'$,

$$\frac{m^p\, e_p^p}{\prod_\mu (e_p/e_\mu)} = m^p\cdot e_1\, e_2 \cdots e_p = \pi'.$$

Let $D$ be the unimodular transformation changing $\Omega$ into $A$. We have

$$\Phi' = \bar{D}^{-1}\,\Phi D^{-1}, \qquad F' = \bar{D}FD.$$

Hence

$$\pi' = D\pi.$$

Finally, if $\Delta$, $\Delta'$ are the determinants formed with the real and imaginary terms of $\Omega$ and $A$, we have

$$\Delta' = D\Delta, \qquad \therefore\ \pi\Delta = \pi'\,\Delta' < 0.$$

Thus, as shown by Bagnera and de Franchis in the case $p = 2$, the pfaffians of the non-degenerate fundamental forms of $\Omega$ all have the sign of $-\Delta$. By applying to $\Omega$ a suitably chosen unimodular transformation, we can arrange matters so that they will all be positive. The dimension of a complete linear system of intermediary functions will then be exactly equal to the corresponding pfaffian.

## § 4. The relation $\rho = 1 + k$. Theorem of Appell-Humbert

69. Let $\Phi_1$ be a principal form, $\Phi_2, \Phi_3, \cdots, \Phi_{k+1}$ other alternate forms of $\Omega$ constituting with the first a set of $k + 1$ independent forms. The conditions that $x\Phi_1 + y\Phi_j$ be principal, ($x$, $y$ are integers), can be expressed in the form of inequalities

$$f_1(x, y) > 0, \qquad f_2(x, y) > 0, \qquad \cdots, \qquad f_n(x, y) > 0,$$

where the $(f)$'s are homogeneous polynomials with integral coefficients. These conditions, which are satisfied for $x = 1$, $y = 0$, will certainly also be when we take $y = 1$, and $x$ great enough. Hence, if $\Phi_j$ is not principal, we can always replace it in the above system by a principal form. This will allow us to assume that all the $(\Phi)$'s are principal forms. To $\Phi_j$ correspond non-degenerate intermediary functions, hence a fundamental form $F_j$, of genus $p$, inverse of $\Phi_j$. Let us represent any alternate form $G$ of genus $p$ by the point $(G)$ of a space $S_{p(2p-1)-1}$ having its coefficients for homogeneous coördinates. The relation between a form and its inverse establishes an involutory birational correspondence between the points of this space, and if $G$ is not degenerate, the point $(G)$ is not fundamental for that correspondence. Let $\Phi$, $F$ be a principal form and the corresponding fundamental form. The $S_k$ which contains the points $(\Phi)$ is transformed by the involution into an algebraic $k$-dimensional variety $W_k$ containing all the $(F)$'s. I say that there is no $W_h$, $h < k$, containing all these last points. For otherwise there would be a variety $W'_h$ containing all the $(\Phi)$'s, hence all the points $\alpha\Phi + \beta\Phi_j$, where $\alpha$, $\beta$ are arbitrary positive integers. In the neighborhood of $(\Phi)$ there would then be found within $W'_h$ a point-set, everywhere dense, not contained in a space of less than $k$ dimensions, which would have as a consequence effectively $h \geqq k$.

It follows that the points $(F)$ are not contained in a space of less than $k$ dimensions. Besides, if $F$, $F'$ are two fundamental forms, $\alpha F + \beta F'$ is another fundamental form ($\alpha$, $\beta$, arbitrary positive integers). Hence if $W_k$ were not an $S_k$, it would have to contain an infinity of straight lines passing through an ordinary point not situated within an $S_k$ which is impossible. It follows that $W_k$ is merely an $S_k$.

Let finally $F'$ be a degenerate fundamental form of genus $p' < p$. There corresponds to it an intermediary function in $p'$ variables. We know that there exists then at least one intermediary function $\phi''$, of genus $p'' = p - p'$, such that $\phi'^{\alpha}\,\phi''^{\beta}$ be of genus $p$, where $\alpha$, $\beta$ are arbitrary positive integers. If $F''$ is the fundamental form corresponding to $\phi''$, $(\alpha F' + \beta F'')$ will be in the space $S_k$ containing the points which correspond to the non-degenerate forms. Hence $(F')$ will also belong to that $S_k$.

Thus the fundamental forms like the Riemann forms give rise to a linear system $\infty^k$.

The modulus generated by the fundamental forms will have a minimum base composed of $k + 1$ forms $F_1, F_2, \cdots, F_{k+1}$. Let $F$ be a fundamental form of genus $p$ corresponding to intermediary functions of order one. The coefficients of $F$ will then be prime to each other, and there exists a minimum base for fundamental forms of which $F$ is an element. Assume then $F = F_1$. The form $F'_j = F_j + xF_1$ ($x$ a positive integer) is not degenerate for $x$

arbitrary, even if $F_j$ is. If $F_j$ is degenerate, we can then replace it by $F'_j$, and we thus see that there exists a minimum base for fundamental forms composed of non-degenerate forms. We shall assume that the above minimum base already has this property.

70. Let us prove now the following important

THEOREM: *The non-degenerate intermediary functions of a complete continuous system of such functions cut out a complete continuous system of hypersurfaces.*

Let us reduce $\Omega$ to a canonical matrix $A$ in such a manner that the intermediary functions in question become $(\theta)$'s belonging to $A$. We have then to prove that the $(\theta)$'s of a given order $m$ cut out a complete continuous system. Let us represent $V_p$, as before, by equations

$$\tau x_j = \theta_n^j \left[ \begin{smallmatrix} g \\ h \end{smallmatrix} \right] (u) \qquad (j = 1, 2, \cdots, s)$$

where $s - 1$ is the dimension of the $(\theta)$'s of order $n$ of given characteristic. We assume that this representation furnishes a non-singular $V_p$ in point to point correspondence without exception with $U_{2p}$.

Let $\{C_m\}$ be the complete system of hypersurfaces of $V_p$ containing as total hypersurfaces those cut out by the $\theta_m$. We have to show that an arbitrary $C_m$ can be cut out by a $\theta_m$.

Let us designate by $|H|$ the system of hyperplane sections. Since $V_p$ is without singularities, an arbitrary hypersurface $lH$ of the complete system $|lH|$, ($l$ great enough), can be cut out in the $S_{s-1}$ containing $V_p$ by a polynomial of order $l$.*

If we substitute in this polynomial the $(\theta)$'s in place of the $(x)$'s, we find that every hypersurface of $lH$ can be cut out by a $\theta_{nl}$. But by means of an ordinary transformation of $V_p$, every linear system of $\{C_{nl}\}$ can be transformed into another containing an arbitrarily assigned linear system, since these systems are all representable by the points of a Picard variety of period matrix isomorphic to $\Omega$, and which is the image of an involution on $V_p$. It follows that all the linear systems in question have the same dimension, equal to that of a $\theta_{nl}$ of given characteristic. This proves the proposition for $\{C_{nl}\}$, when $l$ is above a certain limit.

We conclude from this that every $rC_m$, ($r$ a sufficiently high multiple of $n$), is cut out by a $\theta_{mr}$. In the neighborhood of an arbitrary point $(u^0)$ of $C_m$ we shall have

$$\theta_{mr}(u) = [f(u_1, u_2, \cdots, u_p)]^r \cdot g(u_1, u_2, \cdots, u_p),$$

where $f$ is a polynomial in $u_p - u_p^0$ whose coefficients are holomorphic in $u_1, u_2, \cdots, u_{p-1}$, in the neighborhood of the set of values $u_1^0, u_2^0, \cdots, u_p^0$, and $g$ is holomorphic in the neighborhood of $(u^0)$, with $g(u_1^0, u_2^0, \cdots, u_p^0) \neq 0$.

* Severi, Rendiconti di Palermo, vol. 28 (1909), pp. 33–87.

Hence $\theta_{mr}^{1/r}$ is holomorphic in the vicinity of $(u^0)$ and therefore it is an entire function. Besides, it behaves like a $\theta_m$ with respect to the matrix $A$, hence up to an exponential quadratic factor perhaps, it is a $\theta_m$ which cuts out $C_m$ and the theorem is proved.

71. The preceding theorem will lead us rapidly to the value of $\rho$. Returning to the minimum base $F_1, F_2, \cdots, F_{k+1}$, for the fundamental forms, let $\phi_i$ be the non-degenerate intermediary function corresponding to $F_i$, and $C_i$ the hypersurface which it cuts out. I say that we cannot have a relation such as $\sum \lambda_i C_i = 0$. Indeed, let $\lambda_i = \lambda'_i - \lambda''_i$ where the $(\lambda')$'s and the $(\lambda'')$'s all are positive integers. By definition, there exists a hypersurface $E$ such that $D' = E + \sum \lambda'_i C_i$ and $D'' = E + \sum \lambda''_i C_i$ are total hypersurfaces of the same continuous system. $E$ can be replaced by any hypersurface of a continuous system which contains it. Hence, we can always consider it as being cut out by an intermediary function $\phi$ of order $m$, high enough, belonging to a fundamental form $mF$ of genus $p$. For a suitable choice of $m$, the fundamental forms $mF + \sum \lambda'_i F_i$, $mF + \sum \lambda''_i F_i$ are not degenerate, and the intermediary functions $\phi \Pi_i \phi_i^{\lambda'_i}$, $\phi \Pi_i \phi_i^{\lambda''_i}$ will not be so either. Since the hypersurfaces $D'$, $D''$ which they cut out belong to the same continuous system, this must also be true for the functions in question and their fundamental forms must coincide—that is, we must have

$$\sum \lambda_i F_i = 0.$$

Hence $\lambda_1 = \lambda_2 = \cdots = \lambda_{k+1} = 0$, and the $(C)$'s are effectively independent.

An immediate corollary is that we must have $\rho \geqq k + 1$, and since

$$\rho \leqq k + 1,$$

*it follows that* $k + 1$ *and* $\rho$ *are equal,* a result whose importance is obvious. Incidentally, we have shown that the $(C)$'s form a base for the hypersurfaces of $V_p$.

72. Theorem of Appell-Humbert: *Every hypersurface of* $V_p$ *may be cut out by an intermediary function.**

Let $C$ be an arbitrary hypersurface of $V_p$. We have $\lambda C + \sum \lambda'_i C_i = \sum \lambda''_i C_i$ ($\lambda, \lambda'_i, \lambda''_i$, positive integers). Hence as before there are two hypersurfaces $D'$, $D''$ cut out by non-degenerate intermediary functions, and such that $\lambda C + D'$ and $D''$ belong totally to the same complete continuous system. Hence there are two intermediary functions $\phi$, $\phi'$ cutting out $\lambda C + D'$ and $D'$. The function $\psi = \phi/\phi'$ behaves like an intermediary function relatively to the periods. Besides $D'$ may be taken arbitrarily within its continuous system. Let $D'_1$ be another hypersurface of that system. There exist inter-

* See Humbert, *Journal de mathématiques pures et appliquées*, ser. 4, vol. 9 (1893), pp. 29–170, for the case $p = 2$.

mediary functions $\phi_1$, $\phi_1'$ cutting out $\lambda C + D_1'$ and $D_1'$ and we see at once that $(\phi\cdot\phi_1')/(\phi'\cdot\varphi_1)$ is neither zero nor infinite on any hypersurface of $V_p$. Hence $\psi_1 = \phi_1/\phi_1' = e^{G(u)}\cdot\psi$, where $G$ is an entire function. Since we may always so choose $D_1'$ that $\psi_1$ be holomorphic at a given point $(u^0)$, $\psi$ is an entire function—therefore it is an intermediary function. Thus $\lambda C$ is cut out by an intermediary function $\psi$. We may now show as in No. 70 that $\psi^{1/\lambda}$ is an intermediary function cutting out $C$, even if $\psi$ is degenerate, which proves the theorem.

73. Let $F$ and $C$ be respectively the fundamental form and the hypersurface corresponding to $\psi$. We have $F + \sum \lambda_i' F_i = \sum \lambda_i'' F_i$ ($\lambda_i'$, $\lambda_i''$, positive integers). Just as before there exists here a non-degenerate fundamental form $F'$, such that $F' + F + \sum \lambda_i' F_i = F' + \sum \lambda_i'' F_i$ is not degenerate either. Let $\phi$ be an intermediary function belonging to $F'$. The intermediary functions $\psi\phi\Pi_i\,\phi_i^{\lambda_i'}$, $\phi\Pi_i\,\phi_i^{\lambda_i''}$ belong to the same continuous system of such functions, which leads at once to the relation $C = \sum(\lambda_i'' - \lambda_i')C_i$. Thus $C_1$, $C_2$, $\cdots$, $C_{k+1}$ form a *minimum base*. Incidentally, these hypersurfaces are independent, hence $\sigma = 1$, as we have already found before in a different manner.

74. *Remarks.* I. When $p = 3$, we can arrive much more rapidly at the relation $\rho = 1 + k$. Indeed, we shall show below independently from this relation, that the arithmetic genus of an Abelian variety of genus $p$ is equal to $(-1)^{p-1}$, hence when $p$ is odd it is equal to the geometric genus. It follows in particular that $V_3$ is completely regular, in which case (No. 36) every two-cycle with zero periods for the integrals of the first kind, is algebraic, which gives at once $\rho = 1 + k$.

II. Whatever $\Omega$, the $p$-uple integral of the first kind of $V_p$ has at least one non-zero period since $\Omega$ is an array of rank $p$. We can then apply the theorem of No. 36: Every algebraic surface which is a complete intersection in $V_p$ of hypersurfaces belonging to a sufficiently general linear system has for bases the traces of the bases of $V_p$.

III. Let us mention explicitly that, whatever $p$,

$$\rho_0^2 = \rho_0 = R_2 - \rho = R_2 - (1 + k) = p(2p - 1) - 1 - k.$$

## § 5. Formula for the arithmetic genera

75. In a paper in the Annals of Mathematics of 1916 the author has established this result: If $C = \sum_{i=1}^{s} \lambda_i C_i$, the arithmetic genus of $C$ is given by

$$[C] = [\prod_{i=1}^{s}(1 + C_i)^{\lambda_i} - 1] - \binom{\sum \lambda_i - 1}{p - 1},$$

where at the right one must expand in series, then replace the non-numerical terms of degree $\leqq p$ by the corresponding arithmetic genera. He has besides established a formula according to which $[(lC)^h]$ is a polynomial in $l$ with coefficients of the first degree in $[C^i]$.

Assume now that $C$ is a hypersurface cut out by an intermediary function of non-degenerate form $F$. If $C$ is generic in its continuous system, $V_p$ may be birationally transformed into a variety $V'_p$, non-singular and having for hyperplane sections the complete system $|lC|$ ($l$ sufficiently great). The dimension of the containing space is $\pi l^p - 1$, where $\pi$ is the pfaffian of $F$. On the other hand, the postulation of $V'_p$ with respect to the varieties of order $n$ of its space is $\pi(ls)^p$. Hence (Severi, loc. cit.),

$$\pi(ls)^p = \binom{s+p}{p}[(lC)^p] - \binom{s+p-1}{p-1}([(lC)^p] + [(lC)^{p-1}] - 1)$$
$$+ ([(lC)^{p-1}] + [(lC)^{p-2}])\binom{s+p-2}{p-2}$$
$$+ \cdots + (-1)^p([(lC)] + [V_p]).$$

If in these expressions the quantities $[(lC)^h]$ are replaced by their expressions in terms of genera $[C^i]$ we obtain an algebraic equation verified for an infinite number of sets of values $(s, l)$ which cannot be thus related. The relation obtained must therefore be an identity. Making $l = 1$, we have for every value of $s$

$$\pi s^p = \Big[\binom{s+p}{p}C^p - (C^p + C^{p-1} - 1)\binom{s+p-1}{p-1}$$
$$+ \cdots + (-1)^p(C + V_p)\Big].$$

Moreover we can always find constants $t_i$ such that, whatever $s$,

$$s^p = \sum_{i=0}^{p} t_i \binom{s+i}{i}.$$

By comparing the two values of $s^p$ it is found that

$$[C] = (-1)^{p-1}[\pi(t_1 + t_2 + \cdots + t_p) - 1].$$

Making successively $s = 0$ and $s = -1$ in the preceding formula, we obtain $t_0 + t_1 + \cdots + t_p = 0$, $t_0 = (-1)^p$, whence ultimately $[C] = \pi + (-1)^p$. Finally, substituting $s = 0$ in the formula for $\pi s^p$, we find $[V_p] = (-1)^{p-1}$. *Thus the arithmetic genus of an Abelian variety of rank one and genus $p$ is equal to $(-1)^{p-1}$.* This is in agreement with the value $-1$ already known for the arithmetic genus of hyperelliptic surfaces.

It is interesting to observe that a certain formula predicted by Severi but for which no proof is known as yet, is verified in the case of Abelian varieties.

Designating as before by $i_s$ the number of $s$-uple integrals of the first kind, we must have according to this formula

$$[V_p] = i_p - i_{p-1} + \cdots + (-1)^{p-1} i_1 .$$

Since $i_s = \binom{p}{s}$ and

$$1 - \binom{p}{1} + \binom{p}{2} - \cdots + (-1)^{p-1} \cdot p = (1-1)^p - (-1)^p = (-1)^{p-1},$$

the formula holds here.

76. Let $F_i$ be the fundamental form of $C_i$. The pfaffian $\pi$ will be a polynomial in $\lambda_1, \lambda_2, \cdots$, whose coefficients are simultaneous invariants of the forms $F_i$. Since

$$\pi + (-1)^p = [\prod_i (1 + C_i)^{\lambda_i} - 1] - \binom{\sum \lambda_i - 1}{p - 1},$$

whatever the positive integers $\lambda_i$, this relation is an identity and the coefficients of the powers of the $(\lambda)$'s must be equal on both sides. Comparing those of $\lambda_1, \lambda_2, \cdots, \lambda_s$ we obtain $[C_1 C_2 \cdots C_s] = + G_s$, where $G_s$ is a simultaneous invariant of the fundamental forms, $F_1, F_2, \cdots, F_s$ of which it is not necessary to give here the expression, which is easy enough to obtain. This formula solves the problem of the determination of the arithmetic genus for the complete intersections in $V_p$.

We verify in particular that

$$[C^p] = \pi \cdot p!, \qquad \text{(formula of No. 66; Poincaré)}$$

$$[C^{p-1}] = 1 + \frac{1}{2}\pi (p-1) \cdot p!$$

$$[C^{p-2}] = -1 + \frac{\pi (p-2)(3p-5)}{4!} \cdot p!, \qquad \text{etc.}$$

For $p = 2, [C^2], [C], [C_1 C_2]$ have precisely the values already obtained by Bagnera and de Franchis.

## Chapter II. Abelian Varieties with Complex Multiplication

### § 1. Generalities

77. In the first chapter we have established the identity between the numbers $\rho$ and $1 + k$. The next problem of interest is that of the determination of the numbers $h, k$ for a wide range of cases. The varieties with complex multiplication will furnish them readily.

In order that the transformation $T$

$$(1) \qquad u'_j = \sum_{k=1}^{p} \lambda_{jk} u_k \qquad (j = 1, 2, \cdots, p),$$

define an algebraic correspondence on the Abelian variety of rank one $V_p$, belonging to the matrix $\Omega$, it is necessary and sufficient (Humbert, Scorza) that there exist relations

(2) $$\sum_k \lambda_{jk}\, \omega_{k\mu} = \sum_{\nu=1}^{2p} b_{\nu\mu}\, \omega_{j\nu}$$

$$(j = 1, 2, \cdots, p;\ \mu = 1, 2, \cdots, 2p),$$

where the $(b)$'s are rational numbers. These relations can also be written symbolically:

$$\| \lambda_{jk} \| \cdot \Omega = \Omega \cdot \| b_{\mu\nu} \| .$$

If the so-called "multipliers" as defined below are rational and equal, $T$ is said to be an *ordinary multiplication* of $V_p$, while in the contrary case it is said to be a *complex multiplication* of the variety. If the determinant $|\lambda_{jk}| = 0$ we have to deal with a complex multiplication of an Abelian variety of genus $< p$ contained in $V_p$. We shall assume then in general that this determinant is not zero. Besides, (Frobenius) $|\lambda_{jk}| \cdot |\bar{\lambda}_{jk}| = |b_{\nu\mu}|$, and the rank of the determinant at the right is double that of $|\lambda_{jk}|$. Hence, if one is not zero, neither is the other. *Moreover, the determinant of the $(b)$'s is necessarily positive.* Let us assume that the $(b)$'s are integers. The determinant is then equal to a certain integer $m$ and we have to deal with a correspondence $(1 - m)$. In order that we have a birational transformation, it is necessary and sufficient that the determinant be equal to one.

The *multipliers* of $T$ already alluded to are simply the roots of the equation in $\alpha$,

$$|\lambda_{jk} - \epsilon_{jk} \cdot \alpha| = 0 \qquad (\epsilon_{jj} = 1,\ \epsilon_{jk} = 0 \text{ if } j \neq k),$$

If none of them are zero, we have a multiplication of functions of genus $p$ or *non-singular* multiplication.

According to Frobenius,

$$|\lambda_{jk} - \epsilon_{jk} \cdot \alpha| \cdot |\bar{\lambda}_{jk} - \epsilon_{jk} \cdot \alpha| = |b_{\nu\mu} - \epsilon_{\nu\mu} \cdot \alpha| = F(\alpha),$$

hence the multipliers are roots of an equation of degree $2p$, with rational coefficients, $F(\alpha) = 0$, called the *characteristic equation* of the complex multiplication $T$. A real root must annul each factor on the left side above and is therefore double for $F(\alpha)$. On the contrary, if a multiplier is imaginary its conjugate annuls the second factor on the left and the two may be simple roots of $F(\alpha)$. When the multiplication is singular, $F(\alpha)$ has zero roots and the determinant $|b_{\nu\mu}| = 0$. The characteristic equation whose properties from the point of view that interests us here, have been thoroughly studied by Frobenius, will play a fundamental part in the sequence. The problem which we propose to consider is the following: *Given a certain equation with rational coefficients, to find the most general Abelian varieties possessing a complex multiplication of which this equation is the characteristic equation, and to determine their invariants.*

When the equation is arbitrary, the problem is far from simple, especially since a given Abelian variety may possess multiplications of widely different types. Now we have seen at the beginning of Part II that the problem of the determination of the numbers $h, k$, and therefore also of $\rho$, reduces itself to the following two problems: (*a*) To find the invariants of a pure matrix; (*b*) To find the simultaneous index of two pure matrices. We shall see that every complex multiplication belonging to a pure matrix has a characteristic equation of type $[f(\alpha)]^r = 0$, ($f(\alpha)$ irreducible), and it is the investigation of extensive cases where the characteristic equation is of this type that will occupy us very largely in the sequence.

78. We recall here, first, the point of view and method which Scorza followed in introducing the index $h$. Let $T'$ be another complex multiplication than $T$ and let

$$u'_j = \sum_{k=1}^{p} \lambda'_{jk} u_k, \qquad \sum_{k=1}^{p} \lambda'_{jk} \omega_{k\mu} = \sum_{\nu=1}^{2p} b'_{\nu\mu} \omega_{j\nu}$$
$$(j = 1, 2, \cdots, p;\ \mu = 1, 2, \cdots, 2p)$$

be its equations. Whatever the rational numbers $m, m'$, the equations $u'_j = \sum_k (m\lambda_{jk} + m' \lambda'_{jk}) u_k$ define a new multiplication, since the relations $\sum_k (m\lambda_{jk} + m' \lambda'_{jk}) \omega_{k\mu} = \sum_\nu (mb_{\nu\mu} + m' b'_{\nu\mu}) \omega_{j\nu}$ are verified. This new multiplication will be designated by $mT + m' T'$.

Since the rank of the determinant $|b_{\nu\mu}|$ is equal to the sum of the ranks of the determinants $|\lambda_{jk}|$, $|\bar{\lambda}_{jk}|$, the $(b)$'s cannot all be zero without this being the case for all the $(\lambda)$'s and conversely. If the terms of a multiplication are all zero we shall write $T = 0$. It is clear that, given $4p^2$ multiplications, we can find a linear combination of them identically zero, hence multiplications form a finite modulus, and in fact the base number is $1 + h$.—This fundamental point is proved by Scorza as follows: Consider the elements of any row of $\Omega$ as homogeneous point coördinates in an $S_{2p-1}$. To the $p$ rows correspond $p$ points which define an $S_{p-1}$. The equations (2) show that the projectivity with rational terms $B = \|b_{\nu\mu}\|$, of $S_{2p-1}$, transforms the $S_{p-1}$ into itself. Conversely a projectivity with rational terms having this property defines a multiplication of $V_p$. $B$ is said to be a *Riemann projectivity* of $\Omega$. If $B'$ corresponds thus to $T'$, to the multiplications $mT + m' T'$ and $TT'$ correspond the respective Riemann projectivities $mB + m' B'$ and $BB'$. Hence in particular the base numbers for the projectivities and for the multiplications are the same. Now a bilinear form of $\Omega$ defines a reciprocity of $S_{2p-1}$ with rational terms, or "a rational reciprocity," which transforms each point of the above $S_{p-1}$ into an $S_{2p-2}$ passing through that $S_{p-1}$. Let $C, D$, be two such reciprocities; $C^{-1}D$ is a Riemann projectivity of $\Omega$ and defines therefore a multiplication. Conversely to the Riemann projectivity $B$ corresponds the rational reciprocity $BC$, from which follows readily that there are exactly $1+h$ distinct multiplications of which no linear combination is a zero multiplication. A very simple discussion shows that $1+h \leqq 2p^2$, $1+k \leqq p^2$, and that if $\Omega$ is pure, $1+h \leqq 2p$, $1+k \leqq 2p-1$. We see that if $k > 0$ there is at least one complex multiplication.

We recall again that every Riemann matrix is isomorphic to one of the type

$$\text{(I)} \qquad \left\| \begin{matrix} \omega_1, & 0, & \cdots, & 0 \\ 0, & \omega_2, & \cdots, & 0 \\ \cdot & \cdot & \cdot & \cdot \\ 0, & \cdot & \cdots, & \omega_n \end{matrix} \right\|$$

where the $(\omega)$'s are pure Riemann matrices not isomorphic to each other or else are of the type

$$\text{(II)} \qquad \left\| \begin{matrix} \omega, & 0, & \cdots, & 0 \\ 0, & \omega, & \cdots, & 0 \\ \cdot & \cdot & \cdot & \cdot \\ 0, & \cdot & \cdots, & \omega \end{matrix} \right\|$$

where $\omega$ is pure. Of course we assume that two submatrices $\omega_i$, $\omega_j$, which enter into the type (I) are not composed with isomorphic pure matrices $\omega$ in case they are impure.

By considerations such as the above, Scorza has shown that if two Riemann matrices possess a simultaneous form, they possess isomorphic submatrices, these conditions being moreover necessary. The matrices $\omega_i$ in (I) do not possess therefore any simultaneous forms and if $h_i$, $k_i$ are their invariants we have for $\Omega$, if of type (I),

$$1+h = \sum(1+h_i), \qquad 1+k = \sum(1+k_i),$$

while for a matrix of type (II), with $n$ terms in the main diagonal,

$$1+h = n^2(1+h'), \qquad 1+k = \frac{n(n-1)}{2}(1+h') + n(1+k'),$$

where $h'$, $k'$ are the invariants of $\omega$. Hence, to find $h$, $k$, or $\rho$, it is sufficient to reduce $\Omega$ to the type (I) and to find the invariants $h$, $k$ of certain pure matrices.

By considering the representative spaces, we see at once that:

(*a*) Every Riemann projectivity operates separately on the submatrices $\omega_i$. This is an immediate consequence of the fact that the reduction to the type (I) is unique (Scorza).

(*b*) The characteristic polynomial $F(\alpha)$ corresponding to a Riemann projectivity is a product $f_1, f_2, \cdots, f_n$, where $f_i$ is the characteristic polynomial of the projectivity corresponding to $\omega_i$.

79. Let $B$ be a Riemann projectivity of $\Omega$. Its various powers, $B^0 = 1$, $B, \cdots, B^m, \cdots$ are also such projectivities. The degree of an equation $f(B) = 0$, ($f$ a polynomial with integral coefficients) satisfied by these powers has a minimum $q$, and there exists a unique equation of this degree satisfied by $B$—it is the *minimum equation* of Frobenius.*

The integer $q$ will be called the *degree* of $B$.

The equations $F(\alpha) = 0$, $f(\alpha) = 0$, have the same roots, hence if $f(0) = 0$, $B$ is singular, and conversely. When the roots of $f(\alpha) = 0$ are all simple, $B$ is said to be a *general* projectivity. In the contrary case, it is said to be *special.* A projectivity is general or special according as the space of minimum dimension containing its stationary points is or is not of dimensionality $2p - 1$. We shall see below that if $B$ is special $\Omega$ is impure. For the present, let us show that *if $B$ is general, non-singular, with irreducible minimum equation, its degree $q$ divides* $1 + h$. For, first, there exists a matrix $C$ such that $C^{-1}BC$ is a matrix with terms on one side of the principal diagonal all equal to zero, those of the diagonal itself being the roots of $F(\alpha)$ each taken with its multiplicity. From this it follows readily that the roots of the characteristic equation of $\psi(B)$ ($\psi$ a polynomial with rational coefficients) are the numbers $\psi(\alpha_i)$, where $\alpha_i$ is any root of $F(\alpha)$. Hence in the first place, if $\psi$ is of degree $< q$, $\psi(B)$ is not singular.

Let now $A$ be a projectivity independent of the powers of $B$. I say that the projectivities $1, B, B, \cdots, B^{q-1}, A, AB, \cdots, AB^{q-1}$, are all independent. For otherwise there would have to be a relation $\phi(B) + A\cdot\psi(B) = 0$, where $\phi, \psi$ are of degree $q - 1$ at most. It is easy to see that since $\psi(B) \neq 0$, there exists a projectivity $\chi(B) = \sum a_s B^s$, such that $\psi(B)\cdot \sum a_s B^s = m$, where $m$ is a non-zero integer. Indeed the computation of the coefficients $a_s$ is formally the same as that which presents itself when given an algebraic number $\alpha$ such that $\psi(\alpha) \neq 0$, it is proposed to put $1/(\psi(\alpha))$ in the form $(1/m)\sum a_s \alpha^s$, a problem which is easily solved. We would have then finally $mA = \phi(B)$, contrary to our assumption as to $A$. The $2q$ distinct projectivities $B^s$, $AB^s$ are therefore effectively independent.

Let now $A'$ be a projectivity independent of the preceding. One may show in a similar manner that the projectivities $B^s$, $AB^s$, $A'B^s$ $(s = 0, 1, \cdots, q - 1)$ are independent, etc. Continuing thus there will come a time

* The term is due to Rosati, who in a note of the Torino Atti of May, 1916, investigated this equation more particularly from the point of view of its relations with correspondences on algebraic curves. There is also a recent note by Scorza bearing on the same subject in the Rendiconti dei Lincei of October, 1917.—(Added in 1922:—The results of this note have since been extensively exposed by Scorza in the Palermo Rendiconti, vol. 45 (1921), pp. 1–204. Not a few contacts with this part of our work are of course to be expected).

when we will have exhausted all the independent projectivities. As those obtained will have been grouped in sets of $q$, $1 + h$ is divisible by $q$. In particular, we observe that if $1 + h > q$, then certainly $1 + h \geqq 2q$.

80. We have said that if $B$ is special, $\Omega$ is impure. This point is easy to prove. Let indeed $\phi(\alpha) = 0$ be the equation of which the roots of $F(\alpha)$ are simple roots. $\phi(\alpha)$ is of degree $< q$, hence $\phi(B)$ is a non-zero projectivity with multipliers all zero and therefore $\Omega$, which possesses a singular projectivity, is necessarily impure. Similarly, if $F(\alpha) = F_1(\alpha)\cdot F_2(\alpha)$, where $F_1, F_2$ are irreducible and prime to each other, $\Omega$ is impure. For $F_1(B)$ is a Riemann projectivity of which some multipliers but not all are zero, hence it is a singular projectivity not identically zero, and $\Omega$ is impure.

We thus see that the characteristic equations of pure matrices are all of type $[f(\alpha)]^r = 0$, where $f$ is irreducible of degree $q = 2p/r$, and $f(B) = 0$ is the minimum equation of $B$, so that its degree is also $q$.

It is interesting to notice that the impure matrices reducible to the type (II) of No. 78 possess at least one general projectivity. For if $u_1, u_2, \cdots, u_p$ are the variables corresponding to this matrix itself, its Abelian variety possesses the cyclic birational transformation of order $n$

$$u_i' = u_{p'+i} \qquad (p = n\,p';\ u_{p+i} = u_p).$$

The corresponding Riemann projectivity is cyclic, hence general.

Thus, amongst the matrices which possess a general projectivity are found those of type (II), whether pure or impure.

81. Let then $B$ be a general Riemann projectivity with a characteristic equation $[f(\alpha)]^r = 0$, of degree $q = 2p/r$ with $f$ irreducible. By making a change of variables, equations (1), (2) can be put in the form

$$u_j' = \alpha_j\, u_j; \tag{3}$$

$$\alpha_j\, \omega_{j\mu} = \sum_{\nu=1}^{2p} b_{\mu\nu}\, \omega_{j\nu} \qquad (j = 1, 2, \cdots, p;\ \mu = 1, 2, \cdots, 2p). \tag{4}$$

Let $\gamma_1$ be an arbitrary linear cycle of $V_p$, $\tau_{j1}$ the corresponding period of $u_j$. The equations (3), (4) show that $\tau_{11}\,\alpha_1, \tau_{21}\,\alpha_2, \cdots, \tau_{p1}\,\alpha_p$ form a system of simultaneous periods of $u_1, u_2, \cdots, u_p$, the corresponding cycle $\gamma_2$ being independent of $\gamma_1$. Similarly the quantities $\tau_{11}\,\alpha_1^{i-1}, \cdots, \tau_{p1}\,\alpha_p^{i-1}$ form a system of simultaneous periods belonging to a linear cycle $\gamma_i$ of $V_p$ and since $f(\alpha)$ is irreducible, the cycles $\gamma_1, \gamma_2, \cdots, \gamma_q$ are independent. If $r > 1$ there will be a cycle $\gamma_{q+1}$ independent of the preceding with a period $\tau_{j2}$ for $u_j$. We shall have similarly cycles $\gamma_{q+1}, \cdots, \gamma_{2q}$, the period of $u_j$ relatively to $\gamma_{q+i}$ being $\tau_{j2}\,\alpha_j^{i-1}$. I say that between these $2q$ cycles, there exists no homology. For otherwise there would have to be a relation

$$\frac{\tau_{j2}}{\tau_{j1}} = \frac{m_0 + m_1 \alpha_j + \cdots + m_{q-1} \alpha_j^{q-1}}{m_0' + m_1' \alpha_j + \cdots + m_{q-1}' \alpha_j^{q-1}},$$

where the $(m)$'s and $(m')$'s are integers. From this would follow a relation with integral coefficients,

$$n\tau_{j2} = \tau_{j1} (n_0 + n_1 \alpha_j + \cdots + n_{q-1} \alpha_j^{q-1}),$$

and therefore a homology between the cycles $\gamma_1, \gamma_2, \cdots, \gamma_{q+1}$, which is in contradiction to the assumption as to the independence of $\gamma_{q+1}$ from the $q$ other cycles. This reasoning can be continued and we shall obtain finally $2p$ independent cycles to which will correspond a period matrix $\Omega'$ isomorphic to $\Omega$,

$$\Omega' = \| \tau_{j1}, \tau_{j1} \alpha_j, \tau_{j1} \alpha_j^2, \cdots, \tau_{j1} \alpha_j^{q-1}, \tau_{j2}, \tau_{j2} \alpha_j, \cdots, \tau_{jr} \alpha_j^{q-1} \| \quad (j = 1, 2, \cdots, p),$$

composed with the two arrays*

$$\tau = \| \tau_{j\mu} \|; \qquad \| 1, \alpha_j, \cdots, \alpha_j^{q-1} \| \quad (\mu = 1, 2, \cdots, r; \ j = 1, 2, \cdots, p).$$

82. The projectivity $B$ being still general, assume now that $F(\alpha) = [f(\alpha)]^r \cdot F_1(\alpha)$, where $f(\alpha)$ is irreducible, of degree $q$, and prime to $F_1(\alpha)$. Of the multipliers, $\frac{1}{2}rq$, say $\alpha_1, \alpha_2, \cdots, \alpha_{\frac{1}{2}rq}$, are roots of $f(\alpha)$. Moreover, according to a remark made above, if $g(\alpha)$ is an arbitrary polynomial with rational coefficients, there exists a complex multiplication which may be represented by $g(T)$, with multipliers $g(\alpha_j)$ $(j = 1, 2, \cdots, p)$. There will then be rational numbers $b_{\nu\mu}$ such that

$$g(\alpha_j) \omega_{j\mu} = \sum_{\nu=1}^{p} b_{\nu\mu} \omega_{j\nu} \quad (j = 1, 2, \cdots, p; \ \mu = 1, 2, \cdots, 2p),$$

and the rank of the determinant $|b_{\nu\mu}|$ will be twice the number of expressions $g(\alpha_j)$ that are not zero. Let us take in particular $g(\alpha) = [f(\alpha)]^r$. We shall have

$$\sum_{\nu} b_{\nu\mu} \omega_{j\nu} = 0 \quad (\mu = 1, 2, \cdots, 2p; \ j = 1, 2, \cdots, \tfrac{1}{2}rq).$$

Hence $u_1, u_2, \cdots, u_{\frac{1}{2}rq}$ form a system of reducible integrals of the variety $V_p$ belonging to $\Omega$ and $T$ operates separately upon them. From this, we conclude that $\Omega$ is isomorphic to a matrix

$$\begin{Vmatrix} \Omega_1 & 0 \\ 0 & \Omega_2 \end{Vmatrix}$$

* For $r = 1$, $p$ arbitrary, this reduction has already been indicated by Scorza in his Palermo Rendiconti paper of 1916, p. 24. For $r = p = 2$ it has been repeatedly used by Bagnera and de Franchis.

where $\Omega_1$ is of genus $\frac{1}{2}rq$ and $T$ operates separately upon it in correspondence with the characteristic equation $[f(\alpha)]^r = 0$, and upon $\Omega_2$ with the equation $F_1(\alpha) = 0$. $\Omega_1$ is therefore isomorphic to a matrix of type (II). By reasoning on $\Omega_2$ as we have on $\Omega$ and continuing, we obtain finally the following result: Let $F(\alpha) = [f_1(\alpha)]^{r_1} \cdots [f_n(\alpha)]^{r_n}$ be the decomposition of $F(\alpha)$ into irreducible factors prime to each other. $\Omega$ is isomorphic to a matrix,

$$\left\| \begin{matrix} \Omega_1, & 0, & \cdots, & 0 \\ 0, & \Omega_2, & \cdots, & 0 \\ \cdot & \cdot & \cdot & \cdot \\ \cdot & \cdot & \cdots, & \Omega_n \end{matrix} \right\|,$$

where $\Omega_j$ (which is of type (II)) is transformed separately by $T$ with the characteristic equation $[f_j(\alpha)]^{r_j} = 0$. $\Omega_j$ can be put in the form indicated in No. 81.

Let $h_j, k_j, \rho_j$, be the invariants of $\Omega_j$. We have then at once (Scorza) the relations

$$\begin{aligned} 1 + h &= \sum (1 + h_j) + 2 \sum \lambda_{jk}, \\ 1 + k &= \sum (1 + k_j) + \sum \lambda_{jk}, \\ \rho &= \sum \rho_j + \sum \lambda_{jk}, \end{aligned}$$

where $\lambda_{jk}$ is the simultaneous index of $\Omega_j$ and $\Omega_k$. The last equation just written has already been obtained at the end of Part I.

It follows from our whole discussion that the determination of the invariants of any Riemann matrix is reducible to the following two problems: (*a*) To determine the invariants of a matrix such as $\Omega_j$, that is possessing a general Riemann projectivity with characteristic equation of type $[f(\alpha)]^r = 0$ ($f$ irreducible). (*b*) To determine the simultaneous index of two such matrices.

Let us observe that if $\Omega$ is of type (II), with $T$ not operating separately upon the $(\omega)$'s, and more especially if $\Omega$ is pure, then $n = 1$, and $F(\alpha)$ is the exact power of an irreducible polynomial, whose degree $q$ is what we have called the degree of $T$. As closely related to all this we recall the two notes of Rosati and Scorza already referred to.

We recall again that if the simultaneous index of two matrices $\Omega$ and $\Omega'$ is not zero, they are isomorphic to two matrices, (Scorza)

$$\left\| \begin{matrix} \Omega_1, & 0 \\ 0, & \Omega_2 \end{matrix} \right\|, \qquad \left\| \begin{matrix} \Omega_1, & 0 \\ 0, & \Omega_2' \end{matrix} \right\|.$$

The problem which we propose to attack is the determination of $h$ and $k$ for a wide range of matrices with general projectivities, with special emphasis on pure matrices. Before we enter upon this task we shall add a last remark concerning the characteristic equation.

83. According to Frobenius* if $T$ is a principal transformation, that is, a transformation which maintains identically invariant a principal form of $\Omega$, the multipliers have all the same absolute value, say $\beta$. Assume that $\Omega$ is pure, which requires that $F(\alpha) = [f(\alpha)]^r$, where $f$ is irreducible and of degree $q = 2p/r$. We shall have $\beta^{qr} = |b_{\nu\mu}|$. Moreover $\beta^q$ is a rational number, hence the determinant $|b_{\nu\mu}|$ in this case is the exact power of a rational number. A real multiplier will then be equal to $\pm\beta$. If $q$ is odd the multiplication $T^q$ will possess a rational number for root of its characteristic equation. As the characteristic polynomial of $T^q$ must still be the exact power of an irreducible polynomial, $T^q$ will necessarily be an ordinary multiplication and $\alpha_j^q = \beta^q = m$, where $m$ is a rational number. Hence, finally, $f(\alpha) = \alpha^q - m$. Thus the characteristic equation of a principal transformation of odd degree of a pure matrix is of the form $(\alpha^q - m)^r = 0$, ($m$ a rational number).

84. *Permutable Projectivities.* According to Scorza, the transformation of coördinates

$$x_s = \sum_{j=1}^{p} \omega_{js} X_j + \sum_{j=1}^{p} \overline{\omega}_{js} X_{p+j}$$

reduces the equations of the projectivity $B$ of $S_{2p-1}$ to the form

$$\sigma X'_j = \sum_{h=1}^{p} \lambda_{hj} X_h; \qquad \sigma X'_{p+j} = \sum_{h=1}^{p} \overline{\lambda}_{hj} X_{p+h} \qquad (j = 1, 2, \cdots, p),$$

and since this transformation of coördinates is independent of the projectivity considered, it follows readily that if two projectivities are permutable, the same holds for the corresponding complex multiplications and conversely.

Let now $B_1, B_2, \cdots, B_s$ be permutable projectivities. The corresponding multiplications will also be permutable and the projectivities permutable with the $(B)$'s will be of type $\phi(B_1, B_2, \cdots, B_s)$, where $\phi$ is a polynomial with rational coefficients, of degree $q_i - 1$ at most in $B_i$ ($q_i$ is the degree of $B_i$). The system of projectivities $\phi(B_1, B_2, \cdots, B_s)$ is in every way analagous to the system of numbers of an algebraic domain. In particular, there exists a projectivity $B_0$, of degree $q$, such that any other is a linear combination of its powers.

Among the projectivities of the set, those with integer terms play the same part as its integers for an algebraic domain. Like these integers, they satisfy an equation whose first coefficient is unity, as follows from the fact that the roots of the minimum equation are also roots of the equation $F(\alpha) = 0$ which is of this nature. These roots are therefore algebraic integers and their symmetric functions are ordinary integers. The projectivities here considered possess therefore a minimum base with $q$ terms $B_1, B_2, \cdots, B_q$.

*Journal für die reine und angewandte Mathematik, vol. 95 (1883), pp. 264–297.

A complex multiplication permutable with any other has necessarily all multipliers equal, their common value being an imaginary quadratic number $\alpha$. Replacing if necessary $\Omega$ by a suitably chosen isomorphic matrix, we may assume that $\alpha = i\sqrt{d}$ ($d$ a positive integer). The array $\tau$ can then be reduced to one with $p$ rows and columns

$$\left\| \begin{matrix} 1, & 0, & \cdots, & 0 \\ 0, & 1, & \cdots, & 0 \\ \cdot & \cdot & \cdot & \cdot \\ \cdot & \cdot & \cdots, & 1 \end{matrix} \right\|$$

and $\Omega$ is a matrix with $p$ elliptic integrals, of the type called by Scorza "with *maximum indices*," for which $1 + h = 2(1 + k) = 2p^2$.

In general assume that for some general projectivity $F(\alpha) = [f(\alpha)]^r$ ($f$ irreducible and of degree $q$), and that among the quantities $\alpha_1, \alpha_2, \cdots, \alpha_p$, are found $\frac{1}{2}q$ roots each taken $r$ times, which requires that all roots of $f(\alpha)$ be imaginary. The array $\tau$ is then of the form

$$\left\| \begin{matrix} \tau_1 \\ \tau_1 \\ \cdot \\ \cdot \\ \cdot \\ \tau_1 \end{matrix} \right\|$$

where $\tau_1$ is a square array of the same type as above with $r$ rows and columns, and $\Omega$ is impure of type (II) composed $r$ times with the matrix

$$\| 1, \alpha_j, \cdots, \alpha_j^{q-1} \| \qquad (j = 1, 2, \cdots, \tfrac{1}{2}q).$$

## § 2. Complex multiplication with irreducible characteristic equation.

85. As before, let $F(\alpha) = 0$ be the characteristic equation. We shall assume the roots $\alpha_1, \alpha_2, \cdots, \alpha_{2p}$ arranged in such order that $\bar{\alpha}_j = \alpha_{p+j}$ ($j < p$), and we propose to investigate the matrix

$$\Omega = \| 1, \alpha_j, \cdots, \alpha_j^{2p-1} \| \qquad (j = 1, 2, \cdots, p).$$

In order that it be a Riemann matrix there must exist an alternate Riemann form

$$\sum_{\mu=1}^{2p} c_{\mu\nu} x_\mu y_\nu . \tag{5}$$

This requires that

$$\sum_{\mu, \nu=1}^{2p} c_{\mu\nu} (\alpha_j^{\mu-1} \alpha_k^{\nu-1} - \alpha_j^{\nu-1} \alpha_k^{\mu-1}) = 0 \quad (j, k = 1, 2, \cdots, p), \tag{6}$$

and that the Hermitian form

$$(7) \qquad \sum_{j,k=1}^{p} A_{jk}\, x_j\, \bar{x}_k; \qquad A_{jk} = -\frac{1}{4i}\sum_{\mu,\nu=1}^{2p} c_{\mu\nu}\,(\alpha_j^{\mu-1}\,\alpha_{p+k}^{\nu-1} - \alpha_j^{\nu-1}\,\alpha_{p+k}^{\mu-1}),$$

be definite positive.

Let $G$ be the group of $F(\alpha) = 0$. The equation (6) will be verified when we replace the pair $(\alpha_j, \alpha_k)$ by any pair of roots which may be derived from the pairs taken out of the set $\alpha_1, \alpha_2, \cdots, \alpha_p$ by the substitutions of $G$. On the other hand, in order that (7) be definite, it is necessary that none of the $(A_{ii})$'s be zero. Hence, as a necessary condition among the pairs of roots derived from those with indices both $\leqq p$ must not figure a pair of conjugate roots.

I say that this last condition is sufficient to insure that $\Omega$ be a Riemann matrix. For let $(\alpha_{j_1}, \alpha_{k_1}), \cdots, (\alpha_{j_s}, \alpha_{k_s})$ be the $s$ pairs derived from those of indices $\leqq p$, $(\alpha_{j'_1}, \alpha_{k'_1}), \cdots, (\alpha_{j'_t}, \alpha_{k'_t})$, $(\alpha_1, \alpha_{p+1}), \cdots, (\alpha_p, \alpha_{2p})$, the $p + t$ pairs remaining with $s + p + t = p(2p - 1)$. Finally consider the equations in the unknowns $c_{\mu\nu}$:

$$(8) \qquad \sum_{\mu,\nu=1}^{2p} c_{\mu\nu}\,(\alpha_{j_n}^{\mu-1}\,\alpha_{k_n}^{\nu-1} - \alpha_{j_n}^{\nu-1}\,\alpha_{k_n}^{\mu-1}) = 0 \qquad (n = 1, 2, \cdots, s),$$

$$(9) \qquad \sum_{\mu,\nu=1}^{2p} c_{\mu\nu}\,(\alpha_{j'_n}^{\mu-1}\,\alpha_{k'_n}^{\nu-1} - \alpha_{j'_n}^{\nu-1}\,\alpha_{k'_n}^{\mu-1}) = d_n \qquad (n = 1, 2, \cdots, t),$$

$$(10) \qquad -\frac{1}{4i}\sum_{\mu,\nu=1}^{2p} c_{\mu\nu}\,(\alpha_j^{\mu-1}\,\alpha_{p+j}^{\nu-1} - \alpha_j^{\nu-1}\,\alpha_{p+j}^{\mu-1}) = e_j \qquad (j = 1, 2, \cdots, p).$$

The determinant of the coefficients is up to the sign

$$|\alpha_j^{\mu-1}\,\alpha_k^{\nu-1} - \alpha_j^{\nu-1}\,\alpha_k^{\mu-1}| \qquad (j, k, \mu, \nu = 1, 2, \cdots, 2p;\ j \neq k,\ \mu \neq \nu),$$

or the determinant formed with the minors of order two of the determinant of Vandermonde of $\alpha_1, \alpha_2, \cdots, \alpha_{2p}$, hence equal to a power of that determinant and therefore different from zero. It follows that there will be a unique solution for the $(c)$'s. Whether these coefficients are rational or not, we can consider the corresponding Hermitian form (7). It will be definite positive if the $p$ roots of the equation in $\beta$,

$$|A_{jk} - \epsilon_{jk}\,\beta| = 0 \qquad (\epsilon_{jj} = 1;\ \epsilon_{jk} = 0, j \neq k)$$

are all positive. But when the $(d)$'s are all zero, these $p$ roots are equal to the $(e)$'s, since (7) becomes then

$$\sum_{j=1}^{p} e_j\, x_j\, \bar{x}_j.$$

Hence when we take the $(d)$'s sufficiently small in absolute value and the $(e)$'s positive, the roots in question will all be positive and the Hermitian form will be definite positive.

Let us consider, as we have already had occasion to do, the $(c)$'s as homogeneous point coördinates in an $S_{p(2p-1)-1}$. The point whose coördinates satisfy the $s$ equations (8) is contained in a *rational* $S_{p(2p-1)-s-1}$, by which we mean that the space contains $p(2p-1)-s$ independent points with rational coördinates. We can see this at once by observing that equations (8) are simply permuted among themselves by the substitutions of $G$ and therefore have as many independent solutions in rational numbers as they have independent solutions in general. Their independence results from the fact that the array of the coefficients is composed of $s$ rows taken from a non-zero determinant.

In the $S_{p(2p-1)-s-1}$ in question there will then be a rational point differing as little as we please from any given point and in particular from the point whose coördinates satisfy the equations (8), (9), (10), where the $(d)$'s are sufficiently small and the $(e)$'s positive. In these conditions $\sum c_{\mu\nu} x_\mu y_\nu$ will be a principal form of $\Omega$, which proves our proposition.

*Thus, given an irreducible equation of degree 2p with roots all imaginary, in order that there correspond to it a Riemann matrix $\Omega$, it is necessary and sufficient that its group does not permute any pair of roots taken from a set of p roots among which none are conjugate to each other, with a pair of conjugate roots.*

When $\Omega$ exists, $1+k$ is obviously equal to the number of independent solutions of the equations (8), that is $1+k = p(2p-1)-s = p+t$.

86. To determine $h$, we must find the number of independent rational solutions possessed by the equations in the unknowns $c_{\mu\nu}$,

$$\sum_{\mu,\nu=1}^{2p} c_{\mu\nu}\, \alpha_j^{\mu-1}\, \alpha_k^{\nu-1} = 0 \qquad (j, k = 1, 2, \cdots, p).$$

This equation must be satisfied when $\alpha_j$, $\alpha_k$ are permuted and also when they are replaced by any one of the pairs of roots $(\alpha_{j_n}, \alpha_{k_n})$, $(\alpha_{k_n}, \alpha_{j_n})$ $(n = 1, 2, \cdots, s)$; $(\alpha_j, \alpha_j)$ $(j = 1, 2, \cdots, 2p)$; whence the equations

$$\sum_{\mu,\nu=1}^{2p} c_{\mu\nu}\, \alpha_{j_n}^{\mu-1}\, \alpha_{k_n}^{\nu-1} = 0, \qquad \sum_{\mu,\nu=1}^{2p} c_{\mu\nu}\alpha_{k_n}^{\mu-1}\, \alpha_{j_n}^{\nu-1} = 0 \qquad (n = 1, 2, \cdots, s),$$

$$\sum_{\mu,\nu=1}^{2p} c_{\mu\nu}\alpha_j^{\mu+\nu-2} = 0, \qquad (j = 1, 2, \cdots, 2p).$$

The array of the coefficients is composed of $2(p+s)$ rows of the determinant

$$\left|\alpha_j^{\mu-1}\, \alpha_k^{\nu-1}\right| \qquad (\mu, \nu, j, k = 1, 2, \cdots, 2p),$$

which is not zero. The above equations are therefore independent and as they are merely permuted by the substitutions of $G$, they will possess as many independent rational solutions as independent solutions in general. Hence, finally $1 + h = 4p^2 - 2p - 2s = 2(1 + k)$. Thus the invariants of $\Omega$ satisfy the relations $1 + h = 2(1 + k) = 2p(2p - 1) - 2s = 2(p + t)$.

87. The preceding result gives rise to some interesting observations. For we have seen that when a matrix is pure, $1 + h \leqq 2p$. Hence, in order that $\Omega$ be pure, it is necessary that $s \geqq p(2p - 1) - p$. On the other hand, the $p$ pairs of roots $(\alpha_j, \alpha_{p+j})$ are not among the pairs $(\alpha_{j_n}, \alpha_{k_n})$ $(n = 1, 2, \cdots, s)$. Hence in order that $\Omega$ be pure, it is necessary that $s = p(2p - 1) - p$, which requires that *the only pairs that may not be derived from those of indices* $\leqq p$ *by the substitutions of* $G$, *are the pairs of conjugate roots.*

Let $z_1 = m_1 \alpha_1 + m_2 \alpha_2 + \cdots + m_{2p} \alpha_{2p}$ be a Galois function of $F(\alpha) = 0$, the $(m)$'s being arbitrary integers, and consider the Galois resolvent $\phi(z) = 0$ of which $z_1$ is a root. The $(\alpha)$'s are rational functions of $z_1$ with real coefficients. Hence $z_1$ is not real, as can also be seen directly. Replacing $z_1$ by $\bar{z}_1$ in these rational functions, we obtain a substitution $S$, of order two of $G$, which has the property of permuting any two conjugate roots. Moreover, if another substitution $S'$ had the same property, we would have $SS' = 1$, hence $S' = S^{-1} = S$ and therefore $S$ is unique.

Let us now return to $\Omega$. In order that it be a Riemann matrix and pure, it is necessary and sufficient: (*a*) That $S$ be permutable with all the substitutions of $G$. (*b*) That among the corresponding roots, $\alpha_1, \alpha_2, \cdots, \alpha_p$, none be conjugate, and moreover that from the pairs $(\alpha_j, \alpha_k)$ it be possible to derive all the pairs of non-conjugate roots by the substitutions of $G$. To begin with, the second condition is obviously necessary. Let us show that so is the first. For $G$ must not permute the pairs of conjugate roots with pairs of roots that are not conjugate. Hence, if $T$ is an arbitrary substitution of $G$ and if $T\alpha_j = \alpha_k$ we must have $TS\alpha_j = S\alpha_k$, whence $S^{-1}TS = T$, and $T$ is effectively permutable with $S$.

Let us show that our two conditions are also sufficient. For, if $T\alpha_j = \alpha_k$, we shall also have $T\bar{\alpha}_j = TS\alpha_j = ST\alpha_j = S\alpha_k = \bar{\alpha}_k$, and the pair $(\alpha_j, \bar{\alpha}_j)$ is permuted with $(\alpha_k, \bar{\alpha}_k)$; hence the pairs of conjugate roots are not permuted with any others by the substitutions of the group. The second condition insures then that $t = 0$, the pairs $(\alpha_j, \bar{\alpha}_j)$ being all alone excluded. The equations for the coefficients of an alternate Riemann form are here

$$\sum_{\mu, \nu=1}^{2p} c_{\mu\nu} (\alpha_j^{\mu-1} \alpha_k^{\nu-1} - \alpha_j^{\nu-1} \alpha_k^{\mu-1}) = 0$$

$$(j, k = 1, 2, \cdots, 2p;\ j \neq p + k).$$

The quantities

$$A_{jj} = -\frac{1}{4i}\sum_{\mu,\nu=1}^{2p} c_{\mu\nu}(\alpha_j^{\mu-1}\alpha_{p+j}^{\nu-1} - \alpha_j^{\nu-1}\alpha_{p+j}^{\mu-1})$$

will not all be zero, for otherwise the $(c)$'s would also be zero. But if one of the quantities $A_{jj}$ is zero, all the others will also be zero for they are transitively permuted by the substitutions of $G$. Hence none of them is zero and the Hermitian form corresponding to the arbitrary Riemann form considered becomes here $\sum_{j=1}^{p} A_{jj}\, x_j\, \bar{x}_j$, and is always of genus $p$. Hence finally every Riemann form of $\Omega$ is of genus $p$ and this matrix is actually pure. This completes the proof.

Observe that the invariants of the pure matrix will be given by

$$1 + h = 2(1 + k) = 2p,$$

$h$ having the maximum value which it may have for such a matrix.

88. A particularly interesting case is that where $p$ is prime. Then $\Omega$, which is of type (II), is certainly pure, whence the following result: Let $F(\alpha) = 0$ be an irreducible equation of degree $2p$ ($p$ prime), with roots all imaginary. Either the group of the equation permutes only among themselves the pairs of conjugate roots, or else in every set of $p$ roots there are at least two which are permutable with a pair of conjugate roots.

89. Let us endeavor to characterize $F(\alpha)$ when its group is permutable with $S^*$ or, which is the same, when there is a corresponding pure matrix $\Omega$. The group $G$ merely permutes the equations in the unknowns $d$,

$$\bar{\alpha}_j = d_0 + d_1\alpha_j + \cdots + d_{2p-1}\alpha_j^{2p-1} \qquad (j = 1, 2, \cdots, 2p),$$

and hence these equations possess a rational solution which besides is unique. This means that $\bar{\alpha}_j = Q(\alpha_j)$, where $Q$ is a polynomial with rational coefficients independent of $j$. Thus every root can be expressed as a rational function, the same for all of them, of its conjugate. Conversely if $\alpha_j = Q(\alpha_j)$, whatever $j$, $S$ and $G$ are permutable. For denoting still by $z_1$ the Galois function, we shall have first

$$\alpha_j = R_j(z_1), \qquad \bar{\alpha}_j = Q[R_j(z_1)].$$

We may then obtain every substitution of $G$ by replacing $z_1$ by a suitably chosen root $z_2$ of the Galois equation. If $T\alpha_j = \alpha_k$, we shall have

$$\alpha_k = T\alpha_j = R_j(z_2), \qquad T\bar{\alpha}_j = Q[R_j(z_2)] = Q(\alpha_k) = \bar{\alpha}_k,$$

that is, $TS\alpha_j = ST\alpha_j$, or $TS = ST$.

* From this property follows already that the solution of the characteristic equation is reducible to that of an equation of degree $p$ followed by a quadratic, but this result is insufficient for some applications which we have in view.

Thus, in order that $S$ and $G$ be permutable, it is necessary and sufficient that whatever $j$, $\bar{\alpha}_j = Q(\alpha_j)$, where $Q$ is a polynomial with rational coefficients independent of $j$.

But we can go still further. Indeed, the $p$ points of an $S_{2p-1}$ whose non-homogeneous point coördinates are $\alpha_j + \bar{\alpha}_j$, $\alpha_j^2 + \bar{\alpha}_j^2$, $\cdots$, $\alpha_j^{2p-1} + \bar{\alpha}_j^{2p-1}$ are certainly distinct, for if two of them had their first two coördinates equal, $F(\alpha)$ would have a double root. Given arbitrary integers $d_1, d_2, \cdots, d_{2p-1}$, the distances from these points to the corresponding hyperplane

$$d_1 x_1 + d_2 x_2 + \cdots + d_{2p-1} x_{2p-1} = 0$$

are therefore all different. Whence the following conclusion: If we set

$$\alpha_j' = d_1 \alpha_j + d_2 \alpha_j^2 + \cdots + d_{2p-1} \alpha_j^{2p-1},$$

the quantities $\alpha_j' + \bar{\alpha}_j'$ are all distinct. In other words $\Omega$ possesses a Riemann projectivity with irreducible characteristic equation and such that if its multipliers are still denoted by $\alpha_1, \alpha_2, \cdots, \alpha_p$, the $p$ sums $2\zeta_j = \alpha_j + \bar{\alpha}_j$ are all distinct. If we adjoin $\zeta_j$ to the domain of rationality, the pair of conjugate roots $\alpha_j$, $\bar{\alpha}_j$ is determined in one and only one way, and their product $\alpha_j \bar{\alpha}_j$ is therefore a rational function of $\zeta_j$. On the other hand the $(\zeta)$'s are transitively permuted by the substitutions of $G$, hence they satisfy an irreducible equation $\psi(\zeta) = 0$ of degree $p$, and the characteristic equation is of the form

$$\prod_{j=1}^{p} (\alpha^2 - 2\zeta_j \alpha + R(\zeta_j)) = 0,$$

where $R$ is a polynomial of degree $p - 1$ with rational coefficients and such that $\zeta_j^2 < R(\zeta_j)$ $(j = 1, 2, \cdots, p)$. Of course these conditions are not sufficient to make certain that $\Omega$ is pure. For example, take the case $p = 2$. The characteristic equation is of the form

$$(\alpha^2 - 2\zeta_1 \alpha + m\zeta_1 + n)(\alpha^2 - 2\zeta_2 \alpha + m\zeta_2 + n) = 0,$$

where $m, n$ are rational and $\zeta_1$, $\zeta_2$ are the two roots of an equation

$$a\zeta^2 + b\zeta + c = 0$$

with coefficients $a, b, c$, integers and $a > 0$. We must have $\zeta_j^2 < m\zeta_j + n$ $(j = 1, 2)$, which is equivalent to the inequalities

$$a(an + c)^2 - b(an + c)(am + b) + c(am + b)^2 > 0,$$
$$(am + b)(2na^2 - mab + 2ac - b^2) > 0.$$

Moreover, in order that $\Omega$ be pure, it is necessary that the pairs of conjugate roots be transitively permuted. This requires that $\alpha_1 \alpha_2 + \bar{\alpha}_1 \bar{\alpha}_2$ be irrational, a condition which is besides sufficient. A very simple calculation shows that this is equivalent to requiring that

$$a[a(an+c)^2 - b(an+c)(an+b) + c(an+b)^2]$$

be not the square of a rational number.

When these conditions are satisfied, the matrix

$$\begin{Vmatrix} 1, & \alpha_1, & \alpha_1^2, & \alpha_1^3 \\ 1, & \alpha_2, & \alpha_2^2, & \alpha_2^3 \end{Vmatrix}$$

is a Riemann matrix and pure.

It is scarcely possible to go further when the equation $F(\alpha) = 0$ is arbitrary and so we shall pass on to the investigation of the particularly important case where it is Abelian, when the preceding results may be completed and much increased in precision.

# ON CERTAIN NUMERICAL INVARIANTS OF ALGEBRAIC VARIETIES WITH APPLICATION TO ABELIAN VARIETIES

## § 3. The characteristic equation is irreducible and Abelian

90. The first problem which presents itself is that of the determination of the indices. This determination will result from the following theorem:

*In order that the matrix*

$$\Omega \equiv \| 1, \alpha_j, \alpha_j^2, \cdots, \alpha_j^{2p-1} \| \qquad (j = 1, 2, \cdots, p),$$

*where the* $(\alpha)$*'s are roots of an irreducible Abelian equation* $F(\alpha) = 0$ *with* $\bar{\alpha}_j = \alpha_{p+j}$, *be impure, it is necessary and sufficient that the group* $G$ *of the equation contain a subgroup* $G'$ *maintaining the set* $\alpha_1, \alpha_2, \cdots, \alpha_p$ *invariant.*

In order that $\Omega$ be impure it is necessary and sufficient that there exist a pair of non-conjugate roots $(\alpha_m, \alpha_n)$ which cannot be deduced from a pair of roots with indices $\leqq p$ by any substitution of $G$. We shall have $\alpha_n = U\alpha_m$, where $U$ is a well-defined substitution of $G$. If $\alpha_j$ is a root of index $< p$, there will exist a substitution $T$ such that $\alpha_m = T\alpha_j$. The root $T^{-1}\alpha_n$ must not be one of the roots $\alpha_1, \alpha_2, \cdots, \alpha_p$, hence it is the conjugate $S\alpha_k$ of a root $\alpha_k$ of index $k < p$, that is

$$S\alpha_k = T^{-1}\alpha_n = T^{-1}U\alpha_m = T^{-1}UT\alpha_j = U\alpha_j,$$

and hence $\alpha_k = S^{-1}U\alpha_j = SU\alpha_j$. This shows that the set $\alpha_1, \alpha_2, \cdots, \alpha_p$ is transformed into itself by $SU$ which is not the substitution unity, if $U \neq S$, as is actually the case since $\alpha_m \neq \bar{\alpha}_n$. The condition is therefore necessary.

On the other hand, if a substitution $T$ (which cannot be $S$) maintains invariant the set $\alpha_1, \alpha_2, \cdots, \alpha_p$, I say that the pair of roots $(\alpha_j, \alpha_k)$ $(j, k \leqq p)$ cannot be permuted with a pair $(\alpha_h, ST\alpha_h)$. To begin with, this is equivalent to affirming that it cannot be permuted with $(\alpha_j, ST\alpha_j)$. For if $\alpha_j = T'\alpha_h$, the pair $(\alpha_h, ST\alpha_h)$ can be permuted with $(\alpha_j, T'ST\alpha_h)$ or $(\alpha_j, ST\alpha_j)$. Assume then that there exists a substitution $U$ of $G$ permuting $(\alpha_j, \alpha_k)$ with this last pair. Since $G$ is Abelian it possesses no other substitution than the identity maintaining a root invariant. Now $U$ is certainly not the identity for then we would have $\alpha_k = ST\alpha_j$, which is impossible for $ST$ permutes the set $\alpha_1, \alpha_2, \cdots, \alpha_p$ with $\alpha_{p+1}, \alpha_{p+2}, \cdots, \alpha_{2p}$, and hence $k > p$ contrary to our assumptions. We must therefore have $ST\alpha_j = U\alpha_j$, $\alpha_j = U\alpha_k$, and consequently $U = ST$, $\alpha_j = ST\alpha_k$, which leads to a similar contradiction: $j > p$. It follows that $G$ may not contain any substitution such as $U$. There is then a pair of conjugate roots which may not be derived from those of indices $\leqq p$ by substitutions of $G$, and $\Omega$ is impure, as was to be proved.

91. Let $n$ be the order of $G'$, assumed now to be the maximum subgroup maintaining invariant the set of multipliers $\alpha_1, \alpha_2, \cdots, \alpha_p$. $G'$ does not contain $S$, hence the product of $G'$ by $S$ is a subgroup of order $2n$ of $G$. It follows that $2n$ divides the order $2p$ of $G$, and therefore $n$ divides $p$. The roots $\alpha_1, \alpha_2, \cdots, \alpha_{2p}$ can be subdivided into $2p'$ sets of $n$ roots, each set being composed of the roots derived from one of them by the substitutions of $G'$. Of these sets, $p'$, say $(\alpha_1, \alpha_2, \cdots, \alpha_n)$, $(\alpha_{n+1}, \cdots, \alpha_{2n})$, $\cdots$, $(\alpha_{(p'-1)n+1}, \cdots, \alpha_{p'n})$, form the set of $p$ multipliers.

Let us designate by $A_1, A_2, \cdots, A_{p'}$ these sets of roots, and by $A_{p'+j}$ the set formed by the conjugate roots of those which compose $A_j$. Any substitution of $G$ which does not belong to $G'$ merely permutes the $(A)$'s among themselves. Moreover there exists none maintaining invariant the set $(A_1, A_2, \cdots, A_{p'})$ for otherwise $G'$ would not be the maximum subgroup maintaining invariant the set of multipliers $(\alpha_1, \alpha_2, \cdots, \alpha_p)$. The group $G'' = G/G'$ of permutations of the $(A)$'s is also Abelian and we can apply to $G''$ and the $(A)$'s the same reasoning as before to $G$ and the $(\alpha)$'s. In particular the pai s $(A_j, A_{p'+j})$ are the only pairs which may not be derived from the pairs $(A_j, A_k)$ of indices $j, k \leqq p'$, by the substitutions of $G''$. Let then $\alpha_j, \alpha_k$ be a pair of roots and $A_{j'}, A_{k'}$ the $(A)$'s to which they belong. If the difference $j' - k' \neq \pm p'$, there exists a substitution of $G''$ permuting the pair $(A_{j'}, A_{k'})$ with a pair $(A_{j''}, A_{k''})$ of indices $j'', k'' \leqq p'$. This substitution permutes $(\alpha_j, \alpha_k)$ with a pair of roots of indices $\leqq p$ and $(\alpha_j, \alpha_k)$ is not an excluded pair. Moreover, in order that $(\alpha_j, \alpha_k)$ be not excluded, it is necessary that $(A_{j'}, A_{k'})$ be permutable with a pair $(A_{j''}, A_{k''})$, $(j'', k'' \leqq p')$, which requires that $j' - k' \neq \pm p'$. Hence, the only pairs excluded are those for which $j' - k' = \pm p'$ and these pairs are all effectively excluded. Their number is $p + t = n^2 p' = np$, and therefore finally

$$1 + h = 2(1 + k) = 2np.$$

*Remark:* Let there be a set of roots $\alpha_1, \alpha_2, \cdots, \alpha_{p+qn}$, composed of $p' + q$ groups $A_i$, say $A_1, A_2, \cdots, A_{p'+q}$. Then from the pairs of roots $(\alpha_j, \alpha_k)$ of indices $j, k \leqq p + qn$, we may certainly derive all pairs of roots without exception. This is an immediate consequence of the fact that among the $(A)$'s of the above set there are at least two whose indices differ by $p'$.

92. We shall now endeavor to obtain the structure of $\Omega$ when it is impure. For this purpose let

$$\beta_{q,\,s} = \alpha^{s}_{(q-1)n+1} + \alpha^{s}_{(q-1)n+2} + \cdots + \alpha^{s}_{qn}$$

and observe that $\Omega$ is isomorphic to a matrix $\Omega'$ whose $p'$ first rows are

$$\begin{array}{llll} \beta_{1,\,1}, & \beta_{1,\,2}, & \cdots, & \beta_{1,\,2p} \\ \beta_{2,\,1}, & \beta_{2,\,2}, & \cdots, & \cdots \\ \cdot & \cdot & \cdot & \cdot \\ \beta_{p',\,1}, & \cdots, & \cdots, & \beta_{p',\,2p}. \end{array}$$

The $(\beta)$'s are invariant under the substitutions of $G'$ and those belonging to the same index $s$ are permuted among themselves by the substitutions of $G''$. Consider, then, the equations in the unknowns $n_i$

$$n_1\beta_{s,\,1} + n_2\beta_{s,\,2} + \cdots + n_{2p}\beta_{s,\,2p} = 0 \qquad (s = 1, 2, \cdots, 2p').$$

The array of the coefficients is composed of $2p'$ rows derived from the determinant which may be formed with the periods of $\Omega'$ and their conjugates, hence the array is of rank $2p'$ and these equations are independent. Since $G$ merely permutes them among themselves, they possess $2(p - p')$ independent rational solutions, and $\Omega'$ and hence $\Omega$ is isomorphic to a matrix

$$\left\| \begin{matrix} \omega, & 0 \\ 0, & \Omega_1 \end{matrix} \right\|;$$

$$\omega \equiv \| \beta_{j,\,h_1}, \beta_{j,\,h_2}, \cdots, \beta_{j,\,h_{2p'}} \| \qquad (j = 1, 2, \cdots, p'),$$

the indices $h_1, h_2, \cdots, h_{2p'}$ being chosen so that there exists no relation

$$d_1\beta_{j,\,h_1} + d_2\beta_{j,\,h_2} + \cdots + d_{2p'}\beta_{j,\,h_{2p'}} = 0$$

with integer coefficients $d_i$. For every other $\beta_{j,\,s}$ there will exist a relation

$$\beta_{j,\,s} = e_1\beta_{j,\,h_1} + e_2\beta_{j,\,h_2} + \cdots + e_{2p'}\beta_{j,\,h_{2p'}} \qquad (j = 1, 2, \cdots, 2p'),$$

where the $(e)$'s are rational. Hence an arbitrary rational function of the $(\alpha)$'s belonging to the group $G''$ is of the form

$$\beta_1 = c_1\beta_{1,\,h_1} + c_2\beta_{2,\,h_1} + \cdots + c_{2p'}\beta_{2p',\,h_1},$$

where the $(c)$'s are rational numbers and the $2p' - 1$ conjugate values $\beta_2, \beta_3, \cdots, \beta_{2p'}$ of the algebraic domain determined by the multipliers are given by

$$\beta_j = c_1\beta_{j,\,h_1} + c_2\beta_{j,\,h_2} + \cdots + c_{2p'}\beta_{j,\,h_{2p'}}.$$

We then have $\beta_{j,\,h_s} = g_s(\beta_j)$, where the $(g)$'s are polynomials of degree $2p' - 1$ at most with rational coefficients and $\omega$ is therefore isomorphic to

$$\| g_1(\beta_j), g_2(\beta_j), \cdots, g_{2p'}(\beta_j) \| \qquad (j = 1, 2, \cdots, p'),$$

and finally to

$$\| 1, \beta_j, \beta_j^2, \cdots, \beta_j^{2p'-1} \| \qquad (j = 1, 2, \cdots, p').$$

Now $\beta_1, \beta_2, \cdots, \beta_{2p'}$ are roots of an irreducible Abelian equation of order $2p'$ whose group is $G''$ and no subgroup of $G''$ maintains the set $(\beta_1, \beta_2, \cdots, \beta_{p'})$ invariant, this always because $G'$ is the maximum subgroup of $G$ maintaining invariant the set $(\alpha_1, \alpha_2, \cdots, \alpha_p)$. Hence $\omega$ is pure and its indices $h', k'$ are given by $1 + h' = 2(1 + k') = 2p'$. As to $\Omega$, it is isomorphic to a matrix

$$\left\| \begin{matrix} \omega, & 0, & \cdots, & 0 \\ 0, & \omega, & \cdots, & \cdot \\ \cdot & \cdot & \cdot & \cdot \\ \cdot & \cdot & \cdots, & \omega \end{matrix} \right\|$$

with $n$ terms in the principal diagonal, as results from the fact that the reduction to the type (II) is unique. For its invariants we have then

$$1 + h = n^2(1 + h') = 2n^2 p';$$

hence

$$1 + h = 2(1 + k) = 2n^2 p' = 2np,$$

as we have already shown in a different manner in No. 91.

93. The multiplications of $\Omega$ can be found without difficulty and in fact in two different ways. For let $\alpha_{in+1}, \alpha_{in+2}, \cdots, \alpha_{(i+1)n}$ be again a set of multipliers permuted by the subgroup $G'$ of order $n$. The system of numbers $\lambda_{jk}$ given by the array

$$\left\| \begin{matrix} E_1, & 0, & \cdots, & 0 \\ 0, & E_2, & \cdots, & 0 \\ \cdot & \cdot & \cdot & \cdot \\ \cdot & \cdot & \cdots, & E_{p'} \end{matrix} \right\|,$$

$$E_{i+1} \equiv \left\| \begin{matrix} g_1(\alpha_{in+1}), & g_2(\alpha_{in+1}), & \cdots, & g_n(\alpha_{in+1}) \\ g_n(\alpha_{in+2}), & g_1(\alpha_{in+2}), & \cdots, & g_{n-1}(\alpha_{in+2}) \\ \cdot & \cdot & \cdot & \cdot \\ \cdot & \cdot & \cdot & \cdot \\ g_2(\alpha_{(i+1)n}), & g_3(\alpha_{(i+1)n}), & \cdots, & g_1(\alpha_{(i+1)n}) \end{matrix} \right\|,$$

where the $(g)$'s are arbitrary polynomials of degree $2p - 1$ with rational coefficients, defines a complex multiplication, and as we have here a linear system with $2np = 1 + h$ parameters, every multiplication is of this type.

We may also obtain the multiplications in a different manner. The multiplications of the matrix $\Omega'$ to which $\Omega$ was reduced in No. 92, which depends upon those obtained by transforming the submatrix $\omega$ into itself or into another like it, are of the general type $\| E_{\mu\nu} \| \; (\mu, \nu = 1, 2, \cdots, n)$, where the $(E)$'s are matrices representing multiplications of $\omega$, that is of the type

$$E_{\mu\nu} \equiv \left\| \begin{array}{cccc} g_{\mu\nu}(\beta_1), & 0, & \cdots, & 0 \\ 0, & g_{\mu\nu}(\beta_2), & \cdots, & 0 \\ \cdot & \cdot & \cdot & \cdot \\ \cdot & \cdot & \cdots, & g_{\mu\nu}(\beta_{p'}) \end{array} \right\|,$$

where the $(g)$'s are arbitrary polynomials of degree $2p' - 1$ with rational coefficients. As we have here again $2n^2 p' = 1 + h$ distinct multiplications of $\Omega'$, all its multiplications are of this type.—On the other hand if $\Lambda$, $\Lambda'$ are corresponding multiplications of $\Omega$, $\Omega'$, we have $\Lambda' = M\Lambda M^{-1}$, where $M$ is a definite square matrix of order $p$. Hence having obtained all the multiplications of one of them, we may say that we also possess all those of the other.

It is of interest to consider the multipliers $\eta_i$ of $\Lambda$, which are the same as those of $\Lambda'$. The latter being assumed as the multiplication written above, on forming the equation in the $(\eta)$'s, we see at once that it is obtained by multiplying the left-hand sides of the $p'$ equations of degree $n$

$$\phi(\eta;\, \beta_i) = |g_{\mu\nu}(\beta_i) - \epsilon_{\mu\nu}\eta| = 0$$
$$(i = 1, 2, \cdots, p'; \quad \mu, \nu = 1, 2, \cdots, n; \quad \epsilon_{\mu\nu} = 1; \quad \epsilon_{\mu\nu} = 0, \mu \neq \nu),$$

where $\phi$ is a polynomial in $\eta$ and $\beta_i$ with rational coefficients, and moreover the conjugates $\eta_{p+1}, \eta_{p+2}, \cdots, \eta_{2p}$ of the $(\eta)$'s satisfy the equations $\phi(\eta;\, \beta_{p'+i}) = 0$. Two cases may now present themselves: (*a*) These $2p'$ equations have all the same roots. Then $\phi(\eta;\, \beta_i) = 0$ reduces to an equation $\phi(\eta) = 0$ in $\eta$ alone, of degree at most $n$. We can take for $\phi$ an arbitrary polynomial with rational coefficients since $\Omega'$ is equivalent to a matrix composed with arrays

$$\tau \equiv \| \tau_{j1}, \tau_{j2}, \cdots, \tau_{j,2p'} \|, \quad \| 1, \eta_j, \eta_j^2, \cdots, \eta_j^{n-1} \| \qquad (j = 1, 2, \cdots, p),$$

where $\eta_j$ is an arbitrary root of $\phi(\eta) = 0$, and the array $\tau$ is composed of the matrix $\omega$ superposed $n$ times. Generally speaking we shall obtain all the multiplications of the nature here considered if we succeed in forming all the polynomials with rational coefficients $g_{\mu\nu}(\beta_i)$ such that the coefficients of the polynomial $\phi(\eta;\, \beta_i)$ in $\eta$ are independent of $\beta_i$, that is, are rational numbers. (*b*) Among the $2p'$ equations in the multipliers $\eta_i$ there are at least two whose roots are not the same. In this case the adjunction of one of the $(\beta)$'s to the domain of rationality brings about the reduction of the equation in the multipliers if that equation is not reducible already. In particular, if the new multiplication does not transform into themselves certain submatrices and if it is of degree $> n$, the equation in the multipliers is necessarily reducible in the domain of rationality $K(\beta_i)$, hence in the domain $K(\alpha_i)$.

94. To complete this investigation it is necessary to compare from the point

of view of isomorphism two matrices

$$\Omega \equiv \| 1, \alpha_{j_n}, \alpha_{j_n}^2, \cdots, \alpha_{j_n}^{2p-1} \| ; \quad \Omega' \equiv \| 1, \alpha_{h_n}, \alpha_{h_n}^2, \cdots, \alpha_{h_n}^{2p-1} \|$$

$(n = 1, 2, \cdots, p)$, corresponding to the same equation $F(\alpha) = 0$.

In order that $\Omega$, $\Omega'$ be isomorphic there must exist a simultaneous bilinear form

$$\sum c_{\mu\nu}\, x_\mu\, y_\nu .$$

By reasoning as at the beginning of § 2, we see that we must have

$$\sum_{\mu,\nu} c_{\mu\nu} (\alpha_j^{\mu-1} \alpha_h^{\nu-1} - \alpha_j^{\nu-1} \alpha_h^{\mu-1}) = 0$$

whenever $(\alpha_j, \alpha_h)$ is a pair of roots derived from the pairs $(\alpha_{j_m}, \alpha_{h_n})$ $(m, n = 1, 2, \cdots, p)$, by the substitutions of the group $G$. If the pairs thus derived include all pairs of roots, the system of equations will have as many independent equations as there are unknowns $c_{\mu\nu}$, and these unknowns will all be zero so that $\Omega$, $\Omega'$ will not be isomorphic.

Thus as a condition for isomorphism, we find that it must not be possible to derive from the pairs $(\alpha_{j_m}, \alpha_{h_n})$ all the possible pairs of roots. Let $(\alpha_q, \alpha_r)$ be an excluded pair and $U$, $T$ substitutions of $G$ such that $\alpha_r = U\alpha_q$, $\alpha_{j_m} = T\alpha_q$. It must not be possible for $T$ to permute $\alpha_r$ with an $\alpha_{h_n}$, hence $T$ must permute it with an $\alpha_{h_n}$, and therefore $T\alpha_r = S\alpha_{h_n}$. It follows that $TU\alpha_q = S\alpha_{h_n} = U\alpha_{j_m}$, which shows that $SU$ permutes the set $(\alpha_{j_1}, \alpha_{j_2}, \cdots, \alpha_{j_p})$ with the set $(\alpha_{h_1}, \alpha_{h_2}, \cdots, \alpha_{h_p})$. Thus in order that $\Omega$ and $\Omega'$ be isomorphic, it is necessary that there exist a subgroup of $G$ permuting these two sets.

Conversely let $T$ be a substitution permuting these two sets. No pair $(\alpha_{j_1}, \alpha_{h_1})$ may be permuted with $(\alpha_m, ST\alpha_m)$. For let $U$ be the substitution of $G$ such that $U\alpha_{j_1} = \alpha_m$. We must have $U\alpha_{h_1} = ST\alpha_m$, hence $\alpha_{h_1} = ST\alpha_{j_1}$, from which would follow that $\alpha_{h_1}$ belongs to both sets $(\alpha_{h_1}, \alpha_{h_2}, \cdots, \alpha_{h_p})$, $(\bar\alpha_{h_1}, \alpha_{h_2}, \cdots, \bar\alpha_{h_p})$—an impossibility. Hence $\Omega$, $\Omega'$ certainly possess a simultaneous form. If they are pure this is sufficient to insure their isomorphism. Assume that they are impure, and let $\omega$, $\omega'$ be the corresponding pure matrices after the manner of No. 92, $\beta_j$, $\beta_j'$ the quantities analogous to those already designated by similar letters corresponding to these two matrices. It is seen at once that since the sets $(\alpha_{j_1}, \alpha_{j_2}, \cdots, \alpha_{j_p})$, $(\alpha_{h_1}, \alpha_{h_2}, \cdots, \alpha_{h_p})$ are permuted by some substitution of $G$, the subgroups which maintain them invariant coincide. Hence $\omega$, $\omega'$ are of the same genus $p'$, and the $(\beta)$'s and $(\beta')$'s can be taken as roots of one and the same irreducible equation of degree $2p'$. Moreover to

$$\beta_j = \alpha_{n(j-1)} + \alpha_{n(j-1)+1} + \cdots + \alpha_{nj}$$

corresponds

$$\beta_j' = T\alpha_{n(j-1)} + T\alpha_{n(j-1)+1} + \cdots + T\alpha_{nj} .$$

Hence the sets $(\beta_1, \beta_2, \cdots, \beta_{p'})$, $(\beta_1', \beta_2', \cdots, \beta_{p'}')$ are permuted by a substitution of the group of the equation in the $(\beta)$'s and therefore $\omega$, $\omega'$ are isomorphic, from which follows that this is also the case for $\Omega$, $\Omega'$. Thus *in order that $\Omega$, $\Omega'$ be isomorphic, it is necessary and sufficient that there exist a substitution of the group $G$ permuting their sets of multipliers.*

*Corollary. The number of essentially distinct matrices belonging to $F(\alpha)$ is equal to its number of classes of transitively permutable sets of $p$ roots.*

This ends the discussion of the case where $F(\alpha)$ is irreducible, Abelian.

## § 4. Characteristic equation of type $[f(\alpha)]^r = 0$, $r > 1$. (*a*) Generalities

95. We assume of course $f(\alpha)$ irreducible and shall denote its degree as previously by $q$. The matrix $\Omega$ is composed with two arrays

$$\tau \equiv \| \tau_{j1}, \tau_{j2}, \cdots, \tau_{jr} \|, \quad \| 1, \alpha_j, \alpha_j^2, \cdots, \alpha_j^{q-1} \| \qquad (j = 1, 2, \cdots, p;\ qr = 2p),$$

where the $(\alpha)$'s are roots of $f(\alpha) = 0$. We recall that if $\alpha_j$ is real it must be double root of $f = 0$, hence $r$ must be even.

In order that $\Omega$ be a Riemann matrix it is necessary that there exist rational numbers $c_{\mu\nu}^{mn}$, such that

$$\sum_{m,n}^{1\ldots q} \sum_{\mu,\nu}^{1\ldots r} c_{\mu\nu}^{mn} (\alpha_j^{m-1} \alpha_h^{n-1} \tau_{j\mu} \tau_{h\nu} - \alpha_j^{n-1} \alpha_h^{m-1} \tau_{j\nu} \tau_{h\mu}) = 0 \qquad (j, h = 1, 2, \cdots, p). \tag{11}$$

Let us set

$$\gamma_{\mu\nu}(\alpha_j, \alpha_h) = \sum_{m,n} c_{\mu\nu}^{mn} \alpha_j^{m-1} \alpha_h^{n-1} \qquad (\gamma_{\mu\nu}(x, y) = -\gamma_{\nu\mu}(y, x)).$$

We may say that the array $\tau$ will possess the bilinear forms

$$\sum_{\mu,\nu} \gamma_{\mu\nu}(\alpha_j, \alpha_h) x_\mu y_\nu \qquad (j, h = 1, 2, \cdots, p), \tag{12}$$

in the sense that the particular form corresponding to $(\alpha_j, \alpha_h)$ must vanish when we replace in it the $(x)$'s by the elements of the $j$-th row and the $(y)$'s by those of the $h$-th column. We shall say for the sake of simplicity that this form (12) is a bilinear form of $\Omega$.

In order that $\Omega$ be a Riemann matrix, it is necessary besides that

$$\sum_{j,h}^{1\ldots p} A_{jh} x_j \bar{x}_h; \quad A_{jh} = \frac{-1}{2i} \sum \gamma_{\mu\nu}(\alpha_j, \bar{\alpha}_h) \tau_{j\mu} \bar{\tau}_{h\nu} \tag{13}$$

be a positive definite Hermitian form. We shall assume this condition fulfilled for the present and will return to it later. The relations thus imposed upon the elements of $\tau$ do not determine them completely and they will depend in general upon a certain number of essential parameters. We have thus classes of Riemann matrices and our object will be to determine the invariants of the most general matrix of a given class.

96. Assume then that there exists a single form (12). What are the invariants of $\Omega$? The numbers $c_{\mu\nu}^{mn}$ corresponding to a given Riemann form must always satisfy the relations (11) and as the $\gamma_{\mu\nu}(\alpha_j, \alpha_h)$ are uniquely determined up to a factor of proportionality necessarily rational in $\alpha_j, \alpha_h$, we have

$$\sum_{m,n}^{1\cdots q} c_{\mu\nu}^{mn} \alpha_j^{m-1} \alpha_h^{n-1} = \phi_{jh}(\alpha_j, \alpha_h)\gamma_{\mu\nu}(\alpha_j, \alpha_h) \qquad (j \neq h;\, j, h = 1, 2, \cdots, p),$$

where $\phi_{jh}$ is a polynomial with rational coefficients. The left-hand side must be changed in sign when we interchange at the same time $j, h$ and $\mu, \nu$. Hence

$$\phi_{jh}(\alpha_j, \alpha_h) = \phi_{jh}(\alpha_h, \alpha_j) = \phi_{hj}(\alpha_h, \alpha_j),$$

that is, $\phi_{jh}$ must be symmetrical in $\alpha_j$ and $\alpha_h$. Moreover if the group $G$ of $f(\alpha) = 0$ permutes the pairs $(\alpha_j, \alpha_h)$ and $(\alpha_{j'}, \alpha_{h'})$, then it is necessary that $\phi_{jh}(x, y) = \phi_{j'h'}(x, y)$. Finally if there are equal roots among the multipliers $\alpha_1, \alpha_2, \cdots, \alpha_p$, say $\alpha_j = \alpha_{j'}$—which is certainly the case if $r > 2$ —and if moreover the quantities $\gamma_{\mu\nu}(\alpha_j, \alpha_j)$ are not all zero, we will have to consider a unique function $\phi^j(\alpha_j) = \phi_{jj'}(\alpha_j, \alpha_{j'})$. This will certainly occur when the $(\alpha)$'s are all real, for then in order that (13) be definite none of the quantities $A_{jj}$ must be zero, hence the expressions $\gamma_{\mu\nu}(\alpha_j, \bar{\alpha}_j) = \gamma_{\mu\nu}(\alpha_j, \alpha_j)$ must not all be zero either.

Given the $(\gamma)$'s and the $(\phi)$'s, the numbers $c_{\mu\nu}^{mn}$, when $\mu, \nu$ are assigned, satisfy a system of non-homogeneous linear equations whose number is determined thus:

(*a*) Among the multipliers $\alpha_1, \alpha_2, \cdots, \alpha_p$ two at least are equal. Let then $s$ be the number of pairs of distinct roots derived from the pairs $\alpha_j, \alpha_h$; $j, h < p$, by the substitutions of $G$. When $\mu \neq \nu$ there are $2s + q$ equations and when $\mu = \nu$ there are $s$ of them.

(*b*) The multipliers $\alpha_1, \alpha_2, \cdots, \alpha_p$ form without repetition the totality of the roots of $f(\alpha) = 0$. This occurs only when $p = q$, $r = 2$. The number of equations is then $q(q-1)$ or $\frac{1}{2}q(q-1)$ according as $\mu \neq \nu$ or $\mu = \nu$.

But as we have seen in § 2, the left-hand sides of the equations in the $(c_{\mu\nu})$'s, when these unknowns are considered as variables, are linearly independent. Hence when the $(\gamma)$'s and the $(\phi)$'s are given we have a solution with

$$r\left[\binom{q}{2} - s\right] + 2\binom{r}{2}\left[\binom{q}{2} - s\right] = r^2\left[\binom{q}{2} - s\right]$$

arbitrary parameters in the first case and with $q = p$ of them in the second.

It remains to determine the number of systems

$$\phi_{jh}^1, \phi_{jh}^2, \cdots, \phi_{jh}^n; \phi^1(\alpha_j), \phi^2(\alpha_j), \cdots, \phi^n(\alpha_j)$$

such that there exist no relations

$$\sum_{t=1}^{n} \lambda_t \phi_{jh}^t(\alpha_j, \alpha_h) = 0, \quad \sum_{t=1}^{n} \lambda_t \phi^t(\alpha_j) = 0,$$

where the $(\lambda)$'s are rational numbers and where we must replace in the second relation $\alpha_j$ successively by all the roots of $f(\alpha) = 0$, and in the first $(\alpha_j, \alpha_h)$ by all the pairs of distinct roots, derived from those for which $j, h \leqq p$, by the substitutions of $G$. Moreover, in case (*b*) or in case (*a*) when the $\gamma_{\mu\nu}(\alpha_j, \alpha_j)$ are all zero we must make $\phi(\alpha_j) \equiv 0$. Between the symmetrical functions $\alpha_j^m \alpha_h^n + \alpha_j^n \alpha_h^m$ of two assigned roots there must exist a certain number $t_i$ of linear homogeneous relations with integral coefficients, and these relations will still be satisfied by the symmetrical functions of two roots $T\alpha_j$, $T\alpha_h$, where $T$ is an arbitrary substitution of $G$. The number $t_i$ of linearly independent symmetrical functions characterizes therefore not so much the pair $(\alpha_j, \alpha_h)$ as the set of pairs of roots which are transitively permutable with it by the substitutions of $G$,—and to each set of transitively permutable pairs of roots corresponds such an integer. Let finally $\epsilon$ be a number $= 0$ if $\phi(\alpha_j) \equiv 0$ and $= +1$ if $\phi(\alpha_j) \not\equiv 0$. A very simple discussion shows then that in case (*a*)

$$1 + k = \epsilon q + \sum t_i + r^2 \left[ \binom{q}{2} - s \right],$$

while in case (*b*)

$$1 + k = \sum t_i + p.$$

The determination of $h$ can be made in a similar manner. If we assume the array $\tau$ as general as possible, there will be no other relations between its elements than those which follow from the existence of (11). A non-alternate bilinear form must then correspond to the relations

$$\sum_{m,n}^{1\ldots q} \sum_{\mu,\nu}^{1\ldots r} c_{\mu\nu}^{mn} \alpha_j^{m-1} \alpha_h^{n-1} \tau_{j\mu} \tau_{h\nu} = 0 \qquad (j, h = 1, 2, \cdots, p),$$

whence we will derive as before

$$\sum_{m,n} c_{\mu\nu}^{mn} \alpha_j^{m-1} \alpha_h^{n-1} = \phi_{jh}(\alpha_j, \alpha_h) \gamma_{\mu\nu}(\alpha_j, \alpha_h), \tag{14}$$

where $\phi_{jh}$ is no more constrained to be symmetrical. We shall now have to consider as distinct the pairs of roots $(\alpha_j, \alpha_h)$ and $(\alpha_h, \alpha_j)$, and in place of $t_i$ we shall be led to introduce an integer $t_i'$ to denote the number of products $\alpha_j^m \alpha_h^n$ between which there exists no linear relation with integral coefficients.

Finally, we have

$$1 + h = \epsilon q + \sum t'_i + 2r^2\left[\binom{q}{2} - s\right]$$

in case ($a$), $\epsilon$ having the same meaning as previously, while in case ($b$)

$$1 + h = \sum t'_i + p.$$

If there exists a non-identically zero solution of the equations (11) and (14) and the $(\gamma)$'s are not all zero, it will be necessary to add to $k$ and $h$ respectively the numbers

$$r^2\left[\binom{q}{2} - s\right], \qquad 2r^2\left[\binom{q}{2} - s\right]$$

in case ($a$), and the same number $p$ in case ($b$).

*Remark:* It is not difficult to extend these formulas to the case where $\Omega$ possesses several bilinear forms of the type considered. However, the results obtained are not simple and it is preferable to establish them directly in the few cases where this extension will be needed.

97. As an application of the preceding considerations, consider the case where $f(\alpha) = 0$ is as general as possible and of degree $q > 2$. The group $G$ is then the symmetrical group, hence doubly transitive. There will then be a single integer $t_i = t$ and a single integer $t'_i = t'$. To find $t'$ observe that the equations in the unknowns $d_{mn}$,

$$\sum_{m,\,n}^{1\cdots q} d_{mn}\,\alpha_j^{m-1}\,\alpha_h^{n-1} = 0 \qquad (j, h = 1, 2, \cdots, q;\ j \neq h), \tag{15}$$

possess $q$ independent rational solutions. Hence there are

$$q^2 - q = q(q-1) = t'$$

products $\alpha_j^m\,\alpha_h^n$ between which there exist no linear relations with integral coefficients, and therefore $1 + h = q(q-1) + \epsilon q$. The number of distinct symmetrical functions formed with the products

$$\alpha_j^m\,\alpha_h^n \qquad (m, n < q) \qquad \text{is equal to} \qquad \tfrac{1}{2}q(q-1) + q = \tfrac{1}{2}q(q+1).$$

But between them there are as many linear relations with integral coefficients as there are independent solutions of the equations in the unknowns $d_{mn}$,

$$\sum_{m,\,n}^{1\cdots q} d_{mn}\,(\alpha_j^{m-1}\,\alpha_h^{n-1} + \alpha_j^{n-1}\,\alpha_h^{m-1}) = 0 \qquad (j, h = 1, 2, \cdots, q;\ j \neq h).$$

But these equations are linear combinations of the equations (15), and the manner in which they have been obtained shows that they are independent like the equations (15) themselves. They possess therefore

$\frac{1}{2}q(q+1) - \frac{1}{2}q(q-1) = q$ independent solutions and hence

$$1 + k = t = \frac{q(q+1)}{2} - q = \frac{q(q-1)}{2}.$$

98. We shall now examine more closely the *conditions of existence of* $\Omega$. Let us return to the form (12). We may define two polynomials with integral coefficients

$$\gamma'_{\mu\nu}(x, y) = \sum_{m,n}^{1\cdots q} d_{\mu\nu}^{mn} x^{m-1} y^{n-1},$$
$$\gamma''_{\mu\nu}(x) = \sum_{m=1}^{q} d_{\mu\nu}^{m} x^{m-1}$$

by the relations

$$\gamma'_{\mu\nu}(\alpha_j, \alpha_k) = \gamma_{\mu\nu}(\alpha_j, \alpha_h) \qquad (\alpha_j, \alpha_k \text{ are any two different roots});$$
$$\gamma'_{\mu\nu}(\alpha_j, \alpha_j) = 0; \qquad \gamma''_{\mu\nu}(\alpha_j) = \gamma_{\mu\nu}(\alpha_j, \alpha_j) \qquad (j = 1, 2, \cdots, 2p).$$

Consider the elements of a given row of $\tau$ as homogeneous point coördinates in an $S_{r-1}$—this is the generalization of Scorza's habitual point of view. To the $r_j$ rows corresponding to $\alpha_j$ ($r_j$ is the number of times that $\alpha_j$ is found among the multipliers), say

$$\begin{matrix} \tau'_{11}, & \tau'_{12}, & \cdots, & \tau'_{1r} \\ \tau'_{21}, & \cdot & \cdots, & \cdot \\ \cdot & \cdot & \cdot & \cdot \\ \tau'_{r_j1}, & \cdot & \cdots, & \tau'_{r_jr} \end{matrix}$$

will correspond an $S_{r_j-1}$. The equations (11) can be interpreted as follows:

(*a*) $S_{r_j-1}$ and $S_{r_k-1}$ must be conjugate spaces relatively to the reciprocity

$$\sum_{\mu,\nu}^{1\cdots r} \gamma'_{\mu\nu}(\alpha_j, \alpha_k) x_\mu y_\nu = 0.$$

(*b*) Any two points of $S_{r_j-1}$ are conjugate relatively to the linear complex

$$\sum_{\mu,\nu}^{1\cdots r} \gamma''_{\mu\nu}(\alpha_j) x_\mu y_\nu = 0.$$

(*c*) In the Hermitian form (13) let us annul all variables corresponding to roots other than $\alpha_j$. The remaining expression is, with a change in the indices of the $(x)$'s,

$$-\frac{1}{2i} \sum_{h,k}^{1\cdots r_j} \sum_{\mu,\nu}^{1\cdots r} \gamma_{\mu\nu}(\alpha_j, \bar{\alpha}_j) \tau_{h\mu} \tau_{k\nu} x_h \bar{x}_k.$$

It must be positive for all non-zero values of the $(x)$'s. Now it may be written

$$-\frac{1}{2i} \sum_{\mu,\nu}^{1\cdots r} \gamma_{\mu\nu}(\alpha_j, \bar{\alpha}_j) \sum_{h=1}^{r_j} \tau'_{h\mu} x_h \sum_{k=1}^{r_j} \bar{\tau}'_{k\nu} \bar{x}_k,$$

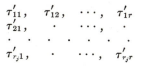

which shows that this amounts to requiring that the Hermitian form

$$-\frac{1}{2i}\sum_{\mu,\nu}^{1\cdots r}\gamma_{\mu\nu}(\alpha_j,\bar{\alpha}_j)\,x_\mu\,\bar{x}_\nu$$

be positive for all points of $S_{r_j-1}$.

99. The condition ($c$) may be interpreted in two distinct ways according as $\alpha_j$ is real or complex:

($c'$) If $\alpha_j$ is real it is equivalent to demanding that the real complex

$$\sum \gamma''_{\mu\nu}(\alpha_j)\,x_\mu\,y_\nu = 0$$

contain no real straight line resting on $S_{r_j-1}$.

($c''$) If $\alpha_j$ is complex the Hermitian form

$$-\frac{1}{2i}\sum \gamma'_{\mu\nu}(\alpha_j,\bar{\alpha}_j)\,x_\mu y_\nu$$

must be positive at all points of $S_{r_j-1}$.

The conditions ($a$), ($b$), ($c'$), ($c''$) are necessary but may not be sufficient. Hence if there exist Riemann matrices for which these conditions but no others are fulfilled, these matrices will certainly be as general as possible of their type, and their invariants will be as small as possible.

Returning to ($c''$), denote for the present by $S_{r'_j-1}$ the space corresponding to $\alpha_j$. The Hermitian form above may also be written

$$\frac{1}{2i}\sum \gamma'_{\mu\nu}(\alpha_j,\bar{\alpha}_j)\,x_\mu\,\bar{x}_\nu = -\frac{1}{2i}\sum \gamma'_{\mu\nu}(\alpha_j,\bar{\alpha}_j)\,\bar{x}_\mu\,x_\nu .$$

This shows that it must be positive at all points of $S_{r_j-1}$ and negative at all those of $\bar{S}_{r'_j-1}$, conjugate space of $S_{r'_j-1}$.

Now it is known from classical theorems on Hermitian forms that by means of a transformation

$$x'_\mu = \sum_{\nu=1}^{r}\lambda_{\mu\nu}\,x_\nu;\qquad \bar{x}'_\mu = \sum_{\nu=1}^{r}\bar{\lambda}_{\mu\nu}\,\bar{x}_\nu \qquad (\mu = 1, 2, \cdots, r)$$

our form may be reduced to the type

$$\sum a_\mu\,x'_\mu\,\bar{x}'_\mu ,$$

where the $(a)$'s are real. If we consider the transformation on the $(x)$'s as applied to the points of $S_{r_j-1}$ and of $\bar{S}_{r'_j-1}$, and that on the $(\bar{x})$'s to the points of $S_{r'_j-1}$, we see that the mutual relations of the three spaces $S$, $S'$, $\bar{S}'$, transformed of the three preceding, are again the same as before in regard to the new Hermitian form. Now however we have the advantage that the new form takes the same sign at $S'$ and $\bar{S}'$, so that we have this situation: The spaces $S$ and $S'$ are conjugate with respect to the reciprocity obtained by making the

Hermitian form vanish, and the form takes the sign + on the one and the sign − on the other. We shall see later that this requires that $r = r_j + r'_j$. Hence the transformed form must not be degenerate and this holds also for the original one.

100. Let us make a few remarks concerning the equation $f(\alpha) = 0$ when it has mixed real and complex roots. We may show that in this case the integers $r_j$ cannot in general be taken arbitrarily. For assume that the group $G$ of the equation is doubly transitive, that is, permutes transitively all the pairs of roots. If one of the reciprocities

$$\sum \gamma'_{\mu\nu}(\alpha_j, \alpha_k) x_\mu y_\nu = 0$$

is degenerate so will all the others be, since when the determinant

$$|\gamma'_{\mu\nu}(\alpha_j, \alpha_k)| \qquad (\mu, \nu = 1, 2, \cdots, r)$$

is zero for a pair of distinct roots, it will be zero for all of them. But as we have just seen this reciprocity is certainly not degenerate for $\alpha_j = \bar{\alpha}_k$, hence it is never degenerate when $\alpha_j \neq \alpha_k$.

Assume now $\alpha_j$ real, $\alpha_k$ still complex. $S_{r_j-1}$ and $S_{r_k-1}$ are conjugate relatively to the non-degenerate reciprocity

$$\sum \gamma'_{\mu\nu}(\alpha_j, \alpha_k) x_\mu y_\nu = 0,$$

hence $r_j - 1 + r_k - 1 = r - 2$. But $r_j = \frac{1}{2}r$, hence $r_k = \frac{1}{2}r$ also. Thus all the $(r)$'s must have the value $\frac{1}{2}r$, that is, among the multipliers must be found every root of $f(\alpha) = 0$ taken the same number of times.

Let us return to the Hermitian form

$$-\frac{1}{2i}\sum \gamma'_{\mu\nu}(\alpha_j, \alpha_j) x_\mu \bar{x}_\nu.$$

It will have one sign on $S_{r_j-1}$ and another on $\bar{S}_{r'_j-1}$. Hence, as we shall see below, the equation of degree $r$ in $\delta$,

$$|\gamma'_{\mu\nu}(\alpha_j, \alpha_j) - \epsilon_{\mu\nu}\cdot\delta| = 0 \qquad (\epsilon_{\mu\mu} = 1;\ \epsilon_{\mu\nu} = 0,\ \mu \neq \nu),$$

must have $\frac{1}{2}r$ positive roots and $\frac{1}{2}r$ negative roots. There is here considerable restriction imposed upon the form (12).

An interesting case of mixed equation has been investigated by Frobenius, —the case of the characteristic equation of a principal transformation. Frobenius does not give himself an arbitrary equation $f(\alpha) = 0$ but assumes a definite Riemann projectivity and starts from the equation

$$|b_{\mu\nu} - \epsilon_{\mu\nu}\cdot\alpha| = 0,$$

showing that, given such an equation, there always corresponds to it a suitable Riemann matrix.

101. Before continuing, let us examine rapidly the lower limits which may be assigned to $h$ and $k$ in some simple cases. First, if the multiplication has real multipliers, we will have necessarily $\epsilon = 1$, on account of ($b$) and ($c'$), hence $1 + h \geqq q$, $1 + k \geqq q$. Assume the multipliers all imaginary, hence $q = 2q'$. According to ($a$) and ($c''$) the expressions $\gamma_{\mu\nu}(\alpha_j, \alpha_j)$ must not all be zero. The group $G$ of $f(\alpha) = 0$ may permute transitively a pair $(\alpha_j, \bar{\alpha}_j)$ with other pairs of roots.—Let $(\alpha_1', \alpha_1''), \cdots, (\alpha_{q''}', \alpha_{q''}'')$ be one of the sets thus transitively permuted and including a pair of conjugate roots. We have $\sum q'' \geqq q'$ and the $(q'')$'s are all $> 1$, if $q > 2$, else $f(\alpha)$ would be reducible. There can be no such relation as

$$\sum_{m,n}^{1\ldots q''} c_{mn}\, \alpha_j'^{\,m-1}\, \alpha_h''^{\,n-1} = 0 \qquad (j, h = 1, 2, \cdots, q''),$$

since the Vandermonde determinants of the $(\alpha')$'s and $(\alpha'')$'s are not zero. Similarly there can be no such relations as

$$\sum_{m,n}^{1\ldots q''} c_{mn}\,(\alpha_j'^{\,m-1}\, \alpha_h''^{\,n-1} + \alpha_j'^{\,n-1}\, \alpha_h''^{\,m-1}) = 0 \qquad (j, h = 1, 2, \cdots, q'').$$

Hence,

$$\sum t_i \geqq \sum \frac{q''(q''-1)}{2} \geqq \sum \frac{(q''-1)^2}{2} + \tfrac{1}{2}\sum q'',$$

and therefore if $q > 2$, $1 + k > \frac{1}{4}q$. Moreover since there are obviously $q$ independent multiplications—namely $q$ powers of the given one—we have $1 + h \geqq q$. These limits are obviously correct a fortiori if $\Omega$ is impure with submatrices invariant under the Riemann projectivity considered. In particular $k = 0$, $h > 0$ is possible only if $f(\alpha) = 0$ is a quadratic equation with imaginary roots. We shall return to this later.

### § 5. The characteristic equation is of type $[f(\alpha)]^r = 0$. ($b$) Two important special cases

102. *Real multipliers** $(r = 2p' > 2)$. When the multipliers are real, the integers $r_j$ are all equal to $p'$ and the only conditions to be satisfied are ($b$) and ($c'$). By a slight change in our notation we may say that it is first necessary that the points of $S_{r_j-1}$ be conjugated to each other with respect to the linear complex

$$\sum_{\mu,\nu}^{1\ldots 2p'} \gamma_{\mu\nu}(\alpha_j)\, x_\mu\, y_\nu = 0. \tag{16}$$

Next at all points of $S_{r_j-1}$ we must have

$$-\frac{1}{2i}\sum \gamma_{\mu\nu}(\alpha_j)\, x_\mu\, \bar{x}_\nu > 0.$$

* (Added in 1922.) See *Comptes Rendus du Congrès de Strasbourg* (1921).

Consider the matrix

$$\tau_j \equiv \| \tau^j_{h1}, \tau^j_{h2}, \cdots, \tau^j_{h,\,2p'} \| \qquad (j = 1, 2, \cdots, p'),$$

formed with the rows of $\tau$ which correspond to $\alpha_j$. The last condition just stated is equivalent to requiring that the Hermitian forms

$$(17) \qquad \sum_{h,k}^{1\cdots p} A^j_{hk}\, x_h\, \bar{x}_k, \qquad A^j_{hk} = -\frac{1}{2i} \sum_{\mu,\nu}^{1\cdots 2p'} \gamma_{\mu\nu}(\alpha_j)\, \tau^j_{h\mu}\, \bar{\tau}^j_{k\nu}$$

be definite positive. Under these conditions I say that (16) defines a principal Riemann form of $\Omega$. For we may determine a Riemann form

$$\sum \delta_{\mu\nu}(\alpha_j, \alpha_k)\, x_\mu\, y_\nu,$$

with coefficients $\delta_{\mu\nu}$ satisfying the relations

$$\begin{aligned} \delta_{\mu\nu}(\alpha_j, \alpha_k) &= 0 \qquad\qquad (\alpha_j \neq \alpha_k), \\ \delta_{\mu\nu}(\alpha_j, \alpha_j) &= \gamma_{\mu\nu}(\alpha_j), \end{aligned}$$

since the equations for the coefficients of the polynomials $\delta_{\mu\nu}$ always have at least one rational solution. On the other hand the Hermitian form corresponding to this Riemann form reduces to the sum of the forms (17),—it is therefore definite positive.

We may remark that in all cases if (16) corresponds to a non-principal-alternate form, the corresponding Hermitian form is the sum of the $q$ forms (17), hence its genus is the sum of their genera.

Let us return to $\tau_j$. It is clear that the conditions imposed upon this matrix are identical with the conditions imposed upon a Riemann matrix of genus $p'$, except that the coefficients of the principal alternate form which occurs in the definition of these matrices are only subjected to being numbers of the algebraic domain $K(\alpha_j)$ but not necessarily rational numbers. We may say that $\tau_j$ is a Riemann matrix belonging to this algebraic domain.

103. By a transformation of coördinates with coefficients rational with respect to those of the complex (16), that is, with coefficients belonging to the domain $K(\alpha_j)$, we may reduce the equation of this complex to the form

$$\sum_{\mu=1}^{p'} e_\mu(\alpha_j)\,(x_\mu\, y_{p'+\mu} - x_{p'+\mu}\, y_\mu),$$

where the $(e)$'s are numbers of the same domain.* This is equivalent to applying a certain transformation of isomorphism upon $\Omega$. By following up this transformation with $q$ others applied each to the rows of $\tau_j$, we shall reduce $\tau_j$ to the form

$$\left\| \begin{matrix} (e_1(\alpha_j))^{-1}, & 0, & \cdots, & 0, & a^j_{11}, & a^j_{12}, & \cdots, & a^j_{1p'} \\ 0, & (e_2(\alpha_j))^{-1}, & \cdots, & 0, & a^j_{21}, & a^j_{22}, & \cdots, & \cdot \\ \cdot & \cdot & \cdot & \cdot & \cdot & \cdot & \cdot & \cdot \\ 0, & \cdot & \cdots, & (e_{p'}(\alpha_j))^{-1}, & a^j_{p'1}, & \cdot & \cdots, & a^j_{p'p'} \end{matrix} \right\|$$

* See for example Bertini, *Lezzioni sulla geometria proiettiva degli iperspazi*, page 106.

and if we set $a^j_{hk} = a'_{hk} + ia''_{hk}$ it is necessary that the quadratic forms $\sum a''_{hk} x_h y_k$ be all definite positive.

The class of matrices thus obtained depends upon $\frac{1}{2}qp'(p'+1) = \frac{1}{4}p(r+2)$ continuous parameters, and $qp' = p$ arbitrary integers—the integers which enter into the composition of the $(e)$'s, where we may assume the coefficients of the powers of the $(\alpha)$'s equal to integers.

The invariants of the most general matrices of the class here considered are given by

$$1 + h = 1 + k = q,$$

since with the notations of § 4, the $(\phi_{jh})$'s are zero, $s = \binom{q}{2}$ and the expressions $\phi(\alpha_j)$ are not zero.

104. Most of the properties of ordinary Riemann matrices belong also to those of the domain $K(\alpha_j)$. Let us indicate a few of them as well as their corollaries for $\Omega$.

Assume that there exist $1 + k_1$ linearly independent complexes such as (16), say

$$\sum_{\mu,\nu}^{1\cdots 2p'} \gamma^s_{\mu\nu}(\alpha_j) x_\mu y_\nu = 0 \qquad (\gamma^s_{\mu\nu} = -\gamma^s_{\nu\mu}),$$

belonging to $\Omega$. The equations in the coefficients $c^{mn}_{\mu\nu}$ of the alternate forms of $\Omega$, analogous to the equations (14), become

$$\sum_{m,n}^{1\cdots q} c^{mn}_{\mu\nu} \alpha_j^{m-1} \alpha_h^{n-1} = 0 \qquad (\alpha_j \neq \alpha_h)$$

$$\sum_{m,n}^{1\cdots q} c^{mn}_{\mu\nu} \alpha_j^{m+n-2} = \sum_{s=1}^{1+k_1} \phi_s(\alpha_j) \gamma^s_{\mu\nu}(\alpha_j),$$

where the $(\phi)$'s are polynomials with rational coefficients. They show that

$$1 + k = q(1 + k_1).$$

Similarly if there are $1 + h_1$ reciprocities

$$\sum_{\mu,\nu}^{1\cdots 2p'} \gamma^s_{\mu\nu}(\alpha_j) x_\mu y_\nu = 0,$$

then $1 + h = q(1 + h_1)$.

The matrix $\Omega$ will be pure if it does not possess any other bilinear forms of the type just defined (this we have tacitly assumed in the above formulas)—and if moreover none of these forms are degenerate.

We may apply to the matrices $\tau_j$ a transformation defined by the symbolic matrix-equation:

$$\tau'_j = \tau_j \cdot \| \beta_{\mu\nu}(\alpha_j) \| \qquad (\mu, \nu = 1, 2, \cdots, 2p'),$$

where the $(\beta)$'s are numbers of the domain $K(\alpha_j)$. Then we may apply to $\tau_j$ a linear transformation of the rows. If we do this simultaneously for every

$\tau_j$ we have a transformation of isomorphism of $\Omega$. The matrices $\tau_j$ must be considered as impure if, for every $j$,

$$\tau'_j \equiv \begin{Vmatrix} \tau_1^j, & 0, & \cdots, & 0 \\ 0, & \tau_2^j, & \cdots, & 0 \\ \cdot & \cdot & \cdot & \cdot \\ 0, & \cdot & \cdots, & \tau_n^j \end{Vmatrix}$$

where $\tau_h^j$ is a matrix with $p'_h$ rows and $2p'_h$ columns, $p'_h$ being independent of $j$. In this case $\Omega$ is isomorphic to a matrix of type (I), the submatrices $\omega_i$ being invariant under the Riemann projectivity considered. Such a condition will certainly arise if $1 + h_1 > 2p'$. In all cases, as Scorza showed for ordinary Riemann matrices, we have

$$1 + h_1 \leqq 2p'^2, \qquad 1 + k_1 \leqq p'^2,$$

hence

$$1 + h \leqq pr, \qquad 1 + k \leqq \tfrac{1}{2}pr.$$

When the indices $h_1$, $k_1$ have their maximum values, $\Omega$ is isomorphic to a matrix composed with an array $\tau$ such that

$$\tau_j \equiv \begin{Vmatrix} 1, & \delta(\alpha_j), & 0, & 0, & \cdots, & 0 \\ 0, & 0, & 1, & \delta(\alpha_j), & \cdots, & 0 \\ \cdot & \cdot & \cdot & \cdot & \cdot & \cdot \\ 0, & 0, & \cdot & \cdot & 1, & \delta(\alpha_j) \end{Vmatrix}$$

where $\delta(\alpha_j)$ is a quadratic number of the domain $K(\alpha_j)$ (Scorza).

The numbers $h_1$, $k_1$ may have lacunary values and they can be obtained as those of $h$, $k$ have been obtained by Scorza.

If $h_1 > 0$, then there exists at least one complex multiplication $T'$ permutable with the multiplication $T$ whose multipliers are the $(\alpha)$'s. There will then be a multiplication of degree $q' \geqq 2q$ permutable with both $T$ and $T'$ and we can investigate $\Omega$ by taking this multiplication as a starting point. But there may be advantage in considering directly the Riemann projectivities of $\tau_j$, and if $\Omega$ is pure $\tau_j$ may be composed with two matrices

$$\| \theta_{k1}^j, \quad \theta_{k2}^j, \quad \cdots, \quad \theta_{kr}^j \|, \qquad \| 1, \quad \beta_{jk}, \quad \beta_{jk}^2, \quad \cdots, \quad \beta_{jk}^{q'-1} \|$$

where the numbers $\beta_{jk}$ are roots of an irreducible equation $f(\beta) = 0$ of degree $q' = 2p'/r'$ with coefficients in the same domain as already considered.

Let $\Omega'$ be another Riemann matrix of genus $p$ possessing a Riemann projectivity having also $f(\alpha) = 0$ for characteristic equation and assume $\Omega$ and $\Omega'$ both pure. They will be isomorphic if there exists a simultaneous form

$$\sum \gamma_{\mu\nu}(\alpha_j) x_\mu y_\nu$$

which vanishes when the $(x)$'s are replaced by the elements of any row of $\tau_j$

and the $(y)$'s by those of any row of $\tau'_j$ which corresponds to $\tau_j$ for $\Omega'$, this for $j = 1, 2, \cdots, p'$. If there exist $\lambda_1$ such forms, the simultaneous index of $\Omega, \Omega'$ is $\lambda = q\lambda_1$.

In concluding, let us remark that the arithmetic properties of alternate forms belonging to ordinary Riemann matrices can be extended at once to those which we have here considered. For these propositions can in general be derived by purely arithmetic methods and without mentioning Abelian varieties at all. For example, the theorems given by Cotty in his thesis, on hyperelliptic surfaces and the corresponding forms, can be at once extended to $\tau_j$ when $r = 2p' = 4$.

105. *Real multipliers* $(r = 2p' = 2)$. When $r = 2$ and $\alpha_1, \alpha_2, \cdots, \alpha_p$ include all the roots of $f(\alpha) = 0$ they must necessarily be real, at least when the array $\tau$ is as general as possible. For if

$$\sum_{\mu, \nu}^{1,2} \gamma_{\mu\nu}(\alpha_j, \alpha_k) x_\mu y_\nu \tag{18}$$

is an alternate form of $\Omega$, it is necessary that $\gamma_{\mu\nu}(\alpha_j, \alpha_k) = 0$ if $\alpha_j \neq \alpha_k$, hence $\gamma_{\mu\nu}(\alpha_j, \bar{\alpha}_j) = 0$ if $\alpha_j \neq \bar{\alpha}_j$, and these matrices exist only if the $(\alpha)$'s are all real. They are then of the type of matrices which we have just investigated. We may always arrange matters so that no $\tau_{j1}$ will be zero. If we divide then every term in the $j$-th row of $\Omega$ by $\tau_{j1}$, the array $\tau$ will assume the form

$$\| 1, \tau_j \| \qquad (j = 1, 2, \cdots, p).$$

The arbitrary constants $\tau_j$ must be imaginary else $\Omega$ would have a row with elements all real. If the form (18) is an alternate form of $\Omega$,

$$\gamma_{\mu\nu}(\alpha_j, \alpha_k) = -\gamma_{\nu\mu}(\alpha_k, \alpha_j), \qquad \gamma_{\mu\nu}(\alpha_j, \alpha_k) = 0 \text{ if } \alpha_k \neq \alpha_j;$$
$$\gamma_{\mu\nu}(\alpha_j, \alpha_j) \neq 0;$$

and hence $1 + k = q$. If on the contrary (18) is not an alternate form of $\Omega$, and if the parameters $\tau_j$ remain entirely arbitrary it is necessary that

$$\gamma_{\mu\mu}(\alpha_j, \alpha_k) = 0 \quad (\mu = 1, 2); \quad \gamma_{12}(\alpha_j, \alpha_k) = \gamma_{21}(\alpha_j, \alpha_k) = 0 \text{ if } \alpha_j \neq \alpha_k;$$

$$\gamma_{12}(\alpha_j, \alpha_j) = -\gamma_{21}(\alpha_j, \alpha_j);$$

whence again $1 + h = q = p = 1 + k$.

Assume that (18) is an alternate form of $\Omega$. The polynomial

$$\gamma(\alpha) = \gamma_{12}(\alpha, \alpha)$$

is after all merely an arbitrary polynomial of degree $q - 1$. The Hermitian form corresponding to (18) is

$$-\frac{1}{2i}\sum_{j=1}^{p} \gamma(\alpha_j)(\bar{\tau}_j - \tau_j) x_j \bar{x}_j = \sum_{j=1}^{p} \gamma(\alpha_j) \tau''_j x_j x_j \qquad (\tau_j = \tau'_j + i\tau''_j).$$

We can always take $\gamma(\alpha)$ such that the coefficients will all be positive, hence whatever the parameters $\tau_j$, $\Omega$ is always a Riemann matrix. If we assume, as we have done so far, that $f(\alpha)$ is irreducible, then when one of the numbers $\gamma(\alpha_j)$ is zero they will all be. Since moreover none of the $\tau_j''$ can be zero, whenever $\Omega$ possesses no other alternate forms than those of the above type, the Hermitian form is always of genus $p$ and $\Omega$ is pure.

Can the matrix be pure and possess a new Riemann projectivity? In order that this be the case it must possess a bilinear form other than of the above type, say

$$\sum_{\mu,\nu}^{1,2} \beta_{\mu\nu}(\alpha_j, \alpha_k) x_\mu y_\nu,$$

imposing some relations between the parameters $\tau_j$. Two cases are possible: (a) The expressions $\beta_{\mu\nu}(\alpha_j, \alpha_k)$ $(\alpha_j \neq \alpha_k)$ are all zero, but the $\beta_{\mu\nu}(\alpha_j, \alpha_j)$ are not. There will then be no new alternate form, hence $1 + h = 2p$, $1 + k = p$. The new complex multiplication is permutable with the former. The matrix $\Omega$ which is still pure possesses a Riemann projectivity with irreducible characteristic equation and we fall back upon the case studied in § 2. This is a consequence of the fact that the reciprocity

$$\sum_{\mu,\nu}^{1,2} \beta_{\mu\nu}(\alpha_j, \alpha_k) x_\mu y_\nu = 0$$

leaves the point $(1, \tau_j)$ invariant, hence the parameters $\tau$ are then quadratic conjugate numbers of the real domains $K(\alpha_j)$. The matrix $\Omega$ is then isomorphic to a matrix with array $\tau$ of type $\| 1, \sqrt{R(\alpha_j)} \|$, where $R(\alpha_j)$ is an integer of the same domain, negative together with all its conjugates. (b) The expressions $\beta_{\mu\nu}(\alpha_j, \alpha_k)$ $(\alpha_j \neq \alpha_k)$ are not all zero. In this case $1 + h = q + q(q-1) = p^2 > 2p$ if $p > 2$. Hence $\Omega$ is impure if $p > 2$. Thus if $r = 2$ and if the multipliers are all real, $\Omega$ if pure cannot possess more than two other Riemann projectivities, one of degree two and the other of degree $2p$, permutable with the given one.

106. *Imaginary multipliers.* We shall first consider a question concerning Hermitian forms.

Given the Hermitian form of genus $r$

$$\sum_{\mu=1}^{r} a_\mu x_\mu \bar{x}_\mu,$$

let $S_{r'-1}$, $S_{r''-1}$ be two spaces of $S_r$ such that $r' + r'' = r$ and conjugated with respect to the reciprocity

$$\sum a_\mu x_\mu y_\mu = 0.$$

Under what conditions can the Hermitian form be positive at all points of the first space and negative at all points of the second? The equations of $S_{r'-1}$ can be put in the form

$$x_{r'+\nu} = \sum_{\mu=1}^{r'} b_{\nu\mu}\, x_\mu \qquad (\nu = 1, 2, \cdots, r''),$$

and it will be necessary that

$$\sum_{\mu=1}^{r'} a_\mu\, x_\mu\, \bar{x}_\mu + \sum_{\nu=1}^{r''} a_{r'+\nu} \left( \sum_{\mu=1}^{r'} b_{\nu\mu}\, x_\mu \right) \left( \sum_{\mu=1}^{r'} \bar{b}_{\nu\mu}\, \bar{x}_\mu \right) > 0 .$$

This expression can certainly be made negative if it is possible to annul all the terms whose coefficients $a_\mu$ are positive without annulling the others. Let $s'$ be the number that are positive and $s''$ the remaining. We shall have $s'$ linear equations in $r'$ unknowns and since they must not have any solutions, it is necessary that $r' \leqq s'$. Similarly we must have $r'' \leqq s''$, but $r' + r'' \geqq s' + s''$; hence $r' = s'$, $r'' = s''$. These conditions are besides sufficient. To see it we may consider the form

$$\sum_{\mu=1}^{r'} \bar{x}_\mu\, x_\mu - \sum_{\nu=1}^{r''} x_{r'+\nu}\, \bar{x}_{r'+\nu}$$

and the two conjugate spaces

$$S_{r'-1}; \quad x_{r'+\nu} = \lambda_\nu\, x_1 \qquad (\nu = 1, 2, \cdots, r''),$$

$$S_{r''-1}; \quad x_2 = x_3 = \cdots = x_{r'} = 0, \qquad x_1 = \sum_{\nu=1}^{r''} \lambda_\nu\, x_\nu ,$$

where the $(\lambda)$'s satisfy the conditions

$$\sum \lambda_\nu \bar{\lambda}_\nu < 1 .$$

On $S_{r'-1}$ the Hermitian form becomes

$$(1 - \sum \lambda_\nu \bar{\lambda}_\nu)\, x_1\, \bar{x}_1 + \sum_{\mu=2}^{r'} x_\mu\, \bar{x}_\mu$$

and its sign is $+$. On $S_{r''-1}$ it reduces to

$$\begin{aligned} &\sum_{\nu=1}^{r''} \lambda_\nu\, x_{r'+\nu} \sum_{\nu=1}^{r''} \bar{\lambda}_\nu\, \bar{x}_{r'+\nu} - \sum_{\nu=1}^{r''} x_{r'+\nu}\, \bar{x}_{r'+\nu} \\ &\quad = |\sum \lambda_\nu\, x_{r'+\nu}|^2 - \sum x_{r'+\nu}\, \bar{x}_{r'+\nu} \\ &\quad\quad < (\sum |\lambda_\nu\, x_{r'+\nu}|)^2 - \sum |x_{r'+\nu}|^2 < \sum |\lambda_\nu|^2 \sum |x_{r'+\nu}|^2 - \sum |x_{r'+\nu}|^2 \\ &\quad = (-1 + \sum \bar{\lambda}_\nu \lambda_\nu) \sum |x_{r'+\nu}|^2 < 0 , \end{aligned}$$

as was desired. This shows also that there is an infinity of pairs of spaces answering the question.

107. We now pass to the study of matrices with imaginary multipliers. We shall limit ourselves to the case *where the group $G$ of the equation $f(\alpha) = 0$ is permutable with the unique operation that permutes each root with its conjugate.* I say that in this case if there exists a form (12) satisfying only the conditions

(a) and $(c'')$, $\Omega$ is a Riemann matrix. For let $\sum \beta_{\mu\nu}(\alpha_j, \alpha_k) x_\mu y_\nu$ be a form answering the question. We may determine polynomials with rational coefficients and of degree $q - 1$ in $\alpha_j, \alpha_k$, say $\gamma_{\mu\nu}(\alpha_j, \alpha_k)$, by the relations

$$\gamma_{\mu\nu}(\alpha_j, \alpha_k) = 0 \qquad \text{if} \qquad \alpha_j \neq \alpha_k;$$
$$\gamma_{\mu\nu}(\alpha_j, \bar{\alpha}_j) = \beta_{\mu\nu}(\alpha_j, \bar{\alpha}_j).$$

That this is possible in view of the property assumed for the group of $f(\alpha) = 0$ follows from § 2, and we see then that the form

$$(19) \qquad \sum \gamma_{\mu\nu}(\alpha_j, \alpha_k) x_\mu y_\nu$$

is a Riemann form of $\Omega$. For, the spaces $S_{r_j - 1}, S_{r_k - 1}$ will be conjugated with respect to the proper reciprocity, in particular $S_{r_j-1}, S_{r'_j-1}$ with respect to

$$(20) \qquad \sum \gamma_{\mu\nu}(\alpha_j, \bar{\alpha}_j) x_\mu y_\nu = 0$$

and the Hermitian forms

$$(21) \qquad -\frac{1}{2i} \sum \gamma_{\mu\nu}(\alpha_j, \bar{\alpha}_j) x_\mu \bar{x}_\nu$$

will be positive at all points of $S_{r_j-1}$ and negative at all points of $S_{r'_j-1}$. As the Hermitian form belonging to (19) is here the sum of the forms (21) it is definite positive, and $\Omega$ is effectively a Riemann matrix.

108. To establish the existence of matrices of the type considered it is sufficient to show that we can take polynomials with rational coefficients $\gamma_{\mu\nu}(\alpha_j, \bar{\alpha}_j)$, such that when the Hermitian form

$$-\frac{1}{2i} \sum \gamma_{\mu\nu} x_\mu \bar{x}_\nu$$

is reduced to the type

$$\sum \gamma_\mu x_\mu \bar{x}_\mu,$$

$r_j$ coefficients $\gamma_\mu$ are positive and $r'_j$ negative. This is equivalent to requiring that of the $r$ roots of the equation in $\gamma$ (all real, as is well-known),

$$\left| -\frac{1}{2i} \gamma_{\mu\nu}(\alpha_j, \bar{\alpha}_j) - \epsilon_{\mu\nu} \cdot \gamma \right| = 0 \qquad (\epsilon_{\mu\mu} = 1;\ \epsilon_{\mu\nu} = 0,\ \mu \neq \nu),$$

$r_j$ be positive and $r'_j$ negative. Indeed, we can then determine polynomials with rational coefficients $\gamma'_{\mu\nu}(\alpha_j, \alpha_k)$ such that

$$\gamma'_{\mu\nu}(\alpha_j, \alpha_k) = 0 \quad \text{if} \quad \alpha_j \neq \alpha_k; \qquad \gamma'_{\mu\nu}(\alpha_j, \bar{\alpha}_j) = \gamma_{\mu\nu}(\alpha_j, \bar{\alpha}_j),$$

and also spaces $S_{r_j}, S_{r'_j-1}$ satisfying the conditions (a), $(c'')$ relatively to

$$\sum \gamma'_{\mu\nu}(\alpha_j, \alpha_k) x_\mu y_\nu,$$

and there will therefore exist a Riemann matrix belonging to $f(\alpha) = 0$ and to the distribution $\alpha_1, \alpha_2, \cdots, \alpha_p$ of its roots.

Now let us take $rq$ numbers $\eta_{j\mu}$ such that with $\alpha_{j'} = \overline{\alpha}_j$,

$$\eta_{j,1} = \eta_{j,2} = \cdots = \eta_{j,r_j} = +1; \qquad \eta_{j,r_j+1} = \cdots = \eta_{j,r} = -1;$$
$$\eta_{j',1} = \eta_{j',2} = \cdots = \eta_{j',r_j} = -1; \qquad \eta_{j',r_j+1} = \cdots = \eta_{j',r} = +1,$$

and consider the equations

$$-\frac{1}{2i}\sum_{m,n}^{1\cdots q} c_{\mu\nu}^{mn}(\alpha_j^{m-1}\overline{\alpha}_j^{n-1} - \overline{\alpha}_j^{m-1}\alpha_j^{n-1}) = \eta_{j\mu} \qquad (j = 1, 2, \cdots, q;\ \mu = 1, 2, \cdots, r),$$

of which we shall take a type solution in real numbers $c_{\mu\nu}^{mn}$. Designate then by $-\frac{1}{2i}\,\delta_{\mu\mu}(\alpha_j, \overline{\alpha}_j)$ the left-hand sides. The equation in $\gamma$

$$\left| -\frac{1}{2i}\,\delta_{\mu\nu}(\alpha_j, \overline{\alpha}_j) - \epsilon_{\mu\nu}\cdot\gamma \right| = 0 \qquad (\epsilon_{\mu\nu} = \delta_{\mu\nu} = 0,\ \mu \neq \nu;\ \epsilon_{\mu\mu} = 1),$$

has $r_j$ positive and $r_j'$ negative roots, this for $j = 1, 2, \cdots, q$, all equal to unity in absolute value. Let us take finally polynomials $\gamma_{\mu\nu}(\alpha_j, \overline{\alpha}_j)$ with rational coefficients differing as little as we please from those of the $(\delta)$'s. The Hermitian form

$$-\frac{1}{2i}\sum \gamma_{\mu\nu}(\alpha_j, \overline{\alpha}_j)\, x_\mu\, \bar{x}_\nu$$

will obviously answer the question. The existence of our matrices is therefore proved and at the same time we have given a construction for them.

To obtain $h$, $\tau$ being assumed as general as possible, we remark that the number of independent relations

$$\sum_{m,n}^{1\cdots q} b_{mn}\,\alpha_j^{m-1}\,\overline{\alpha}_j^{n-1} = 0 \qquad (j = 1, 2, \cdots, q)$$

is equal to the number of independent solutions for the $(b)$'s, that is, to $q(q-1)$. Hence $t' = q$. Similarly for $k$ we must consider the number of relations

$$\sum_{m,n}^{1\cdots q} b_{mn}(\alpha_j^{m-1}\overline{\alpha}_j^{n-1} + \overline{\alpha}_j^{m-1}\alpha_j^{n-1}) = 0 \qquad (j = 1, 2, \cdots, \tfrac{1}{2}q).$$

This number is $\frac{1}{2}q^2$, hence

$$t = \frac{q(q+1)}{2} - \tfrac{1}{2}q^2 = \tfrac{1}{2}q.$$

It follows that

$$1 + h = 2(1+k) = q + 2r^2\left[\binom{q}{2} - s\right].$$

This assumes $r > 2$. When $r = 2$ and the numbers $r_j$ are not all equal to unity, nothing is changed. If they are all equal to unity we must add $q$ alternate forms to $h$ and $k$ (the same forms as occur in the case of real multipliers), and since $s = \binom{q}{2}$, we have

$$1 + h = 2q, \qquad 1 + k = \tfrac{3}{2}q.$$

*Remark:* When $f(\alpha) = 0$ is Abelian, it becomes possible to determine $s$. For it is certainly possible to derive from the set $(\alpha_1, \alpha_2, \cdots, \alpha_p)$ all the pairs of roots by the operations of $G$ if the number of distinct multipliers exceeds $\frac{1}{2}q$ (No. 91). Hence when this occurs, as it does in the most general case, we have $s = \binom{q}{2}$, and therefore $1 + h = 2(1 + k) = q$, with the added condition that if $r = 2$ the multipliers must not all be distinct. When they are we are thrown back on the formulas already derived for this special case.

Assume now that there are exactly $\frac{1}{2}q$ distinct multipliers. Then (No. 84) $\Omega$ is impure of type (II) composed $r$ times with

$$\omega \equiv \| 1, \alpha_j, \cdots, \alpha_j^{q-1} \| \qquad (j = 1, 2, \cdots, \tfrac{1}{2}q)$$

which is itself pure or impure according as the subgroup $G'$ of $G$ which maintains invariant the set $(\alpha_1, \alpha_2, \cdots, \alpha_p)$ does or does not reduce to the identity. Let $n$ be the order of $G'$. The invariants of $\omega$ are given by $1 + h' = 2(1 + k') = nq$ and those of $\Omega$ by $1 + h = 2(1 + k) = nr^2 q = nrp$.

109. The form in the left-hand side of (20) can be reduced to one of the same nature with $\gamma_{\mu\nu}(\alpha_j, \bar{\alpha}_j) = 0 \quad (\mu \neq \nu)$, by a transformation of variables

$$x'_\mu = \sum_\nu b_{\mu\nu}(\alpha_j, \bar{\alpha}_j) x_\nu, \qquad y'_\mu = \sum_\nu b_{\mu\nu}(\bar{\alpha}_j, \alpha_j) y_\nu,$$

where the $(b)$'s are as before polynomials with rational coefficients. Now according to what has been stated in No. 89, these equations can be put in the simple form

$$x'_\mu = \sum_\nu e_{\mu\nu}(\alpha_j) x_\nu, \qquad y'_\mu = \sum_\nu e_{\mu\nu}(\bar{\alpha}_j) y_\nu,$$

where the $(e)$'s are still polynomials with rational coefficients. This is all equivalent to stating that $\Omega$ can be transformed into an isomorphic matrix for which (20) is replaced by $\sum_{\mu=1} \gamma_{\mu\mu}(\alpha_j, \bar{\alpha}_j) x_\mu y_\mu$. Moreover

$$\gamma_{\mu\mu}(\alpha_j, \bar{\alpha}_j) = (\alpha_j - \bar{\alpha}_j) \sum c_{\mu\mu}^{mn} (\alpha_j^m \bar{\alpha}_j^n + \bar{\alpha}_j^m \alpha_j^n)$$

and contains therefore $\frac{1}{2}q$ arbitrary coefficients. Hence $\Omega$ depends upon $\frac{1}{2}rq = p$ arbitrary integers and upon $\frac{1}{2}\sum_{j=1}^{j=q} r_j r'_j$ continuous essential parameters, namely the parameters which determine the position of one of the spaces $S_{r_j-1}$, $S_{r'_j-1}$ for each pair of roots $(\alpha_j, \bar{\alpha}_j)$.

When there are no other alternate forms than those derived from (19), $\Omega$ is pure since the Hermitian form belonging to each alternate form is the sum of $q$ forms of genera $r_j$, with independent variables, hence its genus is $\sum r_j = p$. This is still true if there are $1 + k_0$ forms of the type in question, such that the corresponding forms (20)

$$\sum \gamma^{s}_{\mu\nu}(\alpha_j, \bar{\alpha}_j)\, x_\mu\, y_\nu \qquad (s = 1, 2, \cdots, 1 + k_0)$$

are linearly independent and moreover do not possess any combination with coefficients polynomials in $\alpha_j$, $\bar{\alpha}_j$, degenerate. In all cases if $r > 2$, or $r = 2$, and the multipliers are not all distinct, $1 + k = \frac{1}{2}q(1 + k_0)$, and in the same conditions if there are $h_0$ non-alternate forms of the same type,

$$1 + h = q(1 + h_0).$$

When the multipliers are all distinct $1 + k$ is the same, but $1 + h = q(2 + h_0)$.

Let us consider a little more closely the case just mentioned where $\Omega$ possesses two bilinear forms of type (19) with forms (20) independent. Let $A_j$, $B_j$ be the reciprocities determined by two of them between the spaces $S_{r_j-1}$, $S_{r'_j-1}$. Then $A_j^{-1}\cdot B_j$ is a projectivity transforming $S_{r_j}$ into itself. It is defined by equations such as $x'_\mu = \sum_\nu b_{\mu\nu}(\alpha_j, \bar{\alpha}_j)\, x_\nu$, which as before can be put in the form $x'_\mu = \sum_\nu e_{\mu\nu}(\alpha_j)\, x_\nu$, and it is readily seen that these relations define, as in the case of real roots, a complex multiplication permutable with the multiplication whose multipliers are the $(\alpha)$'s.

*Remark:* Let us assume that the integers $r_j$ are all equal and that moreover $\Omega$ possesses an alternate form (19) such that $\gamma_{\mu\nu}(\alpha_j, \alpha_k) = 0$ if $\alpha_j \neq \alpha_k$. It will then be possible to apply with scarcely any change everything that has been said in the case of real multipliers. However $h$ and $k$ will not be the same. Let there be $k_1$ forms of the nature in question and such that the alternate forms

$$\sum \gamma^{s}_{\mu\nu}(\alpha_j, \alpha_j)\, x_\mu\, y_\nu \qquad (s = 1, 2, \cdots, k_1),$$

$$\gamma^{s}_{\mu\nu}(\alpha_j, \alpha_j) = -\gamma^{s}_{\nu\mu}(\alpha_j, \alpha_j)$$

are linearly independent. We will then have

$$1 + h = \tfrac{1}{2}(1 + k_0 + 2k_1) + r^2\left[\binom{q}{2} - s\right].$$

Similarly if there are $h_1$ non-alternate forms, then

$$1 + h = q(1 + h_0 + h_1) + 2r^2\left[\binom{q}{2} - s\right].$$

110. Can $\Omega$ possess a new bilinear form

$$\sum \delta_{\mu\nu}(\alpha_j, \alpha_k)\, x_\mu\, y_\nu \tag{22}$$

without possessing other alternate forms than those derived from (19)? In order that this be the case, it is necessary that the following relations be verified:

$$\delta_{\mu\nu}(\alpha_j, \alpha_k) - \delta_{\nu\mu}(\alpha_k, \alpha_j) = 0,$$

$$(\alpha_j - \alpha_k)[\delta_{\mu\nu}(\alpha_j, \alpha_k) + \delta_{\nu\mu}(\alpha_k, \alpha_j)] = 0 \quad \text{if} \quad \alpha_j \neq \bar{\alpha}_k,$$

$$\delta_{\mu\nu}(\alpha_j, \bar{\alpha}_j) - \delta_{\nu\mu}(\bar{\alpha}_j, \alpha_j) = \phi(\alpha_j, \bar{\alpha}_j)\cdot\gamma_{\mu\nu}(\alpha_j, \bar{\alpha}_j),$$

$$(\alpha_j - \bar{\alpha}_j)[\delta_{\mu\nu}(\alpha_j, \bar{\alpha}_j) + \delta_{\nu\mu}(\bar{\alpha}_j, \alpha_j)] = \psi(\alpha_j, \bar{\alpha}_j)\cdot\gamma_{\mu\nu}(\alpha_j, \bar{\alpha}_j).$$

From the first two we conclude that

$$\delta_{\mu\nu}(\alpha_j, \alpha_k) = 0 \quad \text{if} \quad \alpha_j \neq \alpha_k \text{ or } \overline{\alpha}_k,$$
$$\delta_{\mu\nu}(\alpha_j, \alpha_j) = \delta_{\nu\mu}(\alpha_j, \alpha_j),$$

and from the last two that

$$\delta_{\mu\nu}(\alpha_j, \overline{\alpha}_j) = \chi(\alpha_j, \overline{\alpha}_j)\cdot\gamma_{\mu\nu}(\alpha_j, \overline{\alpha}_j).$$

This shows that we may combine the two forms (19) and (22) so as to obtain a form

$$\sum \zeta_{\mu\nu}(\alpha_j, \alpha_k)\, x_\mu\, y_\nu$$

such that $\zeta_{\mu\nu}(\alpha_j, \alpha_k) = 0$ if $\alpha_j \neq \alpha_k$, while

$$\sum \zeta_{\mu\nu}(\alpha_j, \alpha_j)\, x_\mu\, y_\nu = 0$$

represents a quadric of $S_{r-1}$ which must contain the space $S_{r_j-1}$. If this quadric were degenerate $\Omega$ would possess a degenerate bilinear form and would therefore be impure. Limiting ourselves to the case of pure matrices, we must then have $r_j - 1 \leqq \frac{1}{2}(r-2)$ when $r$ is even and $r_j - 1 \geqq \frac{1}{2}(r-3)$ when it is odd (Bertini). Similar limits hold of course for $r_j' - 1$. But $r_j + r_j' = r$, hence $r$ must be even and we must have besides $r_j = r_j' = \frac{1}{2}r$. Thus $r$ must be even, and among the multipliers each root must be taken the same number of times. In this case the solution exists actually as can be shown by the following choice of bilinear forms: For (19) we take a form such that $\gamma_{\mu\nu}(\alpha_j, \alpha_k) = 0$ if $\alpha_j \neq \alpha_k$ and

$$-\frac{1}{2i}\sum \gamma_{\mu\nu}(\alpha_j, \overline{\alpha}_j)\, x_\mu\, y_\nu = \sum_{\mu=1}^{r/2} (x_\mu\, y_\mu - x_{\frac{r}{2}+\mu}\, y_{\frac{r}{2}+\mu});$$

for the quadric (22) we take

$$\sum_{\mu=1}^{r/2} x_\mu\, x_{\frac{r}{2}+\mu} = 0;$$

and for the spaces $S_{r_j-1}$, $S'_{r_j-1}$,

$$S_{r_j-1}\colon x_{\frac{r}{2}+\mu} = 0, \qquad S'_{r_j-1}\colon x_\mu = 0$$

$$(\mu = 1, 2, \cdots, r/2), \qquad (j = 1, 2, \cdots, q/2), \qquad \alpha_h \neq \alpha_k \text{ if } h, k \leqq q/2.$$

If there are two non-degenerate reciprocities such as (22), the product of one by the inverse of the other defines a new multiplication permutable with the first. $\Omega$ possesses then a complex multiplication of degree $q' \geqq 2q$, and we have $1 + k > \frac{1}{4}(2q) = \frac{1}{2}q$, hence there must be a new alternate form. In other words, the existence of two such reciprocities increases necessarily the index of singularity, $k$. When there is only one we have

$$1 + k = \tfrac{1}{2}q; \qquad 1 + h = 2q.$$

111. As an application we shall establish the existence of non-singular Abelian varieties with complex multiplication whose existence has recently been announced without proof by Scorza.*

According to No. 101 we must have $q = 2$ and the characteristic equation of the complex multiplication in question must be of the type

$$(a\alpha^2 + b\alpha + c)^q = 0, \qquad b^2 < 4ac,$$

with $p > 2$. We then have $1 + h = 2$, $1 + k = 1$ and it is sufficient to construct the matrices such as those of Nos. 107, 108 corresponding to this complex multiplication. We may observe that among the multipliers one of the roots may be taken any number of times $< p$. When $p$ is even there exist varieties with two complex multiplications such as those of No. 111, and for them $1 + h = 4$, $1 + k = 1$. According to what we have just seen, if there is one more reciprocity of this type we have certainly $k > 0$. The matrices in question are therefore the only non-singular matrices with complex multiplication.

This result may be extended to varieties with a complex multiplication whose multipliers are real and to multiplications permutable with these. The integers $h_0$, $k_0$ having always the same meaning as previously we find that for $k_1 = 0$, $h_1$ can only take the values $0, 1, 3$. Finally it is easy to construct the corresponding matrices—by merely replacing everywhere the ordinary domain of rationality by the domain $K(\alpha_j)$—but we shall not dwell on this any further.

This ends the discussion of complex multiplications. We shall proceed to make a rapid application to the classification of pure matrices for $p = 2$ or $3$.

## §6. Pure matrices of genus two or three

112. *Matrices of genus two.* If a matrix of genus two is singular and pure, it must possess a Riemann projectivity whose characteristic equation is of type $[f(\alpha)]^r = 0$, ($f(\alpha)$ irreducible). Since $rq = 4$, $r$ can only have the values $1, 2$. Let first $r = 1$. We have seen in No. 89 that $\Omega$ is then reducible to the type

$$\left\| \begin{matrix} 1, & \alpha_1, & \alpha_1^2, & \alpha_1^3 \\ 1, & \alpha_2, & \alpha_2^2, & \alpha_2^3 \end{matrix} \right\|,$$

where $\alpha_1$, $\alpha_2$ are roots of an equation of degree four

$$(\alpha^2 - 2\zeta_1\alpha + m\zeta_1 + n)(\alpha^2 - 2\zeta_2\alpha + m\zeta_2 + n) = 0$$

with $m$, $n$ rational and $\zeta_1$, $\zeta_2$ roots of an equation

$$a\zeta^2 + b\zeta + c = 0,$$

*Comptes Rendus, October, 1917. (Added in 1922: The proof has since been supplied in his memoir of the Palermo Rendiconti, vol. 45 (1921), p. 185. It appears to be decidedly different from ours.)

and we have indicated there the conditions that $a, b, c, m, n$ must satisfy. As to the invariants of $\Omega$, they are $1 + h = 2(1 + k) = 4$.

113. Let now $r = 2$ and assume the multipliers real. The matrix $\Omega$ is isomorphic to a matrix composed with two arrays

$$\left\| \begin{matrix} 1, & \tau_1 \\ 1, & \tau_2 \end{matrix} \right\|, \quad \left\| \begin{matrix} 1, & \sqrt{d} \\ 1, & -\sqrt{d} \end{matrix} \right\|,$$

where $d$ is a positive integer not a perfect square. We know that $\Omega$ exists provided $\tau_1, \tau_2$ are imaginary. Just by way of illustration we give the calculation, which is here very simple. There will be an alternate form

$$\gamma_{12}(\sqrt{d}, -\sqrt{d})\, x_1 y_2 + \gamma_{21}(\sqrt{d}, -\sqrt{d})\, x_2 y_1$$

with

$$\gamma_{12}(x, y) = -\gamma_{21}(y, x) = \lambda(xy + d) + \mu(x + y),$$

so chosen as not to impose any condition on $\tau_1$ and $\tau_2$. This form contains two arbitrary parameters, hence $1 + k = 2$, and similarly $1 + h = 2$ as we already knew. The corresponding Hermitian form is, up to the factor $\sqrt{d}$,

$$\tau''(\mu + \lambda\sqrt{d})\, x_1 \bar{x}_1 - \tau_1''(\mu - \lambda\sqrt{d})\, x_2 \bar{x}_2; \qquad \tau_j = \tau_j' + i\tau_j''.$$

We must therefore have

$$\tau_1''(\mu + \lambda\sqrt{d}) > 0, \qquad \tau_2''(\mu - \lambda\sqrt{d}) < 0.$$

We can always assume that one of the numbers $\tau_1''$, $\tau_2''$ is positive—say $\tau_1''$. The existence of $\Omega$ is certain for we can take $\lambda$, $\mu$ positive and such that $\mu - \lambda\sqrt{d}$ has the sign of $-\tau_2''$. Moreover the matrix is pure because the Hermitian form is of genus two provided that $\lambda$ and $\mu$ are not both zero. The sets of integers $(\lambda, \mu)$ satisfying the above equalities define the principal forms and therefore also the systems of Abelian functions corresponding to the matrix.

If there exists a non-alternate bilinear form imposing some relations upon $\tau_1, \tau_2$, these numbers are conjugate quadratic numbers of the domain $K(\sqrt{d})$, and $\Omega$ possesses a Riemann projectivity with irreducible characteristic equation. It is therefore of the type considered in No. 112.

114. Let us pass to the case of complex multipliers. We have now two arrays

$$\left\| \begin{matrix} 1, & \tau_1 \\ 1, & \tau_2 \end{matrix} \right\|, \quad \left\| \begin{matrix} 1, & i\sqrt{d} \\ 1, & -i\sqrt{d} \end{matrix} \right\|,$$

where $d$ is as before a positive integer not a perfect square. From the existence of an alternate form, follows a bilinear relation between $\tau_1$ and $\tau_2$, whose coefficients belong to the domain $K(i\sqrt{d})$. If $\Omega$ is pure we can always reduce it to an isomorphic matrix composed with two similar arrays such that the relation in question is then $c_{11} - c_{22}\,\tau_1\,\tau_2 = 0$, where $c_{11}$, $c_{22}$ are integers. The corresponding Hermitian form is

$$c_{11}\,x_1\,\bar{x}_1 - c_{22}\,x_2\,\bar{x}_2 .$$

It must be positive for $x_1 = 1$, $x_2 = \tau_1$, and negative for $x_1 = 1$, $x_2 = \tau_2$. It follows that $c_{11}$, $c_{22}$ must have the same signs and that their ratio must be included between $\tau_1\,\bar{\tau}_1$ and $\tau_2\,\bar{\tau}_2$, numbers which must not be equal. This is equivalent to the sole condition $|\tau_1| \neq |\tau_2|$. The invariants of $\Omega$ are $1 + h = 4$, $1 + k = 3$. As they have the maximum value for a pure matrix, $\Omega$ cannot acquire any other bilinear forms. This completes the discussion of the case $p = 2$.

We see that for a pure matrix of genus 2, the only possible combinations $(h, k)$ are $(0, 0)$, $(1, 1)$, $(3, 2)$, and $(3, 1)$. The corresponding matrices depend respectively upon 3, 2, 1, 0 continuous parameters.

115. *Matrices of genus three.* When $p = 3$, $qr = 6$; hence $r = 1$, 2, or 3. Let first $r = 1$, and therefore $q = 3$. The characteristic equation is

$$\prod_{j=1}^{3} (\alpha^2 - 2\zeta_j\,\alpha + \lambda\zeta_j^2 + \mu\zeta_j + \nu) = 0,$$

where the $(\zeta)$'s are roots of an irreducible equation $\zeta^3 + p\zeta^2 + q\zeta + r = 0$. Let $T$ be an operation of the group $G$ of the characteristic equation, permuting cyclically the pairs $(\alpha_1, \bar{\alpha}_1)$, $(\alpha_2, \bar{\alpha}_2)$, $(\alpha_3, \bar{\alpha}_3)$. It corresponds to the cyclic operation of order 3 that the group of the equation in $\zeta$ always contains. If we observe that $T$ is permutable with the binary operation $S$ of $G$ permuting pairs of conjugate roots, we see that with a suitable choice of notation $T$ has one of the following two forms:

$$(\alpha_1, \alpha_2, \alpha_3)\cdot(\bar{\alpha}_1, \bar{\alpha}_2, \bar{\alpha}_3); \qquad (\alpha_1, \alpha_2, \alpha_3, \bar{\alpha}_1, \bar{\alpha}_2, \bar{\alpha}_3).$$

Hence $G$ always contains a cyclic subgroup of order 6, of the powers of $TS$ in the first case and of the powers of $T$ in the second. This is sufficient to allow us to affirm, as in the case of an Abelian equation, that there exists a pure matrix corresponding to the above equation since it will always be possible to choose three roots $\alpha_1$, $\alpha_2$, $\alpha_3$ of which none are conjugate to each other and such that from the pairs $(\alpha_1, \alpha_2)$, $(\alpha_2, \alpha_3)$, $(\alpha_3, \alpha_1)$ we may deduce, by the operations of the group, every pair of non-conjugate roots. As to the invariants, they will have the values $1 + h = 2(1 + k) = 6$.

116. Let us assume now $r = 2$. The matrix $\Omega$ is then composed with the arrays

$$\| \tau_{j1}, \tau_{j2} \|, \qquad \| 1, \alpha_j, \alpha_j^2 \| \qquad (j = 1, 2, 3),$$

the multipliers being the roots of an equation of the third degree. I say that in order that $\Omega$ be pure, these roots must all be real. For if $\sum \gamma_{\mu\nu}(\alpha_j, \alpha_k)\,x_\mu\,y_\nu$

is a principal form, the corresponding Hermitian form will be definite only if its coefficients $A_{jj}$ are not all zero, which requires that the quantities $\gamma_{\mu\nu}(\alpha_j, \bar{\alpha}_j)$ $(\mu, \nu = 1, 2, 3, 4)$ be not all zero. Hence, if there is a pair of imaginary multipliers, the expressions $\gamma_{\mu\nu}(\alpha_j, \alpha_k)$ $(\alpha_j \neq \alpha_k)$, $(\mu, \nu = 1, 2, 3, 4)$ are not all zero. Moreover as is well known the equation in the multipliers is not Abelian. But the group of a non-Abelian irreducible equation of degree three is of order six and permutes transitively the three pairs of roots. Hence, if $\gamma_{\mu\nu}(x, y)$ does not vanish for one pair of distinct roots, it does not vanish for any other pair. Now, there are three distinct quantities of type $(\alpha_j^m \alpha_h^n + \alpha_j^n \alpha_h^m)$ $(m, n = 0, 1, 2)$ between which there is no relation with integral coefficients. Hence (No. 96) $1 + k = 3 + 3 = 6 > 2p - 1$ and $\Omega$ if it exists at all is necessarily impure.

When $\Omega$ is pure the three multipliers must then be real. We know then that $\Omega$ always exists provided that the ratios $\tau_{j2}/\tau_{j1}$ are not real. When they are arbitrary, we have $1 + h = 1 + k = 3$. The existence of a new alternate form either will bring us back to the case of No. 115 or else will make $1 + k$ take the value $6 > 2p - 1$, hence $\Omega$ will be impure. The same will hold for the non-alternate forms.

117. Let finally $r = 3$, $q = 2$. Since $r$ is odd, the multipliers must be imaginary. The matrix $\Omega$ is isomorphic to a matrix composed with two arrays,

$$\tau \equiv \| \tau_{j1}, \tau_{j2}, \tau_{j3} \|, \qquad \| 1, \alpha_j \| \qquad (\alpha_1 = \alpha_2 = -\alpha_3 = i\sqrt{d}),$$

where $d$ is again a positive integer and not a perfect square. The determinants of order two, derived from the first two rows of $\tau$, cannot all be zero, for otherwise $\Omega$ would have two rows with proportional terms. Finally if $\tau_{31} = \tau_{32} = 0$, $\Omega$ contains an elliptic submatrix. Hence if $\Omega$ is pure, we can always replace $\tau$ by an array of the type

$$\begin{Vmatrix} 1, & 0, & \tau_1 \\ 0, & 1, & \tau_2 \\ 1, & \tau_4, & \tau_3 \end{Vmatrix}.$$

Since we have made no transformation upon the columns, we may take as form (19)

$$\gamma_{11}(\alpha_j, \bar{\alpha}_j) x_1 y_1 + \gamma_{22}(\alpha_j, \bar{\alpha}_j) x_2 y_2 + \gamma_{33}(\alpha_j, \bar{\alpha}_j) x_3 y_3.$$

The numbers $\frac{-1}{2i}\gamma_{jj}(i\sqrt{d}, -i\sqrt{d}) = a_j$ are real numbers of the domain $K(i\sqrt{d})$—they are therefore rational numbers and we can without inconvenience assume that they are integers. In order that $\Omega$ be a Riemann matrix it is necessary (*a*) that $a_1 x_1 y_1 + a_2 x_2 y_2 + a_3 x_3 y_3$ vanish when the $(x)$'s are replaced by the elements of the first or second rows and the $(y)$'s by those of the third; (*b*) that $a_1 x_1 \bar{x}_1 + a_2 x_2 \bar{x}_2 + a_3 x_3 \bar{x}_3$ be positive at the point $(\lambda, \mu, \lambda\tau_1 + \mu\tau_2)$ whatever $\lambda$, $\mu$, and negative at the point $(1, \tau_4, \tau_3)$.

We have, therefore, the relations

$$a_1 + a_3 \tau_1 \tau_3 = 0, \qquad a_2 \tau_4 + a_3 \tau_2 \tau_3 = 0,$$

then the inequalities

$$a_1 + a_2 \tau_4 \bar{\tau}_4 + a_3 \tau_3 \bar{\tau}_3 < 0$$

$$(a_1 + a_3 \tau_1 \bar{\tau}_1)\lambda\bar{\lambda} + a_3 \tau_1 \bar{\tau}_2 \lambda\bar{\mu} + a_3 \bar{\tau}_1 \tau_2 \bar{\lambda}\mu + (a_2 + a_3 \tau_2 \bar{\tau}_2)\mu\bar{\mu} > 0.$$

Hence the roots of the equation in $\xi$

$$\begin{vmatrix} a_1 + a_3 \tau_1 \bar{\tau}_1 - \xi, & a_3 \tau_1 \bar{\tau}_2 \\ a_3 \bar{\tau}_1 \tau_2, & a_2 + a_3 \tau_2 \bar{\tau}_2 - \xi \end{vmatrix} = 0$$

must both be positive. Moreover the coefficient of $\lambda\bar{\lambda}$ in the Hermitian form in $\lambda$, $\mu$ must be positive, which gives us finally the inequalities

$$\begin{aligned} &a_1 + a_3 \tau_1 \bar{\tau}_1 > 0, \\ &a_1 + a_2 + a_3(\tau_1 \bar{\tau}_1 + \tau_2 \bar{\tau}_2) > 0, \\ &a_1 a_2 + a_3(a_2 \tau_1 \bar{\tau}_1 + a_1 \tau_2 \bar{\tau}_2) > 0. \end{aligned}$$

Let us set

$$a_1 = -ma_3, \qquad a_2 = -na_3, \qquad \tau_j = R_j e^{i\phi_j}.$$

The array $\tau$ assumes then the form

$$\tau \equiv \begin{Vmatrix} 1, & 0, & m/\tau_3 \\ 0, & 1, & n\tau_4/\tau_3 \\ 1, & \tau_4, & \tau_3 \end{Vmatrix},$$

and our inequalities reduce to

$$\begin{aligned} &ma_3(m - R_3^2) > 0, \\ &a_3(m^2 + n^2 R_4^2 - (m+n)R_3^2) > 0, \\ &a_3(m + nR_4^2 - R_3^2) > 0, \qquad mn < 0. \end{aligned}$$

Let us consider $m$, $n$ as rectangular point coördinates and draw the curves representing the functions in the left-hand sides of these inequalities (ellipse with axes parallel to the coördinate axes and straight lines):
The point $(m, n)$ can only be in one of the two regions, I, II, III. If it is in one of the first two regions, we must take $a_3 > 0$, and if it is in the third, we must take $a_3 < 0$. One of these regions always exists, hence $\tau_3$, $\tau_4$ can take arbitrary values, zero excepted. Under these conditions, we shall have $1 + k = 1$, $h = 2$, and the matrix is not singular—it is the simplest matrix of this type of genus $p > 1$.

If the matrix possesses a new non-degenerate bilinear form

$$\sum \gamma_{\mu\nu}(\alpha_j, \alpha_k) x_\mu y_\nu$$

it is impossible that $\gamma_{\mu\nu} = 0$ for $\alpha_j \neq \alpha_k$ if $\Omega$ is pure. Indeed then, if $\alpha_j = \alpha_k$, either the equation obtained, $\sum \gamma_{\mu\nu}(i\sqrt{d}, i\sqrt{d}) x_\mu y_\nu = 0 \quad (\alpha_k = \alpha_j)$, represents a linear complex, necessarily degenerate since the containing space is of odd dimensionality, or else

$$\sum [\gamma_{\mu\nu}(i\sqrt{d}, i\sqrt{d}) + \gamma_{\nu\mu}(i\sqrt{d}, i\sqrt{d})]\, x_\mu\, x_\nu = 0 \qquad (\alpha_j = \alpha_k)$$

is a conic containing all the points of the line that joins $(1, 0, n/\tau_3)$ to $(0, 1, n\tau_4/\tau_3)$, conic necessarily degenerate. In all cases the reciprocity $\sum \gamma_{\mu\nu}(\alpha_j, \alpha_j)\, x_\mu\, y_\nu$ is degenerate and $\Omega$ is impure.

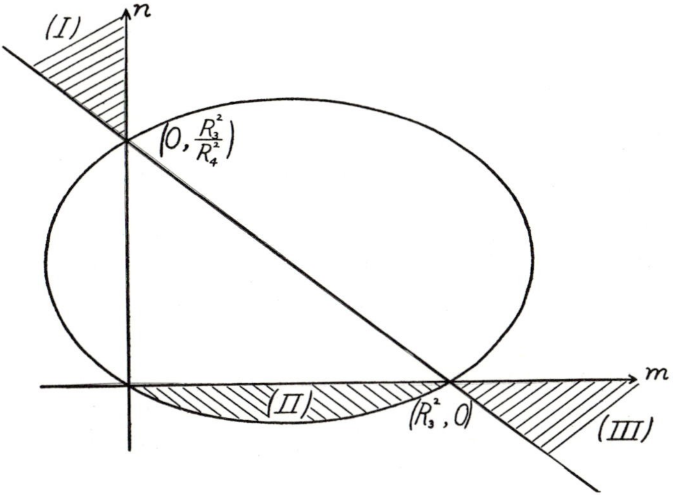


FIG. 1.

If the expressions $\gamma_{\mu\nu}(\alpha_j, \alpha_k)$ are not all zero, $\Omega$ will possess two reciprocities of the type

$$\sum \gamma_{\mu\nu}(\alpha_j, \alpha_k)\, x_\mu\, y_\nu = 0,$$

$$\gamma_{\mu\nu}(\alpha_j, \alpha_j) = 0; \qquad \gamma_{\mu\nu}(\alpha_j, \alpha_k) \neq 0 \text{ if } \alpha_j \neq \alpha_k.$$

If $\Omega$ is pure they are not degenerate and multiplying one of them by the inverse of the other we obtain then a new multiplication permutable with the one whose multipliers are the $(\alpha)$'s. In that case $\Omega$ must possess a Riemann projectivity of degree $\geqq 2q$, hence of degree 6, and we fall back on the type of No. 115.

Thus for a pure matrix of genus three the only possible combinations for the invariants $h$, $k$ are $(0, 0)$, $(0, 1)$, $(1, 1)$, $(2, 5)$. The corresponding matrices depend upon 6, 2, 3, 0 essential parameters.

## CHAPTER III. ABELIAN VARIETIES WITH CYCLIC GROUPS AND VARIETIES OF RANK > 1 IMAGES OF THEIR INVOLUTIONS

### § 1. Varieties of rank one with cyclic group

118. The study of hyperelliptic surfaces with cyclic group and more generally with finite group of birational transformations has already been made by Enriques, Severi, Bagnera and de Franchis. We only propose here to give a few new properties and to calculate the invariants of some simple varieties of genus $p > 2$ and rank $> 1$.

Every birational transformation of finite order $T$ of an Abelian variety of rank one, $V_p$, whose Riemann matrix has been put in a suitable form, will be given by equations such as

$$(1)\qquad \begin{aligned} u_i' &= u_i + \alpha_i && (i = 1, 2, \cdots, p');\\ u_{p'+j} &= \epsilon_j\, u_{p'+j} + \alpha_{p'+j} && (j = 1, 2, \cdots, p - p'), \end{aligned}$$

where the $(\epsilon)$'s are roots of unity other than one. A very interesting case and one to which we shall largely limit the discussion is that where the equations of $T$ are

$$(2)\qquad u_j' = \epsilon^{n_j}\, u_j \qquad (j = 1, 2, \cdots, p),$$

the $(\epsilon^n)$'s being primitive roots of a binomial equation $x^m = 1$. It will be particularly the case if $\Omega$ is pure. We shall assume $m > 2$. The numbers $\pm n_j$ will form a complete set of residues prime to $m$ taken $r$ times so that if $2\mu = \varphi(m)$, where $\varphi$ is the Euler function, then $r = p/\mu$.

We have shown how we may derive from $\Omega$ a matrix $\Omega'$ composed with two arrays

$$\tau \equiv \| \tau_{j1}, \tau_{j2}, \cdots, \tau_{jr} \|, \qquad \| 1, \epsilon^{n_j}, \cdots, \epsilon^{(2\mu-1)n_j} \| \qquad (j = 1, 2, \cdots, p).$$

To this matrix corresponds an Abelian variety of rank one, $V_p'$, in correspondence $(n, 1)$ with $V_p$ if the array $\tau$ is suitably chosen. $V_p'$ is the image of an ordinary involution of order $n$ on $V_p$ and possesses a cyclic birational transformation expressed by the same equations (2). Now it is easy to show that from the point of view which will occupy us, we may replace $V_p$ by $V_p'$, that is, $\Omega$ by $\Omega'$, and this we propose to do in the sequence. We shall then assume that $\Omega$ coincides with $\Omega'$.

When $\tau$ is as general as possible and the $(\epsilon^n)$'s are not all different, the invariants of $V_p$ are given by $1 + h = 2(1 + k) = 2\mu$, while if these multipliers are different $(r = 2)$, they are given by $1 + h = 4\mu$, $1 + k = 3\mu$.

119. The determination of the total group of birational transformations of a $V_p$ is of considerable interest. For $p = 2$, this has already been done by Scorza in his Palermo Rendiconti memoir. We shall show how the solution of this question is related in an important case to the problem of the determination of the units of an algebraic domain.

Let $G$ be a permutable subgroup of the group formed by the products of the complex multiplications, such that no Riemann projectivity corresponding to one of its operations transforms separately a submatrix of $\Omega$ if $\Omega$ is impure. We shall endeavor to characterize the operations of $G$ which lead to birational transformations of $V_p$.

Let $T$ be a birational transformation belonging to $G$ and

$$B \equiv \| b_{\nu\mu} \| \qquad (\nu, \mu = 1, 2, \cdots, 2p)$$

its Riemann projectivity. We know that the determinant $|b_{\nu\mu}| = \pm 1$. Moreover there exists a minimum base $B_1, B_2, \cdots, B_q$ for the projectivities whose terms are integers and we have $B = \sum \lambda_j B_j$, where the $(\lambda)$'s are integers. If $\alpha_1^j, \alpha_2^j, \cdots, \alpha_p^j$ are the multipliers of $B_j$ we have for those of $B$

$$\alpha_s = \sum_j \lambda_j \alpha_s^j \qquad (s = 1, 2, \cdots, p)$$

and their norm is equal to unity. They are therefore conjugate units of a certain algebraic domain, and more especially $\alpha_s$ is a unit contained in the modulus $(\alpha_1^s, \alpha_2^s, \cdots, \alpha_q^s)$. We may remark in passing that each number of this modulus determines a Riemann projectivity whose terms are integers. Moreover the modulus is an *order* in the sense of Dedekind. For, if $\alpha_s'$ and $\alpha_s''$ belong to it, there correspond to them projectivities with integral terms $B'$, $B''$, and $\alpha_s' \alpha_s''$ which is multiplier of the projectivity with integral terms $B'B''$ belongs actually to the same modulus. Thus $\alpha_s$ is a unit contained in a certain order of algebraic numbers and conversely if $\alpha_s$ is such a unit, it defines a projectivity with integral terms whose determinant is unity, hence a birational transformation of $V_p$. Thus the projectivities corresponding to the birational transformations, *and therefore the birational transformations themselves, are combined like units in an order of an algebraic domain.* From a classical theorem due to Dirichlet, follows then that there exists a finite number of permutable birational transformations $T_1, T_2, \cdots, T_\nu$, $\nu < p/r$, such that any other is given by a relation

$$T = T_0 T_1^{n_1} T_2^{n_2} \cdots T_\nu^{n_\nu},$$

where $T_0$ is an arbitrary cyclic transformation of the system and $\nu$ is the number of distinct real multipliers increased by half the number of imaginary multipliers.

Let $\beta_1^j, \beta_2^j, \cdots, \beta_s^j$ be a minimum base for the integers of the domain $K(\alpha_j)$. The solution of the problem which we are considering is related to that of the following: To solve in integers $x_j$ the equation

$$\prod_j \sum_k \beta_k^j x_k = \pm 1.$$

We see that the group of birational transformations of $G$, like $G$ itself, depends solely upon the equation $f(\alpha) = 0$ but not upon the array $\tau$ provided that $r \neq 2$.

In the particularly interesting case where the roots of $f(\alpha) = 0$ are all imaginary, each being given by the same rational function of its conjugate, the problem is somewhat simplified. Let us set $\alpha_j + \bar{\alpha}_j = 2\zeta_j$, $\alpha_j \bar{\alpha}_j = \eta_j$, and assume the multipliers so chosen that the $(\zeta)$'s are all distinct. The norm of $\eta_j$ is obviously *one* and the problem is therefore reduced to finding the real units of the real domain $K(\zeta_j)$ defined by an irreducible equation of degree $p/r$, $\phi(\zeta) = 0$, (No. 89).

*Remarks:* (I) There exist no cyclic birational transformations permutable with a given transformation $T$ of order $q^m$, ($q$ prime), other than the powers of $T$ itself. This follows from the fact that if $\epsilon = e^{2\pi i/q^m}$, the algebraic domain $K(\epsilon)$ contains no other roots of unity than the powers of $\epsilon$.

(II) With $T$ still of order $q^m$, ($q$ prime), and if $r > 1$, we may apply certain considerations of No. 90. If the numbers $\epsilon^{n_j}$ are invariant under a subgroup of order $n$ of the equation in the $q^m$-th primitive roots of unity, the period matrix is composed with $n$ submatrices of genus $p' = p/n$. If there exists a complex multiplication with multipliers of degree $> n$, the corresponding irreducible equation must be reducible in the domain $K(\epsilon)$. But according to a well-known theorem on the cyclic units of a cyclic domain of degree $q^m$, if these multipliers are roots of unity they must be themselves powers of $q$. It follows that if the order $\nu$ of a birational transformation is such that $\varphi(\nu) > n$, then $\nu = q^a \leqq q^m$. In particular, if $m = 1$, then $\nu = q$.

120. As an application, we shall establish, in a different manner, the results obtained by Scorza on the birational transformations of pure hyperelliptic surfaces when the multipliers are roots of an irreducible equation of degree four. Let $u_1' = \alpha_1 u_1$, $u_2' = \alpha_2 u_2$ be the equations of the birational transformation. The expressions $\alpha_1 + \bar{\alpha}_1$ and $\alpha_1 \bar{\alpha}_1$ are integers of a quadratic domain $K(\sqrt{d})$ where $d$ is a positive integer, not a perfect square, and $\alpha_1 \bar{\alpha}_1$ is a unit of this domain. Hence $\alpha_1$ satisfies an equation

$$\alpha^2 + (m + n\sqrt{d})\alpha + t + u\sqrt{d} = 0,$$

where $m, n, t, u$ are integers or halves of integers and $t^2 - du^2 = \pm 1$. The number $\alpha_2$ satisfies the equation obtained when $\sqrt{d}$ is replaced by $-\sqrt{d}$.

In order that these two numbers be imaginary, it is necessary that

$$(m + n\sqrt{d})^2 < 4(t + u\sqrt{d}), \qquad (m - n\sqrt{d})^2 < 4(t - u\sqrt{d}).$$

Hence $(m^2 - dn^2)^2 < 16$, and therefore $|\text{norm}\,(m + n\sqrt{d})| = |\beta| = 1, 2,$ or 3. The surface possesses the birational transformations whose multipliers are $\alpha_1^n, \alpha_2^n$, $n$ being an arbitrary integer. Hence, also $|\text{norm}\,(\alpha_1^n + \bar{\alpha}_1^n)| = 1,$

2, or 3. Now if we set

$$(m + n\sqrt{d})^2(t - u\sqrt{d}) = \xi + \eta\sqrt{d},$$

and observe that $\xi^2 - d\eta^2 = \beta^2$, we obtain by an easy computation

$$\text{norm}\,(\alpha_1^3 + \overline{\alpha}_1^3) = \beta(\beta^2 - 6\xi + 9) = \beta',$$
$$\text{norm}\,(\alpha_1^4 + \overline{\alpha}_1^4) = (\beta^2 - 4\xi + 4)^2 - 4(2\xi^2 - \beta^2 - 4\xi + 4) = \beta''.$$

Hence if $\beta = \pm 3$ we must also have $\beta' = \pm 3$, $18 - 6\xi = \pm 1$, which is impossible since $2\xi$ is an integer. Similarly, if $\beta = \pm 2$, $\beta' = \pm 2$ and $13 - 6\xi = \pm 1$, which requires that $\beta' = \beta$, $\xi = 2$. But in that case $\beta'' = 0$, which is impossible since $d$ is not a perfect square. It follows necessarily that $\beta = \pm 1$. We conclude from this that

$$(5 - 4\xi)^2 - 4(2\xi^2 - 4\xi + 3) = \beta'',$$

or

$$\xi^2 - 3\xi + \tfrac{1}{8}(17 - \beta'') = 0,$$

whence

$$\xi = \frac{1}{2}\left[3 \pm \sqrt{9 - \frac{17 - \beta''}{2}}\right].$$

For $\beta'' = 1$, we obtain $\xi = 1$ or $2$. But for $\xi = 1$, $\beta' = \pm 4$, and for $\xi = 2$, $\beta' = \pm 2$. Hence $\beta'' = -1$, $\xi = 3/2$, and therefore $d = \dfrac{\xi^2 - 1}{\eta^2} = \dfrac{5}{(2\eta)^2}$, and finally $\eta = \pm \frac{1}{2}$, $d = 5$. Ultimately, then, we obtain

$$(m + n\sqrt{5})^2 = (t + u\sqrt{5})\left(\frac{3 \pm \sqrt{5}}{2}\right),$$

$$t + u\sqrt{5} = (m + n\sqrt{5})^2\left(\frac{3 \pm \sqrt{5}}{2}\right).$$

The quantity $\delta = \frac{1}{2}(1 + \sqrt{5})$ is the fundamental unit of the domain $K(\sqrt{5})$. Hence

$$m + n\sqrt{5} = \delta^{\nu}, \qquad t + u\sqrt{d} = \delta^{2(\nu \pm 1)},$$

and therefore $\alpha$ satisfies one of the two equations

$$\alpha^2 + \delta^{\nu}\alpha + \delta^{2(\nu \pm 1)} = 0.$$

If we observe that for $\nu = 1$ the second has for root $\epsilon = e^{2\pi i/5}$, we find for $\zeta$, if we set $\mu = \nu - 1$, the two values $\epsilon\delta^{\mu}$, $(\epsilon + \epsilon^{-1})^{\mu}\left(\dfrac{2\epsilon - 1 + \delta}{2}\right)$, or else, since $\delta = \epsilon + \epsilon^{-1}$, the values

$$\epsilon(\epsilon + \epsilon^{-1})^{\mu}, \qquad (\epsilon + \epsilon^{-1})^{\mu}\left(\frac{3\epsilon + \epsilon^{-1} - 1}{2}\right).$$

Having taken for $\alpha_1$ one of these values we obtain $\alpha_2$ by replacing everywhere

$\epsilon$ by $\epsilon^2$ or $\epsilon^3$. The corresponding matrices are isomorphic to

$$\left\| \begin{matrix} 1, & \alpha_1, & \alpha_1^2, & \alpha_1^3 \\ 1, & \alpha_2, & \alpha_2^2, & \alpha_2^3 \end{matrix} \right\| .$$

For $\mu = 0$ and $\alpha$ of the first type, we have the matrix

$$\left\| \begin{matrix} 1, & \epsilon, & \epsilon^2, & \epsilon^3 \\ 1, & \epsilon^2, & \epsilon^4, & \epsilon \end{matrix} \right\|$$

to which they are all isomorphic and it corresponds to the surface of Jacobi-Humbert investigated by Scorza or the Jacobi surface of the curve

$$y^2 = x(x^5 + 1),$$

which possesses a cyclic transformation of order 5.

*Remark:* We have assumed everywhere that $d$ is not a perfect square. In the contrary case $t + u\sqrt{d} = \pm 1$, $m + n\sqrt{d} = \mu$, integer, and $\alpha_1$ is a root of $\alpha^2 + \mu\alpha \pm 1 = 0$. The surface possesses then a birational transformation with multipliers of degree two.

### § 2. Multiple points of abelian varieties of rank > 1

121. Let us designate by $W_p$ the Abelian variety of rank $m$, image of the cyclic involution (1) or (2). The multiple points of $W_p$ are images of the coincidence points of the involution on $V_p$. If the transformation $T$ corresponds to the equations (1) we shall obtain the coincidence points if we obtain the solutions of the following equations in the unknown integers $x_\mu$,

$$u_i + \sum_{\mu=1}^{2p} x_\mu \, \omega_{i\mu} = u_i + \alpha_i \qquad (i = 1, 2, \cdots, p');$$

$$u_{p'+j} + \sum_{\mu=1}^{2p} x_\mu \, \omega_{p'+j,\mu} = \epsilon_j \, u_{p'+j} + \alpha_{p'+j} \qquad (j = 1, 2, \cdots, p - p').$$

If $p' < p$, there is no solution when the $(\alpha)$'s of index $i \leqq p'$ are not all zero and there is an infinity of them when they all are. In this last case the problem is reduced to the determination of the coincidence points of an involution on a $V_{p-p'}$. Let us assume then that $p' = 0$, or that the equations of $T$ are of the form

$$u_i' = \epsilon_i \, u_i + \alpha_i \qquad (i = 1, 2, \cdots, p).$$

The coincidence points are all given by the formula

$$u_i = \frac{-\alpha_i}{\epsilon_i - 1} + \sum_{\mu=1}^{2p} \frac{x_\mu \, \omega_{i\mu}}{\epsilon_i - 1} \qquad (i = 1, 2, \cdots, p).$$

Let $(x_1^0, x_2^0, \cdots, x_{2p}^0)$ be a solution in integers. If we add to the $(x^0)$'s a solution of the system

$$\sum_{\mu=1}^{2p} \frac{x_\mu \, \omega_{i\mu}}{\epsilon_i - 1} = \sum_{\mu=1}^{2p} y_\mu \, \omega_{i\mu} \qquad (i = 1, 2, \cdots, p),$$

the $(y)$'s being also integers, we shall always obtain the same coincidence point. Now, let

$$\epsilon_i \omega_{i\mu} = \sum_{\nu=1}^{2p} a_{\nu\mu} \omega_{i\nu} \qquad (i = 1, 2, \cdots, p; \mu = 1, 2, \cdots, 2p),$$

be the Riemann p ojectivity belonging to $T$. The above system can be written

$$\sum_{\mu=1}^{2p} x_\mu \omega_{i\mu} = - \sum_{\mu=1}^{2p} y_\mu \omega_{i\mu} + \sum_{\mu,\nu}^{1\cdots 2p} y_\mu a_{\nu\mu} \omega_{i\nu} \qquad (i = 1, 2, \cdots, p),$$

and therefore

$$x_\mu = - y_\mu + \sum_{\nu=1}^{2p} a_{\mu\nu} y_\nu \qquad (\mu = 1, 2, \cdots, 2p).$$

Let us set

$$a_{\mu\mu} - 1 = b_{\mu\mu}, \qquad a_{\mu\nu} = b_{\mu\nu} \quad \text{if} \quad \mu \neq \nu,$$

and let $B_{\mu\nu}$ be the coefficient of $b_{\mu\nu}$ in the expansion of the determinant

$$B = |b_{\mu\nu}| \qquad (\mu, \nu = 1, 2, \cdots, 2p),$$

whose value is

$$B = \prod_{i=1}^{p} (\epsilon_i - 1)(\epsilon_i^{-1} - 1) > 0.$$

The $(x)$'s satisfy the congruences

$$\sum_{\mu=1}^{2p} B_{\mu\nu} x_\nu \equiv 0, \text{ mod. } B \qquad (\nu = 1, 2, \cdots, 2p)$$

which possess $B^{2p-1}$ distinct solutions in numbers included between zero and $B$.* Thus to each of the $B^{2p}$ sets of values of the $(x)$'s all included between zero and $B$—and it is not necessary to consider any others—will correspond $B^{2p-1}$ others giving the same coincidence point of the involution, hence the same multiple point of $W_p$. Therefore the number of these multiple points is precisely equal to $B$.

The value of $B$ is easy to compute. Let $F(\alpha) = 0$ be the characteristic equation of the complex multiplication belonging to $T$. We have obviously $B = F(1)$. For the Kummer surface $F(\alpha) = (\alpha + 1)^4$, $B = 16$. For the hyperelliptic surface of rank 3 belonging to the matrix

$$\left\| \begin{matrix} 1, & \epsilon, & \tau, & \epsilon\tau \\ 1, & \epsilon^2, & \tau', & \epsilon^2\tau' \end{matrix} \right\|, \qquad \epsilon = e^{2\pi i/3},$$

$F(\alpha) = (1 + \alpha + \alpha^2)^2$, $F(1) = 3^2 = 9$, which is actually the number found by Bagnera, de Franchis, Enriques, and Severi.

122. When $m$ is arbitrary, the singular points are of different nature according as the corresponding points of $V_p$ are coincidence points for all the powers

* Krazer, *Lehrbuch der Theta Functionen*, p. 57.

of $T$ or for some of them only. We shall examine in detail the case of $m = q$, prime. The case of $m$ arbitrary can be treated in very much the same way. We may also take the numbers $\alpha_i$ all equal to zero and we then have for the equations of $T$ $u'_i = \epsilon^{n_i} u_i$ $(i = 1, 2, \cdots, p)$, where the numbers $\pm n_i$ form $2p/(q-1)$ times a complete set of residues modulo $q$. The neighborhoods of the multiple points are transitively permuted by a finite group of birational transformations of $W_p$ which corresponds to a group of ordinary transformations of $V_p$ (Bagnera and de Franchis). Hence the neighborhood of any one of them is equivalent to the same number $\kappa$ of infinitesimal hyper-surfaces and, from the point of view of Analysis Situs, they behave alike. It is therefore sufficient to consider the multiple point corresponding to the coincidence point $u_1 = u_2 = \cdots = u_p = 0$.

The groups of points of the involution in the neighborhood of this coincidence point are in one-to-one correspondence with the groups of points of the involutions determined in an $S_p$ by the homogeneous transformation of coördinates

$$\zeta x'_i = \epsilon^{n_i} x_i, \qquad \zeta x'_{p+1} = x_{p+1} \qquad (i = 1, 2, \cdots, p)$$

in the neighborhood of the coincidence point $O$ $(0, 0, \cdots, 0, 1)$. Designate by $M_p$ the image of this involution of $S_p$ and by $s$ the number of distinct exponents modulo $q$ among the $(n)$'s. In the neighborhood of the multiple point $O'$ of $M_p$ transformed of $O$, there will be $s$ distinct branches corresponding to the $s$ infinitesimal varieties of coincidence of the involution in the neighborhood of $O$. Let us apply a quadratic birational transformation to the space containing $M_p$ transforming it into a variety $M'_p$, this in such a manner that $O'$ becomes a hyperplane $H$. The $s$ branches just mentioned become, as far as their parts in the neighborhood of $O'$ are concerned, the neighborhoods of $s$ linear spaces of $H$, of which we shall designate any one by $K''$. To $K''$ corresponds on $M_p$ an infinitesimal variety $K'$ very near $O'$, and on $V_p$ an infinitesimal variety of coincidence $K$ very near $O$. The $p-1$ dimensional elements of $V_p$ passing through $K$ undergo an involution whose representative equations with suitably chosen homogeneous parameters $x_i$ are of the type

$$\zeta x'_i = \epsilon^{n_i - n_h} \cdot x_i \qquad (i = 1, 2, \cdots, t')$$
$$\zeta x'_{t'+l} = x_{t'+l} \qquad (l = 1, 2, \cdots, t),$$

where $n_h$ is one of the indices $n$, perfectly defined when $K$ is known. This involution is of the same type as previously except that we have now only $s-1$ distinct groups of exponents. In the neighborhood of $K''$ there will then be $(s-1)$ distinct branches of $M'_p$. This reasoning may be continued until we isolate elements in the vicinity of which there is only one branch. We shall then have $\kappa = s!$ In particular, if the $(n)$'s form a complete system of residues, then $\kappa = (q-1)!$ The singular points are therefore equivalent to

$B\kappa = Bs!$ infinitesimal algebraic hypersurfaces and therefore $\rho = [\rho] + Bs!$ We recall that when any invariant is written within square brackets, its value is assumed taken disregarding infinitesimal cycles (See No. 11).

*Remark:* The preceding discussion, followed by a reasoning analogous to that in No. 51, may lead to a proof that $W_p$ is birationally transformable into a non-singular variety contained in a suitable space.

## § 3. Indices of connectivity of varieties of rank > 1

123. Let us assume the birational transformation of type (2). The matrix $\Omega$ is then equivalent to the matrix $\Omega'$ of No. 118. Designate by $\delta_h^\mu$ the linear cycle of $V_p$ corresponding to the period $\epsilon^{nh}\tau_{j\mu}$, by $(\delta_{h_1}^{\mu_1}, \delta_{h_2}^{\mu_2}, \cdots, \delta_{h_s}^{\mu_s})$ the $s$-cycle corresponding to the linear cycles in parenthesis, and by $(\bar{\delta}_{h_1}^{\mu_1}, \bar{\delta}_{h_2}^{\mu_2}, \cdots, \bar{\delta}_{h_s}^{\mu_s})$ the corresponding cycle on $W_p$. On this last variety

$$(\bar{\delta}_{h_1}^{\mu_1}, \cdots, \bar{\delta}_{h_s}^{\mu_s}) \sim (-1)^n (\bar{\delta}_{k_1}^{\nu_1}, \cdots, \bar{\delta}_{k_s}^{\nu_s})$$

if the sets of integers $(\mu)$, $(\nu)$ differ only by their order and if moreover, when $\mu_d = \nu_{d'}$, then $h_d = k_{d'} + \nu$. The integer $n$ indicates the number of inversions when we pass from one of the systems of superscripts to the other. The cycles $(\bar{\delta}_{h_1}^{\mu_1}, \bar{\delta}_{h_2}^{\mu_2}, \cdots, \bar{\delta}_{h_s}^{\mu_s})$ form therefore a base for the finite cycles on $W_p$. It is not in general possible to give an exact formula for $[R_s]$, but we can give a geometrical process to obtain this number. Let first $p = \mu$, $s = 2$. We mark on a circle the vertices $1, 2, \cdots, m$ of a regular $m$-sided polygon and denote by $r_1, r_2, \cdots, r_{2\mu}$ the numbers which are prime to $m$ and $< m$. We join the vertices $r_i$ to each other and the number of segments of distinct length thus obtained is equal to $[R_2]$. Similarly $[R_s]$ is the number of distinct incongruent convex polygons having for vertices $s$ of the points $r_i$. If $p > \mu$, $r > 1$, we take $rn$ division points and use the points $r_i, r_i + m, \cdots, r_i + (r-1)m$ as the vertices of convex polygons.

124. *Case of $m = q$, odd prime.* We can then obtain simple formulas for the indices. Let first $2p = q - 1$. An $s$-sided convex polygon will be determined by $s$ integers whose sum is $q$, say $h_1, h_2, \cdots, h_s$. Two convex polygons that correspond to partitions $h_1, h_2, \cdots, h_s$, and $h_1', h_2', \cdots, h_s'$, of $q$ will be congruent if $h_i' - h_i = h_k' - h_k$ $(i, k = 1, 2, \cdots, s)$. But in all there are $\binom{q-1}{s-1}$ arrangements of $s$ integers yielding a sum $q$. Taking account of the possibility of permuting cyclically the $(h)$'s, we obtain $[R_s] = \frac{1}{s}\binom{2p}{s-1}$. This number is actually an integer for $\binom{q}{s} = \frac{q}{s}\binom{q-1}{s-1}$ is an integer and $s$ is prime to $q$. In particular $[R_2] = p$. We can similarly obtain $[R_s]$ whatever $r$, but we will merely indicate here the formula, easy to obtain, $[R_2] = \frac{1}{2}r^2(q-1)$.

125. *Formula for $R_2$.* If we assume that the infinitesimal algebraic cycles in the vicinity of the multiple points are all independent, we have $R_2 = [R_2] + B\kappa$. To prove that this is actually the case, it is sufficient to show that the infini-

tesimal hypersurfaces in the neighborhood of these points are algebraically distinct. We shall merely give some rapid indications on this question:— $A$, $B$ being any two of them we take suitable multiples $C$, $D$ of the hyperplane sections passing the one through $A$ and the other through $B$, then two hypersurfaces, $C_1$, $D_1$, of the same systems infinitely near $C$, $D$ respectively, but passing through neither $A$ nor $B$. We have $[C^i D^{p-i}] = [C_1^i D_1^{p-i}]$, $i > 0$, hence at once $[A^i B^{p-i}] = 0$. By considering then two hypersurfaces through $A$, it may be shown that $[A^p] \neq 0$. Finally $[A^i H^{p-i}] = 0$, $p > i > 0$, for every $H$ of $W_p$ not passing through the multiple points. By a reasoning of Severi's follows readily that the infinitesimal hypersurfaces of $W_p$ are algebraically distinct. Moreover, if $H_1, H_2, \cdots, H_{[\rho]}$ form a base for the hypersurfaces of $W_p$ when we neglect the multiple points, there is no relation between the $(H)$'s and the infinitesimal hypersurfaces. These facts have been established by other methods for the case $p = 2$ by Severi, Bagnera, and de Franchis.

126. Let us pass now to the determination of the invariants $[\sigma_s]$, $\sigma_s$, $\sigma$. Let $\Delta$ be a non-zero $s$-cycle of $V_p$ such that no sum of less than $m$ of the cycles $T^k \Delta$ be $\sim 0$ but $\sum_0^{m-1} T^k \Delta \sim 0$. If $\Delta_1$ is a cycle which is not homologous to the cycles $T^k \Delta_1$ $(0 < k < m)$, we can take $\Delta = T\Delta_1 - \Delta_1$. Let $\delta$ be the cycle corresponding to $\Delta$ on $W_p$. If $k\delta$ bounds on $W_p$, $\Delta + T\Delta + \cdots + T^{k-1} \Delta$ must bound on $V_p$. Hence according to the assumption made, we must have $k = m$. Moreover, we have actually $m\delta \sim 0$ and therefore *$\delta$ is a zero divisor for the $s$-cycles of $W_p$*. Are we really dealing with actual zero divisors? This is certainly the case if $W_p$ is without multiple points, that is, if there are no coincidence points on $V_p$. Then, if $m = q$, prime, $\sigma_s$ is of the form $m^a$ and in particular $\sigma = \sigma_1 = q^r$; $r = 2p'/(q-1)$, $p - p'$ being the genus of the submatrix maintained invariant by $T$.

Let us return now to the case where there is a finite number of coincidence points. I say that then $\sigma_1 = \sigma_{2p-2} = \sigma = 1$. For, if $A$ is a finite hypersurface of $V_p$, algebraically distinct from its transformed by the powers of $T$, $B$ a variable hypersurface of the same continuous system as $A$, while $A'$, $B'$ are the corresponding hypersurfaces of $W_p$, then $A' - B'$ is to be considered as a divisor of zero and we obtain all those divisors in this manner since they are all algebraic. Now, let $B$ approach a coincidence point. $B'$ will approach a singular point, and at the limit $A' - B'$ will have become a zero divisor *if we consider the singular points as ordinary points*, but certainly not if we bring into play the infinitesimal hypersurfaces in the neighborhood of the singular point, for then $A' - B'$ will be equal to a sum of such hypersurfaces. Hence on the variety with ordinary singularities birationally equivalent to $W_p$, $A' - B'$ will not be a zero divisor and we shall have $\sigma = 1$. This is in agreement with the results of Severi, Bagnera, and de Franchis.

### § 4. Integrals of the first kinds. Invariants $\rho$, $[\rho]$, $\rho_s$

127. When $W_p$ is the image of an involution generated by a transformation (1), it possesses $p'$ simple integrals of the first kind, hence if $p' = 0$, $W_p$ is regular.

Let us assume then $W_p$ regular. When we express the coördinates in terms of the $(u)$'s, a $k$-uple integral of the first kind must assume the form

$$\sum A_{j_1 j_2 \cdots j_k} \int\int \cdots \int du_{j_1}\, du_{j_2} \cdots du_{j_k},$$

and remain invariant when we apply $T$. We may assume $T$ in the form (2) since the $(\alpha)$'s have nothing to do with the question. We must then have $n_{j_1} + n_{j_2} + \cdots + n_{j_k} \equiv 0$, mod. $m$, unless $A_{j_1 j_2 \cdots j_k} = 0$. Thus, every integral of the first kind is a linear combination of the integrals invariant under $T$. The converse proposition is obvious.

I say now that *if $T$ is of the form* (2), *does not maintain invariant any reducible system of integrals of the first kind, and maintains invariant a k-uple integral of the first kind* $(k > 1)$, *then $W_p$ does not contain any congruence of spaces.**

By *congruence* we mean a system of algebraic manifolds such that one and only one passes through a given point of $W_p$. For there will be an invariant integral such as $\int\int \cdots \int du_1\, du_2 \cdots du_k$. The variety $W_p$ contains a ruled $k$-dimensional variety $M'_k$, locus of straight lines of which one goes through every point of the variety. Such a variety possesses no $k$-uple integrals of the first kind—this can be proved as done by Picard for $k = 2$. Hence on $M'_k$ the above integral reduces to a constant. It follows that on $M'_k$, and therefore on the corresponding variety $M_k$ of $V_p$, several of the differentials $du$, for example $du_1$, $du_2$, $\cdots$, $du_l$, vanish, and hence at once that $u_1$, $u_2$, $\cdots$, $u_l$ form a system of reducible integrals invariant by $T$.† This contradiction proves our theorem.

128. When $\Omega$ is of the type of No. 118 with $r > 1$ and $T$ does not transform into itself any system of reducible integrals, there will certainly be invariant integrals. For the $(n)$'s are certainly not composed with $r$ times the same set of $\mu$ exponents. There will then certainly be two whose sum is $m$ and therefore at least one double integral will be invariant. In particular, *if $\Omega$ is pure and $r \geqq 2$, $W_p$ does not possess any congruence of spaces.*

129. Let us show that *what we have just stated still holds if $r = 1$ and $m$ is a prime number, $q > 7$.* For since the characteristic equation corresponding to $T$ is irreducible there will then be no system of reducible integrals invariant by $T$. Everything reduces therefore to establishing that among the $(n)$'s we can always find a set $n_1$, $n_2$, $\cdots$, $n_k$ whose sum is divisible by $q$. The

* For $p = 2$ this has been proved by Enriques and Severi.

† *Castelnuovo,* Rendiconti dei Lincei (1905).

condition $k > 2$ must be added as follows from the fact that none of the $(n)$'s may be conjugate to each other.

If $p < n_j < 2p + 1 = q$, we can replace $n_j$ by $-q + n_j$. Hence, modulo $q$, $n_1, n_2, \cdots, n_p$ are nothing more nor less than the numbers $1, 2, \cdots, p$ affected with an arbitrary combination of signs and we have to prove the following: Whatever these signs, we can always form, with some of the numbers of the set, a sum divisible by $q$.

Now, we may at once verify the following: If the integers $1, 2, 3, 4, 5$ are not all affected with the same sign, we may form with them a zero sum except when 1 or 2 are taken with one sign and the four others with the opposite sign. In these two cases, it is easy to verify that we may form a sum equal to $\pm 11$. Hence the theorem is true for $q = 11$. Assume $q > 11$. We can change all the signs, hence we may assume 3 taken positively. Let then $\alpha$ be the first integer $> 5$ taken negatively. If $\alpha = 6$, we have the following three possibilities as to the signs of the first six integers and for which there may be doubt:

$$1, 2, 3, 4, 5, -6; \qquad -1, 2, 3, 4, 5, -6; \qquad 1, -2, 3, 4, 5, -6.$$

In the first two cases $2 + 4 - 6 = 0$ and in the third $-2 + 3 + 5 - 6 = 0$. If $\alpha > 6$ and $\alpha$ is even, in the only doubtful case where $3, 4, 5$ are all taken with the same sign $+$, the combination $\frac{1}{2}\alpha - 1$, $\frac{1}{2}\alpha + 1$, $-\alpha$ gives a zero sum, while if $\alpha$ is odd, the combination $\frac{1}{2}(\alpha + 1)$, $\frac{1}{2}(\alpha - 1)$, $-\alpha$ yields the same result. The theorem is therefore proved.

130. *Numbers* $\rho$, $[\rho]$, $\rho_0$. Let us take again for matrix $\Omega$ the matrix of No. 118. In the notations used above $[\rho]$ is the number of algebraic cycles of $V_p$ of type

$$\sum_{\mu, \nu, s} a_{\lambda, \nu, s} \sum_{t=1}^{2\mu} (\delta_\lambda^t, \delta_\nu^{s+t}).$$

The period of $\int\int du_j\, du_k$ with respect to this cycle is

$$\sum a_{\lambda, \nu, s} \sum_t (\tau_{j\lambda}\, \tau_{k\nu}\, \epsilon^{n_j(t-1)+n_k(s+t-1)} - \tau_{j\nu}\, \tau_{k\lambda}\, \epsilon^{n_j(s+t-1)+n_k(t-1)}),$$

and it is zero as we might expect when $n_j + n_k \not\equiv 0$, mod. $m$, whereas for $n_j \equiv -n_k$, mod. $m$, it has the form

$$\sum \gamma_{\lambda\nu}(\epsilon^{n_j}, \epsilon^{-n_j})\, \tau_{j\lambda}\, \tau_{k\nu} \qquad (\gamma_{\lambda\nu}(x, y) = -\gamma_{\nu\lambda}(y, x)),$$

where $\gamma_{\lambda\nu}$ is a polynomial with integral coefficients. Hence $[\rho] = (1 + k_0)$ where $k_0$ is the number defined in No. 107. As to the other invariants, we have

$$\rho_0 = [R_2] - [\rho], \qquad \rho = [\rho] + B \cdot \kappa.$$

In particular, if $m = q$, prime, and if $\tau$ is general, then $[\rho] = 1$,

$\rho_0 = \frac{1}{2}r^2(q-1) - 1$. If $r = 1$, we never have $n_j + n_k \equiv 0$, mod. $m$, all the periods of the nature in question are zero, hence $[\rho] = [R_2]$, $\rho_0 = 0$. Thus the variety of rank $m$ corresponding to the matrix

$$\| 1, \epsilon^{n_j}, \cdots, \epsilon^{(2p-1)n_j} \| \qquad (j = 1, 2, \cdots, p)$$

has neither double integrals of the first kind nor double integrals of the second kind.

## Chapter IV. A class of algebraic curves with cyclic group and their Jacobi varieties

### § 1. Integrals of the first kind of the curve $y^q = \prod_i (x - a_i)^{\alpha_i}$ ($q$ odd prime)

131. The object of this chapter is the investigation of the curves

$$(1) \qquad y^q = \prod_{i=1}^{r+2} (x - a_i)^{\alpha_i} \qquad (q \text{ odd prime})$$

characterized by the possession of a cyclic group of genus zero. Their importance consists in that their Jacobi varieties are the most interesting example of the varieties discussed in Chapter II. Furthermore they belong to a class much studied by various authors especially in regard to the presence of reducible systems of integrals of the first kind. We believe that the results here given constitute the most far-reaching investigation along that line. The restriction of $q$ to odd primes is of course narrowing, yet it is amply compensated by the greater elegance of the results obtained. Very likely much of the discussion that follows holds for any $q$, and perhaps even when the group is of genus other than zero.

132. Let $p$ be the genus of (1) which we shall call $C_p$ in the sequence, and $T$ the cyclic transformation,

$$x = x', \qquad y = \epsilon y'; \qquad \epsilon = e^{\frac{2\pi i}{q}}.$$

$T$ leaves no rational point function on $C_p$ invariant other than those rational in $x$ alone. $R(x)$ being such a function, any other can be expressed as a rational function of $R(x)$, if and only if $x$ itself can be, which requires that $R(x) = (ax + b)/(cx + d)$. Hence $x$ is characterized by being invariant under $T$, and, up to a projective transformation, by the fact that any other point function on $C_p$ invariant under $T$ is rational in $x$. Thus $x$ is determined by properties invariant under any birational transformation. The anharmonic ratios of the values of $x$ at the critical points, or coincidence points of $T$, are therefore invariant relatively to birational transformations, and as they completely determine $C_p$ those which are functionally independent among them can be taken as the independent moduli of the curve.

The point $A_i(a_i, 0)$ is *critical* for the function $y(x)$ if $\alpha_i$ is not divisible by $q$. If $r' + 2$ is the number of critical points, *$C_p$ depends upon $r' - 1$ moduli.*

Let $C'_p$ be a curve

$$y^q = \prod_i (x - a'_i)^{\alpha'_i}$$

birationally equivalent to $C_p$ and this in such a manner that $T$ becomes a transformation of similar form say

$$x = x', \qquad y = \epsilon^k y'$$

for $C'_p$. Let

$$x' = R_1(x, y), \qquad y' = R_2(x, y)$$

be the transformation changing $C_p$ into $C'_p$. Since $R_1$ and $R_2^q$ are invariant under $T$, $R_1$ must be rational in $x$, and as $x$ must be rational in $R_1$ we have $R_1(x) = (ax + b)/(cx + d)$. The conditions relatively to

$$R_2(x, y) = R_2(x, \prod_i (x - a_i)^{\frac{\alpha_i}{q}})$$

require that $R_2$ be of the form $y^n \cdot R(x)$, where $n$ is an integer prime to $q$. Hence the transformation from $C_p$ to $C'_p$ must be of the form

$$x' = \frac{ax + b}{cx + d}, \qquad y' = y^n \cdot R(x). \tag{2}$$

Conversely if $C_p$ can be changed into $C'_p$ by a transformation of the form (2), (2) is birational. For let $\lambda$, $\mu$ be two integers such that $n\lambda + q\mu = 1$. Then

$$y = y^{n\lambda + q\mu} = y'^{\lambda} \cdot S_1(x) = y'^{\lambda} \cdot S_2(x'),$$

where $S_1$, $S_2$ are rational functions. This is sufficient to show that $x$, $y$ are rational in $x'$, $y'$.

It is readily seen that a suitable transformation (2) will reduce $C_p$ to

$$y^q = \prod_i (x - a_i)^{\alpha'_i},$$

where the $(\alpha')$'s are subjected to the sole condition of being congruent modulo $q$ to the numbers $n\alpha_i$, $n$ being an arbitrary integer. In particular the $(\alpha')$'s may thus be replaced by their least residues modulo $q$, and then the $(a)$'s remaining will all be critical points. We shall assume that this has already been done, the number of critical points being $r + 2$, so that $C_p$ depends upon $r - 1$ moduli.

If we examine the behavior of $y$ at infinity we find that there is a critical point there corresponding to an exponent $-\sum \alpha_i$. Hence the sum of the exponents corresponding to *all* critical points is divisible by $q$.

133. Let us assume then that $a_{r+2} = \infty$ so that the equation is in the form

$$y^q = \prod_{i=1}^{r+1} (x - a_i)^{\alpha_i}. \tag{3}$$

As the sum of the $(\alpha)$'s for all critical points is divisible by $q$, we may assume that the $r+1$ exponents here indicated are prime to $q$ as well as their sum and are their own least positive residues modulo $q$. Each critical point counts for $q-1$ branch points of the Riemann surface representing the function $y(x)$ and there are no others. As the surface is $q$-sheeted, we have $(r+2)(q-1) = 2(p+q-1)$, therefore $p = \frac{1}{2}r(q-1)$.

Any integral of the first kind is of the form

$$\sum_{n=0}^{q-1} \int \frac{R_n(x)}{y^n} dx,$$

where the $(R)$'s are rational functions. If we apply $T^k$ it becomes

$$\sum_n \epsilon^{nk} \int \frac{R_n(x)}{y^n} dx.$$

Summing with respect to $k$ it is found that

$$\int \frac{R_n(x)}{y^n} dx$$

is of the first kind. Hence at once $R_0 \equiv 0$. Moreover by considering what happens at the critical points we find readily that this last integral is of the form

$$u = \int \frac{\prod_{i=1}^{r+1} (x - a_i)^{\beta_i}}{y^n} \phi(x)\, dx$$

where $\phi$ is a polynomial. Any other integral is then linearly dependent upon those such as $u$—a result due to Königsberger.

134. In order that $u$ be of the first kind the following inequalities must be satisfied:

$$(4) \qquad \begin{cases} \beta_i - \dfrac{n\alpha_i}{q} > -1 & (i = 1, 2, \cdots, r+1) \\ \sum \dfrac{n\alpha_i}{q} \sum \beta_i - r' > 0, \end{cases}$$

$r' - 1$ being the degree of $\phi$. To take the $(\beta)$'s as small as possible is tantamount to changing the degree of $\phi$. Denoting as usual by $[m]$ the least integer contained in the positive number $m$, let us take then $\beta_i = [n\alpha_i/q]$ so that all but the last of (4) will be satisfied. The second may be written

$$r' < \left(\left[\sum \frac{n\alpha_i}{q}\right] - \sum \beta_i\right) + \left(\sum \frac{n\alpha_i}{q} - \left[\sum \frac{n\alpha_i}{q}\right]\right).$$

The second parenthesis is positive and $< 1$,

$$\therefore \quad \left[\sum \frac{n\alpha_i}{q}\right] - \sum\left[\frac{n\alpha_i}{q}\right] \geqq r' > 0 .$$

Conversely if this inequality is satisfied $u$ is of the first kind whatever the polynomial $\phi$ of degree $r' - 1$. For it is only necessary to verify the last of (4). Now since neither $n$ nor $\sum \alpha_i$ are divisible by $q$,

$$\sum \frac{n\alpha_i}{q} - \sum \beta_i - r' > \left[\sum \frac{n\alpha_i}{q}\right] - \sum\left[\frac{n\alpha_i}{q}\right] - r' \geqq 0,$$

as was to be proved.

Thus for every $n$ there are $r'$ integrals of the first kind, where

$$r' = \left[\sum \frac{n\alpha_i}{q}\right] - \sum\left[\frac{n\alpha_i}{q}\right].$$

134. When $C_p$ is transformed by $T$ its Jacobi variety undergoes the transformation $u_j' = \epsilon^{n_j} \cdot u_j \quad (j = 1, 2, \cdots, p)$, where $u_1, u_2, \cdots, u_p$ are $p$ independent integrals of the first kind such as $u$, with $n_j$ corresponding to $n$. The corresponding value $r_j$ of $r'$ denotes the number of times that the multiplier $\epsilon^{n_j}$ is repeated. But if $r_j'$ corresponds to $\epsilon^{-n_j} = \epsilon^{q-n_j}$ and if the $(n)$'s are all included between *one* and $q$, then

$$r_j + r_j' = \left[\sum \frac{n_j\,\alpha_i}{q}\right] + \left[\sum \frac{(q - n_j)\,\alpha_i}{q}\right] - \sum\left[\frac{n_j\,\alpha_i}{q}\right] - \sum\left[\frac{(q - n_j)\,\alpha_i}{q}\right] = r.$$

Hence the complex multiplication corresponding to $T$ is of the type which we have studied. The multipliers are the roots of an irreducible equation of degree $q - 1$ whose $r$th power is the characteristic equation. These multipliers, it is scarcely necessary to point out, are all imaginary, each being a rational function of its conjugate—in fact its inverse. This of course does not allow us to apply without a preliminary discussion the formulas of Chapter II, since the Riemann matrices are by no means the most general of their type. It is nevertheless remarkable that if the critical points are arbitrary the formulas for the indices of the Jacobi variety are the same as for the most general Abelian variety of similar type.

The integer $r$, equal to the number of critical points decreased by two, will play the same part as in Chapter II, and we observe at once that if $r = 1$, one of the two numbers $r_j$, $r_j'$ is always zero.

135. Let now $\omega$ be any period of $u$. We have

$$\omega = \sum_{\mu=0}^{r} \epsilon^{n\nu_\mu}\,(1 - \epsilon^{n\nu_{\mu+1}}) \cdot \int_{x_0}^{a_{\mu+1}} du \qquad (\nu_{\mu_0} = 0).$$

Also

$$\epsilon^{n\nu_\mu}(1 - \epsilon^{n\nu_{\mu+1}}) = (1 - \epsilon^n)\cdot g_\mu(\epsilon^n),$$

where $g_\mu$ is a polynomial with integral coefficients. Hence

$$\omega = (1 - \epsilon^n) \sum_{\mu=0}^{r} g_\mu(\epsilon^n)\cdot \int_{x_0}^{a_{\mu+1}} du.$$

On the other hand

$$\tau_\mu = -(1 - \epsilon^n)\cdot \int_{x_0}^{a_\mu} du - \epsilon^n(1 - \epsilon^{-n})\cdot \int_{x_0}^{a_{\mu+1}} du = (1 - \epsilon^n)\cdot \int_{a_\mu}^{a_{\mu+1}} du$$

is a period, and so is $\epsilon^{nh}\tau_\mu$ as well. The $(\gamma)$'s in the sequence designating polynomials with integral coefficients, we have

$$\omega - \gamma_r(\epsilon)\tau_r = (1 - \epsilon^n) \sum_{\mu=0}^{r-1} g'_\mu(\epsilon^n)\cdot \int_{x_0}^{a_{\mu+1}} du$$

$$(g'_\mu = g_\mu, \qquad \mu < r - 1; \qquad g'_{r-1} = g_{r-1} + g_r).$$

This reasoning may be continued until we have finally

$$\omega - \sum_{\mu=1}^{r} \gamma_\mu(\epsilon^n)\tau_\mu = \gamma_0(\epsilon^n)(1 - \epsilon^n) \int_{x_0}^{a_1} du.$$

The left-hand side is a period, hence the other side must be one also. But the corresponding circuit in the $x$ plane can only surround the critical point $a_1$, hence it is a zero cycle of $C_p$ and $\gamma_0 = 0$. Another way of seeing it is to remark that the second side of the equation like the first must be independent of $x_0$ and as it is zero for $x_0 = a_1$, it must vanish identically. We have then $\omega = \sum_{\mu=1}^{r} \gamma_\mu(\epsilon^n)\tau_\mu$. If in place of $n$ we had $n_j$, $\epsilon^n$ would have to be replaced by $\epsilon^{n_j}$, whence follows that the period matrix is equivalent to one composed with the arrays,

$$\tau = \| \tau_{j\mu} \|; \qquad \| 1, \epsilon^{n_j}, \epsilon^{2n_j}, \cdots, \epsilon^{(q-2)n_j} \|$$
$$(j = 1, 2, \cdots, p;\ \mu = 1, 2, \cdots, r),$$

where

$$\tau_{j\mu} = \int_{a_\mu}^{a_{\mu+1}} du.$$

*Remark:* It is clear that instead of taking for limits of integration two $(a)$'s of consecutive indices we could take any two $(a)$'s provided that the quantities $\tau_{j\mu}$ are not related by a linear equation

$$\sum_\mu \gamma'_\mu(\epsilon^{n_j})\tau_{j\mu} = 0 \qquad (j = 1, 2, \cdots, p).$$

136. *Weierstrass points.* We recall that a point $A$ of an algebraic curve of genus $p$ is said to be a *Weierstrass point* if the excluded orders of infinity for rational point functions of which it is the sole pole do not form the series

$1, 2, \cdots, p$. According to Weierstrass the number of such points is finite and $> 0$ if $p > 0$ (lacunary theorem). Let $t$ be a variable such that a certain region around the origin in the $t$ plane be in point-to-point correspondence with the vicinity of $A$ on the curve. We may choose $p$ integrals of the first kind $u_1, u_2, \cdots, u_p$, such that near $A$

$$u_i = t^{s_i} \mathfrak{P}_i(t),$$

where $s_1 < s_2 < \cdots < s_p$, and the $(\mathfrak{P})$'s are holomorphic functions near $t = 0$ and do not vanish there. Of course the choice of the $(u)$'s is not unique but the $(s)$'s are perfectly determined and in fact they are the excluded orders for $A$, which is then a Weierstrass point if and only if

$$m = \sum s_i - \frac{p(p+1)}{2} > 0.$$

The integer $m$ is called the *order* of the point. It is at once obvious that this order is invariant under birational transformations and hence that a birational transformation of the curve into itself simply permutes these points. Hurwitz has shown* that if the curve is not hyperelliptic $m < \frac{1}{2}p(p-1)$, whereas for every Weierstrass point of hyperelliptic curves $m = \frac{1}{2}p(p-1)$.

137. Let us now return to the curves which we are investigating and choose for the $r_j$ integrals belonging to $n_j$ the following:

$$\int \frac{(x-a_i)^{\gamma_j - 1} \prod_{i=1}^{r+1} (x-a_h)^{\beta_h^j}\, dx}{y^{n_j}} \qquad (\gamma_j = 1, 2, \cdots, r_j),$$

and for the variable $t$ relatively to $(a_i, 0)$,

$$t = (x-a_i)^{\frac{1}{q}}.$$

To show that it is suitable observe that in the neighborhood of the point $(a_i, 0)$, $y$ is holomorphic in $t$, and in fact

$$y = t^{\alpha_i} \mathfrak{P}(t^q),$$

where $\mathfrak{P}(z)$ is holomorphic in $z$ near the origin and does not vanish there. Hence if $e, e'$ are integers such that $e\alpha_i + e'q = 1$ (we shall frequently write $e \equiv 1/\alpha_i$, mod. $q$, when $e$ satisfies such a relation), we have

$$t = y^e (x-a_i)^{e'} \mathfrak{P}(x-a_i).$$

It is then verified at once that the integral written above has a development near $t = 0$, of the form

$$t^{(\beta_i^j + \gamma_j)q - n_j \alpha_i} \mathfrak{P}_1(t^q),$$

*Mathematische Annalen, vol. 41, (1893).

where $\mathfrak{P}_1$ behaves like $\mathfrak{P}$. That this function contains only terms in $t^q$ is seen at once by observing that the effect of $T$ on both $t$ and the integral is merely to multiply them by powers of $\epsilon$.

The exponents $(\beta_i^j + \gamma_j)q - n_j \alpha_i$ form the set of integers $s_1, s_2, \cdots, s_p$ for $C_p$ and the point $A_i$.

We may apply to $C_p$ a birational transformation such that $\alpha_i$ be replaced by unity and $n_j$ by $n_j' \equiv n_j \alpha_i$, mod. $q$, $0 < n_j' < q$. Then $\beta_j$ will be replaced by $[n_j'/q] = 0$, and the set of the $(s)$'s becomes the set

$$\gamma_j q - n_j' \qquad (\gamma_j = 1, 2, \cdots, r_j; j = 1, 2, \cdots, q-1).$$

They certainly do not form the set $1, 2, \cdots, p$ if $r > 1$ and the numbers $r_j$ are not all equal. Hence *the critical points are all Weierstrass points when* $r > 1$, *unless the numbers* $r_j$ *are all equal.*

*Assume now* $r = 1$. Then $\gamma_j = 0$, or $1$, and we have to deal with the set $q - n_j'$ $(j = 1, 2, \cdots, p)$ which is up to a factor mod. $q$ the set of the $(n)$'s. Hence if the set of $(n)$'s is not congruent mod. $q$ to the set $1, 2, \cdots, p$, the three critical points are Weierstrass points.

138. We shall now apply similar considerations to the proof of a very useful theorem:—*If two curves,* $C_p$, $C_p'$,

$$y^q = \prod_i (x - a_i)^{\alpha_i}, \qquad y^q = \prod_i (x - a_i')^{\alpha_i'},$$

*are birationally equivalent through a transformation* $S$, *this transformation will be of type* (2) *if and only if it transforms some critical point into another.*

The condition is obviously necessary. Let us show that it is sufficient. Assume then that $S$ transforms $A_i(a_i, 0)$ into $A_i'(a_i', 0)$, these two points being both at finite distance, which is not a restriction. Let $u_1, u_2, \cdots, u_p$ be the integrals of No. 137 corresponding to $A_i$ and consider in particular

$$u_p = t^{s_p} \mathfrak{P}(t^q).$$

It is determined up to a constant factor, for the development of $c_1 u_1 + c_2 u_2 + \cdots + c_p u_p$ begins with a term in $t^{s_i}$ if $c_i$ is the first coefficient $c$ not zero. Hence it begins with a term in $t^{s_p}$ only if all the $(c)$'s except $c_p$ are zero.

Let now $(x, y)$, $(x', y')$ be two corresponding points of $C_p$, $C_p'$ through $S$ and set $x' - a_i' = t'^q$. The two variables $t$, $t'$ are obviously holomorphic functions of each other near the origin. Hence

$$t' = t \cdot \mathfrak{P}(t).$$

($\mathfrak{P}$ here as in the sequence designates a function of the type already described.) In making this substitution in $u_p$ it becomes

$$u_p' = t'^{s_p}(a + bt'^q \cdots).$$

This shows that the integral corresponding to $s_p$ for $A_i$ on $C_p$ corresponds to it also for $A'_i$ on $C'_p$, as might indeed be expected. Now

$$u_p = \int \frac{(x-a_i)^\delta R(x)}{y^n}\,dx, \qquad u'_p = \int \frac{(x'-a'_i)^{\delta'} R'(x')}{y'^{n'}}\,dx'$$

where $R$, $R'$ are rational functions which neither vanish nor become infinite at $a_i$, $a'_i$ respectively, and moreover

$$\frac{\delta - n\alpha_i}{q} = \frac{\delta' - n'\alpha'_i}{q} = \frac{s_p}{q} - 1.$$

Near $x = a_i$, $x' = a'_i$, we have then a relation

$$(x-a_i)^{\frac{s_p}{q}-1}\mathfrak{P}(x-a_i)\,dx = (x'-a'_i)^{\frac{s_p}{q}-1}\mathfrak{P}'(x'-a'_i)\,dx'.$$

When $x$ approaches $a_i$, $x'$ must approach $a'_i$; hence if we integrate, the constants of integration must be made equal to zero. Integrating, then raising to the power $q/s_p$, we obtain

$$(x-a_i)\mathfrak{P}_1(x-a_i) = (x'-a'_i)\mathfrak{P}'_1(x'-a'_i).$$

This shows that $x'$ is holomorphic in $x$ in the vicinity of $x = a_i$. But there is a relation

$$x' = R(x, y) = \sum_{s=0}^{q-1} R_s(x)\,y^s,$$

where the $(R)$'s are rational functions. If the $(R)$'s other than $R_0$ were $\neq 0$, $x'$ would have an algebraic critical point at $a'_i$, hence $x' = R_0(x)$. Similarly $x$ is rational in $x'$, and therefore $x' = (ax+b)/(cx+d)$. But also

$$y' = R'(x, y) = \sum_{s=0}^{q-1} R'_s(x)\,y^s.$$

Now $y'^q$ is rational in $x'$ and therefore in $x$, which requires that of the $(R')$'s only one, and that one of index $> 0$, does not vanish. The theorem is therefore proved.

*Remarks:* I. If the equation of $S$ for $y$ is

$$y' = y^\lambda \cdot R(x),$$

it is seen that its effect upon the integral $u_j$ corresponding to $n_j$ is to transform it into an integral corresponding to $\lambda n_j$, that is, all the $(n)$'s are multiplied by $\lambda$, mod. $q$. But $s_p \equiv -n_p\alpha_i \equiv -n'_p\alpha'_i$, mod. $q$,

$$\therefore \quad \frac{n'_p}{n_p} \equiv \frac{\alpha_i}{\alpha'_i} \equiv \lambda, \text{ mod. } q.$$

II. The equations of the transformation $T'$ of $C'_p$ corresponding to $T$ on $C_p$

are $x = x'$, $y = \epsilon^\lambda y'$, that is, by virtue of $S$ the cyclic transformations apparent for each curve are transformed into powers of each other. If we recall then the invariant meaning given to $x$ in No. 132 relatively to birational transformations we see at once that $S$ transforms the set of $(A)$'s into the set of $(A')$'s. Hence *if a birational transformation changes a single point $A_i$ into an $A'_i$ it does so for all of them.*

139. A particularly interesting case is that where $C_p$ and $C'_p$ coincide. According to the remark just made, *a birational transformation of $C_p$ into itself either permutes the set of $(A)$'s with a similar set having not a single point in common with the first and corresponding to a cyclic transformation of order $q$ not a power of $T$, or else it maintains the set in question completely invariant.* Assume then $S$ to be of the second type and let $\nu$ be its order. Its equations for a suitable choice of $x$ will assume the form

$$x' = \eta x, \qquad y' = y^\lambda R(x); \qquad (\eta = e^{\frac{2k\pi i}{\nu}}; \qquad \lambda^\nu \equiv 1, \text{ mod. } q).$$

Incidentally if $\lambda > 1$, $\nu$ is *not prime* to $q - 1$. If $a_i$ is a critical point, $\eta^s a_i$ is one also, and the corresponding exponent $\alpha$ can be taken to be $\lambda^s \alpha_i$. Also if one of the points $x = 0$, $x = \infty$, is critical, it will be maintained invariant by $S$, hence then (Remark I, No. 138), $\lambda = 1$. There are two distinct possibilities:

(*a*) $\lambda \not\equiv 1$, mod. $q$. $C_p$ can be reduced to

$$y^q = \phi(x)\cdot[\phi(\eta^{-1}x)]^\lambda \cdots [\phi(\eta^{1-\nu}x)]^{\lambda^{\nu-1}}$$

where $\phi$ is a polynomial with $\phi(0) \neq 0$. The critical points of this curve are not completely arbitrary. As our investigation will be limited to curves with critical points as arbitrary as possible for each type the only curve of this nature that we shall have to consider is the curve with three critical points, and $\nu = 3$, reducible to:

$$y^q = (x-1)(x-\eta)^\lambda(x-\eta^2)^{\lambda^2}, \qquad (\lambda^3 = 1, \text{ mod. } q;\ \eta = e^{\frac{2\pi i}{3}}),$$

and which presents itself only when $q - 1$ is a multiple of *three.* It may be found directly that the equations of $S$ are

$$x' = \eta x, \qquad y' = \eta^{\frac{1+\lambda+\lambda^2}{q}} y^\lambda.$$

(*b*) $\lambda \equiv 1$, mod. $q$. $C_p$ can then be reduced to

$$y^q = x^a \prod_i (x^\nu - a_i^\nu)^{\alpha_i}.$$

The only curve of this type with arbitrary critical points is reducible to one of the two hyperelliptic curves

$$y^q = \frac{x^2 - a^2}{x^2 - b^2}, \qquad y^q = x^2 - a^2.$$

For both curves, $S$ is the ordinary binary transformation characteristic of hyperelliptic curves. This transformation is permutable with $T$.

*Remark:* It is easy to show that if $S$ maintains every $A$ invariant then it is a $T^k$. For its equations must be

$$x' = x, \qquad y' = y \cdot R(x).$$

By direct substitution in the equation of $C_p$ we find at once $R^q = 1$, hence $R = \epsilon^k$ which proves our assertion.

140. We can now show that if a transformation $T'$ of order $q$ has also one of the $(A)$'s for coincidence point then $T' = T^k$. In other words *a single coincidence point determines the rest of them and the corresponding cyclic group of order $q$ as well.* For if we choose $x$ properly the equations of $T'$ will be

$$x' = \epsilon^n x, \qquad y' = y^\lambda \cdot R(x).$$

Since $T'$ is of order $q$ we must have $\lambda^q \equiv 1$, mod. $q$, and therefore according to Fermat's theorem, $\lambda \equiv 1$, mod. $q$, and we may take in fact $\lambda = 1$. $C_p$ will then be reducible to the form

$$y^q = x^\alpha \prod_i (x^q - a_i^q)^{\alpha_i},$$

with $\alpha$ not divisible by $q$, if we assume as we may that the common coincidence point of $T$ and $T'$ is at the origin. By direct substitution it is found that we must have $R(x) = \epsilon^{n\alpha/q}$ with the condition that $n\alpha \equiv 0$, mod. $q$, if $T'^q = 1$. Hence we may take $n = 0$, and the equations of $T'$ are finally $x' = x$, $y' = \epsilon^k \cdot y$ which shows that $T' = T^{-k}$ as was to be proved.

As a corollary if $S$ permutes the $(A)$'s among themselves, $STS^{-1}$ being a transformation such as above is a $T^k$, hence $ST = T^k \cdot S$. From this follows readily $S^j \cdot T = T^{k^j} \cdot S^j$. This shows that the group generated by the products of the powers of $S$ and $T$ is of order $q\nu$, where $\nu$ is the order of $S$. To obtain the exact relations existing between the transformations observe that $S^\nu \cdot T = T = T^{k^\nu} \cdot S^\nu$, hence $k^\nu \equiv 1$, mod. $q$. Moreover $k \not\equiv 1$, mod. $q$, else $S$ would be permutable with $T$ and this can only be when $C_p$ is hyperelliptic and $S$ is its ordinary binary transformation. This results from the fact proved in Chapter III, that the only cyclic transformations of the Jacobi variety of $C_p$ permutable with $T$ is the ordinary transformation, $u_i' = -u_i$ $(i = 1, 2, \cdots, p)$, to which corresponds only the ordinary binary transformation present in hyperelliptic curves, and none at all if the curve is not hyperelliptic. This case being set aside for the present, we must take for $\nu$ a root of the congruence other than one.

This being assumed, we find first

$$(S^j T^i)^n = T^l \cdot S^{nj}; \qquad l = ik^j \frac{(k^{nj} - 1)}{(k^j - 1)}.$$

It follows at once that in all cases $S^j T^i$ is either of order $\nu$ or of order $q$. It is certainly of order $\nu$ when: (*a*) there are no relations $S^\alpha = T^\beta$, $\alpha < \nu$; (*b*) $j \neq 0$; (*c*) $k$ is a primitive root of its congruence.

141. Torelli has proved a proposition of which we shall make some very interesting applications:* *Two algebraic curves of same genus with identical period matrices as to sets of normal integrals of the first kind are birationally equivalent.*

It follows at once that if the period matrices of two sets of integrals of the first kind are equivalent the curves are still birationally equivalent. However of more interest to us is this immediate corollary: If a given algebraic curve possesses two distinct sets of integrals of the first kind with *identical* period matrices it possesses a birational transformation into itself by which

---

*Rendiconti dei Lincei (1913). [Added in 1922: Rosati has shown in the Palermo Rendiconti, vol. 44 (1920), that Torelli's theorem requires that the periods correspond to retrosections.—As a matter of fact, in the cases of interest in the sequence this condition is actually fulfilled, and moreover Torelli's theorem is not essential as I proceed to show.

Let $y_1$, $y_2 = T \cdot y_1$, $y_3 = T^2 \cdot y_1$, $\cdots$, be the determinations of $y$. To each pair of loops $aa_1$, $aa_{\mu+1}$, of the $x$ plane, corresponds in a well-known manner a cycle $\delta_k^\mu$ of $C_p$, running entirely (save perhaps in the vicinity of $a_1$ and $a_{\mu+1}$) in the sheets $y_k$, $y_{k+1}$ ($y_{q+1} = y_1$), of the Riemann surface of $y(x)$, and $\delta_k^\mu = T \cdot \delta_{k-1}^\mu$. The $2p$ cycles $\delta_k^\mu$ ($\mu = 1, 2, \cdots, r$; $k = 1, 2, \cdots, q-1$) constitute precisely a fundamental system corresponding to the period matrix of No. 135.—Any birational transformation of the Jacobi variety, to occur later, shall have the property that *it induces permutations of the* ($u$)*'s and of the cycles* $\delta_1^\mu, \delta_2^\mu, \cdots$, *for each* $\mu$, *of common order* $\neq 0$, *leaving the period matrix invariant.* Up to an ordinary transformation, this completely characterizes $U$, and I say that *correspondingly there is a birational transformation of* $C_p$ *into itself, which is not a power of* $T$.

Let $C'_p = U \cdot C_p$. To the coincidence points $A_i$ ($a_i$, 0), of $T$ on $C_p$, correspond those $A'_i$ of $T' = UTU^{-1}$ on $C'_p$, the indices being so chosen that if $\alpha'_i$ is the same for $A'_i$ as $\alpha_i$ for $A_i$, the ratios $\alpha'_i/\alpha_i$ are all congruent mod. $q$. This being done, we readily define a birational transformation $S$ of $C_p$ into $C'_p$ such that $A'_i = S \cdot A_i$, and that the cycles $S\delta_1^\mu$, $S\delta_2^\mu$, $\cdots$, are cyclically permuted by $T'$. Observe that any birational transformation of order $q$ of $C'_p$, with the ($A'$)'s for coincidence points, is necessarily a $T'^h$ (No. 140). Now the ($u$)'s may be defined, up to a constant factor, by the nature of their expansions at a critical point, whence follows readily that their transformed by $U$ on $C'_p$ play the same part for $T'$ as the ($u$)'s for $T$ (i.e., $T'$ merely multiplies them by constants). On considering the periods with respect to the cycles $U \cdot \delta_k^\mu$, and recalling the assumptions as to $U$, we see that they cannot be the same as the cycles $S \cdot \delta_k^\mu$ or $T'^h S \cdot \delta_k^\mu$, of same indices, for the first are not cyclically permuted by $T'$, while the last are. Hence $U \neq ST'^h$, and $S^{-1}U$ is a birational transformation of $C'_p$ into itself not a $T'^h$, and as there is a similar one for $C_p$, our assertion is proved.

We may also verify that the Rosati condition is fulfilled. For each of the sets of $2p$ cycles $U \cdot \delta_k^\mu$, $S \cdot \delta_k^\mu$ ($k < q$), of $C'_p$, has this property: If the Riemann surface of $C'_p$ is cut open along them, it is reduced to a two-cell, and these cells are reducible to each other by a homeomorphism, wherein the cuts $U\delta_k^\mu$, $S\delta_k^\mu$ correspond to one another. Moreover there are integrals of the first kind with identical period matrices with respect to these cycles. From this follows readily enough the existence of two distinct sets of retrosections, with identical canonical period matrices, hence Rosati's condition is verified, and as previously $C'_p$ possesses a birational transformation into itself.

Observe that from the discussion of No. 141 it will appear that unless the curve is hyperelliptic, the order of the birational transformation of the curve is the same as that of $U$.]

the integrals of one system are transformed into those of the other. If then $u_1, u_2, \cdots, u_p$ and $u'_1, u'_2, \cdots, u'_p$ are the two sets of integrals, $M(x, y)$ any point of the curve, $M'(x', y')$ its transformed through $S$, we have

$$u'_j(x', y') = u_j(x, y) + \delta_j \qquad (j = 1, 2, \cdots, p),$$

where the $(\delta)$'s are constants. Since the two sets of integrals are distinct the algebraic correspondence defined by these relations is *singular* and therefore $S \neq 1$.

In particular if the period matrix is changed identically into itself by a certain equivalence the curve possesses a corresponding birational transformation.

We have already remarked that the transformation of the Jacobi variety defined by

$$u_j(x', y') = -u_j(x, y) + \delta_j \qquad (j = 1, 2, \cdots, p)$$

determines a birational transformation of the curve only when the latter is hyperelliptic. If $u_j(x', y') = u_j(x, y) + \delta_j$ were to determine a birational transformation of the curve, $M$ and $M'$ would be distinct points of a linear series of order and dimension one and therefore $p = 0$. Hence, except for the ordinary transformation possessed by hyperelliptic curves, to each birational transformation $U$ of a curve of genus $p > 0$ corresponds a non ordinary transformation of its Jacobi variety.—Under the following form this result will be found very useful: *If to $U$ of $C_p$, non hyperelliptic, corresponds an ordinary transformation of the Jacobi variety, $U$ reduces to the identity.*

142. Returning to the $C_p$ which we are investigating we shall show below that when $r > 1$, and the $(a)$'s are arbitrary, then $1 + h = q - 1$, where $h$ is the index of multiplication of the Jacobi variety. Exception must be made of the case where $r = 2$ and the integers $r_j$ are all equal to unity. Save in that special case which shall be fully examined in its place, the complex multiplications of the variety are then all linearly dependent upon those determined by the powers of $T$, and therefore (No. 119) the only non ordinary cyclic transformations of the variety are the powers of $T$. We shall see that if $C_p$ is hyperelliptic we are in the exceptional case. Thus, save in the exceptional case, if $r > 1$ the total group of $C_p$ is the obvious cyclic group of order $q$, for otherwise there would have to exist non ordinary cyclic transformations of the Jacobi variety other than the powers of $T$.

It follows that when the points $A_i\ (a_i, 0)$ are arbitrary and $r > 1$ there is no analogous set distinct from them on $C_p$, hence any transformation of the curve into itself must maintain them invariant and therefore it is of type (2).

143. Let us show that when $r > 1$ it is not possible that of the two numbers $r_j, r_{q-j}$, one be always zero. For with a suitable choice of integrals of the

first kind the period matrix would take the form

$$\left\| \begin{matrix} \omega, & 0, & 0, & \cdots, & 0 \\ 0, & \omega, & \cdot & \cdots, & \cdot \\ \cdot & \cdot & \cdot & \cdot & \cdot \\ 0, & \cdot & \cdot & \cdots, & \omega \end{matrix} \right\|,$$

$$\omega = \| 1, \epsilon^{n_j}, \cdots, \epsilon^{(q-2)n_j} \| \qquad \left(j = 1, 2, \cdots, \frac{q-1}{2}\right),$$

and depends upon no arbitrary constants whereas as we know it should depend upon $r - 1 > 0$ of them.

144. *Hyperelliptic curves.* The curve $C_p$ being assumed hyperelliptic let $S$ be its ordinary binary transformation. It transforms the point $(x, y)$ of $C_p$ into a point $(x', y')$ such that

$$\int \prod_i (x - a_i)^{\beta_i - \frac{n\alpha_i}{q}} dx$$

changes in sign when $x$ is replaced by $x'$. Dividing two such differential expressions, then raising to the $q$th power we obtain a relation

$$\prod_i (x - a_i)^{\delta_i} = \prod_i (x' - a_i)^{\delta_i} \qquad (\delta_i \text{ integer})$$

It shows that when $x$ approaches one of the $(a)$'s, $x'$ approaches another. Hence $S$ is of type (2) and by a proper choice of $x$ we shall have merely $x' = -x$. The equations of the curve will then be

$$y^q = x^\alpha \prod_i (x^2 - b_i^2)^{\alpha_i}$$

As we must have $y' = y^n \cdot R(x)$, we find by direct substitution $n = 1$, $R = (-1)^\alpha$, and $S$ will have for equations

$$x' = -x, \qquad y' = (-1)^\alpha \cdot y.$$

The only hyperelliptic curves with unrestricted critical points are reducible to one of the following two:

$$y^q = x^2 - a^2,$$
$$y^q = \frac{x^2 - a^2}{x^2 - b^2}.$$

In the first case where $r = 1$, we have $\alpha_1 = \alpha_2 = 1$ and the integers $r_j$ and $n_j$ are determined by $[2n_j/q] - r_j \geqq 0$ which shows that $n_j$ must be $> q/2$ in order that $r_j = 1$. Hence $n_j$ takes the values $p + 1, p + 2, \cdots, 2p$ and the period matrix is at once seen to be equivalent to

$$\| 1, \epsilon^j, \epsilon^{2j}, \cdots, \epsilon^{(q-2)j} \| \qquad (j = 1, 2, \cdots, p).$$

For $r = 2$ the curve may be reduced to such a form that $\alpha_1 = \alpha_2 = 1$,

$\alpha_3 = q - 1$. Hence

$$\left[\frac{n_j(q+1)}{q}\right] - \left[\frac{n_j(q-1)}{q}\right] - r_j = 1 - r_j \geqq 0.$$

Hence $n_j$ can take all the values $1, 2, \cdots, p$ and to each corresponds exactly *one* integral, the period matrix being composed with:

$$\| \tau_{j1}, \tau_{j2} \|, \qquad \| 1, \epsilon^j, \cdots, \epsilon^{(q-2)j} \| \qquad (j = 1, 2, \cdots, p).$$

## § 2. Curves with three critical points

145. Our next object is the study of the $C_p$ whose $r = 1$, and which is reducible to

$$y^q = (x - a_1)^{\alpha_1}(x - a_2)^{\alpha_2}.$$

Since $2p = q - 1$, we may assume $q > 3$, for, if $q = 3$, we have to deal with elliptic curves which offer no great interest.

Let $g$ be a primitive root of $q$ and set $n_j \equiv g^{e_j}$, mod. $q$. The period matrix is equivalent to:

$$\| \epsilon^{g^{e_j}}, \epsilon^{g^{e_j+1}}, \cdots, \epsilon^{g^{e_j+2p-1}} \| \qquad (j = 1, 2, \cdots, p).$$

In conformity with Chapter II, § 3, we must first ask ourselves if there exists a subgroup of order $\nu = p/p'$ of the cyclic group of the equation $(x^q - 1)/(x - 1) = 0$, maintaining invariant the multipliers $\epsilon^{g^{e_j}}$. This problem is equivalent to that of determining if there exists a factor $p'$ of $p$ such that if we add $2p'$ to one of the $(e)$'s we get another, mod. $(q - 1)$. Assume that such is the case. The period matrix may be put in the form

$$\| \epsilon^{g^{e_j}}, \epsilon^{g^{e_j+2p'}}, \cdots, \epsilon^{g^{e_j+2(\nu-1)p'}}, \epsilon^{g^{e_j+1}}, \epsilon^{g^{e_j+1+2p'}} \cdots \| \qquad (j = 1, 2, \cdots, p).$$

Let us range the integrals of the first kind in such an order that $u_{j+kp'}$ $(k \leqq \nu - 1)$ corresponds to $e_j + 2kp'$. We recognize then that under the assumptions made if we permute cyclically the columns of order $k\nu + 1$, $k\nu + 2, \cdots, (k+1)\nu$ $(k = 0, 1, \cdots, 2p' - 1)$, in the period matrix in the form just written, and permute at the same time cyclically the rows $j, j + p', \cdots, j + (\nu - 1)p'$ $(j = 1, 2, \cdots, p')$, the matrix is unchanged. Hence (No. 141), $C_p$ possesses a birational transformation $S$ of order $\nu$. The genus of the involution defined by a given point and its $(\nu - 1)$ transformed through the powers of $S$ is easy to obtain. For if $\eta = e^{2\pi i/\nu}$, $S$ multiplies by $\eta^k$ the $p'$ integrals

$$u_j + \eta^k u_{j+p'} + \cdots + \eta^{(\nu-1)k} u_{j+(\nu-1)p'} \qquad (j = 1, 2, \cdots, p').$$

Hence the multipliers of the multiplication defined by $S$ on the Jacobi variety are composed of the powers of $\eta$, unity included, taken each $p'$ times. There are then exactly $p'$ integrals of the first kind maintained invariant by $S$, which shows that $p'$ is the required genus. Since the involution defined by $T$ is of genus *zero*, no power of $S$ can be a $T^k$. Let now $(x, y)$ be any point of $C_p$, and $(x', y')$ its transformed by $S$. This transformation permutes $u_j$ with $u_{j+p'}$, whence at least three relations such as

$$(x - a_1)^{\gamma_1}(x - a_2)^{\gamma_2}\,dx = (x' - a_1)^{\gamma'_1}(x' - a_2)^{\gamma'_2}\,dx'.$$

If we divide two of them by each other we obtain a relation

$$(x - a_1)^{\delta_1}(x - a_2)^{\delta_2} = (x' - a_1)^{\delta'_1}(x' - a_2)^{\delta'_2},$$

the exponents being rational. From this follows that if $x$ tends towards one of the critical points, $x'$ must tend towards another.

Let us denote as before the critical points by $A_1, A_2, A_3$ and assume first that $S$ maintains one of them, say $A_1$, invariant. Then the integer $\lambda$ of No. 139 must be equal to *unity*, mod. $q$, and the curve is reducible to the form $y^q = x^a(x^2 - a^2)$. It is therefore hyperelliptic. Let $S'$ be its ordinary binary transformation. $SS'$ maintains the $(A)$'s invariant, hence it is a $T^k$ and therefore $S = T^k S'$, which is absurd for $T^k S'$ determines an involution of genus *zero* while $S$ determines one of genus $p' > 0$.

We conclude then that $S$ permutes cyclically the critical points, and that $\lambda^3 \equiv 1$, mod. $q$. $C_p$ is therefore reducible to the curve

$$y^q = (x - 1)(x - \eta)^{\lambda}(x - \eta^2)^{\lambda^2}; \qquad \eta = e^{\frac{2\pi i}{3}}$$

of No. 139, the transformation $S$ being of order *three*, with equations

$$x' = \eta x, \qquad y' = \eta^{\frac{1+\lambda+\lambda^2}{q}}\cdot y^{\lambda}.$$

This curve which we shall denote by $C_{3p'}$ exists only if $q - 1$ is divisible by three. When $q > 3$, the exponents $1, \lambda, \lambda^2$ are distinct and the curve is certainly not hyperelliptic.

Referring to Chapter II, § 3, we may affirm that the invariants of the Jacobi variety when $r = 1$ are given by

$$1 + h = 2(1 + k) = 2p = q - 1$$

unless we deal with $C_{3p'}$ when we have

$$1 + h = 2(1 + k) = 6p = 3(q - 1).$$

In this last case the period matrix is isomorphic to one of type

$$\left\| \begin{matrix} \omega, & 0, & 0 \\ 0, & \omega, & 0 \\ 0, & 0, & \omega \end{matrix} \right\|,$$

where $\omega$ is pure and of genus $p' = p/3$.

146. The curve $C_{3p'}$ may be reduced to a particularly interesting form. For since $\frac{1}{3}(q-1)$ is an integer, $q$ is decomposable in a unique manner in the domain $K(\sqrt{3})$, hence there are two relatively prime positive integers $m, n$, such that $q = m^2 - mn + n^2$. The curve represented in homogeneous coordinates by $x^m y^n + y^m z^n + z^m x^n = 0$ is invariant under the collineation group of order $3q$ generated by the two operations

$$x' = y, \qquad y' = z, \qquad z' = x;$$
$$x' = \epsilon^m x, \qquad y' = \epsilon^n y, \qquad z' = z.$$

Let $\gamma$, $\delta$ be two integers such that $m\gamma + n\delta = 1$. If we apply to $x^m y^n + x^n + y^m = 0$ the birational transformation

$$x = x'^{\delta} y'^{m}, \qquad y = x'^{-\gamma} y'^{n}$$

whose inverse is

$$x' = x^n y^{-m}, \qquad y' = x^{\gamma} y^{\delta}$$

we obtain

$$y'^q = -x'^{-\delta m - \gamma(m-n)} \cdot (1 + x').$$

To prove that this is the curve $C_{3p'}$ we observe first that

$$m^2 - mn + n^2 = q = q(m\gamma + n\delta),$$
$$\therefore \quad m(m - n - q\gamma) = -n(n - q\delta)$$

and since $m$ and $n$ are relatively prime, there is an integer $t$ such that

$$m - n - q\gamma = tn, \qquad n - q\delta = -tm,$$

whence

$$\gamma = \frac{m - n(1+t)}{q}, \qquad \delta = \frac{n + tm}{q};$$
$$\therefore \quad -m\delta - \gamma(m - n) = -(1 + t).$$

The three exponents $\alpha_i$, corresponding to the curve obtained, can therefore be taken to be $1, t, -(1+t)$. Now $nt \equiv m - n$, mod. $q$,

$$\therefore \quad n(1 + t + t^2) \equiv m^2 - mn + n^2 \equiv 0, \text{ mod. } q.$$

But $n$ may be assumed $< m$, and the relation which defines them shows that $n^2$ is less than $q$ and therefore prime to it. It follows that $1 + t + t^2$ is divisible by $q$, or $t^3 \equiv 1$, mod. $q$, so that $t$ is either the integer $\lambda$ or $\lambda^2$, mod. $q$, for $t \equiv 1$, mod. $q$, would mean that either $q$ is 3, or $n^2$ is divisible by $q$. The curve which we have obtained corresponds then to the three exponents $1, \lambda, \lambda^2$ and in the case $r = 1$, this suffices to establish that it is birationally equivalent to $C_{3p'}$.

The two simplest curves $C_{3p'}$ have been considered at length by Klein $(q = 7)$, and Virgil Snyder $(q = 13)$, under the form

$$x^3 y + y^3 z + z^3 x = 0$$
$$x^4 y + y^4 z + z^4 x = 0$$

and the invariants of their Jacobi variety have been directly computed by Scorza.*

147. The consideration of the Weierstrass points will be useful for the problem which will occupy us next. Let $s_i^h$ be the value of the integer $s_i$ of No. 136 for $A_h$. We find at once

$$s_i^1 + s_i^2 + s_i^3 = \left( \beta_1^i + \beta_2^i + \beta_3^i - n_i \frac{(\alpha_1 + \alpha_2 + \alpha_3)}{q} \right) q + 3q .$$

In expressing that $u_i$ is finite at infinity we find that the first term at the right $< -q$, hence the left side is $< 2q$. But $\alpha_1 + \alpha_2 + \alpha_3$ is divisible by $q$. Hence this left side, which is positive, is a multiple of $q$, and therefore is exactly equal to $q$, that is $s_i^1 + s_i^2 + s_i^3 = q$. Hence, if $m_h$ is the order of the point $A_h$,

$$\begin{aligned} m_1 + m_2 + m_3 &= \sum_{i=1}^{p} (s_i^1 + s_i^2 + s_i^3) - \frac{3p(p+1)}{2} \\ &= pq - \frac{3p(p+1)}{2} = \frac{p(p-1)}{2} . \end{aligned}$$

148. We have seen that for the curves other than $C_{3p'}$, $1 + h = q - 1 = 2p$. Hence (No. 142) the total group of these curves is the cyclic group of order $q$ generated by $T$, except in the hyperelliptic case when it is of order $2q$. It remains to examine $C_{3p'}$. We propose to show that *when $q > 7$ the total group of $C_{3p'}$ is the group of order $3q$ generated by the cyclic operations $T$ of order $q$ and $S$ of order* 3.

We have seen that $ST = T^\mu S$, where $\mu^3 \equiv 1$ but $\mu \not\equiv 1$, mod. $q$, and therefore the group generated by $S, T$, is indeed of order $3q$. Moreover either $\mu$ or $\mu^2$ is the integer repeatedly denoted by $\lambda$. Since there are no relations $S^\alpha = T^\beta$, $S^i T^k$ $(i = 1, 2)$ is of order three (No. 140).

149. Let $\sum$ be a birational transformation of $C_{3p'}$ into itself, not an $S^i T^k$, and let $\nu$ be its order. It induces in the Jacobi variety one of equal order, $\sum'$, for when either one is the identity so is the other (No. 141). Hence among the multipliers of $\sum'$ is found at least one primitive $\nu$-th root of unity. On the other hand these multipliers satisfy the characteristic equation, which is of degree $q - 1$, and as we know (No. 93), is reducible in $K(\epsilon)$, its left side becoming there the product of $2p'$ cubic factors.—Now the primitive $\nu$-th roots satisfy an equation whose degree is the Euler function $\varphi(\nu)$, and which is irreducible in the rational domain. Hence the roots of this equation satisfy the characteristic equation as well, and $\varphi(\nu) \leqq q - 1$. $\therefore$ $\nu = q, 2q$, or else it is prime to $q$. But in this last case the equation in the $\nu$-th primitive

* See Klein—Fricke, *Vorlesungen über elliptische Modulfunctionen*, vol. 1, p. 701; Snyder, American Journal of Mathematics, vol. 30, (1908); Scorza, Atti dell' Accademia Gioenia (1917).

roots is irreducible in $K(\epsilon)$ (Kronecker). Hence one of the cubic factors above has for roots all the primitive $\nu$-th roots, whence then $\varphi(\nu) \leqq 3$, $\nu = 2, 3, 4, 6$. Thus finally $\nu$ can only have the values $2, 3, 4, 6, q, 2q$.

The case $\nu = 2q$ may be excluded at once for then the multipliers are necessarily the quantities $-\epsilon^{n_j}$ and $\sum'^q$ is the ordinary multiplication present only when the curve is hyperelliptic. There remain then only the possibilities $\nu = 2, 3, 4, 6$ or $q$. Through $\sum$ the points $A_1, A_2, A_3$ are transformed into new points $A_1', A_2', A_3'$, for by the discussion of No. 140 it is readily seen that $C_{3p'}$ possesses no other transformations than an $S^i\, T^k$ leaving the set of $(A)$'s invariant. These new points are coincidence points for the transformation of order $q$, $\sum T \sum^{-1}$. Let $A_i^k$ be the transformed of $A_i'$ by $T^{k-1}$, or of $A_i$ by $T^{k-1} \sum$. The $(A^k)$'s are also coincidence points for a certain transformation of order $q$, and I say that an $A^k$ can be neither an $A^h$ nor an $A$. For otherwise $\sum^{-1} T^{k-h} \sum$ would transform an $A$ into another, and therefore $\sum^{-1} T^{k-h} \sum = S^j\, T^k$. But if $j \neq 0$ this is impossible for the two transformations are of different orders $q$, and three, while if $j = 0$, it leads to $\sum T^k \sum^{-1} = T^{k-h}$, again impossible whether $k \neq h$, as the two substitutions have then different coincidence points, or still worse $k = h \neq 0$. Thus the existence of $\sum$ brings as a corollary that of $(q + 1)$ sets of three Weierstrass points. The sum of the orders of the $(A^k)$'s is obviously equal to the sum of the orders of the $(A)$'s, that is, to $p(p-1)/2$. The sum of the orders of all the Weierstrass points thus obtained is $(q+1)p(p-1)/2 = p(p^2-1)$. But this is the sum of the orders of all the Weierstrass points (Hurwitz), hence $C_{3p'}$ can have no others. Thus the Weierstrass points are grouped in sets of three and any birational transformation $\sum$ merely permutes the sets among themselves. This follows from the fact that if $\sum$ permutes for example $A_1$ with $A_1^1$, $A_1^1$ is critical for a definite transformation of order $q$, $\sum T \sum^{-1}$, whose cyclic group is uniquely defined by $A_1^1$, and hence also the corresponding set of three coincidence points to which $\sum A_2$ and $\sum A_3$ necessarily belong. This set can only be $A_1^1, A_2^1, A_3^1$. Henceforth the set of $(A^k)$'s will be denoted by $(A^k)$.

150. I say that $S$ permutes at least two sets $(A^k)$. For if it maintained them all fixed, $ST$ would permute $q$ of them cyclically which is impossible as its order is three. Assume then that $S$ permutes $(A^{k+1})$ with some other set. The transformation $U = \sum^{-1} T^{-k} S T^k \sum$ which is of order three permutes cyclically $(A)$ with two other sets, say $(A^1)$, $(A^2)$ in the order named. On the other hand the number of sets is $q + 1$ which is prime to three, hence $S$ must maintain fixed at least one set other than $(A)$, say $(A^{h+1})$. Then $T^{-h} S T^h$ is a transformation of order three maintaining invariant $(A)$ and $(A^1)$ and it may be assumed without inconvenience to be $S$ itself. Finally if $V = UT$, we have $V(A) = UT(A) = (A^1)$, $V(A^1) = UT(A^1) = U(A^2)$

$= ('A)$. It follows that $V^2$ maintains $(A)$ invariant and can only be of order one, three, or $q$ and $V$ therefore of order $2, 3, 6$, or $q$ since $2q$ is excluded. It is also obviously not of order 3 or $q$. Hence $V$ is of order 2 or 6.

It is at once verified that the effect of $V$ on the point $A_1$ is expressed by one of the relations

$$VSA_1 = SVA_1, \qquad VSA_1 = S^2 VA_1.$$

Hence either $VS = SVT^k$, or $VS = S^2 VT^k$. If $k \neq 0$ we must have one of the relations

$$V^{-1} S^{-1} VS = T^k, \qquad V^{-1} S^{-2} VS = T^k$$

which is impossible for $T^k$ maintains $(A)$ alone fixed, while the other substitutions maintain both $(A)$ and $(A')$ fixed. Thus, *if there is a transformation* $\sum$ *of* $C_{3p'}$, *not an* $S^i T^k$, *there is one* $V$ *of order* 2 *or* 6, *such that* $VT^{-1}$ *is of order* 3, *and that moreover either* $VS = SV$, *or* $VS = S^2 V$.

151. Let us examine now the effect of the birational transformations $S^i T^k$ on the integrals of the first kind. It is seen at once that they transform a linear combination $au_j + bu_{j+p'} + cu_{j+2p'}$ into one of the same type. We have there $3q$ birational transformations and if we range the integrals in suitable order, their transformations will be represented by arrays such as

$$\left\| \begin{matrix} B_1, & 0, & \cdots, & 0 \\ 0, & B_2, & \cdots, & 0 \\ \cdot & \cdot & \cdot \quad \cdot & \cdot \\ \cdot & \cdot & \cdots, & B_{p'} \end{matrix} \right\|$$

where $B_j$ is itself an array of one of the three following types

$$\left\| \begin{matrix} \epsilon^{n_j}, & 0, & 0 \\ 0, & \epsilon^{\lambda n_j}, & 0 \\ 0, & 0, & \epsilon^{\lambda^2 n_j} \end{matrix} \right\|, \quad \left\| \begin{matrix} 0, & \epsilon^{n_j}, & 0 \\ 0, & 0, & \epsilon^{\lambda n_j} \\ \epsilon^{\lambda^2 n_j}, & 0, & 0 \end{matrix} \right\|, \quad \left\| \begin{matrix} 0, & 0, & \epsilon^{n} \\ \epsilon^{\lambda n_j}, & 0, & 0 \\ 0, & \epsilon^{\lambda^2 n_j}, & 0 \end{matrix} \right\|,$$

corresponding to $T^j$, $ST^j$, $S^2 T^j$, respectively. The array of the $(B)$'s represents also a complex multiplication. Each set contains $q$ multiplications of which $q - 1$ are independent and between the three sets of $q - 1$ independent multiplications thus obtained there are no linear relations. Since $1 + h = 3(q - 1)$, these multiplications form a base. Therefore any multiplication can be represented by an array such as that of the $(B)$'s with now any one of the $(B)$'s of the form

$$\left\| \begin{matrix} g_1(\epsilon), & g_2(\epsilon), & g_3(\epsilon) \\ g_3(\epsilon^{\lambda}), & g_1(\epsilon^{\lambda}), & g_2(\epsilon^{\lambda}) \\ g_2(\epsilon^{\lambda^2}), & g_3(\epsilon^{\lambda^2}), & g_1(\epsilon^{\lambda^2}) \end{matrix} \right\|,$$

the $(g)$'s being polynomials with rational coefficients. For example, $B_j$ is obtained when $\epsilon$ is replaced by $\epsilon^{n_j}$.

152. Let us assume in particular that we are dealing with the complex multiplication defined by the birational transformation $V$ of No. 150, and let us represent the typical $B_j$ just written by

$$\left\| \begin{matrix} a_1, & a_2, & a_3 \\ b_1, & b_2, & b_3 \\ c_1, & c_2, & c_3 \end{matrix} \right\| .$$

If we express the fact that $VS = SV$ or $S^2 V$ we find that this array is necessarily of one of the two forms*

$$\left\| \begin{matrix} a, & b, & c \\ c, & a, & b \\ b, & c, & a \end{matrix} \right\| , \qquad \left\| \begin{matrix} a, & b, & c \\ b, & c, & a \\ c, & a, & b \end{matrix} \right\| .$$

Finally the multipliers are roots of one of the two equations

$$\left\| \begin{matrix} a - x, & b, & c \\ c, & a - x, & b \\ b, & c, & a - x \end{matrix} \right\| = 0, \qquad \left\| \begin{matrix} a - x, & b, & c \\ b, & c - x, & a \\ c, & a, & b - x \end{matrix} \right\| = 0.$$

153. What are these multipliers? *When $V$ is of order two,* they can only be $+1$ and $-1$, taken one $p'$ times and the other $2p'$ times in the set of $p$ multipliers, or else $-1$ taken $3p'$ times. But this last case may be set aside as it presents itself only for hyperelliptic curves. Similarly I say that $+1$ cannot be taken $2p'$ times. For then $V$ would leave invariant $2p'$ integrals of the first kind, hence it would determine an involution of order *two* and genus $2p'$ on $C_{3p'}$, involution which must have $2p - 2(2 \cdot 2p' - 2) = 2 - 2p' \geqq 0$ coincidences, which is impossible unless $p' = 1$, $q = 7$. Hence except for this last case when $C_{3p'}$ is Klein's quartic, if $V$ is of order *two*, there are $p'$ multipliers equal to $+1$, and $2p'$ equal to $-1$. The numbers $a, b, c$ above are numbers of the domain $K(\epsilon)$, and when we replace $\epsilon$ by $\epsilon^k$ we do not change the roots of the equations at the end of No. 152, if they are rational or roots of unity of degree prime to $q$. Hence these equations will all have $-1$ for double root and $+1$ for simple root. The sum of these roots is $-1$ and their product $+1$.

154. *When $V$ is of order* 6, the multipliers can only be $1, \eta^{\frac{1}{2}}, \eta^{-\frac{1}{2}}$, or $-1, \eta^{\frac{1}{2}}, \eta^{-\frac{1}{2}}$, taken each $p'$ times $(\eta = e^{2\pi i/3})$. The first possibility can again be set aside, for then there would be on the curve an involution of order 6 and genus $p'$, with $2p - 2 - 6(2p' - 2) = 10 - 6p'$ coincidences, which requires here also $p' = 1$, $q = 7$. The second and only possible set of multipliers are the distinct roots of $x^3 + 1 = 0$. Their sum is *zero* and their product $-1$.

* See Klein-Fricke, loc. cit. p. 704, for a similar computation corresponding to $q=7$, $p'=1$.

155. We can now show that to $V$ there cannot correspond a type array

$$\left\| \begin{matrix} a, & b, & c \\ c, & a, & b \\ b, & c, & a \end{matrix} \right\|$$

for any $B$. For if $V^2 = 1$, the condition as to the sum of the multipliers gives at once $3a = -1$. On the other hand $VT^{-1}$ must be of order three. But to this transformation corresponds an array of $(B)$'s of which a typical one is

$$\left\| \begin{matrix} a\epsilon, & b\epsilon, & c\epsilon \\ c\epsilon^{\lambda}, & a\epsilon^{\lambda}, & b\epsilon^{\lambda} \\ b\epsilon^{\lambda^2}, & c\epsilon^{\lambda^2}, & a\epsilon^{\lambda^2} \end{matrix} \right\|.$$

The corresponding multipliers which can only be $1, \eta, \eta^2$ are roots of

$$\left| \begin{matrix} a\epsilon - x, & b\epsilon, & c\epsilon \\ c\epsilon^{\lambda}, & a\epsilon^{\lambda} - x, & b\epsilon^{\lambda} \\ b\epsilon^{\lambda^2}, & c\epsilon^{\lambda^2}, & a\epsilon^{\lambda^2} - x \end{matrix} \right| = 0$$

$$\therefore \quad a(\epsilon + \epsilon^{\lambda} + \epsilon^{\lambda^2}) = 1 + \eta + \eta^2 = 0$$

and as $3a = -1$, we must have $\epsilon + \epsilon^{\lambda} + \epsilon^{\lambda^2} = 0$ which cannot be since $q$ has been assumed $> 3$ and $\epsilon$ is a root of the irreducible equation

$$1 + x + x^2 + \cdots + x^{q-1} = 0.$$

If $V^6 = 1$, we must have first $3a = 0$. Hence the array corresponding to $V$ is of the type

$$\left\| \begin{matrix} 0, & b, & c \\ c, & 0, & b \\ b, & c, & 0 \end{matrix} \right\|.$$

Its square

$$\left\| \begin{matrix} 2bc, & c^2, & b^2 \\ b^2, & 2bc, & c^2 \\ c^2, & b^2, & 2bc \end{matrix} \right\|$$

must be a matrix whose cube is unity. Hence again $6bc = 0$, and either $b$ or $c = 0$. Assume for example $b = 0$. Expressing the fact that the multipliers of $V$ have $-1$ for product we have $c^3 = -1$, hence $c = -1$, $\eta^{\frac{1}{2}}$ or $\eta^{-\frac{1}{2}}$. The values $\eta^{\pm\frac{1}{2}}$ are to be set aside for $c$ must be a number of the domain $K(\epsilon)$ and cannot be a root of unity of degree prime to $q$. As to $-1$ it must be set aside for then $VS$ would be a birational transformation with multipliers all $-1$, and $C_{3p'}$ would be hyperelliptic. Our affirmation is therefore proved.

156. Let us now consider the array

$$\left\| \begin{matrix} a, & b, & c \\ b, & c, & a \\ c, & a, & b \end{matrix} \right\|.$$

Assume first $V^6 = 1$. The sum of the multipliers is zero and their product $-1$, hence

$$a + b + c = 0, \qquad \begin{vmatrix} a, & b, & c \\ b, & c, & a \\ c, & a, & b \end{vmatrix} = -1$$

which cannot be since the determinant just written has $a + b + c$ for factor.

It remains to consider $V^2 = 1$. We have then $a + b + c = -2 + 1 = -1$. Moreover since the square of our array is the identical substitution we must have

$$bc + ca + ab = 0,$$
$$\therefore \quad a(-1 - a) + bc = 0, \qquad a = bc - a^2,$$

and similarly $b = ca - b^2$, $c = ab - c^2$. Finally since $VT^{-1}$ is of order *three*, the roots of

$$\begin{vmatrix} a\epsilon - x, & b\epsilon, & c\epsilon \\ b\epsilon^{\lambda}, & c\epsilon^{\lambda} - x, & a\epsilon^{\lambda} \\ c\epsilon^{\lambda^2}, & a\epsilon^{\lambda^2}, & b\epsilon^{\lambda^2} - x \end{vmatrix} = 0$$

are $1, \eta, \eta^2$. Their sum and the sum of their double products must be zero, hence $a\epsilon + b\epsilon^{\lambda} + c\epsilon^{\lambda^2} = 0$,

$$\epsilon^{(1+\lambda)}(ac - b^2) + \epsilon^{(\lambda+\lambda^2)}(bc - a^2) + \epsilon^{(1+\lambda^2)}(ab - c^2) = 0.$$

By means of the relations above and since $1 + \lambda + \lambda^2 \equiv 0$, mod. $q$, this last equation reduces to $a\epsilon^{-1} + b\epsilon^{-\lambda} + c\epsilon^{-\lambda^2} = 0$, from which follows

$$\frac{a}{\epsilon^{\lambda-\lambda^2} - \epsilon^{\lambda^2-\lambda}} = \frac{b}{\epsilon^{1-\lambda} - \epsilon^{\lambda-1}} = \frac{c}{\epsilon^{\lambda^2-1} - \epsilon^{1-\lambda^2}}.$$

The denominators are double the imaginary parts of powers of $\epsilon$ not divisible by $q$, hence they are not zero. Substituting in $bc + ca + ab = 0$, we have:

$$\epsilon^{\lambda^2-\lambda} + \epsilon^{\lambda-\lambda^2} - (\epsilon^3 + \epsilon^{-3}) + \epsilon^{\lambda-1} + \epsilon^{1-\lambda} - (\epsilon^{3(1+\lambda)} + \epsilon^{-3(1+\lambda)})$$
$$+ \epsilon^{1-\lambda^2} + \epsilon^{-1+\lambda^2} - (\epsilon^{3(1+\lambda^2)} + \epsilon^{-3(1+\lambda^2)}) = 0.$$

If we compare the first exponent to the others mod. $q$, and remember the congruence satisfied by $\lambda$, we find that it cannot be equal to any of them unless $\lambda = -2, -4/5$ or $-1/5$, mod. $q$, and since $\lambda^3 \equiv 1$, mod. $q$, it is easily seen that $q$ must divide one of the integers $9$, $126 = 2 \cdot 3^2 \cdot 7$, $189 = 3^3 \cdot 7$ and therefore must be 7. For $q > 7$ the equation which $\epsilon$ must satisfy is certainly not its irreducible equation since it has less than $q - 1$ terms, or has $q - 1$ terms perhaps for $q = 13$, but with coefficients certainly not all equal.

The only possibility is then $q = 7$, $p' = 1$, $p = 3$, when $C_{3p'}$ is Klein's quartic. The relation just obtained is identically verified and in fact this curve possesses as shown precisely by Klein a binary transformation such as $V$ and its group is the classical $G_{168}$ generated by $S$, $T$, and $V$. For further details and in particular for the equations of the group the reader is referred to his work (loc. cit.). *This ends the proof of the theorem of No. 148.*

157. The properties of the curves whose discussion is now complete may be summarized thus: *The curves of minimum genus possessing a cyclic group of order $q$, odd prime, constitute a finite number of birationally distinct families. Their total group is the cyclic group of order $q$ with the exception of: (a) A hyperelliptic family present, whatever $q$, and whose group is of order $2q$ cyclic. (b) A family present only when $q-1$ is divisible by three whose group is of order $3q$, unless $q=7$ when it is the classical $G_{168}$. The indices of the Jacobi variety are given by*

$$1+h=2(1+k)=2p=q-1$$

*and the variety is pure except in the case (b) when*

$$1+h=2(1+k)=6p=3(q-1),$$

*and the variety is impure.*

*Remark:* The variety of rank $q$ corresponding to the group of order $q$ has no congruence of spaces for $q>7$, or for $q=7$ and Klein's quartic, as then the product of the multipliers is $+1$ and the Jacobi variety possesses a triple integral of the first kind (No. 127). For $q=5$ the first variety is ruled as shown by Enriques and Severi. The only doubtful case is then that of the hyperelliptic curve present for $q=7$.

## § 3. Curves with four critical points

158. Let us consider first two hypergeometric integrals

$$\int_g^h (x-a_1)^{b_1}(x-a_2)^{b_2}(x-y)^{b_3}\,dx,$$

$$\int_g^h (x-a_1)^{b_1'}(x-a_2)^{b_2'}(x-y)^{b_3'}\,dx.$$

We assume that they have a meaning whatever the limits $g, h$, which requires that

$$b_i>-1, \qquad b_i'>-1 \qquad (i=1,2,3),$$
$$b_1+b_2+b_3<-1, \qquad b_1'+b_2'+b_3'<-1.$$

Let us denote by $\omega_1$, $\omega_1'$ the two integrals corresponding to $g=a_1$, $h=y$ and by $\omega_2$, $\omega_2'$ those corresponding to $g=a_2$, $h=y$. Can there exist a bilinear relation

$$\gamma_{11}\,\omega_1\,\omega_1'+\gamma_{12}\,\omega_1\,\omega_2'+\gamma_{21}\,\omega_2\,\omega_1'+\gamma_{22}\,\omega_2\,\omega_2'=0,$$

where the $(\gamma)$'s are independent of $y$? To answer the question set first

$z = \omega_1/\omega_2$, $z' = \omega_1'/\omega_2'$. As is well known $z$ satisfies Schwarz's differential equation *

$$\frac{2\frac{dz}{dy}\frac{d^3z}{dy^3} - 3\left(\frac{d^2z}{dy^2}\right)^2}{2\left(\frac{dz}{dy}\right)^2} = \frac{1-\lambda}{2(y-a_1)^2} + \frac{1-\mu}{2(y-a_2)^2} + \frac{\lambda+\mu-\nu-1}{2(y-a_1)(y-a_2)},$$

$$\lambda = (b_1 + b_3 + 1)^2, \qquad \mu = (b_1 + b_2 + 1)^2, \qquad \nu = (b_2 + b_3 + 1)^2.$$

If the bilinear relation in question exists, $z'$, a linear fractional function of $z$, must satisfy the same differential equation, and therefore

$$(b_i + b_k + 1) = \pm (b_i' + b_k' + 1) \qquad (i \neq k;\ i, k = 1, 2, 3).$$

By considering the various combinations of signs we find the following four possible types of relations:

$$(a) \qquad b_i = b_i' \qquad (i = 1, 2, 3),$$
$$(b) \qquad 1 + b_i = -b_i' \qquad (i = 1, 2, 3),$$
$$(c) \qquad 1 + b_i = -b_j', \qquad 1 + b_j = -b_i', \qquad -1 + b_k = b_1' + b_2' + b_3',$$

in which case $b_i, b_j, b_i', b_j'$ are $> 0$ and $b_k, b_k' < 0$.

$$(d) \qquad b_i = b_j', \qquad b_j = b_i', \qquad -b_k' = b_1 + b_2 + b_3 + 2.$$

The case (*a*) is to be rejected for then $z$ would be a constant, and when any of the three sets of relations (*b*), (*c*) and (*d*) are satisfied the bilinear relation certainly exists as is known from the theory of Schwarz's differential equations. To calculate the coefficients we recall that according to Picard the group of the hypergeometric differential equation satisfied by $\omega_1, \omega_2$ possesses the two fundamental substitutions †

$$(\omega_1, \omega_2;\ \eta_1\omega_1, (\eta_1 - \eta_3)\omega_1 + \omega_2);$$
$$(\omega_1, \omega_2, \omega_1 + (\eta_2 - \eta_3^{-1})\omega_2, \eta_2\omega_2);$$
$$\eta_1 = e^{-2\pi i(b_1+b_3)}, \qquad \eta_2 = e^{2\pi i(b_2+b_3)}, \qquad \eta_3 = e^{-2\pi i b_3}.$$

By accenting all quantities we obtain similarly the group corresponding to $\omega_1', \omega_2'$. The bilinear relation whose existence is discussed must be invariant when corresponding substitutions are applied to the $(\omega)$'s and the $(\omega')$'s. This yields six equations of the first degree between the $(\gamma)$'s. It is found that they are sufficient to determine their ratios provided we have

$$\frac{(\eta_1 - \eta_3)(\eta_2 - \eta_3^{-1})}{(1-\eta_1)(1-\eta_2)} = \frac{(\eta_1' - \eta_3')(\eta_2' - \eta_3'^{-1})}{(1-\eta_1')(1-\eta_2')}.$$

* See Picard *Traité d'Analyse*, vol. 3, second edition, p. 333.

† Picard, loc. cit. p. 324.

This is at once verified for (b). As for (c) or (d) we have either

$$\eta_1' = \eta_1, \qquad \eta_2' = \eta_2, \qquad \eta_3' = \eta_1 \eta_2^{-1} \eta_3^{-1},$$

or

$$\eta_1' = \eta_2, \qquad \eta_2' = \eta_1, \qquad \eta_3' = \eta_1 \eta_2^{-1} \eta_3^{-1},$$

and by means of one or the other of these relations the equation of condition is readily verified, as was to be expected.

159. We are now ready to undertake the investigation of the curve $C_p$ with four critical points. One of these being assumed for the present at infinity, the equation of the curve will be

$$y^q = (x - a_1)^{\alpha_1} (x - a_2)^{\alpha_2} (x - a_3)^{\alpha_3}$$

with the $(\alpha)$'s positive and $< q$, their sum being prime to $q$. The genus is $p = q - 1$ and $C_p$ depends upon a single modulus—the anharmonic ratio of the critical $(x)$'s. We shall assume this modulus arbitrary and propose to determine above all the invariants of the Jacobi variety and next the total group of our curves.

To $n_j$ corresponds a number of integrals of the first kind given by

$$r_j = \left[ n_j \frac{(\alpha_1 + \alpha_2 + \alpha_3)}{q} \right] - \left[ \frac{n_j \alpha_1}{q} \right] - \left[ \frac{n_j \alpha_2}{q} \right] - \left[ \frac{n_j \alpha_3}{q} \right],$$

and they are of the form

$$\int (x - a_1)^{b_1^j} (x - a_2)^{b_2^j} (x - a_3)^{b_3^j} dx,$$

the $(\beta)$'s having the same meaning as previously. If $r_j = 1$ we shall take

$$b_i^j = \beta_i^j \frac{- n_j \alpha_i}{q} \qquad (i = 1, 2, 3),$$

while if $r_j = 2$ there will be another integral which we shall choose so that it differ from the first only in that $b_1^j = \beta_i^j - (1/q) n_j \alpha_i + 1$. We can of course choose two other analogous integrals by permuting $1, 2, 3$ but they are linear combinations of the two just considered. We recall that the period matrix $\Omega$ is composed with the arrays

$$\tau = \| \tau_{j1'}, \tau_{j2} \|, \qquad \| 1, \epsilon^{n_j}, \cdots, \epsilon^{(q-2)n_j} \|, \qquad (j = 1, 2, \cdots, p),$$

where

$$\tau_{j\mu} = \int_{a_\mu}^{a_{\mu+1}} du_j.$$

160. Since $\Omega$ is a Riemann matrix there certainly exists an alternate bilinear form

(6) $$\sum_{\mu, \nu}^{1\,2} \gamma_{\mu\nu} (\epsilon^{n_j}, \epsilon^{n_k}) x_\mu y_\nu$$

in the sense of Chapter II. The $(\gamma)$'s are such that

$$\gamma_{\mu\nu}(x, y) = -\gamma_{\nu\mu}(y, x);$$
$$\gamma_{\mu\nu}(\epsilon^{n_j}, \epsilon^{n_k}) = 0 \quad \text{if} \quad n_j + n_k \not\equiv 0, \quad \text{mod. } q.$$

The ratios $\tau_{j1}/\tau_{j2}$ are certainly variable for otherwise our curves would all have the same period matrix and would not depend upon a variable modulus. It follows that the form (6) is unique.

Can there exist forms of any other type? Whether it is alternate or not assume that such a form leads to an actual relation between the periods of $u_j$ and $u_k$, that is, that the quantities $\gamma_{\mu\nu}(\epsilon^{n_j}, \epsilon^{n_k})$ are not all zero. Then one of the sets of relations (*b*), (*c*), (*d*) of No. 158 must be satisfied. (*b*) leads to

$$n_j \alpha_i \equiv -n_h \alpha_i, \quad \text{mod. } q \qquad (i = 1, 2, 3, 4);$$
$$\therefore \quad n_j + n_h \equiv 0, \quad \text{mod. } q,$$

that is, the $(\gamma)$'s are all zero except when $\epsilon^{n_j}$ and $\epsilon^{n_k}$ are conjugate. The calculation of No. 158 shows that the bilinear relation existing must be unique and as in the case considered (6) already is one there can be no other.

The conditions (*c*) lead to, for example,

$$n_j \alpha_1 = -n_h \alpha_2, \qquad n_j \alpha_2 \equiv -n_h \alpha_1, \text{ mod. } q.$$
$$n_j \alpha_3 = -n_h(\alpha_1 + \alpha_2 + \alpha_3) \equiv n_h \alpha_4, \text{ mod. } q.$$

Hence mod. $q$, either

$$\alpha_1 \equiv \alpha_2, \qquad n_j \equiv -n_h,$$

or

$$\alpha_1 \equiv -\alpha_2, \qquad n_j \equiv n_h, \qquad \alpha_3 \equiv -\alpha_4.$$

In the first case we are in the same situation as before. In the second $C_p$ is birationally equivalent to

$$(7) \qquad y^q = \frac{x - a_1}{x - a_2}\left(\frac{x - a_3}{x - a_4}\right)^{\alpha}, \qquad 0 < \alpha < q.$$

We can take $\alpha_1 = 1$, $\alpha_2 = q - 1$, $\alpha_3 = \alpha$, and find that

$$r_j = \left[\frac{n_j(q + \alpha)}{q}\right] - \left[\frac{n_j}{q}\right] - \left[\frac{n_j(q - 1)}{q}\right] - \left[\frac{n_j \alpha}{q}\right] = 1$$

which shows that for each value of $n_j$ there is only one integral and a bilinear relation such as contemplated in the second case cannot exist. The discussion of conditions $(d)$ is the same as for $(c)$ and the answer negative as well. *Hence $\Omega$ possesses no other bilinear form than* (6).

161. To determine $h$, $k$ we must find out if there are other curves than (7) for which $r_j = 1$ whatever $j$. We shall see that this is not the case.

If $r_j = 1$ whatever $j$, $\Omega$ is composed with arrays

$$\| \tau_{j1}, \tau_{j2} \|, \qquad \| 1, \epsilon^j, \epsilon^{2j}, \cdots, \epsilon^{(q-2)j} \| \qquad (j = 1, 2, \cdots, q-1 = p).$$

The calculation outlined in No. 158 leads to the relation

$$\tau_{q-j,\,2}\,\tau_{j1} - \epsilon^j\,\tau_{q-j,\,1}\,\tau_{j,\,2} = 0.$$

$$\therefore \quad \frac{\tau_{q-j,\,2}}{\tau_{q-j,\,1}} = \frac{\epsilon^j\,\tau_{j,\,2}}{\tau_{j,\,1}} = \epsilon^j \cdot \tau_j$$

and therefore $\Omega$ can be put in the following form, where only the $j$-th and $(q-j)$-th rows are written:

$$\Omega \equiv \left\| \begin{array}{cccccccccc} 1, & \epsilon^j, & \epsilon^{2j}, & \cdots, & \epsilon^{(q-2)j}; & \tau_j, & \epsilon^j\,\tau_j, & \cdots, & \tau_j\,\epsilon^{(q-2)j} \\ \cdot & \cdot & \cdot & \cdot & \cdot & \cdot & \cdot & \cdot & \cdot \\ 1, & \epsilon^{-j}, & \epsilon^{-2j}, & \cdots, & \epsilon^{-(q-2)j}; & \tau_j\,\epsilon^j, & \tau_j, & \cdots, & \tau_j\,\epsilon^{-(q-3)j} \\ \cdot & \cdot & \cdot & \cdot & \cdot & \cdot & \cdot & \cdot & \cdot \end{array} \right\|.$$

Denote respectively by $\delta_\mu$, $\overline{\delta_\mu}$, the cycles corresponding to the periods $\epsilon^\mu$, $\tau_1\,\epsilon^\mu$, in the first line, the indices being taken mod. $q$. If we apply to the integrals and the cycles the binary permutations

$$u_j' = u_{q-j} \qquad\qquad (j = 1, 2, \cdots, p)$$
$$\delta_\mu' \sim \delta_{-\mu}; \qquad \overline{\delta_\mu'} \sim \delta_{1-\mu} \qquad (\mu = 0, 1, 2, \cdots, q-1)$$

$\Omega$ is unchanged. Hence $C_p$ possesses a corresponding binary transformation $U$ whose genus is $\frac{1}{2}p$ for it maintains invariant the integrals $u_j + u_{q-j}$ and no others.

Let $(x, y)$ be any point of $C_p$, $(x', y')$ its transformed by $U$. The integrals of the first kind are all of the form

$$\int \frac{\prod\limits^{(i)} (x - a_i)^{\gamma_i}\,dx}{y^n}.$$

By reasoning as in No. 145 we find therefore again that $U$ merely permutes the critical points. The equations of $U$ are of type (2) and it is not the ordinary transformation present in hyperelliptic curves since its genus is not zero. Besides $U$ must permute every critical point for if it maintained two of them fixed, the anharmonic ratio of the $(a)$'s would not be variable as was assumed. If then $U$ permutes $A_1$ with $A_2$ and $A_3$ with $A_4$, we must have (No. 138)

$$\frac{\alpha_2}{\alpha_1} \equiv \frac{\alpha_1}{\alpha_2}, \qquad \frac{\alpha_4}{\alpha_3} \equiv \frac{\alpha_3}{\alpha_4}, \qquad \text{mod. } q$$

and as we may always take $\alpha_1 = 1$, it is seen at once that we fall back upon (7), as was to be proved.

In accordance then with Chapter II, *if there are four arbitrary critical points*, *the Jacobi variety is pure, its indices being given by*

$$(1 + h) = 2(1 + k) = p = q - 1$$

*unless we deal with* (7) *when the variety is impure and*

$$1 + h = 2(q - 1) = 2p, \qquad 1 + k = 3(q - 1).$$

162. Let us investigate more closely the question of birational transformations. We know that if $C_p$ is not of type (7) its total group is cyclic of order $q$ (No. 142). It remains to consider the curves (7).

In the first place if $\alpha = 1$, the curve is hyperelliptic. The ordinary transformation then present shall be denoted by $S$, and the binary transformation of genus $p'$, as before, by $U$. We propose to show that *in general $C_p$ possesses no other transformations than $S$, $T$, $U$ and the products of their powers.*

In the first place the birational transformations $T^k$, $T^k U$, define the two multiplications

$$u_j' = \epsilon^{kj} u_j, \qquad u_j' = \epsilon^{kj} u_{q-j} \qquad (j = 1, 2, \cdots, p),$$

and among those $2q$ multiplications exactly $2(q-1) = 1 + h$ are linearly independent. If we write the integrals in suitable order the most general multiplications will be represented by an array

$$\left\| \begin{matrix} B_1, & 0, & \cdots, & 0 \\ 0, & B_2, & \cdots, & 0 \\ \cdot & \cdot & \cdot & \cdot \\ 0, & \cdot & \cdots, & B_{\frac{1}{2}p} \end{matrix} \right\| ; \qquad B_j = \left\| \begin{matrix} g(\epsilon^j), & h(\epsilon^j) \\ h(\epsilon^{-j}), & g(\epsilon^{-j}) \end{matrix} \right\|,$$

where $g$, $h$ are polynomials with rational coefficients. The array $B_j$ may be written more simply

$$\left\| \begin{matrix} g, & h \\ \bar{h}, & \bar{g} \end{matrix} \right\|.$$

We first remark that if $C_p$ is hyperelliptic $A_1$ may be permuted with the three other $(A)$'s by one of the three transformations $S$, $U$, $SU = US$, while if $C_p$ is not hyperelliptic, $A$ may be permuted with only one of these points, and that by means of $U$. But as a matter of fact the two cases, hyperelliptic and the other, need not be separated in the discussion.

163. Let us now assume that there exists a transformation $V$ of $C_p$ not a product of powers of those already known. The binary transformation $W = V^{-1} U \cdot V$ may or may not permute the $(A)$'s among themselves. We propose to show that in both cases we are led to an impossibility.

To begin with if $W$ permutes the $(A)$'s merely among themselves it must belong to one of the three types $UT^k$, $ST^k$, $UST^k$. But if $T^k \neq 1$, these transformations are of order $2q$ at least as seen by a reference to their complex multiplications. Hence necessarily $T^k = 1$, and $UV = VU$, $VS$ or $VUS$.

If $V$ corresponds to the array of the $(B)$'s, we obtain by comparison of the arrays for $UV$ with those of the other three substitutions, one of the systems of relations

$$\begin{array}{lll} (a) & g = \bar{g}, & h = \bar{h}, \\ (b) & g = -\bar{h}, & \\ (c) & g = -\bar{g}, & h = -\bar{h}. \end{array}$$

The equation in the multipliers of $V$ is

$$\begin{vmatrix} g - x, & h \\ \bar{h}, & \bar{g} - x \end{vmatrix} = 0.$$

This equation must be reducible in $K(\epsilon)$, whence, as in Nos. 149, 151, the only possible systems of multipliers $(1, -1)$, $(\eta, \eta^{-1})$, $(i, -i)$, $(\eta^{\frac{1}{2}}, \eta^{-\frac{1}{2}})$, $(\epsilon^k, \epsilon^{-k})$, $(-\epsilon^k, -\epsilon^{-k})$, corresponding to the orders $\nu = 2, 3, 4, 6, q, 2q$ for $V$.

164. In the case of relations ($a$), $g$ and $h$ are real and the equation in the multipliers is $x^2 - 2gx + g^2 - h^2 = 0$. Let $x_1, x_2$ be its roots. When $x_1 = 1$, $x_2 = -1$, we have $g = 0$, $h = \pm 1$, hence $V = U$ or $SU$.* In the other cases $h$ takes the values $\pm i$, $\pm \frac{i\sqrt{3}}{2}$, $\pm \frac{(\epsilon^k - \epsilon^{-k})}{2}$, all imaginary whereas $h$ must be real.

In the case of relations ($b$) the equation in the multipliers is

$$x^2 - (g + \bar{g})x = 0$$

and must be rejected since no multipliers can be zero.

Finally in the case of relations ($c$), the equation is $x^2 + h^2 = 0$. Since $h$ is purely imaginary, $h^2 = x_1 x_2$ is a negative real number, and therefore necessarily $x_1 = 1$, $x_2 = -1$, $h = \pm i$. But $h$ is a number of the domain $K(\epsilon)$, and therefore cannot be a root of unity of degree prime to $q$. Hence $V$ cannot correspond to any relations such as those just discussed.

165. Let us assume now that the binary transformation $W$ of No. 163 permutes the $(A)$'s with points of another similar set. The multipliers of $W$ must be $1, -1$, since it is not $S$, and if it corresponds to the array of $(B)$'s, we must have $g + \bar{g} = 0$. $g$ is therefore purely imaginary. If we set $\epsilon^k + \epsilon^{-k} = \eta_k$, $\epsilon^k - \epsilon^{-k} = \zeta_k$, the $(\eta)$'s are real and the $(\zeta)$'s pure imaginary, hence

$$g = \sum_{i=1}^{\frac{q-1}{2}} a_i \zeta_i \qquad (a_i \text{ rational number}).$$

To $TW$ corresponds an array of $(B)$'s with $g\epsilon$ in place of $g$. More exactly, to

* For by the remark at the end of No. 141 it is sufficient to show that to $V^{-1}U$ or $V^{-1}SU$ correspond ordinary transformations of the Jacobi variety, and this is readily done.

the $B_j$ of this array corresponds $g(\epsilon^j)\cdot\epsilon^j$, but if $\epsilon$ designates a suitably chosen $q$-th root of unity, any one of these numbers can be set in the form $g\epsilon$. Now if we set $\eta_0 = 2$, it is at once verified that the real part of $\zeta_i\,\epsilon = \frac{1}{2}\zeta_i(\eta_1 + \zeta_1)$ is equal to $\frac{1}{2}\zeta_i\,\zeta_1 = \frac{1}{2}(\eta_{i+1} - \eta_{i-1})$. Hence the double of the real part of $g\epsilon$, or

$$g\epsilon + \bar{g}\epsilon^{-1} = \sum_{i=1}^{\frac{q-1}{2}} a_i(\eta_{i+1} - \eta_{i-1}).$$

Finally since $\eta_j = \eta_{q-j}$, we can write

$$g\epsilon + \bar{g}\epsilon^{-1} = \sum_{i=1}^{\frac{q-1}{2}} a'_i(\eta_{2i} - \eta_{2(i-1)}),$$

where the $(a')$'s are merely the $(a)$'s written in different order. If we write the equation in the multipliers of $TW$ we see at once that this expression can only have the values $0, \pm 1, \pm(\epsilon^k + \epsilon^{-k})$.

In the first case the $(a')$'s and therefore the $(a)$'s must all be zero. The second leads to

$$a'_1 = \pm\tfrac{1}{2}, \qquad a'_1 = a'_2 = \cdots = a'_{\frac{q-1}{2}},$$

hence $g\epsilon + \bar{g}\epsilon^{-1} = \pm\frac{1}{2}\eta_{\frac{q-1}{2}} \pm 1 \neq \pm 1$, and therefore cannot occur at all.

In the third case we have, $s$ being a suitable integer between 0 and $q-1$, $g\epsilon + \bar{g}\epsilon^{-1} = \pm\eta_{2s}$;

$$\therefore\quad a'_s - a'_{s+1} = \pm 1, \qquad a'_i = a'_s, \qquad i < s; \qquad a'_{s+i} = a'_{s+1}.$$

Hence then $g\epsilon + \bar{g}\epsilon^{-1} = -2a'_s \pm \eta_{2s} + 2a'_{s+1}\eta_{q-1}$ which can only be if $a'_{s+1} = a'_s = 0$ and we have here again a contradiction.

The only possibility is then $g = 0$. But in this case $UW$ corresponds to a multiplication of type $u'_j = \lambda_j u_j$, where $u_j$ is the integral which $T$ multiplies by $\epsilon^j$. This multiplication is permutable with $T$, cyclic, and therefore, by a previous remark, $W = US^i\,T^k$ or $UT^k$ and hence must permute the $(A)$'s among themselves, contrary to the assumptions made.

This completes the proof of the non existence of any transformations of $C_p$ other than those belonging to the group generated by $S$, $T$ and $U$.

166. It is at once verified that $T^k\,U = UT^{-k}$, hence every substitution of the group is of the form $S^i\,U^j\,T^k$ if $C_p$ is hyperelliptic, or $U^j\,T^k$ if it is not. Hence *if $C_p$ is not hyperelliptic, that is, if the integer $\alpha$ in* (7) *is* $> 1$, *the group is of order* $2q$, *while if the curve is hyperlliptic the order is* $4q$.

Observe that the curve can be reduced to the form

$$y^q = \left(\frac{x-a}{ax-1}\right)\left(\frac{x+a}{ax+1}\right)^{\alpha},$$

the equations of $U$ being then $x' = -x$, $y' = y^{-1}$. For $\alpha = 1$ we have, besides, $S$, whose equations are $x' = -x$, $y' = y$. The parameter $a$ may be taken as modulus of the curve.

*Remark:* For particular values of the modulus of which the curves with four critical points depend, there will be other birational transformations different from those considered. For example

$$y^q = (x - 1)(x - \eta)(x - \eta^2), \qquad \eta = e^{\frac{2\pi i}{3}},$$

possesses an obvious cyclic transformation of order three.

### § 4. Curves with $r + 2 > 4$ critical points.

167. We shall take the equation in the form

$$y^q = \prod_{i=1}^{r+1} (x - a_i)^{\alpha_i}, \tag{8}$$

the $(\alpha)$'s of index $< r + 2$ and their sum being prime to $q$, with $a_{r+2}$ at infinity.

We may choose for the $r_j$ integrals of the first kind corresponding to $n_j$, the set

$$\int \frac{(x - a_1)^{\gamma_j} \prod\limits^{(i)} (x - a_i)^{\beta_i^j}\, dx}{y^{n_j}} \qquad (\gamma_j = 1, 2, \cdots, r_j).$$

Now the sums $\alpha_i + \alpha_j$ cannot all be divisible by $q$. For then if $r$ is odd, $\sum_{i=1}^{r+1} \alpha_i$ would be divisible by $q$, while if $r$ is even

$$2 \sum \alpha_i = \alpha_1 + \alpha_2 + \sum_{i=2}^{r+1} \alpha_i + (\alpha_1 + \sum_{i=3}^{r+1} \alpha_i)$$

would be also, leading again to the divisibility of the same sum by $q$.

Let us observe that since $r > 2$ we may arrange matters so that there will be at least three sums $\alpha_\mu + \alpha_{\mu+1}$ not divisible by $q$. For the worst conditions obtained are when:

(*a*) $r = 3$ and we have a set of five exponents corresponding to the five critical points, such as: $\alpha, -\alpha, \alpha, \alpha', \alpha''$. Then $\alpha + \alpha' + \alpha'' \equiv 0$, mod. $q$, and neither $\alpha'$ nor $\alpha''$ are $\equiv -\alpha$. One of them however may be $\equiv \alpha$. In this case the exponents in suitable order form a set such as $\alpha, \alpha, \alpha, \alpha'$, $-\alpha$, and the sums $2\alpha, 2\alpha, \alpha + \alpha'$ are not divisible by $q$.

(*b*) $r = 4$ and the set of exponents is $\alpha, -\alpha, \alpha, -\alpha, \alpha - \alpha$. In another order they form the set $\alpha, \alpha, \alpha, -\alpha, -\alpha, -\alpha$, for which the assertion is at once verified.

Let then $\alpha_1 + \alpha_2 \not\equiv 0$, mod. $q$, and make $a_1$ tend towards $a_2$. Some of the integrals of the first kind will preserve their character for the curve $C_{p'}$ limit of $C_p$, where $p' = \frac{1}{2}(r-1)(q-1)$. I say that the integrals of the first kind corresponding to $n_j$ for $C_{p'}$ will be linearly dependent upon these. For the limiting integrals which preserve their character form the set

$$\int (x - a_1)^{b'_1} \prod_{i=3}^{r+1}{}' (x - a_i)^{b_i}\, dx,$$

where the $(b)$'s have the same value as above, and

$$b'_1 = \frac{n_j(\alpha_1 + \alpha_2)}{q} - \left[\frac{n_j(\alpha_1 + \alpha_2)}{q}\right] + \gamma.$$

We must determine the range of the positive integer $\gamma$. Recalling that

$$\beta_i^j = \frac{-n_j\alpha_i}{q} + \left[\frac{n_j\alpha_i}{q}\right] < 0$$

we see that if the integer $[-\beta_1^j - \beta_2^j]$, which $< 2$, is equal to unity, we must take $\gamma \geqq 1$, while if $[-\beta_1^j - \beta_2^j] = 0$ we must have only $\gamma \geqq 0$. On the other hand the upper limit of $\gamma$ is $r'_j$ corresponding to $r_j$ for $C_p$. But $r'_j = r_j - [-\beta_1^j - \beta_2^j]$, and hence in all cases $\gamma$ takes $r_j$ consecutive values. As the limiting integrals are obviously independent our assertion is proved.

168. I say that if the array $\tau$ possesses the bilinear form

$$\sum \gamma_{\mu\nu}(\epsilon^{n_j}, \epsilon^{n_k})\, x_\mu\, y_\nu$$

then the coefficients $\gamma_{\mu\nu}$ vanish when $\epsilon^{n_j}$, $\epsilon^{n_k}$ are not conjugate, that is when $n_j + n_k \not\equiv 0$, mod. $q$. Since this has been proved for $r = 2$, we may assume it correct for a curve with less than $r + 2$ critical points, then prove it for a curve with that number of critical points.

Assume then first that the theorem is untrue. If the $(\gamma)$'s are not zero for a particular pair of exponents $n_j$, $n_k$ not congruent mod. $q$, they are not zero for any pair $n$, $cn$, where $c \equiv n_k/n_j$, mod. $q$, and hence if we replace in $\sum \gamma_{\mu\nu}(\epsilon^n, \epsilon^{cn})\, x_\mu\, y_\nu$ the $(x)$'s by the elements of a row of $\tau$ corresponding to $\epsilon^n$ and the $(y)$'s by the elements of a row corresponding to $\epsilon^{cn}$, then the form vanishes identically. Let us assume that there is at least one case where the corresponding integers $r_j$, $r_k$ are both $> 1$—for example let this be so for $n_j$, $n_k$ themselves. The critical points being ranged in proper order there will be three distinct sums $\alpha_\mu + \alpha_{\mu+1}$, $\alpha_{\mu'} + \alpha_{\mu'+1}$, $\alpha_{\mu''} + \alpha_{\mu''+1}$, not divisible by $q$. If $a_\mu$ is made to tend towards $a_{\mu+1}$ there will be two integrals of the first kind corresponding to $n_j$ and $n_k$ preserving their character for the limiting curve $C_{p'}$. Let $\tau_{j\nu}$, $\tau_{k\rho}$ $(\nu, \rho = 1, 2, \cdots, r)$ be the corresponding elements of the array $\tau$. We have then $\sum \gamma_{\nu\rho}(\epsilon^{n_j}, \epsilon^{n_k})\, \tau_{j\nu}\, \tau_{k\rho} = 0$. Passing to the limit, $\tau_{j\nu}$, $\tau_{k\rho}$ become similar periods of the limiting integrals, say

$\tau'_{j\nu}, \tau'_{k\rho}$, except that $\tau'_{j\mu} = \tau'_{k\mu} = 0$. Hence then

$$\sum_{\nu, \rho \neq \mu} \gamma_{\nu\rho}(\epsilon^{n_j}, \epsilon^{n_k}) \tau'_{j\nu} \tau'_{k\rho} = 0,$$

and there will be such a relation for every pair of integrals of $C_{p'}$ corresponding to $n_j, n_k$, and even more generally to $cn_j, cn_k$. According to the assumptions we have therefore

$$\gamma_{\nu\rho}(\epsilon^{n_j}, \epsilon^{n_k}) = 0 \qquad (\nu, \rho \neq \mu.)$$

If we reason similarly when $\mu$ is replaced first by $\mu'$, then by $\mu''$, we find successively

$$\gamma_{\mu\rho}(\epsilon^{n_j}, \epsilon^{n_k}) = \gamma_{\rho\mu}(\epsilon^{n_j}, \epsilon^{n_k}) = 0, \qquad \rho \neq \mu';$$
$$\gamma_{\mu\mu'}(\epsilon^{n_j}, \epsilon^{n_k}) = \gamma_{\mu'\mu}(\epsilon^{n_j}, \epsilon^{n_k}) = 0,$$

which completes the proof in the case considered.

169. The above may cease to apply: (*a*) When the integers $r_j$ corresponding to the sequence of exponents $l, cl, c^2 l, c^3 l, \cdots$, are alternately $+1$ and $r-1$, and the limiting curve $C_{p'}$ possesses no integrals corresponding to the exponents for which $r = 1$. But in this case the numbers $r_j$ corresponding to $C_{p'}$ are all $r-1$ which is impossible if $r > 2$. The proof can therefore still be applied as between integrals corresponding to two exponents $c^s l$, $c^{s+1} l$, leading to $\gamma_{\mu\nu}(\epsilon^{c^s l}, \epsilon^{c^{s+1} l}) = 0$. But since the $(\gamma)$'s are polynomials with rational coefficients, it follows that $\gamma_{\mu\nu}(\epsilon^{n_j}, \epsilon^{cn_j}) = \gamma_{\mu\nu}(\epsilon^{n_j}, \epsilon^{n_k}) = 0$. (*b*) When $n_j = n_k$, $r = 3$. In this case we are not certain that to a bilinear relation between the periods of $C_p$ corresponds after passing to the limit a similar one for $C_{p'}$, at least if all the numbers $r_j$ of this last curve are equal to unity. Generally speaking let $\tau'_{j1}, \tau'_{j2}, \tau'_{j3}$ and $\tau''_{j1}, \tau''_{j2}, \tau''_{j3}$ be two lines of $\tau$ corresponding to $n_j$. By assumption then $\sum \gamma_{\mu\nu}(\epsilon^{n_j}, \epsilon^{n_j}) \tau'_{j\mu} \tau''_{j\nu} = 0$ with the relations $\gamma_{\mu\nu}(\epsilon, \epsilon) = -\gamma_{\nu\mu}(\epsilon, \epsilon)$, for otherwise there would exist a bilinear relation between the elements of every row of $\tau$, the impossibility of which may be established as in No. 168. We have then in

$$\sum \gamma_{\mu\nu}(\epsilon^{n_j}, \epsilon^{n_j}) x_\mu y_\nu = 0$$

the equation of a plane "line complex."—We recall that in a space of an even number of dimensions linear complexes are degenerate. Hence there must exist a unique point $(\tau_1, \tau_2, \tau_3)$ of the plane, conjugate of both $(\tau'_1, \tau'_2, \tau'_3)$ and $(\tau''_1, \tau''_2, \tau''_3)$, relatively to the above complex. On the other hand when the moduli of $C_p$ vary these points remain conjugate with respect to the same complex. But when the $(a)$'s describe closed paths the $(\tau')$'s and $(\tau'')$'s are transformed by a certain discontinuous projective group whose fundamental operations have been given by Picard.* Since the point $(\tau_1, \tau_2, \tau_3)$ is unique it must be maintained invariant by this group. However a glance at

*Annales de l' École Normale (1885).

the equation of its operations show that they leave no point invariant. We have therefore an impossibility, hence $\gamma_{\mu\nu}(\epsilon^{n_j}, \epsilon^{n_j}) = 0$.

170. It follows that *for the curves with more than four critical points in general position the period matrix possesses a minimum number of bilinear forms and consequently for the Jacobi variety* $1 + h = 2(1 + k) = q - 1$.

171. CONCLUSION. Our whole discussion may be summarized thus: The curves possessing a cyclic group of order $q$ odd prime, of genus zero, with arbitrary coincidence points, do not possess in general any other birational transformations and their Jacobi varieties are pure with

$$1 + h = 2(1 + k) = q - 1.$$

Exception must be made for the curves birationally equivalent to the following:

(I) $$y^q = x^2 - a^2.$$

The group is then of order $2q$, cyclic, the rest being as in the general case.

(II) $$y^q = (x - 1)(x - \eta)^\lambda (x - \eta^2)^{\lambda^2}$$

$$\left(\frac{q-1}{3} \text{ integer}; \quad \eta = e^{\frac{2\pi i}{3}}; \quad \lambda^3 \equiv 1 \text{ mod. } q\right).$$

The group is of order $3q$ if $q > 7$, of order 168 if $q = 7$ (curve reducible to Klein's quartic). The Jacobi variety is impure with

$$1 + h = 2(1 + k) = 6p = 3(q - 1).$$

(III) $$y^q = \frac{(x - a)}{(ax - 1)}\left(\frac{x + a}{ax + 1}\right)^\alpha.$$

The order of the group is $2q$ if $\alpha^2 \not\equiv 1$, mod. $q$, $4q$ in the opposite case. The Jacobi variety is impure with

$$1 + h = 2(q - 1), \qquad 1 + k = 3/2(q - 1).$$

# III, IV

# III

## INTERSECTIONS AND TRANSFORMATIONS OF COMPLEXES AND MANIFOLDS*

### Introduction

In writing this paper my first objective has been to prove certain formulas on fixed points and coincidences of continuous transformations of manifolds. To this proof for orientable manifolds without boundary is devoted most of the second part, the remainder of which is taken up by a study of product complexes in the sense of E. Steinitz, as they are the foundation on which the proof rests. With suitable restrictions the formulas derived are susceptible of extension to a wider range of manifolds, but this will be reserved for a later occasion. It may be stated that our formulas include and completely generalize the early results due to Brouwer and whatever has been obtained since along the same line.† No such generality would have been possible without that powerful instrument, the product complex.

The principle of the method is best explained by means of a very simple example. Let $f(x)$ and $\varphi(x)$ be continuous and uni-valued functions over the interval 0, 1, and let their values on the interval also lie between 0 and 1. It is required to find the number of solutions of $f(x)=\varphi(x)$, $0\leqq x\leqq 1$.

Graphically the problem is solved by plotting the curvilinear arcs

$$y=f(x), \quad y=\varphi(x), \quad 0\leqq x\leqq 1,$$

and taking their intersections. A slight modification of the functions may change the number of solutions, even make them become infinite in number. However, the difference between the numbers of *positive* and *negative* crossings of sufficiently close polygonal approximations to the arcs is a fixed number, their Kronecker index. Its determination is then a partial answer to the question, and indeed seemingly the only possible general answer.

---

* Presented to the Society under somewhat different title at the Chicago Meeting of April 13, 1923, and the Southwestern Section Meeting of December 1, 1923; received by the editors in November, 1924.

† A good bibliography is found in Kérékjárto's recently published volume *Topologie*, Berlin, J. Springer. For a list of the most recent titles see a paper by J. W. Alexander, these Transactions, vol. 25 (1923), p. 173, to which must be added my notes in the Proceedings of the National Academy of Sciences, vol. 9 (1923), p. 90, vol. 11 (1925), pp. 287, 290, summarizing the results of the present paper.

The two complexes whose product is taken in this case are the unit segments on the $x$ and $y$ axes, their product being the square whose sides they are. Replace the unit segments by two identical manifolds of $n$ dimensions, $M_n$ and $M_n'$, the square by the $M_{2n}$ image of their pairs of points (product of the two), the arcs by manifolds on $M_{2n}$ and the exact situation of Part II is obtained.*

In all questions of the above type, the Kronecker index plays then an essential part. In order to put everything on a solid basis, it seemed vital to discuss thoroughly this index. There is in existence an excellent treatment of it by Hadamard,† leaving little to be desired for euclidean spaces, but distinctly insufficient for general manifolds. Then the Kronecker index is only a special topic in the more interesting and far reaching theory of the intersection of complexes on a manifold,‡ needed in any case, to some slight extent, for a good treatment of the index itself. To this theory is devoted most of Part I, of which the chief result is as follows: given several complexes on an orientable manifold $M_n$, which do not intersect on each other's boundaries nor on that of $M_n$, there exists a well defined cycle of $M_n$, their *intersection*. It is well defined in this sense: no matter how the complexes are approximated by means of straight complexes, the cycle intersection of the latter remains homologous to itself. If the approximating complexes intersect in isolated points there is a definite Kronecker index independent of the mode of approximation.

The independence from covering complexes and related modes of defining straightness has presented some of our most serious difficulties. It is a little surprising that the necessity of freeing the Kronecker index from this vitiating circumstance has never been considered in the literature. That the wider problem has not been attacked is natural enough since intersections of general complexes have been studied but very little if at all.§

* This concept appeared first, applied to the special case of algebraic correspondences, in Severi's paper in the Torino Memorie, vol. 54 (1904). Needless to say, Severi did not suspect the analysis situs aspect of the problem, hence did not and could not derive the Hurwitz coincidence formulas. See in this connection Enriques and Chisini, *Lezioni sulla Teoria Geometrica delle Equazioni*, vol. 3, p. 427, also Chisini, Istituto Lombardo Rendiconti, ser. 2, vol. 7 (1924), p. 481. Their work is anticipated by my first Note, which seems to have escaped their notice.

† Note to Tannery's *Introduction à la Théorie des Fonctions*. See also the first chapter of my recent Borel Series Monograph, *L'Analysis Situs et la Géométrie Algébrique*, and my paper in the 1921 Transactions which both contain important applications of the index to algebraic geometry, and finally a very interesting paper by Veblen that has just appeared in these Transactions, vol. 25 (1923); results are recalled in Part I, §7 of this paper and derived anew in Part II.

‡ Considered and actually applied, I believe, for the first time in my Monograph.

§ Some very important results along that line have been obtained of late by J. W. Alexander. See Proceedings of the National Academy of Sciences, vol. 10 (1924), pp. 99, 101, 493.

## PART I. QUESTIONS OF INTERSECTION

### §1. PRELIMINARIES

1. In notation and terminology, we shall follow essentially Veblen's *Colloquium Lectures on Analysis Situs*. We shall assume the reader fairly familiar with this fundamental work and briefly refer to it as "Coll. Lect." The following designations will recur with special frequency in our paper: $S_n$ = euclidean $n$-space; $E_n = n$-cell; $C_n = n$-dimensional complex; $M_n$ = manifold of $n$ dimensions; $\Gamma_n$ (also $\gamma_n$, $\delta_n$ in Part II) $= n$-cycle.* The various numerical invariants, the signs $\sim$, $\equiv$, for congruence or homology, and also the definition of orientation are as in Chapter I of my Borel Series Monograph, *L'Analysis Situs et la Géométrie Algébrique*. In Part II we shall introduce the sign $\approx$ for homologies with division allowed, that is with zero-divisors neglected.

2. Our ordinary complexes shall be restricted to Veblen's *regular* type. Such a $C_n$ is the homeomorph of an $n$-dimensional polyhedron $\Pi_n$ whose faces are all *simplicial* cells (interiors of simplexes) no two intersecting. The cells of $\Pi_n$ define those of $C_n$. We shall apply the term rectilinear segment, polygonal or polyhedral configuration, etc., to $C_n$, as if it were $\Pi_n$ itself, meaning thereby the images of the $\Pi_n$ configurations. Distances on $C_n$ shall also be measured by reference to $\Pi_n$, which for the purpose is assumed immersed in some $S_n'$, $n' \leqq n$. By a *subcomplex* of $C_n$ we shall mean one made up with cells of $C_n$.

$C_n$ may be subdivided into new complexes, and this can be carried out so that the cells of the new complex be of diameter $<\epsilon$, assigned. Of importance in this connection is the method of *regular* subdivision (Coll. Lect., p. 89).

3. We shall define manifolds in accordance with a suggestion due to Veblen (Coll. Lect., p. 92). It amounts essentially to demanding of a $C_n$ defining an $M_n$ that its cells be grouped about any particular one much as if they were all immersed in an $S_n$. It will be found worth while to examine the matter at closer range.

Let $E_k$, $k<n$, be any simplicial cell on $C_n$, $E_{k+1}$ another incident with it (i. e. with $E_k$ on its boundary). We define a set $\{e\}$ of elements such that (*a*) to every $E_{k+1}$ corresponds one and only one $e$; (*b*) to $e$ corresponds together with a given $E_{k+1}$ all others having such a cell in common with it; (*c*) two elements of the set are said to tend towards one another if and

---

* Or set of oriented $n$-circuits in Veblen's terminology.

only if there are corresponding cells whose vertices opposite to $E_k$ tend towards one another. This last condition gives a definition of continuity for $\{e\}$.

Now $C_n$ is said to define an $M_n$ *without boundary* if for every $E_k$, $0 \leq k \leq n-1$, the set $\{e\}$ is homeomorphic to the boundary of an $(n-k)$-cell. The $M_n$ *with boundary* is then defined as in Coll. Lect., p. 88.

To verify the manifold condition we proceed much as in Coll. Lect., pp. 88, 90. We shall show in §3 (No. 14, Lemma II) that there exists a subdivision $C_n'$ of $C_n$ of which $E_k = A_0 A_1 \cdots A_k$ is a cell, the $A$'s being its vertices. (Incidentally this method, now quite customary, of naming a cell by its vertices will prove very convenient.) Let $E_h = A_0 A_1 \cdots A_h$ be any cell of $C_n'$ incident with $E_k$. Then $A_{k+1} \cdots A_h$ is also a cell of $C_n'$, and the totality of such cells gives rise to a $C_{n-k-1}$ homeomorphic to $\{e\}$. Hence the manifold condition is equivalent to demanding that all these complexes be homeomorphic to cell boundaries. As the complexes for a given $E_k$ are all homeomorphic the verification is really independent of the particular $C_n'$ chosen.

It is as yet unknown whether for a given $M_n$ the manifold condition is verified simultaneously for all defining complexes. We shall therefore agree to consider only defining complexes such that the cells of $M_n$ for which the condition is verified have for logical sum one and the same point set.

4. The *orientation* of a simplicial cell is best defined by the order of naming the vertices (Monograph, p. 13). The oriented $C_n$ is a complex as previously understood plus an assigned orientation for each $n$-cell. With the orientation of $E_n$ there is attached one for its boundary $(n-1)$-cells. $M_n$ is called *orientable* if its defining $C_n$ can be so oriented that every non-bounding $E_{n-1}$ receives opposite orientations from its two adjacent $E_n$'s. This property is independent of the particular defining $C_n$. For match the complex with a copy of itself so that corresponding boundary points coincide. There will result a set of $n$-circuits, orientable or not at the same time as $C_n$ itself. From the known independence for the circuit (Coll. Lect., pp. 100-102) follows that for $M_n$.

5. It is frequently convenient to orient $M_n$ by means of a special $n$-cell used as indicatrix, thus: on an $E_n = A_0 A_1 \cdots A_n$ of the defining complex we choose another $E_n' = B_0 B_1 \cdots B_n$ reducible to the first by an affine transformation of the common $S_n$ with coincidence of vertices in the order named, and instead of assigning the order of the $A$'s, we do it for the $B$'s.

Let $x_1, x_2, \cdots, x_n$ be cartesian coördinates on $S_n$, the origin being $B_0$. Then $E_n'$ is completely defined if we give ourselves the matrix $H'$ of the coördinates of the $B$'s, it being understood that the $i$th row corresponds to $B_i$. Let $E_n''$ be another indicatrix with the same first vertex $B_0$, and let $H''$ be the corresponding matrix. Then $E_n''$ is the indicatrix of $+M_n$ or $-M_n$ according as the signs of the determinants of $H'$, $H''$ (certainly $\neq 0$) are or are not the same.

**Remark.** Whenever we derive from $C_n$ a new complex $C_n'$, subdivision of the first, we shall agree to orient each $n$-cell of $C_n'$ so that it constitutes an indicatrix of the $n$-cell of $C_n$ that carries it.

6. With Veblen we shall call *singular* $k$-cell on $C_n$ a point set $\bar{E}_k$, of $C_n$, uniform and continuous image of an ordinary cell $E_k$ which we may as well assume simplicial. Any statement concerning $\bar{E}_k$, in particular regarding its orientation or boundary cells, is to be interpreted by reference to $E_k$.

Let $E_k^1, \cdots, E_k^p$ be sensed cells on $C_n$, and let $E_{k-1}^1, \cdots, E_{k-1}^q$ be their bounding $(k-1)$-cells, all cells being possibly (but not necessarily) singular. We shall extend the term "$k$-complex on $C_n$" to cover a symbol

$$C_k = \sum x_i E_k^i$$

where the $x$'s are arbitrary integers. The points of $C_k$ are those of the $E_k$'s whose $x$ coefficient is not zero plus their limit points. To the cells correspond Poincaré congruences

$$E_k^i \equiv \sum y_{ij} E_{k-1}^j \ ,$$

and for $C_k$ by definition

$$C_k \equiv \sum x_i y_{ij} E_{k-1}^j \ .$$

The $C_{k-1}$ at the right is the *boundary* of $C_k$. If it reduces to zero, $C_k$ is a *k-cycle*. The fact that $C_{k-1}$ is a boundary is also expressed by the homology

$$C_{k-1} \sim 0 \qquad (\text{mod } C_n) \ .$$

**Remark.** The boundary (sensed) of an $n$-cell is a cycle; the verification is immediate. Hence, by summation the boundary of a $C_n$ is also a cycle.

## §2. Intersection of cells

7. Let $E_h$, $E_k$, $E_n$ be simplicial cells, the first two on the third. $E_h$ and $E_k$ may intersect in various ways. Assume that in the $S_n$ of $E_n$ the spaces $S_h$ and $S_k$ which carry the other cells are linearly independent, so that their intersection is an $S_l$, $l = h + k - n$. Grant furthermore that $l \geqq 0$, so that $S_l$ is an actual space (possibly a point) and also that the cells themselves intersect. This intersection will consist of an element of $S_l$, bounded by a convex polyhedron, and therefore constitutes an $l$-cell $E_l$. If we assume the linear spaces of the boundaries of $E_h$ and $E_k$ also as independent as possible

from those of the cells themselves, no bounding $(l-1)$-cell of $E_l$ (necessarily polyhedral like $E_l$ itself) will be an $(l-1)$-cell of the boundary of $E_h$ or $E_k$, but will be merely on one of their bounding $(h-1)$-cells or $(k-1)$-cells respectively, and this we shall assume for the present. $E_l$ with its bounding polyhedron constitutes an $M_l$ decomposable into simplexes, for example by the regular subdivision process. To this $M_l$ we now propose to assign an orientation corresponding to given orientations of the other cells as follows. Let $E_n' = A_0 A_1 \cdots A_n$ be a small simplicial cell such that $E_l' = A_0 A_1 \cdots A_l$ lies on $E_l$, $E_h' = A_0 A_1 \cdots A_h$ on $E_h$, and $E_k' = A_0 A_1 \cdots A_l A_{h+1} \cdots A_n$ on $E_k$. Let $\alpha_s E_s'$, $\alpha_s = \pm 1$, be the indicatrix of $E_s$ $(s = h, k, l, n)$. Then $\alpha_l$ is to be determined by the relation

$$\alpha_h \cdot \alpha_k \cdot \alpha_l \cdot \alpha_n = +1 \, .$$

The cell so sensed shall be designated by $E_h \cdot E_k$.*

The relation between the $\alpha$'s shows that if one of the cells $E_h$, $E_k$, $E_n$ is inverted so is $E_l$. Furthermore if $E_h$ and $E_k$ are permuted the only indicatrix changed is $E_n'$, whose vertices undergo $(n-h)(h-l) = (n-h)(n-k)$ transpositions. Hence

$$E_h \cdot E_k = (-1)^{(n-h).(n-k)} E_k \cdot E_h = -(-E_h) \cdot E_k = -E_h \cdot (-E_k) \, .$$

8. We have tacitly assumed throughout that $l > 0$. With suitable conventions we may let it take any value whatever. The case $l < 0$ may be dismissed at once; we simply write, then,

$$E_h \cdot E_k = 0.$$

Let now $l = 0$. Then there is a unique point of intersection constituting a zero-cell $E_0$. It is this point with the value of $\alpha_0$ attached which we designate by $E_h \cdot E_{n-h}$. The value of $\alpha_0$ is called the *Kronecker index* or simply index of $E_h$ and $E_{n-h}$ and denoted by $(E_h \cdot E_{n-h})$. The above symbolic relation still holds, and we derive from it and our discussion

$$(E_h \cdot E_{n-h}) = (-1)^{(n+1)h} (E_{n-h} \cdot E_h) = -(-E_h \cdot E_{n-h}) = -(E_h \cdot -E_{n-h}) \, ,$$

which may also be obtained directly by means of the indicatrices.

To make our conventions complete, when the cells do not intersect, we shall write

$$E_h \cdot E_k = 0 \, , \qquad (E_h \cdot E_{n-h}) = 0 \, .$$

The case $h = 0$, $k = n$, is not exceptional. We then have a point $E_0 = A$ and an attached unit $\alpha_0$ in place of $\alpha_h$. The point is on the "intersecting"

* In previous papers the same notation was used, sometimes with, sometimes without the "dot." In this paper it has been essential to use the dot throughout, for in Part II another "product" symbol comes in, whose meaning is wholly different and which will be written as a "cross" product.

cell $\bar{E}_n$ (in place of $E_k$) and $\alpha_k$ is replaced by $\bar{\alpha}_n$ whose value is $+1$ if $E_n$ is sensed like $E_n$, $-1$ otherwise. The index is now

$$(E_0 \cdot E_n') = \alpha_0 = \alpha_n \cdot \bar{\alpha}_n \cdot \bar{\alpha}_0 .$$

Its sign is that of $\bar{\alpha}_0$ if $E_n$ and $\bar{E}_n$ are sensed alike, its opposite otherwise.

9. Our next task is to determine the boundary congruences. We first assume that $E_h \cdot E_k$ *has no boundary* $(l-1)$*-cell on the boundary of* $E_n$. This is indeed the general case, but the exception here referred to is of importance later. Let the boundary congruences for $E_h$ and $E_k$ be

$$E_h \equiv \sum E_{h-1}^i , \qquad E_k \equiv \sum E_{k-1}^j .$$

The boundary of $E_h \cdot E_k$ is then the sum of the cells $E_h \cdot E_{k-1}^j$, $E_{h-1}^i \cdot E_k$, affected with signs that are to be determined.

Let for example $E_h$ actually intersect $E_{k-1}^j$ and choose $E_n'$ with the vertices of

$$E_{k-1}' = A_0 A_1 \cdot \cdot \cdot A_{l-1} A_{h+1} \cdot \cdot \cdot A_n \text{ on } E_{k-1}^j .$$

As $A_l$ must be transposed $l$ times to come to first place, $(-1)^l \cdot \alpha_k \cdot E_{k-1}'$ is an indicatrix of $E_{k-1}^j$. Hence, $\alpha_{k-1}$ corresponding to $E_{k-1}^j$ as $\alpha_k$ to $E_k$, we have

$$\alpha_{k-1} = (-1)^l \alpha_k .$$

Therefore

$$\alpha_{l-1} = (-1)^l \cdot \alpha_l$$

corresponds to $E_h \cdot E_{k-1}$ as $\alpha_l$ to $E_h \cdot E_k$ itself, so that $E_h \cdot E_{k-1}$ has for indicatrix $(-1)^l \cdot \alpha_l \cdot A_0 A_1 \cdot \cdot \cdot A_{l-1}$, which is the precise indicatrix that it should have as a boundary cell of $E_h \cdot E_k$, since $A_l$ must be transposed $l$ times to be brought to first place, and since $\alpha_l A_0 A_1 \cdot \cdot \cdot A_l$ is the indicatrix of $E_h \cdot E_k$.

We conclude then that in the boundary congruence for $E_h \cdot E_k$ we must affect $E_h \cdot E_{k-1}$ with the sign $+$. Similarly $E_k \cdot E_{h-1}$ must be affected with the sign $+$ in the congruence for $E_k \cdot E_h$, hence $(-1)^{(n-h+1)(n-k)} E_{h-1}^i \cdot E_k$ with the sign $+$ in the congruence for $(-1)^{(n-h)(n-k)} E_h \cdot E_k$, from which at once

$$E_h \cdot E_k \equiv (-1)^{(n-k)} \sum E_{h-1}^i \cdot E_k + \sum E_h \cdot E_{k-1}^j .$$

In the exceptional case at first excluded, $E_l$ will have some bounding $E_{l-1}$'s on some bounding $E_{n-1}$ of $E_n$. This will be due to the fact that for example $E_{h-1}^p$ and $E_{k-1}^q$ will both lie in $E_{n-1}$. It is found by considering now intersections in $E_{n-1}$ that $E^{p}_{h-1} \cdot E^{q}_{k-1}$ is positively related to $E_h \cdot E_k$. The method is the same as above: we merely assume in our indicatrix $A_l$ exterior to $E_{n-1}$ and the rest goes through about as before. We shall then have the general congruence

$$(9.1)\quad E_h \cdot E_k \equiv (-1)^{n-k} \sum E^{i}_{h-1} \cdot E_k + \sum E_h \cdot E^{j}_{k-1} + \sum E^{p}_{h-1} \cdot E^{q}_{k-1} \, .$$

We have here a case where it would be distinctly worth while to have a more complicated notation to indicate in which complex intersections are taken. Such instances are comparatively rare and the doubt will always readily be cleared up by reference to the context.

10. **Fundamental theorem on Kronecker indices.** Just as before, the case where $l=1$, and the cells at the right are points (zero-cells) offers no exception. $E_h \cdot E_k$ is then a rectilinear one-cell $E_1$ and the three sums at the right reduce to two terms corresponding to the initial and terminal points of $E_1$, the sensed intersection. To each of these terms corresponds a Kronecker index, computed either as to $E_n$ or as to one of its bounding $(n-1)$-cells. I say, and this is our theorem, that *in all cases the sum of these two-indices is zero* so that they are units of opposite signs.

11. Let then

$$E_1 = E_h \cdot E_{n-h+1} \equiv (-1)^{h-1} E'_{h-1} \cdot E_{n-h+1} + \cdots ,$$

where we have not written the term that we do not wish to discuss. Assume first that the term written represents the initial point $A_0$ of $E_1$ and take it for vertex of same name of the indicatrix previously considered. Here then $\alpha_l = \alpha_1 = +1$. The situation is as follows:

| | | |
|---|---|---|
| $A_0 A_2 \cdots A_h$ | is indicatrix for | $-\alpha_h \cdot E'_{h-1}$ , |
| $A_0 A_1 A_{h+1} \cdots A_n$ | " " " | $\alpha_{n-h+1} \cdot E_{n-h+1}$ , |
| $A_0 A_2 \cdots A_h A_1 A_{h+1} \cdots A_n$ | " " " | $(-1)^{h-1} \cdot \alpha_n E_n$ . |

The Kronecker index for the point $(-1)^{h-1} \cdot E'_{h-1} \cdot E_{n-h+1}$ is then the number $\beta$ defined by the condition

$$(-1)^{h} \cdot \alpha_h \cdot \alpha_{n-h+1} \cdot (-1)^{h-1} \cdot \alpha_n \cdot \beta = 1 \, .$$

We have also

$$\alpha_h \cdot \alpha_{n-h+1} \cdot \alpha_n \cdot \alpha_1 = \alpha_h \cdot \alpha_{n-h+1} \cdot \alpha_n = 1 \, .$$

Hence finally $\beta = -1$. If $A_0$ were the terminal point of $E_1$, we would have merely $\alpha_1 = -1$, the rest being the same, hence $\beta = +1$. Thus we see that the Kronecker indices for the end points have the same signs that the points receive in the boundary congruence for $E_1$. The other two cases (where $A_0$ is $E_h \cdot E_{n-h}$ or $E_{h-1} \cdot E_{n-h}$ and on an $E_{n-1}$) lead to exactly the same conclusion; the proofs, essentially similar, are omitted here. The sum of the indices is therefore always zero as was to be proved.

12. The extension to the intersection of $s$ cells $E_i$, $E_h$, $\cdots$, $E_k$ on $E_n$, goes through with ease. It is denoted by $E_i \cdot E_h \cdots E_k$, a symbol which obeys the associative but in general not the commutative law. A similar remark holds for the index $(E_i \cdot E_h \cdots E_k)$, which exists only when $i+h+\cdots+k=n(s-1)$. The boundary congruences can be written down at once.

## §3. Intersections of polyhedral complexes and their Kronecker indices

13. The complexes which are to occupy us in the rest of Part I shall all be immersed in a connected, orientable and oriented manifold $M_n$ with an assigned defining complex $C_n$. We shall assume throughout that intersecting complexes have non-intersecting boundaries and no common points on the boundary $C_{n-1}$ of $M_n$. Of several intersecting complexes so restricted let one, say $C_h$, have points on $C_{n-1}$. We may subdivide $C_h$ into $C_h'$ with cells so small that those $h$-cells which have points of $C_{n-1}$, or whose boundary has some, carry no points of the intersecting complexes on themselves or on their boundary. Let $\overline{C}_h$ be the complex sum of these cells plus their boundaries. As far as the intersection with the other complexes is concerned, $C_h'$ may be replaced by $C_h'-\overline{C}_h$ which carries no points of $C_{n-1}$. A similar remark applies in case there are points common to some, but not all, the boundaries of the intersecting complexes. Henceforth it shall then be understood once for all that

I. *intersecting complexes have no points on the boundary of $M_n$;*

II. *their boundaries do not meet.*

Our general plan is as follows. We shall first define the intersection of a still narrower class of polyhedral complexes, and then approximate general complexes by means of these. But before defining our special polyhedral complexes, we must prove two lemmas.

14. Lemma I. *Any polyhedral $C_h$ is a sum of simplicial cells.*

Each $h$-cell of $C_h$ is a sum of a finite number of polyhedral regions of a certain $S_h$. Each region is decomposable into a sum of convex polyhedral $h$-cells.* Remove these from $C_h$ and let $C_{h-1}$ be the remaining complex. The lemma is true for $h=1$. Grant it for the dimensionality $h-1$; $C_{h-1}$ can then be decomposed into a sum of simplicial cells. Select a point on each convex $h$-cell and join it by rectilinear segments to the simplicial

* This has been proved by Veblen and others. For detailed references see Coll. Lect., p. 83.

cells of $C_{h-1}$ on the boundary of its $h$-cell. There will follow the requisite decomposition of $C_h$.

**Remark.** A region of the initial decomposition of $C_h$ may lie on several cells of $C_n$. If so the boundaries of the latter decompose it into regions each of which lies on a unique cell. Then $C_h$ will appear as a sum of simplicial cells, each also on a unique cell of $C_n$. Whenever we shall consider in the sequel a polyhedral $C_h$, arising in some manner in the course of the discussion, we shall assume that it *has been decomposed into a sum of simplicial cells each on a unique cell of* $C_n$. Strictly speaking, the initial complex is thus replaced by a subdivision and should be designated by a new notation, but it will simplify matters a good deal to avoid this.

LEMMA II. *There exists a subdivision* $C_n'$ *of* $C_n$ *with* $C_h$ *as a sub-complex.*

Decompose $C_h$ as just stated into a sum of simplicial cells, any one, say $E_k$, on an $E_n$ of $C_n$ or on its boundary. The $S_k$ of $E_k$ is the intersection of certain $S_{n-1}$'s of the $S_n$ of $E_n$. Extend these $S_{n-1}$'s as far as possible on the simplex of their $E_n$. There will result a decomposition of $C_n$ into a new complex $\bar{C}_n$ with $C_h$ as a subcomplex. Apply now Lemma I to $\bar{C}_n$ and $C_n'$ follows.

15. We now seek to define the intersection of two polyhedral complexes $C_h$, $C_k$ and its boundary congruences. Let

$$C_h = \sum E_h^i \, ; \qquad C_k = \sum E_k^j \, ;$$

$$E_h^i \equiv \sum E_{h-1}^{ip} \, ; \qquad E_k^j \equiv \sum E_{k-1}^{jq} \, .$$

We impose the following restrictive conditions:

(*a*) *Intersecting h- and k-cells are on one and the same n-cell of* $C_n$ *and in general position as understood in* §2. *Their intersection is then an l-cell, where, as before,* $l = h + k - n$.

(*b*) *Let* $E_{h-1}^{ip}$ *intersect* $E_{k-1}^{jq}$ *on an* $E_{l-1}$. *Then both are on an* $E_{n-1}$ *of* $C_n$ *and* $E_{l-1}$ *is not on the boundaries of* $C_h$ *and* $C_k$.

When these two conditions are satisfied, the intersection, to be denoted by $C_h \cdot C_k$, is a $C_l$ defined by the relation

$$C_h \cdot C_k = \sum E_h^i \cdot E_k^j \, ,$$

it being understood that, whenever $E_h^i$ and $E_k^j$ do not intersect, $E_h^i \, . \, E_k^j = 0$.

The symbols $C_h \cdot C_k$ obeys the distributive law, as follows at once from the definition. Thus:

$$(C_h' + C_h'') \cdot C_k = C_h' \cdot C_k + C_h'' \cdot C_k \, ,$$

it being granted of course that both $C_h'$ and $C_h''$ satisfy conditions (*a*) and (*b*) as to $C_k$. Similarly for $C_k$ while the effect of permuting the $C'$'s or inverting them is as in No. 7 for the cells.

The boundary congruences will present no great difficulty. From (9.1) follows

$$(15.1) \qquad C_h \cdot C_k \equiv (-1)^{n-k} \sum E_{h-1}^{ip} \cdot E_k^j + \sum E_h^i \cdot E_{k-1}^{jq} + \sum E_{h-1}^{ir} \cdot E_{k-1}^{js},$$

the meaning of each sum being readily apprehended by reference to (9.1). I say that the terms in the third sum cancel each other. Indeed let $E_h^i$ and $E_k^j$ be on the cell $E_n$ and give rise to the term $E_{h-1}^{ir} \cdot E_{k-1}^{js}$, intersection of $E_{h-1}^{ir}$, $E_{k-1}^{js}$, situated in a bounding cell $E_{n-1}$ of $E_n$. We assume the cells $E_{h-1}^{ir}$, $E_{k-1}^{js}$, $E_{n-1}$ positively related to $E_h^i$, $E_k^j$, $E_n$, and $E_{h-1}^{ir}$, $E_{k-1}^{js}$ is the intersection of the two sensed cells, oriented as indicated in §2.

According to (*b*) there exist $E_n'$, $E_h^{i'}$, $E_k^{j'}$ of $C_n$, $C_h$, $C_k$, with $E_{n-1}$, $E_{h-1}^{ir}$, $E_{k-1}^{js}$ on their boundaries and negatively related to them. There is a cell labelled $E_{h-1}^{i'r'} = -E_{h-1}^{ir}$ and one labelled $E_{k-1}^{j's'} = -E_{k-1}^{js}$. Indeed according to (*b*) the cells of $C_h$ adjacent to $E_{h-1}^{ir}$ can be grouped in pairs oppositely related, and we may assume that we have such a pair in $E_h^i$, $E_h^{i'}$. Similarly for $E_k^j$ and $E_k^{j'}$.

There are now several possibilities. It may be that $E_{n-1}$ separates $E_h^i$ and $E_h^{i'}$ (that is, one of them is on $E_n$, the other on $E_n'$) but not $E_k^j$ and $E_k^{j'}$. Then the third sum in (15.1) contains these terms pertaining to the couples considered above and no others:

$$E_{h-1}^{ir} \cdot E_{k-1}^{js} + E_{h-1}^{ir} \cdot E_{k-1}^{j's'}.$$

They represent intersections on $E_{n-1}$ and as $E_{k-1}^{j's'} = -E_{k-1}^{js}$ their sum is zero.

A second possibility is that $E_{n-1}$ separates the two pairs of $h$- and $k$-cells. Then in the sum in question there correspond the terms

$$E_{h-1}^{ir} \cdot E_{k-1}^{js} + E_{h-1}^{i'r'} \cdot E_{k-1}^{j's'}.$$

The first intersection is taken on $E_{n-1}$, the second on $-E_{n-1}$ (that is in the scheme of §2, $E_{n-1}$ must now be replaced by $-E_{n-1}$, the reason being that boundaries of cells on $E_n'$ are now involved and $E_n'$ is negatively related to $E_{n-1}$). When intersections are referred to $E_{n-1}$, the second term must be written

$$-E_{h-1}^{i'r'} \cdot E_{k-1}^{j's'} = -(-E_{h-1}^{ir}) \cdot (-E_{k-1}^{js}) = -E_{h-1}^{ir} \cdot E_{k-1}^{js}$$

and the sum is again zero.

Finally we must consider the case where $E_{n-1}$ separates none of the two pairs of cells. Then the terms to be considered are now

$$\overset{ir}{E}_{h-1} \cdot \overset{js}{E}_{k-1} + \overset{ir}{E}_{h-1} \cdot \overset{j's'}{E}_{k-1} + \overset{i'r'}{E}_{h-1} \cdot \overset{js}{E}_{k-1} + \overset{i'r'}{E}_{h-1} \cdot \overset{j's'}{E}_{k-1} ,$$

the intersections being all referred to $E_{n-1}$. It is immediately verified that the fourth term is the same as the first, and the other two terms its negative. The sum is then again zero. This completes the proof of our assertion.

16. The boundary of $C_h$ is a cycle $\Gamma_{h-1}$, and that of $C_k$ is a cycle $\Gamma_{k-1}$. We have

$$C_h \equiv \Gamma_{h-1}, \quad C_k \equiv \Gamma_{k-1} .$$

The first two sums in (15.1) are respectively $\Gamma_{h-1} \cdot C_k$ and $C_h \cdot \Gamma_{k-1}$. Hence in the last analysis we have this fundamental congruence:

$$C_h \cdot C_k \equiv (-1)^{n-k} \, \Gamma_{h-1} \cdot C_k + C_h \cdot \Gamma_{k-1} . \tag{16.1}$$

17. A series of important corollaries follows at once from the preceding discussion.

I. A complex $C_h$ is called a *generalized manifold* if every non-bounding $E_{h-1}$ of it is incident with just two $h$-cells. Orientability is defined for it as for an ordinary manifold. From our discussion we obtain the following: *if $C_h$ and $C_k$ are orientable generalized manifolds so is $C_h \cdot C_k$.*

II. *If one of the complexes is a cycle the boundary of $C_h \cdot C_k$ is the intersection of this cycle with the other complex or its opposite. If both are cycles so is their intersection.* In symbols,

$$C_h \cdot \Gamma_k = (-1)^{n-k} \, \Gamma_{h-1} \cdot \Gamma_k ; \quad \Gamma_h \cdot C_k \equiv \Gamma_h \cdot \Gamma_{k-1} ; \quad \Gamma_h \cdot \Gamma_k \equiv 0 . \tag{17.1}$$

III. *When the boundary of each complex does not meet the other, $C_h \cdot C_k$ is a cycle.*

IV. *Let $\Gamma_h$ bound $C_{h+1}$ satisfying our restrictive conditions as to its intersection with $C_k$ and not intersecting the boundary of $C_k$. Then $\Gamma_h \cdot C_k$ is a bounding cycle also.* This can be read off from (17.1).

18. **Kronecker index.** This time the dimensions are $h$ and $n-h$. We make the same assumptions as previously, and, in addition, agree that for two non-intersecting cells the index is zero. Then

$$(C_h \cdot C_{n-h}) = \sum (\overset{i}{E}_h \cdot \overset{j}{E}_{n-h}) .$$

From No. 8 follows the distributive law for indices. The result of permuting the two complexes or of changing the sign of one is as for cells and need not be written down.

From No. 10 and (16.1) it follows that if

$$C_h \equiv \Gamma_{h-1}\,, \quad C_{n-h+1} \equiv \Gamma_{n-h}\,,$$

then here

(18.1) $$(C_h \cdot \Gamma_{n-h}) = (-1)^h \cdot (\Gamma_{h-1} \cdot C_{n-h+1})\,,$$

a formula of great importance later (§6), as it transforms an index corresponding to dimensionalities $h$, $n-h$, to one with $h$ replaced by $h-1$.*

19. From the theorem of No. 10, together with (16.1), now follows

*Let $\Gamma_h$ bound a $C_{h+1}$ not intersecting the boundary of $C_{n-h}$, a condition that disappears if we deal with a $\Gamma_{n-h}$. Let furthermore the usual restrictions as to intersecting complexes, $C_{h+1}$ and $C_{n-h}$ or $\Gamma_{n-h}$, hold. Then*

$$(\Gamma_h \cdot C_{n-h}) = 0\,, \quad (\Gamma_h \cdot \Gamma_{n-h}) = 0\,.$$

Observe that owing to the distributive law, it is not necessary that $\Gamma_h \sim 0$, but merely $\approx 0$, for then $t\Gamma_h$ bounds and the multiples of the indices are zero; hence also the indices themselves. This result will have important applications in Part II.

20. The extension to several intersecting complexes offers no particular difficulties. The symbols follow the associative and distributive laws, but not in general the commutative law.

## §4. Approximation of complexes

21. A first, but somewhat inelastic, approximation to a general complex $C_k$ by a polyhedral $C_k'$ will be obtained by direct application of processes due to Alexander (these Transactions, vol. 16 (1915), p. 148) and Veblen (Coll. Lect., pp. 95, 118). $C_k$ appears then as a subcomplex of a subdivision of $C_n$ with cells of suitably small diameter. There are two associated complexes $C_{k+1}$ and $C_k^0$ such that

(21.1) $$C_{k+1} \equiv C_k - C_k' + C_k^0\,,$$

and therefore

(21.2) $$C_k' \sim C_k + C_k^0\,.$$

$C_k^0$ appears only when $C_k$ is not a cycle. Our $C_{k+1}$ is the same as Veblen's $B_{i+1}$. When $C_k$ is not a cycle, the boundary cells of $C_{k+1}$ which join boundary cells of $C_k$ and $C_k'$ are also part of the boundary of $C_{k+1}$, and their sum is precisely $C_k^0$.

* An analogous formula for cells was given by Veblen in the Transactions paper already quoted. p. 542.

In the light of this and upon examining the construction, we find by the simplest continuity considerations that it may be carried out so that the following statements hold.

(*a*) *$C_k'$ and $C_{k+1}$ are both as near as we please to $C_k$.*

(*b*) *The complex $C_k^0$ and the boundary of $C_k'$ are as near as we please to the boundary of $C_k$.*

(*c*) *$C_k'$ includes any particular polyhedral boundary subcomplex of $C_k$.* This is proved by a very transparent application of Lemma II, No. 14.

22. From the preceding method of approximation we may derive this interesting result: Let the approximation be made by means of cells of $C_n$. *If the points of a cell of $C_k$ are sufficiently near those of $E_h$ of $C_n$, then its approximation is $E_h$ itself or a cell on its boundary. The cell of $C_k$ need only be within a certain distance $\delta$ of $E_h$ in order that this be true.*

From this we have the following

THEOREM. *To every polyhedral complex $C_k$ corresponds a positive number $\delta$ such that every cycle $\Gamma_h$ whose points are all within $\delta$ of $C_k$ is homologous to a cycle $\Gamma_h'$ on $C_k$.*

For $C_k$ is a subcomplex of a subdivision $C_n'$ of $C_n$ (Lemma II, No. 14) and in this case $C_h^0$ corresponds to $C_k^0$ of No. 21, for $\Gamma_h$ is absent. We can also affirm that $\Gamma_h - \Gamma_h'$ will bound, by (*a*), a $C_{h+1}$ whose points are as near as we please to $C_k$ at the same time as those of $\Gamma_h$.

Incidentally, since $C_k$ can have no cycle of more than $k$ dimensions, we have this very interesting result: *A non bounding cycle cannot be homologous to a cycle as near as we please to a complex of smaller dimensionality.*

23. The approximations which we have obtained so far are not flexible enough for our purpose, which demands the approximation of two or more complexes at the same time by others with a well-defined intersection. This will be based upon the all-important

THEOREM. *Let $C_k$ be a subcomplex of $C_n$. By subtraction of bounding $k$-cycles, it may be reduced to another complex with the same boundary, whose $(k-i)$-cells not on the boundary of $M_n$ nor on its own are on cells of at least $n-i$ dimensions of $C_n$.*

Let $C_n'$ be a regular subdivision of $C_n$ and $C_k'$ the corresponding subdivision of $C_k$. Any vertex of $C_n'$ shall be affected with an upper index, such as $A_i^p$ to indicate the dimensionality $p$ of the cell of $C_n$ which carries it.

Let us attach to any cell $E_r = A_0 A_1 \cdots A_r$ of $C_n'$ a symbol $(p_0, p_1, \cdots, p_r)$ to describe its type. Observe that *the $p$'s are all distinct*, for $p_0 = p_1 = p$ would mean that two points on distinct $p$-cells of $C_n$ are joined by a rectilinear segment wholly on a $p$-cell.

Another and more significant property is that *the highest $p$ indicates the dimensionality of the cell of $C_n$ which carries $E_r$.* For then $E_r$ has points on such a cell (in the vicinity of the corresponding $A^p$); hence, due to the mechanism of regular subdivision, it lies entirely on it.

Our theorem will then be proved if we can show that *$C_k$ is reducible to a complex of which every $E_{k-i}$ not on its boundary nor on that of $M_n$ has in its symbol an integer $\geqq n-i$.*

24. (*a*) For $n=1$, the reduction is immediate. Then $k=0$ alone needs to be considered. $C_0$ is a sum of points each with an assigned sign. Any such point $A$, say affected with $+$, may be reduced to any point $B$ of $C_1$ by adding the end points of a polygonal line $AB$. These end points constitute the bounding $\Gamma_0$ of the theorem.

(*b*) Let now $i=0$ and $E_k$ be as yet not reduced. Its symbol is then of type $A_0^{p_0} \cdots A_k^{p_k}$ with all the $p$'s less than $n$. The cell lies therefore on the boundary of an $n$-cell of $C_n$ on which there is a vertex $A_{k+1}^n$. Let $\Gamma_k$ be the boundary of the simplex $A_{k+1}^n A_0^{p_0} \cdots A_k^{p_k}$ which is positively related to $E_k$; $C_k'-\Gamma_k$ is a complex which has the same structure as $C_k$ except that $E_k$ has been replaced by $k+1$ cells of same dimensionality in every one of whose symbols appears $A_{k+1}^n$, so that they are on $n$ cells of $C_n$. This carries out the reduction for $i=0$.

(*c*) Assume that the process goes through for any $M_{n'}$, $n'<n$, and also for all cells of more than $k-i=m$ dimensions of $C_k'$. I say that it goes through for all dimensionalities.

Consider an unreduced $E_m=A_0^{p_0} \cdots A_m^{p_m}$ of $C_k'$, the $p$'s being then all $<n-i=n-k+m$. To any $E_h=A_0^{p_0} \cdots A_m^{p_m} \; A_{m+1}^{p_{m+1}} \cdots A_h^{p_h}$ incident with $E_m$ corresponds $E_{h-m+1}=A_{m+1}^{p_{m+1}} \cdots A_h^{p_h}$ which we sense so that if the first set of $A$'s is, as we shall assume, an indicatrix of $E_h$, the last is one for $E_{h-m-1}$. The totality of these cells is a subcomplex of $C_k$ which is an $M_{n-m-1}$ homeomorphic to the boundary of a cell (No. 3). The incidence relations (boundary congruences) between corresponding cells are formally identical.

Let $q_0, q_1, \cdots, q_{n-m-1}$ be the set of integers in increasing order which together with $p_0, \cdots, p_m$ constitute the set $0, 1, \cdots, n$. The manifold $M_{n-m-1}$ carries two defining complexes. The first $C_{n-m-1}$ has the points $A^{q_0}$ for vertices, the second $C_{n-m-1}'$ the remaining points $A^q$. In fact $C_{n-m-1}'$ is a regular subdivision of $C_{n-m-1}$, its vertices $q_i$ being on $i$-cells of it. To show this it will suffice to examine the relation between those which correspond to $q_0$ and $q_1$. Let, for example, the sequence of the $p$'s and $q$'s in increasing order read $p_0, p_1, q_0, p_2, p_3, q_1, \cdots$, so that $p_0=0$, $p_1=1$, $q_0=2$, $\cdots, q_1=5, \cdots$. Consider now $E_2=A^0 A^1 A^3$ of $C_n'$. It is on a certain three-cell of $C_n$ on which it is incident with exactly two cells $A^0A^1A_1^2A^3$ and

$A^0A^1A_2^2A^3$ of $C_n'$. Hence $E_{m+1}=A^0A^1A^3 \cdots A^5A^{p_4}A^{p_5} \cdots A^{p_m}$ whose symbol is $(p_0, \cdots, p_3, q_1, p_4, \cdots, p_m)$ is incident with the two $(m+2)$-cells obtained by placing first $A_1^2$, then $A_2^2$ between $A^1$ and $A^3$. It follows that $A^3$ is on the one-cell $A_1^2A^3+A^3A_2^2$ of $C_{n-m-1}$. With the $q$'s the statement is that $A^{q_1}$ is on the one-cell $A_1^{q_0}A^{q_1}+A^{q_1}A_2^{p_0}$. The same reasoning applies to the other $q$ vertices.

Observe that since the $p$'s are all $< n-k+m$ the $q$'s must include all integers from $n-k+m$ to $m$. Hence $q_{n-m-1}=n$, $q_{n-m-2}=n-1, \cdots, q_{n-k-1}=n-k+m$.

25. Let now $(p_0, \cdots, p_m; q_0', \cdots q_{k-m-1}')$ be the symbol for any $k$-cell of $C_k'$ incident with $E_m$. The reduction being achieved by assumption for all cells of more than $m$ dimensions, $(p_0, \cdots, p_m; q_i')$, the symbol of an $(m+1)$-cell, must possess an integer $\geqq n-k+m+1$ which can only be $q_i'$. Similarly $(p_0, \cdots, p_m; q_i', q_j')$ must include at least one integer $\geqq n-k+m+2$, and this can only be $q_i'$ or $q_j'$, etc. Finally, then, among the $k-m$ integers $q_i'$ there must be one at least equal to each integer of the sequence $n-k+m+1, n-k+m+2, \cdots, n$. As they are all distinct and $\leqq n$ they constitute that very sequence and our cell has then the symbol $(p_0, \cdots, p_m; n-k+m+1, \cdots, n)$. The symbol of the corresponding cell of $C_{n-m-1}'$ is $(n-k+m+1, \cdots, n)$.

26. Let $\bar{C}_k$ be the subcomplex of $C_k'$ which is the sum of its cells incident with $E_m$. Since the latter is not on the boundary of $C_k'$, on any $E_{k-1}$ of $\bar{C}_k$, there are as many positively related incident $k$-cells as negatively related. Hence the complex of $C_{n-m-1}'$ which corresponds to $\bar{C}_k$ is a cycle $\Gamma_{k-m-1}$. Since $C_{n-m-1}'$ is homeomorphic to the boundary of a cell, $\Gamma_{k-m-1}$ bounds on the complex, and in fact bounds a subcomplex $C_{k-m}$ of $C_{n-m-1}'$ (Coll. Lect., pp. 95, 118). Furthermore since the reduction to be proved applies by assumption to an $M_{n-m-1}$, $C_{k-m}$ may be so reduced without changing its boundary that in the symbol of its $s$-cells not on the boundary $\Gamma$ there appears $q_{n-k+s-1}$ or a higher $q$. But since the cells of $\Gamma$ already satisfy this condition, it holds for all cells of $C_{k-m}$ without exception. From this we conclude immediately that its $(k-m)$-cells have all the same symbol:

$$(q_{n-k-1}, q_{n-k}, \cdots, q_{n-m-1}) \equiv (n-k+m, n-k+m+1, \cdots, n).$$

That it has this last simple form follows from the remark at the end of No. 24.

27. To $C_{k-m}$ corresponds a subcomplex $C_{k+1}$ of $C_n'$ whose boundary is a $\Gamma_k$. The cells of this cycle incident with $E_m$ constitute $\bar{C}_k$ so that $E_m$ is not a cell of $C_k''=C_k'-\Gamma_k$. However, among the new cells of $C_k''$ are found those on $k$-cells of $\Gamma_k$ not incident with $E_m$ and we must examine these.

The symbol for any $(k+1)$-cell of $C_{k+1}$ is $(p_0, \cdots, p_m; n-k+m, \cdots, n)$. The new $k$-cells introduced have then a symbol such as $(p_0, \cdots, p_{s-1},$

$p_{s+1}, \cdots, p_m; n-k+m, \cdots, n)$. In the symbol of any cell of $m+j$ dimensions of its boundary will then appear $n-k+m+j$ or a higher integer. The new cells of $m$ or more dimensions have then the desired behavior. Thus $E_m$ and its incident cells have been replaced by a set of cells which fulfil all requirements. *The proof of our theorem is therefore complete.*

**Remark.** We have incidentally obtained the following interesting proposition: *Every cycle of a complex defining a manifold without boundary is homologous to a cycle on its dual.* More precisely *every $\Gamma_k$ of $C_n$ is homologous to a cycle whose k-cells have all the same symbol* $(n-k, n-k+1, \cdots, n)$.

28. We return to our approximation problem. The reduction of $C_k$ has been obtained by adding the boundaries of $(k+1)$-cells incident with its $k$-cells and belonging to $C_n$. Let us replace $C_n$ by a subdivision $\bar{C}_n$ with cells $<\epsilon$ and $C_n'$ by a regular subdivision of $\bar{C}_n$. If we make the same reduction, we shall merely add to $C_k$ the boundaries of $(k+1)$-cells within a distance of $\epsilon$ from the complex.

Applying this reduction to the approximating complex $C_k'$ we find that we do not thereby disturb (21.1) or (21.2). The complex $C_{k+1}$ is simply increased by cells as near as we please to $C_k'$, hence to the approximated complex.

29. The essential property of $C_k$ is that *the $S_{k-i}$ of any $E_{k-i}$ of the complex has the maximum degree of generality relatively to the space of the cell $C_n$ that carries it.* We mean thereby that, by modifying the complex without changing its cellular structure, $S_{k-i}$ may be brought into coincidence with an arbitrary neighboring $S'_{k-i}$. The weakest case is when $E_{k-i}$ lies on an $E_{n-i}$ of $C_n$ with its vertices on $(n-k)$-cells of the boundary of $E_{n-i}$. Let $A$ be a vertex of $E_{k-i}$ on the cell $E_{n-k}$; $S'_{k-i}$ will intersect the latter at a point $B$ very near $A$. Impress upon $A$ the rectilinear displacement $AB$, and similarly for the other $k-i$ vertices of $E_{k-i}$, leaving the remaining vertices of $C_k'$ unchanged. There results an obvious deformation of $C_k'$, into say $C_k''$, whereby $S_{k-i}$ is brought into coincidence with $S'_{k-i}$ thus proving our assertion.

The displacement of $C_k'$ may be so carried out, and in a continuous way, that every point will describe a rectilinear path. Their locus is a $\bar{C}_{k+1} \equiv C_k'' - C_k'$. By adding this to (21.1) we see that $C_k''$ may take the place of $C_k'$ with $C_{k+1}+\bar{C}_{k+1}$ in place of $C_{k+1}$. If $S'_{k-i}$ is sufficiently near $S_{k-i}$ conditions $(a)$, $(b)$, $(c)$ of No. 21 may still be fulfilled. The reasoning in case $E_{k-i}$ lies on a cell of more than $n-i$ dimensions is the same.

## §5. Intersections of general complexes

30. To arrive at something significant, we must narrow down the problem once more. We replace then the second condition of No. 13 by the somewhat more sweeping

III. *Intersecting complexes must not meet the boundaries of one another.*

To complexes so restricted we shall ascribe definite cycles or Kronecker indices. Indices and cycles are fixed, in the sense that the former remain the same when we vary the approximating complexes or even the $C_n$ by means of which we construct them, while the cycles remain homologous to themselves. We shall at first maintain $C_n$ fixed and merely vary the polyhedral approximations, then examine the effect produced by a change of $C_n$.

31. Starting first with two complexes $C_h$, $C_k$ always restricted as in No. 13, we approximate them as closely as we please by $C_h'$, $C_k'$ constructed as in No. 32 with the system of relations

$$C_{h+1} \equiv C_h - C_h' + \overset{0}{C}_h \,; \qquad C_h' \sim C_h + \overset{0}{C}_h \,; \tag{31.1}$$

$$C_{k+1} \equiv C_k - C_k' + \overset{0}{C}_k \,; \qquad C_k' \sim C_k + \overset{0}{C}_k \,; \tag{31.2}$$

the various complexes have the same meaning as those of similar designation in §4. If the approximation is sufficiently fine, $C_h'$ and $C_k'$ will also fulfil the restrictive conditions I, III of Nos. 13 and 30. This we assume henceforth for all our approximating complexes.

It follows at once from No. 29 that we may so choose $C_h'$ and $C_k'$ that they satisfy the two conditions of No. 14 for a well defined intersection $C_h' \cdot C_k'$. Since the boundary of each complex does not meet the other, the intersection will be an $l$-cycle. It remains to be shown that this cycle is independent of the approximating complexes.

32. As a preliminary step let $\bar{C}_h'$ be another approximation whose intersection with $C_k'$ is well defined. I say that

$$C_h' \cdot C_k' \sim \bar{C}_h' \cdot C_k' \,. \tag{32.1}$$

We have now congruences such as (21.1)

$$\bar{C}_{h+1} \equiv C_h + \overset{0}{\bar{C}}_h - \bar{C}_h' \,, \tag{32.2}$$

$$C_{h+1} \equiv C_h + \overset{0}{C}_h - C_h' \tag{32.3}$$

with $\overset{0}{C}_h$, $\overset{0}{\bar{C}}_h$ very near the boundary of $C_h$ and $C_{h+1}$, $\bar{C}_{h+1}$ very near the complex itself. Their approximation is in fact assumed such throughout that none of these complexes meets the boundary of $C_k'$. We have then from (32.2) and (32.3)

$$\bar{C}_{h+1} - C_{h+1} \equiv C_h' - \bar{C}_h' - (\overset{0}{C}_h - \overset{0}{\bar{C}}_h) \,.$$

To the complex at the left we may apply everything said previously for $C_k$ with the following result. There exists a polyhedral complex $C_{h+1}'$ very near it, having a well defined intersection with $C_k'$ and whose boundary is $C_h' - \bar{C}_h'$ plus a certain $h$-complex very near $C_h^0 - \bar{C}_h^0$ and therefore not meeting $C_k'$. Hence by (16.1)

$$C_{h+1}' \cdot C_k' \equiv (C_h' - \bar{C}_h') \cdot \bar{C}_k' \sim 0 \qquad (\text{mod } M_n),$$

from which (32.1) follows.

33. Let now $\bar{C}_h'$, $\bar{C}_k'$ be two approximations with a well defined intersection. In order to show that the intersections are independent of the particular polyhedral approximations provided they have a well defined intersection, we must prove that

$$(33.1) \qquad \bar{C}_h' \cdot \bar{C}_k' \sim C_h' \cdot C_k' .$$

By a slight displacement such as is used in No. 29, we may replace $C_h'$ by a complex $C_h''$ with a well defined intersection with both $C_k'$ and $\bar{C}_k'$. All that is necessary is to replace throughout $C_k'$ by $C_k' - \bar{C}_k'$. We have then according to the preceding number

$$C_h' \cdot C_k' \sim C_h'' \cdot C_k' ; \quad \bar{C}_h' \cdot \bar{C}_k' \sim C_h'' \cdot \bar{C}_k' ;$$

$$C_h'' \cdot C_k' \sim C_h'' \cdot \bar{C}_k' ,$$

whence (33.1) follows.

Regarding the Kronecker indices, Corollary IV at the end of §3 yields at once for $h+k=n$

$$((C_h' - \bar{C}_h' - (C_h^0 - \bar{C}_h^0)) \cdot C_k') = 0 ,$$

and as $C_k'$ does not intersect $C_h^0 - \bar{C}_h^0$,

$$(C_h' \cdot C_k') = (\bar{C}_h' \cdot C_k')$$

and the rest is as before.

34. The extension to more than two intersecting complexes is easy. With obvious notations we must show that

$$C_h' \cdot C_k' \cdots C_l' \sim \bar{C}_h' \cdot \bar{C}_k' \cdots \bar{C}_l' .$$

Introduce $C_h''$ in general position as to $C_k'$, $\bar{C}_k'$, $\cdots$, $C_l'$, $\bar{C}_l'$. As above it may be shown that in the homology $C_h'$ and $\bar{C}_h'$ may both be replaced by $C_h''$, the process continuing in an obvious way. The treatment for indices is the same.

35. As is natural we denote the cycles and indices defined by means of our approximations as $C_h \cdot C_k \cdots C_l$, or $(C_h \cdot C_k \cdots C_l)$. These symbols have the same properties as those for polyhedral complexes them-

selves. Those pertaining to permutation of complexes hold obviously, others less so. We shall examine them in turn, with particular reference to our future needs.

I. *Associative law.* The scheme of the proof is sufficiently illustrated with three complexes and if we show that

$$C_h \cdot (C_k \cdot C_l) \sim C_h \cdot C_k \cdot C_l \, .$$

By definition $C_k \cdot C_l$ and $C_h \cdot C_k \cdot C_l$ are the polyhedral cycles $C_k' \cdot C_l'$ and $C_h' \cdot C_k' \cdot C_l'$. In the approximation that leads to the determination of $C_h \cdot (C_k' \cdot C_l')$, the cycle in parentheses can be taken as its own approximation. Hence the cycle at the left in the homology is by definition $C_h' \cdot (C_k' \cdot C_l')$ and we are back to the case of polyhedral complexes in general position, for which the law holds.

II. *Distributive law.* We wish to show that, say,

$$C_h \cdot (C_k + \bar{C}_k) \sim C_h \cdot C_k + C_h \cdot \bar{C}_k \, .$$

On examining the two successive approximations of §4 it will be seen that $C_k' + \bar{C}_k'$ is an approximation for $C_k + \bar{C}_k$. If each of the two primed complexes has a well defined intersection with $C_h'$ so has their sum. Hence the left side is by definition $C_h' \cdot (C_k' + \bar{C}_k')$. As the terms at the right are also defined by means of the primed symbols, we are again back to the case of polyhedral complexes where the law holds.

The two preceding proofs hold without modification for the Kronecker index.

III. *If $C_h, C_k, \cdots, C_l$ do not actually have a common point, then*

$$C_h \cdot C_k \cdots C_l \sim 0 \, , \textit{ or } (C_h \cdot C_k \cdots C_l) = 0 \, .$$

For then the primed complexes may be taken without any common point and everything is once more reduced to the known case of polyhedral complexes.

IV. *Let $C_h$ bound $\bar{C}_{h+1}$ such that it is the only one of the set $\bar{C}_{h+1}, C_k, \cdots, C_l$ whose boundary may have points in common with the other complexes. Then*

$$C_h \cdot C_k \cdots C_l \sim 0 \textit{ or } (C_h \cdot C_k \cdots C_l) = 0 \, .$$

From (31.1) follows $\bar{C}_{h+1} - C_{h+1} \equiv C_h'$, for $C_h^0$ is now absent since $C_h$ is a cycle. Moreover $C_h$ being a subcomplex of $\bar{C}_{h+1}$ the sequence $C_h, C_k, \cdots, C_l$ behaves like that of the statement. Let us approximate $\bar{C}_{h+1} - C_{h+1}$ by, say, $C_{h+1}'$, in our usual manner, which may be done without changing $C_h$ since it is on the boundary (property (*c*), No. 21). Since $C_{h+1}$ is very near $C_h'$

it will be seen that the primed sequences corresponding to the two considered above behave as they do. The generalization of (16.1) gives here

$$C'_{h+1} \cdot C_k' \cdots C_l' \equiv C_h' \cdot C_k' \cdot \cdots \cdot C_l'$$

from which IV follows.

V. *Let* $\Gamma_h \approx 0$ (*i. e., some multiple of* $\Gamma_h \sim 0$). *Then*

$$\Gamma_h \cdot \Gamma_k \cdots \Gamma_l \approx 0 \text{ or } (\Gamma_h \cdot \Gamma_k \cdots \Gamma_l) = 0 .$$

This is an immediate corollary of IV. In both IV and V it is of course not at all necessary that the complex or cycle singled out be the first.

VI. *In* $C_h$ *we may suppress any subcomplex not intersecting* $C_k, \cdots, C_l$, *without affecting intersection cycle or index.*

This is an immediate but important corollary of II and III.

36. Before we proceed with a thorough examination of the effect of passing from $C_n$ to a new defining $\bar{C}_n$, let us observe that *instead of approximating to* $C_h$ *by means of* $\bar{C}_n$ *it is sufficient to do this for* $C_h'$. Indeed, let $\bar{C}_h'$ be the approximation to $C_h'$ by means of $\bar{C}_n$. We shall have a congruence such as (21.1):

$$\bar{C}_{h+1} \equiv C_h' - \bar{C}_h' + \bar{C}_h^0 ,$$

with $\bar{C}_{h+1}$ very near $C_h'$ and $\bar{C}_h^0$ very near its boundary. Add this congruence to the first of (31.1) (which is the same as (21.1) with $k$ replaced by $h$):

$$(C_{h+1} + \bar{C}_{h+1}) = C_h - \bar{C}_h' + (C_h^0 + \bar{C}_h^0) .$$

This is analogous to (21.1) with $\bar{C}_h'$ as the approximation to $C_h$. As the congruences such as (21.1) plus the structure of the approximating complexes themselves were alone used in defining the intersection cycles and indices and deriving their properties, our assertion is proved.

## §6. Proof that intersection cycles and indices are invariant when the defining complex is changed*

37. For indices the invariance is conditioned upon a certain simple sense convention. Let $C_n$, $\bar{C}_n$ be any two defining complexes. By practically the same reasoning as Veblen's in *Colloquium Lectures*, pp. 101, 102, we may show that one of the two complexes $C_n \pm \bar{C}_n$ is a bounding cycle, but not both (loc. cit., p. 120). We shall assume in the future that *the complexes are so oriented that* $C_n - \bar{C}_n$ *is the bounding cycle.* Once any particular complex has been assigned an orientation, a definite one follows for the rest.

* A first type of proof is outlined in my second Proceedings note. Just as it appeared in print I discovered the much simpler treatment embodied in this section.

We must examine if the present convention agrees with our previous mode of sensing a subdivision $C_n'$ of $C_n$, where it will be remembered a cell and its subdivisions were always so sensed as to have a common indicatrix. It is clearly sufficient to consider the case where $C_n$ is a simplex with a simplicial subdivision:

$$C_n' = \overset{1}{\Sigma}_n + \cdots + \overset{p}{\Sigma}_n .$$

Let the first $q$ simplexes, but no others, have an $(n-1)$-simplex on a given simplex $\Sigma_{n-1}$ of $\Sigma_n$, namely $\overset{i}{\Sigma}_{n-1}$ for $\overset{i}{\Sigma}_n$, with $A$ and $A^i$ as the vertices of $\Sigma_n$ and $\overset{i}{\Sigma}_n$ not on $\Sigma_{n-1}$ or $\overset{i}{\Sigma}_{n-1}$. We take the orientations such that $A$ followed by the vertices of $\Sigma_{n-1}$ corresponds to the same as $A_i$ followed by those of $\overset{i}{\Sigma}_{n-1}$.

In order that $\Sigma_n - C_n'$ be a cycle, it is necessary and sufficient that the boundaries on $\Sigma_{n-1}$ cancel, or that

$$\Sigma_{n-1} - (\overset{1}{\Sigma}_{n-1} + \cdots + \overset{q}{\Sigma}_{n-1})$$

be a cycle. This reduces the verification from $n$ to $n-1$, hence ultimately to $n=1$ for which it is immediate.

38. Let us return to two arbitrary defining complexes $C_n$, $\bar{C}_n$. Our customary approximations, when applied to $\bar{C}_n$ by means of $C_n$, cannot go farther than what is yielded by the Alexander-Veblen process. In this case it comes down to this: We subdivide $C_n$ and $\bar{C}_n$ into $C_n'$ and $\bar{C}_n'$, then establish a correspondence $T$ whereby to each cell of $\bar{C}_n'$ is assigned a unique one of $C_n'$. The $n$-cells of $C_n'$ are each covered positively by $k_2$ cells corresponding to positive cells of $\bar{C}_n'$ and negatively by $k_1$ such cells. Furthermore it is a property of the correspondence that $C_n' - T(\bar{C}_n')$ bounds, hence $k_2 - k_1$ is fixed for all $n$-cells of $C_n$, else this complex would have boundary cells exterior to the boundary of $M_n$ and would not be a cycle. Moreover

$$T(\bar{C}_n') \sim \bar{C}_n' \sim \bar{C}_n \sim C_n \sim C_n' ;$$

therefore

$$C_n' - T(\bar{C}_n') = (1+k_1-k_2)C_n' \sim 0 .$$

When a subcomplex of a $C_n$ bounds it bounds also such a subcomplex (Coll. Lect. p. 118). But $C_n'$ has no $(n+1)$-cells, hence this homology can only be true if $1+k_1-k_2=0$, $k_2-k_1=1$. *Thus $T(\bar{C}_n')$ covers $C_n'$ exactly once.* The importance of this will appear later.

39. **Invariance of the index.** It will be established by means of (18.1) that

$$(C_h \cdot \Gamma_{n-h}) = (-1)^h \cdot (\Gamma_{h-1} \cdot C_{n-h+1}), \tag{39.1}$$

where the conditions of No. 15 must be satisfied. Here the intersecting complexes are the two extremes, the cycles being their boundaries. We have, then, given $C_h$, $C_{n-h}$, their approximations $C_h'$, $C'_{n-h}$ by means of a first defining complex $C_n$, and $\bar{C}_h'$, $\bar{C}'_{n-h}$ by a second $\bar{C}_n$, the latter being, if we wish, approximations to $C_h'$ and $C'_{n-h}$ rather than to the given complexes (No. 37). We denote the indices as to $C_n$ by the usual round parentheses, those as to $\bar{C}_n$ by square parentheses, and our object is to show that

$$(C_h' \cdot C'_{n-h}) = [C_h' \cdot C'_{n-h}] \ . \tag{39.2}$$

Assuming first that neither $h$ nor $n-h$ are zero, we shall replace (39.2) by a similar formula with $h-1$ in place of $h$. This will allow us to reduce everything to the case $h=0$ for which the proof is simple.

40. Suppose that there exists a polyhedral $C'_{n-h+1}$ (until further notice straightness and the like are defined by reference to $C_n$) whose boundary

$$\Gamma'_{n-h} = C'_{n-h} + C''_{n-h}$$

where $C'_h$ and $C''_{n-h}$ do not meet. Then first

$$(C'_h \cdot C'_{n-h}) = (C'_h \cdot \Gamma'_{n-h}) \ .$$

Also assuming all due conditions satisfied, and denoting by $\Gamma'_{h-1}$ the boundary of $C'_h$, we find from (39.1)

$$(C'_h \cdot C'_{n-h}) = (-1)^h \cdot (\Gamma_{h-1} \cdot C'_{n-h+1}) \ . \tag{40.1}$$

The necessary conditions are fulfilled with ease. According to our very construction of approximating complexes, $C'_h$ and $C'_{n-h}$ intersect in a finite number of points, any one of which, say $A_i$, is on cells of maximum dimensionality of the two complexes and of $C_n$. Let in particular $E^i_{n-h}$ be that of $C'_{n-h}$. We can construct a simplicial $E^i_{n-h+1}$ with $E^i_{n-h}$ on its boundary and positively related to it. Let this last cell count $m_i$ times for $C'_{n-h}$. We choose

$$C'_{n-h+1} = \sum m_i E^i_{n-h+1} \ ;$$

$C''_{n-h}$ is now what is left of the boundaries of the cells at the right when the cells $m_i E^i_{n-h}$ are removed from them. We may therefore manifestly so choose the cells $E^i_{n-h+1}$ that $C_{n-h}$ will contain no $A$ point.

Without prejudice to what precedes, the cells of $C''_{n-h}$ may be brought as near as we please to $C'_{n-h}$, hence we may so construct $C'_{n-h+1}$ that $C''_{n-h}$ intersects $C_h$ in a finite number of points (none an $A$), each on $h$-, $(n-h+1)$- and $n$-cells of $C'_{\bar{h}}$, $C'_{n-h+1}$ and $C_n$. Let $B$ be such a point. Remove from $C_h'$

a simplicial cell $E_h$ containing $B$. If it is sufficiently small, it will contain no $A$ and its bounding $(h-1)$-cells will intersect $C'_{n-h+1}$ each at a single point of the cell that carries $B$. Operate similarly for all $B$'s and let the new complex still be denoted by $C'_{h-1}$ and its boundary by $\Gamma'_{h-1}$.

If the cells of $C'_{n-h+1}$ and those removed from the initial $C_h'$ are adequately small, the intersection of the new $C'_h$ with $C'_{n-h+1}$ will be a sum of isolated rectilinear segments each on an $n$-cell of $C_n$, and as between the two complexes the various restrictions of §3 are verified. Since the new $C_h'$ does not meet $C''_{n-h}$, it gives rise to the same index $(C'_h \cdot C'_{n-h})$ as the initial one, so that we are assured that (40.1) holds with $C_h'$ as the new approximation to $C_h$ and $\Gamma'_{h-1}$ as its boundary.

41. At this stage we introduce the second defining complex $\bar{C}_n$. All approximations by means of it are to be denoted by the same symbols barred.

We first construct $\bar{\Gamma}'_{h-1}$, $\bar{C}'_{n-h}$, $\bar{C}''_{n-h}$. On examining our approximating processes, it is seen that the sum of the last two complexes is a suitable $\bar{\Gamma}'_{n-h}$. The cycle $\bar{\Gamma}'_{h-1}$ bounds $C'_h+C^0_h$, where the second complex, introduced by the approximation, is as near as we please to $\Gamma'_{h-1}$. Since $C'_h$ and $C'_{n-h}$ were constructed in general relative position, the second does not meet the boundary $\Gamma'_{h-1}$ of the first. With approximations sufficiently fine, it will not meet $C^0_h$ either. Hence (No. 35, VI)

$$[(C'_h+C^0_h) \cdot C'_{n-h}] = [C'_h \cdot C'_{n-h}] \, .$$

Therefore in place of approximating $C_h'$ we may approximate $C_h'+C^0_h$. To avoid more symbols than necessary, we shall denote that approximation by $\bar{C}_h'$.

Similarly $\bar{\Gamma}'_{n-h}$ bounds a complex $C'_{n-h+1}+C^0_{n-h+1}$, the latter very near the cycle and therefore not intersecting $\Gamma'_{h-1}$, hence, as above, its approximation may take the place of that of $C'_{n-h+1}$ and will then be denoted by $\bar{C}'_{n-h+1}$.

As previously, we may choose the approximations $\bar{C}'_h$ and $\bar{C}'_{n-h+1}$ such as to fulfil the conditions of §3 (Nos. 13, 15), for a well defined intersection. Even for $h=1$ is this the case. $\bar{C}_1'$ may then be so chosen that the points of $\Gamma_0'$ are approximated by any in their vicinity. We have therefore in all cases where $h>0$

$$[\bar{C}'_h \cdot \bar{C}'_{n-h}] = (-1)^h [\bar{\Gamma}'_{h-1} \cdot \bar{C}'_{n-h+1}] \, .$$

By the definition of the index, each expression defines that of the approximated complexes, which are here the primed complexes. Hence

$$[C'_h \cdot C'_{n-h}] = (-1)^h \cdot [\Gamma'_{h-1} \cdot C'_{n-h+1}] \, .$$

42. From (40.1) and (41.1) follows that in place of (39.2) we need only prove

$$(\Gamma'_{h-1} \cdot C_{n-h+1}) = [\Gamma'_{h-1} \cdot C_{n-h+1}] ,$$

of the same type but with $h$ replaced by $h-1$. Proceeding thus if necessary we shall reduce $h$ to zero. The last step will consist in replacing $C_1'$ by its set of terminal points $\Gamma_0'$, the signs affixed to them being $+$ or $-$ according as they are positively or negatively related to the complex. $\bar{\Gamma}_0'$ will then consist of points in the same number as those of $\Gamma_0'$, very near to them and with same signs attached.

For the other complex, we now have a $C_n'$ and by subdividing $C_n$ if necessary, we may assume that $C_n'$ is a subcomplex of it. Furthermore, (No. 41), the points of $\Gamma_0'$ may all be chosen on $n$-cells of $C_n'$.

Owing to the distributive law, we may assume in the last analysis that we have a unique point $A$ with a definite sign affixed, and a unique cell $E_n$ of $C_n$ carrying the point $A$. We have seen that the sign of the point is to be independent of the approximation; let us assume that it is $+$. In the other possible case, the reasoning would be the same with perhaps some signs changed. We must then show that

$$(A \cdot E_n) = +1 = [A \cdot E_n] .$$

Since $A$ is not on $C_n - E_n$, we may add the last complex to $E_n$ without affecting the indices. We have to prove then, that

$$[A \cdot C_n] = +1 .$$

We may choose $A$ on $E_n$ not a cell of less than $n$ dimensions of $\bar{C}_n$ without affecting our indices. Let $A$ be its own approximation by means of $\bar{C}_n$. The approximation of $C_n$ by means of $\bar{C}_n$ will be the sum of the cells of a certain subdivision of $\bar{C}_n$ (No. 40), which is the same as $\bar{C}_n$ itself. Finally then

$$[A \cdot C_n] = [A \cdot \bar{C}_n] = +1 ,$$

which completes the proof of the invariance of the index of two complexes.

43. The extension to the index of several complexes can be made along the same lines. The essential thing is to obtain a relation analogous to (40.1). It will be sufficient to derive it for three complexes $C_h$, $C_k$, $C_l$, $h+k+l=2n$, with the usual conditions as to their boundaries. Their approximations will then satisfy conditions analogous to those of No. 15. With obvious notations, let $C'_{h+1}$ have for boundary $\Gamma'_h = C'_h + C''_h$ with $C''_h$ not intersecting $C_k'$ nor $C_l'$. As a matter of fact, it would be sufficient to have it not intersecting $C_k' \cdot C_l'$. Then

$$(C_h' \cdot C_k' \cdot C_l') = (\Gamma_h' \cdot C_k' \cdot C_l') . \tag{43.1}$$

Also by (16.1) and with $\Gamma'_{k-1}$ the boundary of $C_k'$,

$$C'_{h+1} \cdot C'_k \equiv (-1)^{n-k} \cdot \Gamma'_h \cdot C_k + C'_{h+1} \cdot \Gamma'_{k-1} ,$$

it being granted that conditions of No. 15 are duly fulfilled. If we now further assume that the boundary of $C_l'$ does not meet the complex at the left, then (No. 19)

$$(-1)^{n-k}(\Gamma'_h \cdot C'_k \cdot C'_l) + (C'_{h+1} \cdot \Gamma'_{k-1} \cdot C'_l) = 0 .$$

Hence by (43.1)

$$(C'_h \cdot C'_k \cdot C'_l) = (-1)^{n-k-1}(C'_{h+1} \cdot \Gamma'_{k-1} \cdot C'_l) .$$

Similarly, with notations whose meaning is transparent,

$$C'_h \cdot C'_k \cdot C'_l = (-1)^{n-l-1} \cdot (C'_h \cdot C'_{k+1} \cdot \Gamma'_{l-1}) .$$

Thus we can raise both $h$ and $k$ at the expense of $l$, until $h=k=n$, $l=0$. From this point on the details of the discussion differ in no material way from the case of two complexes and need not be given here.

We have then proved that *the index of any number of complexes, when any exists, is independent of the defining $C_n$ of $M_n$ and the related straightness.* It has of course the various properties given in No. 36.

44. **Invariance of the intersection cycle.** The reasoning is exactly the same whatever the number of complexes, so we need only take two, $C_h$, $C_k$. This time, with our previous notations, we have to show that

$$C'_h \cdot C'_k \sim \bar{C}'_h \cdot \bar{C}'_k .$$

The cycle $C_h' \cdot C_k'$ is carried by a $C_l$, of the simplest type of No. 2, sum of the distinct simplicial $l$-cells of a decomposition of the cycle into such cells. Let these cells be $E_l^1, \cdots, E_l^p$, oriented each in a definite way. We shall have

$$C_l = \sum E_l^i ; \quad C'_h \cdot C'_k = \sum t_i E_l^i ,$$

where the $t$'s are non-zero integers.

Since the barred complexes are as near as we please to the non-barred, $\bar{C}_h' \cdot \bar{C}_k'$ is as near as we please to $C_l$. Hence (No. 22) the latter carries a $\Gamma_l$, subcomplex of a sufficiently fine subdivision, such that $\bar{C}_h' \cdot \bar{C}_k' - \Gamma_l$ bounds a $C_{l+1}$ whose points are as near as we please to $C_l$. But $\Gamma_l$ is homologous, mod $C_l$, to a subcomplex of $C_l$ itself (Coll. Lect., p. 120). Hence we may assume that $\Gamma_l$ itself is such a subcomplex without weakening the assertion as to $C_{l+1}$. We shall have

$$\bar{C}_h' \cdot \bar{C}_k' \sim \sum s_i E_l^i = \Gamma_l .$$

Owing to our usual approximating procedure, each $E_l^i$ is on an $n$-cell of $C_n$. Hence, just as if $C_n$ were an $S_n$, we can construct a simplicial $E_{n-l}$ that intersects $C_l$ at a unique point on $E_l^i$ in such manner that

$$(E_l^i \cdot E_{n-l}) = +1 ; \qquad (E_l^j \cdot E_{n-l}) = 0 , \qquad j \neq i .$$

By taking $E_{n-l}$ sufficiently small, we may dispose of it so that its boundary does not meet $C_l$, while the cell itself does not meet the boundaries of $C_h'$ or $C_k'$. This is due to the fact that the latter do not intersect $C_l$ while $E_{n-l}$ is very near to this complex.

Let us now approximate by means of $\bar{C}_n$. We have two congruences such as in No. 31:

$$\bar{C}_{h+1} \equiv C_h' - \bar{C}_h' + \bar{C}_h^0 ,$$

$$\bar{C}_{k+1} \equiv C_k' - \bar{C}_k' + C_k^0 .$$

When the approximation is carried sufficiently far,

(*a*) the boundary of $E_{n-l}$ will not meet $C_{l+1}$;

(*b*) it will not go through any intersection of $\bar{C}_{h+1}$ with $C_k'$ or $\bar{C}_k'$, nor of $\bar{C}_{k+1}$ with $C_h'$ or $\bar{C}_h'$;

(*c*) $E_{n-l}$ will not meet $\bar{C}_h^0$ nor $\bar{C}_k^0$.

Since $C_{l+1}$ is as near as we please to $C_l$, (*a*) follows from the fact that the boundary does not intersect $C_l$. Now were the first of (*b*) untrue, since $\bar{C}_{h+1}$ is as near as we please to $C_h$, it would only mean that the boundary of $E_{n-l}$ goes through a point of $C_h' \cdot C_k'$, and hence meets $C_l$ which carries this cycle, and this is not the case. Similarly for the rest of (*b*). As to (*c*), it follows at once from the fact that the cell does not meet the boundaries of $C_h'$ and $C_k'$ to which $\bar{C}_h^0$ and $\bar{C}_k^0$ are very near.

Owing to (*a*) and to IV and I of No. 36, we have

$$((\Gamma_l - \bar{C}_h' \cdot \bar{C}_k') \cdot E_{n-l}) = 0 ;$$

therefore

$$\sum s_j (E_l^j \cdot E_{n-l}) = s_i = (\bar{C}_h' \cdot \bar{C}_k' \cdot E_{n-l}) .$$

Also

$$(C_h' \cdot C_k' \cdot E_{n-l}) = \sum t_j (E_l^i \cdot E_{n-l}) = t_i .$$

Again by IV and the first of (*b*),

$$((C_h' - \bar{C}_h' + \bar{C}_h^0) \cdot C_k' \cdot E_{n-l}) = 0 .$$

From ($c$) we have

$$(\bar{C}_h^0 \cdot C_k' \cdot E_{n-l}) = 0 .$$

Therefore

$$(C_h' \cdot C_k' \cdot E_{n-l}) = (\bar{C}_h' \cdot C_k' \cdot E_{n-l}) .$$

Similarly from the last of ($b$) and ($c$),

$$(\bar{C}_h' \cdot (C_k' - \bar{C}_k' + \bar{C}_k^0) \cdot E_{n-l}) = 0 .$$

Therefore

$$(\bar{C}_h' \cdot C_k' \cdot E_{n-l}) = (\bar{C}_h' \cdot \bar{C}_k' \cdot E_{n-l}) .$$

By comparing we have at once

$$t_i = (C_h' \cdot C_k' \cdot E_{n-l}) = (\bar{C}_h' \cdot \bar{C}_k' \cdot E_{n-l}) = s_i .$$

From this follows finally

$$C_h' \cdot C_k' \sim \bar{C}_h' \cdot \bar{C}_k' ,$$

and our invariance proof is complete.

45. Intersection cycles or Kronecker indices are then the same whatever the defining complex of $M_n$ from which they are derived, provided as regards indices that various complexes have their orientations suitably related. Cycles and indices have all the properties established in No. 35 and the effect of a permutation of complexes in the symbols is the same as for cells.

## §7. Fundamental sets on an orientable $M_n$ without boundary

46. Let $\Gamma_h^i (i = 1, 2, \cdots, p_h)$, $\Gamma_{n-h}^j (i = 1, 2, \cdots, p_{n-h})$ be two fundamental sets (No. 7) for the cycles of the same dimensions, and consider the matrix

$$\| (\Gamma_h^i \cdot \Gamma_{n-h}^j) \| . \tag{46.1}$$

Any other fundamental set, say for the dimensionality $h$, can be derived from a given one by applying to its cycles a transformation of determinant unity. This is at once derivable from the result established by Veblen, loc. cit., p. 117. By such a transformation we mean that every cycle of the new set will be homologous to a sum of cycles of the old, the matrix of the coefficients of the homologies being $\pm 1$.

According to the distributive law for indices, if we change fundamental sets the above matrix is merely multiplied to the right or to the left by a matrix of determinant $\pm 1$. *Therefore the invariant factors of* (46.1) *are independent of the two fundamental sets*, and in fact *they are all equal to unity*, an important result just proved by Veblen,* which we shall also derive later in a new way (No. 65). It follows from this and from Poincaré's theorem on the equality of the numbers $R_h$, $R_{n-h}$, that we may select our sets so that

(*a*) for $j > R_h$, $\Gamma_h^j$ and $\Gamma_{n-h}^j$ are zero-divisors;

(*b*) if $h \neq n/2$ or if $h = n/2$ and $n/2$ is even,

$$(\Gamma_h^i \cdot \Gamma_{n-h}^i) = \begin{Bmatrix} +1 & \text{if } h \leqq n/2 \\ (-1)^{(n+1)h} & \text{if } h > n/2 \end{Bmatrix} i \leqq R_h \,,$$

all other indices being zero;

(*c*) if $h = n/2$ and $n/2$ is odd, when (46.1) is an alternate matrix,

$$(\Gamma_{n/2}^{2i-1} \cdot \Gamma_{n/2}^{2i}) = -(\Gamma_{n/2}^{2i} \cdot \Gamma_{n/2}^{2i-1}) = +1 \,,\; i \leqq \tfrac{1}{2}R_{n/2}$$

with all other indices again zero. In this case, of course, *since the matrix is alternate and of rank $R_{n/2}$ this last integer is necessarily even*, a generalization of the well known result for two-dimensional manifolds.

The possibility of choosing the sets as above is based upon well known theorems on the reduction of matrices with integer terms.†

Fundamental sets of the type just described will be called *canonical.*

## Part II. Transformation of manifolds

### §1. Product complexes‡

47. Let $E_p$, $E_q$ be two cells, $A$ a point of $E_p$ or of its boundary, $B$ a similar point for $E_q$. Consider the set of couples $A$, $B$ which by definition vary continuously if either $A$ or $B$ so varies. I say that the set is an $E_n$, $n = p+q$, plus its boundary. This follows at once from the fact that if $x_1, x_2, \cdots, x_n$ are cartesian coördinates for an $S_n$, $E_p$, $E_q$ with their boundaries, then the sets in question are respectively homeomorphic to the following three sets:

$$0 \leqq x_i \leqq 1, \quad i \leqq p; \quad x_{p+j} = 0\,;$$
$$x_i = 0, i \leqq p; \quad 0 \leqq x_{p+j} \leqq 1\,;$$
$$0 \leqq x_i \leqq 1, \quad i = 1, 2, \cdots, n\,.$$

---

* These Transactions, vol. 25 (1923), p. 540. See in the same connection my Monograph already quoted, p. 13. In two recent notes of the Proceedings of the National Academy of Sciences, vol. 10 (1924), pp. 99–103, J. W. Alexander generalizing this notion has been led to a new set of topological invariants.

† See the expository paper by Veblen and Franklin in the Annals of Mathematics, ser. 2, vol. 23.

‡ The term and corresponding notation are due to E. Steinitz, Sitzungsberichte der Berliner mathematischen Gesellschaft, vol 7 (1908). See also in this connection H. Tietze, Abhandlungen des mathematischen Seminars zu Hamburg, vol. 2 (1923), p. 37, H. Künneth, Mathematische Annalen, vol. 90 (1923), p. 65, and my paper in these Transactions, vol. 22 (1921), p. 362 (p. 76 of the present volume).

48. Let $\Pi_p, \Pi_q$ be polyhedra in spaces $S_r, S_t$, by means of which straightness and distances are defined for $E_p$ and $E_q$ and let $x_1, \cdots, x_r$ and $y_1, \cdots, y_t$ be cartesian coördinates for $S_r$ and $S_t$. We may refer an $S_{r+t}$ to the set of cartesian coördinates $x_i, y_j$. Then if a point $(x)$ describes $\Pi_p$ and a point $(y)$ describes $\Pi_q$ the corresponding point $(x, y)$ of $S_{r+t}$ describes a $\Pi_n$ homeomorphic to $E_n$ plus its boundary. The geometry of this is very elementary and we leave it to the reader. This $\Pi_n$ shall be chosen as basis for straightness and distances on $E_n$ and its boundary.

The *point* $A, B$ of $E_n$ or its boundary shall be designated henceforth by $A \times B$. Let $AA_1 \cdots A_p$, $BB_1 \cdots B_q$ be indicatrices of $E_p$ and $E_q$. We shall agree to sense $E_n$ by the indicatrix

$$A\times B\ A_1\times B \cdots A_p\times B\ A\times B_1 \cdots A\times B_q ,$$

and the cell so sensed shall be denoted by $E_p\times E_q$ and called *product* of the two cells, its *factors*.* The notation $A\times B$ merely corresponds to $p=q=0$. At once, we have

$$E_p\times E_q = -(-E_p)\times E_q = -E_p\times(-E_q) = (-1)^{pq} E_q\times E_p .$$

Let $a^i_{p-1}, b^j_{q-1}$ be the boundary cells of $E_p$ and $E_q$, so sensed that

$$E_p \equiv \sum a^i_{p-1}, \qquad E_q \equiv \sum b^j_{q-1} .$$

Then

$$E_p\times E_q \equiv \sum \epsilon_i a^i_{p-1}\times E_q + \sum \eta_j E_p\times b^j_{q-1} , \tag{48.2}$$

where $\epsilon_i$ and $\eta_j$ are $\pm 1$. To determine their actual value assume that $BB_1 \cdots B_{q-1}$ is on $b^j_{q-1}$. Then $(-1)^q \cdot BB_1 \cdots B_{q-1}$ is an indicatrix of $b^j_{q-1}$ as boundary cell of $E_q$. That of $E_p\times b^j_{q-1}$ is

$$(-1)^q A\times B\ A_1\times B \cdots A_p\times B\ A\times B_1 \cdots A\times B_{q-1} .$$

Comparing with the indicatrix of $E_p\times E_q$ we find $\eta_j = (-1)^q$. If we interchange $E_p$ and $E_q$ we shall find by applying this very relation $(-1)^q\ \epsilon_i = (-1)^q$, hence $\epsilon_i = 1$, so that finally (48.2) becomes

$$E_p\times E_q \equiv \sum a^i_{p-1}\times E_q + (-1)^p \sum E_p\times b^j_{q-1} , \tag{48.3}$$

which describes the boundary of the product.

49. Let now $C_p$, $C_q$ be any complexes, $a^i_p$ a generic $p$-cell of the first, $b^j_q$ one of the second. We define their product by the relation

$$C_p\times C_q = \sum a^i_p\times b^j_q .$$

* Concerning this symbolic product see the footnote to No. 7.

By applying (48.3) to each term at the right we find

$$C_p \times C_q \equiv \sum a_{p-1}^{i} \times b_q^{j} + (-1)^p \sum a_p^{i} \times b_{q-1}^{j} \tag{49.1}$$

where only the boundary $(p-1)$- and $(q-1)$-cells appear at the right with the very orientation that they possess in the boundary congruences of the complexes. Hence if $C_{p-1}$ and $C_{q-1}$ are these boundaries,

$$C_p \times C_q \equiv C_{p-1} \times C_q + (-1)^p C_p \times C_{q-1} \, . \tag{49.2}$$

From (49.2) we have the following:

I. *The product is orientable* (in the sense that a manifold is) *if and only if each factor is.*

II. *The product of two manifolds is a manifold.*

III. *The product of two complexes is without boundary if, and only if, the complexes themselves have none.* Observe that all this can be extended to singular complexes, hence

IV. *The product of two cycles of the factors is a cycle of the product.*

From polyhedra of the factors we derive as in No. 48 one for the product, hence corresponding definitions of straightness and distances.

50. Every $\Gamma_k$ of $C_p \times C_q$ is homologous to a polyhedral cycle $\Gamma_k'$ (Coll. Lect., p. 120). Let $\alpha$ be a point of the cell $a_p^i \times b_q^j$ not on $\Gamma_k'$. Draw a rectilinear segment from $\alpha$ to every point $\beta$ of $\Gamma_k'$ on the cell, to its intersection at $\gamma$ with the boundary. The set of all segments $\beta\gamma$ constitutes a polyhedral $C_{k+1}$. The effect of subtracting its boundary from $\Gamma_k'$ is to reduce the latter to a similar cycle without any points on $a_p^i \times b_q^j$. On proceeding thus with all such cells, then with the $C_{n-1}$ made up with the cells of less than $n$ dimensions of $C_p \times C_q$, and so on, we shall reduce $\Gamma_k'$ to a homologous polyhedral cycle whose points are all on cells of at most $k$ dimensions of the complex. From the theorem of the Colloquium Lectures just recalled, as applied to the $C_k$ made up of the cells of at most $k$ dimensions of our complex, follows that the reduced cycle, which we still call $\Gamma_k'$, is a sum of $k$-cells of $C_p \times C_q$. This may also be shown by remarking that if it has points on an $E_k$ the whole cell must belong to it, else it would have a boundary on it.

We conclude then that *every cycle of a product is homologous to a sum of products of cells of its two factors.*

Suppose that such a $\Gamma_k$ bounds on $C_p \times C_q$. We may apply to the complex that it bounds the identical reasoning, the operations made not affecting the boundary, and we conclude that *if a cycle, sum of products of cells of the factors, bounds on the product, it bounds a complex which is also such a sum.*

Henceforth we shall use the notation $\gamma_k$, $\delta_k$ for the $k$-cycles of $C_p$ and $C_q$, keeping $\Gamma_k$ itself for those of the product $C_p \times C_q$.

51. As shown by Veblen* every $E_k$ of $C_p$ is expressible as a sum of multiples of certain cycles $\gamma_p^i$ and certain cells $a_k^j$. Of course no sum of these particular $a$'s is a $\gamma$, hence every $\gamma_k$ of $C_p$ is a sum of multiples of the cycles $\gamma_k^j$ alone. It is not ruled out that some of the latter bound, but their set may be so chosen as to include a *fundamental set* for the $k$-cycles. There exists of course a similar set $\delta_k^i$, $b_k^j$ for $C_q$.

52. THEOREM. *The k-cycles which are products of cycles taken from fundamental sets of the factors constitute a fundamental set for the product.*

According to No. 50, for any $\Gamma_k$ we have

$$\Gamma_k \sim \sum \gamma_\mu^i \times \delta_{k-\mu}^j + \sum \gamma_\mu^i \times b_{k-\mu}^j + \sum a_\mu^i \times \delta_{k-\mu}^j + \sum a_\mu^i \times b_{k-\mu}^j .$$

We must express the fact that the right side has no boundary. The boundary of $a_\mu^i \times b_{k-\mu}^j$ is a sum of terms $\gamma_{\mu-1} \times b_{k-\mu}^j$, $a_\mu \times \delta_{k-\mu-1}$ where $\gamma_{\mu-1}$, $\delta_{k-\mu-1}$ are sums of cycles described in the preceding number. The boundaries of an $a \times \delta$ or of a $\gamma \times b$ are both of type $\gamma \times \delta$. Hence the boundaries of the terms in the fourth sum can only cancel each other, which compels the sum to be similar to the third. Hence for the cycle we have an homology

$$\Gamma_k \sim \sum \gamma_\mu^i \times \delta_{k-\mu}^j + \sum \gamma_\mu^i \times b_{k-\mu}^j + \sum a_\mu^i \times \delta_{k-\mu}^j .$$

We may assume that the second sum cannot be split into two, one of which is a cycle, for it would then be of the same type as the first and could be merged with it. Similarly for the third sum. From this follows that if the second sum is absent so is the third and conversely. Also since the boundaries of the terms in the last two sums must cancel each other, any term $\gamma_\mu^i \times b_{k-\mu}^j$ in the second sum contains only such a $\gamma$ as may come from the boundaries of terms in the third. $\gamma_\mu^i$ bounds then a certain $C_{\mu+1}$ of $C_p$. Let

$$b_{k-\mu}^j \equiv \sum \epsilon_{js} \delta_{k-\mu-1}^s .$$

At once,

$$C_{\mu+1} \times b_{k-\mu}^j \equiv \gamma_\mu^i \times b_{k-\mu}^j + (-1)^{\mu+1} \sum \epsilon_{js} C_{\mu+1} \times \delta_{k-\mu-1}^s \sim 0 \qquad (\text{mod } C_p \times C_q).$$

On subtracting this bounding cycle from the second expression for $\Gamma_k$ the term $\gamma_\mu^i \times b_{k-\mu}^j$ will have been replaced by a sum of terms that will go

---

* Coll. Lect., p. 116. The result in question, not stated explicitly by him, is really derived in the course of the discussion which leads to the theorem there stated.

into the first or the third sum. Therefore by a previous remark, the third must then also disappear. Hence

$$\Gamma_k \sim \sum \gamma_\mu^i \times \delta_{k-\mu}^j .$$

If $\gamma_\mu^i$ bounds $C_{\mu+1}$ of $C_p$, then $\gamma_\mu^i \times \delta_{k-\mu}^j$ bounds $C_{\mu+1} \times \delta_{k-\mu}^j$; hence in the expression of $\Gamma_k$ only the terms for which neither $\gamma$ nor $\delta$ bound need to be preserved, which proves our theorem.

It is not difficult to show that *if two factor cycles do not bound their product does not bound,* but as this is unnecessary for the sequel we omit the proof. A corollary is that *no cycle of the fundamental sets obtained bounds on* $C_p \times C_q$.

Of more import is the following observation. If $\gamma_\mu$ is a zero-divisor for $C_p$, $\gamma_\mu \times \delta_{k-\mu}$ is a zero-divisor or else bounds and similarly with $\delta$ in place of $\gamma$. For if $t\gamma_\mu$ bounds $C_{\mu+1}$, $(t\gamma_\mu) \times \delta_{k-\mu}$ bounds $C_{\mu+1} \times \delta_{k-\mu}$. Hence *the theorem that we have proved holds even when fundamental sets with respect to the operation* $\approx$ *take the place of the others.* For we may now eliminate from the fundamental sets of the product all cycles of which a factor is a zero-divisor.*

53. **Product of orientable manifolds.** We know already (No. 49) that if $M_p$, $M_q$ are orientable manifolds so is $M_p \times M_q$. We are now especially concerned with the question of Kronecker indices.

Let $E_h$, $E_{p-h}$ be simplicial cells on $M_p$ and suppose that they intersect at $A$. Let $E'_k$, $E'_{q-k}$ and $B$ be similar elements for $M_q$. Then $E_h \times E_k'$ and $E_{p-h} \times E'_{q-k}$ intersect at $A \times B$ on $M_p \times M_q$ and the question is *to determine their index in terms of* $(E_h \cdot E_{p-h})$, *and* $(E'_k \cdot E'_{q-k})$.

Let the $A$'s and $B$'s of No. 48 now serve to determine indicatrices for the manifolds, as previously for the cells. As a matter of fact, the $M$'s might simply be $E$'s without changing anything.

We may now assume that the indicatrices are

$$AA_1 \cdots A_h\,, \quad AA_{h+1} \cdots A_p \quad \text{for } E_h \text{ and } E_{p-h}\,,$$
$$BB_1 \cdots B_k\,, \quad BB_{k+1} \cdots B_q \quad \text{for } E_k \text{ and } E_{q-h}\,.$$

Hence for the above named two-cell products the indicatrices

$$A \times B \; A_1 \times B \cdots A_h \times B \; A \times B_1 \cdots A \times B_k\,,$$
$$A \times B \; A_{h+1} \times B \cdots A_p \times B \; A \times B_{k+1} \cdots A \times B_q\,.$$

* From this follow the results of my Transactions paper and those of Künneth, concerning connectivity and torsion indices.

By comparing with the indicatrix of $M_p \times M_q$ and applying the rule for the determination of the index we obtain the following formula:

$$(53.1) \quad (E_h \times E'_k \cdot E_{p-h} \times E'_{q-k}) = (-1)^{k(p-h)} (E_h \cdot E_{p-h})(E'_k \cdot E'_{q-k}) .$$

It holds even for $k=0$, $h=p$, when it yields the following result, verifiable directly with ease:

$$(53.2) \qquad (E_p \times B \cdot A \times E_q) = +1 .$$

The product $E_p \times B$ simply denotes an $E_p$ on the product cell, and similarly for $A \times E_q$.

Let now $C_h$, $C_{p-h}$ be complexes on $M_p$ whose points of intersection are neither on their boundaries nor on that of the manifold, and let $C'_k$, $C'_{q-k}$ be analogous for $M_q$. Polyhedral approximations to $C_h$ and $C_k'$ have for product such an approximation to $C_h \times C_k'$, and similarly for the other two complexes. If the approximations to $C_h$ and $C_{p-h}$ and those to $C'_k$ and $C'_{q-k}$ intersect at isolated ordinary points of the complexes, the approximations to the products will behave likewise. Hence from the definition of the index of two polyhedral complexes with well defined intersections, the extension to arbitrary complexes, and (53.1), (53.2), we have

$$(53.3) \quad (C_h \times C'_k \cdot C_{p-h} \times C'_{q-k}) = (-1)^{k(p-h)} (C_h \cdot C_{p-h})(C'_k \cdot C'_{q-k}) ;$$

$$(53.4) \qquad (M_p \times B \cdot A \times M_q) = +1 .$$

The products in (53.4) represent complexes homeomorphic to $M_p$ or $M_q$ on $M_p \times M_q$. The approximations to $A$ and $B$ in that case consist in choosing points of their vicinity situated on $p$- and $q$-cells of the covering complexes of the manifolds which serve to define straightness and distances on them.

Two complexes $C_h$, $C_{p-h-i}$ on $M_p$, without points on its boundary, may always be approximated by two that do not intersect. Hence if $C'_k$, $C'_{q-k+i}$ are on $M_q$, without points on its boundary, there are non-intersecting polyhedral approximations to $C_h \times C_k'$ and $C_{p-h-i} \times C'_{q-k+i}$. Therefore

$$(53.5) \qquad (C_h \times C'_k \cdot C_{p-h-i} \times C_{q-k+i}) = 0 .$$

54. THEOREM. *Let $M_p$, $M_q$ be without boundary and let $\gamma_k^i$, $\delta_k^j$ be the cycles of their canonical fundamental sets. Then except for the signs their products constitute such a set for $M_p \times M_q$.*

This follows at once from the relations derived from (53.3), (53.4) (53.5):

$$(54.1)\quad (\overset{i}{\gamma}_\lambda\times\overset{j}{\delta}_{k-\lambda}\cdot\overset{i}{\gamma}_{p-\lambda}\times\overset{j}{\delta}_{q-k+\lambda})=(-1)^{(k-\lambda)(p-\lambda)}\cdot(\overset{i}{\gamma}_\lambda\cdot\overset{i}{\gamma}_{p-\lambda})(\overset{j}{\delta}_{k-\lambda}\cdot\overset{j}{\delta}_{q-k+\lambda})\ ;$$

$$(54.2)\quad (\overset{i}{\gamma}_\lambda\times\overset{j}{\delta}_{k-\lambda}\cdot\overset{i}{\gamma}_{p-\mu}\times\overset{j}{\delta}_{q-k+\mu})=0\ ,\qquad \lambda\neq\mu\ .$$

They indicate also the manner in which the cycles are to be associated.

Of great importance for transformations is the special case $p=q=k=n$. We have then

$$(54.3)\quad (\overset{i}{\gamma}_\lambda\times\overset{j}{\delta}_{n-\lambda}\cdot\overset{i}{\gamma}_{n-\lambda}\times\overset{j}{\delta}_\lambda)=(-1)^{n(\lambda+1)}\cdot(\overset{i}{\gamma}_\lambda\cdot\overset{i}{\gamma}_{n-\lambda})\cdot(\overset{j}{\delta}_\lambda\cdot\overset{j}{\delta}_{n-\lambda})\ ,$$

and all other indices vanish.

55. Formulas (53.3), (53.4) are special cases of a more general one corresponding to cycles that are intersections of complexes, which is derived by similar considerations. As we shall not use it later we merely give it here without proof. If $C_l$ and $C_\lambda$ of $M_p$ do not intersect each other's boundary, and have no intersection on the boundary of the manifold, if also $C'_m$, $C'_\mu$ behave likewise relatively to $M_q$, the formula in question is as follows:

$$C_l\times C'_m\cdot C_\lambda\times C'_\mu\sim(-1)^{(p-l)(q-\mu)}\cdot C_l\cdot C_\lambda\times C'_m\cdot C'_\mu \qquad (\text{mod } M_p\times M_q)\ .$$

## §2. Transformations of a manifold without boundary

56. Let $M_n$ be the manifold, $M_n'$ another copy of it, $T$ a transformation of $M_n$ into itself or part of itself, which we subject to this sole condition of a very general nature: *If $A$ is any point of $M_n$, $B$ the image of any transform $A'$ of it on $M_n'$, the set of all points $A\times B$ is an n-cycle $\Gamma_n$ on $M_n\times M_n'$.* The inclusiveness of the class of transformations so defined becomes apparent when we remark that all continuous one-valued transformations (for each $A$ only one $A'$ varying continuously with $A$) belong to it. The non-singular $C_n$ without boundary of which $\Gamma_n$ is then the image in the sense of No. 6 is $M_n$ itself. More generally $k$-valued continuous transformations are also of our type; the corresponding $C_n$ is then $kM_n$.

Much of the rest of this paper will center around the determination of certain Kronecker indices and it becomes essential to define all orientations involved. Let $C_n$ cover $M_n$ and let $C_n'$ be its image covering $M_n'$. Suppose that $\Pi_n$ is a polyhedron which associated with $C_n$ serves to define straightness and distances on $M_n$. The polyhedron $\Pi_n$ has the same cell structure as $C_n'$ and we shall agree to use it, associated with $C_n'$, to define straightness and distances on $M_n'$. Then any rectilinear segment of $M_n$ has for image a similar one on $M_n'$ and both have the same length. To an indicatrix $E_n$ on $M_n$ will correspond an $E_n'$ on $M_n'$. We shall name the vertices of $E_n'$ in the same order as the corresponding vertices of $E_n$, and use the simplex so sensed as indicatrix for $M_n'$. In accordance with our previous conventions the orientation of $M_n \times M_n'$ is now perfectly determined. Owing to its importance for the sequel, it is well perhaps to characterize it more geometrically. Through any point $A \times B$ of the manifold there pass $M_n \times B$ and $A \times M_n'$. Let $B$ be the image of $A$. To $E_n$ with $A$ as its first vertex corresponds $E_n'$ with $B$ as its first vertex. Let $\bar{E}_n$ be the image of the first on $M_n \times B$, $\bar{E}_n'$ that of the second on $A \times M_n'$. The $E_{2n}$ indicatrix of the product manifold has its vertices named in the following order: $A \times B$, the other vertices of $\bar{E}_n$, the other vertices of $\bar{E}_n'$.

There remains the orienting of $\Gamma_n$. If one cycle is suitable for $T$ so is its opposite. Of the two we shall select the one such that

$$(\Gamma_n \cdot A \times M_n') = \alpha_0 \geqq 0 \ ,$$

a perfectly definite condition since the integer is a Kronecker index of cycles.

57. Let us apply to $M_n$ and $M_n'$ the very notations of No. 54, so that $\delta_k^i$ is now the cycle of $M_n'$ that corresponds to $\gamma_k^i$ on $M_n$. We shall then have

$$\Gamma_n \sim \sum \overset{ij}{\epsilon_\mu} \cdot \overset{i}{\gamma_\mu} \times \overset{j}{\delta}_{n-\mu} \qquad (\text{mod } M_n \times M_n') \ . \tag{57.1}$$

The $\gamma$'s and $\delta$'s are, it will be recalled, cycles of fundamental sets. In particular $\gamma_0$, $\delta_0$ are merely points of $M_n$, $M_n'$ while $\gamma_n$, $\delta_n$ are the manifolds themselves. The number of cycles of the fundamental sets is *one* for these two extreme cases.

The $\epsilon$'s are important characteristic integers of $T$. It will be remembered that two transformations are said to be of the same *class* if they belong to one and the same continuous family of transformations. Whenever $T$ varies within its class, $\Gamma_n$ is continuously deformed (a condition that might serve to define the class) and $\Gamma_n$ remains homologous to a fixed cycle, so that the $\epsilon$'s are unchanged. Hence *to a given class of transformations corresponds a fixed set of $\epsilon$'s*. Whether the converse holds or not is as yet unknown.

58. We must now introduce the notion of the *transform of a cycle* of $M_n$ by $T$. The necessity of defining such a notion with precision arises from the fact that while $T$ is a point transformation a cycle is not a mere point set, but consists of a set of cells each taken with a certain multiplicity. The cycle is then really a symbol attached to a set of cells, and what is meant by its transform is by no means evident a priori.

Let $\gamma_\mu$, $0 \leqq \mu \leqq n$, be any cycle of $M_n$. If it is not polyhedral we approximate it by one that is and that furthermore has been reduced in accordance with the theorem of No. 23. Then $\gamma_\mu \times M_n'$ behaves likewise and we may also approximate $\Gamma_n$ in the same manner so that the two have a well defined intersection $\Gamma_n \cdot \gamma_\mu \times M_n' = \Gamma_\mu$. (We economize in notations by designating the approximations like the cycles themselves.) To every point $A \times B$ of $\Gamma_\mu$ corresponds a unique $B$ on $M_n'$ varying continuously when $A \times B$ so varies on $\Gamma_\mu$. Hence $B$ gives rise to a cycle $\bar{\delta}_\mu$ on $M_n'$. It is a singular complex of $M_n'$, continuous image of $\Gamma_\mu$ in the sense of No. 6. The image $\bar{\gamma}_\mu$ of $\bar{\delta}_\mu$ on $M_n$ is by definition the transform of $\gamma_\mu$ by $T$. It will be observed that $\bar{\delta}_\mu$ is *polyhedral*, hence so is $\bar{\gamma}_\mu$. For $\bar{\delta}_\mu$ is represented on the copy $A \times M_n'$ of $M_n'$ by the cycle $A \times \bar{\delta}_\mu$ which is the locus of all intersections of $A \times M_n'$ with all manifolds $M_n \times B$ that meet $\Gamma_n \cdot \gamma_\mu \times M_n'$. In this fashion, however, $A \times \bar{\delta}_\mu$ appears as an intersection of polyhedral complexes, and it is therefore also polyhedral. If $A \times B\ A_1 \times B_1 \cdots A_\mu \times B_\mu$ is an indicatrix of $\Gamma_\mu$ then $BB_1 \cdots B_\mu$ is an indicatrix of $\bar{\delta}_\mu$.

The preceding definition is justified by the fact that for $\mu = 0$, when $\gamma_0$ consists of a finite number of points, the correct transforms of these points are obtained. Furthermore (No. 62) when $T$ is the identity all cycles are left invariant, which is as it should be with any properly selected definition.

Let the cycle transformations be represented by homologies

$$\bar{\gamma}^i_\mu \sim \sum a^{ij}_\mu \gamma^j_\mu \qquad (\mathrm{mod}\ M_n). \tag{58.1}$$

For any given $T$ one is far more likely to possess information concerning the $a$'s than concerning the $\epsilon$'s. Hence the interesting and important problem arises *to determine the $\epsilon$'s in terms of the $a$'s*. This problem shall be solved partly in a more narrow form. Let us drop zero divisors throughout. In accordance with No. 52, with the $R$'s denoting the connectivity indices of $M_n$, and assuming that among the cycles $\gamma^i_\mu$ the first $R_\mu$ are independent, we shall now have in place of (57.1) and (58.1), the following relations whose appearance is the same:

$$\Gamma_h \approx \sum \epsilon^{ij}_\mu \cdot \gamma^i_\mu \times \delta^j_{n-\mu} \qquad (\mathrm{mod}\ M_n \times M'_n); \tag{58.2}$$

$$\bar{\gamma}^i_\mu \approx \sum a^{ij}_\mu \gamma^j_\mu \qquad (\mathrm{mod}\ M_n), \tag{58.3}$$

where as before $\mu$ runs from 0 to $n$, but, for each $\mu$, $i$ and $j$ run from 1 to $R_\mu = R_{n-\mu}$. The problem that we shall completely solve is *the determination of the $\epsilon$'s within this range in terms of the $a$'s similarly limited.* This will suffice to provide all that is needed in our applications to fixed points and coincidences.

59. **An important formula.** The solution of our problem is based upon the following formula, whose proof will now occupy us:

(59.1) $$(\Gamma_n \cdot \gamma_\mu \times \delta_{n-\mu}) = (-1)^\mu \cdot (\bar{\gamma}_\mu \cdot \gamma_{n-\mu}),$$

with indices computed at the left as to $M_n \times M_n'$ and at the right as to $M_n$. In place of (59.1) it will be found more convenient to prove the equivalent

(59.2) $$(\Gamma_n \cdot \gamma_\mu \times \delta_{n-\mu}) = (-1)^\mu \cdot (\bar{\delta}_\mu \cdot \delta_{n-\mu}),$$

where the last index is computed as to $M_n'$, and to this we now turn our attention.

60. By means of our usual approximations as applied to $\Gamma_n$, $\gamma_\mu$ and $\delta_{n-\mu}$, we may so arrange matters that all cycles are polyhedral and have well defined intersections. Let $A \times B$ be an intersection of $\Gamma_n$ with $\gamma_\mu \times \delta_{n-\mu}$. Then $B$ is an intersection of $\bar{\delta}_\mu$ with $\delta_{n-\mu}$. Therefore we need only to show that the contributions of the two points to their respective indices have the ratio $(-1)^\mu$. Owing to the distributive law we may assume that all complexes are cells and that the points count for $\pm 1$ in the indices, or what is the same, that the manifolds are linear spaces and $T$ a projectivity. The method of matrices (No. 4) will be found most convenient for our purpose.

61. We begin with *the contribution of* $A \times B$ *to the left side of* (59.2). Let $A$, $B$, and $A \times B$ be origins of cartesian axes for the spaces $M_n$, $M_n'$, and their product, the coördinates being $x_1, \cdots, x_n$ for $M_n$, $y_1, \cdots, y_n$ for $M_n'$, $x_1, \cdots, x_n, y_1, \cdots, y_n$ for $M_n \times M_n'$. To the points $(x)$, $(y)$ of $M_n$, $M_n'$ corresponds the point $(x, y)$ of the product.

We can assume for $M_n$ a matrix-indicatrix in the sense of No. 4, of type

(61.1) $$\begin{Vmatrix} X_\mu & , & 0 \\ 0 & , & X_{n-\mu} \end{Vmatrix}$$

where the $X$'s denote square matrices of order equal to the index with determinants equal to $+1$, and the zeros matrices whose terms all vanish. The meaning of (61.1) is clear. If $A_i$ is the point of $M_n$ whose coördinates are the terms in the $i$th row, then $AA_1 \cdots A_n$ is an indicatrix of $M_n$. We may so select it that $AA_1 \cdots A_\mu$ and $AA_{\mu+1} \cdots A_n$ are indicatrices of $\gamma_\mu$ and $\gamma_{n-\mu}$. This implies a definite orientation for them, but if we invert $\gamma_\mu$ or $\gamma_{n-\mu}$ we also invert $\bar{\delta}_\mu$ or $\delta_{n-\mu}$ and therefore (59.2) is unaffected, so that there is really no restriction. In the same sense as for $M_n$, we may say that the matrices

$$\|X_\mu, 0\|, \quad \|0, X_{n-\mu}\|$$

define indicatrices of $\gamma_\mu$ and $\gamma_{n-\mu}$.

To the images of the transforms of $M_n, \gamma_\mu, \gamma_{n-\mu}$, by $T$, on $M_n'$, correspond matrix-indicatrices

$$\left\| \begin{matrix} Y_\mu, & Y \\ Y', & Y_{n-\mu} \end{matrix} \right\|, \quad \| Y_\mu, Y \|, \quad \| Y', Y_{n-\mu} \|$$

where $Y$, $Y'$ are rectangular arrays and the two other $Y$'s square arrays of order equal to their index. The first corresponds to $\pm M_n' = \epsilon M_n'$, where the sign is plus if $T$ maintains the indicatrix on $M_n$, minus if it inverts it. The second and third matrices correspond to $\bar{\delta}_\mu$ and $\delta_{n-\mu}$. Since (61.1) defines an indicatrix of $M_n'$ when each row is considered as the coördinates of a $y$ point, and since its determinant, product of $|X_\mu|$ and $|X_{n-\mu}|$, is $+1$, $\epsilon$ has the sign of the determinant

$$\left| \begin{matrix} Y_\mu & , & Y \\ Y' & , & Y_{n-\mu} \end{matrix} \right| .$$

From the definition of $\Gamma_n$ we conclude that

$$\left\| \begin{matrix} X_\mu, & 0 & , & Y_\mu & , & Y \\ 0, & X_{n-\mu} & , & Y' & , & Y_{n-\mu} \end{matrix} \right\|$$

in a matrix-indicatrix of $\theta\Gamma_n$, $\theta = \pm 1$. To determine the sign of $\theta$ observe that since $A \times M_n'$ has for matrix-indicatrix

$$\left\| \begin{matrix} 0 & , & 0 & , & X_\mu & , & 0 \\ 0 & , & 0 & , & 0 & , & X_{n-\mu} \end{matrix} \right\| ,$$

the contribution of $A \times B$ to $(\theta\Gamma_n \cdot A \times M_n')$ has the sign of the determinant

$$\left| \begin{matrix} X_\mu & , & 0 & , & 0 & , & 0 \\ 0 & , & X_{n-\mu} & , & 0 & , & 0 \\ 0 & , & 0 & , & X_\mu & , & 0 \\ 0 & , & 0 & , & 0 & , & X_{n-\mu} \end{matrix} \right| = +1 .$$

Hence $\theta$ represents the contribution of the point to the integer $\alpha_0 = (\Gamma_n \cdot A \times M_n')$ of No. 56. For example, in the case of the identical transformation, every point of intersection of the two cycles must contribute $+1$, therefore $\theta = +1$. This is also true for a continuous 1 to $n$ transformation, where the contributions to the index have a constant sign. That $\theta$ is not $-1$ in these two cases comes from the condition $\alpha_0 > 0$ by which we have definitely oriented $\Gamma_n$.

But this is a digression from our main topic to which we now return. It is readily seen that

$$\left\| \begin{matrix} X_\mu, & 0, & 0, & 0 \\ 0, & 0, & 0, & X_{n-\mu} \end{matrix} \right\|$$

is a matrix-indicatrix for $\gamma_\mu \times \delta_{n-\mu}$. As $M_n \times M_n'$ has for matrix-indicatrix

$$\left\| \begin{matrix} X_\mu, & 0, & 0, & 0 \\ 0, & X_{n-\mu}, & 0, & 0 \\ 0, & 0, & X_\mu, & 0 \\ 0, & 0, & 0, & X_{n-\mu} \end{matrix} \right\|$$

whose determinant is $+1$, we conclude from the definition of the Kronecker index that the contribution of $A \times B$ to $(\theta\Gamma_n \cdot \gamma_\mu \times \delta_{n-\mu})$ has the sign of the determinant

$$\left| \begin{matrix} X_\mu, & 0, & Y_\mu, & Y \\ 0, & X_{n-\mu}, & Y', & Y_{n-\mu} \\ X_\mu, & 0, & 0, & 0 \\ 0, & 0, & 0, & X_{n-\mu} \end{matrix} \right| = (-1)^\mu |Y_\mu| .$$

The contribution of $A \times B$ to $(\Gamma_n \cdot \gamma_\mu \times \delta_{n-\mu})$ has then the sign of $(-1)^\mu \theta |Y_\mu|$.

62. Let us now examine *the contribution of B to* $(\bar{\delta}_\mu \cdot \delta_{n-\mu})$. Suppose that we have found for $\Gamma_n \cdot \gamma_\mu \times M_n'$ a matrix-indicatrix

$$\| Z, Z' \|,$$

where $Z$, $Z'$ are matrices with $\mu$ rows and $n$ columns. Then $Z'$ will define an indicatrix for $\bar{\delta}_\mu$, referred of course to the $y$ coördinates. For any row $x_1, \cdots, x_n, y_1, \cdots, y_n$ represents a point $A' \times B'$ such that $B'$ is the point of $M_n'$ with the $y$ coördinates. From this and the remark, No. 58, as to relations between indicatrices on $\Gamma_n$ and $\bar{\delta}_\mu$, follows the property of $Z'$.

Now $\gamma_\mu \times (\epsilon M_n')$ has for matrix-indicatrix

$$\left\| \begin{matrix} X_\mu, & 0, & 0, & 0 \\ 0, & 0, & Y_\mu, & Y \\ 0, & 0, & Y', & Y_{n-\mu} \end{matrix} \right\|$$

which we may replace by

$$\left\| \begin{matrix} X_\mu, & 0, & Y_\mu, & Y \\ 0, & 0, & Y_\mu, & Y \\ 0, & 0, & Y', & Y_{n-\mu} \end{matrix} \right\|,$$

derived from it by addition of rows, as this change of indicatrix merely corresponds to an affine transformation of determinant $+1$ applied to the $S_{\mu+n}$ involved, and is therefore permissible. The first row defines now an indicatrix of $\zeta\theta\Gamma_n \cdot \gamma_\mu \times (\epsilon M_n')$, where $\zeta = \pm 1$. From our mode of defining sensed intersections (No. 12) it follows that $\zeta$ has the same sign as the determinant

$$\begin{vmatrix} X_\mu, & 0, & Y_\mu, & Y \\ 0, & X_{n-\mu}, & Y', & Y_{n-\mu} \\ 0, & 0, & Y_\mu, & Y \\ 0, & 0, & Y', & Y_{n-\mu} \end{vmatrix} = \begin{vmatrix} Y_\mu, & Y \\ Y', & Y_{n-\mu} \end{vmatrix},$$

that is, the sign of $\epsilon$. As both are unity in absolute value, $\zeta = \epsilon$, so that actually the matrix

$$\| X_\mu \quad , \quad 0 \quad , \quad Y_\mu \quad , \quad Y \|$$

defines an indicatrix of $\epsilon\,\theta\Gamma_n \cdot \gamma_\mu \times (\epsilon M_n') = \theta\Gamma_n \cdot \gamma_\mu \times M_n'$. According to what has been said at the beginning of this number, then

$$\| Y_\mu \quad , \quad Y \|$$

defines an indicatrix for $\theta\bar{\delta}_\mu$.

Since $M_n'$ has a matrix-indicatrix of determinant $+1$, the contribution of $B$ to $(\theta\bar{\delta}_\mu \cdot \delta_{n-\mu})$ has the sign of the determinant

$$\begin{vmatrix} Y_\mu, & Y \\ 0, & X_{n-\mu} \end{vmatrix},$$

since the last line is a matrix-indicatrix for $\delta_{n-\mu}$. As the determinant has for value $|Y_\mu|$, the contribution of $B$ to $(\bar{\delta}_\mu \cdot \delta_{n-\mu})$ has the same sign as $\theta|Y_\mu|$. Its ratio to that of $(\Gamma_n \cdot \gamma_\mu \times \delta_{n-\mu})$ is then $(-1)^\mu$, which suffices to prove (59.1).

**Remark.** One incidental result follows readily from our discussion. The indicatrix of $\bar{\delta}_\mu$ coincides with the image of the transform of that of $\gamma_\mu$ or with its opposite according as $\theta$ is positive or negative. Upon translating this back to $M_n$ itself, we find the following result. *If $A \times B$ contributes $+1$ to $(\Gamma_n \cdot A \times M_n')$ then the indicatrix of the transform of any cycle through $A$ is the transform of the indicatrix; it is the opposite of it if the contribution is $-1$.* For example in the case of a continuous sense preserving 1 to $n$ transformation, the indicatrix of the transform is always the transform of the indicatrix. In particular for the identical transformation the two coincide, and $\bar{\gamma}_\mu$ not only coincides with $\gamma_\mu$ element for element, but also with preservation of orientation. This justifies in a sense our convention as to sensing $\Gamma_n$ and our definition of the transform of a cycle (No. 58).

63. It is important to emphasize the fact that the extreme values $\mu=0$, $n$ are not exceptional at all. For $\mu=0$, we have already imposed

$$(\Gamma_n \cdot A \times M_n') = \alpha_0 . \tag{63.1}$$

The interpretation given for this integer in No. 60 fits in perfectly with our discussion. Let $\alpha_n$ be the similar integer for $T^{-1}$. Its interpretation is then this:* An arbitrary $E_n$ of $M_n$ is covered with a certain number of cells of $T \cdot M_n$. Among these there will be, say, $k'$ positive cells, $k''$ negative cells of $M_n$ and $\alpha_n = k' - k''$. Here of course $\alpha_n$ may well be negative. Thus if $T$ is a homeomorphism inverting the indicatrix, $\alpha_n = -1$. An example of this is the symmetry of a sphere with respect to a diametral plane. It is important to remember that the $T$ considered is not the initial transformation but one that corresponds to a polyhedral approximation of its $\Gamma_n$.

If we compute the index as to $M_n' \times M_n$ we obtain then

$$(\Gamma_n \cdot B \times M_n) = (\Gamma_n \cdot M_n \times B) = \alpha_n .$$

If a manifold is inverted all corresponding indices must be changed in sign, an immediate corollary of their definition. As $M_n \times M_n' = (-1)^n M_n' \times M_n$, when the index is computed with reference to $M_n \times M_n'$,

$$(\Gamma_n \cdot M_n \times B) = (-1)^n \alpha_n . \tag{63.2}$$

This is what (59.1) becomes for $\mu = n$. Both (63.1) and (63.2) can also be derived with ease by means of matrix-indicatrices. Indeed this would merely involve repeating the discussion of Nos. 60, 61.

Observe that the notation $\alpha_0$, $\alpha_n$ is in accord with the meaning given to $\overset{ij}{\alpha_\mu}$ in No. 57. They correspond to the homologies describing the behavior of the zero-cycle (a point) and the $n$-cycle ($M_n$ itself). Explicitly for $\mu=0$, $n$, $i$ and $j$ can only be unity, and $\overset{11}{\alpha_0} = \alpha_0$, $\overset{11}{\alpha_n} = \alpha_n$.

64. We now pass to the actual determination of the $\epsilon$'s. If we remember that in calculating Kronecker indices zero-divisors may be dropped, we see that in applying (59.1) we may substitute from formulas (58.2) and (58.3) in place of (57.1) and (58.1). Let us then substitute in (59.1) for $\Gamma_n$ its expression as given by (58.2), for $\gamma_\mu$, $\delta_{n-\mu}$ the cycles $\overset{h}{\gamma_\mu}$, $\overset{k}{\delta}_{n-\mu}$, a nd for $\overset{-h}{\gamma_\mu}$ its expression (58.3). We obtain

$$\sum_{i,j=1}^{R_\mu} \overset{ij}{\epsilon}_{n-\mu}(\overset{i}{\gamma}_{n-\mu} \times \overset{j}{\delta_\mu} \cdot \overset{h}{\gamma_\mu} \times \overset{k}{\delta}_{n-\mu}) = (-1)^\mu \sum_{j=1}^{R_\mu} \overset{hj}{\alpha_\mu}(\overset{j}{\gamma_\mu} \cdot \overset{k}{\gamma}_{n-\mu}) ,$$

* This is the same as Brouwer's *degree* of $T$. See Mathematische Annalen, vol. 71 (1911), p. 105. He limits himself to the case where every point has a unique transform, which is a special case of transformations with $a_0=1$.

all other terms having dropped out in accordance with (54.1). Transforming the left side by means of (53.3) and (53.4) this becomes

$$(-1)^{\mu(n+1)} \sum_{i,j=1}^{R_\mu} \overset{ij}{\epsilon}_{n-\mu}(\overset{h}{\gamma}_\mu \cdot \overset{j}{\gamma}_{n-\mu})\,(\overset{j}{\gamma}_\mu \cdot \overset{k}{\gamma}_{n-\mu}) = \sum_{j=1}^{R_\mu} \overset{hj}{a}_\mu\,(\overset{j}{\gamma}_\mu \cdot \overset{k}{\gamma}_{n-\mu}).$$

For each $\mu$ we have here a set of $R_\mu^2$ linear equations in the $B_\mu$ unknowns $\overset{ij}{\epsilon}_{n-\mu}$. The determinant of their coefficients is a power of the determinant

$$|(\overset{i}{\gamma}_\mu \cdot \overset{j}{\gamma}_{n-\mu})|,$$

whose value is actually $\pm 1$, as may be deduced from Veblen's theorem mentioned in Part I, §6. These equations may therefore be solved even explicitly if need be, a task which presents little interest. However, if, *as we shall assume from now on, the fundamental sets on $M_n$ are canonical the solution can be carried through in an instant.*

For if we substitute this time the indices as given in Part I, §6, and as they must be applied here, we have at once

(*a*) *$\mu \neq \frac{1}{2}n$ or $= \frac{1}{2}n =$ an even integer.*

Then

$$\overset{hk}{\epsilon}_{n-\mu} = \overset{hk}{a}_\mu, \qquad \mu > \tfrac{1}{2}n;$$

$$\overset{hk}{\epsilon}_{n-\mu} = (-1)^{\mu(n+1)}\,\overset{hk}{a}_\mu, \qquad \mu \leqq \tfrac{1}{2}n.$$

(*b*) *$\mu = \frac{1}{2}n =$ an odd integer.*

Then

$$\overset{2h-1,k}{\epsilon}_{n/2} = -\overset{2h,k}{a}_{n/2}, \qquad \overset{2h,k}{\epsilon}_{n/2} = \overset{2h-1,k}{a}_{n/2}.$$

We obtain for $\Gamma_n$ the following relations:

(*a*) $n \not\equiv 4$, mod 2.

$$\Gamma_n \approx \sum_{0 \leqq \mu < n/2} \sum_{i,j=1}^{R_\mu} \overset{ij}{a}_{n-\mu} \cdot \overset{i}{\gamma}_\mu \times \overset{j}{\delta}_{n-\mu}$$

$$+ \sum_{\mu \geqq n/2} (-1)^{(n+1)\mu} \sum_{i,j=1}^{R_\mu} \overset{ij}{a}_{n-\mu} \cdot \overset{i}{\gamma}_\mu \times \overset{j}{\delta}_{n-\mu}.$$

(*b*) $n \equiv 4$, mod 2.

$$\Gamma_n \approx \sum_{0 \leqq \mu < n/2} \sum_{i,j=1}^{R_\mu} (\overset{ij}{a}_{n-\mu} \cdot \overset{i}{\gamma}_\mu \times \overset{j}{\delta}_{n-\mu} + (-1)^\mu\, \overset{ij}{a}_\mu \cdot \overset{i}{\gamma}_{n-\mu} \times \overset{j}{\delta}_\mu)$$

$$+ \sum_{h=1}^{\frac{1}{2}R_{n/2}} \sum_{k=1}^{R_{n/2}} (\overset{2h-1}{a}_{n/2} \cdot \overset{2h}{\gamma}_{n/2} \times \overset{k}{\delta}_{n/2} - \overset{2h,k}{a}_{n/2} \cdot \overset{2h-1}{\gamma}_{n/2} \times \overset{k}{\delta}_{n/2}).$$

65. **A new proof of Veblen's theorem on fundamental sets.** We refer to the theorem of Part I, §6, according to which the invariant factors of the matrix

$$\| (\overset{i}{\gamma}_{\mu} \cdot \overset{j}{\gamma}_{n-\mu}) \|$$

for any two fundamental sets are all unity. Let them be in any case $e_1$, $e_2, \cdots, e_{R_\mu}$. That their number is $R_\mu$ follows from other considerations that need not detain us.* The reduction to canonical sets will be accomplished as before, only now

$$(\overset{i}{\gamma}_{\mu} \cdot \overset{i}{\gamma}_{n-\mu}) = \pm e_i ,$$

the same indices being zero as previously. Therefore we have now

$$\overset{hk}{\epsilon}_{n-\mu} \cdot e_h e_k = \pm \overset{hk}{\alpha}_{\mu} \cdot e_k .$$

Hence $e_h$ is a factor of $\overset{hk}{\alpha}_{\mu}$, and in particular of $\overset{hh}{\alpha}_{\mu}$ *for every* $T$. But for the identical transformation $\overset{hh}{\alpha}_{\mu}=1$, hence $e_h=1$ which is precisely Veblen's theorem.

## §3. Coincidences and fixed points of transformations of a manifold

66. A *coincidence* of two transformations $T$, $T'$ is a pair of points $A$, $A'$, of $M_n$, such that $A'$ is a transform of $A$, by both $T$ and $T'$. A *fixed point* of $T$ is a coincidence for $T$ and the identical transformation. Let $\Gamma_n$, $\Gamma_n'$ be the cycles corresponding to $T$ and $T'$, and let $B$ be the image of $A'$ on $M_n'$; $A \times B$ is an intersection of $\Gamma_n$ with $\Gamma_n'$ and conversely to such an intersection corresponds a coincidence of $T$ and $T'$. The determination of the *number of coincidences and fixed points* is then reduced to a question of intersections of cycles. The actual numbers are not definite, may even vary for transformations of the same class, become infinite, and so on. Not so, however, with the attached Kronecker index $(\Gamma_n \cdot \Gamma_n')$, whose determination alone is usually possible. Its interpretation is simple enough. If we consider suitable approximations to $T$ and $T'$ there will be only a finite number of intersections. Some of these, say $k'$, shall be counted positively, others, say $k''$, counted negatively according to a definite rule, and the difference $k'-k''$ is independent of the mode of approximation, in a sense sufficiently clear in the light of Part I, §5. From what has been established there, it follows that *the number of coincidences to be obtained in the sequence is an actual topological invariant of the transformations involved*, the same being

* See p. 301; also Veblen's paper on this question quoted in Part I, §6.

true for *the number of fixed points* counted in an analogous manner to the coincidences.

67. We shall now endeavor to characterize topologically the coincidences according to the signs of their contributions to $(\Gamma_n \cdot \Gamma_n')$. We select two polyhedral approximations that intersect in a finite number of points only, in such a way that if $A \times B$ is one of them it has neighborhoods on the cycles and on $M_n \times M_n'$ that are interiors of simplexes. We continue to call the approximations $\Gamma_n$, $\Gamma_n'$. Each intersection $A \times B$ counts for $\pm 1$ in computing the index, and as far as its neighborhood alone and those of $A$ and $B$ on $M_n$ and $M_n'$ are concerned, everything is as if all complexes involved were linear spaces. The matrix-indicatrix method will then be again the most convenient.

To $\Gamma_n$ and $\Gamma_n'$ we may make correspond matrices

$$\| X_n , Y_n \| , \| X_n' , Y_n' \| .$$

Regarding $X_n$ and $X_n'$ we may replace them by any others whose determinants have the same signs. This is an immediate corollary of the definition of the indicatrix. Let us assume our coördinate system such that the unit matrix of order $n$, $I_n$, corresponds to an indicatrix for $M_n$ as referred to the $x$ coördinates. Then $I_n$ will play the same part for $M_n'$ and the $y$ coördinates, $I_{2n}$ for $M_n \times M_n'$ and the coördinates $x, y$. Under the circumstances, as we have shown in No. 60, if $\theta$, $\theta'$ denote the contributions of $A \times B$ to the indices $(\Gamma_n \cdot A \times M_n')$, $(\Gamma_n' \cdot A \times M_n')$, then the determinants in question have the sign of $\theta$ and $\theta'$. We conclude that there will be suitable indicatrices for $\theta\Gamma_n$ and $\theta'\Gamma_n'$ of type

$$\| I_n , Y_n \| \quad , \quad \| I_n , Y_n' \| .$$

Now the contribution of $A \times B$ to $(\theta\Gamma_n \cdot \theta'\Gamma_n')$, in absolute value equal to unity, has the sign of

$$\begin{vmatrix} I_n , & Y_n \\ I_n , & Y_n' \end{vmatrix} ,$$

since $I_{2n}$ is a matrix-indicatrix for $M_n \times M_n'$. This determinant is equal to $|Y_n' - Y_n|$, and therefore the contribution of $A \times B$ to $(\Gamma_n \cdot \Gamma_n')$ has the sign of $\theta\theta'|Y_n' - Y_n|$. The determinant factor is certainly $\neq 0$, else $A \times B$ would not be an isolated intersection of the $\Gamma$'s. Furthermore owing to the degree of arbitrariness in the choice of these cycles we may always assume that the equation of the $n$th degree

$$f(t) = | Y_n' - tY_n | = 0$$

(characteristic equation of $Y_n' - tY_n$) has only distinct roots. We know already that $f(1) \neq 0$.

Let $e = \pm 1$ be the same integer as in No. 60 ($+1$ if $T$ maintains an indicatrix of $M_n$ at $A$, $-1$ if it inverts it). Then as shown there

$$\epsilon \, | Y_n | > 0 .$$

Since the determinant $Y_n$ is the leading coefficient of $f(t)$, $f(1)$ has the sign of $(-1)^\nu \epsilon$, where $\nu$ is the number of real roots of $f(t)$ that are less than unity. Therefore the contribution of $A \times B$ to $(\Gamma_n \cdot \Gamma_n')$ *is equal to* $(-1)^\nu \epsilon \theta \theta'$.

68. It is decidedly desirable and worth while to find a geometric interpretation for $\nu$. The transformation $T$ acts as an affine transformation whereby the vector $V$ whose $x$ components are $\xi_1, \xi_2, \cdots, \xi_n$ is changed into a vector whose components are defined by the matrix product.

$$\| \xi_1 , \xi_2 , \cdots , \xi_n \| \cdot Y_n ,$$

and similarly for $T'$ and $Y_n'$. Let $t_i$ be a real root of $f(t)$. Then there is a $V_i$ whose transforms by $T$ and $T'$ are collinear and in the ratio $1 : t_i$. Let $V_1, V_2, \cdots, V_\nu$ be the similar vectors for all real roots $<1$. The vectors

$$u_1 V_2 + \cdots + u_\nu V_\nu ; \qquad 0 \leqq u_i \leqq 1 , \; \sum u_i = 1$$

fill up a simplex $\Sigma_\nu^0$ of $M_n$ with a vertex at $A$. Its transforms by $T$ and $T'$ are simplexes $\Sigma_\nu$, $\Sigma_\nu'$ with the common vertex $A'$, the second being entirely on the first, with all cells through $A'$ situated in the same linear spaces, and $\nu$ is the largest integer for which such simplexes exist. This is the desired interpretation for $\nu$.

69. We now proceed with the determination of $(\Gamma_n \cdot \Gamma_n')$.

Referring to No. 54, and recalling the special relations for canonical sets, we find that the only indices $(\gamma_\lambda^i \times \bar{\delta}_{n-\lambda}^j \cdot \gamma_\mu^h \times \delta_{n-\mu}^k) \neq 0$ are those for which $\lambda + \mu = n$ and whose values are given below.

(*a*) $\mu \neq \frac{1}{2} n$ *or else* $= \frac{1}{2} n$ *an even integer.*

$$\begin{aligned}(\overset{i}{\gamma}_\mu \times \overset{j}{\delta}_{n-\mu} \cdot \overset{i}{\gamma}_{n-\mu} \times \overset{j}{\delta}_\mu) &= (-1)^{n(\mu+1)} (\overset{i}{\gamma}_\mu \cdot \overset{i}{\gamma}_{n-\mu}) (\overset{j}{\gamma}_\mu \cdot \overset{j}{\gamma}_{n-\mu}) \\ &= (-1)^{n(\mu+1)} ; \quad 1 \leqq i , j \leqq R_\mu ,\end{aligned}$$

for with canonical systems, the two indices dropped are equal, their common value being $\pm 1$.

(*b*) $\mu = \frac{1}{2} n$ *an odd integer.* For entirely similar reasons the indices below alone are left, with values as given:

$$(\overset{2h-1}{\gamma} \times \overset{2k-1}{\delta} \cdot \overset{2h}{\gamma} \times \overset{2k}{\delta}) = (\overset{2h}{\gamma} \times \overset{2k}{\delta} \cdot \overset{2h-1}{\gamma} \times \overset{2k-1}{\delta}) = -1 ,$$

$$(\overset{2h-1}{\gamma} \times \overset{2k}{\delta} \cdot \overset{2h}{\gamma} \times \overset{2k-1}{\delta}) = (\overset{2h}{\gamma} \times \overset{2k-1}{\delta} \cdot \overset{2h-1}{\gamma} \times \overset{2k}{\delta}) = +1 .$$

To facilitate the reading of these formulas we have omitted the lower indices whose common value is $\frac{1}{2}n$.

70. The application to $(\Gamma_n \cdot \Gamma_n')$ is immediate. Let, throughout, the $\eta$'s and $\beta$'s play the same part for $\Gamma_n'$ or $T'$ as the $\epsilon$'s and $\alpha$'s for $\Gamma_n$ or $T$. The distributive law for indices gives

$$(\Gamma_n \cdot \Gamma_n') = \sum_{i,j,\mu} \overset{ij}{\epsilon}_{n-\mu} \overset{hk}{\eta}_{\mu} \ (\overset{i}{\gamma}_{\mu} \sim \overset{j}{\delta}_{n-\mu} \cdot \overset{h}{\gamma}_{n-\mu} \times \overset{k}{\delta}_{\mu}) \, .$$

Therefore for $n \not\equiv 2$, mod 4, and with $\mu$ replaced by $n-\mu$,

$$(70.1) \qquad (\Gamma_n \cdot \Gamma_n') = \sum_{\mu=0}^{n} (-1)^{n\mu} \sum_{i,j=1}^{R_\mu} \overset{ij}{\epsilon}_{n-\mu} \overset{ij}{\eta}_{\mu} \, ,$$

while for $n \equiv 2$, mod 4, after some simplifications,

(70.2) $(\Gamma_n \cdot \Gamma_n') =$ (same sum as in (70.1) except that $\mu$ does not take the value $\frac{1}{2}n$)

$$+ \sum_{h,k=1}^{\frac{1}{2}R_{n/2}} \Big( \overset{2h-1,2k}{\epsilon_{n/2}} \overset{2h,2k-1}{\eta_{n/2}} + \overset{2h,2k-1}{\epsilon_{n/2}} \overset{2h-1,2k}{\eta_{n/2}} - \overset{2h-1,2k-1}{\epsilon_{n/2}} \overset{2h,2k}{\eta_{n/2}} - \overset{2h,2k}{\epsilon_{n/2}} \overset{2h-1,2h-1}{\eta_{n/2}} \Big) \, .$$

In terms of the $\alpha$'s and $\beta$'s, by means of No. 64 we find, if $n \not\equiv 2$, mod 4,

$$(70.3) \qquad (\Gamma_n \cdot \Gamma_n') = \sum_{\mu=0}^{n} (-1)^{\mu} \sum_{i,j=1}^{R_\mu} \overset{ij}{\alpha}_{\mu} \overset{ij}{\beta}_{n-\mu} \, ,$$

and if $n \equiv 2$, mod 4,

(70.4) $(\Gamma_n \cdot \Gamma_n') =$ (same sum as in (70.3) except that $\mu$ does not take the value $\frac{1}{2}n$) .

$$+ \sum_{h,k=1}^{\frac{1}{2}R_{n/2}} \Big( \overset{2h-1,2k}{\alpha_{n/2}} \overset{2h,2k-1}{\beta_{n/2}} + \overset{2h,2k-1}{\alpha_{n/2}} \overset{2h-1,2k}{\beta_{n/2}} - \overset{2h-1,2k-1}{\alpha_{n/2}} \overset{2h,2k}{\beta_{n/2}} - \overset{2h,2k}{\alpha_{n/2}} \overset{2h-1,2k-1}{\beta_{n/2}} \Big) \, .$$

71. To obtain the number of invariant points of $T$, always counted with a certain sign, it is convenient to replace throughout $T$ by the identical transformation, whose cycle we denote by $\Gamma_n^0$, and $T'$ by $T$. The homologies for the identical transformation are (No. 62)

$$\overset{h}{\bar{\gamma}}_{\mu} \sim \overset{h}{\gamma}_{\mu} \, ,$$

hence $\alpha_\mu^{ii}=1$, $\alpha_\mu^{ij}=0$ if $i\neq j$. On replacing the $\alpha$'s by these values and the $\beta$'s by the $\alpha$'s both (70.3) and (70.4) reduce to the very simple formula

$$(\overset{0}{\Gamma}_n \cdot \Gamma_n) = \sum_{\mu=0}^{n} (-1)^\mu \sum_{i=1}^{R_\mu} \alpha_\mu^{ij} . \tag{71.1}$$

Remarkably enough *this formula is still correct if the fundamental sets cease to be canonical.* For if $\gamma_\mu'^i$ is a new fundamental set, we have

$$\gamma_\mu'^i \approx \sum_j A_{ij} \gamma_\mu^j \qquad (\text{mod } M_n) ,$$

where the $A$'s are integers whose determinant

$$|A_{ij}| = \pm 1 .$$

Hence the transformation matrix of the $\gamma'$'s that takes the place of

$$||\alpha_\mu^{ij}||$$

is obtained by transforming the latter by means of the matrix of the $A$'s, an operation which leaves unchanged the sum of the terms in the principal diagonal, which is the term corresponding to $\mu$ in (71.1).

A particularly noteworthy case is when the effect of $T$ on any cycle is merely to increase it by a zero-divisor, which includes as a special case deformations. Then *all* the $\alpha$'s in (71.1) are equal to $+1$ and

$$(\Gamma_n \cdot \Gamma_n) = \sum_{\mu=0}^{n} (-1)^\mu R_\mu .$$

This expression is the well known Euler-Poincaré *characteristic number of* $M_n$ (difference between the number of cells of even dimensionality of any covering $C_n$ and the number of the rest). Since $R_\mu = R_{n-\mu}$ its value is zero for $n$ odd, hence this very neat proposition: *for a $T$ of the preceding type, in particular of the same class as the identity, the number of invariant points counted with their signs is equal to the Euler-Poincaré characteristic. It is zero when $n$ is odd.*

Thus for $n=2$ the number is $2-2p$ as found by Birkhoff, who however confined himself to analytic transformations.

Fixed points for transformations of hyperspheres have proved of importance in many questions. There are then no cycles of dimensions other than 0 or $n$ so that (71.1) becomes

$$(\overset{0}{\Gamma}_n \cdot \Gamma_n) = \alpha_0 + (-1)^n \alpha_n ,$$

a result obtained by Brouwer for $\alpha_0 = 1$.

Finally formula (71.1) for $n=2$ includes practically all related results obtained in recent years by various authors (Brouwer, Nielsen, Kérékjárto, Alexander).

72. Another class of applications is to *coincidences of algebraic correspondences on algebraic curves.* Owing to a proposition which I proved in my Monograph, p. 19, the Kronecker indices give then the *exact* number of intersections. We should therefore expect to be able to identify (70.4) with the well known coincidence formula due to A. Hurwitz.* Observe that his $h'_{ki}$, etc., are the same as the $h_{ki}$, etc., not for $C'$ but for $C'^{-1}$ and should be replaced as indicated in his footnote; for example, $h'_{ik}$ by $G'_{ik}$ and so on. With these changes and a comparison of notations the identification becomes complete.

Thus *the coincidence formulas may be obtained without making use of any function theory.* The classical Hurwitz relations between the periods are simply the analytical translation of the transformations on the linear cycles of $M_n$.

* Mathematische Annalen, vol. 28 (1887), p. 578, formula 35. See in this connection Chisini, Istituto Lombardo Rendiconti, ser. 2, vol. 7 (1924), p. 481.

# IV

## MANIFOLDS WITH A BOUNDARY AND THEIR TRANSFORMATIONS*

This is the continuation of the paper which appeared in the January, 1926, number of these Transactions.† Its chief object is to extend to an $M_n$ with a boundary the results already obtained for transformations of manifolds without boundary. To these already treated in full we devote a few pages chiefly to elucidate and simplify certain points of importance for the extension. We have succeeded in deriving coincidence and fixed points formulas for the two types of transformations that are alone amenable to anything like a general treatment and extended the formulas of this and the preceding paper to transformations between two different manifolds with or without a boundary. As an incidental acquisition there should be pointed out some highly interesting topological propositions obtained in Parts II, III. Of importance also is the fact that by means of ample use of matrices we have been able to put all coincidence formulas of this and the previous paper in very simple and manageable form.

### I. General remarks on manifolds

**1. A theorem on intersecting complexes.** Let $C_h$, $C_k$, $M_p$ be two complexes and a manifold on $M_n$, all polyhedral and with $C_k$ a sub-complex of $M_p$. We wish to prove that *the intersections* $C_h \cdot C_k$ *taken on* $M_n$ *and* $(C_h \cdot M_p) \cdot C_k$ *taken on* $M_p$ *coincide and when* $h+k=n$ *the related Kronecker indices are equal.* We assume that as regards all intersections to be considered the restrictions of Tr., No. 15, are fulfilled. The problem is then reduced at once to the case where the complexes are simplexes and the manifolds

* Presented to the Society, October 30, 1926; received by the editors in November, 1926. See also Proceedings of the National Academy of Sciences, vol. 12 (1926), p. 737.

† Referred to in the sequel as Tr. Unless otherwise stated, the notations, terminology, assumptions, etc., of that paper will apply directly here. The only changes will be actually in the definition of an $M_n$, and using for cycles besides $\Gamma$, also $\gamma$ as in Tr., Part II, and later other letters for special types. Since we shall make considerable use of matrices we may as well give our notations here. All our matrices will have integer terms. Any matrix will be designated as its generic element with position indices omitted. The transverse of a matrix $m$ will be called $m'$; when $m$ is square its determinant is denoted by $|m|$ and the sum of the terms in its principal diagonal is called its *trace*.

linear spaces. The natural procedure is as in Tr., Part I, §2, to compare indicatrices. The indicatrix of Tr., No. 7, is now so chosen that $A_0A_1 \cdots A_q$, $q=p+h-n$, be on $C_h \cdot M_p$, $A_0A_1 \cdots A_qA_{h+1} \cdots A_r$, $r=p+q-n$, be on $M_p$ and that multiplied respectively by $\alpha_q$, $\alpha_p$, they constitute indicatrices of their complexes. Let also $\alpha_l' A_0A_1 \cdots A_l$ be an indicatrix of $(C_h \cdot M_p) \cdot C_k$ when the intersection is taken on $M_p$. Then the relations

$$(1.1) \qquad \alpha_h\alpha_k\alpha_l\alpha_n = \alpha_h\alpha_p\alpha_q\alpha_n = \alpha_q\alpha_k\alpha_l'\alpha_p = 1$$

define the orientations of $C_l = C_h \cdot C_k$ and $C_h \cdot M_n$ taken on $M_n$, and that of $(C_h \cdot M_p) \cdot C_k$ taken on $M_p$. Since the $\alpha$'s are all $\pm 1$, from (1.1) follows at once

$$(1.2) \qquad \alpha_k\alpha_l' = \alpha_p\alpha_q = \alpha_h\alpha_n = \alpha_k\alpha_l,$$

therefore

$$(1.3) \qquad \alpha_l' = \alpha_l.$$

The cells of the complexes to be compared are the same, and by (1.3) similarly oriented. Therefore

$$(1.4) \qquad C_h \cdot C_k = (C_h \cdot M_p) \cdot C_k,$$

as we wished to prove. As we know from Tr., No. 8, when $l=0$, we merely have a Kronecker index to consider and then

$$(1.5) \qquad (C_h \cdot C_{n-h}) = ((C_h \cdot M_p) \cdot C_{n-h}),$$

which may also be established directly as above by comparison of indicatrices.

Of particular interest is the case when the $C$'s are cycles. The passage through suitable approximations, as in Tr., Part I, §4, will enable us to drop all restrictions as to them provided their intersection does not meet the boundary of $M_p$. For the latter being polyhedral, the approximations can always be so carried out that $C_l$ remains on it.

2. In the applications that we have especially in view, $p=n-1$, $k=n-h$, so that we deal with a Kronecker index. Furthermore $M_{n-1}$ will there be a subcomplex of the defining $C_n$ of $M_n$ and the preceding result does not apply outright. The extension is, however, easy on this basis: $C_h$ and $C_{n-h}$ are assumed as general as possible and hence their intersections are isolated points, of which any one, say $A$, is on an $E_{n-1}$ of $C_n$. As far as the contribution of $A$ to the index is concerned, only the two $n$-cells $E_n$, $E_n'$ of $C_n$ incident with $E_{n-1}$ are involved. The situation is then the same as if $M_n$ were reduced to $E_n + E_n'$ + their boundaries. But this system is obviously

homeomorphic with preservation of structure* to a similar one in $S_n$. Practically this means that we may assume that $M_n$ is an $S_n$. On the latter, however, we can always choose a defining $C_n$ with $A$ on an $n$-cell of it. Then $M_p$ will not be a subcomplex of it, or rather its cells through $A$ will not be cells of $C_n$. Therefore we are back to the case already considered and the conclusion is the same.

3. **On the Kronecker index.** Until further notice we assume that $M_n$ is *without boundary*. In an important paper Veblen has had occasion to define so-called *intersection numbers* for associated complexes of special type on $M_n$.† Since we have applied some of his results it is important to show that his numbers are merely the Kronecker indices of the complexes.

We begin with Poincaré congruences

$$C_k \equiv \Gamma_{k-1}, \qquad C_{n-k+1} \equiv \Gamma_{n-k}, \tag{3.1}$$

where the cycles have no common points. They give rise as in Tr., No. 18, to the relation

$$(C_k \cdot \Gamma_{n-k}) = (-1)^k (\Gamma_{k-1} \cdot C_{n-k+1}), \tag{3.2}$$

valid without any other restrictions than the one just stated concerning the cycles. That follows at once by passing to polyhedral approximations for which (3.2) holds. Since the indices are defined in each case by means of the approximations the relation is valid for the initial complexes.

* Two sets of simplicial cells, $\{e\}$, $\{e'\}$, are said to have the same structure, whenever to each $e$ there corresponds one and only one $e'$ of same dimensionality and conversely, and when furthermore the incidence relations between any two $e$'s and the corresponding $e'$'s are the same.

† These Transactions, vol. 25 (1923), pp. 540–550. Substantially the same results, derived in similar fashion, were also obtained simultaneously but independently by Hermann Weyl, Revista de Matematica Hispaño Americana, 1923. Regarding the index, I recently received a communication from Weyl in which he points out that, unknown to me, he had proved its independence from the defining $C_n$: (a) for $n=2$ in Jahresbericht der Deutschen Mathematiker-Vereinigung, vol. 25 (1916), p. 225, also Note to the second edition of *Die Idee der Riemannschen Flächen*; (b) for $(\Gamma_1 \cdot \Gamma_{n-1})$ and any $n$ in the Revista paper. In the same paper he also points out that for $\mu = \frac{1}{2}n$ even, it may not be possible to have a canonical set whose matrix of indices is the identity. The bearing on the coincidence formulas in Tr. is that one must have two associated sets as when $\mu \neq n/2$, with $T$ operating on one and $T'$ on the other. Let these sets be $\gamma^i$, $\gamma'^i$, with $\gamma'^i \approx \Sigma g_{ij}\gamma^j$. All matrices $g$ corresponding to such a pair of associated sets are of the form $pgp'$, where $p$ is an arbitrary square matrix of order $R_{n/2}$ with $|p| = \pm 1$. The properties of the whole class of such matrices are invariants of $M_n$, another form of a remark made by Weyl, loc. cit.

In a letter received in early December, Weyl communicated to me substantially the same derivation as mine of the matrix formulas (10.3) and (39.1) of this paper from those of Tr., Part II (see also the note in the Proceedings of the National Academy of Sciences, for December, 1926). He has thus confirmed my results, at an important point.

4. Let us now adopt the notations of Tr., No. 23. We designate furthermore by $\overline{C}_n$ the dual of $C_n$, whose cells are subcomplexes of $C_n'$ also. Any one $\overline{E}_{n-k}$ is a sum of cells of $C_n'$ of type $(k, p_1, \cdots, p_i)$, with a common vertex $A^k$, and the $p$'s all$>k$. The $(n-k)$-cells of the sum are of type $(k, k+1, \cdots, n)$. A similar statement holds for the cell $E_k$ of $C_n$ that carries $A^k$, except that now the types are $(q_1, \cdots, q_j, k)$, $q_i<k$ (Coll. Lect.,* p. 89).

Let $A^{k-1}$ be any vertex on a cell $E_{k-1}$ of the boundary of $E_k$. It is on a cell $\overline{E}_{n-k+1}$ of $\overline{C}_n$, which as before is the sum of certain cells of $C_n'$ with $A^{k-1}$ for vertex. We assume $E_{k-1}$ positively related to $E_k$, hence a positive cell of its boundary $\Gamma_{k-1}$, and $\overline{E}_{n-k+1}$ so sensed that $\overline{E}_{n-k}$ is a positive cell of its boundary $\Gamma_{n-k}$. Now the two cycles have no common points, since these would have to be vertices of $C_n'$, while the vertices they carry are of the incompatible types $A^{k+i}$, $A^{k-i}$, $i\neq 0$. Hence (3.2) is applicable here and

$$(E_k \cdot \overline{\Gamma}_{n-k}) = (-1)^k(\Gamma_{k-1} \cdot \overline{E}_{n-k+1}). \tag{4.1}$$

$E_k$ and $\overline{\Gamma}_{n-k}$ meet at $A^k$, and nowhere else. For all cells of $C_n'$ of which the cycle is made up are of type $(k, p_1, \cdots, p_r)$, $p_i>k$. Therefore (Tr., No. 23), they are on cells of $k$ dimensions of $C_n$, unless they are merely vertices and of the same type as $A^k$. In that case they are on $k$-cells with one and only one such vertex on each $k$-cell. Hence the $k$-cell $E_k$ of $C_n$ can only meet $\overline{\Gamma}_{n-k}$ at a single point which can only be $A^k$. Therefore

$$(E_k \cdot \overline{\Gamma}_{n-k}) = (E_k \cdot \overline{E}_{n-k}). \tag{4.2}$$

Due to the symmetrical relation of $C_n'$ to the dual complexes, it is not necessary to repeat the discussion for the second index in (4.1) and we infer at once

$$(\Gamma_{k-1} \cdot \overline{E}_{n-k+1}) = (E_{k-1} \cdot \overline{E}_{n-k+1}). \tag{4.3}$$

Therefore in place of (2.1)

$$(E_k \cdot \overline{E}_{n-k}) = (-1)^k(E_{k-1} \cdot \overline{E}_{n-k+1}). \tag{4.4}$$

Now this is precisely the relation proved by Veblen for his intersection numbers in §4 of his paper. Therefore in proving that they are merely Kronecker indices, say for a given $k$, the latter may be increased or decreased by one unit. Hence we may assume $k=n$. But for this special value Veblen's definition reduces essentially to ours. *Therefore his numbers are Kronecker indices for every k.*

---

* *Analysis Situs*, by Oswald Veblen (The Cambridge Colloquium, Part II), will be referred to as Coll. Lect. throughout the present paper.

5. We are then justified in taking over Veblen's theorems bodily. Of particular significance are these two:

I. *In order that* $\gamma_\mu \approx 0$ *on* $M_n$ *without boundary (i.e., that it be a zero-divisor or bounding cycle) it is necessary and sufficient that* $(\gamma_\mu \cdot \gamma_{n-\mu}) = 0$ *whatever* $\gamma_{n-\mu}$.

II. *Let* $\gamma_\mu^i, \gamma_{n-\mu}^j$ *run respectively through the elements of two fundamental sets for their dimensions. Then the rank of*

$$\| (\gamma_\mu{}^i \cdot \gamma_{n-\mu}^j) \| \tag{5.1}$$

*is* $R_\mu$, *and its invariant factors are all unity.*

These propositions are proved by Veblen only when in each pair one cycle is a subcomplex of $C_n$, the other one of the dual $\overline{C}_n$. As every cycle is homologous to a subcomplex of $\overline{C}_n$ (Tr., No. 27, Remark), and as the index is invariant with respect to homology, the two propositions are true without restrictions.

In our previous paper we actually made use only of the first part of II, that is, of the fact that (5.1) is of rank $R_\mu$, and proved the other part directly. Of I also, the necessary condition alone was needed and established directly in Part I. The only question that could have been raised is then as to whether the $R$'s in the formulas are the connectivity numbers and not merely the ranks of the matrices (5.1) and we have just answered it in the negative.

6. Let us return to the fundamental set $\gamma_\mu^1, \gamma_\mu^2, \cdots, \gamma_\mu{}^{R_\mu}$ relative to the operation $\approx$ as considered in Tr., p. 37. It has the property that any $\gamma_\mu$ of $M_n$ is a combination of the cycles of the set plus a zero-divisor. Hence a fundamental set as to $\sim$ is obtained by merely adding zero-divisors to the $\gamma$'s. For example, the set in Coll. Lect., p. 117, is of this very nature, and so are the canonical sets of our first paper, but we need not limit ourselves to these.

Let $\gamma_{n-\mu}^1, \gamma_{n-\mu}^2, \cdots, \gamma_{n-\mu}^{R_\mu}$ correspond in analogous fashion to the dimensionality $n-\mu$. As is well known the number of the cycles is again $R_\mu$ (Poincaré). Complete the two sets by zero-divisors so as to have fundamental sets for $\sim$, then form (5.1). By §3, the matrix will merely consist of the square array

$$L_\mu = \| (\gamma_\mu{}^i \cdot \gamma_{n-\mu}^j) \| \qquad (i,j = 1,2, \cdots, R_\mu) \tag{6.1}$$

bordered with rows and columns of zeros. Hence the determinant $|L_\mu| = \pm 1$, for in absolute value it is the product of the invariant factors of $L_\mu$, all equal to one.

7. **Continuous transformations.** First observation of very general nature: All of Tr. Part I, goes through whether the cells of $C_n$ are simplicial or merely convex (i.e., corresponding to convex polyhedral regions on the representative polyhedron $\Pi_n$). The bearing of this becomes clear when we remember that even when the cells of $C_p$, $C_q$ are simplicial, those of $C_p \times C_q$ are merely convex. Hence it is proper to apply to a product manifold all the approximation work (loc. cit.). Of course the more general type of $C_n$, practically the same as Veblen's, possesses a regular subdivision of the restricted type (Coll. Lect., p. 85).

This important point settled let us return to the situation of Tr., Part II, §2, particularly as regards the definition of $T\gamma_\mu = \bar{\gamma}_\mu$. It is obtained by means of an approximation $\bar{\Gamma}_n$ to $\Gamma_n$, defining cycle of the transformation $T$. The question arises, however, whether $T\gamma_\mu$ is unique. To show that such is the case let a second defining complex $C_n^0$ lead to $\Gamma_n^0, \gamma_\mu^0, \bar{\gamma}_\mu^0$. We do not exclude the possibility that $C_n^0$ coincides with $C_n$. Then by Tr., Part I, §6, if the approximations are sufficiently close,

$$\bar{\Gamma}_n^0 \cdot \gamma_\mu^0 \times M_n' \sim \bar{\Gamma}_n \cdot \gamma_\mu \times M_n' \qquad (\mathrm{mod}\ M_n \times M_n'). \tag{7.1}$$

Let the difference of the two sides bound $C_{\mu+1}$. When the point $A \times B$ describes it, $B$ describes on $M_n'$ a singular image $C'_{\mu+1}$ of the complex, and when $A \times B$ describes the boundary of $C_{\mu+1}$, $B$ describes $\bar{\delta}_\mu - \bar{\delta}_\mu^0$. Hence this last cycle bounds on $M_n'$, and $\bar{\gamma}_\mu - \bar{\gamma}_\mu^0$ therefore bounds on $M_n$, that is $\bar{\gamma}_\mu \sim \bar{\gamma}_\mu^0$ (mod $M_n$), which proves the uniqueness of $T\gamma_\mu$.

8. Throughout our first paper we have assumed that an oriented zero-cell is considered as a cycle of zero dimensions. (See in particular No. 63.) This justifies, for example, the theorem of No. 52, our assertion as to the Euler characteristic, No. 71, etc. Veblen in Coll. Lect., p. 110, does not consider such cycles. The advantage of our procedure is readily perceived. Two points similarly oriented on the same connected piece of a $C_n$ constitute homologous cycles, for their difference is the boundary of an obvious one-cell. Hence $R_0$ is the number of distinct connected pieces of $C_n$, and if it is an $M_n$ without boundary, $R_0 = R_n$, and Poincaré's duality formula holds without exception. His formula for the Euler characteristic also takes the simpler form, assumed in Tr., No. 71,

$$\sum(-1)^i \alpha_i = \sum(-1)^i R_i. \tag{8.1}$$

The convention of Tr., No. 58, reduces to the following for $\mu = 0$: The transform of $\gamma_0 = A$, point of $M_n$, is $\alpha_0 A$, where $\alpha_0$ is as in Tr., No. 56.

Incidentally the orientation of $\Gamma_n$, defined (loc. cit.) by $\alpha_0 \geqq 0$, is indeterminate when $\alpha_0 = 0$. In that case we orient the cycle arbitrarily. This has no great importance, as a change of orientation merely changes the sign of certain Kronecker indices, whose absolute value, however, is alone of interest.

What matters chiefly is to make sure that the fundamental formula (59.1) holds without exception. The proof goes through in fact even for $\mu = 0$, $n$, but the direct verification for both cases is very simple. The case $\mu = n$ has already been considered in No. 63, but there is a simpler and more direct verification, as we shall presently see.

Let first $\mu = 0$. Denote again by $\theta$, as in Tr., No. 61, the contribution of a certain point $A \times B$ to $\alpha_0$. With the same notations as there used, except that the cycle is $\overline{\Gamma}_n$, if it carries the cell $E_n = A \times B \cdots A_n \times B_n$ then $\theta E_n$ is its indicatrix as cell of $\overline{\Gamma}_n$. It follows $\alpha_0 = \sum \theta$, where the sum is extended to all points $A \times B$ of $\overline{\Gamma}_n$ corresponding to a fixed $A$. If $\overline{T}$ is the transformation defined by $\overline{\Gamma}_n$, we may think of them as the points $A \times B$ whose $B$ is the image of some $\overline{T}A$. The verification of (59.1) for $\mu = 0$ requires that

$$(8.2) \quad (\Gamma_n \cdot A \times M_n') = (\overline{\Gamma}_n \cdot A \times M_n') = (\overline{T}A \cdot M_n') = \alpha_0 (A \cdot M_n') = \alpha_0,$$

which is in accordance with the formula in Tr., No. 56.

Let now $\mu = n$, and denote by $\epsilon$ the same integer as in Tr., No. 61; $\epsilon B B_1 \cdots B_n$ is then the indicatrix of $M_n'$ at $B$ and $\epsilon$ has the sign of $|Y_n|$. Hence as at the end of No. 61, $A \times B$ contributes $(-1)^n \epsilon\theta$ to $(\overline{\Gamma}_n \cdot M_n \times B)$ and therefore

$$(8.3) \qquad (\overline{\Gamma}_n \cdot M_n \times B) = (-1)^n \sum \epsilon\theta.$$

Here the sum is extended to all points of $\Gamma_n$ with a fixed $B$, or to all points $A \times B$ of the cycle such that among the points $\overline{T}A$ there is one whose $M_n'$ image is $B$.

Now the image of $\overline{T}C_n$ is a polyhedral $\gamma_n$ on $C_n$. We may so subdivide $C_n$, say into $C_n^0$, that $\overline{T}C_n$ be a subcomplex of the subdivision. Then any particular cell $E_n$ of $C_n^0$ will count say $k'$ times positively and $k''$ times negatively among the cells of $\overline{T}C_n$, and we shall have from the above, if $B$ is on $E_n$,

$$(8.4) \qquad k' - k'' = \sum \theta\epsilon = \alpha_n ;$$

$$(8.5) \qquad \overline{T}M_n = \alpha_n \cdot M_n .$$

The verification of (8.2) for $\mu = n$ requires here that

$$(8.6)\qquad (\bar{\Gamma}_n \cdot M_n \times B) = (\bar{\Gamma}_n \cdot M_n \times B) = (-1)^n(\bar{T}M_n \cdot A) = (-1)^n \alpha_n (M_n \cdot A) = (-1)^n \alpha_n,$$

which comparison with (8.3) shows is correct.

9. **Invariant form for coincidence and fixed point formulas.** The formulas given in Tr. are rather involved and furthermore depend upon a special choice of fundamental sets. It so happens that by making use of matrices there can be derived formulas independent of the particular fundamental sets relative to the operation $\approx$ that may be chosen. To begin with, the first formula on p. 43, Tr., reads*

$$(9.1)\qquad (-1)^{\mu(n+1)} L_\mu \epsilon_{n-\mu} L_\mu = \alpha_\mu L_\mu,$$

where $L_\mu$ is as defined in §6 of the present paper. Since its determinant is not zero, this gives at once

$$(9.2)\qquad (-1)^{\mu(n+1)} L_\mu \epsilon_{n-\mu} = \alpha_\mu,$$

and then

$$(9.3)\qquad \epsilon_{n-\mu} = (-1)^{\mu(n+1)} L_\mu^{-1} \alpha_\mu.$$

This solves then the problem of expressing the $\epsilon$'s in terms of the $\alpha$'s for any choice of fundamental sets as to $\approx$, and not merely for the canonical sets. The explicit formulas of No. 64, Tr., for canonical sets follow from these simply by replacing the $L$'s by the form corresponding to each $\mu$.

10. For the coincidence formula the starting point is as in Tr., No. 70, the relation independent of the choice of sets

$$(10.1)\qquad (\Gamma_n \cdot \Gamma_n') = \sum \epsilon_{n-\mu}^{ij} \eta_\mu^{hk} (\gamma_{n-\mu}^i \times \delta_\mu^j \cdot \gamma_\mu^h \times \delta_{n-\mu}^k).$$

By applying Tr., (53.3), and then replacing afterwards the $\delta$'s by $\gamma$'s, this becomes

$$(10.2)\qquad (\Gamma_n \cdot \Gamma_n') = \sum (-1)^\mu \epsilon_{n-\mu}^{ij} \eta_\mu^{hk} (\gamma_{n-\mu}^i \cdot \gamma_\mu^h)(\gamma_\mu^j \cdot \gamma_{n-\mu}^k) = \sum (-1)^\mu \text{ trace } \epsilon_{n-\mu}' L_{n-\mu} \eta_\mu L_\mu'.$$

By (9.2) or (9.3) applied to both transformations, and recalling that $(ab)' = b'a'$, we have

$$(10.3)\qquad \epsilon_{n-\mu}' L_{n-\mu} \eta_\mu L_\mu' = (L_\mu^{-1}\alpha_\mu)' \beta_{n-\mu} L_\mu' = (L_\mu \beta_{n-\mu}' L_\mu^{-1} \alpha_\mu)'.$$

---

* In that formula $\gamma_{n-\mu}^j$ must be replaced by $\overset{i}{\gamma}_{n-\mu}$.

Since transposed matrices have equal traces, (10.2) may be replaced by

$$(10.4)\qquad (\Gamma_n \cdot \Gamma_n') = \sum (-1)^\mu \operatorname{trace} L_\mu \beta_{n-\mu}' L_\mu^{-1} \alpha_\mu .$$

This is the final form that we need for the coincidence formula. It is manifestly more compact and clear than what is found in No. 70, Tr.,* and has also the requisite invariant form most appropriate for manifolds with boundary. For the fixed point formula we choose here the second transformation as the identity. Then the $\beta$'s are all unit matrices and (10.4) reduces to

$$(10.5)\qquad (\Gamma_n \cdot \Gamma_n^0) = \sum (-1)^\mu \operatorname{trace} \alpha_\mu .$$

This is (71.1), Tr., with the two cycles interchanged. However, with the situation chosen there, $\mu$ should be replaced by $n-\mu$ in (71.1), or, what is equivalent, the two cycles interchanged on the left. Here again, as in loc. cit., for a transformation of the same class as the identity, or more generally for one which merely adds zero-divisors to any cycle,

$$(10.6)\qquad (\Gamma_n^0 \cdot \Gamma_n^0) = \sum (-1)^\mu R_\mu ,$$

the Euler characteristic.

11. While dealing with transformations, let us bring out the following interesting property: *The transform of a zero-divisor or cycle is also a zero-divisor or cycle.* In signs, if $\gamma_\mu \approx 0$, also $\bar{\gamma}_\mu = T\gamma_\mu \approx 0$. For then $\gamma_\mu \times \delta_{n-\mu}$ is also a zero-divisor or cycle for $M_n \times M_n'$, hence, by §3 and Tr. (59.1),

$$(11.1)\qquad 0 = (\Gamma_n \cdot \gamma_\mu \times \delta_{n-\mu}) = (-1)^\mu (\bar{\gamma}_\mu \cdot \gamma_{n-\mu})$$

for every $\gamma_{n-\mu}$, from which at once $\bar{\gamma}_\mu \approx 0$.

## II. Manifolds with a boundary

12. We propose to modify somewhat the definition of manifolds of our earlier paper. The difference, however, pertains only to the boundary and since it has played no direct part there, all results so far obtained will continue to hold. Let $C_n$ be a complex. Consider the star of cells whose center is a given $E_k$ of $C_n$. Between its cells there take place the same incidence relations as between the elements of a certain $C_{n-k-1}$: the $h$-cells of the latter correspond to the $(h+k+1)$-cells of $C_n$ incident with $E_k$, or the star of cells of center $E_k$. Now $C_n$ defines a manifold $M_n$ when $C_{n-k-1}$ is homeomorphic

* In the first formula of No. 70 replace $\sim$ by $\times$.

(a) with the boundary of an $E_{n-k}$ (sphere of $S_{n-k}$);

or

(b) with an $E_{n-k-1}$ plus its boundary.

$E_k$ is an *interior* or a *boundary* cell of $M_n$ according as (a) or (b) is fulfilled. The set $F_{n-1}$ of all boundary cells is the boundary of $M_n$.

Due to (b), the $C_{n-k-2}$ image of the star of $F_{n-1}$ with center $E_k$ assumed on $F_{n-1}$ is homeomorphic with the boundary of an $E_{n-k-1}$. Hence (a) holds for every cell of $F_{n-1}$, which is thus, itself, an $M_{n-1}$ without boundary.

The manifold conditions may be replaced by others equivalent and often more convenient. Instead of considering the $h$-cells of the star of center $E_k$ as $(h-k-1)$-cells of a certain complex, let us think of them as $(h-k)$-cells of a new system $s_{n-k}$ with a unique zero-cell corresponding to $E_k$. Then in place of (a) and (b) we may obviously impose the following conditions:

(a′) when $E_k$ is an interior cell, $s_{n-k}$ is homeomorphic with an $E_{n-k}$;

(b′) when $E_k$ is on the boundary, $s_{n-k}$ is homeomorphic with a star of cells of $S_{n-k}$ that constitutes an $E_{n-k}$ plus an $E_{n-k-1}$ on its boundary, or what is the same thing, with a hemispherical region of $S_{n-k}$ plus its flat base.*

The conditions here imposed for an $M_n$ are more stringent for the boundary than Veblen's (Coll. Lect., p. 88). They are, however, in the nature of a certain homogeneity requirement along the boundary and entirely similar to what is imposed on the interior. It will also be observed that they do not demand that the cells of $C_n$ be *simplicial*, but merely that they be *convex*.

13. *When $C_n$ satisfies the manifold conditions so does any subdivision of it, $C_n'$.* This is proved in outline as follows. With each flat $E_k$ on $C_n$ we associate a system such as $\{e\}$ of Tr., No. 3, where $e$ represents a class of incident $(k+1)$-cells on the same half $S_{k+1}$. The system $\{e\}$ is the same for two flat $k$-cells with a $k$-subcell in common. The $(h+k+1)$-cells incident with $E_k$ may be considered as $h$-cells made up with the similar $(k+1)$-cells as points, that is with the $e$'s as points. Hence when $E_k$ is a cell of any particular subdivision $C_n'$ of $C_n$, $\{e\}$ is homeomorphic with the $C_{n-k-1}'$ similar to $C_{n-k-1}$ of §12. It is then sufficient to show that when $C_n$ behaves as desired, every $\{e\}$ obeys conditions (a), (b).

Let then $E_k$ carry $E_{k-1}$, with a corresponding system $\{e'\}$. When $\{e\}$ behaves as desired so does $\{e'\}$. For to each $e$ there corresponds a one-cell

* One is tempted to replace (b′) by the simpler "$s_{n-k}$ is an $E_{n-k}$ plus an $E_{n-k-1}$ on its boundary." Unfortunately to show that this is equivalent to (b′) we need the following theorem: Two $h$-cells can be homeomorphically transformed into one another in such manner that two $(h-1)$-cells of their boundaries are similarly transformed. For $h=2$ this goes back to the Jordan curve theorem, but beyond that there is no proof.

of $e''$s with fixed end points (images of $E_k$ itself). Hence the relation between the two systems is like that between a sphere or hemisphere in $S_{n-k}$ (a sphere when $E_k$ is an interior cell, a hemisphere otherwise) and its locus when the space, immersed in an $S_{n-k+1}$, rotates through an angle $\pi$ around a diameter. It follows that for our purpose $\{e\}$ may replace $\{e'\}$ and $E_k$ replace $E_{k-1}$. Ultimately, then, we shall merely have to consider some cell of $C_n$ itself; that is, as asserted, the correct behavior of $\{e\}$ follows from that of every $C_{n-k-1}$ attached to the cells of $C_n$.

14. From (a) and (b) as applied to $(n-1)$-cells it follows that every interior $(n-1)$-cell of $C_n$ separates two $n$-cells, and every boundary $(n-1)$-cell is on a unique $n$-cell of $C_n$; $F_{n-1}$ is then the sum of all $(n-1)$-cells on a unique $n$-cell of $C_n$. *We assume again that $M_n$ is orientable.* Then $F_{n-1}$ will also be orientable. For let us orient $C_n$ and then sense each $E_{n-1}$ of $F_{n-1}$ positively in relation to the $E_n$ that it bounds. Between the $n$- and $(n-1)$-cells incident with a given $E_{n-2}$ of $F_{n-1}$ we can write down the same Poincaré congruences as for the one- and zero-cells of a polygonal line. Hence the two end $(n-1)$-cells, which are those of $E_{n-1}$ incident with $E_{n-2}$, are oppositely related to $E_{n-2}$, and $F_{n-1}$ is oriented. Its orientation as thus fixed shall be preserved throughout. It corresponds to the congruence $M_n \equiv F_{n-1}$.

15. **The auxiliary manifold** $V_n$. We assume henceforth that $M_n$ has a boundary $F_{n-1}$. Take, then, another copy $\overline{M}_n$ of the manifold and piece the two together along corresponding boundary points. *The new configuration $V_n$ so obtained is an $M_n$ without boundary.* (Any element of $\overline{M}_n$ corresponding to a given one of $M_n$ will be called its conjugate and designated by the same letter barred.) If $C_n$ is the basic defining complex of $M_n$, we use $\overline{C}_n$ for $\overline{M}_n$ and $C_n+\overline{C}_n$ for $V_n$. Then if $E_k$ is the cell of $M_n$ in §12, when it is not on $F_{n-1}$, the complex $C_{n-k-1}$ plays the same part for it relative to $V_n$ as to $M_n$. Hence it behaves then according to (a), and similarly for $\overline{E}_k$ and $\overline{C}_{n-k-1}$. However, when $E_k$ is on $F_{n-1}$, in place of $C_{n-k-1}$ we have $C_{n-k-1}+\overline{C}_{n-k-1}$. As this set is composed of two $(n-k-1)$-cells pieced together along their boundaries, it is homeomorphic to the boundary of an $E_{n-k}$. This is seen at once by referring to the piecing together of two hemispheres in $S_{n-k}$ into a sphere of that space. Hence $E_{n-k}$ behaves again in accordance with (a), which proves our assertion as to $V_n$.

If $E_n$ is an $n$-cell of $C_n$, $E_n'$ its indicatrix, we sense $\overline{E}_n$ of $\overline{M}_n$ by $-\overline{E}_n'$, hence $V_n = M_n - \overline{M}_n$. The importance of $V_n$ is due to the fact that the solution of the coincidence problem for pairs of transformations of $M_n$ will be reduced to the same problem for pairs of associated transformations of $V_n$.*

* This or a similar procedure has been followed by other authors dealing with this question. See, for example, Brouwer, Comptes Rendus, vol. 168 (1919), p. 1042; Alexander, these Transactions, vol. 23 (1922), pp. 89–95.

It is evident that its topological properties are really inherent properties of $M_n$ itself. We shall be particularly concerned with the study and disposition of its fundamental sets.

16. **Fundamental sets for $V_n$.** As several distinct types of cycles will have to be considered we shall avoid excess of indices by using not only $\Gamma$, $\gamma$ but also $G$, $\Delta$, $D$ to designate them.

Let $\Gamma_\mu{}^1, \cdots, \Gamma_\mu{}^p$ be a fundamental set for the $\mu$ cycles of $M_n$, and consider all possible homologies

$$(16.1) \qquad \sum t_i \Gamma_\mu^i \sim \text{a cycle of } F_{n-1} \qquad (\text{mod } M_n).$$

By paraphrasing a well known process* we may readily establish that these homologies are sums of multiples of a finite number of the same type which constitute a fundamental set for them. The members of the fundamental set and also the cycles can then be combined in such a fashion as to have a new fundamental set of $\Gamma$'s (for which we keep the same designation as above) with fundamental homologies

$$(16.2) \quad \theta_i \Gamma_\mu{}^{r_\mu^1+i} \sim \text{a cycle of } F_{n-1} \qquad (\text{mod } M_n) \qquad (i = 1, 2, \cdots, p - r_\mu^1).$$

The operations referred to correspond to elementary transformations on the matrix of the coefficients of the fundamental homologies. The first $r_\mu^1$ cycles are not related by any homology such as (16.1) and in particular they are entirely independent; the remaining cycles have some non-zero multiple homologous to a cycle on the boundary. Between the cycles $\Gamma_\mu{}^{r_\mu^1+i}$ there may exist homologies mod $M_n$. Reducing those as above, we shall replace the cycles by a new set $G_\mu^j$, $j=1, 2, \cdots, p-r_\mu^1$, whose first say $r_\mu^2$ elements are independent while the others are $\approx 0$. Of course $r_\mu^1 + r_\mu^2 = R_\mu$, the $\mu$th connectivity index of $M_n$. The cycles $\Gamma_\mu^i$, $i=1, 2, \cdots, r_\mu^1$, $G_\mu^j$, $j=1, 2, \cdots, r_\mu^2$, constitute a fundamental set for $M_n$ relative to the operation $\approx$.

It will be convenient to call a cycle *symmetric* when it is $\approx$ mod $M_n$ to a cycle on $F_{n-1}$. Then the difference between the cycle and its conjugate is $\approx 0$, mod $V_n$.

17. Of no less importance than the preceding are the *skew-symmetric* cycles of $V_n$. We so designate those of type $C_\mu - \overline{C}_\mu$, where $C_\mu$ is on $M_n$. In place of them it would be possible to consider the complexes whose boundary is on $F_{n-1}$ and their properties in regard to what might be called "quasi-homologies" or relations:

$$(17.1) \qquad \sum t_i C^i + \text{a complex of } F_{n-1} \sim 0 \qquad (\text{mod } M_n).$$

* Klein, *Elliptische Modulfunctionen*, vol. 2, p. 543.

The theory resulting therefrom would be largely a paraphrase of the one to be found here, with the mild advantage of being strictly confined to elements of $M_n$ itself.

Since $C_\mu - \overline{C}_\mu$ is a cycle, the boundary of $C_\mu$ is on $F_{n-1}$. Furthermore we can remove from it any $\mu$-cell on $F_{n-1}$ without changing the cycle.

If $\rho_\mu$ is the $\mu$th connectivity index of $V_n$, any $\rho_\mu+1$ skew-symmetric $\mu$-cycles are dependent, so that Klein's reasoning applies and we find a fundamental set $\Delta_\mu^1, \Delta_\mu^2, \cdots, \Delta_\mu^q$, for the type. It is reducible in similar fashion to the above as regards homologies between the intersections with the boundary:

$$\sum t_i \Delta_\mu^i \cdot F_{n-1} \sim 0 \qquad (\text{mod } F_{n-1}), \tag{17.2}$$

with a similar conclusion: The set can be replaced by a new one, for which the same designation is preserved, with fundamental homologies for the type (17.2):

$$\tau_j \Delta_\mu^j \cdot F_{n-1} \sim 0 \qquad (\text{mod } F_{n-1}) \qquad (j = 1, 2, \cdots, s \leqq q). \tag{17.3}$$

In short the first $s$ cycles of the new set intersect the boundary of $M_n$ in zero-divisors or bounding cycles of it, while the remaining $q-s$ intersect it in independent cycles of $F_{n-1}$. By Tr., No. 35, Theorem V, they are independent for $V_n$ as well. There may exist, however, homologies between the first $s$. Reducing again as regards these we finally obtain a fundamental set consisting of the following:

(a) $r_\mu^3$ cycles $\Delta_\mu^1, \Delta_\mu^2, \cdots, \Delta_\mu^{r_\mu^3}$, independent (mod $V_n$), but meeting $F_{n-1}$ in cycles $\approx 0$ (mod $F_{n-1}$);

(b) $r_\mu^4 = q-s$ cycles $D_\mu^1, D_\mu^2, \cdots, D_\mu^{r_\mu^4}$, independent (mod $V_n$) and intersecting $F_{n-1}$ in cycles independent (mod $F_{n-1}$);

(c) a set of at most $q - r_\mu^3 - r_\mu^4$ zero-divisors of $V_n$.

Every skew-symmetric cycle is $\approx$ to a sum of $\Delta$'s and $D$'s (these notations are henceforth reserved for cycles (a) and (b)). Also there can be no homology involving both $D$'s and $\Delta$'s, as we see at once by reference to their intersections with $F_{n-1}$. Since there is none involving each type alone, they constitute $r_\mu^3 + r_\mu^4$ independent cycles of $V_n$, and therefore a fundamental set as regards $\approx$ and skew-symmetric cycles.

Let $\gamma_\mu \approx \Delta_\mu + D_\mu$ be a skew-symmetric cycle, $\Delta_\mu$ and $D_\mu$ being sums of cycles (a) or (b). Let $\delta_{n-\mu}$ be any cycle of $F_{n-1}$. From §1 follows, with indices computed as to $F_{n-1}$,

$$((\gamma_\mu \cdot F_{n-1}) \cdot \delta_{n-\mu}) = ((D_\mu \cdot F_{n-1}) \cdot \delta_{n-\mu}). \tag{17.4}$$

Hence, by §5 and the definition of the $D$'s, in order that $\gamma_\mu \cdot F_{n-1} \approx 0$ (mod $F_{n-1}$) it is necessary and sufficient that $\gamma_\mu \approx \Delta_\mu$ alone.

18. Let $\gamma_{\mu-1}$ be a cycle of $F_{n-1}$ not $\approx 0$ on that manifold, but bounding $C_\mu$ on $M_n$. Then $C_\mu - \overline{C}_\mu$ is a skew-symmetric cycle of type $D$ not $\approx 0$ (mod $V_n$), since its intersection $\gamma_{\mu-1}$ with $F_{n-1}$ is not $\approx 0$ (mod $F_{n-1}$). Hence to a set of $t$ independent $\mu-1$ cycles of $F_{n-1}$ bounding on $V_n$ correspond as many independent skew-symmetric cycles of type $D$, and conversely. Therefore $r_\mu^4$ is the *number of distinct* $\mu - 1$ *cycles* $F_{n-1}$ *that bound on* $M_n$.

Another interesting property is the following: *Every* $\Delta$ *has a multiple which is the sum of a cycle on* $M_n$ *and of a cycle on* $\overline{M}_n$. Let $\Delta_\mu$ be the cycle, $\Delta_\mu'$ a polyhedral approximation of maximum generality intersecting $F_{n-1}$ in $\gamma_{\mu-1}$. By §17,

$$\gamma_{\mu-1} \approx 0 \qquad (\text{mod}\, F_{n-1}). \tag{18.1}$$

Since $\Delta_\mu'$ is of maximum generality it has no $\mu$-cells on $F_{n-1}$, hence we may write

$$\Delta_\mu' = C_\mu - \overline{C}_\mu', \tag{18.2}$$

where the first complex is on $M_n$, the second on $\overline{M}_n$, and both have the common boundary $\gamma_{\mu-1}$. From (18.1) we infer that there exists on $F_{n-1}$

$$C_\mu'' \equiv t\gamma_{\mu-1}, \qquad t \neq 0, \tag{18.3}$$

therefore

$$t\Delta_\mu \sim t\Delta_\mu' = (t\,C_\mu - C_\mu'') - (t\,\overline{C}_\mu' - C_\mu''). \tag{18.4}$$

Each parenthesis at the right is a cycle, the first on $M_n$, the second on $\overline{M}_n$, which proves our assertion.

19. **Theorems.** I. *Every cycle of* $V_n$ *is the sum of a skew-symmetric cycle and of one on* $M_n$.

Let $\gamma_\mu$ be the cycle. It may be assumed polyhedral and the sum of two complexes $C_\mu$ and $\overline{C}_\mu'$, the first on $M_n$, the second on $\overline{M}_n$. But

$$\gamma_\mu = (C_\mu + C_\mu') + (\overline{C}_\mu' - C_\mu'). \tag{19.1}$$

The second parenthesis is a skew-symmetric cycle, the first a complex on $M_n$, the difference of two cycles, hence also a cycle, and the theorem is therefore proved.

II. *A* $\gamma_\mu$ *of* $M_n$ *not* $\approx 0$ (mod $M_n$) *cannot be skew-symmetric.*

For let it be $\approx \delta_\mu$, skew-symmetric. Then

$$\delta_\mu - \gamma_\mu \approx 0 \approx \bar{\delta}_\mu - \bar{\gamma}_\mu \approx + \delta_\mu + \bar{\gamma}_\mu \qquad (\text{mod } V_n), \tag{19.2}$$

Therefore

$$t(\gamma_\mu + \bar{\gamma}_\mu) \sim 0 \ (\text{mod } V_n)\,;\ t \neq 0. \tag{19.3}$$

*First let $\gamma_\mu$ be symmetric,* with $\theta\gamma_\mu$, $\theta \neq 0$, homologous (mod $M_n$) to $\gamma_\mu^0$ on $F_{n-1}$. Then

$$\theta\gamma_\mu \sim \theta\bar{\gamma}_\mu \sim \gamma_\mu^0 \qquad (\text{mod } V_n), \tag{19.4}$$

and therefore by (19.3)

$$2t\gamma_\mu^0 \sim 0 \qquad (\text{mod } V_n). \tag{19.5}$$

There exists then on $V_n$ a

$$C_{\mu+1} \equiv 2t\gamma_\mu^0\,. \tag{19.6}$$

Let $C'_{\mu+1}$ be the subcomplex of $C_{\mu+1}$ that includes all its cells on $M_n$ (logical intersection of $C_{\mu+1}$ and $M_n$) and set $C_{\mu+1} - C'_{\mu+1} = \bar{C}''_{\mu+1}$, complex made up of all $(\mu+1)$-cells of $C_{\mu+1}$ interior to $\bar{M}_n$ plus their boundaries. Since the boundary of $\bar{C}''_{\mu+1}$ can only be on $F_{n-1}$, it coincides with that of $C''_{\mu+1}$. Hence

$$C'_{\mu+1} + C''_{\mu+1} \equiv 2t\gamma_\mu^0\,, \tag{19.7}$$

for $C_{\mu+1}$ has the same boundary as the complex at the left. But the latter is on $M_n$, hence

$$\theta\gamma_\mu \sim \gamma_\mu^0 \approx 0\,;\quad \gamma_\mu \approx 0 \qquad (\text{mod } M_n), \tag{19.8}$$

contrary to assumption:

*Assume now that $\gamma_\mu$ is not symmetric.* It is reducible to a polyhedral cycle whose $\mu$-cells are all interior to $M_n$. The first part of this double assertion is established as in Coll. Lect., pp. 95, 118, the second as in Tr. No. 24, (b). Furthermore a complete proof, independent of the present discussion, is given below (§23). The left member of (19.3), polyhedral and without $\mu$-cells on $F_{n-1}$, will also bound a polyhedral $C_{\mu+1}$ (Coll. Lect., p. 120). Let again the sum of its cells on $M_n$ be called $C'_{\mu+1}$. This last complex has for total boundary $t\gamma_\mu$ plus a cycle $\gamma'_\mu$ on $F_{n-1}$. Therefore

$$t\gamma_\mu \sim -\gamma'_\mu \qquad (\text{mod } M_n), \tag{19.9}$$

and $\gamma_\mu$ is a symmetric cycle, — a new contradiction, and II is proved.

In the second part of the discussion we have established that if (19.3) holds then $\gamma_\mu$ is symmetric. The identical reasoning holds for the more general homology, in which $\gamma_\mu$ is still on $M_n$,

$$t\gamma_\mu + \theta\bar{\gamma}_\mu \sim 0 \qquad (\text{mod } V_n). \tag{19.10}$$

which shows that

III. *If $\gamma_\mu$ of $M_n$ is dependent upon its conjugate $\gamma_\mu$ then it is a symmetric cycle.*

IV. *When $\gamma_\mu$ of $M_n$ is not* $\approx 0$, mod $M_n$, *then it is also not* $\approx 0$, mod $V_n$.

This is a special case of II corresponding to $\delta_\mu = 0$.

20. From the preceding propositions follows at once the all important

THEOREM. *The cycles $\Gamma_\mu^{\alpha_1}$, $G_\mu^{\alpha_2}$, $\Delta_\mu^{\alpha_3}$, $D_\mu^{\alpha_4}$* $(\alpha_i = 1, 2, \cdots, r_\mu^i)$ *constitute a fundamental set for $V_n$ and the operation $\approx$. To obtain a fundamental set for $\sim$ it is only necessary to add zero-divisors of $M_n$ and skew-symmetric zero-divisors of $V_n$.*

From their definition we know that the Γ's and $G$'s are independent, and similarly for the Δ's and $D$'s. From II follows that the four sets are independent in their totality. Then again from I and the fact that the Γ's and $G$'s constitute a fundamental set as to $\approx$ and $M_n$, and similarly the Δ's and $D$'s for the skew-symmetric cycles and $V_n$, the theorem follows in its completeness.

COROLLARY. *The $\mu$th connectivity index of $V_n$ is* $\rho_\mu = r_\mu^1 + \cdots + r_\mu^4$.

21. Our present object is to show that with a suitable choice of associated sets for the dimensionalities $\mu$ and $n-\mu$ certain indices are, or can be made to be, zero, which will naturally lead to the canonical sets.

22. LEMMA. *Let $E_n$ be a cell, $\Gamma_\mu$ a cycle on $E_n$, both polyhedral. Then there exists a polyhedral $C_{\mu+1}$ on $E_n$ bounded by $\Gamma_\mu$.*

Since $\Gamma_\mu$ is on $E_n$, it has no points on the boundary of the cell. Hence there exists a cell $E_n'$ which together with its boundary lies on $E_n$, and also carries $\Gamma_\mu$. Let $\sigma$ be the least distance between points of the boundaries of the cells. Cover $E_n$ with a polyhedral complex whose cells are all of diameter $<\sigma$, and remove all its $n$-cells with a boundary point on the boundary of $E_n$. The $n$-cells that are left together with their boundaries constitute a $C_n$ on $E_n$ and carrying $E_n'$. The cycle $\Gamma_\mu$ bounds on $E_n'$, hence also on $C_n$, and on that complex it bounds a polyhedral $C_{\mu+1}$ (Coll. Lect., p. 120) which proves the lemma.

23. Let $C_\mu$ be a complex on $M_n$ with its boundary on $F_{n-1}$, and let us follow step by step the approximation described in Tr., Part I, §4. We first apply the Alexander-Veblen process, and obtain $C_{\mu+1}$, polyhedral with an

associated congruence

$$C_{\mu+1} \equiv C_\mu - C'_\mu + C^0_\mu . \tag{23.1}$$

By reference to Coll. Lect., pp. 95, 118, we find that we may approximate each vertex of the boundary of $C_\mu$ by a vertex of the boundary of $C_n$, that is, by a point of $F_{n-1}$. Then $C^0_\mu$ and the boundary of $C'_\mu$ will be on $F_{n-1}$. Except for that, the rest of the work of approximation is directly applicable here. However, No. 23, Tr., does not apply to non-boundary cells of $C'_\mu$ on $F_{n-1}$, and necessitates a slight modification.

We take $C_n$ such that $C'_\mu$ is now a subcomplex of it (Tr., No. 14, Lemma II), and as a first move, reduce its boundary on $F_{n-1}$ as described in Tr., No. 23. We thus obtain a polyhedral cycle $\Gamma'_{\mu-1}$ of $F_{n-1}$ whose $(\mu-i)$-cells are all on cells of no less than $n-i$ dimensions of $C_n$. Furthermore $\Gamma_{\mu-1}-\Gamma'_{\mu-1}$ bounds a polyhedral $C''_\mu$, and both new cycles and complex are as near as we please to $C'_\mu$, hence to $C_\mu$. It follows that in (23.1), $C'_\mu$ and $C^0_\mu$ may be replaced by $C'_\mu+C''_\mu$ and $C^0_\mu+C''_\mu$, without altering the situation. Therefore we may start with a complex $C'_\mu$ whose boundary is already reduced as indicated. To extend then the reduction of Tr., No. 23, all we need to do is to replace $C'_\mu$ by a complex whose non-boundary cells are interior to $M_n$, since the reduction in question can be applied to these. The reader will verify with ease that the situation pertaining to (23.1) remains unaffected by any step to be taken presently. Furthermore, the reduction will be more thorough than in Tr., No. 23, in that the new elements introduced first here, then by the process of Tr., No. 23, will be throughout interior to $M_n$. Hence the complex as finally reduced will have all non-boundary cells interior to the manifold.

Let first $E_\mu$ be a simplicial cell of $C'_\mu$ on $F_{n-1}$ and on the boundary of the cell $E_n$ of $C_n$. Draw rectilinear segments from a fixed point of $E_n$ to all points of $E_\mu$. The resulting simplicial cell has for boundary a $\Gamma_\mu$ whose $\mu$-cells other than $E_\mu$ (which is one of them) are on $E_n$. There is a $t\neq 0$ such that $C'_\mu - t\Gamma_\mu \sim C'_\mu$ is a complex which no longer includes $E_\mu$. The cell $E_\mu$ has then been replaced by cells on $E_n$, or interior cells of $M_n$. Thus we can reduce $C'_\mu$ to a similar complex whose $\mu$-cells are interior to $M_n$. Assume then that all cells of more than $h$ dimensions, $h<\mu$, of $C'_\mu$ are interior cells. I say that the reduction can be extended to the $h$-cells as well.

Let $E_h$ of $C'_\mu$ be on $F_{n-1}$. Introduce new interior vertices on $C_n$ so chosen that for the new complex $C'_n$ the star of cells of center $E_n$ carries no boundary cells of $C'_\mu$ on its own. Since the star attached to $C_n$ carries no other boundary cells of $C'_\mu$ than those on $E_h$ or its boundary, the construction offers no difficulty. The related $C_{n-h-1}$ of Tr., No. 3, which is merely the boundary of the star, will then intersect $C'_\mu$ in a $(\mu-h-1)$-cycle on $E_{n-h-1}$. By our lemma this cycle bounds a polyhedral $C_{\mu-h}$ on $E_{n-h-1}$, having then no point on $F_{n-1}$. Let $E_h=A_0 A_1 \cdots A_h$, and let also $A_{h+1} \cdots A_i$ be an arbitrary cell of $C_{\mu-h}$; then denote by $\overline{C}_{\mu+1}$ the complex sum of the cells $A_0A_1 \cdots A_i$. It has in common with $C'_\mu$ all cells incident with $E_h$, and except for these cells and their boundary points it is entirely interior to $M_n$. Let us sense its boundary $\Gamma'_\mu$ so that the indicatrix of any cell is obtained by naming first the vertices of $E_h$, next those of the related cell on $C_{n-h-1}$ so that both sets be vertices of an indicatrix for their cell. Then $C'_\mu$ and $\Gamma'_\mu$ will have the same cells incident with $E_h$ each counted with the same multiplicity for both complexes. Hence $C'_\mu-\Gamma'_\mu \sim C'_\mu$ will have lost $E_h$ without acquiring new cells on $F_{n-1}$. This shows that the reduction can also be extended to $h$ cells of $C'_\mu$, hence to its boundary cells.

Combining the whole discussion we have the important

THEOREM. *Let $C_n$ be an assigned defining complex of $M_n$ and $C_\mu$ a complex on the manifold, whose boundary is on $F_{n-1}$. Then there is a corresponding congruence* (23.1) *with* (a) *$C'_\mu$ polyhedral, as near as we please to $C_\mu$, with its boundary on $F_{n-1}$ and as near as we please to that of $C_\mu$, also with its $\mu-i$ cells on cells of no less than $n-i$ dimensions of $C_n$, the non-boundary cells being interior to $M_n$*; (b) *$C^0_\mu$ on $F_{n-1}$ and as near as desired to the boundary of $C_\mu$.*

COROLLARY I. *Every cycle of $M_n$ is homologous to an interior cycle behaving in accordance with the theorem.*

COROLLARY II. *Every skew-symmetric cycle is homologous to one of form $C'_\mu-\overline{C}'_\mu$, where $C'_\mu$ behaves in accordance with the theorem.*

For let $\gamma_\mu=C_\mu-\overline{C}_\mu$. Reduce $C_\mu$ as above with the congruence (23.1). Then also

$$\overline{C}_{\mu+1} \equiv \overline{C}_\mu - \overline{C}'_\mu + C^0_\mu \;; \tag{23.2}$$

therefore

$$\gamma_\mu - (C'_\mu - \overline{C}'_\mu) \sim 0, \tag{23.3}$$

as was to be proved.

Unless otherwise stated we shall always assume the cycles of $M_n$ and the skew-symmetric cycles reduced as far as allowed by theorem and corollaries. Practically everywhere in the sequel, a cycle of one of these two types may be replaced by one which is $\sim$, mod $M_n$ and $V_n$ respectively, and then if it is not already reduced, we shall be at liberty to reduce it without further discussion.

24. **Theorems on indices.** I. $(\Gamma_\mu^i \cdot G_{n-\mu}^j) = (G_\mu^i \cdot G_{n-\mu}^j) = 0$.

There exists $G_{n-\mu}$ of $F_{n-1}$, $\sim t\, G_{n-\mu}^j$, mod $M_n$, $t \neq 0$. If we replace $G_{n-\mu}^j$ by $G_{n-\mu}$ we merely multiply the indices by $t$. Hence we need only show that

$$(24.1) \qquad (\Gamma_\mu^i \cdot G_{n-\mu}) = (G_\mu^i \cdot G_{n-\mu}^j) = 0,$$

which is obvious since $\Gamma_\mu^i$ and $G_\mu^i$ may be chosen interior to $M_n$ and then they will not meet $G_{n-\mu}$.

Explicitly, and since $(\Gamma \cdot G) = \pm (G \cdot \Gamma)$, *the indices* $(\Gamma \cdot G)$, $(G \cdot \Gamma)$, $(G \cdot G)$ *are all zero.*

II. *The index of two skew-symmetric cycles is zero.*

Let $\delta_\mu = C_\mu - \overline{C}_\mu$, $\delta_{n-\mu} = C_{n-\mu} - \overline{C}_{n-\mu}$ be the two cycles in the reduced form in position of maximum generality. Then for evident reasons of symmetry the two indices $(C_\mu \cdot C_{n-\mu})$ and $(\overline{C}_\mu \cdot \overline{C}_{n-\mu})$ taken with $M_n$ and $\overline{M}_n$ as the carrying manifolds are equal. Hence, taken with $V_n = M_n - \overline{M}_n$ as the carrying manifold, they are opposite. But $C_\mu$ does not meet $\overline{C}_{n-\mu}$ and $C_{n-\mu}$ does not meet $\overline{C}_\mu$. Hence

$$(24.2) \qquad \begin{aligned} (\delta_\mu \cdot \delta_{n-\mu}) &= (C_\mu - \overline{C}_\mu)(C_{n-\mu} - \overline{C}_{n-\mu}) \\ &= (C_\mu \cdot C_{n-\mu}) - (\overline{C}_\mu \cdot \overline{C}_{n-\mu}) = 0. \end{aligned}$$

III. $(G_\mu^i \cdot \Delta_{n-\mu}^i) = 0$.

In the proof $G$ may be replaced by $t\, G$, $t \neq 0$. Since there is a $t\, G$ on $F_{n-1}$, we may assume that $G$ itself is on the boundary. Then (§ 18) there is an $s \neq 0$ such that

$$(24.3) \qquad s\Delta \sim \Delta' + \overline{\Delta}'',$$

where $\Delta'$ is a cycle on $M_n$ and $\overline{\Delta}'$ a cycle on $\overline{M}_n$. By I and since $\Delta'$ and $\Delta''$ depend upon the $G$'s and $\Gamma$'s,

$$(24.4) \qquad (G \cdot \Delta') = (G \cdot \Delta'') = -(G \cdot \overline{\Delta}'') = 0\,;$$

therefore

$$(24.5) \qquad (G \cdot \Delta) = 0 = (\Delta \cdot G).$$

CONCLUSION. *All indices* $(\Gamma \cdot G)$, $(\Delta \cdot G)$, $(\Delta \cdot \Delta)$, $(\Delta \cdot D)$, *and those obtained by permuting the cycles, are zero.*

25. The matrix $L_\mu$ will then have the form

$$L_\mu = \begin{array}{c} \quad \Gamma_{n-\mu} \quad \Delta_{n-\mu} \quad G_{n-\mu} \quad D_{n-\mu} \\ \left\| \begin{array}{cccc} A & B & 0 & C \\ H & 0 & 0 & 0 \\ 0 & 0 & 0 & F \\ M & 0 & N & 0 \end{array} \right\| \begin{array}{c} \Gamma_\mu \\ \Delta_\mu \\ G_\mu \\ D_\mu \end{array} \end{array}$$

where the terms are all matrices; for example, $H = \|(\Gamma_\mu^i \cdot \Delta_{n-\mu}^j)\|$, $\|(\Delta_\mu^i \cdot \Delta_{n-\mu}^j)\| = 0$, etc. We shall now study $L_\mu$ and in particular show that by proper choice of cycles in each group it can be reduced to a much simpler form, with only one indeterminate matrix, namely $A$. As an incidental result we shall obtain very interesting duality theorems regarding the integers $r_\mu$, theorems quite similar to Poincaré's relation for the connectivity numbers of manifolds without boundary.

26. I. *F and N are square.* The permissible operations of adding a multiple of a cycle of a fundamental set to another or permuting two of them, applied to the groups $G_\mu$, $D_{n-\mu}$ will amount to the noted elementary transformations as applied to $F$. Hence $F$ may be reduced to the well known form

$$(26.1)\qquad \left\| \begin{array}{cccccc} e_1 & 0 & \cdots & 0 & \overbrace{0 & \cdots & 0}^{q} \\ 0 & e_2 & \cdots & \cdots & \cdots & \cdots \\ \cdots & \cdots & \cdots & \cdots & \cdots & \cdots \\ 0 & \cdots & \cdots & e_s & 0 & \cdots & 0 \\ 0 & \cdots & \cdots & 0 & 0 & \cdots & 0 \\ \cdots & \cdots & \cdots & \cdots & \cdots & \cdots \\ 0 & \cdots & \cdots & \cdots & \cdots & \cdots & 0 \end{array} \right\| \begin{array}{c} \\ \\ \\ \\ \left.\begin{array}{c} \\ \\ \\ \end{array}\right\} p \end{array}$$

where the $e$'s are the invariant factors of $F$ and there are $p$ rows and $q$ columns of zeros. Assume this done and continue to call the reduced matrix $F$. We must show that $p = q = 0$. Evidently, $p = 0$, for otherwise $L_\mu$, as reduced, would have a whole row of zeros, whereas its determinant is $\pm 1$ (§ 6). Then if $q \neq 0$ there is a $D_{n-\mu}$ such that $(G_\mu \cdot D_{n-\mu}) = 0$ for every symmetrical cycle $G_\mu$ and in particular for every one on $F_{n-1}$. Hence (§§5, 17), $D_{n-\mu}$ is dependent upon the cycles $\Delta_{n-\mu}$, which is untrue. Therefore $q = 0$, $F$ is square and so similarly is $N$. Their determinants are integers and factors of $|L_\mu| = \pm 1$, hence they are both $\pm 1$.

In the canonical form the $e$'s for both matrices will be $+1$. When $\mu \neq \frac{1}{2}n$, they may both be separately reduced to that form. Denoting generically by $I_k$ the unit matrix of order $k$, we have, when the reduction is carried out, $F=I_{r_\mu^2}$, $N=I_{r_\mu^4}$. When $\mu=\frac{1}{2}n$, we have at once, by Tr., No. 8, $N=(-1)^{n/2}F'$. The reduction of $F$ brings about that of $N$. For $\frac{1}{2}n$ even, the final form is as above; for $\frac{1}{2}n$ odd, it is $F=I_{r_\mu^2}=-N$.

As an important result proved incidentally we have $r_\mu^2=r_{n-\mu}^4$ or

II. (First duality theorem.) *The number of distinct symmetric $\mu$ cycles is equal to the number of independent $n-\mu-1$ cycles of $F_{n-1}$ that bound on $M_n$.*

Since $F$ is a unit matrix and since a $\Gamma-G$ is also a $\Gamma$, we can subtract from every $\Gamma_\mu$ of the fundamental set a $G_\mu$ cycle so chosen as to reduce the corresponding row of $C$ to zero. This means that we can so select the set of $\Gamma_\mu$'s as to have $C=0$. If $\mu\neq\frac{1}{2}n$, we may operate similarly on the $\Gamma_{n-\mu}$'s and reduce $M$ to zero, while when $\mu=\frac{1}{2}n$, we shall have $M=\pm C=0$. Hence

III. *The fundamental sets can be so selected as to give $L_\mu$ one of the two forms*

$$\begin{Vmatrix} A & B & 0 & 0 \\ H & 0 & 0 & 0 \\ 0 & 0 & 0 & 1 \\ 0 & 0 & 1 & 0 \end{Vmatrix}, \qquad \begin{Vmatrix} A & B & 0 & 0 \\ H & 0 & 0 & 0 \\ 0 & 0 & 0 & 1 \\ 0 & 0 & -1 & 0 \end{Vmatrix},$$

*according as $\mu$ is not or is $\frac{1}{2}n$ and odd.* To simplify we have merely indicated the unit matrices by 1.

IV. *A is a square matrix.* It may be reduced to the type (26.1) by the two operations of the beginning of this section applied to $\Gamma$'s alone. As reduced to that type we shall show again that $p=q=0$. For evident reasons of symmetry it is sufficient to show that $q=0$. For every $\Gamma_\mu$ we have $(\Gamma_\mu\cdot\Gamma_{n-\mu}^{s+1})=0$. Then if $\delta_{n-\mu}=\Gamma_{n-\mu}^{s+1}-\overline{\Gamma}_{n-\mu}^{s+1}$, we have, for every $G_\mu$,

$$(26.2) \qquad (G_\mu\cdot\delta_{n-\mu}) = (G_\mu\cdot\Gamma_{n-\mu}^{s+1}) - (G_\mu\cdot\overline{\Gamma}_{n-\mu}^{s+1}) = 0,$$

for the last two indices are equal in absolute value and the first is zero (§ 24, Theorem I). Also, as we may assume $\Gamma_\mu$ interior to $M_n$ and $\overline{\Gamma}_{n-\mu}^{s+1}$ interior to $\overline{M}_n$, and hence that the two are without common points,

$$(26.3) \qquad (\Gamma_\mu\cdot\delta_{n-\mu}) = -(\Gamma_\mu\cdot\overline{\Gamma}_{n-\mu}^{s+1}) = 0.$$

And finally, since $\delta_{n-\mu}$ is skew-symmetric,

$$(26.4) \qquad (\Delta_\mu\cdot\delta_{n-\mu}) = (D_\mu\cdot\delta_{n-\mu}) = 0.$$

In short, $(\gamma_\mu\cdot\delta_{n-\mu})=0$ for every $\mu$ cycle of the fundamental set, therefore for every $\mu$ cycle of $V_n$. Hence $\delta_{n-\mu}\approx 0$. There exists, then, a $t\neq 0$ such that $t\Gamma_{n-\mu}^{s+1}\sim t\overline{\Gamma}_{n-\mu}^{s+1}$. Hence (§ 19, Theorem III), $\Gamma_{n-\mu}^{s+1}$ is a symmetric cycle, contrary to assumptions. There exists then no $\Gamma_{n-\mu}$ whose upper index $>s$, which means that $q=0$ and proves IV.

Since $A$ is square, $r_\mu^1 = r_{n-\mu}^1$ and therefore

V. (SECOND DUALITY THEOREM.) *The number of cycles of* $M_n$ *of which no combination is a cycle of* $F_{n-1}$ *is the same for the dimensions* $\mu$ *and* $n-\mu$.

VI. *B and H are square matrices and their determinants are* $\pm 1$. We reduce again say $B$ to the form (26.1) and show that $p=q=0$. If $q \neq 0$ there exists a $\Delta_{n-\mu}$ such that $(\gamma_\mu \cdot \Delta_{n-\mu})=0$ whatever $\gamma_\mu$, hence $\Delta_{n-\mu}$ is a zero-divisor contrary to assumptions, and $q=0$. Assume now $p\neq 0$. Then the number of distinct $\Delta_{n-\mu}$'s is $< r_\mu^1 = r_{n-\mu}^1$, the order of $A$. Consider, however, the cycles $\Gamma_{n-\mu}^i - \overline{\Gamma}_{n-\mu}^i$. Since the $\Gamma$'s may be taken interior to $M_n$, these skew-symmetric cycles do not intersect $F_{n-1}$ and therefore are dependent upon the $\Delta$'s. On the other hand they are independent, or there would exist a $\Gamma_{n-\mu} \approx \overline{\Gamma}_{n-\mu}$, and hence symmetric (§ 19, Theorem III) in contradiction to the assumptions on the $\Gamma$ type. Hence there are at least $r_\mu^1$ distinct $\Delta_{n-\mu}$'s and $p=0$. Therefore $B$ is a square matrix and so is $H$. Here again their determinants are integers and factors of $|L_\mu| = \pm 1$, therefore also $= \pm 1$.

From VI it follows that $r_\mu^1 = r_{n-\mu}^1 = r_\mu^3 = r_{n-\mu}^3$. Hence

VII. (THIRD DUALITY THEOREM.) *The number of distinct skew-symmetric cycles that intersect* $F_{n-1}$ *in zero-divisors or bounding cycles is the same for the dimensions* $\mu$ *and* $n-\mu$ *and equal to the number of distinct cycles of* $M_n$ *of* $\mu$ *or* $n-\mu$ *dimensions that are independent of the cycles of* $F_{n-1}$.

27. It follows from the above that for $\mu \neq \frac{1}{2}n$ we can reduce both $B$ and $H$ to $I_{r_\mu^1}$. For $\mu = \frac{1}{2}n$, $H=(-1)^n B$ and the reduction of $B$ will bring about that of $H$.

For the computation to follow it is advisable to select fundamental sets thus: when $\mu < \frac{1}{2}n$ the four groups of cycles are taken in the order $\Gamma$, $\Delta$, $D$, $G$; when $\mu \geqq \frac{1}{2}n$ in the order $\Gamma$, $\Delta$, $G$, $D$. Then

$$\mu \neq \frac{1}{2}n, \quad L_\mu = \left\| \begin{array}{cc|cc} A & 1 & \multicolumn{2}{c}{\multirow{2}{*}{0}} \\ 1 & 0 & & \\ \hline \multicolumn{2}{c|}{\multirow{2}{*}{0}} & 1 & 0 \\ & & 0 & 1 \end{array} \right\|; \tag{27.1}$$

$$\mu = \frac{1}{2}n, \quad L_\mu = \left\| \begin{array}{cc|cc} \multicolumn{1}{c}{A} & 1 & \multicolumn{2}{c}{\multirow{2}{*}{0}} \\ (-1)^{n/2}\cdot 1 & 0 & & \\ \hline \multicolumn{2}{c|}{\multirow{2}{*}{0}} & 0 & 1 \\ & & (-1)^{n/2}\cdot 1 & 0 \end{array} \right\|$$

where the 1 stands in each case for a unit matrix whose order is not essential (see Remark below). What matters chiefly is that for the dimensions $\mu$ and $n-\mu$ those in the same places are of the same orders. In the upper left corner they are both of order $r_\mu^1$, in the right corner of orders $r_\mu^2$ and $r_\mu^4$. Concerning $A$ we have no available information, but fortunately it disappears entirely from our formulas and therefore need not concern us further.

**Remark.** Consider for a moment two matrices written in the form

$$\alpha = \| \alpha_{ij} \|, \qquad \beta = \| \beta_{ij} \|, \tag{27.2}$$

with elements $\alpha_{ij}$, $\beta_{ij}$ themselves matrices. The ordinary multiplication rule

$$\alpha\beta = \| \sum \alpha_{ik}\beta_{kj} \| \tag{27.3}$$

is directly applicable (taking care not to interchange factors in $\alpha_{ik}\,\beta_{kj}$) provided that (a) the number of columns in $\alpha$ is the same as the number of rows in $\beta$; (b) the sequence of the number of columns for the elements in a row of $\alpha$ is the same for all rows and also the same as for $\beta'$. These two conditions are fulfilled if $\alpha$ and $\beta$ are square with their diagonal elements $\alpha_{ii}$, $\beta_{ii}$ also square and of equal order for the same $i$. In that case not only $\alpha\beta$ but also $\beta\alpha$ may be obtained by the usual rule. This is the precise situation that we shall face throughout, where we shall find products of matrices all of the same structure as $L_\mu$.

Let us recall incidentally that, for example, $\alpha' = \|\alpha_{ji}'\|$, that is, the transposed of $\alpha$ is obtained by interchanging rows and columns and replacing each individual term by its transposed. All this goes back to the rule for matrix multiplication.

## III. Continuous transformations of manifolds with a boundary

28. Just as in the no-boundary case the definition of a continuous transformation is best given by reference to $M_n \times M_n'$, where $M_n'$ is a copy of $M_n$. However, before discussing the transformations we shall show that the product is also a manifold. The $M_n'$ image of any $M_n$ configuration will be denoted throughout by the same letter accented.

Let then $E_h$ be any cell of $C_n$, defining complex of $M_n$, $s_{n-h}$ the corresponding system such as appears in § 12 in connection with conditions (a′), (b′). We have

$$s_{n-h} = E_{n-h} + tE_{n-h-1}, \tag{28.1}$$

where $t=0$ or 1 according as $E_h$ is or is not an interior cell of $C_n$. To any cell $E_k$ of $C_n'$ will similarly correspond

$$s_{n-k} = E_{n-k} + t'E_{n-k-1}, \tag{28.2}$$

where all terms have an obvious meaning. But referring to the definition of the $s$'s we find that $s_{n-h}\times s_{n-k}$ corresponds in a similar manner to $E_h\times E_k$ as a cell of $C_n\times C_n'$. Now

$$\begin{aligned} s_{n-h} \times s_{n-k} = E_{n-h} \times E_{n-k} &+ tE_{n-h-1} \times E_{n-k} \\ &+ t'E_{n-h} \times E_{n-k-1} + tt'E_{n-h-1} \times E_{n-k-1}. \end{aligned} \tag{28.3}$$

Each term represents a cell whose dimensions are the sum of those of the factors. Now $E_h\times E_k$ is a boundary cell only when one of the $t$'s is not zero. When both are zero, (a′) is manifestly satisfied, and when, say, $t=1$, $t'=0$, (b′) is satisfied, as they should be. Let, then, $t=t'=1$. We must show that $E_{n-h-1}\times E_{n-k}+E_{n-h}\times E_{n-k-1}+E_{n-h-1}\times E_{n-k-1}$ is homeomorphic to a $2(n-h-k-1)$-cell. It will be remembered that, by condition (b′), $E_{n-k}$ is homeomorphic to the interior of a hemisphere in $S_{n-k}$ with $E_{n-k-1}$ as the flat base of the hemisphere. It follows that the first term is of the same type for an $S_{2n-h-k-1}$ and similarly for the second with the third as the common flat base. The sum is then homeomorphic to the interior of a sphere in the same space, that is to a cell, which completes the proof.

**Remarks.** I. The same proof holds for a product $M_p\times M_q$, $p\neq q$.

II. Since the factors are orientable this is also true for the product (Tr., No. 49).

29. We now define a continuous transformation $T$ of $M_n$ as in Tr., No. 56, by the condition that the set $\{A\times B\}=K_n$ be a $C_n$ of $M_n\times M_n'$ with its boundary on that of the product. Let $T'$ be another transformation, $K_n'$ its complex. The problem is again to determine $(K_n\cdot K_n')$ (number of signed coincidences) in terms of the transformations induced by $T$, $T'$ on the cycles, or of similar data (i. e., information naturally at hand when $T$, $T'$ are known).

No result of any generality is to be expected unless $K_n$ and $K_n'$ are so restricted that the boundaries of suitably defined approximations do not intersect. Indeed, unless this is so, $(K_n\cdot K_n')$ ceases to be an invariant of classes of transformations. We assume of course throughout that $M_n$ is connected, but $F_{n-1}$ need not be so. Let $F_{n-1}^1, \cdots, F_{n-1}^p$ be its connected parts. By Tr., (49.2), the boundary of $M_n\times M_n'$ is the sum of the products $F_{n-1}^i\times M_n'$, $(-1)^n M_n\times F'^{\,i}_{n-1}$, each of which is a manifold (§ 28). The boundary of each $K$ consists of subcomplexes distributed among some or all of these 2 $p$ manifolds. The least that we can exact is that none carry boundary points of both $K$'s in its interior. However, this is still beyond the reach of the method to be used here, at least in this general form. *We shall therefore restrict our discussion to pairs of transformations such that the boundary of one K is on* $F_{n-1} \times M_n'$, *while that of the other is on* $M_n \times F_{n-1}$, *one of the boundaries being actually interior to the carrying manifold.* This will greatly simplify matters, sufficiently indeed to compensate amply for whatever may be lost in generality. In § 32 we shall give a topological interpretation of these types of transformations by means of certain approximating transformations.

30. We shall reduce the coincidence and fixed points problems for $M_n$ to similar problems for certain associated transformations of $V_n$ that we now define.

Case I: *$T$ is of the first type,* that is with the boundary $\Gamma_{n-1}$ of $K_n$ on $F_{n-1} \times M_n'$. According to § 23, Corollary I, it is homologous thereon to $\Gamma_{n-1}'$ interior to $F_{n-1} \times M_n'$ and satisfying in all respects the conditions there stated. If beyond this point we apply the same reductions as before (loc. cit.) to $K_n$ itself, we shall reduce it to a complex $H_n$ bounded by $\Gamma_{n-1}'$ with its non-boundary cells all interior to $M_n \times M_n'$ and behaving in every respect in accordance with the theorem of § 23. If $C_\mu$ of $M_n$ has its boundary on $F_{n-1}$ we first reduce it as in § 23; then its transform $TC_\mu$ is determined as in Tr., No. 58, with $C_\mu$ in place of $\gamma_\mu$ and $H_n$ in place of $\Gamma_n$ (loc cit.). As a special case $C_\mu$ may be a $\gamma_\mu$; then it is first reduced to the interior of $M_n$ and $T\gamma_\mu$ is then determined in the same way.

Since $V_n \times V_n' = M_n \times V_n' - \overline{M}_n \times V_n'$, it may be derived from $M_n \times V_n'$ as $V_n$ from $M_n$. The two parts of the manifold are now matched along their common boundary $F_{n-1} \times V_n'$, and their points associated in conjugate pairs $A \times B$ and $\overline{A} \times B$. To $K_n$ and $H_n$ there correspond in this fashion associated skew-symmetric cycles $K_n - \overline{K}_n \sim H_n - \overline{H}_n = \Gamma_n$, and the latter will serve to define the transformation $T_1$ of $V_n$ associated with $T$.

Since the $B$ points describe on $V_n'$ the images of the transforms of the loci of the $A$ points, we have, by Tr., No. 58,

$$T_1\overline{C}_\mu = T_1C_\mu = TC_\mu\,; \qquad T_1(C_\mu - \overline{C}_\mu) = 0. \tag{30.1}$$

Furthermore, $T_1\gamma_\mu = T\gamma_\mu$ since $K_n'$ alone comes into play in determining the two transforms. Hence $T_1$ *transforms cycles of* $M_n$ *into cycles of* $M_n$ *and skew-symmetric cycles into bounding cycles.*

Let the transformation matrix for the cycles $\Gamma_\mu$, $G_\mu$ of the fundamental sets be

$$P_\mu = \begin{array}{c} \begin{array}{cc} \Gamma_\mu & G_\mu \end{array} \\ \hline \left\| \begin{array}{cc} P^\mu_{11} & P^\mu_{12} \\ P^\mu_{21} & P^\mu_{22} \end{array} \right\| \begin{array}{c} \Gamma_\mu \\ G_\mu \end{array} \end{array}.$$

It is the transformation matrix for the cycles of $M_n$ and $T$, the transformed cycles being given as $\approx$ to a combination of the initial cycles. The similar matrix for $V_n$ is

$$\mu < \frac{1}{2}n,\ \alpha_\mu = \begin{array}{c} \begin{array}{cccc} \Gamma & \Delta & D & G \end{array} \\ \hline \left\| \begin{array}{cccc} P^\mu_{11} & 0 & 0 & P^\mu_{12} \\ 0 & 0 & 0 & 0 \\ 0 & 0 & 0 & 0 \\ P^\mu_{21} & 0 & 0 & P^\mu_{22} \end{array} \right\| \begin{array}{c} \Gamma \\ \Delta \\ D \\ G \end{array} \end{array};\ \mu \geqq \frac{1}{2}n,\ \alpha_\mu = \begin{array}{c} \begin{array}{cccc} \Gamma & \Delta & G & D \end{array} \\ \hline \left\| \begin{array}{cccc} P^\mu_{11} & 0 & P^\mu_{12} & 0 \\ 0 & 0 & 0 & 0 \\ P^\mu_{21} & 0 & P^\mu_{22} & 0 \\ 0 & 0 & 0 & 0 \end{array} \right\| \begin{array}{c} \Gamma \\ \Delta \\ G \\ D \end{array} \end{array}.$$

31. Case II: *$T$ is of the second type*, or with its boundary on $M_n \times F'_{n-1}$. In this case $H_n$ will have its boundary interior to $M_n \times F'_{n-1}$. Then since $V_n \times V'_n = V_n \times M'_n - V_n \times \overline{M}'_n$, the left side is to be considered as obtained by matching the two manifolds at the right whose common boundary is $V_n \times F'_{n-1}$, and $A \times B$, $A \times \overline{B}$ constitute the conjugate pairs. Again $K_n - \overline{K}_n \sim H_n - \overline{H}_n = \Gamma_n$, cycle which serves to define $T_1$.

We find now that $T_1\, C_\mu$ consists of $C'_\mu = TC_\mu$ and $-\overline{C}'_\mu$, the second complex having a minus sign because its orientation is determined by means of $-\overline{K}'_n$. Also no cell on $\overline{M}_n$ has any transform. It follows that *every cycle of $V_n$ is transformed into a skew-symmetric cycle by $T_1$*. More explicitly, let the cycle first polyhedrally approximated be $\gamma_\mu = C'_\mu + \overline{C}''_\mu$, where the first complex is on $M_n$, the second on $\overline{M}_n$. Then $T_1\, \gamma_\mu = TC'_\mu - (\overline{TC'_\mu})$.

The transformation matrix is given below:

$$\mu < \frac{1}{2}n,\ \alpha_\mu = \begin{array}{c} \begin{array}{cccc} \Gamma & \Delta & D & G \end{array} \\ \hline \left\| \begin{array}{cccc} 0 & R^\mu_{11} & R^\mu_{12} & 0 \\ 0 & Q^\mu_{11} & Q^\mu_{12} & 0 \\ 0 & Q^\mu_{21} & Q^\mu_{22} & 0 \\ 0 & R^\mu_{21} & R^\mu_{22} & 0 \end{array} \right\| \begin{array}{c} \Gamma \\ \Delta \\ D \\ G \end{array} \end{array};$$

$$\mu \geqq \frac{1}{2}n, \quad \alpha_\mu = \begin{array}{c} \Gamma \quad \Delta \quad G \quad D \\ \left\| \begin{matrix} 0 & R_{11}^{\mu} & 0 & R_{12}^{\mu} \\ 0 & Q_{11}^{\mu} & 0 & Q_{12}^{\mu} \\ 0 & R_{21}^{\mu} & 0 & R_{22}^{\mu} \\ 0 & Q_{21}^{\mu} & 0 & Q_{22}^{\mu} \end{matrix} \right\| \begin{matrix} \Gamma \\ \Delta \\ G \\ D \end{matrix} \end{array}.$$

The matrix $Q = \|Q_{ij}^{\mu}\|$ is the transformation matrix for skew-symmetric cycles, and alone will appear in the final formulas.

32. The complex $H_n$ defines for both types an approximation $\overline{T}$ to the given $T$ such that (a) when $T$ is of the first type, $\overline{T}M_n$ is wholly interior to $M_n$; (b) when $T$ is of the second type, $\overline{T}F_{n-1}$ does not exist, i.e., the boundary belongs to the set of points that have no $\overline{T}$ transform. This may be considered as a geometric characterisation of the two types.

33. Let us now assume that $T$ is of type I, $T'$ of type II and represent the cycle and complexes attached to $T'$ by the same letters as for $T$ with primes. We are explicitly assuming (§ 29) that $K$ and $K'$ intersect only in interior points of $M_n \times M_n'$. Hence $(K_n \cdot K_n')$ is perfectly determined and by definition equal to $(H_n \cdot H_n')$ (Tr., No. 35). Therefore, at once,

$$(33.1) \qquad (\Gamma_n \cdot \Gamma_n') = ((H_n - \overline{H}_n) \cdot (H_n' - \overline{H}_n')) = (H_n \cdot H_n') = (K_n \cdot K_n'),$$

for when a term of a pair $H$, $H'$ is barred the two complexes do not meet. Hence *the coincidence problem for $T$, $T'$ is reduced to the same problem for $T_1$, $T_1'$*.

## IV. Coincidence and fixed points formulas

34. We have just shown that these formulas are the same for $M_n$ as for the associated transformations of $V_n$. We apply then (10.4), taking for $T_1$ the transformation corresponding to the $\alpha$'s, for $T_1'$ that corresponding to the $\beta$'s, and we have

$$(34.1) \qquad (K_n \cdot K_n') = (\Gamma_n \cdot \Gamma_n') = \sum (-1)^{\mu} \operatorname{trace} L_\mu \beta_{n-\mu}' L_\mu^{-1} \alpha_\mu.$$

We now assume the fundamental sets in the canonical form of § 27 and carry out the computation. The matrices in the product have the same structure, with four square submatrices in the principal diagonal in the same order. $L_\mu^{-1}$ for $\mu \neq \frac{1}{2}n$ odd, is like $L_\mu$ except that

$$\left\| \begin{matrix} A & 1 \\ 1 & 0 \end{matrix} \right\| \text{ is replaced by } \left\| \begin{matrix} 0 & 1 \\ 1 & -A \end{matrix} \right\|.$$

For $\mu=\frac{1}{2}n$ odd,

$$L_{n/2}^{-1}=\left\|\begin{array}{cc|cc} 0 & -1 & & \\ 1 & A & & 0 \\ \hline & & 0 & -1 \\ & 0 & 1 & 0 \end{array}\right\|. \tag{34.2}$$

By way of illustration, we examine the computation for $\mu<\frac{1}{2}n$. Then, dropping the indices $\mu$ and $n-\mu$ for the present,

$$L\beta'=\left\|\begin{array}{cc|c} A & 1 & \\ 1 & 0 & 0 \\ \hline 0 & & 1 \end{array}\right\|\cdot\left\|\begin{array}{cc|cc} 0 & 0 & 0 & 0 \\ R'_{11} & Q'_{11} & R'_{21} & Q'_{21} \\ \hline 0 & 0 & 0 & 0 \\ R'_{12} & Q'_{12} & R'_{22} & Q'_{22} \end{array}\right\|$$

$$=\left\|\begin{array}{cc|cc} R'_{11} & Q'_{11} & R'_{21} & Q'_{21} \\ 0 & 0 & 0 & 0 \\ \hline 0 & 0 & 0 & 0 \\ R'_{12} & Q'_{12} & R'_{22} & Q'_{22} \end{array}\right\|; \tag{34.3}$$

$$L^{-1}\alpha=\left\|\begin{array}{cc|c} 0 & 1 & \\ 1 & -A & 0 \\ \hline 0 & & 1 \end{array}\right\|\cdot\left\|\begin{array}{cc|cc} P_{11} & 0 & 0 & P_{12} \\ 0 & 0 & 0 & 0 \\ \hline 0 & 0 & 0 & 0 \\ P_{21} & 0 & 0 & P_{22} \end{array}\right\| \tag{34.4}$$

$$=\left\|\begin{array}{cc|cc} 0 & 0 & 0 & 0 \\ P_{11} & 0 & 0 & P_{12} \\ \hline 0 & 0 & 0 & 0 \\ P_{21} & 0 & 0 & P_{22} \end{array}\right\|.$$

Hence

$$L\beta'L^{-1}\alpha=\left\|\begin{array}{cccc} Q'_{11}P_{11}+Q'_{21}P_{21} & \cdot & \cdot & \cdot \\ \cdot & 0 & \cdot & \cdot \\ \cdot & \cdot & 0 & \cdot \\ \cdot & \cdot & \cdot & Q'_{12}P_{12}+Q'_{22}P_{22} \end{array}\right\| \tag{34.5}$$

where the terms not in the main diagonal are omitted and need not be computed, as unnecessary for the trace. It follows that

$$\text{trace}\; L_\mu\beta'_{n-\mu}L_\mu^{-1}\alpha_\mu=\text{trace}\;\sum Q'^{\,n-\mu}_{ij}P^{\mu}_{ij}\;=\;\text{trace}\; Q'_{n-\mu}P_\mu, \tag{34.6}$$

therefore

$$(K_n \cdot K_n') = \sum (-1)^\mu \text{ trace } Q'_{n-\mu} P_\mu \tag{34.7}$$

*is the desired coincidence formula.* Its similarity to (10.4) for the non-boundary case is very striking.

35. The preceding formula corresponds to highly specialized fundamental sets. Let us consider the more general case where they are merely composed of the same four types $\Gamma$, $\Delta$, $G$, $D$, in the same order as before for each $\mu$, but no further specialized than is demanded by the condition that all indices $(\Gamma \cdot D)$, $(D \cdot \Gamma)$ be zero. This amounts to requiring that $C = M = 0$, where the two matrices are as in § 25. The passage from a canonical set to the more general type is by means of transformations operating separately on each of the four types. Then

$$L = \left\| \begin{array}{cc|cc} A & B & & \\ H & 0 & \multicolumn{2}{c}{0} \\ \hline & & F & 0 \\ \multicolumn{2}{c|}{0} & 0 & N \end{array} \right\| \qquad (\mu < \tfrac{1}{2}n),$$

$$= \left\| \begin{array}{cc|cc} A & B & & \\ H & 0 & \multicolumn{2}{c}{0} \\ \hline & & N & 0 \\ \multicolumn{2}{c|}{0} & 0 & F \end{array} \right\| \qquad (\mu > \tfrac{1}{2}n), \tag{35.1}$$

$$= \left\| \begin{array}{cc|cc} A & B & & \\ (-1)^{n/2} B & 0 & \multicolumn{2}{c}{0} \\ \hline & & 0 & F \\ \multicolumn{2}{c|}{0} & (-1)^{n/2} F & 0 \end{array} \right\| \qquad (\mu = \tfrac{1}{2}n),$$

where all terms have the subscript $\mu$ omitted for the sake of simplicity. It turns out that the only terms needed are

$$B_\mu = \| (\Gamma_\mu^i \cdot \Delta_{n-\mu}^j) \|, \qquad F_\mu = \| (D_\mu^i \cdot G_{n-\mu}^j) \|. \tag{35.2}$$

Then, either by computing directly as before or else from (34.7), we derive the equivalent invariant form:

$$\begin{aligned} (K_n \cdot K_n') = {} & \sum (-1)^\mu \text{ trace } (B_\mu Q_{11}'^{n-\mu} B_\mu^{-1} P_{11} \\ & + B Q_{21}' F^{-1} P_{21} + F Q_{12}' B^{-1} P_{12} + F Q_{22}' F^{-1} P_{22}), \end{aligned} \tag{35.3}$$

where the dimensionality indices, the same for all four terms at the right, are written only for the first. This is the desired generalization of (34.7).

36. We pass now to the derivation of the *fixed point formulas*. We assume the fundamental sets general in the sense of § 35.

I. *T is of the first type.* Since the identity is of each of the two types, we now assign it to the second type, approximate its $K$, say $K_n^0$, by $H_n^0$ with its boundary on $M_n \times F'_{n-1}$, and interpret the problem as the determination of $(K_n \cdot K_n^0)$. It is in fact exactly that when $T$ actually reduces $F_{n-1}$ to the interior of $M_n$.

We shall so sense $K_n^0$ that the corresponding integer $\theta$ of § 8, here the same at every point of $K_n^0$, is $+1$. Then it is immediately seen that the corresponding $Q_\mu = I$.

Whence

(36.1) $$Q_{12}^{n-\mu} = Q_{21}^{n-\mu} = 0, \quad Q_{11}^{n-\mu} = Q_{22}^{n-\mu} = 1.$$

Therefore

(36.2) $$(K_n \cdot K_n^0) = \sum (-1)^\mu \operatorname{trace} P_\mu.$$

II. *T is of the second type.* The discussion is now the same, with the identity ascribed to the first type. $H_n^0$ determines a deformation of $M_n$ into an interior part of itself, and $(K_n^0 \cdot H_n)$ is then the number of points where $T$ operates as that infinitesimal deformation. In this case

(36.3) $$P_\mu = 1 = P_\mu^{11} = P_\mu^{22}\,; \; P_\mu^{12} = P_\mu^{21} = 0.$$

Hence

(36.4) $$(K_n^0 \cdot K_n') = \sum (-1)^\mu \operatorname{trace} (B_\mu Q'^{\,n-\mu}_{11} B_\mu^{-1} + F_\mu Q'^{\,n-\mu}_{22} F_\mu^{-1}).$$

But by well known properties of matrices,

(36.5) $$\operatorname{trace} uv'u^{-1} = \operatorname{trace} v' = \operatorname{trace} v.$$

This together with a change of $\mu$ into $n-\mu$ gives

(36.6) $$(K_n^0 \cdot K_n') = (-1)^n \sum (-1)^\mu \operatorname{trace} Q_\mu.$$

37. A particularly interesting case of (36.2) corresponds to the fixed points of a deformation $T$ (infinitesimal or otherwise) of $M_n$ into an interior part of itself. Then the actual number of signed fixed points when finite is given by

(37.1) $$(K_n'^0 \cdot K_n^0) = \sum (-1)^\mu R_\mu,$$

where $K_n'^0$ corresponds to the deformation. Here $R_\mu$ is the sum of the traces of $P_{11}^\mu$ and $P_{22}^\mu$ when they both reduce to the identity, that is, the sum of their orders. This is also the number of distinct $\Gamma$'s and $G$'s in the fundamental set for the dimension $\mu$; hence $R_\mu$ is the $\mu$th connectivity index of $M_n$ itself, and the sum in (37.1) is its Euler characteristic. This shows that (10.6) holds for an $M_n$ with a boundary also, of course with the $K$'s in place of the $\Gamma$'s. Therefore

THEOREM. *For every $M_n$, with or without boundary, the number of signed fixed points of a deformation is the Euler characteristic.*

It is understood of course that when $M_n$ has a boundary the deformation reduces $M_n$ to part of itself. It is necessary to point out that instead of a deformation we may equally well consider a $T$ reducing $M_n$ to an interior part of itself and merely adding zero-divisors to the cycles of the manifold.

The question of the *singular points of a vector distribution* is equivalent to the determination of the fixed points of an infinitesimal deformation. The authors dealing with it, notably Brouwer, Birkhoff, and recently Hopf,* have always restricted their manifolds more than in this paper. Hopf, for example, assumes that at every vertex of the defining $C_n$ of $M_n$ the incident cells constitute a star with the same structure as some star embedded in an $S_n$. Then, whatever the point $A$, there exists a region containing it wherein any two points may be joined by a uniquely defined polygonal line, image of a rectilinear segment in the $S_n$ region. Until it is actually proved, as may be done for $n=2$, 3, that *every star of cells which is an n-cell has the same structure as some star in $S_n$*, our manifolds must be considered as much more general, and our results as having a notably wider range. On the other hand it must be stated that all analytical manifolds are of the more restricted type, so that for various applications the restriction may actually not be important.

38. The fixed point formulas derived from the general coincidence formula may also be obtained directly and in a very simple manner. Indeed, instead of associating with the identity the transformations of §§ 30, 31, we may associate with it the identical transformation for $V_n$ also. The corresponding cycle $\Gamma_n^0$ is on $M_n \times M_n'$ and $\overline{M}_n \times M_n'$, hence at once for type I,

$$(K_n \cdot K_n^0) = (\Gamma_n \cdot \Gamma_n^0), \tag{38.1}$$

and for type II

$$(K_n^0 \cdot K_n') = (\Gamma_n^0 \cdot \Gamma_n'). \tag{38.2}$$

* Mathematische Annalen, vol. 96 (1926), pp. 225–250. He has derived (37.1) for a vector distribution on the general $M_n$ of the type that he considers. His seem to be the only investigations on general manifolds along the line of this paper that are to be found in the literature. In Mathematische Annalen, vol. 95 (1925), he has generalized in an interesting manner the well known topological property of the total Gaussian curvature.

Then from (10.5) the desired results follow easily.

39. **Transformation of one manifold into another.** The transformations of $M_n$ into a new manifold $M_n'$ are treated exactly like those of $M_n$ into itself. First let both be without boundary and let $\bar{L}_\mu$ be the analogue of $L_\mu$ for $M_n'$. The transformation of $M_n$ into $M_n'$ will be defined by cycles $\Gamma_n$, $\Gamma_n'$ on $M_n \times M_n'$ and (59.2), Tr., will hold.

The expression for $(\Gamma_n \cdot \Gamma_n')$ will be again (10.4) with the first factor, $L$, in each term replaced by $\bar{L}$, and the others unchanged or*

$$(39.1) \qquad (\Gamma_n \cdot \Gamma_n') = \sum (-1)^\mu \operatorname{trace} \bar{L}_\mu \beta_{n-\mu}' L_\mu^{-1} \alpha_\mu .$$

When there are boundaries, the types of transformations must again be restricted as previously. If $\bar{B}$, $\bar{F}$ correspond to $\bar{L}$ as $B$, $F$ to $L$, we find here the same coincidence formula (35.3) as before, except that $B$, $F$ are replaced by $\bar{B}$, $\bar{F}$, everything else (notably $B^{-1}$, $F^{-1}$) remaining unchanged. It does not seem necessary to write the formula explicitly.

40. **A different type of coincidence formula.** In these Transactions, vol. 25 (1923), Alexander has derived for 1-$\tau$ transformations of surfaces a formula of a different type from ours. The difference consists in the fact that in place of the transformation matrices there appear everywhere the Kronecker index matrices for the intersections of the $n-\mu$ cycles with the transforms of the $\mu$ cycles. Let us show that such formulas can be derived from those of the present paper.

Let first $M_n$ be without boundary and introduce for $T$ the matrix

$$(40.1) \qquad \xi_\mu = \| (T\gamma_\mu^i \cdot \gamma_{n-\mu}^j) \|$$

with a similar matrix $\eta_\mu$ for $T'$. The problem is to express the number of signed coincidences in terms of the $\xi$'s and $\eta$'s. We have

$$(40.2) \qquad (T\gamma_\mu^i \cdot \gamma_{n-\mu}^j) = \left( \sum \alpha_\mu^{ih} \gamma_\mu^h \cdot \gamma_{n-\mu}^j \right) = \sum \alpha_\mu^{ih} (\gamma_\mu^h \cdot \gamma_{n-\mu}^j),$$

therefore

$$(40.3) \qquad \xi_\mu = \alpha_\mu L_\mu ; \quad \eta_\mu = \beta_\mu L_\mu .$$

From this and the readily verified relation

$$(40.4) \qquad L_{n-\mu}' = (-1)^{\mu(n+1)} L_\mu ,$$

* See footnote, §3.

follows in place of (10.6) the desired formula

$$(\Gamma_n \cdot \Gamma_n') = \sum (-1)^{n\mu} \operatorname{trace} \eta_{n-\mu}' L_\mu^{-1} \xi_\mu L_\mu^{-1}. \tag{40.5}$$

For the fixed points, assume the second transformation to be the identity. Then directly from (10.5) or else from (40.5) and

$$\eta_{n-\mu} = L_{n-\mu} = (-1)^{\mu(n+1)} L_\mu', \tag{40.6}$$

we find the number of signed fixed points

$$(\Gamma_n \cdot \Gamma_n^0) = \sum (-1)^\mu \operatorname{trace} \xi_\mu L_\mu^{-1}. \tag{40.7}$$

Let $n=2$ as in Alexander's paper. The terms corresponding to $\mu=0, 2$ are the same as in (10.6), namely $\alpha_0+(-1)^n \alpha_n = \alpha_0+\alpha_2$. The term $\mu=1$ alone needs to be computed. When the fundamental set is canonical, $L_1$ is the well known matrix $||l_{ij}||$ $(i, j=1, 2, \cdots, 2p=R_1)$ with $l_{2i-1,\, 2i}=-l_{2i,\, 2i-1}=1$ and all other terms zero. Let $\gamma_1, \cdots, \gamma_{2p}$ be the retrosections of the Riemann surface. If we recall that for any two cycles $\gamma$, $\delta$ on the surface, $(\gamma \cdot \delta) = -(\delta \cdot \gamma)$, and observe that $L_1' = -L_1$, we find that (40.7) becomes, with $\bar{\gamma}=T\gamma$ as usual,

$$(\Gamma_2 \cdot \Gamma_2^0) = \alpha_0 + \alpha_2 + \sum (\bar{\gamma}_{2i-1} \cdot \gamma_{2i}) + (\gamma_{2i-1} \cdot \bar{\gamma}_{2i}), \tag{40.8}$$

which generalizes Alexander's formula (4), loc. cit., and except for the notation reduces to it when, as he does, we consider $1-\tau$ transformations $(\alpha_0=1,\ \alpha_2=\tau)$.

The treatment for manifolds with a boundary is along the same line. We find

$$\begin{aligned}(K_n \cdot K_n') = \sum (-1)^{n\mu} \operatorname{trace} (&\sigma_{11}'^{\,n-\mu} B_\mu^{-1} \pi_{11}^{\mu} B_\mu^{-1} \\ &+ \sigma_{21}' N^{-1} \pi_{12} B^{-1} + \sigma_{12}' B^{-1} \pi_{21} N^{-1} \\ &+ \sigma_{22}' N^{-1} \pi_{22} N^{-1}),\end{aligned} \tag{40.9}$$

where the sequence of dimensional indices is the same for all four terms under the sum and therefore has been written down only for the first. As for the various letters their meaning is as follows:

$$\begin{aligned} B_\mu &= \| (\Gamma_\mu^i \cdot \Delta_{n-\mu}^j) \|, \quad N_\mu = \| (G_\mu^i \cdot D_{n-\mu}^j) \|, \\ \pi_{11}^{\mu} &= \| (T\Gamma_\mu^i \cdot \Delta_{n-\mu}^j) \| ; \end{aligned} \tag{40.10}$$

$\pi_{12}$, $\pi_{21}$, $\pi_{22}$ are as $\pi_{11}$ with the pair $\Gamma\Delta$ replaced by $G\Delta$, $\Gamma D$ and $GD$ respectively, while $\sigma_{ij}$ is derived from $\pi_{ij}$, by substituting for $T$ and $\mu$, $T'$ and $n-\mu$ and permuting the cycles of the corresponding pair.

For the fixed points of $T$ of first type, we take $T'=1$, and it turns out that (40.7) holds also provided that $\xi$ and $L$ correspond to Kronecker index matrices for the fundamental sets of $M_n$ alone, that is, for the intersections of the $\Gamma$'s and $G$'s and their transforms. For fixed points of $T'$ of second type we take $T=1$ and find

$$(K_n^0 \cdot K_n') = (-1)^n \sum (-1)^\mu \operatorname{trace} (B_\mu^{-1} \overset{\mu}{\sigma}_{11} + N_\mu^{-1} \overset{\mu}{\sigma}_{22}). \tag{40.11}$$

This completes the derivation of the formulas of Alexander's type from those of this paper. The type to be used in any particular case will depend upon the available data on the transformations.

# V

# V

## L'ANALYSIS SITUS

ET

# LA GÉOMÉTRIE ALGÉBRIQUE

# PRÉFACE.

Les surfaces et les variétés algébriques accusent, au point de vue de l'*Analysis Situs*, des différences profondes avec les courbes, différences dues surtout à ce qu'elles ne sont pas les variétés bilatères les plus générales de leur dimension (quatre pour les surfaces, $2d$ pour les variétés à $d$ dimensions). Effectivement elles contiennent des variétés topologiquement très particulières : leurs sections par d'autres surfaces ou variétés algébriques. Il est donc assez naturel d'en profiter pour l'étude de la variété elle-même, et de plus, de rechercher les propriétés spéciales qui en résultent. Après un Chapitre préliminaire, c'est là la tâche que nous avons poursuivie dans le second, le troisième et dans le début du cinquième. Dans les autres on trouvera diverses applications, notamment à la distribution des courbes des surfaces ou des hypersurfaces des variétés algébriques. C'est en effet, dans cette direction, que l'*Analysis Situs* se montre à son maximum d'utilité.

Parmi les divers écrits qui m'ont influencé pendant ce travail, sont à signaler tout particulièrement : les Mémoires de Poincaré sur l'*Analysis Situs* et sur la distribution des courbes d'une surface algébrique, les travaux de M. Picard, tels qu'ils sont exposés dans le Traité qu'il a écrit en collaboration avec M. Simart, enfin ceux des superbes géomètres de l'École italienne. Pour une bonne part, d'ailleurs, les résultats exposés ici sont inédits quant à la forme ou àu fond. Tel est notamment le cas, pour ceux relatifs aux intersections de cycles, qui paraissent ici pour la première fois.

Cette monographie est basée, en partie, sur une série de conférences faites à Rome, au printemps de 1921, sous les auspices de l'*Institute of International Education*, ainsi que sur les recherches poursuivies, un peu auparavant, sous ceux de l'*American Association for the Advancement of Science*.

Aux nombreux amis qui, des deux côtés de l'Atlantique, m'ont prodigué leur cordial encouragement, je tiens à exprimer ici ma grande reconnaissance. Enfin je veux remercier, d'abord M. Borel, pour m'avoir accordé l'honneur de paraître dans sa Collection, justement célèbre; ensuite la maison Gauthier-Villars, pour avoir bien voulu entreprendre la publication de mon Ouvrage, malgré ces temps si durs, où elle continue à maintenir son rang, hors pair, parmi les éditeurs scientifiques.

# TABLE DES MATIÈRES.

# L'ANALYSIS SITUS

ET

# LA GÉOMÉTRIE ALGÉBRIQUE

## CHAPITRE I.

### PROPRIÉTÉS GÉNÉRALES DES VARIÉTÉS ANALYTIQUES.

### I. — Premières définitions. Connexions des divers ordres (1).

1. Deux variétés sont *homéomorphes*, si elles sont en correspondance biunivoque, continue, sans aucune exception. Tels sont deux segments de ligne, deux surfaces de Riemann de même genre, les régions de l'espace ordinaire limitées par un cube et une sphère, etc. L'*Analysis Situs* a pour but l'étude des propriétés des figures, préservées quand on les remplace par d'autres homéomorphes.

La cellule à $n$ dimensions $E_n$ est une variété homéomorphe à l'intérieur d'une hypersphère dans un espace à $n$ dimensions, $S_n$. (Ces notations : $E_n$, $S_n$, reviendront constamment.) La frontière de $E_n$ est la généralisation du circuit. En particulier, $E_1$ est un segment de ligne, $E_2$ une aire simplement connexe. Une variété à $n$ dimensions est *homogène*, si deux $E_n$ de la variété en conte-

(1) *Voir* à ce sujet : PICARD et SIMART, *Traité des fonctions algébriques de deux variables*, vol. I, Chap. II, ainsi que les Mémoires de POINCARÉ, *Journal de l'École Polytechnique*, 2e série, vol. I, 1895. — HEEGAARD, *Bulletin de la Société mathématique*, vol. XLIV, 1916. — TIETZE, *Monatshefte für Mathematik und Physik*, vol. XIX, 1908. — VEBLEN et ALEXANDER, *Annals of Math.*, 2e série, vol. XIV, 1913. — Enfin l'ouvrage de M. VEBLEN : *Analysis Situs, the Cambridge Colloquium, published by the American Mathematical Society*, New-York, 1922.

nant un point en ont toujours en commun une troisième le contenant également. Une courbe plane en forme de 8, un cône à deux nappes offrent des exemples de variétés qui ne le sont pas.

2. **Les variétés analytiques**, seules a être considérées ici, se définissent ainsi : Soient $x_1, x_2, \ldots, x_{n'}$ un système de coordonnées cartésiennes, relatif à un $S_{n'}$ $(n' \geqq n)$, puis considérons les équations

$$x_i = \varphi(u_1, u_2, \ldots, u_n) \qquad (i = 1, 2, \ldots, n'; n' \gtreqless n), \tag{1}$$

où, dans le champ de variation des $u$, les $\varphi$ sont des fonctions analytiques, à déterminants fonctionnels non tous nuls à la fois, enfin faisons le prolongement analytique des seconds membres, obtenant ainsi une série de nouveaux systèmes analogues, etc. A certains d'entre eux peuvent être adjointes des inégalités

$$\psi(u_1, u_2, \ldots, u_n) > 0.$$

L'ensemble de points ainsi obtenu constitue une variété ou multiplicité *analytique* à $n$ dimensions, $M_n$, dans un $S_{n'}$.

Pour permettre de considérer aussi des points à l'infini, nous introduirons des multiplicités définies en partie par des systèmes d'équations tels que (1), en partie par d'autres tels que

$$x_i = \varphi_i(u_1, u_2, \ldots, u_n) \qquad (1 = 1, 2, \ldots, n_1),$$
$$\frac{1}{x_j} = \varphi_j(u_1, u_2, \ldots, u_n) \qquad (j = n_1 + 1, n_1 + 2, \ldots, n'),$$

où les $\varphi$ ont les mêmes propriétés qu'avant.

On n'exclut pas la possibilité que les variables prennent des valeurs complexes. Seulement il suffira alors que $2n' \geqq n$.

En changeant une des inégalités en égalité on obtient une *frontière* de $M_n$. Nous admettrons que les frontières se composent d'une ou plusieurs variétés analytiques à $n - 1$ dimensions, $M^i_{n-1}$ (1).

3. **Orientation des variétés.** — La notion familière des deux

(1) *Voir* la Note de M. Hadamard à la fin de l'*Introduction à la théorie des fonctions* de Jules Tannery.

sens distincts que l'on peut attribuer à un segment de courbe s'étend aux variétés *bilatères*. Soit d'abord une $E_n$ en forme de pyramide généralisée, $P_n = A_1 A_2 \ldots A_{n+1}$, à faces non nécessairement planes. $P_n$ correspond à la pyramide à faces planes comme par exemple le triangle sphérique au triangle ordinaire. Supposons que l'on nomme ses sommets dans l'ordre $i, j, \ldots, k$. Nous dirons que $P_n$ n'a pas changé ou bien qu'elle est invertie suivant que le nombre de transpositions dans la permutation

$$\begin{pmatrix} 1, 2, \ldots, n+1 \\ i, j, \ldots, k \end{pmatrix}$$

est pair ou impair. La pyramide invertie sera désignée par $-P_n$. Ceci généralise de manière évidente les notions familières pour $n = 1, 2$, où $P_n$ devient un segment ou un triangle.

Traçons maintenant dans une $M_n$ quelconque une petite $P_n$. Est-il possible de faire décrire à un point intérieur à $P_n$ un petit circuit de manière à ramener $P_n$ à $-P_n$? Si oui $M_n$ est *unilatère*, si non *bilatère*. Toute $M_1$ est bilatère, mais il existe des $M_2$ unilatères et le feuillet unilatère de Möbiüs est bien connu.

La petite $P_n$ est appelée *indicatrice* de $M_n$. A l'ensemble ($M_n$ + indicatrice) on peut attacher un sens ou signe et l'on convient qu'en changeant l'orientation de l'indicatrice on remplace $M_n$ par $-M_n$. Une fois $M_n$ orientée on définit le sens d'une frontière $M_{n-1}^i$ comme ceci : On amène $P_n$ au contact avec $M_{n-1}^i$ de façon que la $P_{n-1} = A_2 A_3 \ldots A_{n+1}$ soit dans $M_{n-1}^i$ dont elle deviendra par définition une indicatrice.

*Remarque.* — On peut aussi définir une orientation de $M_n$ par les signes des déterminants fonctionnels des $\varphi$ (Poincaré). Les deux définitions sont équivalentes ([1]).

**4. Congruences, homologies.** — La relation entre $M_n$ et ses frontières orientées sera exprimée par une *congruence*

$$M_n \equiv \sum M_{n-1}^i$$

([1]) *Voir* l'article de M. Heegaard cité plus haut.

ou en particulier quand $M_n$ est fermée (sans frontières)

$$M_n \equiv o.$$

Lorsqu'il s'agit simplement d'exprimer que les $M_{n-1}^i$ forment la frontière totale d'une $M_n$ quelconque dans une $M_{n'}$, on écrira une *homologie*

$$\sum M_{n-1}^i \sim o \quad (\text{mod } M_{n'}),$$

en omettant toutefois la mention (mod $M_{n'}$) quand il n'y aura pas d'équivoque possible

Ces notations et presque toutes les définitions remontent à Poincaré. L'emploi du signe $\equiv$ peut prêter à objection, car la relation qu'il exprime n'est pas transitive. Il n'en est pas moins fort commode et l'on verra qu'il ne cause guère de confusion.

5. Le terme *multiplicité* sera étendu à une somme de multiplicités du type défini plus haut, les frontières étant alors par définition les sommes des frontières. On attribue ainsi un sens à une multiplicité

$$\sum_i \lambda_i M_n^i \quad (\lambda_i \text{ entier}).$$

Grâce à cette convention on peut faire subir aux congruences et aux homologies l'addition, la multiplication par un entier, et *aux congruences seules* la division par un facteur entier commun à tous les coefficients. Ce dernier point se démontre ainsi : la congruence

$$(2) \qquad \lambda \sum \lambda_i M_i^n \equiv \lambda \sum \mu_s M_{n-1}^s$$

provient de congruences que l'on multiplie par des entiers $\lambda\lambda_i$. Si l'on se contentait de multiplier par les $\lambda_i$, on obtiendrait précisément (2) avec le facteur $\lambda$ supprimé.

**Déplacements.** — Nous avons eu l'occasion d'employer des déplacements (n° 3). Les seuls déplacements d'une $M_k$ de type restreint que nous considérerons à l'avenir seront ceux exprimables par des équations analytiques. Leur effet sur $M_k$ sera de lui faire engendrer des $M_{k+1}$. Soit en particulier $M_k \equiv M_{k-1}$ et supposons qu'en vertu d'un déplacement durant lequel les deux multiplicités en engendrent d'autres $M'_{k+1}$, $M'_k$, la première soit amenée en $M''_k$. Évidemment

$$M'_{k+1} \equiv M''_k - M_k + M'_k.$$

Par suite, en supposant toutes les multiplicités contenues dans $M_n$,

$$M''_k - M_k + M'_k \sim 0 \qquad (\text{mod } M_n).$$

En particulier, si $M_k$ est fermée, ses positions extrêmes sont homologues.

6. **Cycles. Connexions des divers ordres.** — Dorénavant nous prendrons pour base de nos considérations une variété à $n$ dimensions $W_n$, fermée, homogène, bilatère. Les $M_k$ fermées qu'elle contient sont ses *cycles à k dimensions*. Le cycle est *linéaire* si $k = 1$, *nul* si $M_k \sim 0$. Plusieurs cycles à $k$ dimensions sont *indépendants* ou *distincts* s'ils ne satisfont à aucune homologie, c'est-à-dire s'il n'y en a pas de combinaison linéaire formant la frontière complète d'une $M_{k+1}$. Le maximum $R_k$ (fini) de cycles à $k$ dimensions distincts est l'*indice de connexion à k dimensions de* $W_n$. Les cycles à $k$ dimensions seront en général désignés par la notation $\Gamma_k$, des indices supérieurs servant à les distinguer quand il y en a plusieurs. D'après le n° 5, quand on déplace un cycle il reste homologue à lui-même.

**Diviseurs de zéro, torsion.** — $W_n$ peut posséder des $\Gamma_k$ tels que $\lambda\Gamma_k \sim 0$ pour $\lambda > 1$ sans que $\Gamma_k \sim 0$. Le cycle est dit alors *diviseur de zéro à k dimensions*, ou plus simplement, diviseur à $k$ dimensions. Le nombre $\sigma_k$ de ces diviseurs est fini; l'entier $\sigma_k$ est l'*indice de torsion à k dimensions*. Raisonnant comme M. Severi dans un cas entièrement semblable relatif aux courbes d'une surface algébrique, considérons $+\Gamma_k$ comme une opération à effectuer sur les cycles de sa dimension. En ce sens les opérations relatives aux diviseurs engendrent un groupe abélien fini $G_k$ dont $\sigma_k$ est l'ordre; c'est le *groupe de la torsion à k dimensions*. $G_k$ possède un certain nombre $\zeta_k$ d'opérations formant base, soient $\Gamma_k^1$, $\Gamma_k^2$, ., ., $\Gamma_k^{\zeta_k}$, c'est-à-dire telles que toute autre soit de la forme

$$\lambda_1\Gamma_k^1 + \lambda_2\Gamma_k^2 + \ldots + \lambda_{\zeta_k}\Gamma_k^{\zeta_k} \qquad (\lambda_i \text{ entier}).$$

Les ordres $t_1$, $t_2$, $t_{\zeta_k}$ des opérations de la base sont les *coefficients de torsion* de Poincaré ([1]), et l'on a $\sigma_k = t_1 t_2 \ldots t_{\zeta_k}$. Le terme *torsion* provient de ce que la présence des diviseurs entraîne une sorte de torsion interne de la variété (Poincaré).

*Remarque.* — Les entiers $R_k$, $\zeta_k$, $\sigma_k$, $t_i$ sont tous invariants par rapport à l'homéomorphisme.

**7. Systèmes fondamentaux de cycles.** — Soit $\Gamma_k^1, \Gamma_k^2, \ldots, \Gamma_k^r$ un système fondamental pour les $\Gamma_k$, c'est-à-dire tel que tout cycle

$$\Gamma_k \sim \sum \lambda_i \Gamma_k^i. \tag{3}$$

L'existence d'un tel système est une conséquence de ce que, $R_k$ étant fini, entre $R_k + 1$ cycles $\Gamma_k$ quelconques il y a toujours une homologie. D'ailleurs nous le vérifierons directement pour les variétés algébriques.

Les $\Gamma_k$ satisfont à $r - R_k$ homologies

$$t_{h1}\Gamma_k^1 + t_{h2}\Gamma_k^2 + \ldots + t_{hr}\Gamma_k^r \sim 0 \qquad (h = 1, 2, \ldots, r - R_k). \tag{4}$$

Considérons maintenant des cycles $\overline{\Gamma}_k^1, \overline{\Gamma}_k^2, \ldots, \overline{\Gamma}_k^r$, tels que

$$\Gamma_k^i \sim \sum_h a_{ih} \overline{\Gamma}_k^h \qquad (i = 1, 2, \ldots, r), \tag{5}$$

le déterminant des $a_{ih}$ étant $\pm 1$, de sorte que leur matrice A définit ce que l'on nomme une transformation *unimodulaire*. Les $\overline{\Gamma}$ constituent un nouveau système fondamental. Au lieu de (4), ils satisferont à un certain système (4′). On peut d'ailleurs remplacer (4′) par tout autre système (4″) obtenu en appliquant à ses premiers membres une transformation unimodulaire B. Or, d'après Frobeniüs ([2]), pour un choix convenable de A, B, (4″) sera de la forme

$$t_k \overline{\Gamma}_k^{R+h} \sim 0 \qquad (h = 1, 2, \ldots, r - R_k). \tag{4″}$$

Parmi les $\overline{\Gamma}$ on peut supprimer les cycles nuls, ce qui revient à

([1]) *Proceedings of the London Mathematical Society*, 1900.

([2]) Le calcul est le même que pour les formes bilinéaires diophantines. (*Voir* E. Cahen, *Théorie des nombres*, t. I. p. 269.)

supposer que les $t$ sont tous $> 1$. Ces entiers ne sont autres alors que les coefficients de torsion et les $\overline{\Gamma}^{R_k+h}$ forment une *base* pour le groupe $G_k$, ou si l'on veut, pour les diviseurs $\Gamma_k$. On aura alors $r = R_k + \zeta_k$ ($r = R_k$ s'il n'y a pas de diviseurs). Quand on ne s'inquiète pas d'avoir le plus petit nombre possible de cycles dans le système fondamental, on peut en prendre un composé de $R_k$ cycles indépendants associés à $\sigma_k - 1$ diviseurs [1].

Remarque. — *Si les cycles d'un système fondamental sont indépendants*, $\sigma_k = 1$.

En effet il n'y a pas alors de diviseurs proprement dits, car de

$$\Gamma_k \sim \sum \lambda_i \Gamma_k^i, \quad t\Gamma_k \sim 0, \qquad t > 1,$$

on tire successivement

$$\sum t\lambda_i \Gamma_k^i \sim 0, \qquad t\lambda_i = 0 = \lambda_i, \qquad \Gamma_k \sim 0.$$

*Ceci se présente en particulier pour les* $W_2$, puisque leurs rétrosections constituent un système fondamental à la propriété requise.

**8. Rappel des théorèmes principaux de Poincaré.** — Relativement à nos $W_n$ il a démontré ces propriétés fondamentales, que nous allons retrouver en faisant l'étude des variétés algébriques.

I. *Les* $R_k$ *équidistants des extrêmes sont égaux*, c'est-à-dire $R_k = R_{n-k}$.

II. *Le groupe de torsion* $G_{n-1}$ *se réduit à l'identité. Les groupes* $G_k$, $G_{n-1-k}$ *sont isomorphes* [2].

$W_n$ étant analytique, tout point en sera contenu dans une cellule $E_n$ de frontière analytique, donc d'après un théorème classique (Borel), il y aura un nombre fini de $E_n$ recouvrant $W_n$ tout entière. Ces cellules peuvent se recouvrir partiellement, mais (on

---

(1) De cette discussion résulte aisément que tout système fondamental peut être dérivé de celui des $\overline{\Gamma}_k$, donc d'un quelconque d'entre eux, par une transformation unimodulaire.

(2) Ici, comme à l'avenir, il sera entendu que l'isomorphisme est holoédrique.

le démontre aisément, et en tout cas nous l'admettrons ici) $W_n$ peut être subdivisée en cellules sans points internes communs, les frontières étant elles-mêmes composées de cellules $E_{n-1}$ aux mêmes propriétés, etc. C'est ce qu'on appelle *réduire* $W_n$ *à un polyèdre généralisé.* Le polyèdre possédera un certain nombre $\alpha_i$ de cellules $E_i$. Poincaré a démontré une formule, généralisation d'un résultat classique d'Euler, qui, avec la convention $R_0 = R_n = 1$ adoptée pour la suite, se met sous la forme très simple

$$\text{III.} \qquad \sum(-1)^i \alpha_i = \sum(-1)^i R_i.$$

Poincaré a d'ailleurs démontré une formule analogue pour les variétés unilatères, mais nous ne nous y arrêterons pas.

9. **Quelques modèles de variétés.** — On sait que pour $n > 2$ on n'a pas réussi jusqu'à présent à obtenir de modèle type pour les $W_n$, analogue, par exemple, au disque troué de Clifford pour les surfaces bilatères. Il est par suite fort utile de considérer des exemples particuliers pour obtenir une idée des circonstances variées qui peuvent se présenter. C'est ainsi que Poincaré en a étudié plusieurs dans son premier Mémoire (¹).

Commençons par un type simple dû à M. Volterra. Dans un espace à $n$ dimensions $S_n$, prenons une hypersphère en contenant à son intérieur $k$ autres, extérieures les unes aux autres, et soit T la partie de $S_n$ limitée par toutes les hypersphères. Prenons le symétrique T′ de T par rapport à un hyperplan ne coupant pas T. Enfin faisons coïncider, par exemple à l'aide d'une déformation dans un $S_{n'}$, $(n' > n)$, les points correspondants des frontières de T et T′. On obtient ainsi une $W_n$ avec, comme on le vérifie aisément, $R_1 = R_{n-1} = k$. Ceci démontre que *la connexion linéaire d'une* $W_n$ *peut être égale à un entier quelconque.*

Prenons en particulier $n = 3$, $k = 2$, et les sphères qui limitent T dans $S_3$ (espace ordinaire), concentriques. Soit P le plan de symétrie considéré ci-dessus, et par le centre commun des deux sphères, passons un plan Q perpendiculaire à P. Opérons maintenant comme précédemment sur l'ensemble T + T′ en rem-

(¹) *Voir* la planche et la Note II à la fin de l'ouvrage.

plaçant toutefois P, en tant que plan de symétrie, par le plan Q. On vérifie aisément qu'une variété homéomorphe à $W_3$ peut être construite en remplaçant partout T par l'espace intérieur à un tore (Volterra).

10. Indiquons maintenant un modèle, intéressant surtout, car c'est le plus simple à nombre $\sigma_1$ quelconque. Il remonte au fond à Poincaré. Soient T, T' deux tores, S, S' leurs espaces intérieurs, $a$, $b$ les circuits méridien et parallèle de T, $a'$, $b'$ ceux de T'. Rendons T simplement connexe par les coupures usuelles suivant $a$ et $b$; puis faisons-en autant pour T', mais cette fois avec des coupures $b'$ et $a'_1 = kb' + a'$. Le circuit $a'_1$ doit être tracé de manière à ne pas avoir de points doubles et à ne couper $b'$ qu'en un seul point, conditions faciles à remplir. Les deux cellules $E_2$ ainsi obtenues ont toutes deux quatre frontières dans la même situation relative. On peut donc établir entre elles un homéomorphisme faisant correspondre $a'_1$ à $a$ et $b'$ à $b$. Faisons coïncider chaque point de T avec le point correspondant de T'. On aura ainsi dérivé de $S + S'$ une certaine $W_3$. Pour avoir toutes les homologies entre ses cycles linéaires, il faut ajouter aux homologies fondamentales déjà existantes pour S, S', $a \sim 0$, $a' \sim 0$ (et il n'y en a pas d'autres), l'homologie $a' + kb' - a \sim 0$ résultant de l'homéomorphisme. Les cycles linéaires de $W_3$ sont tous des combinaisons des cycles $a$, $a'$, $b'$, aux trois homologies fondamentales que nous venons d'écrire. Donc $kb' \sim 0$ sans que $b' \sim 0$, c'est-à-dire $k$ est un coefficient de torsion et, comme il n'y en a pas d'autres, $G_1$ est un groupe cyclique d'ordre $\sigma_1 = k$ entier quelconque.

11. **L'anneau à $n$ dimensions et ses cycles.** — Prenons le cube à $n$ dimensions $U_n$, ou solide défini par les relations

$$0 \leq x_i \leq 1 \qquad (i = 1, 2, \ldots, n),$$

et déformons-le dans un $S_{n+1}$ de manière à en faire coïncider les faces parallèles de chaque dimension. La $W_n$ que l'on obtient ainsi est l'*anneau à n dimensions*. Cette construction généralise de manière évidente celle du tore à partir du carré.

Pour trouver les valeurs de $R_k$, $\sigma_k$, nous nous appuierons sur ce lemme :

*Dans une cellule* $E_n$ *toute* $M_k$ *de frontière dans celle* $W_{n-1}$ *de* $E_n$ *peut être réduite par déformation à* $W_{n-1}$ *sans changement de frontière.*

En effet prenons pour $E_n$ l'intérieur d'une hypersphère, puis, s'il le faut, déformons légèrement $M_k$ de façon qu'elle ne passe pas par le centre. Il suffira alors de faire glisser $M_k$ le long des rayons pour arriver au résultat voulu.

L'application à l'anneau est immédiate. Les seuls cycles à considérer sont ceux provenant des faces de $U_n$. Les faces parallèles à $k$ dimensions donnent lieu à des $\Gamma_k$ homologues, et il n'y a pas d'autres homologies entre les $\Gamma_k$ que celles exprimant ce fait. Donc tout $\Gamma_k$ est homologue à une somme de multiples des $\binom{n}{k}$ cycles dérivés d'un système de faces non parallèles. Par suite, ces derniers cycles constituent un système fondamental. Comme ils sont indépendants, $R_k = \binom{n}{k}$, $\sigma_k = 1$. Les groupes de torsion se réduisent tous ici à l'identité.

## II. — Intersection des variétés.

**12.** Considérons dans $W_n$ deux multiplicités du type restreint du n° 2, $M_h$, $M_k$, se coupant suivant une $M_l$ du même type, avec $l = h + k - n$. Soit $A_1$ un point quelconque de $M_l$ et donnons-nous les indicatrices en $A_1$,

$$P_l = A_1 A_2 \ldots A_{l+1}, \qquad P_h = A_1 A_2 \ldots A_{l+1} A_{l+2} \ldots A_{h+1},$$
$$P_k = A_1 A_2 \ldots A_{l+1} A_{h+2} \ldots A_{n+1}.$$

Nous dirons que $M_h$ et $M_k$ se coupent suivant $+M_l$ ou $-M_l$ suivant que

$$P_n = A_1 A_2 \ldots A_{n+1}$$

est, ou bien n'est pas, une indicatrice de $W_n$. Cette intersection orientée sera désignée par $M_h M_k$. Si $M_h$ et $M_k$ se coupent suivant plusieurs variétés $M_l^i$, chacune orientée *ad libitum*, on écrira

$$M_h \cdot M_k = \sum \pm M_l^i,$$

les signes étant déterminés comme avant. On vérifie de suite que

$$M_k M_h = (-1)^{(n-h)(n-k)} M_h M_k.$$

Pour $l = 0$, donc $k = n - h$, on considérera les points $M_0^i$ comme constituant leurs propres indicatrices $P_0$ ; à part cela, rien ne sera changé. Supposons, en particulier, qu'il y ait $q'$ de ces points affectés du signe $+$, $q''$ affectés du signe $-$. Nous nommerons la différence $q' - q''$ *nombre algébrique de points d'intersection de* $M_h$ *avec* $M_k$, par opposition au nombre ordinaire, ou nombre *arithmétique* $q' + q''$. Le nombre $q' - q''$ sera désigné par la notation $(M_h M_k)$. Dorénavant, sauf avis contraire, c'est toujours du premier qu'il s'agira. Ce nombre est destiné à jouer un rôle des plus importants dans la suite.

Relevons en particulier les relations suivantes :

$$(M_h M_{n-h}) = (-1^h)(M_{n-h} M_h) \qquad (n \text{ pair});$$
$$(M_h M_{n-h}) = (M_{n-h} M_h) \qquad (n \text{ impair}).$$

La considération des nombres algébriques de points d'intersection remonte à Kronecker et à Poincaré ([1]).

**13.** Au fond, nous venons de définir une relation entre les quatre variétés $M_h$, $M_k$, $M_l$, $W_n$, en vertu de laquelle l'orientation de l'une est déterminée par celle des trois autres. Un exemple bien connu d'une telle situation est celui de la détermination d'un sens sur la normale à une surface dans l'espace ordinaire, quand la surface et l'espace ont été orientés de manière définie.

*Remarques.* — I. Dans toute la discussion, on a suppose implicitement que, quand $A_1$ se déplace sur l'intersection $M_l$, $P_n$ ne cesse jamais d'être une indicatrice de $W_n$. Ceci pourrait fort bien se produire en certains points exceptionnels de $M_l$ ; mais puisque cela n'aura lieu dans aucune de nos applications, il est inutile de nous en préoccuper.

---

([1]) HADAMARD, *loc. cit.*

II. L'extension de tout ce qui précède à l'intersection de deux variétés de type non restreint, ou bien à celle de plusieurs variétés est immédiate.

**14.** A l'aide des indicatrices on démontre aisément ces deux théorèmes, d'ailleurs intuitivement vrais :

I. *L'intersection de la frontière de* $M_h$ *avec* $M_k$, *fermée, de type restreint, est, au signe près, frontière de l'intersection.* Le signe ne dépend d'ailleurs que des dimensions.

II. *Pour trouver l'intersection de deux variétés, on peut remplacer l'une d'elles par son intersection avec une variété* W′ *contenue dans* $W_n$ *et analogue à elle.* Ainsi, symboliquement, W′ passant par $M_k$,

$$M_h M_k = \pm M_h(W' M_k).$$

Le signe ne dépend que de l'orientation de W′, et à droite, en prenant l'intersection, on considère les deux variétés comme contenues dans W′.

*Remarque.* — Nous avons ici un cas où il serait opportun d'indiquer la variété où sont placées celles dont on étudie l'intersection. Dans la pratique, on la déterminera aisément par la condition $n = h + k - 1$ qui relie les dimensions. Il semble préférable d'éviter de multiplier les notations, même au prix de légères équivoques.

**15.** De I, on déduit ce résultat d'une importance capitale :

Théorème. — *Si* $\Gamma_h \sim 0$, *aussi* $\Gamma_h \Gamma_k \sim 0$, *quel que soit* $\Gamma_k$. *En particulier si* $k = n - h$, *le nombre* $(\Gamma_h \Gamma_{n-h}) = 0$, *quel que soit* $\Gamma_{n-h}$.

Corollaire. — *Les cycles* $\Gamma_h \Gamma_k$, *ou les nombres* $(\Gamma_h \Gamma_{n-h})$, *sont invariants par rapport à l'homologie*, c'est-à-dire quand on remplace les cycles par d'autres homologues.

On verra combien tout ceci est important dans toutes les applications aux variétés algébriques. A titre d'exemple, soit $\Gamma_1$ un cycle nul d'une $W_2$. On aura $(\Gamma_1 \Gamma'_1) = 0$ pour tout cycle linéaire $\Gamma'_1$. Géométriquement, cela signifie que des deux régions en lesquelles $\Gamma_1$ décompose $W_2$, $\Gamma'_1$, passe autant de fois de la première dans la deuxième que de la deuxième dans la première.

*Remarque.* — Il est bien entendu que l'on ne considère jamais que des cycles se coupant exactement suivant un autre de la dimension voulue. Si cette condition n'est pas remplie, il suffira de déformer légèrement un des cycles pour qu'elle le devienne.

## III. — Intersection des cycles des systèmes fondamentaux. Étude de certains invariants qui s'y rattachent.

16. Soient $\Gamma_k^1, \Gamma_k^2, \ldots, \Gamma_k^q$, et $\Gamma_{n-k}^1, \Gamma_{n-k}^2, \ldots, \Gamma_{n-k}^r$, deux systèmes fondamentaux pour les cycles de leurs dimensions. $\Gamma_k$, $\Gamma_{n-k}$ étant quelconques, on aura

$$\Gamma_k \sim \sum x_i \Gamma_k^i, \qquad \Gamma_{n-k} \sim \sum y_j \Gamma_{n-k}^j.$$

Par suite, grâce à la relation évidente,

$$((M'_k + M''_k)\, M_{n-k}) = (M'_k M_{n-k}) + (M''_k M_{n-k}),$$

on obtient

$$(6) \qquad (\Gamma_k \Gamma_{n-k}) = \sum (\Gamma_k^i\, \Gamma_{n-k}^j) x_i y_j.$$

Ainsi, *à chaque paire d'indices complémentaires, $k$, $n-k$, on peut rattacher une certaine forme bilinéaire à coefficients entiers, par laquelle s'expriment les nombres $(\Gamma_k \Gamma_{n-k})$ dès que l'on sait comment les cycles dépendent de leurs systèmes fondamentaux.*

17. Prenons maintenant deux nouveaux systèmes fondamentaux, $\overline{\Gamma}_k^i$, $\overline{\Gamma}_{n-k}^j$. Il y aura des relations

$$(7) \qquad \overline{\Gamma}_k^i \sim \sum_s a_{is} \Gamma_k^s, \qquad \overline{\Gamma}_{n-k}^j \sim \sum b_{jt} \Gamma_{n-k}^t.$$

D'ailleurs évidemment

$$(8) \qquad (\Gamma_k \Gamma_{n-k}) = \sum (\overline{\Gamma}^i_k \overline{\Gamma}^j_{n-k}) \, x'_i y'_j,$$

$$\Gamma_k \sim \sum x'_i \overline{\Gamma}^i_k, \qquad \Gamma_{n-k} \sim \sum y'_j \overline{\Gamma}^j_{n-k}.$$

Désignons par A, B, les matrices aux coefficients $a$, $b$, des homologies (7), par $\overline{\mathrm{B}}$ la transposée de B, c'est-à-dire la matrice obtenue en y permutant les lignes et les colonnes. On tire des équations

$$(\overline{\Gamma}^i_k \overline{\Gamma}^j_{n-k}) = \sum\nolimits_{s,t} a_{is} \, b_{jt} \, (\Gamma^s_k \Gamma^t_{n-k})$$

la relation symbolique entre matrices

$$\| (\overline{\Gamma}^i_k \overline{\Gamma}^j_{n-k}) \| = \mathrm{A} \, \| (\Gamma^i_k \Gamma^j_{n-k}) \| \, \overline{\mathrm{B}}.$$

Ceci montre que l'on peut passer de (6) à (8) par un changement de variables à coefficients entiers. Mais il est évident que l'on peut passer de la même manière de (8) à (6). Donc, Frobenius (1), *les deux formes bilinéaires ont même diviseur élémentaire* $e_1$, $e_2, \ldots, e_f$, *et même rang* $f$. On peut aussi énoncer ce résultat en remplaçant les formes par les matrices. Ces diviseurs élémentaires sont des *invariants* de $\mathrm{W}_n$ par rapport à l'homéomorphisme, au même titre que les nombres $\mathrm{R}_k$, $\sigma_k$, $t_i$.

Remarquons que, A étant unimodulaire quelconque, les $\overline{\Gamma}^i_k$, définis par les homologies (7) comme plus haut, constituent toujours un système fondamental. En effet, on peut exprimer les $\Gamma^i_k$ en termes des $\overline{\Gamma}^i_k$, donc tout $\Gamma_k$ en termes de ces cycles. Une remarque analogue s'applique pour B et les $\overline{\Gamma}^j_{n-k}$. Avec un choix de A, B, ou si l'on veut, avec des systèmes fondamentaux convenables, on pourra réduire (6) à la forme

$$\sum e_i x_i y_i.$$

Nous dirons alors que les systèmes fondamentaux sont *canoniques*. Supposons que ceux dont nous sommes partis le soient déjà. On

(1) E. CAHEN, *loc. cit.*, p. 273. Le *rang* d'une matrice est l'ordre maximum d'un déterminant non nul que l'on en peut tirer. Le *rang* d'une forme bilinéaire est celui de sa matrice aux coefficients.

aura

$$(\Gamma_k^i \Gamma_{n-k}^i) = e_i \ (i \leq f);$$

$$(\Gamma_k^i \Gamma_{n-k}^j) = 0 \qquad (i \neq j \text{ ou bien } i = j > f).$$

De

$$(\Gamma_k \Gamma_{n-k}) = \sum{}_i e_i x_i y_i,$$

on tire

$$(\Gamma_k^i \Gamma_{n-k}) = e_i y_i; \qquad (\Gamma_k \Gamma_{n-k}^i) = e_i x_i,$$

Donc

$$(\Gamma_k^i \Gamma_{n-k}) = (\Gamma_k \Gamma_{n-k}^i) \equiv 0 \ (\mathrm{mod}\, e_i).$$

Ainsi, propriété caractéristique des systèmes fondamentaux canoniques, *leurs cycles coupent ceux de dimension complémentaire en un nombre de points divisible par un entier bien défini.*

18. Voici un théorème général pour les nombres de points d'intersection des cycles :

*Si* $(\Gamma_k \Gamma_{n-k}) = 0$ *quel que soit* $\Gamma_{n-k}$, $\Gamma_k$ *est un cycle nul ou un diviseur de zéro.*

Ce théorème est fort probablement vrai pour toute $W_n$ quoiqu'on n'en connaisse pas de démonstration générale. Nous en donnerons une pour les surfaces et les variétés algébriques. En se basant sur lui, on montre aisément que $f = R_k$. En effet, on en déduit que $\Gamma_k^{f+i}$ est nul ou diviseur de zéro, puisque

$$(\Gamma_k^{f+i} \Gamma_{n-k}) = y_{f+i} (\Gamma_k^{f+i} \Gamma_{n-k}^{f+i}) = 0, \qquad \Gamma_{n-k} \sim \sum y_i \Gamma_{n-k}^i.$$

D'un autre côté, de

$$\sum_1^f {}_i \lambda_i \Gamma_k^i \sim 0,$$

on tirerait, en considérant l'intersection avec $\Gamma_{n-k}^i$,

$$\lambda_i (\Gamma_k^i \Gamma_{n-k}^i) = \lambda_i e_i = 0, \qquad \text{d'où} \qquad \lambda_i = 0.$$

Les cycles $\Gamma_k^i$ $(i \leq f)$ sont donc indépendants. Enfin $\Gamma_k$, arbitraire, dépend des cycles du système fondamental, donc des $\Gamma_k^i$ $(i \leq f)$. Par suite, $f$ est le nombre maximum de $\Gamma_k$ distincts, d'où $f = R_k$.

Le même raisonnement donne $f = R_{n-k}$, donc $R_k = R_{n-k}$. Toutefois, ceci ne constitue qu'une vérification de la relation de Poincaré, car pour démontrer la propriété admise comme point de départ nous emploierons cette relation même.

**19.** Revenons aux nombres $e_i$. *Pour* $k = 1$ *ou* $n - 1$, *pour les variétés algébriques à d dimensions et* $k = 2$ *ou* $2d - 2$, *les* $e_i$ *sont égaux à* 1. En particulier, pour les surfaces algébriques ces entiers sont tous égaux à 1 [1].

Pour démontrer ce théorème relativement à la valeur $k$, il suffit d'établir que si $(\Gamma_k^i, \Gamma_{n-k})$ est divisible par $e_i$ quel que soit $\Gamma_{n-k}$, $\Gamma_k^i$ est, à un diviseur de zéro près, le multiple d'un certain cycle. En effet, il en résultera que ce dernier n'est pas exprimable en termes du système fondamental canonique, contradiction qui démontrera le théorème. Nous ne traiterons pas le cas d'une $W_n$ générale et de ses cycles $\Gamma_1$, $\Gamma_{n-1}$. Les autres seront discutés aux Chapitre III et V.

Remarquons que pour une $W_2$ les théorèmes de ce numéro et du précédent s'établissent de suite. En effet, soient $\gamma_1, \gamma_2, \ldots, \gamma_{2p}$ des rétrosections de Riemann de la surface, et supposons, comme nous le ferons toujours dans la suite, que $\gamma_i$ et $\gamma_{p+i}$ soient associées, de sorte que

$$(\gamma_i \gamma_{p+i}) = 1, \qquad (\gamma_i \gamma_k) = 0 \qquad (i \neq p \pm k).$$

Pour tout $\Gamma_1$ on peut écrire

$$\Gamma_1 \sim \sum \lambda_i \gamma_i$$

et s'il coupe tout autre en un nombre nul de points,

$$(\Gamma_1 \gamma_i) = \mp \lambda_{i \pm p} = 0,$$

donc $\Gamma_1 \sim 0$. Enfin prenant pour systèmes fondamentaux associés $\gamma_1, \gamma_2, \ldots, \gamma_{2p}$, et $\gamma_{p+1}, \gamma_{p+2}, \ldots, \gamma_{2p}, -\gamma_1, -\gamma_2, \ldots, -\gamma_p$, on aura bien

$$(\Gamma_h^i \Gamma_{n-h}^i) = 1, \qquad (\Gamma_h^i \Gamma_{n-h}^j) = 0 \qquad (i \neq j).$$

---

[1] Pendant la correction des épreuves j'ai appris de M. Veblen qu'il a réussi à démontrer que les $e_i$ sont tous égaux à l'unité *quelle que soit* $W_n$. En vertu de ce résultat fort remarquable ils ne constitueraient pas des invariants topologiques nouveaux.

# CHAPITRE II.

## SURFACES ALGÉBRIQUES. ÉTUDE APPROFONDIE DES CYCLES D'UNE COURBE VARIABLE DANS UN FAISCEAU LINÉAIRE.

### I. — Généralités sur les surfaces algébriques.

1. La surface algébrique est le lieu des points d'un espace à trois dimensions complexes dont les coordonnées satisfont à une équation

$$f(x, y, z) = 0,$$

où $f$ est un polynome. Nous la désignerons par la notation $V_2$, et nous la supposerons irréductible et en position générale par rapport aux axes et à l'infini. L'ensemble de points qui constitue $V_2$ est à quatre dimensions réelles; sa représentation correcte est à l'aide d'une $W_4$. Il est entendu qu'à l'avenir $V_2$ dénotera indifféremment la surface ou bien la $W_4$ qui la représente. De même par *courbe algébrique de* $V_2$, on entendra indifféremment la courbe en tant qu'entité de points satisfaisant à l'équation de la surface, ou bien la $W_2$ qui lui correspond dans la $W_4$.

2. **Systèmes linéaires. Rappel de quelques notions simples.** — Un système linéaire de surfaces découpe sur $V_2$ un *système linéaire de courbes*, et si C en est la courbe générique, on le désigne par $|C|$. Le nombre $r$ de paramètres dont C dépend est la *dimension* du système. Le *faisceau linéaire* est un système à un paramètre. En particulier un faisceau linéaire contenu dans $|C|$ et dont la courbe générique dépend du paramètre $u$ sera désigné par $\{C_u\}$. Les courbes C pouraient être astreintes à passer par des points fixes donnés et à s'y comporter de manière voulue. Ces points sont les *points-bases* de $|C|$.

3. La surface $V_2$ n'est pas nécessairement homogène en tant que $W_4$. Pour qu'il en soit ainsi il faut que ses singularités soient de nature relativement simple. Pour éviter les complications, nous supposerons *la surface non singulière, ou bien la projection d'une surface* $V'_2$ *non singulière située dans un espace de dimension* $> 3$. $V'_2$ est nécessairement homogène, donc $V_2$, qui lui est homéomorphe, le sera aussi. Les singularités de $V_2$ consisteront alors uniquement en une courbe double, à nombre fini de points triples (à la fois pour la surface et la courbe) et de points pinces, ou points où les plans tangents le long de la courbe double coïncident. Toute $V_2$ de cette nature est la projection d'une surface non singulière, donc homogène. Enfin théorème fort intéressant : *Toute surface algébrique est la transformée birationnelle d'une surface non singulière située dans un espace convenable* (Beppo Levi, Chisini).

4. *La surface* $V_2$ *est bilatère.* En effet, soit $m$ le degré de $f(x, y, z)$ en $z$. On peut étendre à $V_2$ la représentation d'une courbe algébrique par un plan recouvert de feuillets. Ici il faudra recouvrir de $m$ feuillets l'espace réel $S_4$ image des valeurs complexes $x, y$. Ces feuillets devront être rejoints le long de la projection de la courbe simple commune aux surfaces $f = 0$, $f'_z = 0$, projection représentée dans $S_4$ par une $W_2$. On pourra toujours choisir les axes de façon que la projection d'un circuit donné $\zeta$ de $V_2$ ne rencontre pas cette $W_2$. Il suffira pour cela qu'aucune droite parallèle à l'axe des $z$ et rencontrant $\zeta$ ne soit tangente à $V_2$.

Supposons en particulier que $\zeta$ invertisse une indicatrice de $V_2$. Sa projection dans $S_4$ fournit un circuit $y$ invertissant une indicatrice aussi. Par déformation on en déduit un circuit fini à la même propriété, et l'on montre aisément que cela est impossible.

5. *Les surfaces algébriques sont-elles topologiquement aussi générales que possible?* Autrement dit, existe-t-il une $V_2$ homéomorphe à toute $W_4$? La réponse ici est négative, contrairement à ce qui a lieu dans le cas analogue des courbes algébriques. Nous verrons en effet que l'indice $R_1$ de toute $V_2$ est nécessairement pair, tandis que pour une $W_4$ générale il peut avoir une valeur quelconque (Chap. I, n° 9).

Ce qui distingue les surfaces algébriques, c'est la présence des cycles *algébriques*, ou cycles $\Gamma_2$ formés par leurs courbes algébriques ([1]). Les propriétés topologiques spéciales que nous allons obtenir reposent en entier sur l'existence d'un système de $\Gamma_2$ se conduisant comme les courbes d'un système linéaire irréductible. Elles appartiendraient donc à toute $W_4$ en possédant un. Nous pouvons maintenant énoncer notre problème.

Problème fondamental. — *Étudier les propriétés topologiques des surfaces algébriques surtout en tant qu'elles se relient à celles de leurs courbes algébriques.*

6. La propriété suivante des cycles algébriques, d'ailleurs fort importante pour la suite, montre à quel point ils sont particularisés.

Théorème. — *Pour une orientation convenable de* $V_2$, *les nombres arithmétique et algébrique de points d'intersection de deux cycles algébriques sont égaux* ([2]).

Orientons pour commencer une section plane H donnée; il en

---

([1]) Nous étendrons plus tard (Chap. IV, § VII), cette définition aux cycles simplement homologues à une somme ou une différence de courbes algébriques.

([2]) Voici une démonstration analytique de cet important théorème. Elle est basée sur une identité assez simple que l'on trouvera développée dans mon Mémoire couronné. Soient, pour simplifier, $f(x, y)$, $\varphi(x, y)$ deux fonctions algébriques de $x = x' + ix''$, $y = y' + iy''$. Si l'on pose $f = f' + if''$, $\varphi = \varphi' + i\varphi''$, on a

$$\frac{D(f', f'', \varphi', \varphi'')}{D(x', x'', y', y'')} = \left|\frac{D(f, \varphi)}{D(x, y)}\right|^2, \tag{1}$$

généralisation d'une relation d'ailleurs classique pour les fonctions d'une seule variable. Or soient F, $\Phi$ les $M_2$ de $S_4$ réel $x', \ldots, y''$, aux équations $f' = f'' = 0$; $\varphi' = \varphi'' = 0$. Il suffit de se reporter aux définitions de Poincaré pour les orientations et les nombres algébriques d'intersection des variétés, telles qu'il les a données dans son Mémoire de l'École Polytechnique (1895), pour reconnaître qu'en vertu de (1) les contributions des divers points de l'intersection de F et $\Phi$ au nombre $(F\Phi)$ sont égales. Ceci revient à démontrer notre théorème pour le plan complexe. L'extension à plus de deux variables, puis à des surfaces ou des variétés algébriques quelconques, est facile.

Remarque non sans utilité, tout ceci est applicable même quand les fonctions ne sont pas algébriques, pourvu que l'on ne considère que leurs intersections contenues dans un domaine où elles sont toutes holomorphes.

résultera une orientation pour toute autre H′ en vertu de ce que $H' \sim H$. Démontrons d'abord le théorème pour H et H′. A cet effet, en un de leurs points d'intersection P, construisons les indicatrices, puis orientons $V_2$ de facon que P compte pour +1 dans (HH′). Faisons ensuite varier le plan de H, sans qu'il devienne jamais tangent à H′, de manière à ramener P à un autre point d'intersection P′. Durant le mouvement, l'orientation de H ne changera pas et elle reviendra à sa position primitive avec la même orientation, puisque le cycle H reste constamment homologue à lui-même. De plus, les indicatrices resteront toutes positives pour leurs variétés respectives, toutes bilatères. Celles qu'elles fourniront en P′ auront donc la même situation relative qu'en P. Par suite, P′ compte aussi pour +1 dans (HH′), ce qui démontre notre affirmation quand il s'agit de deux sections planes.

Pour orienter une courbe algébrique quelconque C, nous ferons comme ceci : Soit H une section plane osculant la courbe au point P. On peut tracer d'une infinité de manières deux petits triangles $\zeta$, $\zeta'$, entourant P sur C et H respectivement, dont le second est une indicatrice de H, alors que le premier lui correspond point par point, biunivoquement, de façon que la distance entre deux points correspondants soit très petite par rapport aux dimensions de $\zeta$ ou $\zeta'$. On nommera les sommets de $\zeta$ dans le même ordre que ceux correspondants de $\zeta'$; $\zeta$ ainsi orienté servira d'indicatrice à C.

Ceci posé, il est clair que l'orientation ainsi attribuée à C ne dépend aucunement du point P. Enfin si D est une courbe algébrique quelconque passant par P, les contributions de P à (CD) et à (HD) sont égales. Donc, pour la détermination de (CD), on peut remplacer en chacun de leurs points d'intersection les courbes C, D par deux sections planes, ce qui suffit pour compléter la démonstration.

*Remarque.* — Bien entendu on suppose C et D non tangentes, ce qui nous suffira pour la suite. L'extension au cas où il y aurait contact ne présente d'ailleurs pas de difficultés sérieuses.

Corollaire. — *Une courbe algébrique* C *ne peut former frontière.*

En effet il en existe toujours une autre D qui la rencontre. Comme (CD) $\neq$ o, C ne peut être un cycle nul.

7. **Transformations birationnelles.** — Les nombres $R_k$, $\sigma_k$, etc. sont bien invariants par rapport à l'homéomorphisme, mais comment se comportent-ils par rapport au groupe fondamental de la Géométrie algébrique, celui des transformations birationnelles? Le nombre $R_1$ est un *invariant absolu*, c'est-à-dire par rapport à toutes les transformations en question. Pour $R_2$, au contraire, cela dépendra de la transformation birationnelle considérée et il pourra fort bien arriver qu'il change si la transformation introduit des courbes *exceptionnelles*, ou courbes transformées de points, car dans ce cas la surface et sa transformée ne sont pas homéomorphes. Sans entrer davantage dans la question, contentons-nous d'affirmer que $R_2$ se comporte comme l'invariant de Zeuthen-Segre (*voir* Chap. III, n° 11).

## II. — Étude des cycles d'une courbe variable dans un faisceau linéaire.

8. Soit $|C|$ un système linéaire, $\infty^2$ au moins, aux courbes génériques irréductibles de genre $p$ et se coupant deux à deux en $m > 0$ points. Nous supposerons explicitement que : ($a$) si $|C|$ est $\infty^r$, il ne contient pas $\infty^{r-1}$ courbes réductibles; ($b$) quand une C acquiert une nouvelle singularité, celle-ci consiste uniquement, en général, en un point double ordinaire, le genre s'abaissant alors de $p$ à $p - 1$.

Toute surface à singularités ordinaires possède de tels systèmes. En effet, lorsqu'elle n'est pas une surface réglée ou de Steiner, les propriétés ($a$), ($b$) sont vérifiées pour le système des sections planes (Castelnuovo), et dans les deux cas exceptionnels par ceux que découpent les surfaces d'ordre $\geqq 3$. D'ailleurs les propriétés ($a$) et ($b$) ne sont pas d'importance fondamentale, et nous ne les imposons que pour faciliter la discussion. En particulier, presque tous nos résultats sont vrais même quand ($a$) n'est pas vérifiée.

9. Prenons donc dans C un faisceau générique $\{C_u\}$. En vertu

de $(a)$ toute $C_u$ sera irréductible et son genre sera $p$, sauf pour certaines valeurs critiques $a_1, a_2, \ldots, a_n$, de $u$ où, d'après $(b)$, il s'abaissera à $p - 1$.

Soient X, Y des coordonnées courantes sur C; $\alpha_1, \alpha_2, \ldots$ les points de branchement de la fonction $Y(X)$; $aa_i$ des coupures

Fig. 1.

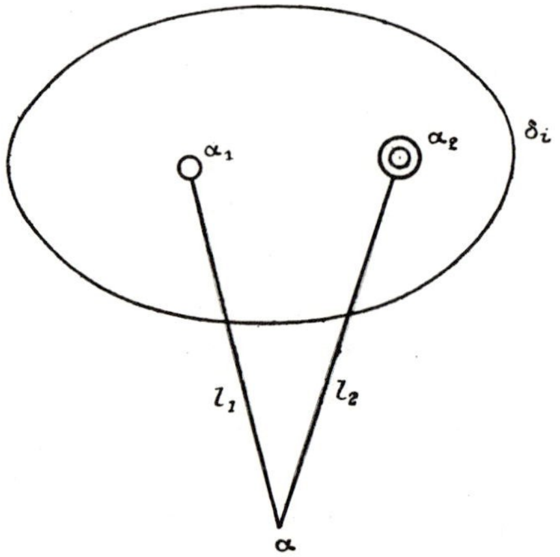


dans le plan $u$. Tant que $u$ décrit un circuit ne traversant pas ces coupures, des lacets $\alpha\alpha_i$ du plan X sont ramenés à leur position initiale, et tout cycle linéaire de $C_u$ à une position homologue. Supposons que $u$ traverse $aa_i$, et soient $\alpha_1, \alpha_2$ les points de branchement de $Y(X)$ en coïncidence pour $u = a_i$. En ce point les deux mêmes valeurs de Y sont permutées, car seulement deux valeurs en deviennent égales au nouveau point double de $C_{a_i}$. Désignons par $l_k$ les lacets $\alpha\alpha_k$. On observe de suite que $\delta_i \sim l_2 - l_1$ est invariant au voisinage de $a_i$, car ce n'est autre que le cycle entourant $\alpha_1$ et $\alpha_2$ dans un des feuillets de la surface de Riemann qui représente la variation de la fonction $Y(X)$. Pour tout autre cycle $\gamma$, on a

$$\gamma \sim t_1 l_1 + t_2 l_2 + l,$$

où $l$ est une somme de lacets $l_k$, $k > 2$, donc invariante au voisinage de $a_i$. Or

$$\gamma \sim (t_1 + t_2) l_1 + t_2 \delta_i + l,$$

et quand $u$ tourne autour de $a_i$, $l_1$ se transforme en $l_2$ (*fig.* 2).

Donc si $\gamma'$ est le transformé de $\gamma$, on a

$$\gamma' - \gamma \sim (t_1 + t_2)(l_2 - l_1) \sim (t_1 + t_2)\,\delta_i.$$

D'ailleurs (*fig.* 1),

$$(l_1 \delta_i) = +1, \qquad (\delta_i \delta_i) = (\delta_i l) = 0;$$

donc

$$(\gamma\,\delta_i) = t_1 + t_2,$$

et par suite finalement

$$(\gamma' - \gamma) \sim (\gamma\,\delta_i).\delta_i.$$

Observons enfin que quand $u$ tend vers $a_i$, $\delta_i$ tend vers un point,

Fig. 2.

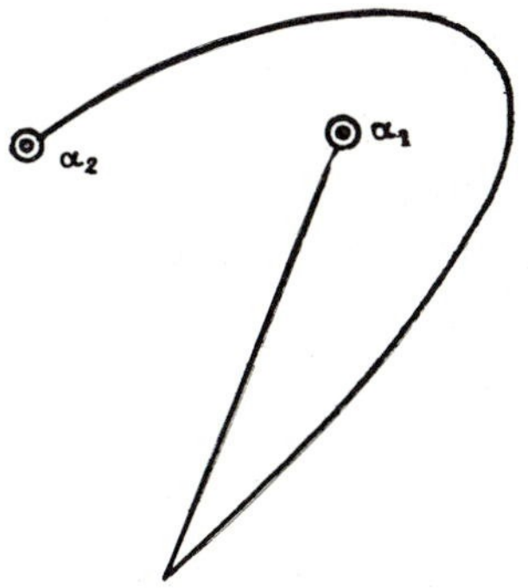


le nouveau point double de $C_{a_i}$, ou, plus exactement, il tend vers un cycle réductible à ce point dans $C_{a_i}$. Nous pouvons donc énoncer :

Théorème fondamental. — *A tout point critique $a_i$ correspond un cycle linéaire $\delta_i$ de $C_u$, invariant dans son voisinage et réductible à un point pour $u = a_i$, Quand $u$ tourne autour de $a_i$, tout autre cycle $\gamma$ s'accroît de $(\gamma\,\delta_i).\delta_i$.*

Au coefficient $(\gamma\,\delta_i)$ de $\delta_i$ près, ce théorème remonte à M. Picard [1]. Il n'est que juste de dire que ce coefficient jouera un rôle capital dans la suite. C'est en effet sa connaissance qui va nous permettre de résoudre la plupart des questions qui vont se

---

[1] *Voir* Picard et Simart, *Traité*, vol. I, p. 95.

présenter, sans avoir recours à des éléments étrangers, tels que la théorie trancendante, comme le faisait d'une manière d'ailleurs si brillante M. Picard. Outre sa valeur didactique, notre méthode va nous fournir des résultats nouveaux ; de plus elle, seule, est pour l'instant susceptible d'être étendue aux variétés algébriques.

*Corollaire.* — Pour que $\gamma$ soit complètement invariant (nous dirons « invariant » tout court), il faut et il suffit que les nombres $(\gamma\,\delta_i)$ soient tous nuls.

10. *Il existe un cycle $\delta'_i$ rencontrant $\delta_i$ en un seul point*, donc s'accroissant de $\delta_i$ quand $u$ tourne autour de $a_i$. Il suffit de prendre

$$\delta'_i = l_1 + l,$$

où $l$ est un chemin permutant les mêmes valeurs de $Y(X)$ que $l_1$ et $l_2$, mais égal à une somme de lacets ne les comprenant ni l'un ni l'autre. Ce chemin existe certainement, car autrement on ne pourrait tracer dans $C_{n_i}$ aucun chemin permutant ces deux valeurs, puisque pour cette courbe les lacets $l_1$, $l_2$ disparaissent. La courbe $C_{n_i}$ serait donc réductible, ce qui est contre nos hypothèses.

Comme corollaire de ce que $(\delta_i\delta'_i) = 1$, $\delta_i$ *n'est pas un cycle nul de* $C_u$, quoiqu'il le soit pour $V_2$.

11. **Lieux de certaines multiplicités** — Il convient d'ouvrir ici une parenthèse pour discuter une question qui reviendra fréquemment. Soit $M_k$ une multiplicité contenue dans $C_u$, ou de frontière dans $C_u$. Quelle est la nature de son lieu quand $u$ varie ? Précisons la question. D'abord on supposera que $M_k$ varie de façon continue avec $u$. Ensuite, une fois pour toutes, on fera varier $u$ de cette manière : on considère la demi-droite

$$\text{argument } (u - a) = \text{constante}, \tag{1}$$

et l'on suppose que sur chacune de ces demi-droites $u$ varie de l'infini à la valeur $a$, puis on fera varier l'argument de $0$ à $2\pi$. Dans le plan $u$ on considérera les coupures $aa_i$ et, en outre, une coupure le long de la demi-droite (1) d'argument zero. Nous appellerons cette dernière coupure la coupure $a\infty$. Dans le cas où la demi-droite en question passerait par un des points critiques, nous la remplacerions par une autre correspondant à un argument $\theta$ quelconque, quitte à faire varier l'argument de (1) de $\theta$ à $\theta + 2\pi$.

Dans les conditions indiquées, on est assuré que les frontières de $M_{k+2}$ lieu de $M_k$ se composent uniquement : ($a$) du lieu de celles de $M_k$; ($b$) des multiplicités suivantes : Soit $u$ un point sur une coupure, et désignons-le par $u'$ ou $u''$, suivant qu'on le considère comme sur un bord ou sur l'autre. La différence entre les $M_k$ relatives à $C_{u'}$ ou $C_{u''}$ est un $\Gamma_k$ dont le lieu, quand $u$ décrit la coupure, est une des multiplicités frontières de $M_{k+2}$. Les différentes frontières ainsi obtenues se rattachent par des éléments de la courbe $C_a$

La nécessité d'introduire la coupure $a\infty$ est due à ce fait : Il n'y a pas de raison pour que, quand $|u - a|$ reste fixe, son argument variant de 0 à $2\pi$, ou bien de $\theta$ à $\theta + 2\pi$ suivant le cas, la position finale de $M_k$ coïncide avec sa position initiale. Leur différence engendre une frontière de type ($b$) quand $u$ décrit la coupure $a\infty$.

**12. Application : Lieu d'un cycle invariant.** — Dans le cas d'un cycle linéaire invariant, on se trouve dans les conditions les plus simples. Si $\gamma$ est invariant, le cycle $\Gamma_1$ relatif à une coupure quelconque est nul pour $C_u$. Soit $M'_2$ la partie de $C_u$ qu'il limite. Quand $u$ décrit la coupure, $M'_2$ engendre une $M'_3$; les $M'_3$ ainsi obtenues, ajoutées au lieu propre de $\gamma$, forment une multiplicité $\Gamma_3$ ne possédant au pis aller qu'une partie de $C_a$ comme frontière. Mais $C_a$, courbe irréductible comme toutes celles de son faisceau, est une $M_2$ connexe. Donc cette frontière ne peut se composer que de $C_a$ tout entière, ce qui est impossible, car $C_a$ est un $\Gamma_2$ non nul (n° 6). Donc enfin $\Gamma_3$ est un cycle à trois dimensions coupant $C_u$ précisément suivant $\gamma$. Il n'est d'ailleurs pas difficile de voir que les $M'_3$ sont réductibles à des multiplicités de dimension moindre. Ceci justifie le terme *lieu de* $\gamma$ appliqué à $\Gamma_3$.

13. THÉORÈME. — *Tout cycle invariant homologue à une somme de* $\delta_i$ (mod $C_u$) *forme frontière sur la courbe* (1).

(1) *Voir* Chapitre IV, n° 7, pour une démonstration transcendante. Celle de mon article des *Annals of Mathematics*, vol. XXI, 1920, est incorrecte comme

On verra combien cette propriété est importante pour la classification de ses cycles.

Soit $\Delta_i$ l'$M_2$ lieu de $\delta_i$ quand $u$ décrit $aa_i$, lieu évidemment homéomorphe à une hémisphère et que nous nommerons l'*onglet* $\Delta_i$. Soit $\gamma$ le cycle de l'énoncé. Si on le suppose placé pour le moment, ainsi que les $\delta_i$, dans $C_a$, on pourra écrire

$$\Delta_i \equiv \delta_i; \qquad \sum t_i\Delta_i \equiv \gamma = \sum t_i\delta_i \sim o \qquad (\mathrm{mod}\, V_2).$$

Faisant varier $u$ et supposant maintenant $\gamma$ dans $C_u$, soit $\Gamma_3$ son lieu. Nous allons montrer que $\Gamma_3 \sim o$ (mod $V_2$), d'où découlera notre théorème, car de $M_4 \equiv \Gamma_3$ on tire de suite (Chap. I. n° **14**)

$$M_4\, C_u \equiv \gamma.$$

**14.** Soit d'abord $C_0$ une courbe quelconque du système $|C|$, n'appartenant pas à $\{C_u\}$. Il existe un faisceau linéaire unique, $\{C'_v\}$, du système, contenant $C_0$ et une $C_u$ quelconque donnée. Nous supposerons la variable $v$ choisie de façon que $C_u$ corresponde à la valeur $v = b$, constante quelconque.

Soient $b_1(u)$, $b_2(u)$, ..., $b_n(u)$ les valeurs critiques de $v$. Faisons tendre $\{C_u\}$ vers $\{C'_v\}$ dans $|C|$, de manière à amener $a$ en $b$ et chaque point $a_i$ au point $b_i$ de même indice. Les $\delta_i$, $\Delta_i$ deviendront des éléments semblables, $\delta'_i$, $\Delta'_i$, pour $\{C'_v\}$, relatifs à certaines coupures $bb_i$ du plan $v$. Considérant cette fois les $\delta'_i$ comme situés dans $C_u = C'_b$, nous aurons

$$\sum t_i\Delta'_i \equiv \gamma = \sum t_i\delta'_i.$$

Aux coupures déjà pratiquées dans le plan $u$, ajoutons-en d'autres $aa'_j$, allant aux points $a'_j$ où plusieurs $b_i$ coïncident. Nous allons rechercher les frontières de l'$M_4$ lieu de $\sum t_i\Delta'_i$ quand $u$ varie. En se reportant au n° **11** on voit qu'il s'agit essentiellement

---

me l'a obligeamment indiqué M. Alexander. C'est une observation qu'il m'a faite qui m'a suggéré l'idée de faire intervenir un second faisceau appartenant à |C|, c'est-à-dire au fond le fait que le système est $\infty^2$ au moins. La simplicité relative de la démonstration transcendante que je possédais depuis longtemps est frappante. Elle est d'ailleurs en relation étroite avec un résultat de M. Picard (*loc. cit.*, vol. II, p. 388).

d'étudier un certain cycle $\Gamma_2$ correspondant à chaque coupure et son lieu quand $u$ la décrit.

15. *Examinons d'abord la coupure* $a\infty$. Je dis d'abord que la frontière correspondante de $M_4$, frontière lieu du $\Gamma_2$ relatif à la coupure, peut être supprimée. Soient en effet $u'$, $u''$ deux points opposés de la coupure. Quand la demi-droite (1) du n° 11 tourne autour de $a$, on peut concevoir qu'un de ses points parte de $u'$ pour aboutir à $u''$, sans jamais traverser les coupures. L'ensemble des trajets des points de (1) donne ainsi lieu à une famille de circuits du plan $u$, tous intérieurs les uns aux autres, partant d'un cercle de rayon infiniment grand pour terminer avec la ligne polygonale formée par les bords de toutes les coupures.

Durant le déplacement de $u$, la figure formée par les lacets du plan $v$ varie, mais reste cependant réductible à sa position primitive par une déformation durant laquelle on ne permet pas aux lacets $bb_i$ de se traverser. Par suite la position finale de cette figure est réductible à sa position initiale par une déformation analogue. Puisque le trajet de $u$ est un chemin n'entourant aucun des points $a_i$, $a'_j$, dans son plan, les points critiques $b_k(u)$ reviennent exactement à leurs positions primitives. On peut donc contrôler la déformation de notre figure pendant que $u$ varie, de manière qu'elle retourne exactement à sa position première. Il y aura alors coïncidence des $M_2$ relatives à $u'$ et $u''$, dont la différence donne lieu à $\Gamma_2$. Cela montre bien que la frontière de $M_4$, relative à $a\infty$, peut être supprimée.

*Passons aux coupures* $aa_i$. Quand $u$ tend vers $a_i$, un des points $b_k$, soit $b_1$, tend vers $b$; quand $u$ tourne une fois autour de $a_i$, $b_1$ en fait autant pour $b$. En faisant la figure on voit de suite que les $M_2$ relatives à deux points opposés, $u'$, $u''$, de $aa_i$, ne diffèrent que par le lieu de $\gamma$ quand $v$ décrit le lacet $bb_1$, c'est-à-dire va de $b$ à $b_1$, puis retourne à $b$, après avoir tourné autour de $b_1$. Ici $\Gamma_2$ se compose de deux multiplicités opposées; donc il peut être supprimé. Ainsi il n'y a pas non plus de frontières de $M_4$ relatives aux coupures $aa_i$.

16. *Examinons finalement les coupures* $aa'_j$. Lorsque $u$ tend vers $a'_j$, certains des points $b_k$, soient pour l'instant deux seule-

ment $b_1$, $b_2$, tendent à coïncider. Soient encore $u'$, $u''$ deux points opposés sur la coupure $aa'_j$. Cette fois $\Gamma_2$ est réductible à un multiple du lieu $\Delta$ de $\delta'_1$, quand $v$ décrit une certaine ligne $b_1 b_2$. D'ailleurs le cycle $\delta'_1$ se réduit à un point aux deux extrémités de la ligne. On le sait pour $b_1$ ; quant à $b_2$, puisque $\Gamma_2 \sim \lambda \Delta$, $\Delta$ doit être fermée, et ceci par une partie de $C'_{b_2}$, ce qui exige que

$$\lambda \delta'_1 \sim 0 \qquad (\mathrm{mod}\, C'_{b_2});$$

donc $\delta'_1$ est homologue au cycle de $C'_v$ qui s'évanouit en $b_2$. Nous ne devons pas en conclure que $\delta'_1 \sim \delta'_2$ dans toute $C'_v$, mais seulement que le cycle de $C'_v$ en coïncidence avec $\delta'_1$ pour $v$ voisin de $b_1$, tend aussi à devenir homologue à $\delta'_2$ quand $v$ tend vers $b_2$ le long de la ligne $b_1 b_2$.

Je dis que l'on peut toujours s'arranger pour que $\Delta$ ne rencontre pas la courbe $C_0$, qui, rappelons-le, appartient à tout $\{C'_v\}$. Soient en effet cette fois X, Y deux coordonnées courantes sur $C'_v$, puis construisons encore la surface de Riemann à plusieurs feuillets, image de la fonction Y(X). Le cycle $\delta'_1$ y entoure, dans l'un des feuillets, deux ou plusieurs points de branchement $\alpha_k$ qui tendent à coïncider quand $u$ tend vers $a'_j$, en même temps que $v$ tend vers $b_1(u)$. D'ailleurs les points de branchement sont les seuls que l'on ne puisse peut-être pas faire éviter au cycle par une déformation convenable. Or, au pis aller, il y a un nombre fini de faisceaux $\{C'_v\}$, donc seulement un nombre fini de valeurs de $u$, pour lesquelles ces points de branchement coïncident, dans une ou plusieurs $C'_v$, avec un des points-bases du faisceau. Faisons donc éviter, aux coupures $aa'_j$, ces valeurs particulières de $u$, puis pour chaque $u$ sur $aa'_j$, à la ligne $b_1 b_2$ le point $v$ correspondant à $C_0$, condition facile à satisfaire. Dans ces circonstances, $\Delta$ ne rencontrera effectivement pas $C_0$.

Dans le cas où plusieurs $b_k$, soient $b_1$, $b_2$, ..., $b_s$, coïncideraient au lieu de deux seulement, il y aurait plusieurs $\Delta$, lieux de certains cycles de $C'_v$ quand $v$ décrit des segments $b_h b_k$ convenables, mais à part cela rien ne serait changé. Nous pouvons donc affirmer, dans tous les cas, que quand $u$ est quelconque sur $aa'_j$, $\Gamma_2$, somme de multiplicités $\Delta$, ne rencontrera pas $C_0$, et il en sera de même pour son lieu $\Gamma'_3$ quand $u$ décrit $aa'_j$, ainsi que pour la somme $\Gamma'_3$ de ces lieux.

17. En définitive, la frontière totale de $M_4$ se compose de $\Gamma'_3$ et du lieu $\Gamma_3$ de $\gamma$ quand $u$ varie. Donc $\Gamma_3 + \Gamma'_3 \equiv 0$. Comme $\Gamma_3 \equiv 0$, aussi $\Gamma'_3 \equiv 0$, c'est-à-dire $\Gamma'_3$ est un cycle.

Faisons maintenant tendre $C_0$, dans $|C|$, vers une $C_u$ quelconque, sans jamais lui faire acquérir de singularités nouvelles. A partir d'un certain moment il arrivera que $\Gamma'_3$, que l'on maintient fixe, coupera la courbe variable suivant un cycle $\gamma'$. Mais comme le genre de la courbe est fixe $\gamma' \sim 0$. Sa position limite $\gamma''$ dans $C_u$ y formera aussi frontière. Mais l'intersection de la frontière de $M_4$ avec $C_u$ y forme frontière, d'où $\gamma + \gamma'' \sim 0$, et enfin $\gamma \sim 0$. *Ceci démontre notre théorème.*

18. Théorème — *Tout cycle de $C_u$ est relié aux $\delta_i$ et aux cycles invariants, par une homologie.*

Soient $\gamma_1, \gamma_2, \ldots, \gamma_{2p}$, $2p$ cycles indépendants de $C_u$, $2p - r$ le rang de la matrice

$$\|(\gamma_i\delta_k)\| \qquad (i = 1, 2, \ldots, 2p;\ k = 1, 2, \ldots, n).$$

En fait, supposons, ce qui ne diminue pas la généralité, le déterminant

$$|(\gamma_i\delta_k)| \neq 0 \qquad (i, k = 1, 2, \ldots, 2p - r).$$

Alors, comme $(\gamma\gamma') = 0$ pour tout $\gamma'$ entraîne $\gamma \sim 0$ (Chap. II, n° 19) :

1° Les cycles $\gamma_1, \gamma_2, \ldots, \gamma_{2p-r}$ sont indépendants et il en est de même de $\delta_1, \delta_2, \ldots, \delta_{2p-r}$.

2° Tout $\delta_i$ dépend de $\delta_1, \delta_2, \ldots, \delta_{2p-r}$. En effet, on peut toujours trouver des entiers $t$, avec $t_0 \neq 0$, tels que

$$(\gamma,\ t_0\delta_i - t_1\delta_1 - t_2\delta_2 \ldots - t_{2p-r}\delta_{2p-r}) = 0$$

pour tout $\gamma$.

3° Pour tout $\gamma$ on peut trouver des entiers $t$, avec $t_0 \neq 0$, tels que

$$(\delta_i,\ t_0\gamma - t_1\gamma_1 - t_2\gamma_2 \ldots - t_{2p-r}\gamma_{2p-r}) = 0,$$

quel que soit $\delta_i$. Le cycle

$$t_0\gamma - t_1\gamma_1 - t_2\gamma_2 \ldots - t_{2p-r}\gamma_{2p-r}$$

sera donc invariant. En particulier, on pourra remplacer ainsi

$\gamma_{2p-r+1}, \gamma_{2p-r+2}, \ldots, \gamma_{2p}$ par autant de cycles invariants, soient $\gamma'_1, \gamma'_2, \ldots, \gamma'_r$.

Ceci posé, toute somme de $\delta_i$ invariante est nulle, donc aucune combinaison des cycles $\delta_1, \delta_2, \ldots, \delta_{2p-r}$ ne peut l'être. Ces $2p-r$ cycles sont donc indépendants des $\gamma'$ et enfin les $2p$ cycles $\delta_1, \delta_2, \ldots, \delta_{2p-r}, \gamma'_1, \ldots, \gamma'_r$ sont indépendants, d'où le théorème s'ensuit.

19. Corollaires. — I. *Tout cycle* $C_u$ *dépend des cycles invariants* $(\text{mod}\, V_2)$.

II. *Le déterminant gauche*

$$|(\delta_i\delta_j)| \neq 0 \qquad (i, j = 1, 2, \ldots, 2p-r).$$

En effet, les entiers $(\gamma'_j\delta_i)$ sont tous nuls. Donc la matrice analogue à celle écrite plus haut, relative aux $2p$ cycles $\delta_1, \delta_2, \ldots, \delta_{2p-r}, \gamma'_1, \ldots, \gamma'_r$, est de même rang que

$$\|(\delta_i\delta_h)\| \qquad (i = 1, 2, \ldots, 2p-r;\ h = 1, 2, \ldots, n),$$

dont le rang doit être $2p-r$. On doit donc pouvoir en tirer un déterminant non nul de cet ordre. Puisque tout $\delta_i$ dépend des $2p-r$ premiers, tout déterminant d'ordre $2p-r$ de la matrice est le produit de celui écrit plus haut par un nombre rationnel. Donc, effectivement, ce déterminant n'est pas nul.

III. *L'entier r est pair*. En effet, $2p-r$, ordre d'un déterminant gauche non nul, est pair, donc $r$ l'est aussi. Soit $r = 2q$. L'entier $q$ est l'*irrégularité* de la surface. C'est un invariant dont l'importance ressortira nettement plus tard.

IV. Puisque les $2p$ cycles écrits plus haut sont indépendants, le déterminant d'ordre $2p$, dont les éléments sont leurs nombres d'intersections mutuelles, n'est pas nul. Comme $(\gamma'_i\delta_j) = 0$, la valeur de ce déterminant est égale au produit

$$|(\delta_i\delta_j)| \cdot |(\gamma'_h\gamma'_k)| \qquad (i, j = 1, 2, \ldots, 2p-r;\ h, k = 1, 2, \ldots, r);$$

d'où

$$|(\gamma'_h\gamma'_k)| \neq 0,$$

résultat utile plus loin.

20. Ajoutons quelques mots sur la variation des lignes joignant entre eux les points-bases $A_1, A_2, \ldots, A_m$ de $\{C_u\}$. Soit $l_j$ une ligne joignant $A_1$ à $A_j$ dans $C_u$. Quand $u$ tourne autour de $a_i$, $l_j$ s'accroît de $(l_j \delta_i)\delta_i$; on le montre comme pour les cycles. Reprenons les cycles indépendants $\gamma_1, \gamma_2, \ldots, \gamma_{2p}$. Je dis que si

$$l = \sum_1^{2p}{}_h \lambda_h \gamma_h + \sum_2^m{}_j \mu_j l_j$$

est une ligne, ou plutôt une somme de lignes, invariante, les coefficients des $l_j$ y sont tous nuls, de sorte que cette somme est un cycle. En effet, autrement le lieu de la somme est une $M_3$ ouverte comme sa trace sur $C_u$. Or la frontière de $M_3$ ne peut que se réduire à $C_u$, ce qui est impossible, puisque le cycle $C_u$ n'est pas nul (nº 6).

Comme conséquence, la matrice

$$\left\| \begin{matrix} (\gamma_1\delta_1) & (\gamma_1\delta_2) & \ldots & (\gamma_1\delta_n) \\ (\gamma_2\delta_1) & \ldots\ldots & \ldots & \ldots\ldots \\ \ldots\ldots & \ldots\ldots & \ldots & \ldots\ldots \\ \ldots\ldots & \ldots\ldots & \ldots & (\gamma_{2p}\delta_n) \\ (l_2\delta_1) & \ldots\ldots & \ldots & \ldots\ldots \\ \ldots\ldots & \ldots\ldots & \ldots & \ldots\ldots \\ \ldots\ldots & \ldots\ldots & \ldots & (l_m\delta_n) \end{matrix} \right\|$$

est de rang $2p - 2q + m - 1$.

# CHAPITRE III.

## TOPOLOGIE DES SURFACES ALGÉBRIQUES.

### I. — Réduction à une cellule ([1]).

**1.** Nous allons faire l'étude des cycles de $V_2$ en pratiquant des coupures la réduisant à une cellule. D'après le lemme du Chapitre I, n° **11**, pour avoir tous les cycles, il suffira de rechercher ceux formés avec les éléments des coupures. Dans cette recherche, la discussion du Chapitre précédent va jouer un rôle essentiel.

Tout d'abord, les coupures $aa_i$, $a\infty$ réduisent le plan $u$ à une cellule $E'_2$. Traçons ensuite sur $C_u$ des rétrosections $\gamma_h$ partant du point $A_1$ pour y retourner et ne se coupant nulle part ailleurs. Pratiquons enfin des coupures le long de ces rétrosections et aussi le long des lignes $l_j = A_1 A_j$, que nous supposerons maintenant tracées de manière à ne se couper mutuellement et à ne rencontrer les $\gamma_h$ qu'au même point $A_1$. Ces conditions sont faciles à réaliser, car les coupures $\gamma_h$ réduisent $C_u$ à une cellule dont les $A_i$ $(i > 1)$ sont des points internes, $A_1$ seul étant sur la frontière. Il est alors évidemment possible de tracer les $l_j$ de la manière voulue.

Nos coupures vont réduire $C_u$ à une cellule $E''_2$, dont les $A_i$ ne sont pas des points internes, condition nécessaire comme on va le voir. Quand $u$ décrit $E'_2$, $E''_2$ engendre une cellule $E_4$ dérivée de $V_2$ à l'aide de coupures à trois dimensions, lieux de celles de $C_u$; c'est ce que nous appellerons « la cellule $V_2$ ».

Si l'on n'avait pas pratiqué les coupures $l_j$ dans $C_u$, on n'aurait

---

([1]) *Voir* Picard et Simart, vol. II, Chap. XI et XII. — Poincaré, *Journal de Mathématiques pures et appliquées*, 5ᵉ série, vol. VIII, 1902; 6ᵉ série, vol. II, 1906. — Lefschetz, *Annals of Mathematics*, vol. XXI, 1920.

pu conclure à l'existence d'une cellule à quatre dimensions, car le lieu de $E''_2$ aurait comme points internes les $A_i$ et l'on n'aurait pas le droit d'affirmer qu'il est homéomorphe à une cellule.

**2.** La frontière totale de la cellule comprend : 1° le lieu de $C_u$ quand $u$ décrit la frontière de $E'_2$, c'est-à-dire les coupures $aa_i$; 2° les lieux $G_h$, $L_j$ des $\gamma_h$ et des $l_j$. Comme éléments à deux dimensions, il y a donc d'abord les courbes $C_a$, $C_\infty$, $C_{a_i}$, ensuite les onglets $\Delta_i$ du Chapitre II, n° **13**. En effet, il y a au moins un cycle linéaire coupant $\delta_i$ en un point, donc non invariant en $a_i$ (Chap. II, n° **10**). Par suite une des rétrosections, soit $\gamma_h$, rencontre $\delta_i$ en un nombre non nul de points. La différence entre les lieux de la coupure $\gamma_h$ quand $u$ décrit les bords de $aa_i$, est une multiplicité déformable en $(\gamma_h\delta_i).\Delta_i$. Cette déformation revient à une déformation continue de la coupure $G_h$.

## II. — Réduction des cycles linéaires à $C_u$. Relation fondamentale $R_1 = 2q$. Cycles à trois dimensions.

**3.** En examinant les éléments linéaires de la frontière de la cellule $V_2$, on voit de suite qu'ils gissent tous dans des $C_u$. Il ne serait pas difficile d'en déduire la réduction à une $C_u$ unique. Il est plus aisé cependant de procéder autrement.

Tout $\Gamma_1$ de $V_2$ se compose de plusieurs circuits, et il suffit d'établir la reduction voulue pour l'un d'eux, soit $\Gamma'_1$. Déformons $\Gamma'_1$ en un circuit simple passant par $A_1$ d'une manière quelconque, avec une $C_u$ unique tangente, qu'il n'y a pas d'inconvénients à supposer être $C_a$. Soit B un point quelconque de $\Gamma'_1$, et dans sa $C_u$ joignons-le à $A_1$ par une ligne $h$. Quand B décrit $\Gamma'_1$, $h$ part d'une position $h'$ dans $C_a$ pour revenir à une position $h''$ dans la même courbe. $h'$, $h''$ sont des circuits de $C_a$, et l'on peut évidemment choisir $h$ de telle manière que l'un d'eux, soit $h''$, se réduise à un point. Alors l'$M_2$ décrite par $h$ aura pour frontière $\Gamma'_1$ et $h'$ et l'on a bien $\Gamma'_1 \sim -h'$. La réduction de $\Gamma'_1$ à $-h'$ se fait d'ailleurs en le glissant le long de $M_2$, car la figure formée par $M_2$, $\Gamma'_1$, $h'$ est homéomorphe à la partie d'une sphère comprise entre deux sections planes tangentes l'une à l'autre. En appliquant le même raisonnement aux autres circuits de $\Gamma_1$, on réduira le cycle entier à $C_a$, qui est d'ailleurs une $C_u$ quelconque.

Corollaires. — I. *Les rétrosections de* $C_u$ *forment un système fondamental pour les cycles linéaires de* $V_2$.

II. *Tout cycle linéaire de* $V_2$ *dépend des cycles invariants de* $C_u$ (mod $V_2$).

4. Pour que la réduction des cycles linéaires à $C_u$ puisse se faire, il suffit que $\{C_u\}$ possède un point-base. Cela revient à exiger que le nombre $(C_u C_u)$, que l'on dénote aussi par $(C_u^2)$ et que l'on appelle *degré* du faisceau, soit positif. D'où ce théorème de M. Severi (1) :

*Pour que tout cycle linéaire de* $V_2$ *soit réductible à une courbe algébrique donnée, il suffit qu'elle appartienne à un faisceau linéaire de degré positif.*

La démonstration de M. Severi est par voie transcendante. Pour les sections planes, M. Picard en avait donné une se rapprochant de la nôtre.

5. Reprenons les cycles invariants indépendants $\gamma'_1, \gamma'_2, \ldots, \gamma'_r$, du Chapitre II, n° 18. $\gamma'_j$ a pour lieu, quand $u$ varie, un cycle à trois dimensions $\Gamma^j_3$, et l'on a

$$(\gamma'_i \Gamma^j_3) = (\gamma'_i \gamma'_j).$$

Donc (Chap. II, n° 19) le déterminant

$$|(\gamma'_i \Gamma^j_3)| \neq 0.$$

Ceci entraîne l'impossibilité d'une homologie

$$\sum t_i \gamma'_i \sim 0 \qquad (\text{mod } V_2),$$

Car on en déduit

$$\sum\nolimits_i t_i (\gamma'_i \Gamma^j_3) = 0 \qquad (j = 1, 2, \ldots, r),$$

(1) *Palermo Rendiconti*, 21e vol., 1906.

d'où, puisque le déterminant ci-dessus n'est pas nul, les $t$ le sont. Ainsi les $\gamma'_i$ sont indépendants $(\bmod V_2)$, et, puisque tout cycle de la surface leur est relié par une homologie,

$$r = 2q = R_1,$$

c'est-à-dire : *L'indice de connexion linéaire est pair et égal au double de l'irrégularité* [1]. Ce théorème fournit une première raison du rôle important joué par $q$ dans la théorie des surfaces algébriques.

6. Passons maintenant aux *cycles à trois dimensions*. Dans la composition d'un $\Gamma_3$ ne peuvent entrer les lieux de $C_u$ quand $u$ décrit les lignes $aa_i$, car on ne peut obtenir ainsi de multiplicités fermées. Par suite,

$$\Gamma_3 \sim \sum_1^{2p}{}_j \lambda_j G_j + \sum_2^{m}{}_k \mu_k L_k \tag{1}$$

Son intersection avec $C_u$

$$\gamma \sim \sum_1^{2p}{}_j \lambda_j \gamma_j + \sum_2^{m}{}_k \mu_k l_k$$

doit être fermée, donc tous les $\mu$ doivent s'annuler, d'où

$$\Gamma_3 \sim \sum_1^{2p}{}_j \lambda_j G_j.$$

D'ailleurs $\gamma$ est invariant, sans quoi $\Gamma_3$ aurait des frontières sommes de $\Delta_i$. Réciproquement, tout cycle invariant a pour lieu un $\Gamma_3$ (Chap. II, n° 12).

*Il y a équivalence entre les homologies*

$$\Gamma_3 \sim 0 \quad (\bmod V_2); \qquad \gamma \sim 0 \quad (\bmod C_u).$$

---

(1) La parité de $R_1$ a été démontrée jusqu'ici uniquement par voie transcendante. Une première démonstration topologique dans mon Mémoire des *Annals of Mathematics* était basée sur des théorèmes imparfaitement démontrés, quoique vrais. *Voir*, pour une démonstration transcendante, PICARD et SIMART, vol. II, p. 423.

En effet, de $M_4 \equiv \Gamma_3$ on déduit $M_4 C_u \equiv \gamma$, donc $\gamma \sim 0 \pmod{C_u}$. D'un autre côté, si $\gamma \sim 0 \pmod{C_u}$, les $\lambda$ sont tous nuls et $\Gamma_3 \sim 0$.

Comme corollaire, *le nombre de cycles $\Gamma_3$ distincts est le même que le nombre r des cycles invariants distincts de* $C_u$, c'est-à-dire $R_3 = r = R_1$. *C'est l'égalité de Poincaré pour les indices de connexion équidistants des extrêmes.*

**7. Système fondamental et torsion pour les cycles à trois dimensions.** — Pour que $\gamma$ soit invariant, il faut et il suffit que

$$\sum_1^{2p}{}_j \lambda_j(\gamma_j \delta_i) = 0 \qquad (i = 1, 2, \ldots, 2p - r).$$

Ce système aura $r$ solutions fondamentales

$$\lambda_1^h, \quad \lambda_2^h, \quad \ldots, \quad \lambda_{2p}^h \qquad (h = 1, 2, \ldots, r),$$

et il n'y a pas d'inconvénient à supposer que les cycles invariants correspondants sont précisément les $\gamma'_h$. Pour toute autre solution, on aura

$$\lambda_j = \sum_1^r{}_h t_h \lambda_j^h \qquad (t_h \text{ entier};\ j = 1, 2, \ldots, 2p),$$

d'où de suite les homologies

$$\gamma \sim \sum_1^r{}_h t_h \gamma'_h \qquad (\operatorname{mod} C_u),$$

$$\Gamma_3 \sim \sum_1^{2p}{}_h t_h \Gamma_3^h \qquad (\operatorname{mod} V_2),$$

dont la première entraîne la deuxième. Par suite, les $\Gamma_3^h$ forment un système fondamental. Leur indépendance même démontre l'absence de diviseurs de zéro. D'ailleurs, directement, $\gamma$ étant encore, ici comme plus loin, trace de $\Gamma_3$, de $t\Gamma_3 \sim 0$, on tire $t\gamma \sim 0$, donc $t = 0$, car $C_u$ n'a pas de diviseurs de zéro, et finalement $\Gamma_3 \sim 0$. Ainsi le groupe de la torsion à trois dimensions se réduit bien à l'identité (*voir* Chap. I, n° 8).

## III. — Intersection des cycles linéaires et à trois dimensions.

8. Nous allons démontrer pour ces cycles les théorèmes annoncés à la fin du premier Chapitre.

I. *Soit d'abord* $(\Gamma_1\Gamma_3)=0$, *pour tout* $\Gamma_3$, et supposons $\Gamma$ réduit à $C_u$. Avec les mêmes notations qu'au n° 7, nous aurons

$$\lambda\Gamma_1 \sim \sum_1^r {}^h \lambda_h \gamma'_h + \sum^{2p-r} {}^k \mu_k \delta_k \qquad (\mathrm{mod}\, C_u),$$

et comme

$$(\Gamma_1\Gamma_3^h)=0 \qquad (\delta_k\gamma_h)=0,$$

on aura

$$\sum \lambda_h(\gamma'_h\gamma'_k)=0 \qquad (k=1,2,\ldots,r).$$

Or le déterminant des coefficients n'est pas nul (Chap. II, n° 19). Donc les $\lambda_h$ le sont, d'où

$$\lambda\Gamma_1 \sim \sum \mu_k\delta_k \sim 0 \qquad (\mathrm{mod}\, V_2).$$

Ceci démontre le premier théorème pour les cycles linéaires. Pour les $\Gamma_3$, la démonstration est immédiate, car de

$$(\Gamma_1\Gamma_3)=(\Gamma_1\gamma)=0,$$

quel que soit $\Gamma_1$, on tire de suite

$$\gamma \sim 0 \qquad (\mathrm{mod}\, C_u);$$

donc $\Gamma_3 \sim 0$.

9. Passons au second théorème : *Les invariants $e_i$ relatifs aux intersections des $\Gamma_1$ et des $\Gamma_3$ sont tous égaux à* 1. D'après le Chapitre I, n° 19, il suffit d'établir que *si*

$$(\Gamma_3\Gamma_1)\equiv 0 \qquad (\mathrm{mod}\, e),$$

*quel que soit* $\Gamma_1$, $\Gamma_3$ *est le multiple d'un certain cycle de* $V_2$.

De l'hypothèse on déduit

$$(\gamma\Gamma_1)\equiv 0 \qquad (\mathrm{mod}\, e)$$

pour tout $\Gamma_1$ de $C_u$, Or, les $\gamma_i$ étant toujours les rétrosections,

$$\gamma \sim \sum \lambda_i \gamma_i \qquad (\text{mod } C_u).$$

En prenant les intersections de $\gamma$ avec les $\gamma_i$, on obtient de suite

$$\lambda_i = (\gamma\gamma_{p+i}), \qquad \lambda_{p+i} = -(\gamma\gamma_i) \qquad (i \leqq p).$$

Donc les $\lambda$ sont tous divisibles par $e$. Soit $\lambda_i = e\lambda'_i$. On a

$$\gamma \sim e\bar{\gamma} = e\sum \lambda'_i \gamma_i \qquad (\text{mod } C_u).$$

Les nombres $(\bar{\gamma}\,\delta_i)$, proportionnels aux $(\gamma\,\delta_i)$, nuls en vertu de l'invariance de $\gamma$, sont nuls eux aussi; donc $\bar{\gamma}$ est invariant. Soit $\bar{\Gamma}_3$ le cycle dont il est la trace. Comme corollaire de l'homologie précédente

$$\Gamma_3 - e\bar{\Gamma}_3 \sim 0 \qquad (\text{mod } V_2),$$

ce qui démontre la proposition.

## IV. — Cycles à deux dimensions.

10. Reprenons la cellule $V_2$. Les multiplicités-coupures $G_j$, $L_k$ sont elles-mêmes des cellules à trois dimensions dont les frontières sont $C_a$, $C_\infty$, les $C_{a_i}$ et les $\Delta_i$. Par suite, en vertu du lemme du Chapitre I, n° 11, tout $\Gamma_2$ est réductible à ces multiplicités.

Je dis que si un $\Gamma_2$ ainsi réduit comprend une $M_2$ en partie dans $C_{a_i}$, il comprend la courbe tout entière. En effet, autrement cette $M_2$ doit s'appuyer par une ligne sur une autre appartenant à d'autres $C_u$ ou à des $\Delta_i$, ce qui est impossible, car elle ne rencontre ces multiplicités qu'en des points isolés ou bien pas du tout. Le même raisonnement s'applique à $C_\infty$ et aux $\Delta_i$ *mais non pas à* $C_a$, à moins toutefois que $\Gamma_2$ ne lui appartienne en entier.

On conclut de ce qui précède que pour tout $\Gamma_2$, on a, en désignant par $(C_a)$ une $M_2$ quelconque contenue dans $C_a$,

$$(1) \qquad \Gamma_2 \sim \sum \lambda_i \Delta_i + (C_a) + \sum \mu_j C_{a_j} + \nu C_\infty.$$

Comme les $C_{aj}$ sont toutes homologues, les trois derniers termes peuvent être mis en un seul $(C_a)$, de sorte que, finalement,

$$\Gamma_2 \sim \sum \lambda_i \Delta_i + (C_a). \tag{2}$$

Considérons de nouveau les $\delta_i$ comme étant dans $C_a$. Des congruences

$$\Delta_i \equiv \delta_i, \qquad \sum \lambda_i \Delta_i + (C_a) \equiv 0$$

on tire

$$(C_a) \equiv -\sum \lambda_i \delta_i, \tag{3}$$

d'où l'homologie

$$\sum \lambda_i \delta_i \sim 0 \qquad (\text{mod } C_a). \tag{4}$$

Ainsi *tout* $\Gamma_2$ *est réductible au type* (2) *avec une homologie* (4) *correspondante.*

Réciproquement à (4) correspond une congruence (3), indiquant que la multiplicité $\Gamma_2$ qui apparaît dans (2) est un cycle.

**Système fondamental.** — (4) est équivalente au système diophantin

$$\sum\nolimits^i \lambda_i (\delta_k \delta_i) = 0 \qquad (k = 1, 2, \ldots, n),$$

qui exprime que la somme à droite dans (4) est un cycle invariant (Chap. II, n° 13). En raisonnant comme pour les $\Gamma_3$ (n° 7), on trouve de suite qu'il y a $\nu = (n - 2p + 2q)$ cycle $\Gamma_2^1, \Gamma_2^2, \ldots, \Gamma_2^\nu$, tels que pour tout autre on ait

$$\Gamma_2 - \sum t_h \Gamma_2^h \sim (C_a).$$

Tout comme plus haut, la multiplicité à droite, partie fermée de $C_a$, en est un multiple, $C_a$ et les $\Gamma_2^h$ forment le système fondamental cherché.

**11.** Le nombre de cycles $\delta_i$ distincts est égal à $2p - 2q$. Donc, comme nous venons de le dire plus haut, il y a $\nu = (n - 2p + 2q)$ homologies (4) distinctes. Mais à (4) peut fort bien correspondre un $\Gamma_2$ nul. On aura alors

$$\sum \mu_j G_j + \sum \nu_k L_k \equiv \Gamma_2.$$

D'ailleurs,

$$G_j \equiv \sum (\gamma_j \delta_i) \Delta_i + (C_a), \qquad L_k \equiv \sum (l_k \delta_i) \Delta_i + (C_a).$$

Par suite,

$$\Gamma_2 \sim \sum (l\, \delta_i) \Delta_i + (C_a), \qquad l = \sum \mu_j \gamma_j + \sum \nu_k l_k.$$

$\Gamma_2$ sera du type (2) aux $\lambda$ non nuls si la ligne $l$ n'est pas invariante. Il faut donc défalquer du nombre de cycles obtenu autant de cycles qu'il y a de lignes $l$ distinctes dont aucune combinaison n'est invariante, soit $2p - 2q + m - 1$ (Chap. II, n° 20). Par suite, il y a $n - (4p + m - 1) + 4q$ cycles $\Gamma_2$ distincts aux $\lambda$ non tous nuls. Quand ces nombres sont nuls, $\Gamma_2$ appartient en entier à $C_a$; donc il comprend la courbe tout entière (n° 10), et le cycle est un multiple du cycle $C_a$, cycle d'ailleurs non nul (Chap. II, n° 6, corollaire). On a alors finalement

$$R_2 = n - (4p + m - 1) + 4q + 1.$$

Le nombre $I = n - m - 4p$ ne dépend pas du système $|C|$. En fait ce n'est autre que le nombre caractéristique de la surface connu sous le nom d'*invariant de Zeuthen-Segre*. On a donc, la formule ne comprenant que des nombres caractéristiques de $V_2$,

$$R_2 = I + 4q + 2, \tag{5}$$

formule donnée correctement pour la première fois par M. Alexander [1]. M. Picard avait obtenu auparavant, par voie transcendante, la valeur $I + 4q + 1$ pour le nombre de cycles finis [2]; à partir de là, il suffit d'établir que tout $\Gamma_2$ dépend des cycles finis et de celui que constitue la courbe à l'infini, pour arriver à (5).

12. **Formule d'Euler-Poincaré.** — La méthode que nous allons suivre pour l'établir est au fond celle dont s'est servi M. Alexander pour obtenir (5).

---

(1) *Rendiconti dei Lincei*, août 1914.
(2) PICARD et SIMART, vol. II, p. 497.

Nous pouvons obtenir une subdivision de $V_2$ en cellules en en faisant autant pour le plan $u$ et la courbe $C_u$ : Pour le premier il suffit de joindre les points $a_i$ aux points $a, \infty$. Quant à $C_u$ on le fera d'une manière quelconque, sauf à prendre soin que les sommets comprennent les $m$ points fixes $A_k$. Le plan $u$ aura $n$ faces, $2n$ arêtes, $n+2$ sommets. Désignons par $\alpha_h$, $\alpha'_h$, $\alpha''_h$ les nombres d'éléments à $h$ dimensions de $V_2$, de $C_u$ et des $C_{a_i}$.

Les $\alpha_4$ éléments à quatre dimensions de $V_2$ proviennent de ceux à deux dimensions de $C_u$ quand $u$ décrit les « faces » (cellules $E_2$) de son plan ; d'ailleurs il n'y en a pas d'autres. Donc

$$\alpha_4 = n\alpha'_2.$$

Les éléments à trois dimensions proviennent des lignes ou des faces de $C_u$ quand $u$ décrit les faces ou les lignes de son plan. Donc

$$\alpha_3 = 2n\alpha'_2 + 2\alpha'_1.$$

Les $\alpha_2$ éléments à deux dimensions sont de trois natures. Il y a d'abord ceux dérivés des $\alpha'_0 - m$ sommets variables de $C_u$ ou de ses $\alpha'_i$ lignes quand $u$ décrit les faces ou les arêtes de son plan. Mais il y a en outre à envisager les faces des courbes $C_a$, $C_\infty$, $C_{a_i}$. Donc

$$\alpha_2 = n(\alpha'_0 - m) + 2n\alpha'_1 + 2\alpha'_2 + n\alpha''_2.$$

De même on a de suite

$$\alpha_1 = 2n(\alpha'_0 - m) + 2\alpha'_1 + n\alpha''_1.$$

Quant a $\alpha_0$ on en obtient l'expression exacte en remarquant que le nouveau point double de $C_{a_i}$ est un point ordinaire, mais que toutefois il compte pour deux dans $\alpha''_0$ comme appartenant à deux branches différentes de $C_{a_i}$ : il suffit donc d'ajouter tous les sommets des courbes $C_a$, $C_\infty$, $C_{a_i}$, et en vue de la remarque ci-dessus, d'en retrancher $n$, pour avoir $\alpha_0$. Il ne faut pas oublier bien entendu que les $m$ points $A_i$ appartiennent à toutes les courbes. Par suite, enfin,

$$\alpha_0 = n(\alpha''_0 - m) + 2(\alpha'_0 - m) + m - n.$$

Rappelons maintenant que pour une $W_2$ de genre $p$,

$$\alpha_0 - \alpha_1 + \alpha_2 = 2 - 2p.$$

Il n'y a qu'à se reporter par exemple au modèle à l'aide d'un disque à $p$ trous pour le voir.

On aura donc ici

$$\alpha'_0 - \alpha'_1 + \alpha'_2 = 2 - 2p, \qquad \alpha''_0 - \alpha''_1 + \alpha''_2 = 2 - 2(p-1),$$

d'où de suite pour $V_2$,

$$\sum(-1)^i \alpha_i = n - m - 4p + 4 = I + 4 = R_2 - 4q + 2 = \sum(-1)^i R_i,$$

qui n'est autre que la formule cherchée.

## V. — Torsion à une et à deux dimensions.

**13.** Reprenons les lignes $l_j = A_1 A_j$ de $C_u$, ainsi que leurs lieux $L_j$, dont soient $\Gamma_2^j$ les frontières. Puisque $\Gamma_2^j$ est nul

$$(\Gamma_2^j C_u) = 0.$$

Mais $\Gamma_2^j$ ne coupe $C_u$ qu'en $A_1$ et en $A_j$; de plus les contributions de ces points au premier membre sont $+1$ pour $A_j$, $-1$ pour $A_1$. En effet, $C_u$ ne coupe $\Gamma_2^j$ que dans la partie de $C_a$ qui lui appartient, car elle ne coupe $L_j$ que suivant la ligne $l_j$. On le voit aussi en remarquant que $C_u$ ne rencontre pas les onglets $\Delta_i$, dont se compose précisément la partie du cycle extérieure à $C_a$. On pourrait se demander si les contributions, quoique de signes opposés, ont précisément les signes indiqués ci-dessus. A supposer qu'il en soit autrement, il suffirait de remplacer $L_j$ par $-L_j$ pour avoir la situation désirée.

Soit maintenant $E_2^j$ une petite cellule de $C_a$ entourant le point $A_j$. Désignons enfin d'une façon générique par $((C_a))$ une partie de $C_a$ ne contenant aucun des points $A_j$. On aura pour un cycle quelconque $\Gamma_2$ et pour $\Gamma_2^j$,

$$\Gamma_2 \sim \sum \lambda'_i \Delta_i + ((C_a)) + \sum_1^m {}_j\, \mu'_j E_2^j,$$

$$\Gamma_2^j \sim \sum \lambda''_i \Delta_i + ((C_a)) + E_2^j - E_2^1 \sim 0.$$

Par suite,

$$\Gamma_2 \sim \sum \lambda_i \Delta_i + ((C_a)) + \mu_1 E_2^1.$$

Comme $C_u$ ne rencontre ni $\Delta_i$, ni $((C_a))$,

$$(\Gamma_2 C_u) = \mu_1(E_2^1 C_u) = \mu_1,$$

d'où finalement

$$\Gamma_2 \sim \sum \lambda_i \Delta_i + ((C_a)) + (\Gamma_2 C_a) E_2^1,$$

relation dont nous ferons un usage répété.

Corollaire. — *Si $\Gamma_2$ est nul ou diviseur de zéro, on a*

$$\Gamma_2 \sim \sum \lambda_i \Delta_i + ((C_a)),$$

car alors un multiple du nombre $(\Gamma_2 C_u)$ est nul, donc ce nombre l'est aussi.

**14.** Théorème. — *Pour avoir tous les diviseurs $\Gamma_2$, il faut et il suffit de rechercher toutes les homologies*

$$t \sum \lambda_i \Delta_i + ((C_a)) \sim 0 \tag{6}$$

*qui cessent d'être vérifiées quand on y remplace t par l'unité.*

D'après le corollaire, une telle homologie est bien vérifiée pour tout diviseur de zéro.

Réciproquement supposons-la vérifiée. On en tire, les $\delta_i$ étant tous dans $C_a$,

$$((C_a)) \equiv - t \sum \lambda_i \delta_i.$$

La somme à droite, prise $t$ fois, sépare $C_a$ en deux parties dont une seule contient tous les A (c'est là précisément le sens de la congruence). Cette somme a donc nécessairement elle-même cette propriété, ce qui revient à dire qu'il y a une congruence

$$((C_a)) \equiv - \sum \lambda_i \delta_i.$$

On a donc bien en

$$\Gamma_2 \sim \sum \lambda_i \Delta_i + ((C_a))$$

un diviseur de zéro, puisque $t\Gamma_2 \sim 0$, sans que $\Gamma_2 \sim 0$.

15. Soit $\overline{\Gamma}_2^h$ la frontière de $G_h$, lieu de la rétrosection $\gamma_h$ de $C_u$. Les homologies cherchées sont toutes de type

$$\sum_1^{2p}{}_h \nu_h \overline{\Gamma}_2^h + \sum_2^m{}_j \nu'_j \Gamma_2^j \sim 0.$$

Je dis que les coefficients $\nu'$ sont tous nuls. En effet, en exprimant les cycles en termes des éléments $\Delta_i$, $((C_a))$, $E_2^j$, l'homologie écrite prend la forme

$$\sum \lambda_i \Delta_i + ((C_a)) + \sum_2^m{}_j \nu'_j (E_2^j - E_2^1) \sim 0,$$

et pour qu'elle soit de la forme voulue, il faut bien que $\nu'_j = 0$.

Écrivons donc les homologies (n° 11)

$$(7) \qquad \overline{\Gamma}_2^h \sim \sum{}_i (\gamma_h \delta_i)\, \Delta_i + ((C_a)) \sim 0,$$

fondamentales pour celles que l'on recherche. En appliquant deux transformations unimodulaires convenables, l'une aux homologies (7), l'autre aux $\Delta$, on obtiendra un nouveau système équivalent à (7), de la forme [1]

$$(8) \qquad t_i \Delta'_i + ((C_a)) \sim 0,$$

dont les homologies sont encore fondamentales pour celles cherchées. Les entiers $t$ sont ici les diviseurs élémentaires de la matrice

$$\| (\gamma_h \delta_i) \| \qquad (h = 1, 2, \ldots, 2p;\ i = 1, 2, \ldots, n).$$

Quant aux $\Delta'$ ce ne sont autres que des sommes de $\Delta$. Donc (n° 14), si $t_i > 1$, il y a un diviseur

$$t_i \Delta'_i + ((C_a))$$

dont l'ordre, en tant qu'opération du groupe des diviseurs, est précisément $t_i$. Ces opérations forment d'ailleurs base pour le groupe. Par suite les $t$ supérieurs à l'unité sont les coefficients de torsion et

$$\sigma_2 = \prod{}_i t_i.$$

(1) E. Cahen, *loc. cit.*, p. 269. Le résultat de Fröbeniüs est immédiatement applicable ici.

16. Passons aux $\Gamma_1$. Puisqu'ils sont tous réductibles à $C_u$ il suffit de considérer les homologies (mod $V_2$) existant entre les cycles de la courbe. Les homologies fondamentales sont

$$(9) \qquad \delta_i \sim 0 \qquad (i = 1, 2, \ldots, n).$$

Exprimons les $\delta$ en termes des rétrosections. Toutefois, pour faciliter l'extension aux variétés algébriques, opérons comme ceci : Aux rétrosections, associons à la manière du Chapitre I, n° 19, le système fondamental

$$\bar{\gamma}_1, \quad \bar{\gamma}_2, \quad \ldots, \quad \bar{\gamma}_{2p},$$

tel que

$$(\gamma_h \bar{\gamma}_h) = 1, \qquad (\gamma_h \bar{\gamma}_k) = 0 \qquad (h \neq k).$$

On aura

$$\delta_i \sim \sum{}^h \beta_{ih} \bar{\gamma}_h; \qquad \beta_{ih} = (\gamma_h \delta_i).$$

Le système (9) peut donc être remplacé par

$$(9') \qquad \sum{}^h (\delta_h \delta_i) . \bar{\gamma}_h \sim 0 \qquad (\text{mod. } V_2).$$

Deux transformations unimodulaires convenablement choisies nous conduiront à un nouveau système fondamental

$$\bar{\gamma}'_1, \quad \bar{\gamma}'_2, \quad \ldots, \quad \bar{\gamma}'_{2p},$$

avec, au lieu de (9'),

$$(10) \qquad t_i \bar{\gamma}'_i \sim 0 \qquad (\text{mod. } V_2),$$

les $t$ désignant toujours les diviseurs élémentaires de la matrice

$$\| (\gamma_h \delta_i) \|$$

déjà considérée plus haut. Les $t$ jouent donc le même rôle ici que pour les $\Gamma_2$, d'où le théorème de Poincaré :

*Les groupes de torsion à une et deux dimensions sont isomorphes. En particulier,* $\sigma_1 = \sigma_2$.

## VI. — Intersection des cycles à deux dimensions.

**17.** Proposons-nous d'abord de trouver $(\Gamma_2\Gamma'_2)$, où

$$\Gamma_2 \sim \sum \lambda_i \Delta_i + ((C_a)) + \sum \mu_j E_2^j,$$

$$\Gamma'_2 \sim \sum \lambda'_i \Delta_i + ((C_a)) + \sum \mu'_j E_2^j.$$

A cet effet, remplaçons $\Gamma'_2$ par le cycle homologue obtenu ainsi : Dans le plan $u$, au lieu de $a$, on prend comme origine des lacets un point voisin $a'$, autour duquel on trace des lacets $a'a_i$, se sui-

Fig. 3.

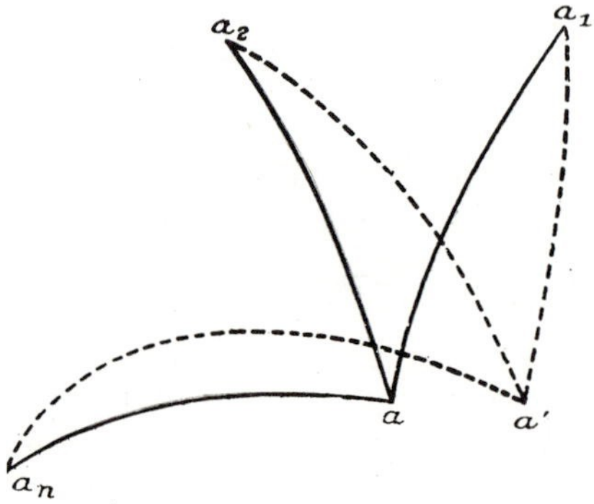


vant dans le même ordre que les $aa_i$ autour de $a$ ; puis $\delta'_i$ étant un cycle qui augmente de $\delta_i$ quand $u$ tourne autour de $a_i$ (Chap. II, n° 10), remplaçons $\Delta_i$ par le lieu $\Delta'_i$ de $\delta'_i$ quand $u$ partant de $a'$ décrit un circuit différant peu de $a'a_i$ (*fig.* 4) ; enfin il faudra remplacer les cellules $E_2^j$ par d'autres analogues $E_2'^j$ situées dans $C_{a'}$. On aura alors

$$\Gamma'_2 \sim \sum \lambda'_i \Delta'_i + ((C_{a'})) + \sum \mu'_j E_2'^j.$$

Or le circuit voisin de $a'a_i$ coupera $aa_i$ en un point unique $b$. Par suite $\Delta'_i$ coupe $\Delta_i$ en un seul point, intersection de $\delta'_i$ avec $\delta_i$ dans $C_b$, d'où $(\Delta_i\Delta'_i) = -1$, comme on le montre sans difficulté [1].

---

[1] On a en tous cas $(\Delta_i\Delta'_i) = \pm 1$ et le choix du signe ne dépend pas de la

Pour trouver le nombre de points communs à $\Delta_i$ et $\Delta_k$, on peut réduire le circuit voisin du lacet $a'a_i$ au lacet lui-même. Comme ce lacet ne coupe le lacet $aa_k$ que si $i > k$, il n'y aura pas inter-

Fig. 4.

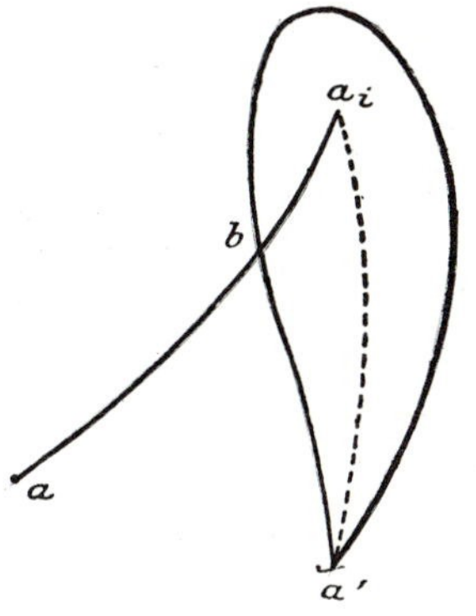


section pour $i < k$, et pour $i > k$ on aura

$$(\Delta_i \Delta'_k) = -(\delta_i \delta_k) = -\alpha_{ik},$$

le signe étant le même qu'auparavant. Les multiplicités $((C_a))$, $((C_{a'}))$ sont sans points communs, donc finalement

$$(11) \qquad (\Gamma_2 \Gamma'_2) = -\sum \lambda_i \lambda'_i - \sum_{i>k}^{i,k} \alpha_{ik} \lambda_i \lambda'_k + \sum \mu_j \mu'_j.$$

Cette formule donne le nombre d'intersections dès que l'on connaît l'expression des cycles en termes des mêmes $\Delta$. Pour les applications, il convient mieux d'en obtenir une autre plus simple. On pourrait la dériver de (11), mais on y arrivera plus vite comme il suit :

Traçons de nouvelles coupures $aa_i$ ne se coupant pas entre elles et ne traversant pas non plus les $aa_i$. Désignons enfin tous les éléments relatifs aux nouvelles coupures en surmontant d'une barre

---

surface envisagée. Considérons donc une quadrique avec pour $\{C_n\}$ un faisceau de sections planes et envisageons sur elle le cycle nul $\Gamma_2^2$ du n° 13. Le choix du signe + donne au lieu de (11) une formule en vertu de laquelle $(\Gamma_2^2 \Gamma_2^2)$ est la somme des carrés de quatre entiers non nuls, alors que ce nombre doit être nul, puisque $\Gamma_2^2 \sim 0$. Ceci impose bien le choix du signe —, comme dans le texte.

ceux correspondants des anciennes. En particulier, nous désignerons dorénavant par $\delta_i$ ce que devient dans $C_a$ le cycle évanouissant en $a_i$ quand $u$ décrit $a_i a$, et par $\bar{\delta}_i$ le cycle analogue pour $C_{\bar{a}}$ et $a_i \bar{a}$. Soit maintenant

$$\bar{\Gamma}_2 = \sum \bar{\lambda}_i \bar{\Delta}_i + ((C_{\bar{a}})) + \sum \bar{\mu}_j \bar{E}_2^j$$

le second cycle. On trouve de suite, les $\Delta'$ ayant le même sens que plus haut,

$$(\bar{\Delta}_i \Delta'_i) = -1, \qquad (\bar{\Delta}_i \Delta'_k) = 0.$$

Par suite, en remplaçant dans l'expression de $\Gamma_2$ les $\Delta$ par les $\Delta'$,

$$(12) \qquad (\Gamma_2 \bar{\Gamma}_2) = -\sum \lambda_i \bar{\lambda}_i + \sum \mu_j \bar{\mu}_j.$$

Les $\bar{\lambda}$ ne sont autres que les coefficients des $\lambda$ dans la première formule dérivée, comme on pourrait le montrer directement.

18. Théorème. — *Si $(\Gamma_2 \bar{\Gamma}_2)$ est nul pour tout $\bar{\Gamma}_2$, $\Gamma_2$ est nul ou diviseur de zéro.*

D'abord, puisque $(\Gamma_2 C_u)$ est nul, on aura (n° 13, corollaire)

$$\Gamma_2 \sim \sum \lambda_i \Delta_i + ((C_a)).$$

Donc, pour tout $\bar{\Gamma}_2$,

$$(13) \qquad \sum \lambda_i \bar{\lambda}_i = 0.$$

Cette relation sera vérifiée pour tous les $\bar{\lambda}$ tels que

$$(14) \qquad \sum \bar{\lambda}_i \bar{\delta}_i \sim 0 \qquad (\bmod C_{\bar{a}}).$$

Pour exprimer que le cycle à gauche est nul, il suffit d'écrire qu'il est invariant, ce qui donne

$$(15) \qquad \sum_i \bar{\lambda}_i (\bar{\delta}_k \bar{\delta}_i) = 0 \qquad (k = 1, 2, \ldots, n);$$

(13) doit être une conséquence des équations (15), d'où $t$ et les $t_k$

étant entiers, avec $t \neq 0$,

$$t\lambda_i = \sum^k t_k(\overline{\delta}_k \overline{\delta}_i) = (\gamma \overline{\delta}_i); \qquad \gamma = \sum t_k \overline{\delta}_k.$$

Par suite,

$$t\Gamma_2 \sim \sum (\gamma \overline{\delta}_i) \Delta_i + ((C_a)).$$

Or, en introduisant les $\overline{\delta}$, l'homologie (7) du nº 15 devient

$$\overline{\Gamma}_2^h \sim \sum (\gamma_h \overline{\delta}_i) \Delta_i + ((C_a)) \sim 0.$$

En effet, en vertu de la manière dont on fait varier $u$ pour engendrer les $G_h$, les $(\gamma_h \delta_i)$ sont supposés calculés en prenant les cycles dans $C_\infty$, $\delta_i$ y étant réduit en faisant décrire à $u$ le prolongement de $aa_i$. Cela revient à calculer $(\gamma_h \overline{\delta}_i)$ pour $\overline{a} = \infty$. Comme ce nombre ne dépend pas de la position de $\overline{a}$ tant qu'il ne traverse pas les coupures $aa_i$, notre affirmation s'ensuit.

Ceci posé, on peut écrire une homologie

$$\gamma \sim \sum_1^{2p}{}^h \theta_h \gamma_h \qquad (\mathrm{mod}\, C_{\overline{a}}).$$

Par suite,

$$t\Gamma_2 - \sum \theta_h \overline{\Gamma}_2^h \sim t\Gamma_2 \sim ((C_a)),$$

et comme $((C_a))$ ne peut être fermée à moins d'être réduite à une dimension moindre que deux, $t\Gamma_2$ est bien un cycle nul, ce qui démontre notre théorème.

19. Théorème. — *Si $(\Gamma_2 \Gamma'_2)$ est divisible par $e$, quel que soit $\Gamma'_2$, il existe un cycle $\overline{\Gamma}_2$ tel que $\Gamma_2 - e\overline{\Gamma}_2$ soit nul ou diviseur de zéro.*

Commençons par prendre $\Gamma_2$ sous la forme du nº 13

$$\sum \lambda_i \Delta_i + ((C_a)) + \mu_1 E_2^1.$$

Ici $\mu_1 = (\Gamma_2 C_a)$ est divisible par $e$, donc la condition satisfaite par $\Gamma_2$ donne la congruence

$$\sum_{1}^{n} {}_i \overline{\lambda_i}\,\overline{\lambda}_i \equiv 0 \qquad (\mathrm{mod}\, e).$$

Je dis que l'*on peut s'arranger pour que e divise tous les* $\lambda$. Tout d'abord les expressions

$$\sum {}_k\, t_k(\overline{\delta}_k \overline{\delta}_i)$$

formées avec les nombres rationnels $t$, si elles sont égales à des entiers, sont précisément les nombres $\lambda$ d'un diviseur de la forme

$$\sum \mu_i \Delta_i + ((C_a)).$$

En effet, en les multipliant par un entier suffisamment élevé $t$, elles deviennent les nombres $(\gamma \overline{\delta}_i)$ relatifs à un certain cycle linéaire $\gamma$, et le cycle à deux dimensions en question, pris $t$ fois, devient le cycle nul

$$\sum (\gamma\, \delta_i)\, \Delta_i + ((C_a)).$$

Donc si l'on pose

$$\nu_i = \lambda_i + \sum_{1}^{n} {}_k\, t_k(\overline{\delta}_k \overline{\delta}_i), \tag{16}$$

les $t$ étant choisis de manière que les $\nu$ soient entiers, on peut remplacer les $\lambda$ par les $\nu$ et il suffit de démontrer que les $\nu$ sont divisibles par $e$. D'ailleurs

$$\sum_{1}^{n} {}_i\, \nu_i \overline{\lambda}_i \equiv 0 \qquad (\mathrm{mod}\, e), \tag{17}$$

comme on le déduit du reste de suite de (15).

Appliquons aux $\overline{\delta}$, aux $\lambda$ et aux $\nu$ une certaine transformation unimodulaire A, aux $t$ et aux $\overline{\lambda}$ la transformation contragrédiente $\overline{A}^{-1}$ [1]. Désignons les éléments transformes en accentuant

[1] Ou transformation inverse de la *transposée* de A. La transposée se dénote par $\overline{A}$, d'où $\overline{A}^{-1}$ pour la contragrédiente.

ceux dont ils dérivent. Les relations (16), (17) seront maintenues, et il suffit de montrer que, pour un choix convenable des $t$, les $\nu'$ sont tous divisibles par $e$.

Choisissons en particulier A de manière que la matrice

$$\| (\bar{\delta}'_k \bar{\delta}'_i) \|$$

soit canonique, ce qui est possible puisqu'elle est alternée ([1]) (Fröbeniüs). Comme elle est de rang $2p - 2q$ (Chap. II, n° 19), on aura alors ([2])

$$(\bar{\delta}'_{2i-1}\bar{\delta}'_{2i}) = -(\delta'_{2i}\delta'_{2i-1}) = g_i > 0 \qquad (i \leqq 2p-2q),$$

et les autres nombres d'intersection sont nuls. Au lieu de (15), (16), on aura maintenant

$$(15') \qquad g_i\bar{\lambda}'_{2i-1} = g_i\bar{\lambda}'_{2i} = 0 \qquad (1 = 1, 2, \ldots, 2p-2q),$$

$$(16') \quad \nu'_{2i-1} = \lambda'_{2i-1} + g_i t'_{2i}; \qquad \nu'_{2i} = \lambda'_{2i} - g_i t'_{2i-1} \qquad (i = 1, 2, \ldots, p-q);$$

$$\nu'_{2p-2q+j} = \lambda'_{2p-2q+j}.$$

Puisque les $g$ ne sont pas nuls, les $\lambda$ d'indice $\leqq 2p - 2q$ le sont et au lieu de (17) on aura

$$(17') \qquad \sum_{2p-2q+1}^{n}{}_i \; \nu'_i\bar{\lambda}'_i \equiv 0 \qquad (\text{mod}\, e).$$

Les $t'$ sont des nombres rationnels soumis à la seule condition de rendre les $\nu'$ entiers. Choisissons-les tels que les $\nu'$ d'indice $\leqq 2p - 2q$ soient nuls. Dans (17'), les coefficients des $\nu'$ qui restent sont arbitraires. Donc cette congruence est identiquement satisfaite, d'où

$$\nu'_{2p-2q+j} \equiv 0 \qquad (\text{mod}\, e).$$

Ainsi on se sera arrangé pour que les $\nu'$ soient nuls ou divisibles par $e$, ce qui démontre notre affirmation.

---

([1]) Si cette matrice n'était pas alternée, il faudrait introduire une seconde transformation unimodulaire B et appliquer $\bar{B}^{-1}$ aux $t$, mais à part cela rien ne serait changé.

([2]) E. CAHEN, *loc. cit.*, p. 280.

En définitive, $\Gamma_2^0$ étant un diviseur de zéro, on pourra écrire

$$_2 + \Gamma_2^0 \sim e\left(\sum \lambda_i \Delta_i + \mu_1 E_2^1\right) + ((C_a)),$$

où les $\lambda$ ne sont pas les mêmes entiers qu'auparavant. Il suffit de répéter le raisonnement du n° **14** pour voir que $((C_a))$ est une partie de $C_a$ prise $e$ fois. Notre homologie peut donc être écrite

$$(18) \qquad \Gamma_2 + \Gamma_2^0 \sim e\overline{\Gamma}_2 = e\left(\sum \lambda_i \Delta_i + ((C_a)) + \mu_1 E_2^1\right).$$

$e\overline{\Gamma}_2$ est fermée, donc $\overline{\Gamma}_2$ l'est aussi. C'est par suite un cycle à deux dimensions et (18) démontre notre théorème.

Corollaire. — *Les invariants $e_i$ relatifs aux intersections des* $\Gamma_2$ *sont tous égaux à l'unité* (*voir* Chap. I, n° **19**).

## VII. — Un exemple : la surface image des couples de points de deux courbes algébriques [1].

**20**. Soient $C_1$, $C_2$ deux courbes algébriques de genres $p_1$, $p_2$, $V_2$ une surface algébrique en correspondance biunivoque et sans aucune exception avec les couples de points des deux courbes. Nous allons en étudier rapidement la topologie, et en particulier vérifier sur elle nos théorèmes d'intersection de cycles. Le caractère très spécial de la surface va nous permettre de le faire de façon particulièrement simple.

Réduisons $C_1$ et $C_2$ à des cellules $E_2^1$, $E_2^2$, par des contours classiques aux rétrosections $K_1$, $K_2$. Il en résultera une réduction de $V_2$ à une cellule $E_4$. Soient M, $m_1$, $m_2$ un point de $V_2$ et les points correspondants des deux courbes. Pour avoir la frontière de $E_4$, on doit faire varier $m_1$ sur $E_2^1$ et $m_2$ sur $K_2$, puis renverser les rôles. On en tire de suite (Chap. I, n° **11**, lemme) que tout cycle de $V_2$ est une somme des cycles obtenus ainsi :

1° *Pour les cycles linéaires*, $m_1$ doit décrire les rétrosections de $C_1$, tandis que $m_2$ est fixe et *vice versa*. Soient $\gamma_i^1$, $\gamma_j^2$ les rétro-

[1] Pour une étude approfondie de ces surfaces, *voir* un Mémoire de M. Severi dans les *Torino Memorie*, vol. LIV, 1903.

sections des deux courbes. Il leur correspond respectivement des cycles linéaires $\Gamma_1^i$, $\Gamma_1^{2p_1+j}$, de $V_2$, constituant un système fondamental.

2° *Pour les cycles à trois dimensions*, $m_1$ doit parcourir les $\gamma_i^1$, tandis que $m_2$ décrit $E_2^2$ et *vice versa*. A $\Gamma_1^{p_1+i}$, $\Gamma_1^{2p_1+p_2+j}$ correspondent ainsi des cycles $\Gamma_3^i$, $\Gamma_3^{2p_1+j}$, et à $-\Gamma_1^i$, $-\Gamma_1^{2p_1+j}$ des cycles $\Gamma_3^{p_1+i}$, $\Gamma_3^{2p_1+p_2+j}$ ($i \leqq p_1$, $j \leqq p_2$). Avec ces notations, en apparence assez particulières, on vérifie de suite que

$$(\Gamma_1^i \Gamma_3^i) = 1, \qquad (\Gamma_1^i \Gamma_3^j) = 0 \qquad (i \neq j).$$

Donc la matrice carrée

$$\| (\Gamma_1^i \Gamma_3^j) \|$$

se réduit à la matrice identique. Ses diviseurs élémentaires sont tous égaux à l'unité et les cycles des systèmes fondamentaux sont tous indépendants, ce qui d'abord permet la vérification de nos théorèmes d'intersection (*voir* Chap. I), et ensuite démontre que

$$R_1 = R_3 = 2p_1 + 2p_2, \qquad \sigma_1 = \sigma_3 = 1.$$

3° *Pour les cycles à deux dimensions*, on fait décrire à $m_1$ et $m_2$ des rétrosections, d'où $4p_1p_2$ cycles $\Gamma_2^{ij}$ ($i = 1, 2, \ldots, 2p_1$; $j = 1, 2, \ldots, 2p_2$), le cycle $\Gamma_2^{ij}$ étant relatif à $\gamma_i^1$ et $\gamma_i^2$. Il y a en outre deux cycles $\Gamma_2^1$, $\Gamma_2^2$, le premier obtenu en maintenant $m_1$ fixe et faisant décrire $E_2^2$ par $m_2$, le second en renversant les rôles. Prenons alors un premier système fondamental

$$\Gamma_2^1, \Gamma_2^2; \Gamma_2^{ik}, \Gamma_2^{p_1+i,\, p_2+k}; \Gamma^{i, p_2+k}, \Gamma^{p_1+i, k}$$
$$(i = 1, 2, \ldots, p_1; k = 1, 2, \ldots, p_2),$$

puis un second obtenu en permutant les deux cycles de chaque couple, y compris le premier. Désignons les cycles du premier système fondamental dans un ordre quelconque par $\Gamma_2^h$, ceux correspondants du second par $\overline{\Gamma}_2^h$ ($h = 1, 2, \ldots, 4p_1p_2 + 2$). On vérifie de suite que

$$(\Gamma_2^h \overline{\Gamma}_2^h) = 1; \qquad (\Gamma_2^h \overline{\Gamma}_2^k) = 0 \qquad (h \neq k).$$

La matrice carrée

$$\| (\Gamma_2^h \overline{\Gamma}_2^k) \|$$

se réduit donc ici aussi à l'identité, d'où vérification immédiate des théorèmes d'intersection. De plus, le rang est égal à l'ordre, donc les cycles $\Gamma_2^h$ sont indépendants et

$$R_2 = 4p_1 p_2 + 2, \qquad \sigma_2 = 1.$$

On a bien effectivement $\sigma_1 = \sigma_2$.

# CHAPITRE IV.

## L'ANALYSIS SITUS ET LES SYSTÈMES DE COURBES D'UNE SURFACE ALGÉBRIQUE.

Dans deux Mémoires très importants (¹), Poincaré a jeté la base d'un traitement nouveau, entièrement analytique de la théorie des surfaces algébriques. Prenant comme point de départ certaines sommes abéliennes, il arrive à caractériser par elles les courbes algébriques d'une surface donnée. Les conditions nécessaires pour qu'à un système de telles sommes corresponde effectivement une courbe algébrique ont été obtenues par lui sous forme purement analytique. Or il se trouve, comme je l'ai montré pour la première fois dans mon Mémoire couronné (²), que ces conditions sont susceptibles d'une interprétation particulièrement simple et élégante à l'aide de la topologie unie aux périodes d'intégrales doubles de première espèce. La voie ainsi ouverte mène à nombre de résultats, aussi intéressants que nouveaux.

On peut entrevoir dans cette direction la possibilité de réunir en une doctrine unique ce que l'on sait aujourd'hui sur les surfaces algébriques, tant au point de vue géométrique qu'au point de vue transcendant. Ce qui va suivre constituerait peut-être le premier chapitre d'une telle théorie.

### I. — Intégrales simples et doubles de première espèce.

1. La définition des intégrales simples ou multiples de fonctions

(¹) *Sur les courbes tracees sur les surfaces algébriques* (*Annales de l'École Normale*, 3ᵉ série, vol. XXVII, 1910; *Sitzungsb. der Berliner math. Gesellschaft*, vol. X, 1911).

(²) *Transactions of the American Math. Society*, 22ᵉ vol., 1921. *Voir* à ce sujet trois Notes de M. Severi, *Rendiconti dei Lincei*, 5ᵉ série, 30ᵉ vol., 1921.

de variables complexes n'est pas à faire (1). Nous allons nous borner à celles de fonctions rationnelles du point sur la surface $V_2$ des Chapitres précédents. Soient $m$ son ordre, $f(x, y, z) = 0$ son équation. On peut envisager deux catégories d'intégrales : les intégrales *simples* ou de *différentielles totales* dues à M. Picard,

$$\int R(x, y, z)\,dx + S(x, y, z)\,dy, \qquad \frac{\partial B}{\partial y} = \frac{\partial S}{\partial x},$$

et les intégrales *doubles*, qui remontent à Nöther et Clebsch,

$$\int\int R(x, y, z)\,dx\,dy.$$

La classification des deux types se fait, à peu près, comme pour les intégrales abéliennes (2); ici encore il y a lieu de considérer des *périodes* relatives à des $\Gamma_1$ pour les intégrales simples, à des $\Gamma_2$ pour les intégrales doubles. Les intégrales de *première espèce*, simples ou doubles, presque seules à entrer en jeu dans la suite, sont toujours définies par la condition d'être partout finies. D'une discussion simple (3) résulte alors, pour les secondes, la forme

$$\int\int \frac{Q(x, y, z)}{f'_z}\,dx\,dy,$$

où Q est un polynome d'ordre $m-4$, adjoint à $V_2$, c'est-à-dire s'évanouissant sur sa courbe double.

2. Le rôle du système $|C|$ du Chapitre II va être joué, dans ce Chapitre, par $|H|$, système des sections planes, qui se conduit comme lui. Le faisceau particulier $\{C_u\}$ sera remplacé par $\{H_y\}$, découpé par les plans $y = \text{const.}$, mais à part cela, nous garderons intactes les notations des deux Chapitres précédents. En particulier, les valeurs critiques $a_j$ correspondront aux plans tangents parallèles au plan $xz$. Le choix de $|H|$ au lieu de $|C|$ ne dimi-

(1) *Voir* Picard et Simart, vol. I, Chap. III.

(2) Nous supposerons connues les propriétés classiques des intégrales abéliennes. *Voir* à ce sujet Appell et Goursat, *Théorie des fonctions algébriques et de leurs intégrales;* E. Picard, *Traité d'Analyse*, vol. II.

(3) Picard et Simart, vol. I, Chap. VII.

nuera en rien la généralité et a pour unique but de faciliter la discussion.

Partons de l'intégrale de première espèce attachée à $H_y$,

$$\int \frac{P(x, y, z)}{f'_z} dx,$$

où P est adjoint à la surface et d'ordre $m - 3$ en $x$, $z$, mais peut être $m - 3 + r$ ($r \gtreqless 0$), en $x$, $y$ et $z$. Je dis d'abord *qu'il y a p intégrales de ce type linéairement indépendantes* (on y considère bien entendu $y$ comme un paramètre). En effet, imposer au polynome le plus général d'ordre $m - 3$ en $x$ et $z$ la condition d'être adjoint à la courbe $f(x, y, z) = 0$ revient à faire satisfaire à ses coefficients un système d'équations linéaires, algébriques par rapport aux coefficients de la courbe, donc par rapport à $y$, invariantes dans leur ensemble quand $y$ varie. Les solutions fondamentales d'un tel système peuvent être choisies composées entièrement de polynomes, ce qui démontre notre affirmation.

*Corollaire.* — Toute intégrale de première espèce de $H_y$ est une combinaison linéaire d'intégrales du type ci-dessus.

3. **Comportement des intégrales de première espèce de $H_y$.** — Les points-bases $A_1, A_2, \ldots, A_m$, du faisceau $\{H_y\}$, sont ici les points à l'infini de ces courbes. Il y a avantage à étudier plus particulièrement les intégrales ayant pour limite inférieure d'intégration l'un des points A. Posons donc

$$u(x, y, z) = \int_{A_1}^{(x, z)} \frac{P(x, y, z)}{f'_z} dx.$$

Cette fonction est évidemment holomorphe, sauf peut-être pour $y$ infinie ou égale à une valeur critique. A l'aide de la transformation homographique

$$x = \frac{x'}{y'}, \qquad y = \frac{1}{y'}, \qquad z = \frac{z'}{y'},$$

on montre aisément que $u$, pour $x$ fixe, est une fonction de $y$ à pôle d'ordre $r - 1$ à l'infini. Examinons ce qui se passe pour $y$ voisin de $a_j$. Supposons, pour simplifier, le nouveau point double de $H_{a_j}$ au point $(o, a_j, o)$. Si $(x, y, z)$ ne tend pas vers ce point quand $y$ tend vers $a_j$, on pourra, en ajoutant à $u$ une période convenable, la rendre holomorphe au voisinage étudié. Ce voisinage est une cellule $E_1$ entourant un point quelconque de $H_{a_j}$, mais ne contenant pas son nouveau point double. Tout revient donc à examiner le comportement de $u$ dans une $E_4$ contenant ce point.

A une fonction holomorphe près, $u(x, y, z)$ se conduit comme

$$\int_{[\alpha, z(\alpha)]}^{(x, y)} \frac{P(x, y, z)}{f'_z}\, dx,$$

avec $|x|$, $|y - a_j|$, $|z|$, $\alpha$, $< \rho$, nombre positif fini, choisi de manière à rendre holomorphe certaines fonctions considérées plus loin.

Appliquons deux fois le théorème sur la représentation d'une fonction au voisinage d'un zéro, en ayant égard au fait que le nouveau point singulier de $H_{a_j}$ est un point double ordinaire. On pourra écrire

$$f(x, y, z) = [z^2 + 2A(x, y).z + B(x, y)].f_1(x, y, z),$$
$$A^2 - B = [x^2 + 2C(y).x + D(y)].\varphi(x, y),$$

où les fonctions nouvelles introduites sont toutes holomorphes quand les variables sont limitées de la manière ci-dessus, avec $\rho$ suffisamment petit, et seules $f_1$ et $\varphi$ ne s'y annulent pas en $(o, a_j, o)$.

Un calcul élémentaire donne alors pour $u$ les deux formes équivalentes suivantes, où les nouvelles fonctions écrites sont encore holomorphes dans les mêmes conditions que plus haut,

$$E(x, y) + F(x, y)\sqrt{x^2 + 2C(x).y + D(y)}$$
$$+ \psi(y)\log\left[x + C(y) + \sqrt{x^2 + 2C(y)x + D(y)}\right],$$
$$E(x, y) + zF(x, y) + \psi(y)\log[z + G(x, y)].$$

*Prenons en particulier pour limite supérieure d'intégration un point sur une courbe algébrique passant par le nouveau point double de* $H_{a_j}$; $u(x, y, z)$ devient alors une fonction de $y$ seule. Pour $\rho$ convenable, $x$ et $z$ seront des séries entières en $(y - a_j)^{\frac{1}{q}}$, $q$ entier, et l'on pourra affirmer que *le produit de u par* $(y - a_j)$ *tend vers zéro avec* $(y - a_j)$. En effet, il ne peut y avoir de doute que pour le produit par le terme logarithmique. Mais, à une fonction holomorphe tendant vers zéro près, ce produit est égal à une constante multipliée par

$$(y - a_j)^s \log(y - a_j) \qquad (s \text{ entier positif}),$$

et l'on sait que cette expression tend bien vers zéro avec $(y - a_j)$. Cette remarque nous sera fort utile plus loin.

4. **Comportement des périodes.** — Appliquons ce qui précède aux périodes de l'intégrale $u$. Soit d'abord $\Omega_j(y)$ celle relative au cycle évanouissant $\delta_j$. Comme il est invariant en $a_j$, $\Omega_j(y)$ est uniforme au voisinage de ce point. De plus, $\Omega_j(y).(y - a_j)$ tend vers zéro avec $y - a_j$, donc $\Omega_j(y)$ est holomorphe au point $a_j$. Soient $\gamma$ un cycle linéaire quelconque de $H_y$, $\omega(y)$ la période correspondante. La fonction

$$\omega'(y) = \omega(y) - \frac{(\gamma\delta_j)}{2\pi i}\Omega_j(y)\log(y - a_j)$$

a exactement les mêmes propriétés que $\Omega_j(y)$. Donc elle est aussi holomorphe en $a_j$. Par suite, au voisinage de ce point,

$$\omega(y) = \omega'(y) + \frac{(\gamma\delta_j)}{2\pi i}\Omega_j(y)\log(y - a_j),$$

résultat obtenu pour la première fois par M. Picard, à l'aide de l'équation différentielle aux périodes (équation de Picard-Fuchs).

L'expression

$$(1) \qquad \omega(y) - \sum \frac{\lambda_j}{2\pi i}\int_a^{a_j} \frac{\Omega_j(Y)\,dY}{Y - y}, \qquad \lambda_j = (\gamma\delta_j)$$

est holomorphe pour toute valeur finie de $y$. En effet, on voit d'abord, comme pour $\Omega_j(y)$ et $\omega'(y)$, qu'elle l'est au voisinage de $a_j$. Il suffit donc d'examiner ce qui se passe au voisinage de $a$. Lorsque $y$ tourne autour de ce point, (1) s'accroît de

$$\sum \lambda_j \Omega_j(y).$$

Cet accroissement n'est autre chose que la période de $u$ par rapport au cycle

$$\gamma' \sim \sum \lambda_j \delta_j = \sum (\gamma\delta_j)\delta_j.$$

Or dans $H_a$ ce cycle représente l'accroissement de $\gamma$ quand $y$ décrit le contour formé par la suite des lacets $aa_j$. Ce contour est déformable, sans traverser les points critiques, en un cercle de grand rayon. Comme $V_2$ est en position générale par rapport à l'infini, $y=\infty$ n'est pas une valeur critique, et l'accroissement correspondant de $\gamma$, qui n'est autre que la position de $\gamma'$ dans une $H_y$ quelconque, est nul. La période correspondante de l'intégrale de première espèce $u$ est donc nulle, ce qui montre que (1) est uniforme au voisinage de $a$. On vérifie de suite que son produit par $y-a$ tend vers zéro avec $y-a$, donc enfin (1) est bien holomorphe en $a$ aussi.

Les intégrales dans (1) tendent vers zéro quand $y$ devient infini. Enfin le point à l'infini est pour $\omega(y)$ un pôle d'ordre $r-1$ (n° 3). C'en est donc un de même nature pour (1) qui est alors nécessairement un polynome de degré $r-1$, $B(y)$. Nous avons par suite pour $\omega(y)$ *la représentation valable dans tout le plan, due à Poincaré*,

$$\omega(y) = \sum \frac{\lambda_j}{2\pi i}\int_a^{a_j} \frac{\Omega_j(Y)}{Y-y}\,dY + B(y).$$

*Corollaire*. — Les périodes de $u$ par rapport aux $2q$ cycles invariants sont des polynomes de degré $r-1$. Pour $r=0$, c'est-à-dire pour P d'ordre $m-3$ en $x$, $y$, $z$, ces périodes sont nulles.

5. Ce qui précède permet de montrer que

$$2\pi i . \psi(y) = \Omega_j(y),$$

où $\psi$ est la fonction du n° 3. En effet, soit $\delta'_j$ un cycle tel que $(\delta_j\delta'_j)=1$. La période correspondante, $\omega(y)$, se conduit au voisinage de $a_j$ comme

$$2G(\beta, y)\sqrt{\beta^2+2C.\beta+D} + 2\psi(y)$$
$$\times \log\left(\beta+\sqrt{\beta^2+2C.\beta+D}\right) + \text{une fonction holomorphe},$$

$\beta$ étant l'un des points de branchement de la fonction $z(x)$ qui coïncident pour $y=a_j$. Or $\beta$ annule le radical. De plus, au voisinage de $a_j$,

$$\beta(y) = (y - a_j)^{\frac{1}{2}} \sum_0^{\infty} {}_s c_s (y - a_j)^{\frac{s}{2}} \qquad (c_0 \neq 0).$$

En effet, puisque les axes sont arbitraires, le nouveau point double de $H_{a_j}$ est un point ordinaire de la courbe simple d'intersection de $f'_z = 0$ avec notre surface; donc $a_j$ est un point de branchement ordinaire pour la fonction algébrique $\beta(y)$, ce qui justifie ce développement.

On aura par suite

$$\omega(y) = \omega'(y) + \psi(y) \log(y - a_j),$$

$\omega'$ étant holomorphe au point $a_j$. En identifiant avec l'expression de M. Picard, on obtient bien

$$2\pi i \psi(y) = \Omega_j(y).$$

On en conclut *l'expression suivante pour* $u$, *valable au voisinage* de $a_j$,

$$u(x, y, z) = F(x, y) + z G(x, y) + \frac{\Omega_j(y)}{2\pi i} \log[z + H(x, y)].$$

6. Tout ce que nous venons de dire s'applique presque sans changement à une intégrale

$$\int \frac{P(x, y, z)}{f'_z} dx,$$

où P est adjoint, d'ordre $> m - 3$, non seulement en $x$, $y$, $z$, mais aussi en $x$, $z$, de sorte que l'intégrale n'est plus de première espèce. En particulier, les périodes ont la forme que nous avons déjà donnée. La seule différence est qu'il peut y avoir $m - 1$ résidus à l'infini, eux aussi polynomes en $y$. Effectivement, dans la discussion, nous ne nous sommes servis nulle part de ce que l'intégrale est de première espèce, mais uniquement de ce qu'elle est finie à distance finie. Au lieu de $A_1$ comme limite inférieure d'intégration, on pourra prendre un point variable d'intersection de $H_y$ avec une courbe algébrique quelconque.

7. En se basant sur ce que nous venons de dire, on montre aisé-

ment [1] qu'il existe une intégrale de différentielles totales *de seconde espèce* (c'est-à-dire se conduisant partout comme une fonction rationnelle), à périodes quelconques par rapport aux cycles invariants. Pour le démontrer on n'a besoin de se servir d'aucune propriété particulière de ces cycles. On en déduit ensuite que s'il y a $r$ cycles invariants, la surface possède $r$ intégrales distinctes de différentielles totales de seconde espèce, c'est-à-dire dont aucune combinaison linéaire n'est une fonction rationnelle.

Si donc on admet le théorème du Chapitre II, n° 13, et les propriétés qu'on en tire pour les cycles de $C_u$, ou bien ici de $H_y$, on en déduit que $V_2$ possède $2q$ intégrales distinctes de seconde espèce. Mais, remarque fort intéressante, la propriété ci-dessus fournit une démonstration transcendante du théorème topologique. Supposons en effet que

$$\gamma \sim \sum \lambda_j \delta_j$$

soit un cycle invariant. Il existe une intégrale de différentielles totales de seconde espèce J à période $+1$ par rapport à $\gamma$. D'un autre côté, J étant de seconde espèce, sa période par rapport à $\delta_j$ ne change pas quand on déforme ce cycle d'une manière quelconque dans $H_y$, en même temps que $y$ varie. Or, de cette manière, on peut réduire $\delta_j$ à un point ordinaire de la surface, le nouveau point double de $H_{a_j}$. La période correspondante de J doit donc être nulle, aussi bien que celle relative à $\gamma$, contradiction évidente qui établit le théorème.

## II. — Les courbes algébriques et leurs sommes abéliennes.

**8. Formules de Poincaré.** — Donnons-nous (nous avons vu qu'il est possible de le faire) $p$ intégrales distinctes de première espèce, telles que $u$ considérée plus haut, soit

$$u_h = \int \frac{P_h(x, y, z)}{f'_z} dx \qquad (h = 1, 2, \ldots, p),$$

où $P_h$ sera maintenant supposé d'ordre $m - 3 + q_h$.

---

(1) Picard et Simart, vol. I, p. 100.

Soit D une courbe algébrique quelconque, coupant $H_y$ aux points $m_1, m_2, \ldots, m_\nu$. D'ailleurs ces points n'ont pas besoin de comprendre la totalité du *groupe* $DH_y$, comme nous désignerons dorénavant l'ensemble des points d'intersection des deux courbes. On pourrait en fait se limiter à l'ensemble des points variables y compris seulement quelques-uns des points fixes. L'essentiel pour la suite, c'est d'avoir sur $H_y$ un groupe de points déterminés rationnellement dans leur ensemble, en fonction de $y$. Ces points fixes, remarquons-le de suite, ne peuvent être que les points $A_i$, pris chacun un certain nombre de fois. Ainsi pour $D = H_y$, on sera amené à considérer comme groupe particulier de points $m_k$, un quelconque des $A_i$.

Ceci étant posé, on peut appliquer sans la moindre modification, aux sommes abéliennes

$$v_h(y) = \sum_1^\nu {}^k \int_{A_1}^{m_k} du_h,$$

le raisonnement du n° 4, relatif au comportement des périodes. On obtient ainsi les *formules de Poincaré*, fondamentales pour ce qui va suivre,

$$(2) \qquad v_h(y) = \sum{}^j \frac{\lambda_j}{2\pi i} \int_a^{a_j} \frac{\Omega_{hj}(Y)}{Y - y}\, dY + B_h(y),$$

où $\Omega_{hj}$ est la période de $u$ par rapport à $\delta_j$ et $B_h$ est un polynome de degré $q_h - 1$. Quant aux $\lambda$, ils ont ici le sens suivant : Lorsque $y$ tourne autour de $a_j$ les chemins d'intégration $A_1 m_k$ sont ramenés à des positions nouvelles ; leur variation totale est un multiple de $\delta_j$. En comparant avec l'accroissement $\lambda_j \Omega_{hj}(y)$ de $v_h(y)$ dans les mêmes conditions, on voit de suite que cette variation est précisément $\lambda_j \delta_j$.

Soit $M_3$ la multiplicité engendrée par les chemins $A_1 m_k$ quand $y$ varie. L'examen de ses frontières donne

$$M_3 \equiv D - \sum \lambda_j \Delta_j - (H_a);$$

on en tire

$$(3) \qquad D \sim \sum \lambda_j \Delta_j + (H_a),$$

homologie qui indique clairement la relation entre les entiers $\lambda$ et la courbe D.

Une conséquence immédiate de (3) est (Chap. III, n° 10),

$$\sum \lambda_j \delta_j \sim 0.$$

Les périodes correspondantes des $u$ doivent être nulles, d'où

$$\sum \lambda_j \Omega_{hj}(y) = 0 \qquad (h = 1, 2, \ldots, p).$$

Comme nous le verrons plus loin (n° 15), ces relations expriment simplement que les $v$ ne dépendent pas de $a$.

9. **Discussion relative aux points à l'infini.** — Tout ce qui précède s'applique à leurs sommes abéliennes, comme nous l'avons déjà remarqué. Ils donnent cependant lieu à quelques observations spéciales, intéressantes.

Tout d'abord, si le groupe de points $m_k$ se réduit à une somme de multiples des $A_i$, il faudra, dans l'homologie (3), remplacer D par zéro. D'ailleurs, réciproquement, supposons que, dans (3), il faille remplacer D par zéro. Il y aura alors une

$$M_3 \equiv \sum \lambda_j \Delta_j + (H_a) \sim 0.$$

Nous avons vu (Chap. III, n° 11) qu'alors $M_3$ est le lieu d'une ligne

$$\gamma + \sum_2^m {}^i\, t_i A_1 A_i,$$

où $\gamma$ est un cycle linéaire convenable de $H_y$. Les $\lambda$ correspondront donc aux sommes abéliennes

$$v_h(y) = \int_\gamma du_h + \sum_2^m {}^i\, t_i \int_{A_1}^{A_i} du_h,$$

donc à

$$t_1 A_1 + t_2 A_2 + \ldots + t_m A_m,$$

où $t_1$ est un entier quelconque.

Je dis que les sommes abéliennes relatives à

$$t_1 A_1 + t_2 A_2 + \ldots + t_m A_m,$$

où l'un au moins des $t$ d'indice $> 1$ n'est pas nul, ne peuvent être égales à des périodes. En effet, nous avons fait correspondre à $A_i (i > 1)$ un cycle nul, frontière du lieu de la ligne $A_1 A_i$,

$$\Gamma_2^i = \sum \lambda_j^i \Delta_j + ((H_a)) + E_2^i - E_2^1.$$

Ici $((H_a))$, d'après nos conventions partie de $H_a$ sans points $A_i$, est simplement une partie finie de la courbe. Quant aux $E_2^i$, ce sont des cellules contenant les $A_i$ et que nous ne pouvons plus qualifier comme très petites. Pour préciser on prendra cette fois les $m$ cellules de $H_a$ représentant l'ensemble des points dont l'$x$ est extérieur à un cercle de grand rayon. On aura ainsi

$$H_a = ((H_a)) + \sum_1^m {}^i E_2^i. \tag{4}$$

Aux périodes correspond un cycle à deux dimensions nul,

$$\Gamma_2 = \sum \lambda_j \Delta_j + ((H_a)).$$

Si notre affirmation n'est pas correcte, les entiers $\lambda$ relatifs à un certain cycle

$$\Gamma_2 + \sum t_i \Gamma_2^i$$

doivent tous être nuls. Ce cycle devra donc se réduire à

$$((H_a)) + \sum_2^m {}^i t_i (E_2^i - E_2^1);$$

par suite, il devra être un multiple de $H_a$. En comparant avec (4), on en déduit

$$t_2 = t_3 = \ldots = t_m = -(t_1 + t_2 + \ldots + t_m) = t,$$

d'où $mt = 0$, $t = 0 = t_i$, ce qui démontre notre affirmation.

### III. — Conditions d'existence des courbes algébriques.

**10**. *Les expressions* (2) *étant données, leur correspond-il une courbe algébrique?* Ou, si l'on préfère, peuvent-elles servir à déterminer rationnellement un groupe unique de points sur $H_y$?

Nous allons chercher à déterminer rationnellement sur $H_y$ un groupe de $p$ points $m_k(x_k, y, z_k)$, tels que

$$(5)\quad \sum_1^p {}_k \int_{A_1}^{m_k} du_h = \sum \frac{\lambda_j}{2\pi i} \int^{a_j} \frac{\Omega_{hj}(Y)}{Y-y}\, dY + B_h(y) \qquad (h = 1, 2, \ldots, p).$$

La discussion va nous fournir des conditions suffisantes pour l'existence d'une solution de ce type spécial. Il se trouvera que ces conditions sont nécessaires pour l'existence d'une solution quelconque, qu'elle soit de type spécial ou non. La question proposée se trouvera ainsi complètement résolue.

**11**. Et d'abord, pour $y_0$ finie, donnée, il ne peut être question de déterminer un groupe de points sur $H_{y_0}$, à partir des sommes abéliennes $v_h(y_0)$, que si les polynomes $P_h(x, y_0, z)$ sont linéairement indépendants (théorème d'inversion). S'ils ne le sont pas, c'est qu'il y a des constantes $c_h$, telles que

$$(6)\qquad \sum c_h P_h(x, y, z) = (y - y_0)^s P(x, y, z).$$

Parmi les polynomes qui figurent à gauche, c'est-à-dire dont le coefficient $c$ n'est pas nul, soit $P_k$ celui de degré maximum. $P(x, y, z)$ est adjoint d'ordre $m-3$ en $x$, $z$, de degré total inférieur à celui de $P_k$. Remplaçons $P_k$ par $P$, puis continuons de même tant qu'il reste des valeurs $y_0$. Ce procédé a certainement un terme, car chaque fois qu'on l'applique, on abaisse le degré d'un des polynomes. Finalement, on en aura donc $p$ qui ne satisferont à aucune relation (6) pour $y$ finie. Nous supposerons qu'ils correspondent précisément aux intégrales $u_h$. Parmi ces polynomes, il y en aura $q'$, soient $P_1, P_2, \ldots, P_q$, de degré total $> m-3$, les autres étant de degré $\leqq m-3$. Pour ces derniers, il faudra dans (5) remplacer les polynomes $B_h$ par zéro. Nous verrons d'ailleurs plus loin que $q' = q$, et que, de plus, les P d'indices $\leqq q'$ sont de degré total $m - 2$.

**12.** Le déterminant fonctionnel des équations (5) par rapport aux $x_k$ (on suppose les $z_k$ fonctions connues des $x_k$) est égal à

$$\left| \frac{P_h(x_k, y, z_k)}{f'_{z_k}} \right|. \tag{7}$$

Lorsque $y$ est arbitraire, les équations (5) peuvent être résolues par rapport aux points $m_k$, et si pour une solution le déterminant s'annule, il y en aura une infinité. Soit $p'$ le rang du déterminant. On sait qu'on pourra alors se donner $p - p'$ des points $m_k$ (nous les ferons coïncider avec $A_1$) et en déduire les autres à l'aide de $p'$ des équations (5) convenablement choisies. Tout reviendra alors à discuter un système de moins de $p$ équations, mais à part cela, rien ne sera changé. Nous supposerons donc que pour $y$ arbitraire, la solution fournie par les équations (5) est unique et nous allons en discuter la nature.

**13.** Il est clair d'abord que toute fonction rationnelle et symétrique des coordonnées des $m_k$, $R(x_1, x_2, \ldots, x_p; z_1, z_2, \ldots, z_p)$ est une fonction uniforme de $y$ dans tout le plan. Pour que les $m_k$ engendrent une courbe algébrique, il faut et il suffit que R soit *méromorphe pour toute valeur finie ou infinie de* $y$. Elle ne peut cesser de l'être que pour $y$ voisin des points critiques, du point $a$, de l'infini, ou enfin des valeurs pour lesquelles le déterminant (7) s'annule. Nous allons examiner ces diverses circonstances en commençant par la dernière.

1° **Pour $y = y_0$, non critique, la solution de (5) fait évanouir le déterminant (7).** — Au voisinage de $y_0$, les équations (5) sont de la forme

$$F(x_1, z_1, \ldots, x_p, z_p; y) = 0,$$

les F étant holomorphes en leurs variables. Les solutions de ces équations sont donc des fonctions *algébroïdes* de $y$ au voisinage de $y_0$, ou, plus précisément, une quelconque des coordonnées $x_k$, $z_k$ est racine d'une équation

$$X^r + c_1(y).X^{r-1} + \ldots + c_r(y) = 0,$$

à coefficients holomorphes au voisinage de $y_0$. La fonction symétrique R est donc certainement méromorphe au point $y_0$.

**14.** 2° **Comportement aux points critiques.** — Soustrayons des seconds membres des équations (5) les périodes par rapport au cycle $\lambda_j \delta'_j$, $\delta'_j$ étant toujours le cycle tel que $(\delta'_j \delta_j) = 1$, et dont la propriété essentielle est de s'accroître de $\delta_j$ quand $y$ tourne autour de $a_j$. Les seconds membres sont alors holomorphes au voisinage de $a_j$, sans que cependant les solutions $x_k$, $z_k$ en soient changées pour cela. Il s'agit de montrer que, quelle que soit la distribution des points $m_k$, pour $(y - a_j)$ suffisamment petit, les solutions des équations (5) restent algébroïdes au voisinage de $a_j$. Supposons ici encore que le nouveau point double de $H_{a_j}$ est $(0, a_j, 0)$. Il n'y a de doute que si pour certains des $m_k$, soient $m_1, m_2, \ldots, m_s$, les $x_k$, $z_k$ sont suffisamment petits. En se reportant à ce que nous avons dit sur le comportement des $u$ au voisinage de $a_j$, on voit que le système (5) peut être remplacé par un autre dont une quelconque des équations aura la forme

$$\sum_{1}^{s}{}^{k} \log[z_k + H(x_k, y)] = \frac{\Phi(x_1, x_2, \ldots, x_p; z_1, z_2, \ldots, z_p; y)}{\Omega_j(y)},$$

où $\Omega_j$ est une quelconque des $\Omega_{hj}$, et $\Phi$ est holomorphe en ses variables quand les quantités $|x_k|$, $|z_k|$, $|y - a_j|$ sont suffisamment petites.

Au pis aller, $p - 1$ des $u$ tendent vers des intégrales de première espèce de $H_{a_j}$ quand $y$ tend vers $a_j$ et une au moins deviendra de troisième espèce. Le nouveau point double est à tangentes distinctes, donc il compte comme deux points distincts de $H_{a_j}$ : ces deux points seront les seuls points logarithmiques de l'intégrale de troisième espèce nouvelle, les périodes logarithmiques correspondantes étant $\pm \Omega_j(a_j)$. Par suite, $\Omega_j(a_j) \neq 0$. Donc en remplaçant au besoin les $u$ par des combinaisons linéaires convenables, on pourra s'arranger pour que $\Omega_{hj}(a_j) \neq 0$ quel que soit $h$. Les fonctions

$$\frac{\Phi}{\Omega_j(y)} = \Psi$$

se conduiront alors comme $\Phi$ dans les mêmes conditions que plus haut. Les équations (5) seront donc équivalentes à un système d'équations de type

$$\prod_{1}^{s}{}_{k}[z_k + \mathrm{H}(x_k, y)] = e^{\Psi},$$

dont les solutions sont bien algébroïdes dans la région considérée.

**15.** 3° **Comportement au point** $a$. — Il sera certainement à souhait si les seconds membres des équations (5) sont indépendants de $a$. En écrivant que leurs dérivées par rapport à $a$ sont nulles, on trouve

$$\sum_j \lambda_j \Omega_j(a) = 0 \qquad (h = 1, 2, \ldots, p). \tag{8}$$

Ceci indique que la période de toute intégrale de première espèce de $\mathrm{H}_a$ par rapport au cycle

$$\sum \lambda_j \delta_j$$

doit être nulle, donc il forme frontière, soit

$$-(\mathrm{H}_a) \equiv \sum \lambda_j \delta_j, \tag{9}$$

et, par suite,

$$\Gamma_2 \sim \sum \lambda_j \Delta_j + (\mathrm{H}_a)$$

est un cycle à deux dimensions. Ces conditions, *suffisantes* pour le comportement désiré au point $a$, sont aussi *nécessaires* (n° 8). Elles expriment simplement que les entiers $\lambda$ déterminent un certain cycle à deux dimensions.

**16.** 4° **Comportement à l'infini.** — Appliquons de nouveau la transformation homographique du n° 3. Une quelconque des équations (5) deviendra (le sens des lettres accentuées est évident)

$$\sum_{1}^{p}{}_{k} \int_{\mathrm{A}'_1}^{(x'_k,\, z'_k)} \frac{\mathrm{P}'_h(x', y', z')}{\mathrm{F}'_{z'}}\, dx'$$
$$= y'^{q_h} \sum \frac{\lambda_j}{2\pi i} \int_a^{a_j} \frac{\Omega_{hj}(\mathrm{Y})}{\mathrm{Y}y' - 1}\, d\mathrm{Y} + y'^{q_h - 1} \mathrm{B}_h\left(\frac{1}{y'}\right) \qquad (h = 1, 2, \ldots, p).$$

Lorsque $q_h \geqq 0$ le second membre est holomorphe au point $y' = 0$. Soit $q_h < 0$. Le polynome $B_h$ disparaît alors, et il faut trouver les conditions sous lesquelles le second membre reste holomorphe. Nous pouvons supposer $a \neq 0$, de sorte que, pour un choix convenable des chemins $aa_j$, $|Y|$ sera toujours supérieur à une certaine limite. Le coefficient de $y'^{q_h}$ est développable en série entière en $y'$. Pour que le second membre soit holomorphe, il faut et il suffit que les $-q_h$ premiers termes de cette série s'annulent. Ceci fournit les relations

$$(10) \qquad \sum{}_j \lambda_j \int_a^{a_j} Y^s \Omega_{hj}(Y)\, dY = 0 \qquad (s = 0, 1, 2, \ldots, -q_h - 1).$$

Le polynome $y^s P_h(x, y, z)$, $(s < -q_h)$ est adjoint d'ordre $m - 3 + q_h + s \leqq m - 4$, donc

$$(11) \qquad \iint \frac{y^s P_h(x, y, z)}{f'_z}\, dx\, dy$$

est de première espèce. *Je dis que* (10) *est sa période par rapport au cycle* $-\Gamma_2$. En effet,

$$\Gamma_2 \sim \sum \lambda_j \Delta_j + ((H_a)) + \sum_1^m{}_i t_i E_2^i.$$

Or d'abord

$$\iint_{\Delta_j} = -\int_a^{a_j} Y^s \Omega_{hj}(Y)\, dY.$$

Ensuite, en tous les points de $((H_a))$ l'intégrale sous forme (11), ou bien sous la forme équivalente

$$\iint \frac{y^s P_h(x, y; z)}{f'_x}\, dy\, dz,$$

a son coefficient différentiel fini, avec $dy = 0$. Donc

$$\iint_{((H_a))} = 0.$$

Enfin, transformons homographiquement $V_2$ de manière que $E_2^i$ devienne une petite cellule $E'_2$, située à distance finie dans la section par le plan $y = 0$. On aura

$$\int\int_{E_2^i} = \int\int_{E_2'} = 0,$$

l'intégrale étendue à $E_2'$ étant la transformée de (11).

On en conclut que la période de l'intégrale (11) par rapport à $-\Gamma_2$ est égale à la valeur de l'intégrale étendue à

$$\sum \lambda_j \Delta_j,$$

c'est-à-dire à l'expression (10), comme nous l'avions affirmé.

D'ailleurs soient

$$\int\int \frac{Q(x, y, z)}{f_z'} dx\, dy \tag{12}$$

une intégrale double quelconque de première espèce,

$$\int \frac{Q(x, y, z)}{f_z'} dx \tag{13}$$

l'intégrale abélienne correspondante attachée à $H_y$, $\Omega_j$ la période de (13) relative à $\delta_j$. Supposons que les $v$ dérivent, à la manière du paragraphe II, d'une courbe algébrique absolument quelconque D. On aura alors

$$\sum^k \int_{A_1}^{(x_k,\, z_k)} \frac{Q(x, y, z)}{f_z'} dx = \sum^j \frac{\lambda_j}{2\pi i} \int_a^{a_j} \frac{\Omega_j(Y)}{Y - y} dY,$$

et, en appliquant la même transformation homographique que plus haut, on trouve aisément que

$$\frac{1}{y'} \sum^j \frac{\lambda_j}{2\pi i} \int_a^{a_j} \frac{\Omega_j(Y)}{Y y' - 1} dY$$

doit être holomorphique au point $y' = 0$, et que par suite

$$\sum^j \lambda_j \int_a^{a_j} \Omega_j(Y)\, dY = 0,$$

c'est-à-dire que la période de l'intégrale arbitraire de première espèce (12) par rapport à $\Gamma_2$ doit être nulle. Comme conclusion de toute cette discussion, nous sommes donc en mesure d'énoncer le théorème d'existence suivant :

THÉORÈME D'EXISTENCE. — *Pour que les fonctions $v$ soient les sommes abéliennes d'une courbe algébrique* D, *il faut et il suffit que les* $\lambda$ *appartiennent à un cycle* $\Gamma_2$ *et que, de plus, les périodes des intégrales doubles de première espèce par rapport à* $\Gamma_2$ *soient nulles. Le cycle* $\Gamma_2 \sim$ D.

*Remarque.* — Aucune condition particulière n'est imposée aux polynomes $B_h$, *donc leurs coefficients sont arbitraires*. D'une façon plus précise, si à un système de valeurs de ces coefficients correspond une courbe algébrique, il en correspondra une aussi à tout autre.

**17.** Il est particulièrement intéressant, eu égard à leur importance, de retrouver les conditions du n° **16**, en se mettant sous un point de vue un peu différent.

S'il correspond aux $v$ une courbe algébrique D, on devra avoir (n° 8)

$$D \sim \sum \lambda_j \Delta_j + (H_\alpha).$$

Dans certains cas exceptionnels, il faudra mettre D = 0, mais alors la période de (12) par rapport au cycle au second membre est nulle, et la condition considérée est bien satisfaite.

Dans le cas général, soit $\varphi(x, y) = 0$ la projection de D. Isolons sur D des cellules entourant les points où $\varphi'_x = 0$, et soit (D) la partie restante. En effectuant s'il le faut une transformation homographique convenable, on pourra s'arranger pour que D — (D) soit une multiplicité finie, très petite, ne comprenant aucun des nouveaux points doubles des $H_{\alpha_j}$. Alors soit sous la forme (12), soit sous la forme équivalente

$$\int\int \frac{Q(x, y, z)}{f'_x} dy\, dz,$$

l'intégrale aura son élément différentiel fini en tous les points de D — (D), et par suite

$$\int\int_{D-(D)}$$

est aussi petite que l'on veut en valeur absolue. D'ailleurs

$$\int\int_{(D)} = \int\int_{(D)} \frac{Q(x, y, z)}{f'_z \varphi'_x} d\varphi\, dy = 0.$$

Donc la période par rapport à D, aussi petite que l'on veut, est nécessairement nulle. Ceci revient à vérifier les conditions du n° **16**.

## IV. — Intégrales de différentielles totales de première espèce.

**18.** Supposons que les $v_h$ satisfassent aux conditions d'existence. Il en sera encore ainsi quand les coefficients des $B_h$ varient, coefficients que nous désignerons dans un ordre quelconque par $\xi_1, \xi_2, \ldots, \xi_{q''}$. Le rang du déterminant fonctionnel (7) du n° **12**, qui varie de façon continue avec les $\xi$, aura une valeur fixe $p' \leqq p$, quand ces $\xi$ ont des valeurs arbitraires, et aux $v_h$ correspondra alors une courbe algébrique D, coupant $H_y$ en $p'$ points variables, dont les fonctions sont les sommes abéliennes.

D'un autre côté, à D correspond tout système de valeurs des $\xi$, obtenu en ajoutant au précédent un système de coefficients appartenant à des périodes *polynomes en y*. Or, il n'y a que $2q$ périodes distinctes de cette nature, les périodes relatives aux cycles invariants (n° **4**, corollaire). Considérons avec Poincaré, dont nous suivons ici le raisonnement, les $\xi$ comme des coordonnées cartésiennes de point d'un $S_{q''}$ complexe. Cet espace, à $2q''$ dimensions réelles, pourra être subdivisé en prismes congruents, chacun limité par $4q$ hyperplans parallèles et contenant un point unique relatif à D. Mais à toute D correspond un système de $\xi$ finis. Donc nos prismes doivent être finis, *ce qui exige* $q'' = q$.

Puisqu'il y a exactement $q'$ intégrales $u$ aux polynomes B non nuls, la série de points découpée par les D sur une $H_y$ fixe quelconque, soit $H_{y_0}$, dépend de $q'$ paramètres arbitraires, les valeurs des quantités $v_1(y_0), v_2(y_0), \ldots, v_q(y_0)$. Ces quantités sont bien arbitraires, puisque les termes constants des polynomes $B_h(y)$ le sont. Le raisonnement que nous venons d'appliquer plus haut à D s'applique à la série de points, d'où cette fois $q = q'$, et enfin $q' = q'' = q$.

Ainsi le nombre de coefficients variables, dans les polynomes $B_h$ non identiquement nuls, est égal au nombre même de ces polynomes. D'où, résultat d'une importance capitale, *ces polynomes se réduisent tous à des constantes, et leur nombre est égal à q*.

Comme corollaire, *il y a tout juste q polynomes* $P_h(x, y, z)$

*d'ordre* $m - 2$ *et les autres sont d'ordre* $m - 3$, au plus, résultat fort intéressant, dû à M. Picard. Sous une autre forme, on peut dire que toute intégrale de première espèce de $H_y$ dépend linéairement de celles du type

$$\int \frac{P(x, y, z)}{f'_z} dx,$$

où P est adjoint à $V_2$, d'ordre $m - 3$ en $x$, $z$ et $m - 2$ en $x$, $y$, $z$.

**19.** Les fonctions symétriques des coordonnées des points variables du groupe $DH_{y_0}$ sont des fonctions $2q$-uplement périodiques de $v_1(y_0)$, $v_2(y_0)$, ..., $v_q(y_0)$, c'est-à-dire, grâce à

$$v_h(y_0) = \sum \frac{\lambda_j}{2\pi i} \int_a^{a_j} \frac{\Omega_{hj}(Y)}{Y - y_0} dY + \xi_h \qquad (h \leq q),$$

des fonctions $2q$-uplement périodiques des $\xi$. Ces fonctions sont attachées à la même matrice aux périodes que les $u$. Puisque dans les expressions ci-dessus les $\lambda$ sont fixes, les seules périodes effectives des fonctions en question sont indépendantes de $y_0$; ce sont donc des périodes par rapport aux cycles invariants, celles par rapport aux $\delta$ étant au contraire toutes nulles. Ainsi *la matrice aux périodes de* $u_1$, $u_2$, ..., $u_q$ *par rapport à* $2q$ *cycles invariants distincts est aussi celle d'un système de fonctions* $2q$*-uplement périodiques de* $q$ *variables*.

En vertu d'un résultat fondamental de la théorie de ces fonctions (Poincaré, Picard, Scorza) [1], on peut alors se donner $q$ intégrales de première espèce $u'_h(y_0)$, ou bien, en remplaçant $y_0$ par $y$, $u'_h(y)$ ($h = 1, 2, \ldots, q$), à périodes, nulles par rapport aux $\delta$, égales à celles, constantes, des $u$ de même indice par rapport aux cycles invariants. Pour les $u_{q+k}$, ce sont, au contraire, les périodes par rapport aux cycles invariants qui sont nulles. Donc $u'_1$, $u'_2$, ..., $u'_q$, $u_{q+1}$, ..., $u_p$ sont linéairement indépendantes. Soit

$$u'_h = \sum_1^p {}^k\, c_{hk}(y)\, u_k \qquad (h = 1, 2, \ldots\, q).$$

[1] *Voir* le Mémoire de M. Scorza dans les *Rendic. del Circolo Mat. di Palermo* de 1916.

Les $c_{hk}$ devront satisfaire au système

$$(14) \qquad \sum_{1}^{p}{}_{k}\, c_{hk}(y)\,\Omega_{kj}(y) = 0 \qquad (j = 1, 2, \ldots, n),$$

et en outre, pour $h$ fixe, à $2q$ équations linéaires, exprimant que les périodes des $u'$ par rapport aux cycles invariants sont constantes. Quand $y$ décrit un chemin fermé quelconque, toute équation (14) est, au pis aller, remplacée par une somme de plusieurs autres; par suite, les $c$ sont uniformes. Mais on peut préciser davantage.

Soient d'abord $\gamma_1, \gamma_2, \ldots, \gamma_p$ les $p$ premières rétrosections d'une courbe de genre $p$. Le déterminant des périodes correspondantes de $p$ intégrales de première espèce linéairement indépendantes n'est pas nul, car par exemple pour les intégrales normales il est $+1$. Ajoutons aux $\gamma$ un petit circuit entourant un point A de la courbe et aux intégrales une de troisième espèce avec A pour point logarithmique : cela reste toujours vrai. On peut même remplacer les $p+1$ intégrales par autant de combinaisons linéaires distinctes sans rien changer.

Prenons $H_y$ pour courbe avec $\gamma_1 = \delta_j$, choix possible à cause du type de $\delta_j$. La première situation ci-dessus se présente pour $y$ non critique et les intégrales $u$, la seconde pour $y = a_j$ et les mêmes intégrales, avec toutefois $p-1$ au lieu de $p$. Donc le déterminant $e(y)$ des périodes relatives aux $\gamma$ est $\neq 0$ pour $y$ non critique et aussi pour $y = a_j$. Exprimons maintenant que les périodes de $u'_h$ relativement aux $\gamma$ sont constantes. Il en résulte pour les $p$ fonctions $c_{hk}$ autant d'équations linéaires à seconds membres constants, avec $e(y)$ pour déterminant des coefficients des inconnues. Par conséquent, ces fonctions sont finies pour $y$ finie, non critique, ainsi que pour $y = a_j$, donc enfin pour toute valeur finie de $y$ sans exception. Il suffit d'écrire les solutions pour voir que $c_{hk}$ se conduit comme un polynome de degré un lorsque $k > q$, comme une constante autrement. C'est donc un polynome de degré un dans le premier cas, une constante dans le second.

Conclusion immédiate de cette discussion : les $u'$ sont de même type que les $q$ premières $u$. D'ailleurs elles sont de première espèce pour toute $H_y$, y compris la courbe à l'infini. Pour cette dernière, c'est l'affaire d'une transformation homographique convenable. Quant aux $H_{a_j}$ cela résulte de suite de ce que les périodes par rapport aux $\delta$ sont nulles, de sorte que quand $y$ tend vers $a_j$ les $u'$ n'acquièrent pas de singularités logarithmiques. Prenons plus spécialement

$$u'_h = \int_{A_1}^{(x,z)} du'_h.$$

Cette fonction est définie en chaque point de $V_2$, du moins à des périodes par rapport aux cycles linéaires de la surface près. En tant que fonction de point sur $V_2$, $u'_h$ est holomorphe partout et ses dérivées partielles par rapport à $x$ ou $y$ sont uniformes, sans singularités essentielles, donc rationnelles. *Par suite* $u'_h$ *est une intégrale de différentielles totales de première espèce de* $V_2$, d'où, avec Poincaré, ce théorème capital en Géométrie algébrique, dû à MM. Castelnuovo, Enriques et Severi :

*Le nombre d'intégrales de différentielles totales de première espèce linéairement indépendantes est égal à l'irrégularité q.*

20. Dorénavant nous prendrons, pour nos $q$ premières intégrales $u$, les intégrales de différentielles totales de la surface. Les sommes abéliennes auront alors la forme

$$v_h(y) = \xi_h \qquad (h = 1, 2, \ldots, q);$$

$$v_{q+k}(y) = \sum \frac{\lambda_j}{2\pi i} \int_a^{a_j} \frac{\Omega_{q+k,j}(Y)}{Y - y}\, dY \qquad (k = 1, 2, \ldots, p - q).$$

## V. — Systèmes linéaires et systèmes continus de courbes algébriques.

21. *Deux courbes algébriques,* C, D, *aux mêmes sommes abéliennes sont contenues totalement dans un même système linéaire.*

Poincaré, à qui ce théorème est dû, impose aux courbes la condition d'avoir même ordre. Cette restriction est inutile, comme nous allons le voir.

Soient $\nu$, $\nu'$ les ordres de C, D avec $\nu \geqq \nu'$. Il existe une fonction rationnelle $R(x, z)$, attachée à $H_y$, ayant pour zéros les points de $CH_y$ et pour pôles ceux de $DH_y$ plus le point $A_1$ compté $\nu - \nu'$ fois. $R(x, z)$ peut être déterminée par des opérations algébriques par rapport à $y$, donc elle sera algébrique en $y$. Comme elle est unique pour $y$ donnée, elle en sera fonction rationnelle aussi. Dénotons-la alors par $R(x, y, z)$. En tant que fonction rationnelle attachée à $V_2$, elle aura comme courbe de zéros C et peut-être certaines courbes $H_y$, soient $H_{y_1}, H_{y_2}, \ldots, H_{y_t}$. De même, outre D, elle aura comme courbes d'infini $H_{y'_1}, H_{y'_2}, \ldots, H_{y'_{t'}}$. Je dis d'abord que $t = t'$. En effet,

$$R(x, y, z) = \frac{A(x, y, z)}{B(x, y, z)} \qquad \text{(A, B polynomes).}$$

Le faisceau linéaire découpé sur $V_2$ par les surfaces $A + kB = 0$ contient comme courbes totales les courbes

$$C + \sum H_{y_r}, \qquad D + \sum H_{y'_s};$$

donc leurs sommes abéliennes sont les mêmes. Par suite, celles pour $(t - t')H_y$, c'est-à-dire en définitive pour

$$(t - t')(A_1 + A_2 + \ldots + A_m),$$

sont egales à des périodes, ce qui entraîne bien $t = t'$ (n° 9). La fonction rationnelle

$$R(x, y, z) . \frac{(y - y'_1) \ldots (y - y'_{t'})}{(y - y_1) \ldots (y - y_t)}$$

aura C comme seule courbe de zéro, et D comme seule courbe d'infini. Ces deux courbes sont donc bien contenues dans un même système linéaire.

Corollaire (Severi). — *Pour que* C, D *soient contenues totalement dans un même système linéaire, il faut et il suffit qu'elles découpent sur une* $H_y$ *générique des groupes de points appartenant à une même série linéaire.*

En effet, si C, D sont dans un même système linéaire, elles ont mêmes sommes abéliennes. Pour $y$ fixe, ces sommes appartiennent aussi aux deux groupes de points découpés sur $H_y$. D'après l'inverse du théorème d'Abel ces deux groupes de points appartiennent bien à une même série linéaire. D'un autre côté, quand cette condition est remplie les sommes abéliennes sont les mêmes; d'après ce qui précède, les deux courbes appartiennent bien alors à un même système linéaire.

**22.** Reprenons la courbe D du n° 18; soit $|D|$ le système *complet* (système le plus ample), d'ailleurs unique, qu'elle détermine. Quand les $\xi$ varient, $|D|$ engendre un système $\sum$, $\infty^q$, de systèmes linéaires de courbes de même ordre que D : Je dis que $\sum$ est complet en tant qu'ensemble de systèmes linéaires. En effet, soient $\sum'$ un système semblable le contenant, $|D'|$ un de ses systèmes linéaires. Les sommes abéliennes de $|D'|$ ne diffèrent de celles de $|D|$ que par les valeurs des $\xi$. Comme il y a un système $|D|$, relatif à tout système de valeurs des $\xi$, il y en a un aux mêmes sommes abéliennes que $|D'|$, donc coïncidant avec lui, puisqu'ils sont tous deux complets. Par suite, $\sum$ contient bien $|D'|$ et enfin $\sum'$.

La courbe particulière D du n° 18, donc aussi $|D|$, peut être déterminée par un certain nombre de paramètres, fonctions $2q$-uplement périodiques des $q$ variables $\xi$. Il en résulte (*voir* plus loin, Chap. VI, n^os^ 11 et 12), que $\sum$ en tant qu'ensemble de systèmes linéaires constitue une variété algébrique. Cette variété, en fait abélienne, est en général désignée sous le nom de *variété de Picard* de la surface. On conclut de ceci que $\sum$ *en tant qu'ensemble de courbes constitue aussi une variété algébrique*.

*Remarque.* — Le système $\sum$, irréductible en tant qu'ensemble de systèmes linéaires, ne l'est pas nécessairement en tant qu'ensemble de courbes. Toutefois, entre deux composantes irréductibles, on peut toujours en intercaler plusieurs autres, telles que chacune ait au moins une courbe en commun avec celles qui la comprennent ([1]).

**23.** Un système complet $|D|$ ne découpe pas nécessairement, sur une courbe algébrique quelconque C, une série linéaire complète. On appelle *défaut* de la série, l'excès, sur sa dimension, de celle de la série complète qui la contient.

Rappelons que, par *série canonique* d'une courbe plane d'ordre $m$, on entend celle découpée par ses adjointes d'ordre $m-3$. Pour une courbe gauche, on la définit en s'en rapportant à une projection plane. Les courbes de $V_2$, qui découpent sur toute D des groupes canoniques, appartiennent à un même système linéaire $|D'|$, dit *adjoint* à $|D|$. Pour $|H|$ cela résulte du corollaire du n° 21 et se démontrerait de même pour tout $|D|$, $\infty^1$ au moins, tel que $(D^2) > 0$.

*Le système* $|D'|$ *découpe sur* D *une série de défaut* $q$. La démonstration qui suit pour $|H'|$ s'étend au cas général. La série canonique complète est découpée sur $H_y$ par les surfaces

$$\sum_1^p {}_h\, c_h P_h(x, y, z) = 0.$$

Toutefois, si les $c$ d'indice $\leqq q$, coefficients des polynomes d'ordre total $m-2$, ne sont pas nuls, les groupes découpés ne sont pas des groupes canoniques, mais bien de tels groupes augmentés des $m$ points à l'infini. Seuls les polynomes $P_{q+k}$ et leurs combinaisons linéaires découpent bien les groupes canoniques. La dimension de la série incomplète correspondante est $p-1-q$, ce qui donne bien $q$ comme défaut.

**24.** En résumé, nous avons trouvé pour l'irrégularité $q$ les propriétés suivantes :

1° $R_1 = 2q$.

2° Il y a $q$ intégrales distinctes de différentielles totales de première et $2q$ de seconde espèce.

---

(1) *Voir* à ce sujet un Mémoire de M. Albanese dans les *Annali di Mat.* de 1915.

3° Les systèmes complets de courbes, suffisamment généraux, sont composés d'$\infty^q$ systèmes linéaires formant un ensemble irréductible.

4° Le défaut de la série découpée sur une courbe d'un système linéaire $|D|$, $\infty^1$ au moins, de degré $(D^2) > 0$, par le système adjoint est égal à $q$.

Plusieurs de ces énoncés peuvent être précisés davantage, mais nous ne nous y arrêterons pas ici (1).

25. **Surfaces régulières.** — On appelle ainsi les surfaces pour lesquelles $q = 0$. Elles jouissent de propriétés particulièrement simples, conséquences de ce que nous venons de dire. Contentons-nous de relever celle-ci : *Sur une surface régulière, tout système complet est linéaire, donc en particulier irréductible.*

## VI. — Équivalence des courbes d'après M. Severi (2) Identité avec leur homologie en tant que cycles.

26. On dit que deux courbes algébriques C, D sont *équivalentes* et l'on écrit $C = D$, s'il en existe une troisième E, telle que $C + E$, $D + E$ appartiennent à un même ensemble irréductible de systèmes linéaires. Nous entendrons par *système continu* complet $\{C\}$ l'ensemble des courbes équivalentes à C.

*Remarque.* — Si C, D appartiennent déjà à un ensemble irréductible de systèmes linéaires, on aura simplement $E = 0$.

Les sommes abéliennes relatives aux courbes de $\{C\}$ (il s'agit de celles pour le groupe total $CH_y$) ne diffèrent que par les $\xi$ et non par les entiers $\lambda$. Bien entendu, on se permet toujours de négliger des périodes. Il en est ainsi pour $C + E$ et $D + E$, donc aussi pour C et D.

---

(1) Pour des renseignements plus complets, notamment sur la bibliographie, *voir* la Note de MM. Castelnuovo et Enriques à la fin du Traité de MM. Picard et Simart.

(2) *Voir* SEVERI, *Math. Ann.*, 1906, où cette notion est introduite pour la première fois sous une forme à peu près équivalente à la nôtre. *Voir* aussi la discussion de M. Albanese dans les *Annali di Mat.* de 1915.

27. Théorème fondamental. — *Deux courbes équivalentes sont homologues en tant que cycles, et réciproquement.*

Comme l'homologie $C \sim D$ ne dépend en rien de E, on voit que cette courbe auxiliaire disparaît finalement de la discussion.

Le théorème direct est immédiat. Les courbes d'un même système linéaire sont homologues; quand ce système varie ses courbes restent homologues à elles-mêmes. Donc de $C = D$ on tire $C + E \sim D + E$, d'où $C \sim D$.

La réciproque est moins facile. Soit donc $C \sim D$, et donnons-nous une courbe K telle que $\{K\}$ soit composé d'$\infty^q$ systèmes linéaires. Les courbes $C' = C + K$ et $D' = D + K$ ont elles-mêmes cette propriété puisque, par exemple, la courbe de $\{C + K\}$ composée de C et de K décrit, en même temps que cette dernière, $\infty^q$ systèmes linéaires distincts. On le voit de suite en s'en rapportant, par exemple, aux sommes abéliennes.

Soit donc

$$C' \sim \sum \lambda_j \Delta_j + (H_a),$$

$$D' \sim \sum \lambda'_j \Delta_j + (H_a),$$

les $\lambda$, $\lambda'$ étant les entiers relatifs aux sommes abéliennes de $C'$ et $D'$. Comme $C' \sim D'$, les entiers $\lambda_j - \lambda'_j$ correspondent aux sommes abéliennes d'un groupe de points $t_1 A_1 + t_2 A_2 + \ldots + t_m A_m$. Les sommes abéliennes d'une courbe quelconque $C''$ de $\{C\}$ ne diffèrent de celles de $D' + t_1 A_1 + \ldots + t_m A_m$ que par les valeurs des $\xi$. Comme $\{C'\}$ contient $\infty^q$ systèmes linéaires, on peut choisir $C''$ de telle manière que les deux systèmes de sommes abéliennes coïncident. Le raisonnement du n° 21 montre alors qu'il existe un faisceau linéaire dont une courbe totale est $C''$, et une autre, $D'$, associée à une certaine courbe $D'_1$. De

$$C'' \sim D' + D'_1, \qquad C'' \sim C' \sim D',$$

on tire $D_1 \sim 0$. Donc $D'_1$ n'existe pas (Chap. II, n° 6, corollaire); $D'$ appartient par suite à $|C''|$, donc enfin à $\{C'\}$. Toutes les conditions pour l'équivalence de C et D sont bien remplies et notre théorème est démontré.

**28. Opérations sur les courbes. Courbes effectives ou virtuelles.** — Nous savons ce que l'on entend par la *somme* de deux courbes C, D. Que doit-on entendre par leur *différence* C — D? S'il existe une courbe E telle que $D + E = C$, il n'y a pas à hésiter, E sera cette différence. Si E n'existe pas, nous conviendrons cependant, avec M. Severi, de considérer le symbole $(C - D)$ comme définissant une certaine courbe dite alors *virtuelle*, par opposition aux courbes ordinaires ou *effectives*. Les courbes virtuelles individualisent un système de sommes abéliennes et un cycle $\Gamma_2$. Que $(C - D)$ soit virtuelle ou effective, $E + (C - D)$, quand elle est effective, est bien définie et coïncide avec $(E - D) + C$, comme on le voit de suite à l'aide des sommes abéliennes. Nous la désignerons par $E + C - D$, ou bien $E - D + C$, en abandonnant les parenthèses. En vertu de ceci, nous pouvons toujours attribuer un sens au symbole

$$t_1 C_1 + t_2 C_2 + \ldots + t_r C_r$$

et à la relation

$$t_1 C_1 + t_2 C_2 + \ldots + t_r C_r = 0. \tag{15}$$

Lorsque les C sont toutes effectives, en posant $t_i = t'_i - t''_i$, avec $t'_i, t''_i > 0$, on aura

$$t'_1 C_1 + \ldots + t'_r C_r = t''_1 C_1 + \ldots + t''_r C_r,$$

relation avec un sens géométrique bien précis. Quand une relation telle que (15) est satisfaite, les C sont dites *algébriquement dépendantes* ou simplement *dépendantes*.

Un corollaire immédiat du théorème fondamental est l'*équivalence entre* (15) *et*

$$t_1 C_1 + t_2 C_2 + \ldots + t_r C_r \sim 0. \tag{16}$$

## VII. — Les cycles algébriques et les nombres σ, ρ.

**29.** Nous étendrons à l'avenir le terme *cycle algébrique* à un $\Gamma_2$ homologue à une courbe algébrique effective ou virtuelle.

*Pour que* $\Gamma_2$ *soit algébrique, il faut et il suffit que les périodes correspondantes des intégrales de première espèce soient nulles. On aura alors*

$$\Gamma_2 \sim C + kH \qquad (C \text{ courbe effective}).$$

La condition est nécessaire, car elle est remplie pour les cycles homologues aux courbes effectives, donc aussi pour ceux homologues aux courbes virtuelles, toutes reliées aux premières par des homologies. Soit maintenant

$$\Gamma_2 \sim \sum \lambda_j \Delta_j + (H_a)$$

un cycle satisfaisant à la condition énoncée. Il existe une courbe C, coupant $H_y$ en $p' \leqq p$ points variables, dont les sommes abéliennes correspondent précisément aux entiers $\lambda$. C pourra passer $t_i$ fois par $A_i$. En désignant toujours par $\Gamma_2^i$ les cycles frontières des lieux des lignes $A_1 A_i$, au groupe total de points $CH_y$ vont correspondre les mêmes entiers $\lambda$ que pour le cycle

$$\Gamma_2 + \sum t_i \Gamma_2^i.$$

Par suite,

$$C \sim \Gamma_2 + \sum t_i \Gamma_2^i + (H_a) \sim \Gamma_2 + (H_a),$$

car $\Gamma_2^i \sim 0.(H_a)$ ici est un cycle, d'où, comme nous l'avons vu maintes fois, à cause de l'irréductibilité de $H_a$ (*voir* Chap. III, n° 10),

$$(H_a) = -kH_a,$$

et finalement

$$\Gamma_2 \sim C + kH,$$

ce qui démontre notre théorème.

30. Voici une application intéressante : La courbe C obtenue au cours de la discussion définit un système $\{C\}$ qui contient $\infty^q$ systèmes linéaires. Supposons, en particulier, que $\Gamma_2$ soit une courbe effective D. On aura

$$D \sim C + kH,$$

donc

$$D = C + kH.$$

La courbe $D + lH$ déterminera, comme C, un système continu

$\{D + lH\}$ contenant $\infty^q$ systèmes linéaires, pourvu que $l \geqq -k$. Ainsi, *étant donnée une courbe algébrique effective quelconque* D, *le système complet* $\{D + lH\}$, *où* $l$ *est suffisamment grand, contient* $\infty^q$ *systèmes linéaires.*

31. Théorème. — *Les ensembles des courbes algébriques ou virtuelles et des cycles à périodes d'intégrales de première espèce nulles se correspondent élément à élément. Les zéros des deux ensembles sont des éléments correspondants.*

Ceci résulte immédiatement du théorème du n° 29.

Corollaires. — I. *Le nombre de courbes algébriques, indépendantes, a un maximum* $\rho \leqq R_2$. L'entier $\rho$ n'est autre que celui introduit par M. Picard dans la théorie des intégrales de différentielles totales de troisième espèce ([1]). Son rapport avec la théorie de l'équivalence des courbes est une des plus belles découvertes de M. Severi ([2]). Enfin le théorème précédent et ceux qui y conduisent ont été démontrés pour la première fois dans mon Mémoire couronné (1^re^ Partie, Chap. II).

II. Soit $\Gamma_2$ un diviseur tel que $\lambda\Gamma_2 \sim 0$. En vertu du théorème sur l'intégrale de Cauchy, étendu par Poincaré aux fonctions de plusieurs variables, les périodes des intégrales doubles de première espèce par rapport à $\lambda\Gamma_2$ sont nulles; donc celles par rapport à $\Gamma_2$ le sont aussi et $\Gamma_2$ est algébrique. Soit $\Gamma_2 \sim C$. On aura

$$\lambda C = 0, \qquad C \neq 0.$$

La courbe C, nécessairement virtuelle, est un *diviseur de zéro* pour les courbes algébriques, par rapport à la relation d'équivalence. Réciproquement, un tel diviseur pour les courbes en constitue un aussi pour les cycles. Avec M. Severi, qui a été le premier à étudier ces diviseurs ([3]), nous en désignerons le nombre par $\sigma - 1$. On aura alors

$$\sigma = \sigma_1 = \sigma_2.$$

---

([1]) Picard et Simart, vol. II, Chap. X.

([2]) Mémoire déjà cité au paragraphe VI.

([3]) *Annales de l'École Normale*, 1909. Il y démontre aussi l'existence d'un *système fondamental* de courbes (*base minima* dans sa terminologie).

Ce que nous venons de dire mérite d'être résumé en un théorème.

THÉORÈME. — *Le nombre* $\rho$ *de M. Picard est le même que celui des* $\Gamma_2$ *distincts à périodes d'intégrales de première espèce nulles. Les diviseurs* $\Gamma_2$ *sont tous algébriques, et*

$$\sigma = \sigma_1 = \sigma_2.$$

## VIII. — Application aux surfaces images des couples de points de deux courbes algébriques.

**32.** Soient $C_1$, $C_2$ les deux courbes; $u_1, u_2, \ldots, u_{p_1}$ les intégrales de première espèce de $C_1$; $v_1, v_2, \ldots, v_{p_2}$, celles de $C_2$; $(\xi_1, \eta_1)$, $(\xi_2, \eta_2)$, $(x, y, z)$ deux points des deux courbes et le point correspondant de la surface, que nous dénotons toujours par $V_2$. Les deux premiers points satisfont respectivement aux équations

$$f_1(\xi_1, \eta_1) = 0, \qquad f_2(\xi_2, \eta_2) = 0$$

des deux courbes et $x$, $y$, $z$ sont rationnelles en $\xi_1, \xi_2, \eta_1, \eta_2$.

Considérons avec M. Picard une intégrale de première espèce de $V_2$,

$$\int\int R(x, y, z)\, dx\, dy. \tag{17}$$

En vertu de ce que nous venons de dire, elle peut se mettre sous la forme

$$\int\int S(\xi_1, \xi_2, \eta_1, \eta_2)\, d\xi_1\, d\xi_2,$$

où S est encore rationnelle. Pour que cette intégrale soit de première espèce, il faut que, quand on maintient $(\xi_2, \eta_2)$ fixe sur $C_2$,

$$\int S(\xi_1, \xi_2, \eta_1, \eta_2)\, d\xi_1$$

devienne une intégrale de première espèce de $C_1$. On devra donc avoir

$$S(\xi_1, \xi_2, \eta_1, \eta_2) = \sum_1^{p_1} {}^i\, A_i(\xi_2, \eta_2) \frac{du_i}{d\xi_1},$$

où les A ne dépendent que de $\xi_2$, $\eta_2$, et ceci rationnellement. De même

$$S = \sum_{2}^{p_2} {}_k\, B_k(\xi_1, \eta_1) \frac{dv_k}{d\xi_2},$$

où les B sont aussi rationnelles. Par suite,

$$\sum_{1}^{p_1} {}_i\, A_i(\xi_2, \eta_2) \frac{du_i}{d\xi_1} = \sum_{1}^{p_2} {}_k\, B_k(\xi_1, \eta_1) \frac{dv_k}{d\xi_2}.$$

Substituons pour $\xi_1$, $\eta_1$ les coordonnées de $p_1$ points de $C_1$ non situés sur une même courbe canonique. On aura ainsi $p_1$ équations linéaires aux A, de déterminant $\neq 0$. En résolvant pour les A, on a donc, les $\alpha$ étant des constantes

$$A_i(\xi_2, \eta_2) = \sum_{1}^{p_2} {}_k\, \alpha_{ik} \frac{dv_k}{d\xi_2}.$$

Par suite,

$$S = \sum \alpha_{ik} \frac{du_i}{d\xi_1} \frac{dv_k}{d\xi_2}.$$

Donc enfin (17) est une combinaison linéaire des intégrales

$$\iint du_i \, dv_k. \tag{18}$$

D'ailleurs (18) est évidemment finie sur $V_2$. De plus, puisqu'à chaque point de $V_2$ correspond un point et un seul de $C_1$ ou $C_2$, les $\xi$, $\eta$ sont rationnelles en $x$, $y$, $z$. Donc (18), qui peut s'écrire

$$\iint \frac{du_i}{d\xi_1} \frac{dv_k}{d\xi_2} \frac{D(\xi_1, \xi_2)}{D(x, y)} \, dx \, dy,$$

est une intégrale double de fonction rationnelle attachée à $V_2$. C'est donc bien une intégrale de première espèce de la surface. Par suite, le nombre de ces intégrales est

$$p_g = p_1 p_2.$$

$p_g$, nombre d'intégrales doubles de première espèce, est le *genre géométrique*. De la relation

$$R_1 = 2p_1 + 2p_2 \quad (\text{Chap. III, n}^\circ\ 20),$$

on tire

$$q = p_1 + p_2.$$

En raisonnant comme ci-dessus, on voit que les $p_1 + p_2$ intégrales $u$, $v$ donnent lieu à $q$ intégrales simples de première espèce de $V_2$. Donc toute autre en est une combinaison linéaire.

Nous avons trouvé (*loc. cit.*) $4p_1p_2 + 2$ cycles à deux dimensions pour $V_2$. Les deux premiers, $\Gamma_2^1$, $\Gamma_2^2$, sont évidemment algébriques. Soient ici encore $\gamma_\mu^1$, $\gamma_\nu^2$, $\Gamma_2^{\mu\nu}$ des rétrosections des deux courbes et le cycle à deux dimensions correspondant,

$$\Omega_1 = \| \omega_{i\mu}^1 \| \qquad (i = 1, 2, \ldots, p_1;\ \mu = 1, 2, \ldots, 2p_1);$$

$$\Omega_2 = \| \omega_{k\nu}^2 \| \qquad (k = 1, 2, \ldots, p_2;\ \nu = 1, 2, \ldots, 2p_2),$$

les matrices aux périodes des intégrales $u$, $v$, les colonnes y correspondant précisément aux rétrosections.

La période de (18), par rapport à $\Gamma_2^{\mu\nu}$, est égale à $\omega_{i\mu}^1 \omega_{k\nu}^2$. Donc, pour que

$$\sum c_{\mu\nu} \Gamma_2^{\mu\nu}$$

soit algébrique, il faut et il suffit que

$$\sum\nolimits^{\mu,\nu} c_{\mu\nu} \omega_{i\mu}^1 \omega_{k\nu}^2 = 0 \qquad (i = 1, 2, \ldots, p_1;\ k = 1, 2, \ldots, p_2),$$

ou bien, si l'on préfère, que la forme bilinéaire à coefficients entiers

$$(19) \qquad \sum c_{\mu\nu} x_\mu y_\nu$$

s'annule quand on y remplace les $x$ par les termes d'une ligne quelconque de $\Omega_1$ et les $y$ par ceux d'une ligne de $\Omega_2$.

Soit $\lambda$ le nombre de formes bilinéaires (19) linéairement indépendantes correspondant aux deux matrices; c'est le nombre que M. Scorza appelle leur *caractère simultané*. On aura alors

$$\rho = \lambda + 2,$$

formule due à M. Severi. Il l'a obtenue en se basant sur la théorie de A. Hurwitz, des correspondances singulières entre les courbes. $\lambda$ n'est autre, en effet, que le nombre de celles qui existent entre $C_1$ et $C_2$.

# CHAPITRE V.

## VARIÉTÉS ALGÉBRIQUES A PLUS DE DEUX DIMENSIONS.

### I. — Propriétés topologiques [1].

**1.** Les généralités relatives aux surfaces s'étendent de suite à une variété algébrique à $d$ dimensions $V_d$. Nous la supposons dans un espace $S_{d+1}$, à singularités *ordinaires*, dont il nous suffira de dire : 1° qu'elles sont les mêmes que pour la projection d'une $V_d$ non singulière d'un $S_{d+1+k}$; 2° que si $V_d$ n'en a pas d'autres, elle est homogène.

La plupart des propriétés topologiques des $V_2$ s'étendent aux $V_d$, d'autant plus que nous avons conduit les démonstrations autant que cela était possible *en vue d'une telle extension*. Toutefois de nouvelles propriétés se présentent, et c'est sur elles que nous insisterons surtout.

**2.** Soit donc dans $V_d$ un système linéaire d'*hypersurfaces* (ce sont les $V_{d-1}$ de $V_d$), $|C|$, $\infty^d$ au moins, tel que :

1° La $V_{d-k}$ commune à $k$ hypersurfaces C arbitraires (nous la désignerons par $C^k$) est elle-même irréductible, à singularités ordinaires, toute singularité nouvelle consistant, en général, en un point double, isolé, ordinaire.

2° $d$ hypersurfaces C ont en commun un nombre de points $\neq 0$. On démontre tout à fait, comme pour les surfaces, que ce nombre de points est égal au nombre algébrique $(C^d)$, pourvu que $V_d$ soit orientée convenablement. Plus généralement, le nombre de points communs aux hypersurfaces $C_1, C_2, \ldots, C_d$ sera alors égal à $(C_1 C_2 \ldots C_d)$.

---

(1) *Voir* Mémoire couronné, 1re Partie, Chap. I.

3° Si $r$ est la dimension du système linéaire découpé par les C sur une $C^k$ générique, ce système ne contient pas $\infty^{r-1}$ $C^{k+1}$ réductibles. Un multiple convenable des sections hyperplanes fournit un exemple d'un système tel que $|C|$.

Prenons dans $|C|$ un faisceau linéaire quelcónque $\{C_u\}$, aux valeurs critiques $a_1, a_2, \ldots, a_n$ (valeurs où $C_u$ acquiert un nouveau point double), puis traçons toujours les coupures $a\infty$, $aa_i$. Désignons enfin par A la $C^2$, de base du faisceau. Si l'on sait réduire $C_u$ à une cellule dont la frontière contienne A, il sera possible, par la construction même qui nous a servi dans le cas des $V_2$, de réduire $V_d$ à une cellule dont la frontière contienne $C_a$. La réduction de $V_d$ se ramène ainsi à celle d'une $V_{d-1}$, donc finalement à celle d'une $V_2$. On peut, par suite, la considérer comme établie.

3. Théorèmes [1]. — I. *Tout* $\Gamma_k$ *de* $C_u$ $(k < d - 1)$ *est invariant.*

II. *Tout* $\Gamma_k$ *de* $V_d$ $(k < d)$ *est homologue à un cycle de* $C_u$.

III. *Pour* $k \leqq d - 2$, $\Gamma_k$ *et le cycle homologue de* $C_u$ *forment frontière, en même temps, dans leurs variétés respectives.*

IV. *Dans les mêmes conditions* $(k \leqq d - 2)$, $V_d$ *et* $C_u$ *ont mêmes indices* $R_k$.

La démonstration par récurrence s'impose ici. Tout d'abord celle de II, pour les $\Gamma_1$ d'une $V_2$ (Chap. III, n° 3), s'étend de suite à $d$ quelconque. Elle est même encore applicable quand la surface possède des points singuliers isolés, pourvu qu'on exclue les cycles qui y passent. Remarquons que pour $d = 2$, les $\Gamma_1$ sont seuls à entrer en jeu.

Par suite, nos théorèmes, en tant qu'ils s'appliquent, sont déjà démontrés pour les surfaces. Supposons-les donc vrais pour une $V_{d-1}$, même avec cette aggravation que II s'applique encore

[1] Ces questions ont déjà été traitées, par voie transcendante, pour les $\Gamma_1$, par MM. Castelnuovo et Enriques dans leur Mémoire des *Annales de l'École Normale*, 3e série, vol. XXII, 1906.

quand la variété a des points singuliers isolés. Je dis qu'ils sont encore vrais pour une $V_d$, II en particulier l'étant dans le cas plus général que nous venons de considérer.

Tout d'abord, pour I, c'est immédiat : $\Gamma_k$ ($k < d - 1$), supposé dans $C_u$, y est alors réduit à A, base de $\{C_u\}$. Quand $u$ varie, le cycle réduit reste homologue à lui-même, puisqu'il est dans une variété fixe. $\Gamma_k$, qui lui est homologue, est bien invariant.

4. Passons à II. On suppose $1 < k < d$, $\Gamma_k$ ne passant pas par les points doubles isolés des $C_{a_i}$, ce qui est légitime puisque $V_d$ est homogène. Le cycle à $k - 2$ dimensions, $\Gamma_k C_u$ est homologue $(\text{mod}\, C_u)$ à un cycle fixe de A. Soit $M_{k-1}$ la multiplicité de $C_u$ dont les deux cycles forment la frontière. Son lieu quand $u$ varie est une $M_{k+1}$ dont la frontière comprend, outre $-\Gamma_k$ et une multiplicité de $C_a$, d'autres telles que celle-ci : $u$ étant quelconque sur $a_i a$, soit $\Gamma^i_{k-1}$ le cycle de $C_u$, différence entre les positions de $M_{k-1}$, suivant que $u$ tend vers sa position d'un côté ou de l'autre de la coupure. Quand $u$ décrit $a_i a$, $\Gamma^i_{k-1}$ engendre une multiplicité $M^i_k$. C'est celle que nous avions en vue.

On peut traiter $M^i_k$, comme plus haut $\Gamma_k$, en réduisant d'abord $\Gamma^i_{k-1}$ à A. En effet, comme $k - 1 < d - 1$, on peut appliquer II à ce cycle de $C_u$. Le cycle, réduit à A, est fixe et devient une multiplicité de dimension moindre, comme $\Gamma^i_{k-1}$ lui-même, lorsque $u$ vient en $a_i$. Donc sa dimension est constamment plus petite que $k - 1$. Ceci montre que $\Gamma^i_{k-1} \sim 0 \ (\text{mod}\, C_u)$. La $M_k$, dont $\Gamma^i_{k-1}$ est la frontière, se réduit à une dimension moindre pour $u = a_i$ ; son lieu $M^i_{k+1}$, quand $u$ décrit $a_i a$, aura donc pour frontière $M^i_k$, ainsi qu'une multiplicité de $C_a$.

On conclut de ceci que $M_{k+1} - \sum M^i_{k+1}$ aura pour frontières uniques $-\Gamma_k$ et une $M_k$ de $C_a$ : $\Gamma_k$ est homologue à cette $M_k$, ce qui démontre II. On voit que cette réduction à $C_a$ n'est pas affectée par la présence de points singuliers isolés extérieurs au cycle. Ceci justifie l'hypothèse faite au début de la discussion.

5. Quant à III, clairement si $\Gamma'_k$ cycle réduit à $C_a \sim 0 \ (\text{mod}\, C_a)$, $\Gamma_k \sim 0 \ (\text{mod}\, V_d)$. Réciproquement de $\Gamma_k \sim 0 \ (\text{mod}\, V_d)$, on tire $\Gamma'_k \sim 0 \ (\text{mod}\, V_d)$. Soit $M_{k+1} \equiv \Gamma'_k$. On peut répéter pour $M_{k+1}$ le

raisonnement que nous venons de faire pour les cycles. Comme $k+1 \leqq d-1$, on trouvera cette fois qu'elle peut être remplacée par une multiplicité de même frontière, située en entier dans $C_a$. Par suite $\Gamma'_k \sim 0 \pmod{C_a}$. Ceci démontre III.

Le théorème IV est un corollaire immédiat de III.

6. **Cycles évanouissants.** — Pour simplifier nous allons nous limiter à une $V_3$. Prenons dans $|C|$ un faisceau $\{C_u\}$, puis appliquons à la surface $C_u$ les notations du Chapitre II. Toutefois, comme faisceau particulier, nous en prendrons un, $\{C_{uv}\}$, découpé par un faisceau quelconque, $\{C_v\}$, de $|C|$, et, au lieu de $a$, $a_j$, nous aurons certaines valeurs, $b$, $b_j(u)$, de $v$.

Pour tout $\Gamma_2$ de $C_u$, on aura donc

$$\Gamma_2 \sim \sum \lambda_j \Delta_j + (C_{ub}).$$

Soit $a_i$ une valeur critique de $u$ pour laquelle $b_1 = b_2$. Supposons les lacets $bb_1$ et $bb_2$ consécutifs, ce qui a pour seul effet de changer peut-être les $\lambda$. On pourra écrire

$$\Gamma_2 \sim \lambda_1 \Delta_1 + \lambda_2 \Delta_2 + M_2,$$

$M_2$ étant invariant quand $u$ tourne autour de $a_i$. Dans les mêmes conditions, au contraire, $\Delta_1$, $\Delta_2$, $\Gamma_2$ seront ramenées à de nouvelles multiplicités $\Delta'_1$, $\Delta'_2$, $\Gamma'_2$, et l'on aura

$$\Gamma'_2 - \Gamma_2 \sim \lambda_1(\Delta'_1 - \Delta_1) + \lambda_2(\Delta'_2 - \Delta_2).$$

Faisons tendre $b$ vers $b_2$. Le second membre devient alors un cycle $\sim \Gamma'_2 - \Gamma_2$, lieu de $\delta_1$ quand $v$ décrit une certaine ligne $b_1 b_2$. Pour que ce lieu soit fermé, il faut que $\delta_1$ soit évanouissant en $b_2$. Donc $\delta_1$ est un multiple de $\delta_2$. Cela veut dire que les deux points de branchement de $C_{uv}$ qui coïncident en $b_1$ en font autant en $b_2$ [1]. Puisque $\delta_1$ et $\delta_2$ sont des circuits entourant ces deux points dans la même surface de Riemann, il faut $\delta_1 \sim \pm \delta_2$. Nous allons voir que le signe doit être $+$. En tout cas il ne dépend ni

[1] On le voit aussi directement en remarquant que pour $u$ voisin de $a_i$ et $v$ de $b_1(a_i) = b_2(a_i)$ il n'y a que deux points de branchement de $C_{uv}$ voisins l'un de l'autre.

de $a_i$, ni de $|C|$, ni de $V_3$, car certainement la situation topologique ne change pas quand on modifie ces divers éléments.

A $\Delta'_2$ correspond un lacet de $b$ à $b_1$, lacet réductible au lacet $bb_1$ primitif, en vertu de l'invariance de $\delta_1$ en $b_2$. On aura ainsi réduit $\Delta_2$ à $\Delta_1$. De même on pourra réduire $\Delta'_1$ à $\Delta_2$. Par suite,

$$\Gamma'_2 - \Gamma_2 \sim (\lambda_2 - \lambda_1)(\Delta_2 - \Delta_1) = (\lambda_2 - \lambda_1).\delta^i_2.$$

*Je dis que $\delta^i_2$ est un cycle*, nécessairement alors évanouissant en $a_i$. Si $\Gamma_2$ n'est pas invariant, $\lambda_2 - \lambda_1 \neq 0$, la multiplicité à droite est un cycle, et $\delta^i_2$ en est un aussi. Ceci entraîne

$$\delta_2 - \delta_1 \sim 0 \quad (\text{mod } C_{ub}),$$

indiquant alors que $+$ était bien le signe à prendre ci-dessus. Réciproquement, si $\delta_2 \sim + \delta_1$, il y a un certain cycle

$$\Delta_2 - \Delta_1 + (C_{ub}).$$

Comme il n'est pas invariant près de $a_i$, $\delta^i_2$ est un cycle. D'après ce que nous avons dit plus haut, nous pourrons remplacer $V_3$ et $|C|$ par un $S_3$ et ses surfaces d'ordre $m > 1$. Il s'agit d'établir que leurs $\Gamma_2$ ne sont pas tous invariants. Or, on peut montrer, indépendamment de notre discussion, que le nombre de $\Gamma_2$ invariants de $C_u$ est égal à l'indice $R_4$ de sa $V_3$. Pour $S_3$, $R_4 = 1$, alors que pour ses surfaces d'ordre $m > 1$, $R_2 > 1$. Donc elles possèdent des cycles non invariants, *et notre affirmation est ainsi démontrée.*

Revenant à la variation de $\Gamma_2$, puisque $(\delta_1 \delta_2) = 0$, la formule pour l'intersection des $\Gamma_2$ d'une $V_2$ (Chap. III, formule 11) donne

$$(\Gamma_2 \delta^i_2) = \lambda_1 - \lambda_2,$$

d'où finalement

$$\Gamma'_2 - \Gamma_2 \sim -(\Gamma_2 \delta^i_2).\delta^i_2.$$

*Ceci exprime l'extension du théorème de M. Picard sur la variation des cycles.* Signalons toutefois cette différence : Ici $(\delta^i_2 \delta^i_2) = 2$, donc $\delta^i_2$ n'est pas invariant, mais au contraire changé en $-\delta^i_2$ quand $u$ tourne autour de $a_i$.

7. On peut maintenant étendre à $V_3$, puis de proche en proche à une $V_d$ quelconque, les résultats relatifs à $V_2$. Nous allons passer

en revue les divers théorèmes, en indiquant les modifications qui se présentent.

THÉORÈMES SUR LES $\Gamma_{d-1}$ DE $C_u$. — I. *A tout point critique $a_i$ correspond un cycle à $d-1$ dimensions, $\delta_i$, de $C_u$, invariant en ce point, tel en outre que quand $u$ tourne autour de lui, tout autre $\Gamma_{d-1}$ s'accroisse* de $(-1)^d.(\Gamma_{d-1}\delta_i).\delta_i$.

II. *$\delta_i$ est invariant en $a_i$ pour $d$ pair, au contraire changée en $-\delta_i$ pour $d$ impair.*

III. *Toute somme de $\delta_i$, invariante, forme frontière sur $C_u$.*

IV. *Tout $\Gamma_{d-1}$ de $C_u$ dépend des $\delta_i$ et des cycles invariants.*

V. *Si $\Gamma_{d-1}$ forme frontière sur $V_d$ il est homologue à une somme de $\delta_i$.*

Ceci se démontre à peu près comme la réduction indiquée plus bas (théorème VI) pour les $\Gamma_a$.

8. THÉORÈME SUR LES $\Gamma_d$ DE $V_d$. — VI. *Soit $\Delta_i$ l'onglet, lieu de $\delta_i$ quand $u$ décrit $a_ia$. Tout*

$$\Gamma_d \sim \sum \lambda_i \Delta_i + M_d, \tag{1}$$

*où $M_d$ est dans $C_a$. De plus les $\delta_i$ y étant aussi pour l'instant, on a la congruence et l'homologie*

$$-M_d \equiv \sum \lambda_i \delta_i; \tag{2}$$

$$\sum \lambda_i \delta_i \sim 0 \qquad (\text{mod } C_a). \tag{3}$$

Pour le montrer appliquons la réduction du n° 4. Avec les mêmes notations, on ne pourra cette fois se débarrasser des $M_d^i$. Toutefois, pour que la frontière de $M_{d+1}$ soit fermée, $\Gamma_{d-1}^i$ devra se réduire à une dimension moindre dans $C_{a_i}$. Donc $\Gamma_{d-1}^i$ est un multiple de $\delta_i$ et $M_d^i$ un multiple de $\Delta_i$, d'où (1) s'ensuit. Quant à (2) et à (3) ce sont des conséquences immédiates de (1).

Réciproquement, à toute homologie (3) correspond un cycle (1).

Quand les $\lambda$ sont nuls, $\Gamma_d$ est en entier dans $C_a$. La multiplicité $M_d$ coupe A suivant un $\Gamma_{d-2}$, car ses frontières, sommes de $\delta_i$, ne rencontrent pas A. Nous nommerons $\Gamma_{d-2}$ le *cycle de* A *correspondant à* $\Gamma_d$. On a d'ailleurs

$$\Gamma_{d-2} \sim \Gamma_d C_u \qquad (\text{mod } C_u).$$

En appliquant de façon répétée le théorème II, n° 3, on peut réduire tout $\Gamma_k$ de $V_d (k < d)$ à une $C^{d-k}$. Soit alors un système de $C^{d-k}$ passant toutes par une $C^{d-k+1}$ donnée. Les éléments de ce système peuvent être représentés par les points d'un $S_k$. Soit $V_{k-1}$ la variété algébrique de cet espace, image des $C^{d-k}$ possédant des singularités nouvelles. Réduisons $S_k$ à une cellule dont la frontière contienne $V_{k-1}$, puis $C^{d-k}$ à une cellule dont la frontière contienne la $C^{d-k+1}$ fixe de son système. Il en résultera une réduction de $V_d$ à une cellule quelque peu plus générale que celle du n° 2. Il suffit de raisonner avec $C^{d-k}$ et cette cellule comme auparavant avec celle rattachée au faisceau $\{C_u\}$ et $V_2$, pour démontrer :

VII. *Les* $\Gamma_k$ *de* $C^{d-k}$ *dépendent des cycles invariants et des cycles évanouissants le long de* $V_{k-1}$.

VIII. *Tout* $\Gamma_k$ *invariant de* $C^{d-k}$ *engendre un* $\Gamma_{2d-k}$ *de* $V_d$, *et il y a équivalence entre les homologies*

$$\Gamma_k \sim 0 \qquad (\text{mod } C^{d-k}); \quad \Gamma_{2d-k} \sim 0 \qquad (\text{mod } V_d).$$

IX. *Il existe un système fondamental pour les* $\Gamma_2$ *quel que soit* $k$.

9. Théorèmes sur les indices $R_k$. — De VIII on tire :

X. *Relation de Poincaré :*

$$R_k = R_{2d-k}.$$

Soit $r_k$ l'indice à $k$ dimensions de $C^{d-k}$. On a

XI.

$$R_d - R_{d-2} = n - 2(r_{d-1} - R_{d-1}) - (r_{d-2} - R_{d-2}).$$

Pour le démontrer on remarque d'abord que l'indice de connexion à $d$ dimensions de $C_u$ est égal à celui à $d-2$ dimensions, c'est-à-dire à $R_{d-2}$ (n° 3, théorème IV). Donc il y a $R_d - R_{d-2}$ cycles $\Gamma_d$ distincts, aux $\lambda$ non nuls. D'un autre côté, (3) est équivalente au système diophantin

$$\sum_i \lambda_i(\delta_k \delta_i) = 0 \qquad (k = 1, 2, \ldots, n),$$

exprimant que le premier membre de l'homologie est invariant. Le rang de la matrice aux coefficients est égal au nombre

$$r_{d-1} - R_{d-1}$$

de $\Gamma_{d-1}$ non invariants, distincts, de $C_u$. Donc il y a $n-(r_{d-1} - R_{d-1})$ systèmes de nombres $\lambda$ distincts. Il s'agit de savoir combien correspondent à des cycles nuls. Soient $D_j$ des $M_{d-1}$ de $C_u$, analogues aux $\Delta$ pour $V_d$, relatives à un faisceau de $C^2$ dont A fait partie, A y jouant le rôle de $C_a$ pour $\{C_u\}$. Soit $M^j_{d+1}$ le lieu de $D_j$ quand $u$ varie. La frontière de $\sum \alpha_j M^j_{d+1}$ est un $\Gamma_d$ nul, aux $\lambda$ non nuls, pourvu que $\sum \alpha_j D_j$ ne soit pas une $M_{d-1}$ invariante. On montre, par une discussion simple, qu'il y a à exclure de ce chef $r_{d-1} - R_{d-1} + r_{d-2} - R_{d-2}$ cycles à $\lambda$ non nuls, de sorte qu'il en reste $n - 2(r_{d-1} - R_{d-1}) - (r_{d-2} - R_{d-2})$. La comparaison avec le nombre déjà obtenu nous donne de suite XI.

XII. *Formule de M. Alexander.* — $I_d$ étant l'invariant de Zeuthen-Segre de $V_d$,

$$R_d = I_d + 2(-1)^d(d-1) + 2\sum_1^d i(-1)^{d-1-i} R_i.$$

Soit $I_k$ l'invariant de Zeuthen-Segre de $C^{d-k}$. Les I seront définis par des relations récurrentes dont la dernière est

$$I_d = n - 2I_{d-1} - I_{d-2},$$

et les autres sont pareilles. On a de suite

$$(R_d - I_d) + 2(r_{d-1} - I_{d-1}) + (r_{d-2} - I_{d-2}) = 2(R_{d-1} + R_{d-2}).$$

Il suffit de multiplier la relation semblable pour $k$ par

$$(-1)^{k-1}.(k+1),$$

puis de sommer par rapport à $k$, afin d'arriver au résultat voulu. De là on déduit ensuite :

XIII. *Formule d'Euler-Poincaré :*

$$(-1)^d I_d + 2d = \sum (-1)^i R_i = \sum (-1)^i \alpha_i.$$

**10. Intersections des cycles.** — XIV. *Pour l'intersection des deux cycles*

$$\Gamma_d \sim \sum \lambda_i \Delta_i + M_d, \qquad (M_d A) = \Gamma_{d-2};$$

$$\Gamma'_d \sim \sum \lambda'_i \Delta_i + M'_d, \qquad (M'_d A) = \Gamma'_{d-2};$$

*on a, en posant*

$$(\delta_i \delta_k) = \alpha_{ik},$$

$$(\Gamma_d \Gamma'_d) = -\sum \lambda_i \lambda'_i - \sum_{\substack{i,k \\ i>k}} \alpha_{ik} \lambda_i \lambda'_k + (\Gamma_{d-2} \Gamma'_{d-2}),$$

*ou encore, avec des notations analogues à celles du Chapitre III*, n° 17,

$$(\bar{\Gamma}_d \Gamma'_d) = -\sum \lambda_i \bar{\lambda}_i + (\bar{\Gamma}_{d-2} \Gamma_{d-2}).$$

XV. *Si* $(\Gamma_k \Gamma_{2d-k})$ *est nul quel que soit* $\Gamma_{2d-k}$, $\Gamma_k$ *est nul ou diviseur de* $V_d$.

Grâce aux théorèmes des n^os 3, 4, 5, il suffit de prendre $k = d$. Supposons le théorème vrai pour les $V_{d-1}$ et étendons-le à $V_d$.

On aura d'abord, $\Gamma_{d-2}$ étant le cycle de A correspondant à $\Gamma_d$ et $\Gamma'_d$ quelconque dans $C_u$,

$$(\Gamma_d \Gamma'_d) = (\Gamma_{d-2} \Gamma'_d) = 0.$$

Donc $\Gamma_{d-2}$ est un diviseur de $C_u$. On peut remplacer $\Gamma_d$ par $t\Gamma_d$, $t$ étant un entier quelconque. En particulier, supposons-le choisi tel que $t\Gamma_{d-2} \sim 0 \pmod{C_u}$. Cela revient à supposer que déjà $\Gamma_{d-2} \sim 0 \pmod{C_u}$. Soit $M_{d+1}$ le lieu de la variété dont il forme frontière, quand $u$ varie. On aura

$$M_{d+1} \equiv \sum \lambda'_i \Delta_i + M^0_d \sim 0, \qquad (M^0_d A) = \Gamma_{d-2},$$

et au cycle

$$\Gamma_d - \sum \lambda'_i \Delta_i - M^0_d \sim \Gamma_d$$

correspond un $\Gamma_{d-2}$ nul de A. Supposons finalement cette condition déjà remplie par $\Gamma_d$. $\overline{\Gamma}_d$ étant quelconque,

$$(4) \qquad -(\Gamma_d \overline{\Gamma}_d) = \sum \lambda_i \overline{\lambda} i = 0,$$

le terme $(\Gamma_{d-2}\overline{\Gamma}_{d-2})$ étant maintenant absent à droite. A partir d'ici la démonstration s'achève comme pour $d = 2$ (Chap. III, n° 18).

*Remarque.* — Le théorème du Chapitre III, n° 19, ne s'étend qu'aux cycles à 1, 2, $2d-2$, $2d-1$ dimensions, les démonstrations se réduisant de suite à celles pour $d=2$; aussi ne nous y arrêterons-nous pas.

**11. Torsion.** — XVI. *Le groupe de torsion à $2d-1$ dimensions se réduit à l'identité; ceux à $k$ et $2d-k-1$ dimensions sont isomorphes. En particulier,*

$$\sigma_{2d-1} = 1, \qquad \sigma_k = \sigma_{2d-1-k}.$$

Le traitement des diviseurs $\Gamma_{2d-1}$ est le même que pour $d=2$. Pour les autres, la démonstration se fait encore par récurrence.

Prenons d'abord $k = d$. Soit pour commencer $\Gamma^0_{d-2}$ un cycle de A, diviseur de $C_u$. Je dis qu'*il existe un diviseur $\Gamma_d$ correspondant de* $V_d$. Soit $t$ l'entier tel que

$$t\Gamma^0_{d-2} \sim 0 \qquad (\mathrm{mod}\ C_u).$$

Le raisonnement du numéro précédent nous fournira un cycle

$$\Gamma'_d \sim \sum \lambda_i \Delta_i + M'_d \sim 0, \qquad (M'_d A) = t\Gamma^0_{d-2}.$$

Donnons-nous-en un autre quelconque

$$\overline{\Gamma}_d = \sum \overline{\lambda}_i \overline{\Delta}_i + \overline{M}_d, \qquad (\overline{M}_d A) = \overline{\Gamma}_{d-2}$$

Puisque $\Gamma'_d$ forme frontière

$$(\Gamma'_d \bar{\Gamma}_d) = -\sum \lambda_i \bar{\lambda}_i + (\Gamma^0_{d-2} \bar{\Gamma}_{d-2}) = 0.$$

Par suite, *quel que soit* $\bar{\Gamma}_d$,

$$\sum \lambda_i \bar{\lambda}_i \equiv 0 \qquad (\text{mod}\, t).$$

Raisonnant alors comme au Chapitre III, n° 19, on en déduit que les $\lambda$ sont tous divisibles par $t$, de sorte que l'on peut écrire

$$\Gamma'_d \sim t \sum \lambda_i \Delta_i + M'_d \sim 0.$$

On en conclut, d'après VI,

$$t \sum \lambda_i \delta_i \sim 0 \qquad (\text{mod}\, C_a).$$

Le cycle à gauche est invariant, donc la somme l'est aussi. Par suite, d'après III, elle forme frontière sur $C_a$. Soit

$$-M''_d \equiv \sum \lambda_i \delta_i.$$

Comme

$$-M'_d \equiv t \sum \lambda_i \delta_i.$$

$M'_d - tM''_d$ est un cycle de $C_a$. Soit $\Gamma''_{d-2}$ la trace de $M''_d$ sur A. $t\Gamma''_{d-2}$, trace du cycle en question, est invariant; donc $\Gamma''_{d-2}$ l'est aussi. Par suite, il existe un cycle $\Gamma''_d$ de $C_a$ dont $\Gamma''_{d-2}$ est la trace sur A. En définitive, $M'_d - t(M''_d - \Gamma''_d)$ coupe A suivant un

$$\Gamma_{d-2} \sim 0 \qquad (\text{mod}\, A),$$

donc

$$M'_d - t(M''_d - \Gamma''_d) \sim 0 \qquad (\text{mod}\, C_a \text{ ou bien } V_d),$$

et finalement

$$\Gamma_d \sim t\left(\sum \lambda_i \Delta_i + M''_d - \Gamma''_d\right) \sim 0.$$

La variété entre parenthèses constitue bien un diviseur du type annoncé, correspondant a $\Gamma^0_{d-2}$.

**12.** Soit maintenant $\Gamma_d$ un diviseur quelconque de $V_d$. Le cycle $\Gamma_{d-2}$ correspondant de A est un diviseur quelconque de $C_u$. Soit $\Gamma_d^0$ le diviseur correspondant obtenu à la manière ci-dessus. A $\Gamma_d - \Gamma_d^0 = \Gamma'_d$ correspond un cycle nul de A. Or la démonstration de notre théorème pour $d = 2$, appliquée ici, fournit ce résultat :

*Le sous-groupe des diviseurs $\Gamma_d$ correspondant à des $\Gamma_{d-2}$ nuls de A est isomorphe à celui des diviseurs $\Gamma_{d-1}$ qui, réduits à $C_u$, n'en sont pas des diviseurs.*

Soit donc $\Gamma_{d-2}^1, \Gamma_{d-2}^2, \ldots, \Gamma_{d-2}^{n'}$ un système fondamental de diviseurs de $C_u$. D'après XVI appliquée à $C_u$ (ceci est licite, puisque nous raisonnons par récurrence), il y en correspond un $\Gamma_{d-1}^1, \Gamma_{d-1}^2, \ldots, \Gamma_{d-1}^{n'}$, pour ses diviseurs $\Gamma_{d-1}$. A $\Gamma_{d-2}^i$ faisons correspondre un diviseur bien défini, $\Gamma_d^i$, de $V_d$. Soient enfin $\Gamma_d^{n'+1}, \Gamma_d^{n'+2}, \ldots, \Gamma_d^n$ et $\Gamma_{d-1}^{n'+1}, \Gamma_{d-1}^{n'+2}, \ldots, \Gamma_{d-1}^n$, les systèmes fondamentaux pour les diviseurs de l'énoncé ci-dessus, $t_i$ le plus petit entier positif tel que

$$t_i \Gamma_{d-1}^i \sim 0, \qquad t_i \Gamma_d^i \sim 0 \qquad (\mathrm{mod}\ V_d).$$

Les deux expressions

$$\sum_1^n {}_i\, \theta_i \Gamma_{d-1}^i, \quad \sum_1^n {}_i\, \theta_i \Gamma_d^i \qquad (\theta_i = 1, 2, \ldots, t_i)$$

représentent une fois et une fois seulement les diviseurs correspondants. Ceci démontre XVI pour $k = d$ ou $d - 1$.

13. Prenons maintenant $k = d - 2$. Il s'agit de comparer les diviseurs $\Gamma_{d-2}, \Gamma_{d-1}$. Pour commencer, tout diviseur $\Gamma_{d+1}$ a, pour trace sur $C_u$, un diviseur $\Gamma_{d-1}$ de $C_u$, diviseur d'ailleurs invariant. Mais tout diviseur $\Gamma_{d-1}$ de $C_u$ est invariant, car les nombres

$$(\Gamma_{d-1}\, \delta_i)$$

sont nuls pour lui. Donc à tout diviseur $\Gamma_{d-1}$ de $C_u$ en correspond un $\Gamma_{d+1}$ de $V_d$. Par suite, les groupes des diviseurs $\Gamma_{d+1}$ de $V_d$ et $\Gamma_{d-1}$ de $C_u$ sont isomorphes. D'un autre côté, $V_d$ et $C_u$ ont mêmes diviseurs $\Gamma_{d-2}$. Le théorème à démontrer se ramène donc au même pour $C_u$, vrai par hypothèse.

Puisque les diviseurs $\Gamma_{d-1-h}$ de $C_u$ sont certainement invariants, cette démonstration s'étend de suite à $k < d - 2$. Le théorème est donc complètement démontré.

## II. — Les hypersurfaces contenues dans les variétés algébriques.

**14.** Pour simplifier nous nous limiterons aux $V_3$, quoique la discussion s'étende de suite aux $V_d$ quelconques.

Soit donc une $V_3$ d'équation

$$f(x, y, z, t) = 0.$$

Au lieu de $|C|$ nous considérerons le système des sections hyperplanes $|H|$, avec $\{H_z\}$ à la place de $\{C_u\}$. Enfin la courbe d'intersection de $H_y$ avec $H_z$ sera désignée par $H_{yz}$.

Parmi les intégrales de première espèce de $H_{yz}$, il y en aura $p - q$ distinctes :

$$u_{q+h} = \int \frac{P_h(x, y, z, t)}{f'_t} dx \qquad (h = 1, 2, \ldots, p - q),$$

où $P_h$ est un polynome adjoint d'ordre $m - 3$ en $x, y, t$, et en fait en $z$ aussi, mais peu importe. Bien entendu $2q = R_1$, indice linéaire commun à $V_3$ et à $H$, et $p$ est le genre des sections planes.

**15.** Soient $\delta_1, \delta_2, \ldots, \delta_{2p-2q}$ un système de cycles linéaires évanouissants, indépendants, de $H_{yz}$ ; $\gamma_1, \gamma_2, \ldots, \gamma_{2q}$ un système de cycles invariants, indépendants. Les $\gamma$ et les $\delta$ forment ensemble $2p$ cycles indépendants de la courbe. Les périodes des intégrales $u_{q+h}$ par rapport aux $\gamma$ sont nulles.

Adjoignons à nos intégrales, $q$ autres, $u_1, u_2, \ldots, u_q$, de manière à en avoir $p$ linéairement indépendantes, et soit

$$\Omega = \| \omega_{i\mu} \| \qquad (i = 1, 2, \ldots, p; \ \mu = 1, 2, \ldots, 2p)$$

la matrice aux périodes. Les colonnes se rapportent dans l'ordre aux cycles $\gamma_1, \ldots, \gamma_{2q}, \delta_1, \ldots, \delta_{2p-2q}$, de sorte que

$$\omega_{h\mu} = 0 \qquad (h = q + 1, \ldots, p; \ \mu = 1, 2, \ldots, 2q).$$

On peut choisir $u_1, u_2, \ldots, u_q$, de manière que $\Omega$ ait la forme

$$\left\| \begin{matrix} \Omega_1, & 0 \\ 0, & \Omega_2 \end{matrix} \right\|,$$

où $\Omega_1$ est une matrice à périodes de genre $q$, $\Omega_2$ la matrice de genre $p-q$ aux périodes des $u_{q+h}$, les zéros représentant des matrices à termes tous nuls. Le choix de $u_1, u_2, \ldots, u_q$ ne peut se faire que d'une seule manière, en ce sens que tout autre système d'intégrales satisfaisant à la condition voulue est composé d'intégrales combinaisons linéaires des précédentes.

Or, comme il est connu, $\Omega$ est de rang $p$, donc $\Omega_1$ est de rang $q$. Par suite il y aura $q$ cycles invariants, soient $\gamma_1, \gamma_2, \ldots, \gamma_q$, par rapport auxquels le déterminant des périodes de $u_1, u_2, \ldots, u_q$, n'est pas nul. Une quelconque de ces intégrales peut donc être définie ainsi : Quand on y fait $y = y_0$, c'est-à-dire quand on la suppose attachée à la courbe fixe $H_{y_0 z}$, elle doit se réduire à une intégrale à périodes données, d'ailleurs arbitraires, par rapport à $\gamma_1, \gamma_2, \ldots, \gamma_q$, et à périodes nulles par rapport aux $\delta$. En se rapportant au traitement des intégrales simples des surfaces, on voit qu'on a ainsi défini $q$ intégrales de différentielles totales de première espèce pour $H_z$.

Supposons en particulier que $H_{yz}$ devienne la courbe à l'infini, soit A, de $H_z$. On aura déterminé, sur toute $H_z$, une intégrale de différentielles totales de première espèce, à périodes données par rapport aux $\gamma$. Le raisonnement du Chapitre IV, nº 19, s'applique maintenant *verbatim*. Il n'y a qu'à prendre pour origine d'intégration un point quelconque à l'infini de $H_{yz}$ et l'on montrera ici encore que $u_1, u_2, \ldots, u_q$ ne sont autres que les valeurs prises par certaines intégrales de différentielles totales de première espèce de $V_3$ quand on maintient $y$ et $z$ fixes. Rappelons que par définition ces intégrales sont du type

$$\int R\,dx + S\,dy + T\,dz,$$

où R, S, T sont des fonctions rationnelles, telles que sous le signe d'intégration on ait une différentielle totale. Ainsi (Castelnuovo-Enriques), $V_3$ *possède $q$ intégrales de différentielles totales de première espèce.*

Les intégrales $u_1, u_2, \ldots, u_q$ adjointes aux $u_{q+h}$ fournissent $p$ intégrales distinctes de première espèce de $H_{yz}$.

16. Soit maintenant D une surface algébrique quelconque. On aura pour les sommes abéliennes sur $H_{yz}$ prises à partir d'un des points à l'infini de la courbe jusqu'aux points variables d'intersection avec D,

$$v_h(y, z) = \xi_h \qquad (h = 1, 2, \ldots, q),$$

$$v_{q+h}(y, z) = \sum \frac{\lambda_k}{2\pi i} \int_a^{a_k(z)} \frac{\Omega_{hk}(Y, z)}{Y - y}\, dz,$$

les notations étant les mêmes qu'au Chapitre IV. Les $\xi$, uniformes, partout finies, sont ici aussi des constantes.

Les $\lambda$ correspondent à un certain cycle *invariant*, algébrique de $H_z$,

$$\Gamma_2 \sim \sum \lambda_k \Delta_k + (H_{az}).$$

Ces conditions *nécessaires* pour qu'aux $v$ corresponde une surface algébrique D sont aussi *suffisantes*, c'est-à-dire *si* $\Gamma_2$ *est invariant, algébrique* (¹), *il lui correspond une surface algébrique* D *coupant* $H_{yz}$ *en des points variables aux sommes abéliennes* $v$.

En effet, $\Gamma_2$ détermine sur chaque $H_z$ un système $\{d\}$ de courbes, dont l'une $d_0$ bien déterminée coupe $H_{yz}$ en un nombre de points variables $\leqq p$, aux sommes abéliennes $v$. $d_0$ est complètement définie par son comportement aux points à l'infini de $H_{yz}$ et par le groupe de ses points à l'infini, puisque de ces derniers on déduit de suite les $\xi$. Ainsi $\{d\}$ étant connu, $d_0$ y est déterminée par des conditions rationnelles par rap ort à $z$. D'ailleurs $\{d\}$ lui-même est défini rationnellement par rapport à $z$, car l'ordre $(d^2)$ de ses courbes est égal à $(\Gamma^2)$, donc bien déterminé. Or les courbes d'un ordre donné d'un $S_3$ se partagent en un nombre fini de familles algébriques (Nœther, Halphen) (²). Si l'on exprime que les courbes

(¹) Il faut en outre que $\Gamma_2$ soit homologue à une courbe effective. Dans le cas contraire on remplacera $\Gamma_2$ par $\Gamma_2 + k\,\mathrm{HH}z$ auquel correspondra une surface effective D lorsque $k$ est suffisamment elevé. A $\Gamma_2$ correspond alors la surface virtuelle $\mathrm{D} - k\mathrm{H}$.

(²) Par toute $d$ il passe, outre $H_z$ une surface d'ordre $\mu$ déterminé, $F_\mu$. On

de l'une d'elles appartiennent à une surface donnée, on impose, aux paramètres dont elles dépendent, des équations rationnelles par rapport aux coefficients de la surface, donc ici par rapport à $z$. Comme $\{d\}$ est unique sur $H_z$, les paramètres correspondants devront être des fonctions rationnelles de $z$ et algébriques d'un certain nombre d'entre eux.

En définitive, l'ensemble des conditions imposées à $d_0$ est rationnel par rapport à $z$. $d_0$ est donc définie par une équation $\varphi(x, y, z, t) = 0$, où $\varphi$ est un polynome en $x$, $y$, $t$, à coefficients rationnels en $z$; on peut le supposer polynome en $z$ aussi. Le lieu D, de $d_0$, intersection partielle ou totale des deux variétés algébriques $f = 0$, $\varphi = 0$, est bien une surface algébrique.

17. On définira, comme pour une $V_2$, les $\Gamma_4$ *algébriques* comme homologues à une somme ou une différence de surfaces algébriques de $V_3$.

*Pour que* $\Gamma_4$ *soit algébrique, il faut et il suffit qu'il coupe* $H_z$ *suivant un* $\Gamma_2$ *algébrique.*

La condition est évidemment nécessaire. Elle est aussi suffisante car quand elle est remplie il y a une surface algébrique D coupant $H_z$ suivant une courbe $d$, et passant, soit $k$ fois, par la courbe à l'infini. Alors $\Gamma_4 + kH$ coupe toute $H_z$ suivant un cycle homologue à $DH_z$. Donc

$$\Gamma_4 + kH \sim D$$

ce qui démontre notre affirmation.

*Deux surfaces coupant* $H_z$ *suivant des courbes d'un même système linéaire sont elles-mêmes contenues dans un même système linéaire* (même démonstration que pour une $V_2$).

L'équivalence des surfaces et les nombres $\rho$, $\sigma$ se définissent comme pour les courbes d'une $V_2$.

---

obtient $\{d\}$ en exprimant que l'intersection de $F_\mu$ et $H_z$ se décompose en deux courbes d'ordre donné, qui se rencontrent en un certain nombre de points, etc. Les paramètres dont $\{d\}$ dépend sont alors clairement algébriques en $z$, donc (voir le texte) rationnels en cette variable.

*Pour que deux surfaces* C, D *soient équivalentes, il faut et il suffit que leurs intersections* $CH_z$, $DH_z$, *avec* $H_z$, *en soient des courbes équivalentes.*

La condition est évidemment nécessaire; montrons qu'elle est suffisante. Soit $g$ une courbe de $H_z$ telle que $CH_z + g$ et $DH_z + g$ appartiennent à un même système continu de courbes à $\infty^q$ systèmes linéaires. En vertu de l'hypothèse, $g$ existe certainement. Alors pour $k$ suffisamment grand $CH_z + g + (kHH_z - g)$ et $DH_z + g + (kHH_z - g)$ seront des courbes effectives ayant la même propriété; c'est-à-dire que $(C + kH)H_z$ et $(D + kH)H_z$ seront contenues totalement dans un même système continu possédant $\infty^q$ systèmes linéaires. Chacun de ces systèmes linéaires est individualisable par une courbe unique, se comportant de manière donnée en certains points; peu importe d'ailleurs, il nous suffit de savoir qu'elle est déterminée rationnellement sur $H_z$. En raisonnant comme plus haut, on voit que cette courbe a pour lieu une surface algébrique coupant $H_z$ en outre suivant la courbe à l'infini comptée un certain nombre de fois. En faisant varier les courbes, on arrive ainsi à un système continu de surfaces, coupant $H_z$ suivant un système continu de courbes, dont une est équivalente à $(C + k'H)H_z$ et une autre à $(D + k''H)H_z$. Chacune de ces surfaces définit à son tour un système linéaire unique, dont un contient C, laissant comme résidu $k'H$, tandis qu'un autre contient D, avec un résidu $k''H$. Par suite,

$$C + k'H = D + k''H.$$

De ceci résulte

$$(CH_z) - (DH_z) \sim (k' - k'').HH_z \sim 0 \qquad (\text{mod } H_z).$$

Mais $HH_z$ ne peut être un cycle nul ou un diviseur de $H_z$. Par suite,

$$k' = k'', \qquad C = D.$$

Corollaires. — I. *Des quatre relations*

$$C = D; \qquad CH_z = DH_z; \qquad C \sim D \quad (\text{mod } V_3); \qquad CH_z \sim DH_z \quad (\text{mod } H_z)$$

*une quelconque entraîne les trois autres.*

En effet, elles sont toutes équivalentes à la seconde.

II. *Les diviseurs $\Gamma_4$ sont tous algébriques et leur groupe est isomorphe à celui des diviseurs pour les surfaces.* En particulier $\sigma_4 = \sigma_1 = \sigma$.

En effet, un diviseur $\Gamma_4$ coupe $H_z$ suivant un diviseur $\Gamma_2$, algébrique comme nous l'avons vu (Chap. IV, n° **31**); donc $\Gamma_4$ l'est aussi. Soient C, D deux hypersurfaces telles que $\Gamma_4 \sim C - D$. Le plus petit entier positif $t$, tel que $t\Gamma_4 \sim 0$, est aussi celui tel que $t(C - D) = 0$, ce qui démontre notre affirmation.

III. *La variété et ses sections hyperplanes ont même nombre $\sigma$.*

Ceci résulte de ce qu'elles ont même nombre $\sigma_1$, égal pour chacune à leur nombre $\sigma$.

*Remarque.* — Tout ce qui précède reste vrai lorsque $|H|$ est remplacé par le système $|C|$ considéré maintes fois.

**18.** La variété peut posséder des intégrales doubles dites « de première espèce », intégrales partout finies, de forme

$$\int\int R\,dy\,dz + S\,dz\,dx + T\,dx\,dy, \tag{5}$$

où R, S, T sont rationnelles et satisfont à la « condition d'intégrabilité »

$$\frac{\partial R}{\partial x} + \frac{\partial S}{\partial y} + \frac{\partial T}{\partial z} = 0.$$

L'intégrale double relative à $H_z$,

$$\int\int T\,dx\,dy, \tag{6}$$

est de première espèce. Pour que $\Gamma_4$ soit algébrique, il faut que la période de toute intégrale (6) par rapport à $\Gamma_4 H_z$ soit nulle, ou bien, ce qui revient au même, que celle de toute intégrale (5) par rapport au même cycle le soit. *Cette condition nécessaire est-elle suffisante?* Nous ne pouvons l'affirmer et pour le moment la question doit rester en suspens. Tout ce que l'on peut dire c'est que *le nombre $\rho$ est au plus égal à celui des $\Gamma_2$ à périodes d'inté-*

*grales doubles toutes nulles.* Ainsi au lieu d'une valeur précise, comme c'était le cas pour les surfaces, nous n'avons plus cette fois qu'un maximum de $\rho$.

*Remarque.* — Tout ce que nous avons dit s'étend à une $V_d$ quelconque. Il suffira de considérer au lieu des $\Gamma_4$ les $\Gamma_{2d-2}$ et au lieu de (5) les intégrales

$$\sum \int\int R_{ik}\, dx_i\, dx_k$$

avec

$$R_{ik} = -R_{ki}, \qquad \frac{\partial R_{ik}}{\partial x_l} + \frac{\partial R_{kl}}{\partial x_i} + \frac{\partial R_{li}}{\partial x_k} = 0.$$

## III. — Un théorème sur les cycles algébriques des surfaces d'une $V_3$. Application aux courbes des surfaces non singulières de l'espace ordinaire.

**19.** $V_3$ étant quelconque, reprenons le système $|C|$ du paragraphe I et faisons-y varier $\{C_u\}$. Soit $r$ la dimension de $|C|$ $(r > 3)$. Les surfaces C, à singularités nouvelles, y formeront un système algébrique $\infty^{r-1}$, $\sum$, que nous supposerons irréductible. Cette condition sera certainement remplie si $r \geqq 4$. En effet, $|C|$ étant irréductible, n'est pas *composé* avec les surfaces d'un système algébrique de dimension moindre, c'est-à-dire que C générique ne se compose pas de plusieurs surfaces C′ appartenant à un système de dimension $\leqq 2$. Donc les C ayant un point singulier en un point de $V_3$ forment un système linéaire $\infty^{r-4}$ et l'ensemble de ces systèmes linéaires est un système $\infty^{r-1}$ composé d'$\infty^3$ systèmes linéaires. Ce système est représentable dans un $S_r$ par une $V_{r-1}$ lieu d'$\infty^r S_{r-4}$. La section de cette $V_{r-1}$ par un $S_4$ de $S_r$ est une $V_3$ birationnellement équivalente à celle donnée, donc irréductible. Par suite, $V_{r-1}$ est elle-même irréductible. Comme ses points correspondent biunivoquement aux éléments de $\sum$, ceci démontre notre affirmation.

**20.** Théorème. — *Si l'un des* $\Gamma_2$ *évanouissants,* $\delta_i$ *de* $C_u$, *est algébrique, ils le sont tous.*

Déplaçons $\{C_u\}$ dans $|C|$ de façon à permuter $C_{a_i}$ avec $C_{a_k}$, ce qui peut se faire grâce à l'irréductibilité du système $\sum$ ci-dessus. Le cycle $\delta_i$ se permutera avec $\delta_k$. Pour préciser, $u$ étant voisin de $a_i$ un certain cycle $\delta_i$ de $C_u$ tend à devenir nul quand $u$ tend vers $a_i$. Ce cycle sera alors devenu pour $u$ voisin de $a_k$ le cycle semblable $\delta_k$ relatif à ce point.

Remarquons que le genre géométrique $p_g$ de $C_u$ ne peut changer que pour un nombre fini de valeurs de $u$, comme il résulte de suite de ce qu'elles seront déterminées par des relations algébriques. Soit $r_2$ l'indice de connexion à deux dimensions d'une C générique. On peut se donner $p_g$ intégrales doubles distinctes de première espèce attachées à $C_u$, à éléments différentiels rationnels par rapport à $u$. Leurs périodes seront des fonctions analytiques de $u$, méromorphes pour toute valeur non critique. Le rang de la matrice aux périodes relatives à $r_2$ cycles distincts est égal à $r_2 - \rho$, où $\rho$ est le nombre de Picard de $C_u$. Ce rang ne peut varier que pour un ensemble dénombrable de valeurs de $u$, puisqu'il appartient à l'ensemble des zéros communs à plusieurs fonctions analytiques (les déterminants de la matrice) se conduisant comme les périodes. Donc le nombre $\rho$ de $C_u$ ne peut varier que pour un ensemble dénombrable de ces surfaces.

On en conclut que les surfaces C, à nombre $\rho$ autre que celui de C générique, se partagent en un ensemble dénombrable de familles $\infty^{r-1}$ au plus.

Revenons au déplacement de $\{C_u\}$ dans $|C|$ considéré plus haut. D'après ce qui précède, pendant qu'il s'accomplit, on pourra faire varier une $C_u$ d'une position voisine de $C_{a_i}$ à une autre voisine de $C_{a_k}$, sans changer $\rho$, donc le nombre de cycles algébriques distincts, à aucun moment du déplacement. Par suite, si $\delta_i$ est algébrique pour $u$ voisin de $a_i$, $\delta_k$ l'est aussi pour $u$ voisin de $a_k$.

Mais si un cycle est algébrique pour $u$ dans une certaine région, il l'est encore pour $u$ arbitraire, car la condition pour qu'il le soit s'exprime par l'évanouissement identique de certaines fonctions analytiques, les périodes des $p_g$ intégrales doubles de première espèce. Donc enfin le cycle $\delta_k$ de $C_a$, évanouissant quand $u$ tend vers $a_k$ le long de $aa_k$, est algébrique pourvu que le cycle semblable $\delta_i$ le soit. Ceci démontre notre théorème.

**21.** Supposons que les $\delta$ ne soient pas algébriques. Alors tout cycle algébrique est invariant, car autrement son accroissement quand $u$ tourne autour de $a_i$ serait un cycle $\lambda\delta_i$ nécessairement algébrique; $\delta_i$ en serait donc un aussi puisque les périodes correspondantes des intégrales de première espèce, tout comme celles relatives à $\lambda\delta_i$, devraient toutes être nulles. Dans ces conditions, les $\Gamma_2$ algébriques de C sont les traces des $\Gamma_4$ algébriques de $V_3$; par suite, les nombres $\rho$ de $V_3$ et de C sont égaux. Ainsi *lorsque les cycles évanouissants de* C *ne sontpas algébriques, ses* $\Gamma_2$ *algébriques sont invariants et traces des* $\Gamma_4$ *algébriques de* $V_3$.

Dans les mêmes conditions, *tout système fondamental pour les courbes de* C *est trace d'un tel système pour les surfaces de* $V_3$.

Supposons au contraire les $\delta$ algébriques. Il résulte alors de l'étude de la topologie de $V_3$ que *tous les* $\Gamma_2$ *de* C *dépendent des cycles invariants et des cycles algébriques.*

**22.** Application a un théorème de Noether ([1]). — *Une surface générique* $F_m$, *d'ordre* $m > 3$, *d'un* $S_3$, *ne possède que des courbes intersections complètes avec d'autres surfaces de l'espace.*

L'espace $S_3$, image des valeurs complexes de trois variables, est une $V_3$ dont la région à l'infini se compose d'un $S_2$ (plan), lui-même homéomorphe à l'image des valeurs complexes de deux variables. Tout $\Gamma_k$ est déformable en une multiplicité en entier à l'infini, ou en définitive dans un $S_2$ quelconque de $S_3$. Si sa dimension $k \leqq 2$, le même raisonnement peut être continué. On en tire de suite $R_{2k+1} = 0$, $R_{2k} = 1$. Par exemple $R_1 = 0$, car tout $\Gamma_1$, finalement réductible à un $S_1$ (droite), y est déformable en un point. Enfin, ceci nous intéresse surtout, tout $\Gamma_4$ étant homologue à un multiple d'un $S_2$, est nécessairement algébrique. Tout cycle invariant de $F_m$ est multiple de celui formé par une section plane H,

([1]) *Voir* Nœther, Mémoire du prix Steiner; G. Fano, *Torino Atti*, 1908, ainsi que mon Mémoire couronné. La démonstration ci-dessus est plus simple, mais ne va pas aussi loin.

car c'est son intersection avec le $\Gamma_4$ fondamental de $S_3$. Les $\Gamma_2$ invariants sont donc tous algébriques.

Le système $|F_m|$ est $\infty^4$ au moins, irréductible; donc on peut appliquer le théorème du n° **20**. Supposons que tous ses $\Gamma_2$ évanouissants soient algébriques. Alors tous les $\Gamma_2$ le seront. Je dis que pour $m > 3$ et $F_m$ arbitraire, cela n'a pas lieu. Il suffit pour cela de montrer qu'il y a toujours une $F_m$ possédant des $\Gamma_2$ non algébriques.

Soient $\varphi(x, y, z)$, $\psi(x, y, z)$ deux polynomes réels de degrés respectifs $m - 2$ et $m$, tels que

$$\varphi(0, 0, 0) > 0, \qquad \psi(0, 0, 0) > 0,$$

mais à part cela absolument quelconques. Soit en outre

$$x^2 + y^2 + z^2 = r^2$$

une sphère réelle, de rayon assez petit pour que $\varphi$ et $\psi$ soient positives à son intérieur et sur elle. La surface $F_m$

$$f(x, y, z) = (x^2 + y^2 + z^2 - r^2)\varphi + \varepsilon\psi = 0,$$

où $\varepsilon$ est un nombre positif très petit, comprend une ovale réelle très voisine de notre sphère et intérieure à elle. Cette ovale est un $\Gamma_2$ de la surface et je dis qu'il n'est pas algébrique. En effet,

$$\iint \frac{dx\,dy}{f'_z} \tag{7}$$

en est une intégrale de première espèce, car il n'y a pas ici de courbe multiple, de sorte que tout polynome est adjoint; si son degré est $\leqq m - 4$, il lui correspond bien une intégrale de première espèce. L'intégrale étendue a $\Gamma_2$ diffère très peu de sa valeur étendue à la sphère, c'est-à-dire de

$$\iint \frac{dx\,dy}{2z\,\varphi(x, y, z)}$$

étendue à la sphère. En coordonnées polaires, cette intégrale devient

$$r\int_0^{\pi}\int_0^{2\pi} \frac{\sin\theta_1\,d\theta_1\,d\theta_2}{\varphi(r\sin\theta_1\cos\theta_2,\ r\sin\theta_1\sin\theta_2,\ r\cos\theta_1)}.$$

Comme $\sin\theta_1$ et $\varphi$ sont positifs sur toute la sphère, l'intégrale est essentiellement positive et il en est de même pour la période de (7) par rapport à $\Gamma_2$. Ce cycle n'est donc effectivement pas algébrique.

Pour $m = 3$, il n'y a pas d'intégrales de première espèce et le procédé ne réussit plus.

On doit donc conclure que, pour $m \geqq 4$, les cycles algébriques de $F_m$ générique sont invariants, donc multiples de la trace H du $\Gamma_4$ fondamental de $S_3$. Ainsi, quel que soit C, courbe de $F_m$, on aura $C = kH$. Or $F_m$ est régulière, puisque son indice $R_1$ est nul comme celui de $S_3$. Par suite, C et $kH$ sont contenues dans un même système linéaire. Prenons pour H la courbe à l'infini. Il y aura une fonction rationnelle attachée à $F_m$,

$$\frac{P(x, y, z)}{Q(x, y, z)} \qquad (\text{P, Q polynomes}),$$

s'annulant sur C et infinie sur $kH$, c'est-à-dire finie à distance finie. P devra donc s'annuler en tous les points de $F_m$ où Q s'annule. D'après l'extension d'un théorème classique de Nœther [1],

$$P = Q.P_1 + P_2 f \qquad (P_1,\ P_2 \text{ polynomes}).$$

Donc, sur $F_m$,

$$\frac{P}{Q} = P_1.$$

Ainsi $P_1$ est un polynome qui a pour zéro simple sur $F_m$ la seule courbe C. Cela revient à dire que C est l'intersection complète de $F_m$ avec la surface $P_1(x, y, z) = 0$. Ceci achève la démonstration.

Je renvoie le lecteur à mon Mémoire couronné pour diverses extensions et applications se rattachant à ce qui précède.

---

(1) Pour la démonstration, *voir* PICARD et SIMART, vol. II, p. 17.

# CHAPITRE VI.

## L'ANALYSIS SITUS ET LES FONCTIONS ABÉLIENNES [1].

### I. — Théorème d'existence des fonctions abéliennes.

1. Étant donnée une matrice à $p$ lignes et $2p$ colonnes

$$\Omega = \| \omega_{j\mu} \| \qquad (j = 1, 2, \ldots, p;\ \mu = 1, 2, \ldots, 2p),$$

existe-t-il des fonctions $2p$-uplement périodiques qui lui appartiennent, c'est-à-dire des fonctions méromorphes de $p$ variables $u_1, u_2, \ldots, u_p$, invariantes quand on ajoute simultanément aux $u$ les éléments d'une ligne quelconque de $\Omega$? La réponse constitue le théorème de Weierstrass, démontré pour la première fois par MM. Poincaré et Picard :

*Pour qu'il existe des fonctions de la nature voulue, il faut et il suffit : 1° qu'il y ait une forme alternée à coefficients entiers*

$$(1) \qquad \sum_{1}^{2p} {}^{\mu,\nu}\, c_{\mu\nu}\, x_\mu\, y_\nu; \quad c_{\mu\nu} = -c_{\nu\mu},$$

*s'annulant quand on y remplace les $x$ et les $y$ par les éléments de deux lignes quelconques de $\Omega$; 2° que si $\xi_\mu + i\eta_\mu$ sont les périodes d'une combinaison linéaire quelconque des $u$,*

$$(2) \qquad \sum c_{\mu\nu}\, \xi_\mu\, \eta_\nu > 0.$$

Nous allons reprendre la question au point de vue de l'*Analysis*

---

[1] *Voir* FROBENIUS, *Journal de Crelle*, vol. 79, 1884. — WIRTINGER, *Monatshefte für Math. und Physik*, vol. VI, 1895. — POINCARÉ, *Acta mathematica*, vol. XXVI, 1902. — LEFSCHETZ, Mémoire couronné, 2e Partie Chap. I.

*Situs*. Outre une précision majeure des résultats connus, nous en obtiendrons de nouveaux, notamment sur la distribution des fonctions périodiques et à multiplicateurs. De plus, ce nouveau traitement se recommande par son élégance et sa simplicité. Les seuls théorèmes d'*Analysis Situs* dont nous nous servirons sont assez élémentaires : théorèmes sur les intersections de cycles, sur les systèmes fondamentaux pour l'anneau et généralisation du résultat du Chapitre II, n° 6.

2. Dans l'$S_{2p}$ réel, où l'on représente les valeurs complexes des $u$, les zéros d'une fonction entière ou méromorphe, $\varphi(u_1, u_2, \ldots, u_p)$ (¹), ou, pour abréger $\varphi(u)$, ont pour image une variété analytique $W_{2p-2}$. Nous allons faire ces hypothèses :

*a*. $W_{2p-2}$ existe et est invariante quand on ajoute des périodes quelconques aux $u$.

*b*. Elle n'est pas *cylindrique*, c'est-à-dire n'est pas le lieu d'un espace linéaire se déplaçant parallèlement à lui-même. En particulier, il n'y a pas de combinaison linéaire des $u$ constante sur $W_{2p-2}$.

Dans ces circonstances, je dis que les conditions classiques d'existence sont satisfaites. Ces hypothèses, un peu plus générales que celles du n° 1, ont l'avantage de couvrir, outre les fonctions périodiques, les fonctions à multiplicateurs (fonctions thêta, fonctions intermédiaires).

3. Les points de $S_p$ définis par

$$u_j = t_1\omega_{j1} + t_2\omega_{j2} + \ldots + t_{2p}\omega_{j2p} \qquad (0 \leqq t_\mu < 1)$$

y constituent un domaine fondamental $U_{2p}$, pour nos fonctions, au même titre que, par exemple, le parallélogramme des périodes pour les fonctions elliptiques. Je dis que $U_{2p}$ est un parallélépipède fini. En effet, dans le cas contraire, les points donnés par les mêmes formules, mais où les $t$ prennent toutes les valeurs entières possibles, forment un ensemble partout dense, contenu dans un certain $S_k$ $(1 \leqq k < 2p)$. Soient $(u)$ et $(u')$ un point de $W_{2p-2}$ et

(¹) POINCARÉ, *Acta mathematica*, 1883, p. 97. — COUSIN, *Ibid.*, 1895, p. 1.

de l'ensemble. D'après $a$, le point $(u+u')$ appartient aussi à $W_{2p-2}$. Ainsi, par $(u)$ passe un $S_k$ contenant un ensemble partout dense qui appartient à la variété. Soient D, $(v)$, $(v')$ une droite quelconque de cet $S_k$ et deux de ses points. Quand la variable $z$ prend toutes les valeurs réelles possibles, le point

$$\frac{v_i + z v'_i}{1+z}$$

décrit D tout entière. En vertu de ce qui précède, tout point de $S_k$, donc aussi de D, est un point limite pour les zéros de $\varphi$. La fonction $\varphi\left(\frac{v+zv'}{1+z}\right)$ est méromorphe en $z$ [1] et très petite pour $z$ réelle et $\neq -1$. Donc elle s'annule identiquement dans ce cas. On en conclut que D appartient à $W_{2p-2}$, et il en est de même pour l'$S_k$ considéré tout entier. Ainsi, par tout point de $W_{2p-2}$, passe un $S_k$ qui lui appartient en entier, ce qui contredit $b$. *Notre affirmation est donc démontrée.*

*Corollaire.* — Désignons par $\bar{x}$ le conjugué d'un nombre $x$. Le déterminant d'ordre $2p$, formé avec $\Omega$ et la matrice aux éléments conjugués,

$$\begin{vmatrix} \omega_{11} & \omega_{12} & \dots & \omega_{1,2p} \\ \dots & \dots & \dots & \dots \\ \bar{\omega}_{11} & \bar{\omega}_{12} & \dots & \bar{\omega}_{1,2p} \\ \dots & \dots & \dots & \dots \end{vmatrix} \neq 0.$$

4. A $U_{2p}$ correspond un anneau à $2p$ dimensions, où $W_{2p-2}$ sera représentée par un $\Gamma_{2p-2}$, C. En effet, $\varphi$ s'annule sur un nombre fini de multiplicités de $U_{2p}$, car autrement il y aurait au moins un point $(u^0)$, du domaine, au voisinage duquel passent une infinité de ces multiplicités; $(u^0)$ serait alors un point singulier essentiel de $\varphi$, alors qu'on l'a supposée méromorphe à distance finie. Quand on déforme $U_{2p}$ en un anneau, ces multiplicités se raccordent (hypothèse $a$, n° 2) et donnent bien lieu à un $\Gamma_{2p-2}$.

On peut étendre aux cycles C les considérations du Chapitre II, n° 6, et leur attribuer une *orientation type*, ce que nous suppose-

---

(1) Ceci résulte de la représentation de la fonction méromorphe $\varphi(u)$ par le quotient de deux fonctions entières (POINCARÉ, COUSIN, *loc. cit.*).

rons fait une fois pour toutes. Ces cycles et leurs intersections seront alors des multiplicités bien définies. En particulier, pour une orientation convenable de l'anneau, que nous supposerons choisie une fois pour toutes, *les nombres arithmétiques et algébriques d'intersections de p tels cycles sont égaux*.

5. Numérotons les arêtes de $U_{2p}$ issues d'un même sommet O, et sur la $i^{\text{ième}}$ marquons, près de O, un certain point $A_i$. A la pyramide à faces planes $P_{2p} = OA_1 A_2 \ldots A_{2p}$ de l'espace des $u$ correspond une $P'_{2p}$ de l'anneau. Nous supposerons l'ordre des $A_i$ tel que $P'_{2p}$ soit une indicatrice de l'anneau. De même à $P_k = OA_{i_1} A_{i_2} \ldots A_{i_k}$ correspond une $P'_k$ dans l'anneau située dans le cycle déterminé par les arêtes correspondantes (Chap. II, n° 12) et qui, par définition, lui servira d'indicatrice. Ce $\Gamma_k$ ainsi orienté sera dénoté par $(i_1, i_2, \ldots, i_k)$.

Il résulte de suite des théorèmes les plus élémentaires sur les intersections des variétés que :

1° Si les nombres $i_1, i_2, \ldots, i_{2p}$ sont tous distincts,

$$(i_1, i_2, \ldots, i_k)(i_{k+1}, i_{k+2}, \ldots, i_{2p}) = (-1)^n,$$

où $n$ est le nombre de transpositions dans la permutation

$$\begin{pmatrix} 1, & 2, & \ldots, & 2p \\ i_1, & i_2. & \ldots, & i_{2p} \end{pmatrix}.$$

2° Si les $i$ ne sont pas distincts,

$$(i_1, i_2, \ldots, i_k)(i_{k+1}, \ldots, i_{2p}) = 0,$$

car alors, en faisant subir au besoin à l'$S_k$ de l'espace des $u$, image du premier cycle, une translation convenable, il ne rencontrera plus l'espace semblable relatif au second.

6. Ceci posé, pour chaque dimension, les cycles du type précédent forment un système fondamental. Il suffit d'ailleurs de prendre pour chaque combinaison d'indices un seul des cycles correspondants. Nous choisirons celui aux $i$ rangés en ordre croissant. Alors

$$C \sim \sum m'_{i_1 i_2} (i_3. i_4, \ldots, i_{2p}),$$

avec $i_1 < i_2$, $i_3 < i_4 < \ldots < i_{2p}$, et pour chaque terme de la somme les $2p$ nombres $i$ sont tous différents. Posons

$$[C(\mu, \nu)] = m_{\mu\nu} = -m_{\nu\mu}.$$

On trouve de suite

$$(-1)^n [C(i_1, i_2)] = (-1)^n m_{i_1 i_2} = m_{i_1 i_2};$$

par suite,

$$C \sim \sum (-1)^n m_{i_1 i_2}(i_3, i_4, \ldots, i_{2p}) \qquad (i_1 < i_2;\ i_3 < i_4 \ldots < i_{2p}).$$

L'ensemble des zéros de la fonction $\varphi(u_1 - g_1, \ldots, u_p - g_p)$ aux constantes $g$ arbitraires [fonction que nous dénoterons plus simplement par $\varphi(u-g)$] remplit encore une multiplicité analytique représentée dans l'anneau par un cycle $C' \sim C$. Je dis qu'en général l'intersection des deux cycles est un $\Gamma_{2p-4}$. En effet, autrement, les deux multiplicités $\varphi(u) = 0$, $\varphi(u-g) = 0$ ont en commun dans $U_{2p}$ une $M_{2p-2}$, ceci *quelles que soient les* $g$. Cette multiplicité doit être une de celles en nombre fini correspondant à C. Comme elle doit varier continûment avec les $g$, elle est fixe. Il y a donc un point $(u^0)$ tel que $\varphi(u^0 - g) = 0$, quelles que soient les $g$. Cela revient à dire que $\varphi(u)$ est identiquement nulle, circonstance à écarter. Notre affirmation est donc correcte et, conformément à nos conventions, nous pouvons parler du cycle CC′, que nous écrirons $C^2$ pour une raison évidente.

On démontre de même l'existence des cycles $C^3, C^4, \ldots, C^{p-1}$, de dimensions $2p-6, \ldots, 2$. Quant à $C^p$, ce sera le groupe de points communs à $p$ cycles C. Leur nombre sera désigné par $(C^p)$. Il n'est autre d'ailleurs que celui des zéros communs à $p$ fonctions $\varphi(u-g)$ situés dans $U_{2p}$. On montre, tout comme nous l'avons fait pour C, que

$$C^k \sim k! \sum (-1)^n m_{i_1 i_2} m_{i_3 i_4} \ldots m_{i_{2k-1} i_{2k}} (i_{2k+1}, i_{2k+2}, \ldots, i_{2p}),$$
$$i_1 < i_2; \qquad i_3 < i_4; \qquad i_{2k-1} < i_{2k}; \qquad i_{2k+1} < i_{2k+2} < \ldots < i_{2p}.$$

Nous retiendrons plus particulièrement

$$C^{p-1} \sim (p-1)! \sum (-1)^n m_{i_1 i_2} m_{i_3 i_4} \ldots m_{i_{2p-3} i_{2p-2}} (i_{2p-1}, i_{2p}),$$
$$(C^p) = p!\, \varpi,$$

ou $\varpi$ est le *déterminant pfaffien* ou simplement *pfaffien* de la forme alternée à coefficients entiers

$$F = \sum_{1}^{2p}{}^{\mu,\nu}\, m_{\mu\nu}\, x_\mu y_\nu.$$

Ce pfaffien, rappelons-le, est au signe près la racine carrée du déterminant de la forme.

La formule donnant ($C^p$) généralise un résultat classique de Poincaré, relatif aux zéros communs à $p$ fonctions thêta.

Soit $M_{\mu\nu}$ le coefficient de $m_{\mu\nu}$ dans le développement du déterminant de F. La forme alternée à coefficients entiers

$$\Phi = \sum M_{\mu\nu}\, x_\mu\, y_\nu$$

est l'*inverse* de F. Des théorèmes les plus simples sur les pfaffiens (¹) on déduit de suite que

$$C^{p-1} \sim \frac{(p-1)!}{2\varpi} \sum M_{\mu\nu}(\mu, \nu).$$

7. Ceci posé, considérons l'intégrale

$$\int\int du_j\, du_k = \int\int \frac{\partial u_j}{\partial \varphi}\, d\varphi\, du_k.$$

On peut supposer qu'en tout point $\varphi = 0$, $\frac{\partial \varphi}{\partial u_j}$ n'est pas nécessairement nulle, car alors les zéros de $\varphi$ seraient tous multiples et on la remplacerait par $\frac{\varphi}{\frac{\partial \varphi}{\partial u_j}}$. Par une discussion très simple, analogue à celle du Chapitre IV, n° 17, on montre alors que, comme $\varphi = 0$ en tout point de $C^{p-1}$,

$$\int\int_{C^{p-1}} du_j\, du_k = 0.$$

En vertu du théorème de Cauchy-Poincaré, que nous avons déjà cité une fois, $C^{p-1}$ peut être remplacé par son expression en fonc-

(¹) *Voir* KOVALEWSKI, *Einführungin die Determinantentheorie*, p. 149.

tion des cycles fondamentaux. Par suite,

$$(3) \qquad \sum M_{\mu\nu}\, \omega_{j\mu}\, \omega_{k\nu} = 0 \qquad (j, k = 1, 2, \ldots, p).$$

Soient ensuite $v = v' + iv''$ une combinaison linéaire quelconque des $u$, $\xi_\mu + i\eta_\mu$ sa $\mu^{\text{ième}}$ période, $x = x' + ix''$ une variable telle qu'en un point arbitraire de $C^{p-1}$ $v$ en soit fonction holomorphe. $x$ pourrait être, par exemple, une des $u$. Soit enfin $E_2$ une petite cellule positive de $C^{p-1}$. Le signe de l'intégrale

$$\int\int_{E_2} dv'\, dv'' = \int\int_{E_2} \left[\left(\frac{\partial v'}{\partial x'}\right)^2 + \left(\frac{\partial v'}{\partial x''}\right)^2\right] dx'\, dx''$$

ne dépend pas de la position de $E_2$ sur $C^{p-1}$ (Riemann), et même, vu nos conventions sur les orientations, il ne dépend ni de $\varphi$, ni même de $\Omega$. On en conclut, en vertu du théorème de Stokes généralisé,

$$(4) \qquad \int\int_{C^{p-1}} dv'\, dv'' = \sum M_{\mu\nu}\, \xi_\mu\, \eta_\nu > 0,$$

car la somme a le signe $+$, quand $\varphi$ est une fonction thêta relative à une matrice aux périodes convenable.

*La relation* (3) *et l'inégalité* (4) *satisfaites par* $\Phi$ *démontrent le théorème de Weierstrass.*

*Définitions.* — Nous dirons avec M. Scorza d'une forme telle que $\Phi$ qu'elle est *principale* pour sa matrice, d'une matrice telle que $\Omega$, qui possède une telle forme, qu'elle est *de Riemann.*

8. L'inégalité (4) a été réduite par M. Scorza à une autre notablement plus suggestive. Posons

$$v = \lambda_1 u_1 + \lambda_2 u_2 + \ldots + \lambda_p u_p.$$

En vertu des identités

$$2\xi_\mu = \sum (\lambda_j\, \omega_{j\mu} + \bar{\lambda}_j\, \bar{\omega}_{j\mu}); \qquad 2i\eta_\mu = \sum (\lambda_j\, \omega_{j\mu} - \bar{\lambda}_j\, \bar{\omega}_{j\mu}),$$

l'inégalité se réduit à

$$(5) \qquad \sum_{j,k}^{p} A_{jk} \lambda_j \bar{\lambda}_k > 0; \qquad A_{jk} = \bar{A}_{kj} = \frac{-1}{4i} \sum M_{\mu\nu}\, \omega_{j\mu}\, \bar{\omega}_{k\nu}.$$

Elle exprime qu'une certaine *forme hermitienne* (forme à indéterminées conjuguées), attachée à la forme principale $\Phi$, est *définie*, positive.

Considérons maintenant la matrice carrée d'ordre $2p$,

$$B = \| b_{rs} \|, \qquad b_{rs} = \omega_{sr}, \qquad b_{r,p+s} = \overline{\omega}_{sr} \qquad (s \leqq p).$$

En vertu des relations

$$\sum\nolimits^{\mu,\nu} M_{\mu\nu}\, \omega_{j\mu}\, \omega_{k\nu} = \sum\nolimits^{\mu,\nu} M_{\mu\nu}\, \overline{\omega}_{j\mu}\, \overline{\omega}_{k\nu} = 0 \qquad (j, k = 1, 2, \ldots, 2p)$$

et en désignant par A la matrice aux coefficients de la forme hermitienne, on a de suite la relation entre matrices,

$$\overline{B}\, \| M_{\mu\nu} \|\, B = \left\| \begin{matrix} O, & A \\ \overline{A}, & O \end{matrix} \right\|,$$

où, rappelons-le, $\overline{A}$ et $\overline{B}$ sont les transposées de A et B.

Or, puisque (5) est définie, A est de rang $p$ et la matrice au second membre de rang $2p$. Donc, des déterminants des matrices à droite, aucun n'est nul. Quant au déterminant de B, cela constitue une nouvelle démonstration du corollaire du n° 3. Mais nous trouvons en outre que le déterminant

$$| M_{\mu\nu} | \neq 0,$$

ce qui montre que la forme $\Phi$, et pareillement *une forme principale quelconque*, *ne peut être dégénérée*.

## II. — Propriétés générales des fonctions périodiques et à multiplicateurs.

9. Dans la première partie de l'exposé suivant, nous nous bornerons à des indications sommaires sur les démonstrations, car elles sont aujourd'hui bien connues des géomètres [1].

I. *Toute matrice de Riemann possède des fonctions à multiplicateurs.* — D'une façon précise, soit $\Phi$ une forme principale. Il existe une fonction entière $\varphi(u)$, à variété de zéros du type du

[1] *Voir* par exemple KRAZER, *Lehrbuch der Thetafunktionen*, Chap. IV.

paragraphe I, telle que

$$(6)\qquad \varphi(u+\omega_\mu) = e^{2\pi i\left(\sum j\, \alpha_{j\mu} u_j + \beta_\mu\right)} \varphi(u),$$

avec

$$(7)\qquad \sum j\, (\alpha_{j\mu}\,\omega_{j\nu} - \alpha_{j\nu}\,\omega_{j\mu}) = m_{\mu\nu} = - m_{\nu\mu},$$

les $\alpha$, $\beta$ étant des constantes et les $m$ des entiers. Une fois l'existence de $\varphi$ admise, on établit (7) en comparant les deux valeurs de $\varphi(u+\omega_\mu+\omega_\nu)$ suivant qu'on ajoute les $\omega_\mu$, $\omega_\nu$ dans un ordre ou dans l'autre.

Pour montrer l'existence de $\varphi$, on choisit un nouveau système de périodes fondamentales (ce qui revient à appliquer aux variables de $\Phi$ une transformation unimodulaire) tel qu'au lieu de $\Phi$ on ait une forme canonique

$$\sum_1^p \mu\, e_\mu (x_\mu y_{p+\mu} - x_{p+\mu} y_\mu).$$

Un changement linéaire et homogène de variables ramènera finalement $\Omega$ à la forme classique

$$\left\| 0, \ldots, 0, \frac{1}{e_\mu}, 0, \ldots, 0, a_{\mu 1}, \ldots, a_{\mu p} \right\|; \quad a_{\mu\nu} = a_{\nu\mu} \quad (\mu = 1, 2, \ldots, p).$$

Les fonctions que l'on recherche ne sont autres alors que les thêtas généralisées, appartenant à la matrice précédente, et dont l'existence est bien connue ([1]). En retournant aux variables $u$, on aura alors une fonction $\varphi(u)$ de la nature voulue.

Les fonctions entières telles que $\varphi$ sont dites *intermédiaires*.

**10.** *Les entiers m relatifs à $\varphi$ sont les mêmes que ceux déjà rencontrés au paragraphe I.* En effet, soit $E_2$ une petite cellule de $U_{2p}$ passant par un point de C, variété des zéros de $\varphi$, telle que $(E_2 C) = +1$. (Le double emploi de la notation C pour la variété et son image dans l'anneau ne causera aucune confusion.)

([1]) Krazer, *loc. cit.*

$\zeta$ étant le contour frontière de $E_2$,

$$\frac{1}{2\pi i}\int_\zeta d\log\varphi = \pm 1,$$

le signe dépendant uniquement de l'orientation de $U_{2p}$. Prenons maintenant dans $U_{2p}$ une $M_2$ quelconque, coupant C aux points $\alpha_1, \alpha_2, \ldots, \alpha_s$, et soit $\eta$ sa frontière. Soit $E_2^h$ une petite cellule positive de $M_2$ contenant $\alpha_h$ en son intérieur, et de frontière $\zeta_h$. On aura

$$\int_{\zeta_h} = \pm\int_\zeta,$$

le signe étant le même que celui de $(E_2^h C)$, c'est-à-dire + si $\alpha_h$ compte pour $+1$ dans $(M_2 C)$, — dans le cas contraire. On en tire

$$\frac{\sum\int_{\zeta_h}}{\int_\zeta} = \frac{\int_\eta}{\int_\zeta} = (M_2 C).$$

Prenons en particulier pour $M_2$ le parallélogramme de $U_{2p}$ relatif au cycle $(\mu\nu)$. Un calcul des plus simples donne de suite, pour le rapport des intégrales, la valeur $m_{\mu\nu}$, d'où $m_{\mu\nu} = [C(\mu, \nu)]$, ce qui démontre notre affirmation.

La forme alternée F, aux coefficients $m$, est dite *forme fondamentale* de la fonction $\varphi$.

**11.** II. *La fonction $\varphi(u)$ appartient à un système linéaire complet* [1], $\infty^{\varpi-1}$, *de fonctions se conduisant de même par rapport à* $\Omega$. $\varpi$ est toujours le pfaffien de F. *Le système linéaire lui-même fait partie d'un système continu* $\infty^p$.

Ces deux propriétés appartiennent aux fonctions thêta dont $\varphi$ provient. La dimension du système linéaire a bien la valeur en question pour ces fonctions, et, comme les transformations unimodulaires ne changent pas le pfaffien, l'énoncé s'ensuit.

III. $p+1$ *fonctions arbitraires périodiques* $f_1(u), f_2(u), \ldots,$

[1] C'est-à-dire contenu dans nul autre semblable, de dimension plus grande.

$f_{p+1}(u)$ *satisfont à une relation algébrique. Toute autre est une fonction rationnelle de celles-ci.*

Considérons les équations

$$(8) \qquad x_1 = f_1(u), \qquad x_2 = f_2(u), \qquad \ldots, \qquad x_{p+1} = f_{p+1}(u).$$

A tout système de valeurs des $x$ correspond un nombre fini de points $(u^i)$ de $U_{2p}$, car les fonctions $f_h(u) - x_h$ sont arbitraires et les cycles, analogues à C, qui leur correspondent, ne se couperont qu'en un nombre fini de points (n° **6**). Les fonctions symétriques élémentaires des expressions $f_{p+1}(u^i)$ sont des fonctions uniformes de $x_1, x_2, \ldots, x_p$ sans singularités essentielles, donc rationnelles. Par suite, $f_{p+1}(u)$ est bien une fonction algébrique des autres $f$.

Il résulte de ce qui précède que les équations (8) sont les équations paramétriques d'une variété algébrique $V_p$. Toute fonction périodique est une fonction uniforme du point sur $V_p$ et n'y a, au pis aller, que des pôles. C'est donc une fonction rationnelle des $x$, c'est-à-dire, en définitive, des $f$.

**12. Variété abélienne attachée à Ω** [1]. — On peut trouver un système de fonctions périodiques $f_1(u), f_2(u), \ldots, f_n(u)$ telles que les équations

$$x_h = f_h(u)$$

représentent une variété algébrique $V_p$, située dans un $S_n$, privée de singularités, homéomorphe à l'anneau. C'est la *variété abélienne* attachée à Ω. Elle n'est d'ailleurs définie qu'à une transformation birationnelle près.

Montrons que $V_p$ existe effectivement. Cela sera certainement le cas si, pour un choix convenable des $f$, on ne peut trouver deux points distincts $(a)$, $(b)$ de U tels que

$$f_h(a) = f_h(b) \qquad (h = 1, 2, \ldots, n).$$

Cela revient à établir que l'on ne peut avoir $f(a) = f(b)$, quelle que soit la fonction périodique $f$. Or, $(v)$ étant quelconque,

---

[1] Les théorèmes qui suivent ont été traités dans mon Mémoire déjà cité. Les démonstrations en ont été, toutefois, quelque peu simplifiées.

$f(u+v)$ est une fonction périodique des $u$ appartenant à $\Omega$. Si la circonstance en question se présente, $f(a+v)=f(b+v)$, quel que soit $(v)$, donc $f(u+a-b)=f(u)$, quel que soit $(u)$. Ceci signifie que $(a-b)$ est un système de périodes simultanées et $(a)$, $(b)$ coïncident.

**13.** *Les fonctions $\varphi$ d'un même système continu découpent un système complet d'hypersurfaces algébriques de* $V_p$.

Soit $C_0$ l'hypersurface découpée par $\varphi$. Les dérivées secondes de $\log\varphi$ sont périodiques, donc rationnelles sur $V_p$ ; par suite, son hypersurface d'infini $C_0$ est bien algébrique.

Soit $|C|$ le système complet déterminé par $C_0$. Il existe une fonction rationnelle $R(x_1, x_2, \ldots, x_n)$ nulle sur C, infinie sur $C_0$, mais nulle part ailleurs sur $V_p$. La fonction

$$\psi(u)=\varphi(u).R[f_1(u), f_2(u), \ldots, f_n(u)]$$

est uniforme, finie à distance finie, donc entière. Le deuxième facteur à droite est une fonction périodique, donc $\psi$, qui découpe C, se conduit comme $\varphi$ par rapport aux périodes. Par suite, elle appartient au même système linéaire (n° **11**, théorème II). Ainsi les fonctions $\varphi$ d'un même système linéaire découpent un système linéaire complet d'hypersurfaces.

En faisant varier les $\varphi$ dans leur système continu, $|C|$ parcourt un système continu $\sum$ contenant $\infty^p$ systèmes linéaires. Mais $R_1=2q=2p$, donc $p$ est l'irrégularité de $V_p$; par suite $\sum$, en tant qu'ensemble de systèmes linéaires, a la dimension maximum. Il coïncide donc avec le système continu complet $\{C\}$, ce qui démontre le théorème.

**14.** Exprimées en termes des $u$, les intégrales simples ou doubles de première espèce dépendent toutes respectivement de celles telles que

$$\int du_h, \qquad \int\int du_h\, du_k,$$

résultat d'ailleurs vrai pour les intégrales de toutes les multiplicités. Il nous suffira pour la suite de montrer que ces intégrales doubles sont bien de première espèce. Dans la relation

$$\int\int du_h\, du_k = \sum \int\int \frac{dx_r\, dx_s}{\dfrac{D(x_r, x_s)}{D(u_h, u_k)}},$$

les dénominateurs à droite sont des fonctions périodiques, donc rationnelles sur $V_p$, et nous avons bien là une intégrale de fonctions rationnelles. D'un autre côté, l'intégrale aux $u$ est de différentielles totales, finie, donc sa transformée, l'intégrale aux $x$, l'est aussi, car ces deux propriétés se préservent dans un changement de variables. Ceci démontre notre affirmation.

### III. — Le nombre $\rho$ pour les variétés abéliennes.

**15.** Voici comment on arrive de suite à une limite supérieure de $\rho$. $\rho$ hypersurfaces algébriques distinctes découperont, sur une $V_2$ convenable de $V_p$, autant de courbes algébriquement distinctes. Les périodes des intégrales doubles

$$\int\int du_j\, du_k$$

par rapport à ces courbes, considérées comme cycles $\Gamma_2$, seront nulles. Il y aura ainsi $\rho$ cycles distincts, à périodes d'intégrales de première espèce toutes nulles. Soit

$$\sum c_{\mu\nu}(\mu, \nu)$$

l'un de ces cycles. La période correspondante sera

$$\sum_{\mu<\nu} c_{\mu\nu}(\omega_{j\mu}\,\omega_{k\nu} - \omega_{k\mu}\,\omega_{j\nu}) = 0.$$

Cela veut dire que la forme alternée à coefficients entiers

$$(9) \qquad \sum c_{\mu\nu}\, x_\mu\, y_\nu \qquad (c_{\mu\nu} = -c_{\nu\mu})$$

s'annule quand on y remplace les $x$ et les $y$ par les éléments de deux lignes quelconques de $\Omega$. En fait, (1) est une forme principale, mais peu importe.

$\Omega$ possédera en général un certain nombre, $1+k$, de formes telles que (9), $k \gtreqless 0$. Le nombre $k$ est dit *indice de singularité* de la matrice (Humbert, Scorza). *D'après ce qui précède*, $\rho \leqq 1+k$.

**16.** Soit $\Phi_1$ une forme principale à coefficients premiers entre eux. Il existe alors un système fondamental de formes alternées dont $\Phi_1$ fait partie ([1]). Soient $\Phi_2, \ldots, \Phi_{k+1}$ ses autres formes. La forme hermitienne, relative à $x\Phi_1 + y\Phi_j$, est à coefficients polynomes en $x$, $y$. Pour qu'elle soit définie positive, donc pour que $x\Phi_1 + y\Phi_j$ soit principale, il faudra et suffira que certaines inégalités de type

$$f(x, y) > 0 \qquad (f, \text{ polynome homogène})$$

soient satisfaites. Or $f(1, 0) > 0$, puisque $\Phi_1$ est principale. Donc si $f$ est de degré $h$, il contient un terme en $Ax^h$ avec $A > 0$. Par suite, $f(\alpha, 1) > 0$ pour $x$ positif, suffisamment grand. On en conclut que $x\Phi_1 + \Phi_j$ est principale pour $x$ suffisamment grand.

Alors, si $\Phi_j$ ne l'est pas, on pourra la remplacer, dans le système fondamental, par la forme que nous venons d'écrire. Nous pourrons donc supposer, comme nous le ferons, que les formes $\Phi_j$ sont toutes principales.

Le système linéaire des formes principales étant $\infty^k$, celui des formes fondamentales, leurs inverses, aura la même dimension. D'un autre côté, si F, F′ sont fondamentales, relatives à des fonctions intermédiaires $\varphi$, $\varphi'$, la forme $\lambda F + \lambda' F'$ ($\lambda$, $\lambda'$ entiers positifs), est fondamentale pour la fonction $\varphi^\lambda \varphi'^{\lambda'}$. Donc *le système des formes fondamentales est linéaire*, $\infty^k$, *comme celui des formes alternées de* $\Omega$.

**17.** Soit donc $F_1, F_2, \ldots, F_{k+1}$ un système fondamental pour les F. Une modification minime du raisonnement du début du numéro précédent permet de montrer que l'on peut les supposer toutes non dégénérées. Soient $\varphi_i$ la fonction intermédiaire relative à $F_i$, $C_i$ l'hypersurface qu'elle découpe sur $V_p$. Je dis que

([1]) Cela résulte de suite d'un théorème de Fröbenius. *Voir* E. CAHEN, *Théorie des nombres*, t. I, p. 168.

$C_1, C_2, \ldots, C_{k+1}$ *sont algébriquement distinctes.* En effet, de

$$t_1 C + t_2 C_2 + \ldots + t_{k+1} C_{k+1} = 0,$$

on tirerait

$$\Gamma_{2p-2} = t_1 C_1 + \ldots + t_{k+1} C_{k+1} \sim 0 \qquad (\mathrm{mod}\ V_p).$$

En se reportant au n° 10, on vérifie que les nombres $(\Gamma_{2p-2}(\mu, \nu))$ sont précisément les coefficients de la forme alternée

$$t_1 F_1 + \ldots + t_{k+1} F_{k+1}.$$

Puisque $\Gamma_{2p-2} \sim 0$, ces nombres sont tous nuls, d'où

$$t_1 F_1 + t_2 F_2 + \ldots + t_{k+1} F_{k+1} \equiv 0,$$

et par suite $t_1 = t_2 = \ldots = t_{k+1} = 0$, puisque les F sont linéairement indépendantes.

On conclut de ceci $\rho \geqq 1 + k$ et, comme $\rho \leqq 1 + k$, on a enfin

$$\rho = 1 + k,$$

c'est-à-dire :

Théorème fondamental. — *Le nombre $\rho$ d'une variété abélienne est égal à l'indice de singularité de sa matrice de Riemann augmenté d'une unité* (1).

18. Soit C une hypersurface quelconque de $V_p$. On aura

$$\lambda C = \sum \lambda_i C_i.$$

Posons $\lambda_i = \lambda''_i - \lambda'_i$ ($\lambda'_i$, $\lambda''_i$ positifs), puis

$$D' = \lambda C + \sum \lambda'_i C_i, \qquad D'' = \sum \lambda''_i C_i.$$

$D''$ est découpée par la fonction intermédiaire

$$\prod_i \varphi_i^{\lambda''_i}.$$

Je dis que l'on peut toujours supposer cette fonction *non dégé-*

(1) Le cas de $p = 2$ a déjà été traité, de manière d'ailleurs entièrement différente, par MM. Bagnera et de Franchis (*Rendic. del Circolo di Palermo*, vol. XXX, 1910).

*nérée* (non réductible à une fonction intermédiaire de moins de $p$ variables). En effet, on peut évidemment remplacer D′, D″ par $D' + xC_1$, $D'' + xC_1$ et, en raisonnant à peu près comme au n° **16**, on montre que, pour $x$ suffisamment grand, la fonction intermédiaire correspondante n'est pas dégénérée.

Ceci posé, puisque $D' = D''$, D′ peut être découpée par une fonction $\psi$ du même système que celle ci-dessus. Or la fonction

$$\chi(u) = \left[\frac{\psi(u)}{\prod_i \varphi_i^{\lambda'_i}}\right]^{\frac{1}{\lambda}},$$

qui découpe C sur $V_p$, se conduit comme une fonction intermédiaire par rapport à $\Omega$; elle est finie à distance finie; donc, pour établir qu'elle est *intermédiaire*, il suffit de démontrer qu'elle est uniforme. Comme elle ne s'annule que sur C, il n'y a de question que pour le voisinage de cette hypersurface. Soit $(u^0)$ un point quelconque de C. D'après le théorème des fonctions implicites

$$[\chi(u)]^\lambda = f(u) f_1(u),$$

où $f_1$ est holomorphe près de $(u^0)$ et ne s'y annule pas. Quant à $f$, c'est un polynome en une des différences $u_i - u_i^0$, par exemple $u_1 - u_1^0$, à coefficients fonctions holomorphes en $u_2$, $u_3, \ldots, u_p$, au même voisinage et s'annulant en $(u^0)$, sauf le premier, qui est égal à l'unité. Mais puisque le second membre découpe C compté $\lambda$ fois, les racines de ce polynome ont toutes la multiplicité $\lambda$, d'où

$$f = [F(u)]^\lambda,$$

où F est de même type que $f$. On en conclut de suite que $\chi(u)$ est holomorphe au point $(u^0)$. C'est donc bien une fonction intermédiaire. Ainsi : *Toute hypersurface algébrique de* $V_p$ *peut être découpée par une fonction intermédiaire.*

Pour $p = 2$, ce théorème a été obtenu par M. Humbert, s'appuyant sur un résultat préliminaire de M. Appell.

17. Soit F la forme fondamentale relative à $\chi(u)$. On aura

$$F \quad \sum t_i F_i.$$

Par suite, en vertu d'une remarque faite plus haut, le cycle

$$C - \sum t_i C_i$$

coupe les cycles $(\mu, \nu)$ en un nombre nul de points. C'est donc un cycle nul ou un diviseur de $V_p$. Mais les nombres $\sigma$ de $V_p$, comme ceux de l'anneau, sont égaux à l'unité. Donc le cycle est nul. De

$$C \sim \sum t_i C_i$$

on tire de suite

$$C = \sum t_i C_i.$$

Les hypersurfaces $C_1, C_2, \ldots, C_{r+1}$ forment donc un système fondamental.

# NOTE I.

## INTÉGRALES DOUBLES DE SECONDE ESPÈCE ET INTÉGRALES SIMPLES DE TROISIÈME ESPÈCE DES SURFACES ALGÉBRIQUES (¹).

Les intégrales doubles de seconde espèce, introduites dans la Science et étudiées de façon fort complète par M. Picard, se rattachent de multiple manière aux théories traitées dans cet Ouvrage. Elles vont nous donner l'occasion de faire de nouvelles applications de l'*Analysis Situs.* En particulier, nous allons mettre en relief un beau théorème reliant la théorie de ces intégrales aux propriétés de certains cycles (n° 15). J'ajoute que le théorème et sa démonstration sont susceptibles d'être étendus aux intégrales de multiplicité quelconque d'une $V_d$.

Une fois la théorie des intégrales doubles bien assise, le théorème fondamental de M. Picard sur le nombre $\rho$ s'obtient si simplement que nous n'avons pu résister à la tentation de le présenter ici.

### I. — Certaines propriétés des cycles à deux dimensions.

1. Théorème. — *Deux cycles* $\Gamma_2$, $\Gamma'_2$ *de* $V_2$ *peuvent être remplacés par d'autres homologues à nombres arithmétique et algébrique d'intersections égaux au signe près.*

Nous supposons que les cycles se coupent en des points ordinaires, sans y être tangents, condition que l'on pourra toujours remplir en déformant légèrement au besoin l'un d'eux.

Je dis d'abord que l'on peut s'arranger pour que l'ensemble des points *ordinaires* de $\Gamma_2$ soit connexe (²). Car autrement, cet ensemble se partage en un nombre fini de régions, dont soient $Q_1$ et $Q_2$ deux quelconques. Dans $Q_i$ prenons une indicatrice $\zeta_i$ formant la frontière d'une cellule $E_2^i$, puis joignons $Q_1$ à $Q_2$ par un petit tube T se raccordant à $\Gamma_2$ suivant les $\zeta$.

(¹) *Voir* Picard et Simart, vol. II, Chap. VII et suivants; Lefschetz, *Annals of Mathematics*, vol. XXI, 1920; *Comptes rendus*, 1923, ainsi qu'un Mémoire en cours de publication au *Journal de Mathématiques*.

(²) Un point d'une $M_2$ est *ordinaire* lorsque deux $E_2$ le contenant en ont toujours en commun une troisième le contenant également.

On aura

$$E_2^1 \equiv \zeta_1, \qquad E_2^2 \equiv \zeta_2, \qquad T \equiv \zeta_1 + \zeta_2.$$

Par conséquent,

$$\Gamma_2 + T - E_2^1 - E_2^2 \equiv 0.$$

La multiplicité à droite est un cycle $\sim \Gamma_2$, car elle lui est réductible par déformation continue. Si l'on prend T suffisamment délié et que l'on évite de lui faire traverser nos cycles, le nouveau cycle coupera toujours $\Gamma'_2$ de manière convenable et il contient une région Q de moins. En poursuivant on en réduira le nombre à une seule, c'est-à-dire que l'ensemble des points ordinaires sera alors connexe.

Soient $A_1, A_2, \ldots, A_n$ les points de l'intersection de $\Gamma_2$, $\Gamma'_2$ contribuant $+1$ à $(\Gamma_2 \Gamma'_2)$, $B_1, B_2, \ldots, B_{n'}$ les autres, de sorte que

$$(\Gamma_2 \Gamma'_2) = n - n'.$$

On suppose $n \leqq n'$, condition en rien restrictive comme on le verra.

On peut joindre $A_i$ à $B_i$ par une ligne composée en entier de points ordinaires. Déformons $\Gamma'_2$ de manière à faire glisser $A_i$ le long de la ligne jusqu'à le faire venir en $B_i$. Nous supposerons qu'en même temps une petite $E_2$ contenant $A_i$ sur $\Gamma'_2$ est ramenée au voisinage de $B_i$ sur le même cycle. Puisque les contributions de $A_i$ et de $B_i$ au nombre $(\Gamma_2 \Gamma'_2)$ sont de signes contraires, l'$E_2$ sera superposée à une cellule orientée négativement pour $\Gamma'_2$, de sorte qn'elle pourra être supprimée dans $\Gamma'_2$. Il en résultera que $A_i$ et $B_i$ seront supprimés en tant qu'intersections des deux cycles. En continuant, on réduira bien le nombre arithmétique d'intersections à

$$n' - n = -(\Gamma_2 \Gamma'_2). \qquad \text{C. Q. F. D.}$$

Corollaire. — *Lorsque* $(\Gamma_2 \Gamma'_2) = 0$, *on peut réduire les cycles à d'autres ne se coupant plus nulle part.*

2. Soient $\Gamma_2^1, \Gamma_2^2, \ldots, \Gamma_2^{\rho_0}$ des cycles distincts en nombre maximum coupant toute courbe algébrique C de $V_2$ en un nombre algébrique nul de points. Complétons le système par $R_2 - \rho_0$ autres cycles, $\Gamma_2^{\rho_0+1}, \ldots, \Gamma_2^{R_2}$, de façon à en avoir $R_2$ de distincts. Toute combinaison des $\Gamma_2^{\rho_0+i}$ coupera au moins une C en un nombre algébrique de points $\neq 0$.

1° On peut trouver $R_2 - \rho_0$ courbes $C_1, C_2, \ldots, C_{R_2-\rho_0}$ telles que le déterminant d'ordre $R_2 - \rho_0$,

$$\text{(1)} \qquad |(\Gamma_2^{\rho_0+h} C_i)| \neq 0,$$

car autrement il y aurait des entiers $\lambda$ non tous nuls tels que

$$\sum \lambda_h (\Gamma_2^{\rho_0+h} C) = 0 = \left[\left(\sum \lambda_h \Gamma_2^{\rho_0+h}\right) C\right],$$

pour toute C, contrairement à la manière de choisir nos cycles.

2° C étant toujours quelconque, il y a des entiers $\lambda$, $\lambda_i$, avec $\lambda \neq 0$, tels que

$$\sum_i \lambda_i(\Gamma_2^{\rho_0+h} C_i) + \lambda(\Gamma_2^{\rho_0+h} C) = 0$$

On en conclut que $C + \sum \lambda_i C_i$ coupe tout $\Gamma_2$ en un nombre algébrique nul de points, d'où (Chap. V, n° 18)

$$t\lambda C + t\sum \lambda_i C_i \sim 0 \qquad (t \neq 0),$$

et enfin (Chap. IV, n° 28)

$$t\lambda C + t\sum \lambda_i C_i = 0. \tag{2}$$

3° D'un autre côté, si l'on avait

$$\sum \lambda_i C_i = 0, \tag{3}$$

on en tirerait

$$\sum \lambda_i C_i \sim 0,$$

auquel cas pour tout $\Gamma_2$

$$\sum \lambda_i(\Gamma_2 C_i) = 0,$$

et par conséquent le déterminant (1) est nul. Comme ceci n'a pas lieu, la relation (3) ne peut exister elle non plus, de sorte que les $C_i$ sont indépendantes au sens de M. Severi. La relation (2) montre que toute autre courbe dépend d'elles, donc le nombre des $C_i$ est égal à $\rho$ (Chap. IV, n° 31). On en conclut :

THÉORÈME. — *Il y a exactement* $R_2 - \rho$ *cycles* $\Gamma_2$ *distincts rencontrant toute courbe algébrique de* $V_2$ *en un nombre algébriquement nul de points.*

*Remarque.* — M. Severi a démontré que le déterminant

$$|(C_i C_k)| \neq 0.$$

D'un autre côté (Chap. IV, n° 28), les $C_i$ sont des $\Gamma_2$ indépendants. Par conséquent on peut prendre, au lieu des $\Gamma_2^{\rho_0+h}$, ces courbes mêmes. On aura alors le système de cycles

$$\Gamma_2^1, \quad \Gamma_2^2, \quad \ldots, \quad \Gamma_2^{\rho_0}, \quad C_1, \quad \ldots, \quad C_\rho,$$

aux mêmes propriétés que celui considéré plus haut.

3. Une autre question fort importante pour la suite est celle ***du nombre de cycles finis à systèmes de $\lambda$ distincts, du type***

$$\sum \lambda_i \Delta_i + ((H_a)). \tag{4}$$

[La notation $((H_a))$ désigne, comme au Chapitre IV, n° 9, une partie finie de $H_a$.]

A (4) correspond l'équivalence

$$-((H_a)) \equiv \sum \lambda_i \delta_i \qquad (\text{les } \delta \text{ sont dans } H_a),$$

ou plus généralement, le sens de $((H_y))$ étant clair,

$$-((H_y)) \equiv \sum \lambda_i \delta_i \qquad (\text{les } \delta \text{ sont dans } H_y). \tag{5}$$

Réciproquement, en raisonnant comme au Chapitre III, n° 10, on montre qu'à (5) correspond toujours un $\Gamma_2$ fini (4). On est donc ramené à déterminer le nombre d'équivalences (5) distinctes.

Reprenons les lignes $l_i$ du Chapitre II, n° 20, joignant l'un des points bases de $\{H_y\}$ (ici les points à l'infini) aux $m-1$ autres. Pour qu'un $\Gamma_1$ de $H_y$ forme la frontière d'une région finie de la courbe, il faut et il suffit d'abord qu'il soit $\sim 0$, puis que

$$(\Gamma_1 l_h) = 0 \qquad (h = 1, 2, \ldots, m-1).$$

En effet, ces relations expriment simplement que des deux régions en lesquelles $\Gamma_1$ partage $H_y$, chacune des $l$ passe un nombre pair de fois de l'une dans l'autre. Les points à l'infini sont alors nécessairement dans la même région.

On conclut de ce qui précède que, pour que (5) soit satisfaite, il faut et il suffit que :

$$\sum \lambda_i (\delta_i \Gamma_1) = 0 \qquad (\text{pour tout } \Gamma_1 \text{ de } H_y), \tag{a}$$

$$\sum \lambda_i (\delta_i l_h) = 0 \qquad (h = 1, 2, \ldots, n-1). \tag{b}$$

D'après la fin du Chapitre II, les équations linéaires aux $\lambda$, auxquelles on est ainsi conduit, ont une matrice aux coefficients de rang

$$2p - 2q + m - 1 = 2p - R_1 + m - 1.$$

Le nombre de solutions distinctes est donc

$$N - (2p - R_1 + m - 1) = R_2 - 1 + 2p - R_1$$

(Chap. III, n° 11). Tel est aussi le nombre d'équivalences (5) distinctes, et par suite enfin le nombre cherché des cycles finis (4) à systèmes de $\lambda$ distincts.

4. D'un autre côté, soit $\Gamma_2$ quelconque. $\Gamma'_2 \sim n\Gamma_2 - (\Gamma_2 H).H$ coupe H en un nombre algébriquement nul de points, donc il est homologue à un cycle fini (n° 1, corollaire). On obtiendra ainsi de suite $R_2 - 1$ cycles finis distincts. Comme $(H^2) = m \neq 0$, H ne peut être homologue à un cycle fini. Par suite, $R_2 - 1$ est le nombre maximum de $\Gamma_2$ finis distincts. On en conclut que les cycles finis (4) en comprennent au moins $2p - R_1$ à systèmes de $\lambda$ distincts, formant cependant frontière sur $V_2$. Il n'est pas difficile de montrer que $2p - R_1$ est précisément le nombre de ces cycles. Soit en effet

$$M_3 \equiv \sum \lambda_i \Delta_i + ((H_a)).$$

$M_3$ coupe une $H_y$ générique suivant un $\Gamma_1$, car $H_y$ n'en rencontre ni les frontières $\Delta_i$ ni $((H_a))$, les premières, puisque $y$ n'est pas en général sur les coupures $aa_i$, la seconde parce que $((H_a)$ ne contient pas les points-bases de $\{H_y\}$.

$M_3$, lieu de $\Gamma_1$, a pour frontières (Chap. III, n° 11)

$$\sum (\Gamma_1 \delta_i).\Delta_i + ((H_a)). \tag{6}$$

Par suite, $\lambda_i = (\Gamma_1 \delta_i)$. On en conclut que le nombre de $\Gamma_2$ considérés ne dépasse pas le nombre de systèmes d'entiers $(\Gamma_1 \delta_i)$ distincts, quand on envisage tous les $\Gamma_1$ de $H_y$, c'est-à-dire $2p - 2q = 2p - R_1$ (Chap. II, n° 18). Ceci suffit pour démontrer que $2p - R_1$ est bien le nombre de ces cycles.

On peut construire de façon fort simple des $\Gamma_2$ réductibles aux précédents par déformation finie. Soit $\overline{\Gamma_2}$ le lieu qui est fini, du $\Gamma_1$ en évidence dans (6), quand $y$ décrit un cercle de grand rayon $\zeta$. Il est clair que $\overline{\Gamma_2} \sim 0$. On peut considérer, si l'on veut, $\overline{\Gamma_2}$ comme un tube dont l'axe est la position limite de $\Gamma_1$ à l'infini. Déformons $\zeta$ sans lui faire traverser les $a_i$ jusqu'à ce qu'il coïncide avec les coupures. $\overline{\Gamma_2}$ sera alors devenu le cycle (6), comme on le voit de suite en considérant le lieu de $\Gamma_1$ quand $y$ décrit successivement les deux lèvres d'une même coupure.

## II. — Généralités sur les périodes et résidus des intégrales doubles.

5. Une période de l'intégrale double

$$\iint R(x, y, z)\, dx\, dy \qquad (\text{R fonction rationnelle}) \tag{1}$$

en est la valeur étendue à un $\Gamma_2$ qui ne rencontre pas les courbes d'infini de l'intégrale [1]. Lorsque $\Gamma_2 \sim 0$, la période est un *résidu* de (1). Soit dans ce cas $M_3 \equiv \Gamma_2$. Si $M_3$ ne rencontre pas les courbes d'infini, le théorème de Cauchy-Poincaré montre de suite que le résidu est nul. Supposons que $M_3$ rencontre une de ces courbes, soit C. Puisque $\Gamma_2$ ne la rencontre pas, on pourra déformer au besoin légèrement $M_3$ de manière que $M_3C$ soit un $\Gamma_1$ et non une $M_2$. Dans $M_3$ isolons le voisinage de ce $\Gamma_1$ par un cycle $\Gamma_2$ infiniment voisin, nécessairement $\sim 0 \pmod{V_2}$. Soit $M'_3$ ce qui reste de $M_3$ quand on y supprime toutes ces parties isolées. Ni $M'_3$ ni ses frontières ne rencontrent les courbes d'infini. De

$$M'_3 \equiv \Gamma_2 - \sum \overline{\Gamma_2}$$

on conclut que le résidu relatif à $\Gamma_2$ est égal à la somme de ceux relatifs aux $\overline{\Gamma_2}$. Il suffira donc de considérer ces résidus spéciaux.

On peut donner une construction fort simple des $\overline{\Gamma_2}$. Supposons que C fasse partie de la courbe $D_{u_0}$ appartenant à un certain faisceau linéaire $\{D_u\}$. En particulier, C peut coïncider avec $D_{u_0}$. Soit $\zeta$ un petit circuit entourant $u_0$ dans son plan, puis considérons la partie de $M_3$ relative aux $u$ intérieurs à $\zeta$. Sa frontière pourra de toute évidence être prise pour un $\overline{\Gamma_2}$. Ce dernier est le lieu d'un $\Gamma_1$ de $D_u$ quand $u$ décrit $\zeta$. Au fond, ce $\overline{\Gamma_2}$ est simplement un tube dont l'axe est la position limite de $\Gamma_1$ dans $D_{u_0}$. On reconnaît là la construction de la fin du n° 4, qui correspond au cas où

$$C = H_\infty, \qquad \{D_u\} = \{H_y\}.$$

$\overline{\Gamma_2}$ n'est autre alors que le cycle fini de ce numéro.

## III. — Intégrales doubles de seconde espèce propres et impropres.

6. Définitions. — Une intégrale double de type

$$(1) \qquad \iint \left(\frac{\partial U}{\partial x} + \frac{\partial V}{\partial y}\right) dx\, dy \qquad (U, V, \text{ fonctions rationnelles})$$

est dite *impropre de seconde espèce*. Toute intégrale double de fonction

[1] Lorsqu'une $M_2$ rencontre uniquement des courbes polaires d'ordre *un* de R, (1) étendue à $M_2$ donne lieu à une intégrale convergente (Picard). Toutefois, cette valeur varie avec l'intersection de $M_2$ et des courbes. L'intégrale étendue à un tel $\Gamma_2$ conduirait donc à des périodes variant continûment avec le cycle. Il n'y a, à ce que je sache, aucune application de telles périodes; c'est pourquoi la restriction imposée pour l'instant à $\Gamma_2$, restriction d'ailleurs levée en partie plus loin (n° 7).

rationnelle est de *seconde espèce* quand elle a cette propriété : à chaque courbe d'infini de l'intégrale en correspond une impropre, telle que la différence entre les deux soit finie au voisinage d'un point arbitraire de la courbe ([1]). Enfin, des intégrales de seconde espèce en nombre quelconque sont *linéairement indépendantes* s'il n'en existe pas de combinaison linéaire à coefficients constants, impropre (sous-entendu ici comme dans la suite « de seconde espèce »).

7. Théorèmes. — I. *Une transformation birationnelle ne change pas la forme des intégrales impropres* ([2]).

En effet, soient $x'$, $y'$, $z'$ les coordonnées courantes sur une surface transformée birationnelle de $V_2$. Le théorème est une conséquence immédiate des identités

$$\iint \frac{\partial u}{\partial x}\,dx\,dy = \iint \frac{D(U, y)}{D(x, y)}\,dx\,dy = \iint \frac{D(U, y)}{D(x', y')}\,dx'\,dy'$$
$$= \iint \left[\frac{\partial}{\partial x'}\left(U\frac{\partial y}{\partial y'}\right) - \frac{\partial}{\partial y'}\left(U\frac{\partial y}{\partial x'}\right)\right] dx'\,dy'$$

et de celles semblables pour $\dfrac{\partial U}{\partial y}$.

II. *Soient* U *une fonction rationnelle*, $\Gamma_2$ *un cycle ne rencontrant ni ses courbes d'infini, ni la courbe simple d'intersection*, D, *de* $V_2$ *avec* $f'_z = 0$. *On a*

$$(2) \qquad \int\!\!\int_{\Gamma_2} \frac{\partial U}{\partial x}\,dx\,dy = \int\!\!\int_{\Gamma_2} \frac{\partial U}{\partial y}\,dx\,dy = 0.$$

Le changement de va·iables $x' = U$, $y' = y$, pour la première intégrale, $x' = x$, $y' = U$, pour la seconde, réduisent de suite le théorème en question à celui de Cauchy-Poincaré.

III. *Les intégrales de seconde espèce n'ont pas de résidus.*

Il suffit évidemment de considérer les intégrales impropres et même les intégrales (3). De plus, pour cycles à résidus, on peut se contenter de prendre ceux désignés par $\overline{\Gamma}_2$ au paragraphe 2. On voit alors de suite qu'en déformant convenablement les axes $\Gamma_1$ de ces tubes, ils rempliront toutes les conditions de l'énoncé de II, d'où III s'ensuit.

Corollaires. — I. *Lorsque* $\Gamma_2 \sim \Gamma'_2$, *les périodes correspondantes d'une intégrale de seconde espèce* J *sont égales.*

---

([1]) Cette définition, légèrement différente de celle de M. Picard, semble quelque peu plus commode. *Voir* Picard et Simart, vol. II, p. 160.

([2]) *Ibid.*, p. 161.

II. *Pour qu'il y ait un cycle* $\sim \Gamma_2$ *par rapport auquel* J *possède une période, il faut et il suffit que* $\Gamma_2$ *coupe toute courbe d'infini de* J *en un nombre algébriquement nul de points.*

Dorénavant, nous n'hésiterons plus à considérer les périodes de J relativement à un $\Gamma_2$ coupant les courbes d'infini, pourvu que les conditions du corollaire II soient remplies. Il s'agira alors toujours de la période relative à un cycle convenable $\sim \Gamma_2$.

THÉORÈME IV. — *Réciproquement, si* J *n'a pas de résidus, elle est de seconde espèce* (1).

Soit $g(x, y) = 0$ la projection d'une courbe d'infini C de J, n'appartenant pas à $\{H_y\}$. (Lorsque C est une $H_y$, il suffit d'échanger les rôles de $x$ et $y$ dans la discussion.) L'intégrale aura la forme

$$\int\int \frac{A}{g^\alpha}\, dx\, dy,$$

où A est une fonction rationnelle ni nulle ni infinie sur C.

Si $\alpha > 1$, on remplace J par l'intégrale aux mêmes résidus

$$J + \int\int \frac{1}{\alpha - 1}\frac{\partial}{\partial x}\frac{A}{g'_x g^{\alpha-1}}\, dx\, dy = \int\int \frac{1}{(\alpha-1)g^{\alpha-1}}\frac{\partial}{\partial x}\left(\frac{A}{g'_x}\right) dx\, dy,$$

de même type, avec $\alpha - 1$ au lieu de $\alpha$. En définitive, on peut donc toujours être ramené à $\alpha = 1$.

Transformons alors $V_2$ birationnellement de manière que la transformée de C sur la nouvelle $V_2$ (pour laquelle nous garderons les mêmes notations que pour l'ancienne) fasse partie de $H_{y_0}$. L'intégrale transformée, qui aura toujours les mêmes résidus que la primitive, sera de la forme

$$(3) \qquad \int\int \frac{A}{y - y_0}\, dx\, dy, \qquad A(x, y_0, z) \neq 0.$$

En se reportant à ce que nous avons dit au n° 5 sur la construction des cycles à résidus, on voit que les résidus de (3) sont les périodes de l'intégrale abélienne attachée à $H_{y_0}$,

$$(4) \qquad 2\pi i \int A(x, y_0, z)\, dx.$$

Réciproquement, toute période de (4) est un résidu de (5), donc aussi de J. Par conséquent, ces périodes sont toutes nulles et (3) se réduit à une

(1) *Loc. cit.*, p. 203.

fonction rationnelle sur C, soit $2\pi i U(x, z)$. On en conclut que

$$\iint \frac{A(x, y, z)}{y - y_0} dx\, dy - \iint \frac{\partial}{\partial x}\left[\frac{U(x, z)}{y - y_0}\right] dx\, dy$$

est finie au voisinage d'un point arbitraire de C, ou plutôt de la transformée de C sur la nouvelle $V_2$. En retournant à la surface originelle, et en se rappelant la propriété d'invariance des intégrales impropres, on voit que J se conduit comme une intégrale de seconde espèce au voisinage de C. Ceci suffit pour démontrer notre théorème.

*Remarque.* — De là on conclut aisément à l'invariance des intégrales de seconde espèce sous les transformations birationnelles.

## IV. — Étude des périodes d'une classe particulièrement simple d'intégrales doubles.

8. Nous allons faire l'étude des périodes des intégrales

$$\iint \frac{P(x, y, z)\, dx\, dy}{f'_z}, \tag{1}$$

où P, ici comme d'ailleurs partout dans la suite, dénote un polynome adjoint. Cette intégrale a comme propriété essentielle d'être finie à distance finie. Plus généralement, nous nous occuperons aussi des intégrales infinies uniquement sur uu certain nombre de courbes $H_y$.

L'étude de (1) se rattache intimement à celle de l'intégrale abélienne attachée à $H_y$,

$$\int \frac{P(x, y, z)\, dx}{f'_z}. \tag{2}$$

**Théorème.** — *Lorsque* P *est quelconque,* (2) *possède* $2p + m - 1$ *périodes cycliques et logarithmiques complètement arbitraires, c'est-à-dire autant que si le polynome n'était pas astreint à être un polynome en* $y$ *également* [1].

Même démonstration qu'au Chapitre IV, n° 2.

---

[1] Je dis en effet que

$$\int \frac{P(x, z)\, dx}{f'_z} \quad (\text{P polynome adjoint à } H_y) \tag{1}$$

*possède* $2p$ *periodes cycliques et* $m - 1$ *logarithmiques à l'infini complètement arbitraires.* Les périodes logarithmiques sont bien arbitraires, car les intégrales normales de troisième espèce avec deux points logarithmiques à l'infini sont

*Remarque.* — Soient $\gamma_1, \gamma_2, \ldots, \gamma_{2p+m-1}$ un système de cycles de $H_y$ correspondant aux périodes en question. Nulle combinaison des $\gamma$ ne peut former frontière pour une partie finie de $H_y$, car alors la période correspondante de toute intégrale (2) correspondante serait nulle (théorème de Cauchy). Mais la théorie élémentaire des surfaces de Riemann nous apprend qu'il y a au plus $2p+m-1$ cycles linéaires ayant cette propriété. Donc $\Gamma_1$ étant quelconque dans $H_y$, il y a un cycle

$$\lambda\Gamma_1+\sum\lambda_i\gamma_i \qquad (\lambda\neq 0)$$

formant frontière pour une partie finie de $H_y$.

Corollaire. — *On peut se donner* $2p+m-1$ *intégrales* (2) *à déterminant de périodes* $\neq 0$. Les cycles correspondants constituent précisément un système tel que les $\gamma$.

9. D'après le paragraphe 1, l'intégrale double (1) possède $R_2-1+2p-R_1$ périodes relatives à des cycles finis

$$\sum\lambda_j\Delta_j+((H_a)). \tag{3}$$

Soit $\Omega_j(y)$ la période de (2) par rapport à $\delta_j$. On montre, comme au Cha-

---

toutes de type (1). Il suffit donc de montrer que les intégrales de seconde espèce (1) ont des périodes arbitraires.

Soit $ax+bz+c=0$ une tangente en un point arbitraire $B_1$ de $H_y$, $B_2, B_3, \ldots, B_m$, ses autres intersections avec la courbe, $b_i$ l'abscisse de $B_i$. Considérons la différence

$$\int\frac{P(x,z)\,dx}{(ax+bz+c)f'_z}-k\,\frac{\prod^i(x-b_i)}{(x-b_1)(ax+bz+c)} = \int\left[\frac{Q(x,z)}{(ax+by+c)}\right]\frac{dx}{f'_z}, \tag{2}$$

où à gauche se trouve l'intégrale normale de seconde espèce relative à $B_1$ et une constante $k$ telle que le tout soit fini en ce point. Ln quantité entre crochets à droite est finie à distance finie, donc (Picard et Simart, vol. II, p. 10) égale à un polynome, nécessairement adjoint. [La tangente (*loc. cit.*) est parallèle à $Oy$, mais on passe de là au cas plus général par une simple rotation d'axes.] L'intégrale à droite dans (2) est alors de forme (1), aux mêmes périodes que l'intégrale normale. Donc les intégrales de seconde espèce (1) ont leurs périodes cycliques aussi arbitraires qu'une combinaison linéaire d'intégrales normales et d'intégrales de première espèce [elles-mêmes aussi de forme (1)]. Ceci suffit pour démontrer notre affirmation.

pitre IV, n° 16, que la période relative à (3) est égale à [1]

$$\sum \lambda_j \int_a^{a_j} \Omega_j(y)\,dy.$$

THÉORÈME. — *Les périodes d'une intégrale double* (1) *arbitraire sont elles-mêmes complètement arbitraires* [2].

En effet, celles de

$$\int\int \frac{\varphi(y)\,\mathrm{P}(x, y, z)\,dx\,dy}{f'_z} \qquad (\varphi \text{ polynome})$$

sont égales à

$$\sum \lambda_j \int_a^{a_j} \varphi(y)\Omega_j(y)\,dy.$$

Si notre théorème tombe en défaut, c'est qu'il y a une relation

$$\sum \mu_j \int_a^{a_j} y^k \Omega_j(y)\,dy = 0,$$

aux $\mu$ non tous nuls, satisfaite pour toutes les valeurs de l'entier $k$. De là on tire de suite

$$\sum \mu_j \int_a^{a_j} \frac{\Omega_j(y)}{y-u}\,dy \equiv 0,$$

quelle que soit $u$. Soit $\mu_j \neq 0$. L'accroissement $2\pi i \mu_j \Omega_j$ de cette expression quand $u$ tourne autour de $a_j$ doit être nul, d'où $\Omega_j(y) = 0$. Mais $\delta_j$ ne fait pas frontière sur $\mathrm{H}_y$ (Chap. II, n° 10), donc, d'après le numéro précédent, la période correspondante d'une intégrale (2) arbitraire ne peut être nulle, contradiction qui démontre notre théorème.

10. THÉORÈME. — *Une intégrale double* J *infinie uniquement sur des* $\mathrm{H}_y$ *et dont toutes les périodes et résidus sont nuls, est impropre de seconde espèce* [3].

Soient

$$\mathrm{J} = \int\int \mathrm{R}(x, y, z)\,dx\,dy, \qquad \mathrm{J}_y = \int \mathrm{R}\,dx,$$

---

(1) Ici, comme au Chapitre IV, n° 15 (*voir* aussi plus loin, n° 10),

$$\sum \lambda_j \Omega_j(y) \equiv 0.$$

(2) PICARD et SIMART, vol. II, p. 355.

(3) *Ibid.*, p. 365. Le cas considéré ici est un peu plus général, mais la démonstration est la même.

$\Omega_j$ la période de $J_y$ relative à $\delta_j$,

$$J_y^h = \int \frac{P_h(x, y, z)}{f'_z}\, dx \qquad (h = 1, 2, \ldots, 2p + m - 1),$$

les intégrales du corollaire au n° 8, à déterminant de périodes non nul, $\Omega_{hj}$ la période de $J_y^h$ relative à $\delta_j$. Pour démontrer le théorème, il suffit d'établir l'*existence de fonctions rationnelles* $c_h(y)$, $U(x, y, z)$, *telles que*

$$J_y - \frac{\partial}{\partial y} \sum c_h J_y^h = U(x, y, z). \tag{4}$$

Soient, dans $H_y$, $\gamma_1, \gamma_2, \ldots, \gamma_{R_1}$ un système de cycles invariants au nombre maximum de $R_1$ (Chap. II et III), et soient $\omega_k$, $\omega_{hk}$ les périodes de $J_y$, $J_y^h$ relativement à $\gamma_k$. Pour que (4) soit vraie, il faut d'abord que les périodes de l'intégrale au premier membre soient nulles. Il en sera bien ainsi lorsque

$$\sum_1^{2p+m-1}{}^h c_h \omega_{hk}(y) = \int_a^y \omega_k(y)\, dy \qquad (k = 1, 2, \ldots, R_1), \tag{5}$$

$$\sum_1^{2p+m-1}{}^h c_h \Omega_{hj}(y) = \int_{a_j}^y \Omega_j(y)\, dy \qquad (j = 1, 2, \ldots, N). \tag{6}$$

En effet, en vertu des relations (6), les périodes relatives aux $\delta$ seront nulles. Ceci entraîne l'invariance de toutes les périodes alors que, grâce aux relations (5), les périodes invariantes elles aussi s'evanouissent. Nous allons voir que les relations (5), (6) suffisent pour déterminer des fonctious $c_h(y)$ convenables.

Au cycle fini (3) correspond l'équivalence

$$-((H_y)) \equiv \sum \lambda_j \delta_j. \tag{7}$$

Par suite,

$$\sum_1^{N}{}^j \lambda_j \Omega_{hj}(y) \equiv 0 \qquad (h = 1, 2, \ldots, 2p + m - 1).$$

Les expressions des périodes de J sont de même type que pour (1). En écrivant qu'elles sont nulles, on obtient

$$-\sum \lambda_j \int_a^{a_j} \Omega_j(y)\, dy = 0 = \sum \lambda_j \int_{a_j}^{y} \Omega_j(y)\, dy.$$

Vu que le nombre de cycles (3) à systèmes de $\lambda$ distincts est égal à

$$R_2 - 1 + 2p - R_1,$$

ce même nombre de relations (6) est une conséquence des

$$N-(R_2-1+2p-R_1)=2p-R_1+m-1$$

autres. Jointes aux équations (5), elles nous en donnent juste le nombre $2p+m-1$ qu'il nous faut pour déterminer les inconnues $c$, bien entendu pourvu que le déterminant des coefficients ne soit pas nul. Or ce déterminant n'est autre que celui des périodes des $J_y^h$ relativement à autant de cycles dont nulle combinaison ne forme frontière pour une partie finie de $H_y$, et il ne sera pas nul.

Ainsi nos relations nous fournissent une solution unique aux inconnues $c$. Remarquons que $\omega_h$, $\omega_{hk}$ sont rationnelles, car elles sont uniformes et régulières à l'infini. De plus, $\int \omega_h(y)\,dy$ l'est également, car ses périodes logarithmiques, s'il y en avait, seraient des résidus de J. Enfin, lorsque $y$ décrit un chemin fermé quelconque, les seconds membres des relations (6) se combinent comme leurs premiers; la vérification en est immédiate. Il suffit alors d'écrire les solutions aux $c$ pour voir que ces fonctions sont uniformes, régulières à l'infini comme les $\Omega$, et partant *rationnelles*.

Sauf à être assuré qu'elle est rationnelle, on pourra déterminer la fonction U. Mais en tout cas

$$R-\frac{\partial V}{\partial y}=\frac{\partial U}{\partial x}, \qquad V=\sum \frac{c_h P_h}{f'_z}, \tag{8}$$

d'où l'on conclut que U est certainement algébrique. Outre $y$, elle peut dépendre de certains paramètres. Maintenant ces derniers fixes, soient $U_1, U_2, \ldots, U_s$ ses déterminations pour $y$ donné. Leur moyenne U satisfait tout aussi bien à (8) et elle est de plus rationnelle en $x, y, z$. Cela revient à démontrer le théorème.

*Remarque.* — Un point peut prêter à objection dans la démonstration précédente : nous avons admis sans plus que les périodes de J ont les mêmes expressions que si elle était finie à distance finie. Cela ne fait aucun doute tant que, parmi ses courbes d'infini, ne se trouvent pas de $H_{a_j}$ (courbes $H_y$ relatives aux valeurs critiques). Montrons que, même dans ce cas, il n'y a que peu de choses à changer.

Soit $\delta'_j$ le cycle de $H_y$ s'accroissant de $\delta_j$ quand $y$ tourne autour de $a_j$ (Chap. II, n° 10). Soit $\Delta'_j$ le lieu de $\delta'_j$ quand $y$ décrit, non plus la coupure $aa_j$, mais un lacet en différant fort peu (Chap. III, n° 17). Les cycles finis sont réductibles au type

$$\sum \lambda_j \Delta'_j + ((H_a)),$$

cycle qui ne rencontre plus les $H_{a_j}$. Il suffira alors de remplacer partout les $\Omega$ par les périodes relatives aux $\delta'$ et les chemins d'intégration $aa_j$ par les lacets; mais, à part cela, le reste de la discussion ira comme auparavant.

## V. — Réduction des intégrales de seconde espèce. Théorème définitif.

11. Théorème. — *Toute intégrale de seconde espèce* J *est réductible par soustraction d'une intégrale impropre à une intégrale infinie uniquement sur des* $H_y$.

Soit $g(x, y) = 0$ la projection d'une courbe d'infini de J n'appartenant pas à $H_y$, et soit

$$\int\int \left(\frac{\partial U}{\partial x} + \frac{\partial V}{\partial y}\right) dx\, dy \tag{1}$$

telle que la différence entre elle et J soit finie au voisinage de C. Les fonctions rationnelles U, V peuvent être mises sous la forme

$$\frac{A(x, y, z)}{g^\alpha h(x, y)},$$

où A et $h$ sont les polynomes ne s'annulant pas identiquement pour

$$g = 0 \quad (^1).$$

On pourra d'ailleurs écrire

$$\frac{1}{g^\alpha h} = \frac{M(x, y)}{g^\alpha k(y)} + \frac{N(x, y)}{h.k(y)} \qquad (\text{M, N, } k \text{ polynomes}).$$

On en conclut que la différence

$$\frac{A}{g^\alpha . h} - \frac{A.M}{g^\alpha . k}$$

est finie au voisinage de C. On pourra donc remplacer U et V par des fonctions semblables, sauf que $h$ ne contient plus $x$. En soustrayant alors (1) de J, on aura remplacé C comme courbe d'infini par un certain nombre de courbes $H_y$. En continuant, on n'aura plus que de telles courbes comme courbes d'infini de l'intégrale réduite. C. Q. F. D.

Corollaire. — J *est réductible à la forme*

$$\int\int \frac{P(x, y, z)}{f'_z} dx\, dy.$$

---

(1) E. Picard, *Traité d'Analyse*, 3e édition, vol. I, p. 69.

En effet, nous pouvons d'abord supposer J infinie uniquement sur des $H_y$. Il existe une intégrale du type précédent aux mêmes périodes que J, soit J'. Pas plus que J, J' ne possède de résidus à l'infini, et, comme elle est finie partout ailleurs, J' est de seconde espèce. J — J' est sans périodes, infinie uniquement sur des $H_y$, donc impropre, et J est bien réductible à J'.

12. Soit C une courbe algébrique d'ordre $\nu$, en position générale par rapport aux axes. Formons une intégrale abélienne attachée à $H_y$,

$$\int V\,dx,$$

à périodes logarithmiques $+1$ relativement aux points de $CH_y$ et $-\nu$ relativement à l'un des points à l'infini. Les conditions imposées à V sont rationnelles par rapport à $y$ dans leur ensemble. Raisonnant alors comme à la fin du n° 10, on montre que cette fonction rationnelle en $x$ et $z$ l'est en $y$ également. Alors

$$\frac{\partial}{\partial y}\int V\,dx$$

est de seconde espèce et, partant, réductible à l'aide d'opérations rationnelles par rapport à $y$ (soustraction de fonctions rationnelles bien déterminées) à une intégrale $J_y$ finie à distance finie. On aura

$$J_y = \frac{\partial}{\partial y}\int V\,dx - U = \int R(x, y, z)\,dx.$$

Par conséquent,

$$J = \int\int R(x, y, z)\,dx\,dy = \int\int\left(\frac{\partial U}{\partial x} + \frac{\partial V}{\partial y}\right)dx\,dy$$

est une intégrale impropre infinie uniquement sur des $H_y$, reliée de façon définie à la courbe C. Il s'agit d'en étudier les périodes.

Soit $\Gamma_2$ un cycle fini (seule espèce de cycle par rapport auquel J puisse avoir des périodes). Réduisons $\Gamma_2$, à la manière du n° 1, à un cycle ne coupant C qu'en $\pm(\Gamma_2 C)$ points. En se rapportant à la discussion (*loc. cit.*), on voit que l'on peut s'arranger pour que ces points soient tous dans $H_{y_0}$ ($y_0$ arbitraire) et que de plus $\Gamma_2$ et $H_{y_0}$ y aient en commun des petites $E_2$ entourant les points. En déformant $\Gamma_2$ on devra prendre soin d'éviter les $H_y$ d'infini de J, chose facile à faire.

Supprimons les $E_2$ dans $\Gamma_2$. La $M_2$ restante aura pour frontières les circuits frontières $\zeta$ des $E_2$. Puisque J est finie sur nos $E_2$,

$$\left|\int\int_{\Gamma_2} R\,dx\,dy - \int\int_{M_2}\right|$$

peut être rendue aussi petite que l'on veut. Or

$$\int\int_{M_2} = \sum \int_\zeta (U\,dy - V\,dx) = -\sum \int_\zeta V\,dx,$$

où il est entendu que les $\zeta$ sont orientés comme des indicatrices de $\Gamma_2$. Ces circuits en seront alors tout aussi bien pour $H_y$ (¹). On a donc

$$\int\int_{M_2} = -\sum \int_\zeta V\,dx = -(\Gamma_2 C).$$

Telle est aussi la valeur de la période de J relativement à $\Gamma_2$, puisqu'elle en diffère d'aussi peu que l'on veut.

*Remarque.* — Au lieu d'une courbe unique, prenons-en plusieurs, $C_1$, $C_2$, ..., $C_r$, puis supposons maintenant que

$$\int V\,dx$$

possède pour points logarithmiques les points $C_k H_y$ avec une période égale à $\mu_k$ On pourra toujours former J et le même raisonnement donne cette fois, pour sa période relativement à $\Gamma_2$, la valeur

$$-\sum \mu_k (\Gamma_2 C_k).$$

**13.** Les $\Gamma_2^j (j = 1, 2, \ldots, \rho_0)$ du paragraphe 1, ne rencontrant toute courbe algébrique qu'en un nombre algébriquement nul de points, sont tous homologues à des cycles finis

$$\sum \lambda_i \Delta_i + ((H_a)).$$

Toutefois, il y aura encore $\rho - 1$ tels cycles, soient

$$\Gamma_2^h \quad (h = 1, 2, \ldots, \rho - 1),$$

rencontrant chacun au moins une courbe en un nombre algébriquement non nul de points. On peut même affirmer que toute combinaison de ces cycles aura toujours cette propriété.

---

(¹) On le démontre en se basant sur ceci : Soient $\Gamma_2$, $\Gamma'_2$, $\Gamma''_2$, passant par un point A ordinaire pour tous trois, les deux premiers y ayant de plus en commun un circuit $\zeta$ passant par A. $\zeta$ sera orientable de manière à être une indicatrice (ou bien l'opposée d'une indicatrice) pour les deux cycles à la fois, pourvu que les contributions de A à $(\Gamma_2 \Gamma''_2)$ et $(\Gamma'_2 \Gamma''_2)$ soient égales. Ici $\Gamma''_2 = C$, $\Gamma'_2 = H_{y_0}$ et $\zeta$ entoure A, mais de faciles considérations de continuité montrent que le raisonnement est toujours valide.

Soient $C_1, C_2, \ldots, C_\rho$ des courbes indépendantes au sens de M. Severi. De la discussion du paragraphe 1, on conclut que la matrice

$$\| (\Gamma_2^{\rho_0+h} C_k) \| \qquad (h = 1, 2, \ldots, \rho - 1;\ k = 1, 2, \ldots, \rho)$$

est de rang $\rho - 1$. On peut par conséquent supposer le déterminant

$$|(\Gamma_2^{\rho_0+h} C_k)| \neq 0 \qquad (h, k = 1, 2, \ldots, \rho - 1).$$

Soit $J_k$ l'intégrale double analogue à celle du numéro précédent relative à $C_k$ ($k \leqq \rho - 1$). Le déterminant précédent est celui des périodes de $J_k$ relativement aux $\Gamma_2^{\rho_0+h}$. En se rappelant la nature de ces intégrales, on conclut: *Il y a une intégrale impropre, infinie uniquement sur des* $H_y$, *à périodes par rapport aux* $\Gamma_2^{\rho_0+h}$ *égales à des constantes quelconques.*

14. Théorème. — *Pour que* J *soit impropre, il faut et il suffit qu'elle soit sans période par rapport aux* $\Gamma_2^j (j \leqq \rho_0)$ (*cycles coupant toute courbe algébrique en un nombre algébriquement nul de points*).

D'après le corollaire II du n° 7, il y a lieu de considérer les périodes par rapport à nos $\Gamma_2$, quelle que soit l'intégrale double envisagée. En vertu du théorème I (*loc. cit.*), quand J est impropre, les périodes sont bien nulles, de sorte que la condition est effectivement *nécessaire.* Elle est aussi *suffisante.* En effet, nous pouvons prendre J sous la forme (1) du paragraphe 4, puisque la réduction s'y fait précisément par soustraction d'une intégrale impropre, donc sans changer les périodes envisagées. Soit alors J', impropre aux mêmes périodes que J relativement aux $\Gamma_2^{\rho_0+h}$, infinie uniquement sur des $H_y$. J — J' est sans périodes, infinie uniquement sur des $H_y$, donc impropre, et il en est de même de J. C. Q. F. D.

15. Théorème définitif. — *Le nombre d'intégrales doubles de seconde espèce linéairement indépendantes est égal au nombre* $\rho_0$ *de cycles distincts rencontrant toute courbe algébrique en un nombre algébriquement nul de points.*

En effet, nous pouvons former $\rho_0$ intégrales distinctes de la forme (1) du paragraphe 4 à résidus tous nuls et à déterminant de périodes par rapport aux $\Gamma_2^j (j \leqq \rho_0)$ non nul (n° 9). Ces intégrales sont de seconde espèce, linéairement indépendantes. Enfin, J étant quelconque de seconde espèce, il y aura une combinaison linéaire J' des intégrales précédentes, aux mêmes périodes qu'elle relativement aux cycles ci-dessus. J — J' sans périodes par rapport à eux est impropre (n° 14). Cela veut dire que J dépend des $\rho_0$ premières intégrales, ce qui démontre le théorème.

Corollaire : Formule de M. Picard. — De ce qui précède, on tire de suite cette formule de M. Picard :

$$\rho_0 = R_2 - \rho = I + 2R_1 - \rho + 2,$$

où, comme toujours, I désigne l'invariant de Zeuthen-Segre.

**16.** Le théorème précédent admet une généralisation complète aux $V_d$. On peut définir pour leurs intégrales $k$-uples ce que l'on entend par intégrales de seconde espèce linéairement indépendantes, et l'on a alors

THÉORÈME. — *Soient $r$ le nombre maximum de $\Gamma_k$ distincts d'une $V_d$ ne rencontrant pas un groupe donné d'hypersurfaces, $\rho_0^k$ le minimum de $r$ quand on envisage tous les groupes possibles. Le nombre d'intégrales $k$-uples de seconde espèce linéairement indépendantes est égal à $\rho_0^k$.*

## VI. — Intégrales simples de troisième espèce.

**17.** Nous avons déjà envisagé au Chapitre IV les intégrales simples

$$(1) \qquad \int U\,dy + V\,dx, \qquad \frac{\partial U}{\partial x} = \frac{\partial V}{\partial y},$$

nous limitant presque exclusivement à celles de première espèce. Une intégrale (1) générale possédera des *courbes logarithmiques*, c'est-à-dire ayant cette propriété : C étant l'une d'elles, $\zeta$ la frontière d'une petite $E_2$ rencontrant C sans contact, l'intégrale étendue à $\delta$ n'est pas nulle. En fait, sa valeur est constante quand $\zeta$ est déformé de manière quelconque sans rencontrer les courbes logarithmiques et cette constante est la *période logarithmique* relative à C. Assujettissons le point $(x, y, z)$ à rester dans une courbe non singulière, D, de l'intégrale. Cette dernière devient alors une intégrale abélienne de troisième espèce attachée à D, aux points logarithmiques CD, avec pour période relativement à chacun d'eux la période même de (1) relativement à C.

L'intégrale (1) est dite *de troisième espèce*, si elle possède effectivement des courbes logarithmiques.

**18.** Soient $C_1, C_2, \ldots, C_r$ les courbes logarithmiques de (1), $\mu_k$ la période relative à $C_k$. Sur une H, l'intégrale en définit une aux points logarithmiques $C_k H$ avec une période correspondante égale à $\mu_k$. Par suite,

$$(2) \qquad \sum \mu_k (C_k H) = 0,$$

d'où immédiatement $r \geqq 2$, du moins quand les $\mu$ ne sont pas tous nuls, c'est-à-dire quand (1) est effectivement de troisième espèce, comme on le supposera dorénavant. On doit à M. Picard un résultat d'une importance capitale en géométrie algébrique, dont la démonstration est l'objet de ce paragraphe :

THÉORÈME DE M. PICARD. — *Il existe toujours une intégrale simple dont* $\rho+1$ *courbes données sont les seules courbes logarithmiques. Par contre, il existe des systèmes de* $\rho$ *courbes ou moins qui ne peuvent être les seules courbes logarithmiques d'une intégrale simple de* $V_2$.

L'entier $\rho$ est bien entendu le même que nous avons désigné précédemment par cette lettre et que l'on nomme d'ailleurs communément *nombre de M. Picard.*

On trouvera la démonstration de ce théorème dans le Traité de MM. Picard et Simard, vol. II, Chap. IX. Nous allons en démontrer un ici un peu plus général, en suivant en quelque sorte la marche inverse de M. Picard. En effet, après avoir obtenu directement le théorème ci-dessus, l'illustre savant s'en est servi pour dériver certains résultats importants sur les intégrales doubles (*loc. cit.*, Chap. X). Ici, au contraire, nous allons faire usage de ce que nous avons appris sur les intégrales doubles pour en déduire très rapidement le théorème que nous avons en vue.

On consultera d'ailleurs avec grand profit sur cette question un fort important Mémoire de M. Severi (*Mathematische Annalen*, 1906), Mémoire que nous avons déjà eu l'occasion de citer. Les travaux de M. Severi ont éclairé d'un jour nouveau la théorie que l'on doit à M. Picard.

19. En se rappelant la définition de $\rho$ donnée au Chapitre IV, on voit que la proposition de M. Picard se ramène de suite à ce théorème :

THÉORÈME. — *La condition nécessaire et suffisante pour qu'il existe une intégrale simple aux courbes logarithmiques* $C_1, C_2, \ldots, C_r$, *est qu'elles soient reliées par une homologie* $(\operatorname{mod} V_2)$, *ou, ce qui revient au même, qu'elles soient dépendantes au sens de M. Severi.*

1° *La condition est nécessaire.* En effet, soient

$$(3) \qquad \int U\,dy - V\,dx, \qquad \frac{\partial U}{\partial x} + \frac{\partial V}{\partial y} = 0$$

l'intégrale de l'énoncé, $\mu_k$ sa période relative à $C_k$ [1]. Clairement, les périodes de

$$\int\int \left(\frac{\partial U}{\partial x} + \frac{\partial V}{\partial y}\right) dx\,dy$$

sont toutes nulles. La remarque à la fin du n° 12 est évidemment appli-

---

[1] On suppose que les C ne comprennent pas la courbe à l'infini. Cette circonstance un peu gênante peut être évitée par une transformation homographique préalable.

cable. En écrivant que l'expression des périodes ainsi obtenue est **nulle**, on a

$$\sum \mu_k(\Gamma_2 C_k) = 0 \tag{4}$$

pour tout $\Gamma_2$ fini. De plus, (2) montre que ceci est encore vrai pour le cycle H. Or tout $\Gamma_2$ dépend de H et des cycles finis. Donc (4) est vraie pour tout $\Gamma_2$. Soient $\Gamma_2^1, \Gamma_2^2, \ldots, \Gamma_2^{R_2}$ des cycles indépendants quelconques. La matrice

$$\| (\Gamma_2^j C_k) \| \qquad (j = 1, 2, \ldots, R_2;\ k = 1, 2, \ldots, r)$$

est de rang $r - 1$ au plus. Puisque ses termes sont tous entiers, il y a des entiers $\lambda$ non tous nuls tels que

$$\sum \lambda_k(\Gamma_2^j C_k) = \left[\left(\sum \lambda_k C_k\right) \Gamma_2^j\right] = 0 \qquad (j = 1, 2, \ldots, R_2).$$

Par conséquent (Chap. III, n° 18),

$$t\sum \lambda_k C_k \sim 0 \qquad (\operatorname{mod} V_2). \tag{5}$$

C. Q. F. D.

2° *La condition est suffisante*. En effet, (5) étant vérifiée, formons l'intégrale double

$$J = \int\int \left(\frac{\partial U}{\partial x} + \frac{\partial V}{\partial y}\right) dx\, dy$$

infinie uniquement sur des $H_y$ et telle que

$$\int V\, dx$$

ait pour points singuliers logarithmiques les points $C_k H$ avec une période égale à $\lambda_k$ (n° 12). En vertu de (5) on aura pour tout $\Gamma_2$

$$\sum \lambda_k(C_k \Gamma_2) = 0, \tag{6}$$

et en particulier pour H

$$\sum \lambda_k(C_k H) = 0. \tag{7}$$

Donc toutes les périodes de J sont nulles. Par conséquent (n° 10),

$$J = \int\int \left(\frac{\partial \overline{U}}{\partial x} + \frac{\partial \overline{V}}{\partial y}\right) dx\, dy, \qquad \overline{V} = \frac{A(x, y, z)}{\varphi(y) f'_z}$$

(A, polynome adjoint; $\varphi$, polynome).

On en tire

$$\frac{\partial(\mathrm{U}-\overline{\mathrm{U}})}{\partial x}+\frac{\partial(\mathrm{V}-\overline{\mathrm{V}})}{\partial y}=0,$$

de sorte que

$$(8)\qquad \int (\mathrm{V}-\overline{\mathrm{V}})\,dx-(\mathrm{U}-\overline{\mathrm{U}})\,dy$$

est une intégrale simple attachée à notre surface. De la forme de $\overline{\mathrm{V}}$ on conclut que (8) n'a, outre les C, que des $\mathrm{H}_y$ pour courbes logarithmiques. Soient $\mathrm{H}_{y_0}$ l'une d'elles, autre que la courbe à l'infini, $2\pi i\alpha$ la période correspondante de (8). Soustrayons de cette dernière

$$\alpha \log(y-y_0)=\int \frac{\alpha}{y-y_0}\,dy.$$

L'intégrale restante acquerra au pis aller la seule courbe à l'infini comme courbe logarithmique nouvelle, et aura perdu $\mathrm{H}_{y_0}$. Supprimons ainsi toutes les courbes logarithmiques de même nature. Soit $\lambda$ la période de l'intégrale finale relativement à la courbe à l'infini. La relation (2) qui lui correspond donne

$$m\lambda+\sum \lambda_k(\mathrm{C}_k\mathrm{H})=0.$$

Par suite, en vertu de (7),

$$m\lambda=0, \qquad \text{donc enfin} \qquad \lambda=0.$$

Ainsi, en réalité, la courbe à l'infini n'est nullement logarithmique pour l'intégrale finale, intégrale qui répond de toute évidence à la question. Ceci achève la démonstration.

# NOTE II.

## SUR LES MODÈLES DE M. VOLTERRA.

Nous avons eu l'occasion, au Chapitre I, n° 9, de décrire certaines variétés dues à M. Volterra. A propos de ces variétés, le grand savant de Rome a construit des modèles fort curieux. La planche ci-après, que nous devons à son obligeance, a pour but de montrer comment deux variétés, définies de manières fort différentes, peuvent n'en être pas moins homéomorphes, comme nous allons l'expliquer rapidement.

La $W_3$ de la planche correspond à la valeur 3 de l'entier $k$ (*loc. cit.*). Voici sa construction exacte : Soit, dans l'espace ordinaire, T la région comprise entre quatre sphères dont trois sont extérieures les unes aux autres, mais intérieures à la quatrième. On suppose de plus que les centres des quatre sphères sont dans un même plan Q. Soit maintenant T′ la région symétrique de T par rapport à un plan P ne rencontrant pas la plus grande sphère et perpendiculaire à Q. Convenons enfin de considérer comme coïncidants les points frontières de T et T′, qui se correspondent dans la symétrie par rapport à P. L'ensemble de T et T′ donne lieu ainsi à la $W_3$ de M. Volterra.

Il est clair que $W_3$ est tout aussi bien symétrique par rapport à Q. Coupons la figure par ce plan, et soient $T_1$, $T_2$ les deux sections de T, $T'_1$, $T'_2$ celles correspondantes de T′. On peut maintenant considérer $W_3$ comme formée de quatre parties, dont chacune est une $E_3$ limitée par quatre hémisphères et une partie plane. Les points des frontières planes doivent être liés par la pensée à leurs symétriques par rapport à Q, les autres comme auparavant. La figure 1 illustre nettement cette situation. Les parties bombées appartenant aux sphères extérieures sont derrière le plan de la figure.

Déformons maintenant $T_1$ en même temps que les parties correspondantes des autres $E_3$, de manière à aplanir les grandes hémisphères et à rendre au contraire les parties planes sphériques. On obtient ainsi la situation représentée par la figure 2. Chaque $E_3$ a pris maintenant la forme d'une hémisphère avec trois cavités dans la partie sphérique. Elles reposent sur le plan de la figure, la partie bombée en avant. Cette fois les parties planes se correspondent par symétrie relativement à P et il en est de même des cavités, alors qu'au contraire les hémisphères se correspondent par symétrie relativement à Q.

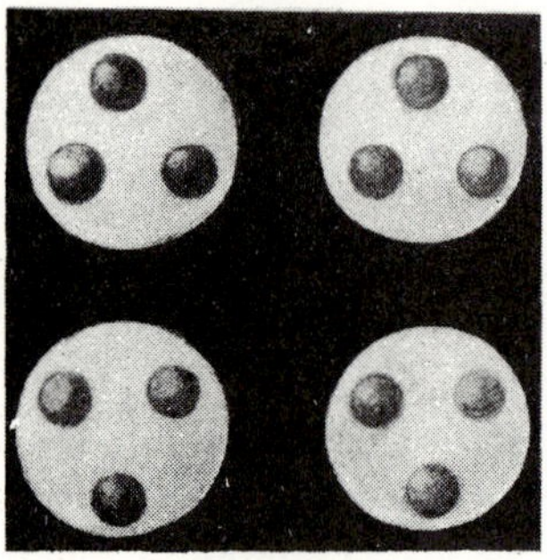

Fig. 1.

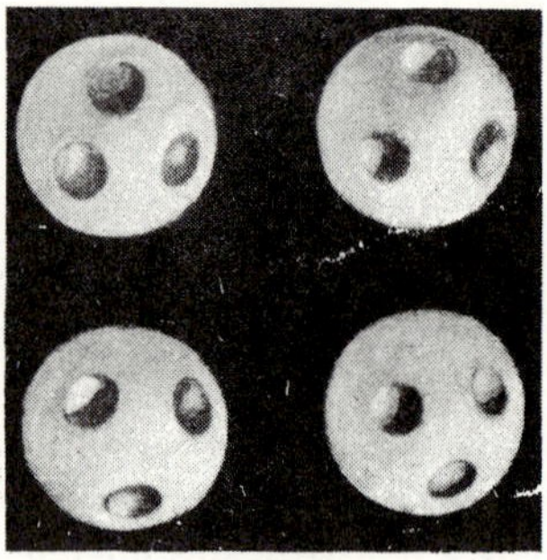

Fig. 2.

Fig. 3.

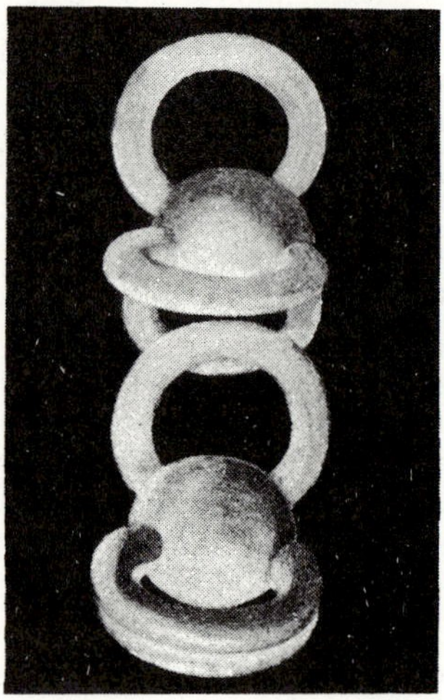

Fig. 4.

Joignons les parties planes à celles qui leur correspondent. $T_1$ et $T'_1$ donneront lieu ainsi à une sphère de centre dans P, sphère dont la surface contient six cavités deux à deux symétriques relativement à P. De même pour $T_2$ et $T'_2$.

Déformons enfin les cavités comme l'indique la figure 3, puis rejoignons chaque partie à sa symétrique (*fig.* 4). Tout cela revient à changer $W_3$ en une variété homéomorphe.

La forme finale obtenue est celle d'une région limitée par une sphère à trois anses et de sa symétrique par rapport à Q. Au fond, ce que l'on a fait, c'est de faire coïncider chaque point frontière de T avec son symétrique par rapport à P sur la frontière de T', quitte à introduire de nouvelles coupures.

# VI - VIII

# VI

## Invariance absolue et invariance relative en géométrie algébrique

1. Le but du présent travail est de montrer comment certaines de nos recherches récentes peuvent servir à jeter quelque lumière sur les types d'invariance que l'on rencontre en géométrie algébrique.

Le groupe fondamental d'une variété algébrique $V_n$ est, comme on le sait, son groupe birationnel. Or parmi ses opérations il y a lieu de distinguer:

a) les transformations qui ne modifient la dimension de nulle $V < V_n$ et n'introduisent, ni ne font disparaître aucune singularité;

b) les transformations à éléments dits fondamentaux, c'est-à-dire dont la dimension est altérée, ou qui sont transformés en nouveaux éléments singuliers.

Certaines fonctions, numériques ou autres, de $V_n$ sont invariantes par rapport à toutes les transformations birationnelles sans exception; d'autres au contraire sont modifiées par les transformations b). Les premières sont nommées invariants absolus, les autres — invariants relatifs. Par exemple les nombres maxima d'intégrales de première espèce des diverses multiplicités, linéairement indépendantes sont des invariants absolus, les nombres de Betti peuvent ne pas l'être. Ceci n'est pas surprenant puisque les transformations b) n'induisent pas un homéomorphisme sur $V_n$. Il y a donc lieu de considérer la transformation topologique associée à une transformation birationnelle donnée, en vue d'en faire ressortir, si possible, les caractères absolument invariants. C'est ce que nous nous proposons de faire, nous bornant, sauf quelques remarques finales, aux surfaces algébriques.

En fait de notations et de terminologie nous suivrons notre ouvrage: „Topology" (American Mathematical Society Colloquium Publications, Vol. XII, New York 1930).

2. Pour éclairer la question il sera bon de considérer quelques exemples.

Premier exemple. Les transformations projectives d'un $S_n$ sont du type a), ses transformation quadratiques du type b).

Deuxième exemple. On prend deux surfaces algébriques $\Phi$, $\Phi'$, de $S_n$, dont la première est un cône de genre $p$, la seconde — une surface réglée dont une section hyperplane coïncide avec une base du cône. Les deux surfaces sont alors birationnellement équivalentes. On supposera qu'à part son sommet $\Phi$ n'a pas de singularités et que $\Phi'$ n'en a aucune — conditions facilement réalisables pour $n$ assez grand et $p$ quelconque. On a de suite pour le premier nombre de Betti (connexion linéaire) $R_1(\Phi') = 2p$.

Par contre tout cycle linéaire de $\Phi$ peut être déformé homotopiquement en son sommet; d'où $R_1(\Phi)=0$. Donc le nombre $R_1$ n'est certes pas un invariant absolu. C'est cependant ce que l'on a toujours admis jusqu'ici. En fait ce n'est pas $R_1$ qui est invariant, mais un autre nombre, comme nous le verrons plus loin.

Troisième exemple. Soit $\Phi$ une quadrique non dégénérée de $S_3$; on prend un point $A$ sur $\Phi$ et l'on projette $\Phi$ de $A$ sur un plan $S_2 \not\ni A$. Il en résulte une transformation birationnelle de $\Phi$ en $S_2$, en vertu de laquelle les deux génératrices $G$, $G$ de $\Phi$ passant par $A$ sont transformées en deux points $g$, $g'$, et $A$ lui-même en la droite $gg'$ toute entière. Par contre en passant de $S_2$ à $\Phi$ par la transformation inverse, $gg' \to A$, $g \to G$, $g' \to G'$. Si l'on considère les nombres $\rho$ de Picard et $R_2$ (deuxième nombre de Betti), on trouve

$$R_2(\Phi)=\rho(\Phi)=2,\quad R_2(S_2)=\rho(S_2)=1.$$

Ainsi ces deux nombres sont modifiés par la transformation; ce sont des invariants relatifs, c'est-à-dire seulement sous les transformations a). Par contre $R_2-\rho=0$, tant pour $\Phi$ que pour $S_2$, et en fait c'est un invariant absolu.

3. Avant d'aborder notre problème nous allons démontrer un lemme topologique qui a d'ailleurs quelque intérêt propre. Il se rapporte à une multiplicité relative $M_n=K_n-L$, où $K_n$ est un complexe simplicial et $L$ un sous-complexe de $K_n$. Nous désignerons génériquement par $\Gamma$ et $G$ les cycles de $K_n$ mod $L$ et ceux du type duel: cycles de $K_n-L$.

Lemme. *Soient $\Gamma_p^i (i=1, 2,\ldots, r)$ des sous-cycles mod $L$ ayant ces deux propriétes: a) ils sont des circuits simples mod $L$; b) ils n'ont en commun deux à deux sur $K_n-L$ que des cellules de dimension $<p-1$. Pour que $G_{n-p}$ soit homologue sur $K_n-L$ à un cycle ne rencontrant pas les $\Gamma_p^i$, il faut et il suffit que les indices de Kronecker $(\Gamma_p^i \cdot G_{n-p})$ soient tous nuls.*

La subdivision indéfinie étant permise nous pouvons choisir $K_n$ simplicial, puis réduire $G_{n-p}$ à un sous-cycle du duel $K^*$ de $K_n-L$. Nous continuerons à nommer ce sous-cycle $G_{n-p}$.

4. Prenons d'abord deux cellules adjacentes $E_p$, $E_p'$ de $\Gamma_p^i$ sur $K-L$, c'est-à-dire incidentes sur la même $E_{p-1}$ de $\Gamma_p^i-L$. Comme $\Gamma_p^i$ est un circuit simple, il n'aura que ces deux $p$-cellules en incidence avec $E_{p-1}$. De plus $E_{p-1}$ n'appartient à aucun $\Gamma_p^j$, $j \neq i$; par suite il n'y aura que les deux cellules $E_p$, $E_p'$ de $\Gamma_p=\Sigma\Gamma_p^i$ en incidence avec $E_{p-1}$.

Passant aux cellules duelles on trouve que $E_{n-p}^{*\prime}$, $E_{n-p}^*$ sont des cellules de la frontière de $E_{n-p+1}^*$ et qu'en dehors d'elles cette frontière ne rencontre pas $\Gamma_p$. Eu égard à nos conventions d'orientation („Topology" p. 137) on aura

$$F(E_{n-p+1}^*)=E_{n-p}^{*\prime}+E_{n-p}^*+C_{n-p}\sim 0 \text{ sur } K_n-L, \tag{1}$$

où $C_{n-p}$ ne rencontre pas $\Gamma_p$.

Je dis que (1) subsiste même quand $E_p$, $E_p'$ sont des cellules non adjacentes de $\Gamma_p^i$. En effet, puisque la somme des $p$- et $(p-i)$-cellules de $\Gamma_p^i-L$ est connexe, il existe une suite de ses cellules, $E_p^1=E_p, E_p^2, \ldots, E_p^{r+1}=E_p'$, dont deux consécutives sont toujours adjacentes. On aura alors

$$E^{*i}_{n-p} + E^{*i+1}_{n-p} + C^{*i}_{n-p} \sim 0 \text{ sur } K_n - L,$$

où $C^{*i}_{n-p}$ ne rencontre pas $\Gamma_p$; en prenant la somme de ces homologies on obtient de nouveau (1).

Ceci posé, désignons pour l'instant par $E^{*ij}_{n-p}$ les duelles des cellules de $\Gamma^i_p - L$. On aura

$$G_{n-p} = \sum x_{ij} E^{*ij}_{n-p} + C_{n-p}, \tag{2}$$

où $C_{n-p}$ ne comprend nulle cellule $E^{*ij}$. On a, eu égard à nos conventions d'orientation:

$$(\Gamma^i_p \cdot G_{n-p}) = \sum_j x_{ij}. \tag{3}$$

Soit maintenant $E^{*i}_{n-p}$ la duelle d'une cellule fixe de $\Gamma^i_p - L$. D'après (1) nous avons les homologies

$$E^{*i}_{n-p} + E^{*ij}_{n-p} + C^{*ij}_{n-p} \sim 0 \text{ sur } K_n - L,$$

où les $C^{*ij}_{n-p}$ ne rencontrent pas les $\Gamma$. Par suite, en vertu de (2) et (3),

$$G_{n-p} \sim - \sum (\Gamma^i_p \cdot G_{n-p}) E^{*i}_{n-p} + C_{n-p} \text{ sur } K_n - L, \tag{4}$$

la chaine $C^{*}_{n-p}$ ne rencontrant pas les $\Gamma$. Donc, si les indices de Kronecker dans la somme sont tous nuls, le cycle à droite ne rencontre pas les $\Gamma$. La condition du lemme est donc suffisante. Elle est aussi nécessaire, car les indices ne changent pas quand on remplace $G_{n-p}$ par un cycle homologue sur $K_n - L$, et, lorsqu'un cycle $G_{n-p}$ ne rencontre pas les $\Gamma$, les indices $(\Gamma^i \cdot G)$ sont tous nuls. Le lemme est donc démontré.

Remarques. I. On démontrerait à peu près de même un lemme analogue où les $\Gamma$ et $G$ seraient des deux types duels $K_n - L^1$ mod $L^2$ et $K_n - L^2$ mod $L^1$ considérés au Ch. III de „Topology", No. 29.

II. Une autre modification qui ne changerait rien à la démonstration serait de prendre pour $G$ une chaine de $K_n - L$ dont la frontière ne rencontre pas les $\Gamma$. Le lemme est donc également applicable à ce cas.

III. On trouvera un résultat particulier analogue dans notre monographie de la collection Borel: „L'Analysis Situs et la Géométrie Algébrique", p. 129, démontré à l'aide de certains types de tubes, les variétés étant toutes analytiques. Le complexe duel $K^*$ dont nous nous sommes servis ici constitute l'instrument idéal pour éviter les considérations assez délicates de tubes et autres configurations compliquées.

5. Reprenons maintenant la surface algébrique $\Phi$. Elle peut posséder un certain nombre de points singuliers isolés et de courbes multiples constituant son ensemble singulier $A_0$. On sait que $\Phi - A_0$ est une $M$ (voir „Topology", Ch. VIII), et que l'on peut définir une fois pour toutes pour elle et pour ses courbes algébriques des 4- et 2-cycles fondamentaux bien précis. Les 2-cycles donnent lieu à un module $a$, ou groupe abélien, que nous nous proposons d'analyser d'un peu plus près. Lorsque $\Phi$ n'a que des singularités ordinaires, voire pas de singularités du tout, on sait déjà d'après nos résultats antérieurs que:

1°. Le groupe des diviseurs de zéro de $a$ est le même que celui de $\Phi$.

2°. Le nombre de Betti $R_2(a) \leqslant R_2(\Phi)$. D'ailleurs $R_2(a)$ n'est autre que le nombre $\rho$ de Picard (nombre de base de Severi).

Puisque le groupe $G_2$ de tous les $\Gamma_2$ de $\Phi$ est abélien, à nombre fini d'éléments générateurs, il en est de même de ses sous-groupes, et en particulier de $a$. On en déduit sans peine qu'il existe un certain nombre de courbes algébriques formant base pour $a$. Une telle base pourrait comprendre certaines courbes de $A_0$ et en outre certaines courbes nouvelles $A_1, \ldots, A_r$. Ici, comme partout dans la suite, $A_1$ dénote indifféremment la courbe ou le $\Gamma_2$ fondamental associé.

Posons $B = A_0 + \ldots + A_r$ et désignons par $B$ le sous-groupe de $G$ composé de la totalité des deux-cycles de $\Phi - B$. Un élément quelconque de $B$ est donc un $\Gamma_2 \subset \Phi - B$, et le zéro du groupe (écrit additivement) est formé par les cycles $\sim 0 \bmod B$ sur $\Phi$. Le n o m b r e d e B e t t i, ou r a n g du groupe, est le nombre maximum d'éléments dont aucune combinaison linéaire n'est $\sim 0$ sur $\Phi$. C'est le nombre que nous avons désigné par $r_2(\Phi - B)$ dans „Topology" (p. 146).

6. T h é o r è m e I. *Le nombre* $r_2(\Phi - B)$ *est un invariant absolu de* $\Phi$.

Je dis d'abord que l'addition d'un nombre fini de courbes algébriques ou de points isolés à l'ensemble $B$ ne modifie pas le sous-groupe $B$. Il suffit évidemment de le démontrer pour l'addition d'une seule courbe algébrique irréductible $A$ ou d'un seul point $P \not\subset B$.

Soit donc $\Gamma_2 \subset \Phi - B$. On peut supposer $\Phi$ recouverte d'un $K_4$ simplicial dont un sous-complexe recouvre $B + A$. D'un autre côté comme certaines courbes de $B$ forment base pour $a$, on a $A \sim 0 \bmod B$ sur $\Phi$. Comme $\Gamma_2 \subset \Phi - B$ il s'ensuit que $(\Gamma_2 \cdot A) = 0$, et, par suite, d'après le lemme, $\Gamma_2$ est homologue sur $\Phi - B$ à un cycle ne rencontrant pas $A$, c'est-à-dire à un cycle de $\Phi -$ $- (B + A)$. On en déduit de suite que $B$ est isomorphe au groupe semblable correspondant à $B + A$. Au point de vue abstrait $B$ n'est donc pas modifié quand on ajoute $A$ à $B$.

Quant à l'addition du point $P$, pour en déterminer l'effet remarquons que l'on peut recouvrir $\Phi$ d'un $K_4$ simplicial, avec $P$ parmi ses sommets, un sous-complexe recouvrant $B$ et tel, de plus, que nulle cellule fermée de sommet $P$ n'ait un sommet sur $B$. Puisque $\Phi - B$ est une $M_4$, l'étoile de $P$ est une $E_4$. Si l'on remplace dans $K_4$ toutes les cellules de l'étoile par cette $E_4$ on obtient un complexe combinatoire $K_4'$. D'aprés ce que nous avons démontré („Topology", Ch. II, p. 107), tout $\Gamma_2$ de $K_4' - B$ est homologue sur $K_4' - B$ à un sous-cycle $\Gamma_2$ de $K_4'$. Comme $\Gamma_2'$ ne passe pas par $P$, nous voyons que l'addition de $P$ ne modifie pas $B$.

Soit maintenant $T$ une transformation birationnelle de $\Phi$ en une nouvelle surface $\Phi'$, et soient $B', \ldots$ des éléments analogues à $B, \ldots,$ pour $\Phi'$. D'après ce qui précède nous pouvons ajouter à $B$ les éléments fondamentaux de $T$ sur $\Phi$, ainsi que les points de $T^{-1}B'$ qu'il ne comprend pas encore, et de même pour $B'$. Ceci étant le cas, $T$ est univoque et continue sur $\Phi - B$; son inverse $T^{-1}$ y est partout déterminée et est univoque et continue sur $\Phi' - B'$. De plus $T(\Phi - B) = \Phi' - B'$, $T^{-1}(\Phi' - B') = \Phi - B$, et $T$ est une transformation homéomorphique entre ces deux ensembles. Comme $\Phi$, $B$, $\Phi'$ $B'$ coïncident avec des complexes, ces ensembles sont localement connexes au sens de „Topology", Ch. II, § 5. On montre alors essentiellement, comme loc. cit., que $r_2(\Phi - B)$ n'est pas modifié par l'homéomorphisme $T$ opérant sur $\Phi - B$. Ceci démontre notre théorème.

7. Théorème II. *Le nombre* $r_2(\Phi - B) = \rho_0$, *nombre maximum d'intégrales doubles de seconde espèce de* $\Phi$ *dont aucune combinaison linéaire n'est impropre.*

Soient $x, y$ deux fonctions rationelles sur $\Phi$, fonctionnellement indépendantes. Une intégrale impropre de seconde espèce de $\Phi$ est une intégrale de forme

$$\int\int\left(\frac{dU}{dx}+\frac{dV}{dy}\right) dx\, dy; \quad U,\ V\text{— rationnelles sur } \Phi. \tag{5}$$

Une intégrale de seconde espèce de $\Phi$ est une intégrale de forme

$$\int\int R\, dx\, dy; \quad R \text{ rationnelle sur } \Phi, \tag{6}$$

qui, à une intégrale (5) près, est finie localement sur un voisinage arbitrairement petit d'un point quelconque $P$ de $\Phi$ (Picard). Le terme voisinage n'est guère indiqué ici avec précision; il nous suffit en fait de savoir que, pour une $\Phi$ sans singularités, le voisinage d'un point $P$ est une petite $E_4$ analytique $\supset P$. Remarquons d'ailleurs que: 1° les intégrales du type (1) sont de formes birationnellement invariantes; 2° nous pouvons définir les intégrales de seconde espèce par la condition d'être de seconde espèce au sens ci-dessus quand on transforme $\Phi$ birationnellement en une surface sans singularités $\Phi'$. L'existence d'une telle $\Phi'$ ayant été démontrée par MM. Beppo Levi, Chisini, Albanese, nous avons une définition précise des intégrales en question, sans avoir à passer par les fourches caudines de voisinages plus ou moins bien définis sur $\Phi$.

8. Passons à la démonstration du théorème II. D'après le théorème I, nous pouvons supposer que $\Phi$ n'a pas de singularités, de sorte qu'elle est une $M$ absolue. Soient alors $A_1, \ldots, A_\rho$ un système maximal de courbes indépendantes en tant que cycles, dans $B$. D'après un théorème connu (Severi; Lefschetz „Topology" p. 389) le déterminant $|(A_i\, A_k)| \neq 0$. Par suite étant donné un $\Gamma_2$ de $\Phi$ on peut trouver des nombres rationnels $x$ tels qu'en posant

$$\Gamma_2' = \Gamma_2 - \sum x_i \cdot A_i,$$

on ait $(\Gamma_2' \cdot A_i) = 0$ pour tous les $i$. Or pour toute courbe algébrique $A$ on a une homologie

$$A \approx \sum y_i\, A_i, \quad y_i\text{— rationnel,}$$

d'où l'on tire

$$(\Gamma_2' \cdot A) = \sum y_i\, (\Gamma_2' \cdot A_i) = 0.$$

Comme on peut recouvrir $\Phi$ d'un $K_4$ simplicial dont un sous-complexe recouvre $B$, notre lemme est applicable à $\Gamma_2'$: le cycle est homologue sur $\Phi$ (donc mod $B$) à un cycle de $\Phi - B$. Nous en concluons que tout 2-cycle de $\Phi$ a un multiple homologue mod $B$ à un cycle de $\Phi - B$. Donc

$$r_2(\Phi - B) = R_2 - \rho = \rho_0,$$

en vertu de résultats déjà connus (Picard, Lefschetz; voir „L'Analysis Situs et la Géométrie Algébrique", p. 145).

Le théorème II, eu égard au théorème I, démontre indirectement l'invariance absolue de $\rho_0$ (sous toutes les transformations birationnelles sans exception).

Il résulte de la discussion précédente que, lorsqu'elle est non-singulière, $\Phi$ ne possède pas de $\Gamma_2$ indépendants des courbes algébriques est des cycles de $\Phi - B$. Ces $\Gamma_2$ seraient les cycles indépendants des courbes et des cycles de $\Phi - B$, dont toutefois aucune combinaison linéaire ne pourrait éviter de rencontrer des courbes et en particulier $B$. Le nombre maximum de tels cycles est l'invariant que nous avons désigné par $t_2(B)$ („Topology", p. 149). D'après ce qui précède on a donc $t_2(B) = 0$.

9. Le cas des cycles linéaires ne présente aucune difficulté nouvelle. Tout d'abord $B$ étant de dimension deux, tout $\Gamma_1$ est homologue sur $\Phi$ à un cycle de $\Phi - B$. Désignons alors génériquement par $\Delta_1$ un cycle de $\Phi - B$ situé sur une courbe algébrique quelconque $A$ et $\sim 0$ sur $A$. Les cycles $\Delta_1$ donnent lieu à un groupe abélien $D_1$. Soit $G_1$ le groupe de tous les cycles linéaires de $\Phi - B$. On a alors:

Théorème. *Le groupe-quotient* $H_1 = G_1 \div D_1$ *(ou additivement* $G_1 - D_1$*) est un invariant absolu de* $\Phi$.

Comme pour la dimension deux il suffit de démontrer que si l'on ajoute à $B$ un point $P$ ou une courbe irréductible $A$ on ne change pas $H_1$. Pour $P$ cela est immédiat, et pour $A$ cela résulte de ce que l'on peut construire une $E_2 \to \Delta_1$, où $E_2$ est sur une courbe algébrique et $(E_2 \cdot A) = 1$.

Comme conséquence du théorème précédent le rang $r$ du groupe $H_1$ et l'ordre $o$ de son groupe des diviseurs de zéro sont des invariants absolus. En s'en rapportant à l'image birationnelle, sans singularités, $\Phi'$ de $\Phi$ on montre que: 1° $\varsigma$ est égal au même nombre pour le groupe de tous les cycles absolus de $\Phi'$ et partant au nombre $\sigma$ de M. Severi; 2° $r = 2q$, $q$ — irrégularité de $\Phi'$, donc de $\Phi$. C'est là la première démonstration par voie purement topologique de l'invariance absolue de ces deux nombres.

10. Cas d'une $V_n$, $n > 2$. Ce qui précède porte à faire croire que pour une $V_n$ quelconque on peut définir des invariants absolus, analogues aux groupes de Betti, de la façon suivante: on se donne un ensemble algébrique $B$ tel que $V_n - B$ soit une $M_{2n}$ avec $G_p$ pour groupe de ses $p$-cycles. Soit $D_p$ le groupe des $p$-cycles des $V_d - B$ de $M_{2n}$ qui sont $\sim 0$ sur $V_d$. Admettons que pour $B$ suffisamment ample, $G_p \div D_p = H_p$ ne change pas quand on ajoute à $B$ des sous-variétés algébriques de $V_n$ en nombre fini. Alors $H_p$ est un invariant absolu topologique de $V_n$.

(Поступило в редакцию 29/V 1932 г.)

# Абсолютная и относительная инвариантность в алгебраической геометрии

**С. Лефшец** (Принстон, САСШ)

*(Резюме)*

Бирациональные преобразования $n$-мерного алгебраического многообразия $V_n$ бывают двух типов — в зависимости от того, всюду ли они регулярны или нет. Бирациональный инвариант многообразия $V_n$ называется *абсолютным*, если он не меняется ни при каком бирациональном преобразовании $V_n$, и *относительным*, — если лишь известно, что он не меняется при преобразованиях первого типа. В настоящей работе дается (впервые) чисто топологическое определение некоторых абсолютных инвариантов алгебраической поверхности с указаниями, относящимися к многообразиям высшего числа измерений.

# VII

## Locally connected sets and their applications

In the last few years an increasing number of investigations have had for their objective the sets which we propose to discuss in some of their phases here. As is well known LC-sets (abridged for locally connected sets) were first considered in connection with the characterization of the so-called continuous curves (Peano or Jordan continua), but it is only through algebraic topology that they have assumed their present development.

Let us first examine and codify, as it were, the various types of spaces falling under the LC category. Taking for the sake of simplicity a compact metric space $\mathcal{R}$, we shall call it $p$-LC whenever for every $\varepsilon$ there is an $\eta(\varepsilon, p)$ such that every singular $p$-sphere $H_p$ of diameter $< \eta$ on $\mathcal{R}$ bounds on $\mathcal{R}$ a singular cell $E_{p+1}$ whose diameter $< \varepsilon$. We shall say that $\mathcal{R}$ is

LC$^p$ whenever it is $q$-LC for every $q \leqslant p$,

weak LC whenever it is LC$^\infty$ and the numbers $\eta(\varepsilon, p)$, $p = 0, 1, \ldots$, have a lower bound $\eta > 0$.

One would be tempted to think that the weak LC type is the only reasonable LC type uniform as to $p$ for $p \to \infty$. Curiously enough this is not the case and a more stringent type is needed. To define it we must first introduce the very useful notion of a s e m i - s i n g u l a r complex $K$ on $\mathcal{R}$. Roughly speaking it is a singular image of an assigned simplicial complex $\mathcal{K}^\circ$ with only part of the singular cells, including all vertices, present. $K$ is the actual singular image of a subcomplex of $\mathcal{K}^\circ$. Its mesh is by definition the maximum diameter of the semi-singular images of the cells of $\mathcal{K}^\circ$ (sum of the faces effectively present). The LC$^p$ condition is equivalent to this: for every $\varepsilon$ there exists an $\eta(\varepsilon, p)$ such that every semi-singular $K_p$ of mesh $< \eta$ may be completed to the proper singular image with mesh $< \varepsilon$. The strong LC or merely LC is characterized by the fact that it fulfils the preceding condition for every $p$, and with a lower bound $\eta > 0$ for the numbers $\eta(\varepsilon, p)$. If $\eta(\varepsilon)$ and $\varepsilon$ have the same upper bound we designate $\mathcal{R}$ as an $\overline{\text{LC}}$. For such a set the semi-singular complex may be completed without restrictions on the meshes.

At this point we come upon the sets called r e t r a c t s by Borsuk. We recall that under his definition $A \subset B$ is a retract (= R) of $B$ whenever there exists a continuous single valued transformation (c. s. v. t.) of $B \to A$, $= 1$ on $A$. The set $A$ is a neighborhood R (= NR) whenever $A$ is an R of a neighborhood of $A$ on $B$. Finally $A$ is an absolute R (= AR) if it is an R for every $B \supset A$ and similarly for ANR. In our work on coincidences and fixed points we already encountered ANR sets and showed that the basic formulas for signed fixed points hold for a c. s. v. t. of $A \to B$ when $B$ is an ANR. We have recently shown that:

ANR $\backsim$ strong LC, AR $\backsim \overline{\text{LC}}$, the characterizations being independent of dimensions.

The chief tool in the proof was this, a noteworthy lemma:

If $A$ is a closed subset of the Hilbert parallelotope $\mathcal{H}$, there is an infinite complex $K \subset \mathcal{H} - A$, finite away from $A$, and a c. s. v. t. of $\mathcal{H} \to K + A, = 1$ on $A$ and such that $\mathcal{H} - A \to K$.

A close parallel to the preceding theory is obtained if we make the following changes:

(a) in the definition of $p$-LC replace $p$-sphere and $(p+1)$-cell by $p$-cycle and $(p+1)$-chain, the bounding being chain-bounding; the new types of sets are described by prefixing H (= in the sense of homology): $p$-HLC, etc.,

(b) replace c. s. v. t. or deformations wherever they occur by chain-deformation $\mathfrak{D}$ or a relation between two chains $C_p$, $C'_p$ wherein there occur two auxiliary chains $\mathcal{D}C_p$, $\mathcal{D}F(C_p)$ ($\mathcal{D}$ raises the dimension by one) such that

$$F\mathcal{D} = \mathfrak{D} - 1 - \mathcal{D}F,$$

(c) replace complexes by certain analogous aggregates of chains (chain-complexes),

(d) H-retraction (in the natural sense) is defined by comparing HLC-sets and subsets only.

The new theory thus obtained is in a sense the abstraction of the LC-theory properly dosed for the purposes of algebraic topology. Both LC- and HLC-sets have in fact the same general homology properties as finite complexes.

There is of course an HLC-theory corresponding to every general class of homology groups. However in view of the recent results of E. Cech and N. E. Steenrod it will probably be unnecessary to consider any other coefficients than integers and real numbers mod 1. On the other hand singular chains, or the different types of abstract chains may well give different results. In any case when the chains are singular

$$\text{LC}^p \supset \text{HLC}^p, \quad \text{LC} \supset \text{HLC}.$$

As we have already observed the LC-sets arise naturally in the study of continuous curves and in the fixed point problem. The HLC-theory is particularly adapted to the critical point problem and the topology of extremals developed by Morse. For in the first place, his topological theorems belong to algebraic topology; in the second place they are obtained by means of the spaces whose elements are broken extremals and these are closely related to the HLC-types. In fact the use of abstract chains in this problem enables one to simplify very greatly the topological work.

There are a number of sets related to the LC- and HLC-types, notably the locally contractible and the locally polyhedral sets. Both occur in the study of extremal spaces and could be used very conveniently in algebraic geometry, where the topology would thus also be considerably simplified. They include the classes of topological manifolds investigated before the recent work of Cech, Wilder and the author. Furthermore, the LC- or HLC-properties are closely related to the generalized manifolds of these authors. We recall finally that local contractibility is at the very root of the method which enabled Lusternik and Schnirelmann to characierize and single out classes of extremals. Their work is in a certain sense parallel to that of Morse, with chain-deformation replaced by certain set-homotopies throughout.

---

## Локально связные множества и их приложения

**С. Лефшец** (Принстон, С. Ш. А.)

*(Резюме)*

Исследуются различные типы многомерной локальной связности в их взаимоотношении с другими понятиями (ретракт, сингулярный комплекс и т. п.).

---

# VIII

## Singular and continuous complexes, chains and cycles [1]

Few topological concepts have caused more misgivings than the so-called singular configurations. They were first introduced by Alexander [2] to prove the topological invariance of the homology characters [3]. In time, however, two distinct types of concepts were described by similar terms, hence causing obvious difficulties. The intuitive value of «singular» paths or «manifolds» of integration, for example, appear to justify a careful examination of the situation. This we have endeavored to do here with a view of making these notions more generally and also more comfortably useful. This has been greatly facilitated by a result just obtained by A. W. Tucker [4] making it possible to eliminate the troublesome degenerate cells from the argument.

We have sharply separated singular and continuous elements from one another and given a theorem which states explicitly how they are related. We have also illustrated their differences by simple examples and indicated some of their applications.

### I. Singular complexes, chains and cycles

**1.** We recall our definition of singular cells on a topological space $\mathfrak{R}$ (**Bull. M. S.**). We suppose that we have a mapping $T$ of a closed oriented geometric simplex $\bar{\sigma}_p$ to a subset $\bar{\alpha}$ of $\mathfrak{R}$, where $\alpha = T\sigma_p =$ the transform of the open simplex. Consider all the triples $(\sigma_p, T, \alpha)$ and define as equivalent two

---

[1] We shall need the following principal references to our other writings which we give together with the abbreviation adopted for them:

Colloquium Lectures: Topology, New York, (1930). (**Topology**);

Annals of Math., **35**, (1934), 118—129. (**Ann. M.**); see 610—621;

Duke Journal, **1**, (1935), 1—19. (**D. J.**); see 525—542;

Bull. Am. Math. Soc., **39**, (1933), 124—129. (**Bull. M. S.**); see 479—484.

[2] J. W. Alexander, Transactions of the American Math. Soc., **16**, (1915), 148—154. See also Veblen's proof of the same theorem in his Colloquium Lectures: Analysis Situs, p. 55, as well as our treatment, **Topology**, p.p. 84—89.

[3] It is not at all our aim to prepare invariance proofs of these characters by the Alexander-Veblen «deformation method», as we believe that chains, singular, continuous (and of other types as well) may be utilized to great advantage in many other questions.

[4] A. W. Tucker, Annals of Mathematics, **34**, (1933), 191—243.

triples $(\sigma_p, T, \alpha)$ and $(U\sigma_p, TU^{-1}, \alpha)$, $U$ affine. This equivalence is reflexive, symmetric and transitive and we may introduce classes of equivalent triples. Such a class is called an *oriented singular* $p$-cell $E_p$. Any particular triple in the class is a *parametrization* of $E_p$, and $\sigma_p$ is an *antecedent* of the singular cell. The set $\alpha$ is designated by $|E_p|$, its diameter by diam $E_p$. The class of the triple $(-\sigma_p, T, \alpha)$ determines an oriented singular cell designated by $-E_p$. Thus a given geometric simplex $\sigma_p$ and mapping $T$: $\bar{\sigma}_p \rightarrow \bar{\alpha}$ may give rise to two oppositely oriented singular $p$-cells $+E_p$ and $-E_p$.

We now define the singular $p$-chains over any group of coefficients $\mathfrak{G}$ as expressions

$$C_p = x_i E_p^i, \quad x_i \varepsilon \mathfrak{G}. \tag{1,1}$$

Here we make the usual convention

$$x(-E_p) = (-x)(+E_p), \tag{1,2}$$

where $(-E_p)$ designates $(+E_p)$ «reversed», or the cell derived from $(+E_p)$ by reversing its antecedent $\sigma_p$. Under these conditions $\{C_p\}$ is an additive Abelian group, the group of the singular $p$-chains over $\mathfrak{G}$.

Since $T$ is defined over $\sigma_p$, if $\sigma_q$ is a face of $\sigma_p$, with $T\sigma_q = \beta$, then $E_q = \{(\sigma_q, T, \beta)\}$ is a singular $q$-cell, called a *face* of $E_p$. Unless otherwise stated we do not exclude $q = p$ and hence consider $E_p$ as a face of itself.

In Tucker's notation [4] we are now free to introduce the geometric incidences and incidence-numbers, between singular cells. For the first we agree that $E_q < E_p$ whenever $E_q$ is a face of $E_p$. This condition is transitive and it furnishes the incidence part of the three conditions for abstract complexes. The other two refer to the boundary operator $F$. We must define it as a homomorphism $\{C_p\} \rightarrow \{C_{p-1}\}$ such that $FF = 0$. Let $\sigma_{p-1}^i$ be the $(p-1)$ faces of $\sigma_p$ and $E_{p-1}^i$ their images. If

$$F(\sigma_p) = \eta_i \sigma_{p-1}^i, \tag{1,3}$$

we define as the boundary of the cell $E_p$ the chain

$$F(E_p) = \eta_i E_{p-1}^i. \tag{1,4}$$

After collecting at the right, the coefficients of the distinct singular $(p-1)$-faces are the incidence-numbers $[E_p : E_{p-1}^i]$. We notice that $F(E_p)$ is independent of the parametrization, hence $F$ is a single-valued function of $E_p$. For if we have another parametrization $(U\sigma_p, TU^{-1}, \alpha)$, the definition of the faces shows that the only alteration in (1,3) is the transcription $\sigma \rightarrow U\sigma$, and that (1,4) is unchanged. We then define $F$ for all chains by linearity. It will be a homomorphism $\{C_p\} \rightarrow \{C_{p-1}\}$ provided that (1,1) always implies

$$(-x) F(E_p) = x F(-E_p), \quad x \varepsilon \mathfrak{G}. \tag{1,5}$$

This is, however, a direct consequence of the fact that the relations (1,3), (1,4) for $(-E_p)$ are the same as those for $(+E_p)$ with all coefficients $\eta$ reversed.

Now $F(E_p)$ is deduced from $F(\sigma_p)$ by the transcription $\sigma \to E$ plus identification of equal singular $(p-1)$-cells. Since the latter have also equal boundaries, from the known relation $FF(\sigma_p)=0$, there follows $FF(E_p)=0$ and finally $FF=0$ for every singular chain.

We have thus shown that the aggregate of all singular cells on $\mathfrak{R}$ with geometric incidences and incidence-numbers as defined has all the properties of an abstract complex [5]. Hence

Theorem I. *The singular cells on* $\mathfrak{R}$ *constitute an abstract complex* $\mathfrak{K}$.

We call $\mathfrak{K}$ the fundamental singular complex of $\mathfrak{R}$. By a singular complex $K$ on $\mathfrak{R}$ we shall understand any finite closed subcomplex of $\mathfrak{K}$.

The cycles and homology groups over $\mathfrak{G}$, whether for $K$ or for $\mathfrak{R}$ are defined in the usual manner, and those of $\mathfrak{R}$ are topological invariants of the space.

We have made no mention of degenerate cells. It has just been proved by A. W. Tucker [6] that the degenerate cycles are all bounding. For this reason it is not necessary to separate them off, as has always been done hitherto. See a more complete discussion in Appendix I.

**2.** Subdivision. Let the closed simplex $\bar{\sigma}_p$ undergo a barycentric subdivision and let $\sigma_{q'}$ be one of the simplexes of the subdivision. If $\sigma_{q'}$ is on a $q$-face $\sigma_q$ of $\sigma_p$ it is to be oriented like $\sigma_q$. The aggregate of singular cells $\{(\sigma_{q'}, T, T\,|\,\sigma_{q'}\,|)\}$ constitutes the first derived or barycentric subdivision of $E_p$. From there we pass to the second, third, ...derived in the obvious way. They are called generically a subdivision of $E_p$. Each induces a similar subdivision on the faces of $E_p$. Hence we may take the successive derived not only of a cell but of a whole complex. Since barycentric subdivision commutes with affine transformation, the derived of $E_p$ are all functions of $E_p$ alone and not of its parametrization.

Let $\overset{i}{E}_q$ be the singular cells which are images of the subdivision of the face $\sigma_q$ of $\sigma_p$. We define a chain-operator $\tau$ by

$$\tau E_q = \sum \overset{i}{E}_q.$$

The same argument as before shows that the operation is independent of the parametrization. We define $\tau$ for the chains by making it homomorphic. Operating with $\tau$ on a chain $C_p$ is called taking the barycentric subdivision or first derived of $C_p$. Similarly $\tau^k C_p$ is the $k$th derived of the chain.

We may prove by direct computation for $\sigma_p$ that $\tau F = F\tau$ (see **Topology**, p. 65). But here also we pass from $\sigma_p$ to the singular cells, by transcription plus identification of equal singular cells. Since this compound operation com-

---

[5] A. W. Tucker, Annals of Mathematics, **37**, (1936), 97.

[6] A. W. Tucker, in the paper immediately following.

mutes with both $F$ and $\tau$, the relation just written holds likewise for $E$ and hence for any chain. As a consequence $\tau^k F = F\tau^k$: *subdivision commutes with* $F$. The practical bearing of such a commutation is that cycles go into cycles, a chain and its boundary into a chain and its boundary, and hence a bounding cycle into a bounding cycle.

**3.** Singular deformations. We must describe in some way an operation which gives exact content to the intuitive notion of geometric deformation. We may think of a deformation as a succession of «bumps» produced in a surface. This suggests that in formalizing the concept we start with the faces of a single group of cells and define their deformations, then the more general type by superposition.

Let $aa'L$ be a closed geometric simplicial complex, the join of $\sigma_1 = aa'$ and $L$ where $L$ does not have $a, a'$ for vertices. We may define two operations $\theta$, $\mathfrak{D}$ on the simplexes of $aL$ as follows

$$\left.\begin{aligned} \theta a\sigma &= a'\sigma, \\ \mathfrak{D}a\sigma &= aa'\sigma, \end{aligned}\right\} \quad \text{for simplexes } a\sigma,$$

$$\left.\begin{aligned} \theta\sigma &= \sigma, \\ \mathfrak{D}\sigma &= 0, \end{aligned}\right\} \quad \text{for } \sigma \in L.$$

They are single-valued operations on the basic chains of $aL$ and we extend them to all the other chains by linearity. We thus obtain chains $\theta C_p$, $\mathfrak{D}C_p$ for every $C_p$ of $aL$. From the two relations

$$F(aa'\sigma) = a'\sigma - a\sigma + aa'F(\sigma),$$

$$F(a\sigma) = \sigma - aF(\sigma),$$

where $\sigma \in L$, we deduce

$$F(\mathfrak{D}a\sigma) = \theta a\sigma - a\sigma - \mathfrak{D}F(a\sigma). \tag{3,1}$$

Likewise when $\sigma$ is a simplex of $L$

$$F(\mathfrak{D}\sigma) = \theta\sigma - \sigma - \mathfrak{D}F(\sigma), \tag{3,2}$$

so that formally (3,2) holds for every simplex of $aL$. Since $\theta$, $\mathfrak{D}$, $F$ are all linear we have then

$$F(\mathfrak{D}C_p) = \theta C_p - C_p - \mathfrak{D}F(C_p) \tag{3,3}$$

valid for every chain of $aL$.

**4.** To transfer the elements of the preceding situation to $\mathfrak{R}$ we consider a mapping of $aa'L$ on $\mathfrak{R}$, and preserve for the images the same symbols as before, except for $E$ everywhere in place of $\sigma$. Let us suppose that we now have two singular closed complexes $K = aL + M$, $K' = a'L + M$, where $M$ contains no cell of the (singular) complex $aL$. We define $\theta$, $\mathfrak{D}$ for all oriented cells of $K$ to oriented cells of $K'$ as follows:

$$\left.\begin{aligned} \theta aE &= a'E, \\ \mathfrak{D}aE &= aa'E, \end{aligned}\right\} \quad \text{for cells } aE \text{ of } K,$$

$$\left.\begin{aligned} \theta E &= E, \\ \mathfrak{D}E &= 0, \end{aligned}\right\} \quad \text{for all other cells of } K.$$

We then extend $\theta$, $\mathfrak{D}$ by linearity to all the chains of $K$. The pair of operators $T=(\theta, \mathfrak{D})$ is called an elementary singular deformation of $K$ into $K'$ over $\mathfrak{R}$.

The argument leading to (3,3) holds here also, since all that is needed is to replace everywhere $\sigma$ by $E$ and $\sigma \varepsilon L$ by $E \varepsilon L + M$ in its derivation.

Suppose that we have several elementary singular deformations: $T_i=(\theta_i, \mathfrak{D}_i)$, $i=1, 2, \ldots, r$, where $T_i$ is from $K_i$ to $K_{i+1}$. If $C_p$ is a chain of $K_1$ and we write $\theta_1 C_p = \overset{1}{C}_p, \ldots, \theta_r \overset{r-1}{C}_p = \overset{r}{C}_p$ we have

$$F(\mathfrak{D}_i \overset{i-1}{C}_p) = \overset{i}{C}_p - \overset{i-1}{C}_p - \mathfrak{D}_i F(\overset{i-1}{C}_p). \tag{4,1}$$

Let us set

$$\theta C_p = \theta_r \theta_{r-1} \ldots \theta_1 C_p, \quad \mathfrak{D} C_p = \sum \mathfrak{D}_i \overset{i-1}{C}_p, \quad \overset{0}{C}_p = C_p.$$

The compound operation $T=(\theta, \mathfrak{D})$ is called a singular deformation of $K$ into $K'$ over $\mathfrak{R}$. If we add up all the relations (4,1) we get once more (3,3). Thus this basic relation is valid for all singular deformations.

We refer sometimes to $\theta$ alone as the singular deformation, and to $\mathfrak{D}C$ as its deformation-chains.

We say that the singular deformation $T$ just considered is the product of the elementary singular deformations $T_i$ and write $T = T_r T_{r-1} \ldots T_1$. This product is manifestly associative, but not commutative. It follows from the associativity that repetition of the same argument for singular deformations, instead of elementary singular deformations, merely yields a singular deformation, so that the class of these operations is closed under the product. The product of any singular deformations is likewise associative but not commutative.

It is convenient sometimes to write (3,3) in the more abstract «operator» symbolism (see **D. J.**):

$$F\mathfrak{D} + \mathfrak{D}F = \theta - 1. \tag{4,2}$$

We refer also the reader to **D. J.** for the concept of chain-deformation, which has increasing applications.

There are two noteworthy properties of singular deformations:

I. *The extreme positions of a deformed cycle are homologous on* $\mathfrak{R}$.

For if $\Gamma_p$ is a cycle then from (3, 3) follows:

$$F(\mathfrak{D}\Gamma_p) = \theta\Gamma_p - \Gamma_p \sim 0. \tag{4, 3}$$

II. $\theta F = F\theta$: *a singular deformation induces a transformation on the chains of* $\mathfrak{R}$ *which commutes with* $F$.

Multiply both sides of (4, 2), first to the right then to the left by $F$. Remembering $FF=0$ we find

$$F\mathfrak{D}F = \theta F - F = F\theta - F, \quad \theta F = F\theta.$$

The operation $T$ on complexes is in general without inverse. Therefore it does not establish a symmetric relationship between $K$ and $TK$, and so does not form a basis for grouping complexes into classes. In many questions, for example in homology theory, what matters is not whether $K$ is deformable into $K'$ or vice versa, but only that one of the two is deformable into the other. This suggests the following definition. We call $K$, $K'$ equivalent and write $K \simeq K'$ whenever there exists a chain $K = K_1, \dots, K_r = K'$, wherein one of two consecutive terms is always deformable into the other. This equivalence is reflexive, symmetric and transitive, and hence it allows for a subdivision into classes. The same equivalence may of course be extended to chains. One might be tempted to consider the equivalence just introduced as not sufficiently restrictive. That this is not the case is shown by the following two examples:

(a) two equivalent cycles are homologous;

(b) if $A$ is a closed subset of $\mathfrak{R}$, then equivalent cycles mod $A$, with induced equivalence (in an obvious sense) for the boundaries on $A$ are homologous mod $A$.

**5.** Singular deformations have a ready application to invariance questions in the theory of geometric complexes. Let us examine briefly these problems.

Let $K$ be a geometric complex and $K'$ the first derived of $K$. Let us send each vertex of $K'$ into a vertex of the simplex of $K$ that carries it. As is well known this induces a simplicial transformation $\theta$: $K' \longrightarrow K$. We have then that much towards making $\theta$ a deformation of $K'$ into $K$. It is not difficult to verify that the appropriate $\mathfrak{D}\sigma'$ may be inserted for every $\sigma'$ of $K'$, and that each is on the simplex $\sigma$ of $K$ that carries $\sigma'$. Therefore $K'$ is deformable into $K$ in such manner that each simplex of $K'$ is merely deformed over the simplex of $K$ that carries it. From this we may deduce that the homology groups are subdivision invariants. This is the so-called combinatorial invariance of these groups. We have given the argument a deformation «slant», but if we omit the mention that $\sigma'$ is on a $\sigma$ of $K$ or deformed over $\sigma$, it becomes purely combinatorial.

Since each $\mathfrak{D}\sigma'$ is on a definite $\sigma$ and may be described purely in terms of that $\sigma$ and invariantly relative to its affine transformations the whole argument may be transferred over to singular complexes or chains. Therefore a subdivision $K'$ of a singular complex $K$ may be deformed into $K$ in the same way as above, each $E'$ of $K'$ remaining on the $E$ of $K$ which carries it, and similarly for the chains. Therefore the same properties hold for the $k$th derived. From this follows that a cycle and its derived are homologous on $\mathfrak{R}$. This is the first step in the proof of the topological invariance of the combinatorial homology groups by the «deformation» method, for it enables one to replace the initial chains by chains of suitably small mesh ready for the application of the deformation theorem (**Topology**, p.p. 86—88).

### II. Continuous complexes, chains and cycles

**6.** An alternate approach to the singular elements consists in defining each by means of an appropriate polyhedral antecedent which forms thus an inherent part of the element. Following Alexandroff-Hopf [7] we shall call these elements «continuous», reserving the term «singular» for those defined previously.

[7] Alexandroff-Hopf, Topologie, (1935), p. 332.

Let $\mathfrak{K}$ be a finite geometric simplicial complex and $T$ a mapping of $\mathfrak{K}$ to a set $\alpha$ on $\mathfrak{R}$. We may paraphrase the definition of singular cells (No. 1) and introduce classes $\{(U\mathfrak{K}, T\overset{-1}{U}, \alpha)\}$, where $U$ is a homeomorphism and coincides with an affine transformation on each cell of $\mathfrak{K}$. Such a class $K$ is called a continuous complex on $\mathfrak{R}$. Any particular triple is a parametrization of $K$, and $\mathfrak{K}$ is an antecedent of $K$.

The transforms $E_p$ of the simplexes $\sigma_p$ of $\mathfrak{K}$ give rise to the continuous $p$-cells of $K$. We extend to these cells the incidences and incidence-numbers which exist in $\mathfrak{K}$. That is to say if $\sigma_p$, $\sigma_q$ are two simplexes of $K$ and $E_p = T\sigma_p$, $E_q = T\sigma_q$, we say that $E_q$ is a face of $E_p$ whenever $\sigma_q$ is a face of $\sigma_p$, and when $q = p - 1$ we choose as incidence-numbers $[E_p : E_{p-1}] = [\sigma_p : \sigma_{p-1}]$. As a consequence if $\sigma^i_p$ has for image $E^i_p$ and $c_p = x_i \sigma^i_p$ is a $p$-chain of $\mathfrak{K}$ over a group $\mathfrak{G}$, then $C_p = x_i E^i_p$ is a continuous $p$-chain of $\mathfrak{R}$ over the same group. Thus a continuos chain $C$ has now a very specific antecedent $c$.

The sets $\alpha$ associated with $K$, $C$ are designated by $|K|$, $|C|$. The mesh of $K$ or $C$ is max diam $|E|$, where $E$ is any cell of $K$ or $C$.

If $K'$, $c'$ are subdivisions of $\mathfrak{K}$ or $c$, then $K' = (\mathfrak{K}', T, |\mathfrak{K}'|)$, $C' = (c, T, |c|)$ are called subdivisions of $K$ or $C$.

The triple $(F(c), T, |TF(c)|)$ determines a well-defined continuous chain $F(C)$, the boundary of $C$. When $F(c) = 0$ likewise $F(C) = 0$ and $C$ is a continuous cycle. Since the algebraic expressions $c$, $C$ differ only in the substitution $\sigma \longrightarrow E$, from $FF(c) = 0$ follows $FF(C) = 0$. That is to say the boundary operator $F$ still satisfies the basic relation $FF = 0$. If $C'$ is the $k$th derived of $C$ then $F(C')$ is the $k$th derived of $F(C)$. Here also if $\tau$ denotes first derivation $\tau^k F = F\tau^k$.

**7.** Continuous deformations may be defined by a mere paraphrase of No. 4, or for that matter of No. 3. Let $\Lambda = aa'L + M$, where $\Lambda$ is a closed geometric simplicial complex and $M$ does not have $a$ for vertex. Then if $\Lambda$ has a continuous image on $\mathfrak{R}$ with $K$, $K'$ as the continuous images of $aL + M$, $a'L + M$, we say that we have an elementary continuous deformation $T$ of $K$ into $K'$ over $\mathfrak{R}$. A succession of such elementary continuous deformations $T_i : K_i \longrightarrow K_{i+1}$ gives rise to a continuous deformation $T = T_r, \ldots, T_1$, also called a homotopy, of the first complex into the last over $\mathfrak{R}$. Of course the complexes $K_i$ must all be maps of the same geometric complex.

Let $T$ deform $K$ into $K'$ and let $c$ be any chain of $\mathfrak{K}$, and let $C$, $TC$, $\mathfrak{D}C$ be the images of $a \times c$, $a' \times c$, $aa' \times c$. We call $\mathfrak{D}C$ the deformation-chain of $C$. From the known boundary formula for Euclidean-products of chains (**Topology**, p. 226) we deduce once more (3, 3) which holds therefore for continuous deformations.

## III. Critical comparison of the two concepts of singular and continuous elements

**8.** First of all the both concepts are closely inter-dependent. The aggregate $\{E\}$ of the cells of a closed continuous complex $K_1$ constitutes a closed singular complex $K_2$. For the $E$'s are singular cells and if $E$ is a member of the aggregate so are all its faces. Furthermore the incidences and incidence-numbers in $K_1$ are those appropriate to the singular cells also. We shall say that the continuous complex coincides with the singular complex $K_2$. It should be observed that, due to the autonomy among the singular cells, $K_1$ determines a unique $K_2$, but the reverse need not be true. Consider for example a figure 8 curve subdivided into arcs by the points $ABCDE$, where $C$ is the double point and $A$, $B$ are on one loop and $D$, $E$ on the other. This graph is a continuous image of a circumference subdivided into 6 arcs in an obvious way, and as such it constitutes a definite $K_1$. It is also the continuous image of two triangles $\alpha\beta\gamma$, $\gamma'\delta\varepsilon$ without common vertices, so mapped onto the figure that $\alpha\rightarrow A$, $\beta\rightarrow B$, $\gamma+\gamma'\rightarrow C$, $\delta\rightarrow D$, $\varepsilon\rightarrow E$ and this is a continuous complex $K_1'$, different from $K_1$. On the other hand the aggregate of 6 one-cells and 5 vertices of the figure make up a singular $K_2$ coincident with both $K_1$ and $K_1'$. Thus $K_2$ does not determine a unique $K_1$ coincident with it.

**9.** *There is always at least one continuous complex $K_2$ coincident with a given singular complex $K_1$.* All that is necessary in fact is to choose for the cells of $K_1$ antecedent geometric simplexes which do not intersect (see Appendix II). Even when $K_1$ consists of a countably infinite set of cells the antecedents may be properly chosen in the Hilbert parallelotope. Their aggregate makes up a geometric complex imaged on $\mathfrak{R}$ into a continuous complex $K_2$ coincident with $K_1$. This $K_2$ is highly disjointed, but this does not matter. Furthermore we may make it more coherent by coalescence of as many faces as possible having the same image in $K_1$ without destroying the polyhedral character of the antecedent complex.

Consider for example three singular cells $\overset{i}{E}_1$, $i=1, 2, 3$, with the same end-points $A$, $B$. We take as antecedents three segments $l_i=\alpha_i\beta_i$, where $\alpha_i$, $\beta_i$ are respective antecedents of $A$, $B$. We may coalesce the three points $\alpha_i$ into a single antecedent $\alpha$ of $A$, but no more. To proceed further we are obliged to subdivide once if $A\neq B$, twice if $A=B$. We may then coalesce further the points $\beta_i$ into a single antecedent of $B$, which will be $\alpha$ or another point $\beta$, accordingly as $A=B$ or $A\neq B$. Then $l_i$ will have for antecedent a broken line made up of four segments in the first case (two subdivisions), of two in the second (a single subdivision).

The preceding example leads to surmising the following noteworthy theorem which is the bridge, as it were, between the singular and continuous elements:

Theorem 2. *A finite singular n-complex $K_n$ always has an abstract isomorph $L_n$ whose elements are true cells and which has a simplicial subdivision $L_n^*$ whose continuous image coincides with the second derived of $K$.*

By «subdivision» of $L$ we merely mean here a simplicial complex $L^*$ such that each cell of $L$ is a sum of simplexes of $L^*$. To be precise if $K=\left\{\overset{i}{E}_p\right\}$, $L=\left\{\overset{i}{\mathfrak{E}}_p\right\}$, the correspondence $\overset{i}{E}_p\longleftrightarrow\overset{i}{\mathfrak{E}}_p$ is one-one. Moreover each $\mathfrak{E}$ is a sub-chain of the simplicial complex $L^*$ and the boundary relations between the $\mathfrak{E}$'s are the same as between the $E$'s. From our theorem will then follow:

Corollary. *Every singular cycle has a subdivision which is continuous, and if it bounds it has a subdivision which bounds a continuous chain.*

Since subdivision permutes with $F$, this shows that «continuous» homologies yield no less than «singular» homologies. Granting of course that the former give rise to sharply defined groups.

**10.** Passing to the proof of our theorem, we notice that it is trivial for $n=0$, so that induction on $n$ is in order. Let then $K_{n-1}$ denote the singular complex left when the $n$-cells of $K_n$ are removed. By hypothesis there are associated complexes $L_{n-1}$, $L^*_{n-1}$ such as demanded by the theorem. We may assume that $L^*$ is a geometric simplicial complex in a space $S_r$, $r=2n+1$, whose vertices are independent in $S_r$ (i. e. have a coordinate-matrix of maximum rank).

Let $E_n$ be any $n$-cell of $K_n$ and let $\sigma_n$ be an antecedent of $E_n$ with $\sigma^i_p$ $(p<n)$, $H_{n-1}$ as its faces and boundary-sphere. We shall designate their second derived by $\sigma'_n, \ldots$ Each $\sigma^i_p$ has for image a specific face $E^i_p$ of $E_n$ and hence it is represented in $L_{n-1}$ by a cell $\mathfrak{E}^i_p$ which in turn is an aggregate of simplexes of $L^*_{n-1}$. Since subdivision is permutable with $F$, the incidence relations between these component aggregates (in $L^*_{n-1}$) are the same as for their images in $K$. Therefore $F(\sigma'_n)$ is already imaged into a cycle $\Gamma_{n-1}$ of $L^*_{n-1}$. All that is necessary then is to add to $L^*_{n-1}$ the images of the cells of $\sigma'_n$, and this so as to have $\sigma'_n$ isomorphic with its images and $F(\sigma'_n)$ imaged into $\Gamma_{n-1}$ in the same manner as before. Since a simplicial transformation is permutable with $F$, it will be sufficient to introduce the images so that the passage from $\bar{\sigma}$ to its image is such a tiansformation.

**11.** Let us choose in $S_r$ one new point for each vertex of $\sigma'_n$, and for all analogous simplexes corresponding to the $n$-cells of $K$. We assume these points so chosen that together with the vertices of $L^*_{n-1}$ they form an independent set of points in $S_r$. Every vertex of $\bar{\sigma}'_n$ has now an image in $S_r$. For each simplex $\sigma$ of $\sigma'_n$ we now introduce the simplex $\zeta$ of $S_r$ whose vertices are the images of the vertices of $\sigma$. Those of $H_{n-1}$ already have such $\zeta$ images in $L^*_{n-1}$. Therefore the tiansformation $\sigma \longrightarrow \zeta$ is simplicial. Let us designate by $\mathfrak{E}_n$ the set of images of the $\sigma$'s of $\sigma'_n$, and write

$$L_n = L_{n-1} = \Sigma\, \mathfrak{E}_n, \quad L^*_n = L^*_{n-1} + \Sigma\, \{\zeta\}.$$

Since no two $\mathfrak{E}$'s corresponding to distinct $E$'s have common simplexes, $L_n$, $L^*_n$ will behave as our theorem demands provided that we can show that the particular aggregate $\{\zeta\}$ corresponding to $E_n$ is isomorphic with $\sigma'_n$. This in turn merely requires that we prove the two properties:

(a) every $n$-simplex of $\sigma'_n$ has a non-degenerate image,

(b) two distinct $n$-simplexes of $\sigma'_n$ always have distinct images.

Indeed from (a), (b) will follow that $\mathfrak{E}_n$ is an $n$-cell with a subdivision isomorphic with $\sigma'_n$.

Suppose that (a) were false. Some one-simplex $\lambda = AB$ of $\sigma'_n$ would then have a degenerate image. Since one of the vertices of $\lambda$ say $A \subset \sigma'_n$, if $B \subset \sigma'_n$, their images are distinct since distinct vertices of $\sigma'_n$ have distinct images. The same conclusion holds if $B \subset H_{n-1}$, since its image is then in $L_{n-1}$ and that of $A$ is not. Therefore $\lambda$ has a non-degenerate image in $S_r$, and (a) is proved.

Consider now (b). If it were false two distinct segments $\lambda = AB$, $\lambda' = A'B'$ of $\sigma'_n$ would have the same image $\alpha\beta$ in $S_r$, the images being non-degenerate, and the notation so chosen that $A, A'$ have coincident images, and so have $B, B'$. Now this can only happen if the points in one of the two pairs, say $A$, $A'$ are in $H_{n-1}$. By a previous remark necessarily $B \subset \sigma'_n$, $B' \subset \sigma'_n$. If $B \neq B'$ then their images are distinct and $\lambda \neq \lambda'$. We must therefore have $B = B'$. That is to say $\lambda$, $\lambda'$ have a common end-point. But in the second derived of $\sigma_n$ there are no one-simplexes $AB$, $BA'$ with $A$, $A' \subset H_{n-1}$ and $B \subset \sigma''_n$. Therefore (b) is true and the proof of our theorem is now complete.

**12.** What we have said so far shows clearly that the continuous elements: complex, chain, cycle, are strongly tied up with definite antecedent polyhedra. They are essentially images of very definite objects, exterior to the basic space $\mathfrak{R}$ and whose structure is not variable. As the objects are deformed, they must remain the images of the initial exterior polyhedra throughout the deformation. On the contrary a singular element is essentially a certain aggregate of images of objects out of a very definite and simple collection: the simplexes of various dimensions. Singular elements are thus in a certain sense much more «intrinsically» related to the space $\mathfrak{R}$ than the continuous elements and they may be deformed much more freely, and without any reference to anything exterior to $\mathfrak{R}$.

We may illustrate the situation by a very simple example. Let $ABC$ be the periphery of a triangle in a plane $\Pi$, and let $\alpha\alpha'$ be a segment in $\Pi$ divided into three parts by points $\beta$, $\gamma$. The cells of $ABC$ and $\alpha\beta\gamma\alpha'$ make up two linear graphs $K_1$, $K_2$. We may consider $K_1$ as a continuous image of $K_2$ in such manner that $\alpha + \alpha' \longrightarrow A$, $\beta \longrightarrow B$, $\gamma \longrightarrow C$, that is to say as a «continuous segment». As such $ABC$ is homotopic to $\alpha\beta\gamma\alpha'$. On the other hand if we consider $ABC$ as a continuous image (homeomorph in fact) of a proper boundary of a triangle it ceases to be homotopic to the segment altogether. Finally if both $ABC$ and $\alpha\beta\gamma\alpha'$ are considered as singular complexes (with the identity as $T$) the second is always singularly deformable into the first.

In short as regards «point-sel» properties the scheme of continuous elements constitutes a much more delicate instrument.

**13.** When we pass however to algebraic properties the situation is reversed. For the addition of continuous chains and cycles gives rise to serious difficulties. Let indeed $c_p = x_i E_p^i$, $D_p = y_i E_p'^i$ be two continuous chains with polyhedral antecedents

$$c_p = x_i \sigma_p^i, \quad d_p = y_i \sigma_p'^i.$$

We may choose the two antecedents in some $(2p+1)$-space and without common points. Together they constitute a single polyhedral chain

$$c_p + d_p = x_i \sigma_p^i + y_i \sigma_p'^i$$

whose continuous image may be described as $C_p + D_p$. So far so good, but what is in fact this sum? Consider for example two triangles $ABC$, $ABD$ and let

$$C_1 = AB + BC + CA, \quad D_1 = BA + AD + DB.$$

Intuitively one would expect the sum to be

$$\Delta_1 = AD + DB + BC + CA.$$

Consider however each chain as its own continuous image. To add them up in the «continuous» sense, we are to consider first two triangles $A'B'C'$, $B''A''D''$ in some three-space, without common points and $C_1$, $D_1$ as the respective images of

$$C_1' = A'B' + B'C' + C'A', \quad C_2' = B''A'' + A''D'' + D''B''.$$

The sum is then to be taken as the continuous image of $C_1' + C_2'$, that is to say as $\Delta_1$ *associated with* $C_1' + C_2'$. Here then the chain-sum is the oriented boundary of a quadrilateral, but it must be considered as the sum of two oriented triangles. This is awkward to say the least. If we deal with the addition of homology classes we may in a measure circumvent these difficulties but it cannot be said that the situation becomes particularly smooth.

On the other hand, since singular chains and cycles are chains and cycles of the fundamental singular complex of $\mathfrak{R}$, their addition offers no difficulty whatever. Thus if we consider $C_1$, $D_1$ as singular chains their sum is $\Delta_1$ without further ado.

To sum up then the singular elements are very well adapted to the treatment of algebraic properties.

In the presence of these complications many topologists are tempted to dispense with one or the other or both of the concepts: singular element, continuous element. A more natural conclusion would be that singular elements are to have the preference in all questions where algebraic considerations predominate, and continuous elements in all those which belong in the main to point-set theory. That both are too valuable and too close to geometric intuition to be left out of the picture is clearly shown by the applications.

Consider for example the Alexander-Veblen deformation theorem (**Topology**, p. 86) which is often used to prove the topological invariance of the homology characters of a geometrie complex. This is essentially a part of algebraic topology and so the singular elements may be allowed full latitude. The deformations considered should be singular deformations, and when they are utilized with a little care the proof offers no difficulty. However in the next section we shall deal with a group of questions where one needs sometime singular elements, sometime continuous elements.

## IV. The LC and HLC properties

**14.** We have introduced several years ago (**Ann. M.**) a concept which has proved useful in many questions, namely the concept of semi-singular complex. There is given a closed simplicial geometric complex $K$ and a closed subcomplex $L$ of $K$ which includes all the vertices of $K$. Let $\mathfrak{K}_1$ be a continuous complex image of $L$ on $\mathfrak{R}$. It constitutes in a sense a partially realized image of $K$ in which every expected cell is represented at least by its vertices. For this reason we have called $\mathfrak{K}_1$ a semi-singular complex. It is more in keeping with the terminology which we have adopted in the present paper to call $\mathfrak{K}_1$ a partially realized continuous complex, or partially realized image of $K$. The problem arises if it is possible to complete $\mathfrak{K}_1$ to a full image of $K$, and in particular: granting that the initial mesh is small, is it possible to complete the realization to a mesh that is also quite small?

The same problem may also be formulated in terms of extension of a function or of a transformation:

Let $f(x)$ be a continuous single-valued function of $L$ to $\mathfrak{R}$. Is it possible to extend it to a function $F(x)$ of the whole of $K$ to $\mathfrak{R}$, where $F(x)=f(x)$ for $x$ on $L$?

**15.** Returning to the first formulation, at this stage we need a little more precision. Given any $\sigma$ of $K$, among $\sigma$ and its faces a certain number, in $L$, are realized and they make up a set $\Phi(\sigma)$. By the mesh of the partial realization $\mathfrak{K}_1$ we understand max diam $\Phi(\sigma)$ for all $\sigma \varepsilon K$.

Definition[8]. The space $\mathfrak{R}$ is said to be $\mathrm{LC}^n$ whenever for every positive $\varepsilon$ there exists a positive $\eta(\varepsilon, n)$ such that every partially realized image of a complex whose dimension $\leqslant n+1$ and whose mesh $<\eta$ may be completed to a full realization whose mesh $<\varepsilon$. Whenever it is possible to choose $\eta(\varepsilon)$ independent of $n$ and fulfilling the same purpose the space is said to be LC.

What gives these spaces much of their importance is that in many respects (homology groups, fixed point problem) they are closely assimilable to finite complexes. There are also close relations to retracts. The abbreviation LC stands for «locally connected». The $\mathrm{LC}^n$ property is in fact equivalent to $p$-locally connected for every $p \leqslant n$. For further details and references we refer the reader to our papers on these questions.

An alternate definition for LC spaces[9]. Let $K$ be a countably infinite geometric simplicial complex which is locally finite, i. e. such that (a) the star of every vertex is finite, (b) if $\sigma^i$, $i=1, 2, \ldots,$ be its simplexes

[8] The definition of $\mathrm{LC}^n$ spaces in **Ann. M.** is different but shown there to be equivalent to the one adopted here.

[9] Given here for the first time.

ranged in some order, then diam $\overset{i}{\sigma} \longrightarrow 0$ with $\frac{1}{i}$. Let a partial realization $\mathfrak{K}_1$ and the sets $\Phi(\sigma)$ be defined as before. Then the LC condition is equivalent to the following: if $\operatorname{diam} \Phi(\sigma^i) \longrightarrow 0$ with $\frac{1}{i}$, then except for a finite subcomplex, $\mathfrak{K}_1$ may be completed to a full realization such that if $E^i$ is the image of $\sigma^i$ then diam $E^i \longrightarrow 0$ with $\frac{1}{i}$.

The equivalence of this definition with our earlier one follows quite readily from the developments of **Ann. M.** It may also be established indirectly by showing that the spaces corresponding to the second definition like those corresponding to the first (see loc. cit.) are absolute neighborhood retracts. If «except for a finite complex» is not needed, then the space is an absolute retract.

**16.** It might be surmised that in the LC definitions continuous complexes could be replaced by singular complexes. This does not seem to be readily manageable because we have to describe explicitly the structure of the expected ultimate complex $\mathfrak{K}$ of which only a certain part $\mathfrak{K}_1$ is at hand. In view of our ignorance of the general (combinatorial) structure of singular complexes this appears to be a serious difficulty. Something could be accomplished by taking instead of arbitrary singular cells, cells which have $p+1$ distinct vertices when their dimension is $p$. The expected complex $\mathfrak{K}$ is then essentially simplicial, although it may occur that distinct $p$-cells have the same $p+1$ vertices. However the structure is now simple enough for all practical purposes. While we have not investigated this type we believe likely that it is identical with the LC (or $\mathrm{LC}^n$) type, as the case may be (see Appendix II).

A generalization of a totally different type leads to the HLC classes of **D. J.** (the abbreviation stands for «locally connected in the sense of homology»). Here we are given an abstract complex $K = \left\{ \sigma_p^i \right\}$ and we are to realize it by chains $\left\{ C_p^i \right\}$ of $\mathfrak{R}$. In keeping with § 11 let the chains be singular, so that we have convenient addition, etc. There is a single-valued transformation $\sigma_p^i \longrightarrow C_p^i$ which preserves boundary relations. A partial realization $\mathfrak{K}_1$ and its mesh are defined as before, and a paraphrase of our earlier definition yields the $\mathrm{HLC}^n$ and HLC classes. By modifying the types of chains envisaged we may obtain widely different types of spaces of the same general category. Roughly speaking on passing from LC to HLC all algebraic properties are preserved, and only the point-retraction properties are lost. These spaces are then particularly interesting to the algebraic topologist, the LC class to the point-set topologist.

## APPENDIX I

### Degenerate singular cells

**17.** While Tucker's recent result [6] will serve largely to eliminate the concept of singular degenerate cell it is not without interest to show how the concept arose.

Let $E_p$ be a singular cell of $\mathfrak{R}$. Whenever it admits a parametrization $(\sigma_p, T, \alpha)$ such that $T = T_1 V$, where $V$ is a singular affine transformation of

$\sigma_p$ into a $\sigma_q$, $q < p$, we say that $E_p$ is degenerate. Degeneracy thus essentially occurs when $E_p$ is the transform of a degenerate $p$-simplex. If $(\sigma', T', \alpha)$ is another parametrization so that $T' = TU$, $U$ non-singular affine, we have $T' = T_1 VU$, where $VU$ is singular affine on $\sigma'$. Hence when $E_p$ is degenerate according to one parametrization it is degenerate according to all others; degeneracy is independent of the parametrization, and is a property of the singular cell alone.

The preceding type of degeneracy, which we adopted for example in **Bull. M. S.**, may be termed geometric. Another type of degeneracy, that may be described as algebraic, arises when the two classes whereby cells are oriented (No. 1) coincide and we have $E_p = -E_p$[10]. Geometric degeneracy implies algebraic degeneracy but the converse need not hold. While subdivision preserves geometric degeneracy, an algebraically degenerate cell need not have a degenerate subdivision. Thus a segment folded in two in the middle, degenerate under this orientation property, would have a non-degenerate subdivision made up of two equal and opposite segments. Geometric degeneracy is the exact analogue of degeneracy as usually understood in a simplex, where it always involves a «collapsing» of dimension. Algebraic degeneracy is the analogue of degeneracy in a simplicial chain when all its terms cancel.

A word to justify the consideration of degenerate cells may not be out of place. In proving the topological invariance of the homology groups of a geometric complex $K$ we were led to consider $p$-cycles made up of $q$-simplexes, $q < p$. These cycles were obviously superfluous, so that in order to eliminate them we took all the singular chains of $K$ modulo the (geometric) degenerate chains (see **Topology**, p. 73). If it could be shown that all degenerate cycles were $\sim 0$ we would not need to pay further attention to degeneracy. This is precisely what Tucker has just succeeded in doing.

## APPENDIX II

### Relations between singular and continuous complexes

**18.** We mentioned in No. 16 that, owing to our ignorance regarding the abstract structure of singular complexes, we could not see readily how to utilize them as a basis for an extension of the $\mathrm{LC}^n$ and LC concepts. On examining further the «antecedent» of Theorem 2, and another to be described presently we have been able to circumvent the difficulty. It turns out, however, that no really new classes of spaces are thus obtained. Let us describe briefly the situation.

An element of an abstract complex shall be called maximal whenever it is not the face of any other element.

Consider now the maximal cells of a singular complex $K$. If $r > 2 \dim K$ we may choose for them antecedent simplexes in an $r$-space $S_r$ whose closures do not meet. The aggregate of these closed simplexes is a closed geometric complex $\mathfrak{K}$ which is manifestly the antecedent of a continuous complex coinci-

[10] See **Bull. M. S.**, also Seifert-Threlfall, Lehrbuch der Topologie, B. G. Teubner, (1934), p. 93.

dent with $K$. Moreover, if $\mathfrak{K}_1$ is another similar complex, there is a homeomorphism $U$: $\mathfrak{K} \longrightarrow \mathfrak{K}_1$ which is affine on each cell of $\mathfrak{K}$. Therefore the continuous complex in question is uniquely determined by $K$.

Let $\mathfrak{K}^*$ be a finite abstract complex whose $p$-elements $E_p$ are true $p$-cells; we assume that their totality makes up a certain topological space $S$ and that in that space, whenever $E$ is a (proper) face of $E'$, then the cell $E$ is a subset of $\overline{E} - E$. In other words $\mathfrak{K}^*$ consists of cells with incidences analogous to those taking place in a simplicial complex. These complexes generalize slightly the complexes considered by Veblen («Analysis Situs», p. 77). It may also be proved as he has done (loc. cit., p. 88) that the first derived $\mathfrak{K}^{*\prime}$ of $\mathfrak{K}^*$ (Veblen's regular subdivision) is simplicial. We may define in an obvious sense a new type of continuous complex $K^*$ on $\mathfrak{R}$ by mapping $\mathfrak{K}^*$ on $\mathfrak{R}$. Let us call $K^*$ a generalized continuous complex (= g. c. complex) on $\mathfrak{R}$. We must, however, allow the antecedent $\mathfrak{K}^*$ to vary under any homeomorphism.

Now referring to Theorem 2 we find that any singular complex $K = \left\{ E_p^i \right\}$ coincides with a g. c. complex whose antecedent $\mathfrak{K}^* = \left\{ \mathfrak{E}_p^i \right\}$ is unique and is (abstractly) isomorphic with $K$.

We have thus obtained two distinct unique antecedents $\mathfrak{K}$, $\mathfrak{K}^*$ for $K$ considered as a continuous complex. We may utilize the one or the other to define the space $\mathfrak{R}$ as $LC^n$, or LC in the singular sense. Let us merely assert, without giving the proof here, that by taking advantage of the mutual relations between $\mathfrak{K}$ and $\mathfrak{K}^*$, it may be shown that these new types of $LC^n$, or LC, are equivalent to those defined in the text. In other words, there exists in fact only one $LC^n$ class and only one LC class.

(Поступило в редакцию 1/IX 1937 г.)

---

## Сингулярные и непрерывные комплексы, цепи и циклы

**С. Лефшец** (Принстон, С.Ш.А.)

*(Резюме)*

Существуют два различных, но тесно связанных между собою топологических понятия: сингулярный элемент — комплекс, цикл и т. д. и непрерывный элемент. В настоящей работе исследуются и сравниваются при помощи примеров оба эти понятия. Доказывается теорема, служащая мостом между обоими понятиями. Работа заканчивается кратким обзором приложений к локально связным пространствам.

# IX

# IX

## Complete families of periodic solutions of differential equations

*To my friend Heinz Hopf on his sixtieth birthday*

1. The following question has been investigated at length by Poincaré especially in connection with his research on the problem of three bodies : — Consider a real differential system

$$\frac{dy}{dt} = Y(y\,;\mu\,;t) \tag{1.1}$$

where $y$ is an $n$-vector and the components of $Y$ are holomorphic at $y = 0$, $\mu = 0$, and continuous and periodic with period $2\pi$ in $T$. Suppose that for $\mu = 0$ there is known a solution $\xi(t)$ with period $2\pi$ in $t$. Does there exist a solution $\xi(t\,;\mu)$ with period $2\pi$ in $t$, holomorphic in $\mu$ about $\mu = 0$ and $\rightarrow \xi(t)$ as $\mu \rightarrow 0$. Poincaré proceeds in this way : — He considers the solution $\xi(t) + x + z(x\,;t\,;\mu)$ with the initial value $\xi(0) + x$ for $t = 0$. In particular $z$ is holomorphic in $x$ and $\mu$ at $x = 0$, $\mu = 0$ and has period $2\pi$ in $t$. Furthermore $z(x\,;0\,;\mu) = 0$. Expressing then the periodicity of the solution there follows a system

$$z(x\,;2\pi\,;\mu) = 0\ . \tag{1.2}$$

If this system has a real solution $x(\mu)$ which $\rightarrow 0$ with $\mu$, there is defined a periodic solution $\xi(t\,;\mu)$ of the desired type. Everything comes down to the determination of the real solutions of a real system (1.2) which actually depend on $\mu$. This problem was only solved by Poincaré in specially simple cases. We propose to give a complete solution of the problem. It will rest upon a rather simple application of Kronecker's method of elimination.

2. Before proceeding let us recall a well known terminology. Let $f(u) = f(u_1, \ldots, u_p)$ be a real or complex function of the indicated variables holomorphic at the origin. We call $f$ a *unit* if $f(u) \neq 0$, a *non-unit* otherwise. If

$$f = u_1^q + f_1(u_2, \ldots, u_p)\, u_1^{q-1} + \cdots + f_q(u_2, \ldots, u_p)\ ,$$

where the $f_h$ are non-units, then $f$ is referred to as a *special polynomial* in $u_1$. Units are written $E$.

3. Returning to our problem let us write (1.2) in the general and explicit form

$$X_{1h}(x_1, \ldots, x_n, \mu) = 0, \quad h = 1, 2, \ldots, n_1 . \tag{3.1}$$

In point of fact here $n = n_1$. However we shall not need to take advantage of this fact and it will make our argument clearer not to assume it. We suppose that the $X_{1h}$ are real non-units and we shall determine all the suitable families of solutions then select the real families among these.

Since we are only interested in solutions depending on $\mu$, if $X_{1h}$ is divisible say by $\mu^h$ we cross this factor out and continue to call the quotient $X_{1h}$. Thus $X_{1h}(x\,;0) \neq 0$. We may now apply a real linear transformation to the $x_j$ and dispose of the situation so that in $X_{1h}$ the variable $x_1$ appears to the power $x_1^{mh}$ where $m_h > 0$ is the lowest degree of any term in the $x_j$ in $X_{1h}$. The Weierstrass preparation theorem yields then

$$X_{1h} = X'_{1h} E(x\,;\mu)$$

where $X'_{1h}$ is a special polynomial in $x_1$ (in the system $x_1, x_2, \ldots, \mu$) of degree $m_h$ in that variable. Notice that its coefficients will all be real since their determination never involves any irrationality. To simplify matters we may therefore suppose that in (3.1) the $X_{1h}$ are already special polynomials in $x_1$.

The $X_{1h}$ may have a common factor $D_1(x_1, \ldots, x_n, \mu)$. It is readily shown to be likewise a special polynomial in $x_1$ and with quotients $X_{1h}/D_1 = X^*_{1h}$. Take any irreducible factor $D^*_1$ of $D_1$. Both $D^*_1$ and $X^*_{1h}$ are again, up to unit-factors, special polynomials in $x_1$. Then $D^*_1 = 0$ represents an irreducible $(n-1)$-dimensional family of solutions depending on $\mu$.

Consider now the system

$$X^*_{1h} = 0 , \quad h = 1, 2, \ldots, n_1 . \tag{3.2}$$

Following Kronecker introduce two linear combinations with arbitrary parameters

$$U = \Sigma\, u_h X^*_{1h} , \qquad V = \Sigma\, v_h X^*_{1h} ,$$

and form the resultant $R(U, V)$ as to $x_1$. We will have

$$R = \Sigma\, W_k(u, v) X_{2k}(x_2, \ldots, x_n, \mu) ,$$

where the $W_k$ are monomials in the $u_h$ and $v_k$. A n. a. s. c. in order that the system (3.2) possess a solution in $x_1$ is that

$$X_{2k}(x_2, \ldots, x_n, \mu) = 0 , \quad k = 1, 2, \ldots, n_2 . \tag{3.3}$$

This system is wholly analogous to (3.1) but with one variable less. We reason then with (3.2) as with (3.1), and so on, and the argument manifestly terminates.

The ultimate result may be described as follows: — There may exist for each $k<n$ and each $\mu$ sufficiently small a certain number of $n-k$ dimensional families of solutions, each represented in suitable coordinates by a system

$$X_h(x_h, \ldots, x_n, \mu) = 0 , \quad h = 1, 2, \ldots, k , \tag{3.4}$$

where $X_h$ is a polynomial in $x_k$ with non-unit non-leading coefficients, and $X_k$ is special in $x_k$. Furthermore $X_k$ is irreducible as a polynomial in $x_k$, and the $X_h$, $h>k$, are irreducible in a similar sense.

One may even proceed further. Let $d_h$ be the degree of $X_h$ in $x_h$ and let $d = d_1 d_2 \ldots d_k$. Choose $k$ real constants $c_1, \ldots, c_k$ such that the $d$ values of $c_1 x_1 + \cdots + c_k x_k$ are distinct. Upon making the change of variables

$$x_h \to x_h, h<k ; \quad c_1 x_1 + \cdots + c_k x_k \to x_k ,$$

we will have in place of (3.4) a system in which the equation $h$ $(h<k)$ will be of degree one in $x_h$. Hence (3.4) assumes the form

$$\left.\begin{aligned} &X_k(x_k, \ldots, x_n, \mu) = 0 , \\ &A_0(x_{k+1}, \ldots, x_n, \mu)\, x_{k-h} - A_h(x_k, \ldots, x_n, \mu) = 0 \end{aligned}\right\} \tag{3.5}$$

where $X_k$ may in fact be taken to be a real special polynomial in $x_k$ and the $A_j$ are non-units and real.

4. Passing now to the problem of the reality of the solutions we must distinguish three types of solutions or points.

I. The points where the Jacobian matrix $J$ of the left-hand sides in (3.5) is of maximum rank $k$ at the same time as $A_0 \neq 0$. These are the *ordinary points*. If $M$ is their set and $P \in M$ then there is a neighborhood $U$ of $P$ in $M$ which is a complex analytical cell of real dimension $2(n-k+1)$. This follows at once from the fact that about $P$ one may express the coordinates and $\mu$ as power series in $n-k+1$ of them.

II. The points where $J$ is of rank $<k$. These are the *singular points* of $M$ and we denote their set by $S$.

III. The points where $A_0 = 0$ are *exceptional points* and we denote their set by $E$.

Each of the three types may yield real points. If $X_k = 0$ has a real solution in $x_k$ for $x_{k+1}, \ldots, x_n, \mu$ arbitrary real and small then the system (3.5) represents a continuous family of real periodic solutions, of dimension $n - k$ for each small real $\mu$. This will certainly occur if $X_k$ is of odd degree in $x_k$.

Regarding the singular points let $Z_j(x; \mu)$, $j = 1, 2, \ldots, \nu$ be the minors of order $n - k + 1$ of $J$. The set $S$ is then defined by the system

$$X_k = 0\ , \qquad A_0 x_{k-h} - A_h = 0\ , \qquad Z_j = 0\ . \tag{4.1}$$

This may be subjected to the same treatment as (3.4). It will yield a finite number of families of complex dimension $< n - k + 1$ whose real points are to be found.

For the exceptional points the argument is the same save that (4.1) is replaced by

$$X_k = 0\ , \qquad A_0 = A_1 = \cdots = A_{k-1} = 0\ . \tag{4.2}$$

It is clear from the preceding argument that the complete determination of all real periodic solutions may be accomplished in a finite number of steps.

5. As a mild application let us determine the families of periodic solution of period $2\pi$ of the system

$$\begin{aligned} \frac{dx_1}{dt} &= -x_2 + \mu g_1(x_1, x_2, \sin t, \cos t, \mu)\ , \\ \frac{dx_2}{dt} &= x_1 + \mu g_2(x_1, x_2, \sin t, \cos t, \mu) \end{aligned} \tag{5.1}$$

where the $g_i$ are polynomials in the indicated variables. If we set $x = x_1 + ix_2$, $g = g_1 + ig_2$ then (5.1) assumes the form

$$\frac{dx}{dt} - ix = \mu g(x, \bar{x}, e^{it}, e^{-it}, \mu)$$

with $g$ still a polynomial. To simplify matters we shall assume that $g$ does not contain $\bar{x}$ so that the system to be treated is

$$\frac{dx}{dt} - ix = \mu g(x, e^{it}, e^{-it}, \mu) \tag{5.2}$$

with $g$ a polynomial in the indicated variables. This system has recently been discussed by Friedrichs (Symposium on non-linear circuit analysis, Brooklyn Polytechnic Institute).

We are then looking for solutions $x(t, \mu)$ of period $2\pi$ of (5.2) which as $\mu \to 0$ tend to a solution of the first approximation

$$\frac{dx}{dt} - ix = 0 . \tag{5.3}$$

Let $\xi$ be the initial value of $x$ so that (5.3) has the general solution $\xi e^{it}$ and (5.2) a general solution of the form

$$x = \xi e^{it} + \mu A_1(\xi, t) + \cdots . \tag{5.4}$$

The substitution of (5.4) in (5.2) yields a simple recurrent system for the $A_h$ together with $A_h(\xi, 0) = 0$. As a consequence the periodicity condition assumes the general form

$$F_0(\xi) + \mu F_1(\xi) + \cdots = 0 \tag{5.5}$$

where the $F_j$ are polynomials and $\neq 0$. If $\xi(\mu)$ is a solution and $\xi(0) = \xi_0$, then $\xi_0$ must be a root of the equation

$$F_0(\xi) = 0 . \tag{5.6}$$

Let $\xi_0$ be a root of order $p$. We have then from the preparation theorem

$$\begin{aligned} &F_0(\xi) + \mu F_1(\xi) + \cdots \\ &= \{(\xi - \xi_0)^p + f_1(\mu)(\xi - \xi_0)^{p-1} + \cdots + f_p(\mu)\}\, E(\xi - \xi_0, \mu) , \end{aligned}$$

where the $f_h(\mu)$ are non-units. Hence the solution of (5.5) for $\xi(\mu)$ such that $\xi(0) = \xi_0$, reduces to that of

$$(\xi - \xi_0)^p + f_1(\mu)(\xi - \xi_0)^{p-1} + \cdots = 0 . \tag{5.7}$$

The required solutions may be obtained in a systematic manner by the Puiseux process. In the present case there will be $s$ so-called circular systems each consisting of $q$ conjugate sets

$$\xi - \xi_0 = \mu^{r/q} E(\mu^{1/q}) , \qquad r > 0 . \tag{5.8}$$

The values $\xi(\mu)$ defined by (5.8) correspond to a single periodic family $x\big(\xi(\mu), t\big)$ such that $x\big(\xi(\mu), 0\big) = \xi_0$ and $\Sigma q = p$. Thus we have obtained a complete solution of our problem.

# X

# X

## ON SINGULAR CHAINS AND CYCLES

1. *Introduction.* The theory of the topological invariance of the absolute or relative combinatorial characters of a complex, as developed in our Colloquium Lectures on *Topology* (Chapter II), was based, following Alexander and Veblen, upon the concept of singular chain. Our presentation, and indeed any known to us, appears to give rise to many misconceptions which it is proposed to clear up in the present note. Unless otherwise stated the notations are those of *Topology.*

2. *Singular Cells.* Let $\mathfrak{R}$ be a topological space and let $e_p$ be a simplicial oriented cell such that there exists a continuous single-valued transformation (=c.s.v.t.) $T$ of the point set $e_p$ into a subset $E_p$ of $\mathfrak{R}$, where $E_p = Te_p$. The symbol $(e_p, T, E_p)$, associated with the set $\overline{E}_p$ is called a *singular oriented p-cell on* $\mathfrak{R}$. If $e_p'$ is another $e_p$, there exists a barycentric transformation $U$ of $\bar{e}_p'$ into $\bar{e}_p$: $U\bar{e}_p' = \bar{e}_p$. If we set $T' = TU$, it is evident that $(e_p', T', E_p)$ defines also a singular oriented $p$-cell on $\mathfrak{R}$. We shall agree to consider it as identical with the first:

$$(e_p', T', E_p) = (e_p, T, E_p). \tag{1}$$

This has the advantage of freeing the notion of singular cell from a too narrow connection with a specific image $e_p$.

3. *Singular Chains.* The singular $p$-chain $C_p$ on $\mathfrak{R}$ is now defined as the association of a symbol

$$C_p = \sum t_i(e_p^i, T^i, E_p^i) \tag{2}$$

with coefficients $t$ belonging to one of the three rings (rational numbers, integers, integers mod $m$) considered in *Topology*, together with the set of all sets $\overline{E}_p^i$ corresponding to $t$'s$\neq 0$. As a special case the $e$'s might be cells of a finite complex $k$ such that there exists a c.s.v.t. $T$ of $k$ into a subset of $\mathfrak{R}$. Then the chain symbol may take the form

$$C_p = \sum t_i(e_p{}^i, T, E_p^i), \tag{3}$$

and $C_p$ may be considered as the image of the subchain

$$c_p = \sum t_i e_p^i \tag{4}$$

of $k$, but that is not essential. In this instance we might have represented $C_p$ by the symbol $(\sum t_i e_p^i, T, \sum t_i E_p^i)$ analogous to the cell symbol. Observe also that we may find for any chain (2) an equivalent representation (3). For we may take the cells $e_p^i$ to be simplexes in some $S_n$ whose closures do not meet, then define $T$ as coincident with $T^i$ on $e_p^i$. The closure of the sum of the cells $e_p^i$ will then be $k$, and (2) will assume the form (3).

If we have several singular chains $C_p^i$, then $\sum s_i C_p^i$, where the coefficients $s_i$ belong to the same ring as those of the chains, defines a $p$-chain which is called the linear combination with coefficients $s_i$ of the chains $C_p^i$. We have thus moduli of singular $p$-chains wholly analogous to the moduli of subchains of a complex.

4. *Boundary Relations.* Returning to $(e_p, T, E_p)$, let the boundary relations for $e_p$ be

$$e_p \rightarrow \sum \eta_i e_{p-1}^i = F(e_p). \tag{5}$$

Since $T$ is a transformation of $\bar{e}_p$ into $\overline{E}_p$, it transforms $e_{p-1}^i$ into a subset $E_{p-1}^i$ of $\overline{E}_p$ and hence $(e_{p-1}^i, T, E_{p-1}^i)$ is a singular $(p-1)$-cell on $\mathfrak{R}$. The singular $(p-1)$-chain

$$F(e_p, T, E_p) = \sum \eta_i (e_p^i, T, E_{p-1}^i) \tag{6}$$

is called the boundary of $(e_p, T, E_p)$ and we write here also

$$(e_p, T, E_p) \rightarrow F(e_p, T, E_p). \tag{7}$$

The boundary of the chain (2) is now by definition

$$F(C_p) = \sum t_i F(e_p^i, T^i, E_p^i), \tag{8}$$

for which we also write

$$C_p \rightarrow F(C_p). \tag{9}$$

Owing to (1) this boundary depends solely on $C_p$ but not on the particular transformations $T^i$ that occur in (8). Let $C_p$ be in the special form (3) with an associated non-singular image (4), and let the boundary relation for $c_p$ be

$$c_p \rightarrow \sum s_i e_{p-1}^i. \tag{10}$$

Then we have well defined singular cells $(e_{p-1}^i, T, E_{p-1}^i)$ and we find that

$$(11) \qquad C_p \to \sum s_i(e_{p-1}^i, T, E_{p-1}^i),$$

which may be described by the statement: the boundary of a singular image of a chain is the singular image of the boundary of the chain.

5. *Degenerate Case.* Let $\mathfrak{R}$ undergo a c.s.v.t. $U$ into a new space $\mathfrak{R}'$. Then the singular cell $(e_p, T, E_p)$ will go over into a singular $p$-cell $(e_p, UT, UE_p)$ and $C_p$ into

$$(12) \qquad UC_p = \sum t_i(e_p^i, UT^i, UE_p^i).$$

To different representations of the same singular $p$-cell on $\mathfrak{R}$ there will merely correspond different representations of the same singular $p$-cell on $\mathfrak{R}'$, and to $F(C_p)$ there will now correspond $F(UC_p)$. In particular also if $C_p \equiv 0$, likewise $UC_p \equiv 0$.

The preceding observations have an immediate application to degenerate cells. Let $(e_p, T, E_p)$ be a singular cell on $\mathfrak{R}$, and let us suppose that there exists a simplex $\sigma_q$, $q<p$, and two c.s.v.t.'s $T'$, $T''$, such that $T'$ is a simplicial transformation of $e_p$ into $\sigma_q$ and that $T'' \cdot T' = T$. The cell $(e_p, T, E_p)$ is called a *singular degenerate $p$-cell* on $\mathfrak{R}$ and chains made up exclusively of such cells are called *degenerate chains.* If $\mathfrak{R}' = U\mathfrak{R}$ as above, the degenerate cells and chains of $\mathfrak{R}$ go over into degenerate cells and chains of $\mathfrak{R}'$.

According to *Topology*, Chapter II, No. 2, $F(e_p, T', \sigma_q)$ is a degenerate $(p-1)$-chain, and hence when $(e_p, T, E_p)$ is degenerate so is its boundary. Hence this holds likewise as regards degenerate $p$-chains. Let us agree to consider all degenerate chains as identically zero. By the observation just made degenerate chains will then completely disappear from all boundary relations.

6. *Homologies.* From the preceding section it appears clearly that when $\mathfrak{R}$ and $\mathfrak{R}'$ are homeomorphic, the homeomorphism between them associates respectively to one another their moduli of $p$-chains, of bounding $p$-chains and their degenerate $p$-chains. These are therefore topological and the homology characters derived from the moduli are topological invariants.

Regarding these homologies, we introduce them exactly as for complexes. In particular if $A$ intersects $B$ in a set $A \cdot B$ closed relatively to $B$, the neglect of the singular cells $(e_p, T, E_p)$, such that $E_p \subset A$, leads to the characters of $B$ mod $A$.

7. *Invariance of the Combinatorial Homology Characters.* Suppose now that $\mathfrak{R}$ itself is a finite simplicial complex $K$ and let $\epsilon_p^i$ designate its cells. They can be considered as the singular cells $(\epsilon_p^i, 1, \epsilon_p^i)$ and it is readily seen that the formal singular boundary relations involving these cells alone are the same as the combinatorial relations between the cells $\epsilon_p^i$ themselves. Therefore whenever only singular cells of this type are involved, the singular boundary relations (7) for $K$ are reduced to the combinatorial relations.

The invariance of the combinatorial homology characters of $K$ is established by identifying them with the corresponding topological characters. The steps in the proof are as follows.

(a) Let $C_p$ be a singular chain on $K$ which we assume henceforth in the simplified form (11) with $T$ and the non-singular prototype $c_p$ fixed. There exists an $\eta > 0$ depending on $K$ but not on $C_p$, such that when mesh $C_p < \eta$, the chain can be homotopically deformed into a subchain $C_p'$ of $K$, the deformation keeping each cell on the closure of the cell of $K$ that carries it. This is the deformation theorem (*Topology*, p. 86). It implies (loc. cit., p. 78) that there are deformation chains, all singular, indicated by $\mathfrak{D}$, such that

$$\begin{aligned} \mathfrak{D}C_p &\rightarrow C_p' - C_p - \mathfrak{D}F(C_p), \\ C_p' &= \sum s_i E_p^i. \end{aligned} \tag{13}$$

(b) If mesh $C_p > \eta$ the chain has a subdivision chain $C_p$ whose mesh is suitable. Subdivision is defined as in *Topology* (p. 85), by reference to a subdivision of $c_p$.

(c) Suppose that $C_p$ possesses certain cells (not necessarily $p$-cells) which belong to $K$ and whose sum is therefore a subcomplex $K_q$ of $K$. Then the subdivision and deformation in (b) may be so chosen as to leave $K_q$ fixed point for point. The proof indicated in *Topology* (p. 87, Remark I) only shows that $C_p$ may be so modified as to leave the cells of $K_q$ invariant individually but not point for point. The more accurate result, which is of interest for its own sake, is proved as follows. We show by induction as in *Topology* (p. 86) that the deformations there indicated leave $K_q$ invariant point for point provided that any cell of $C_p$ without vertices on $K_q$ is of diameter $<\eta$, and that a cell having a face in common with $K_q$ has all its points not farther than $\eta$ from that face. If $C_p$ does not fulfill these conditions we find by reference to $c_p$ that a suitable subdivision of $C_p$ without new vertices on $K_q$ will behave as required. For if $c_p$ is a subchain, say of $k$, there is a subcomplex $k'$ of $k$ such that $T\cdot k'=K_q$. We can then apply to $k$ a series of subdivisions differing from regular subdivisions only in so far that no new vertices are ever introduced on $k'$. Given any $\zeta$ we can thus obtain a subdivision $c_p^*$ of $c_p$ whose cells fulfill relatively to $k'$ and $\zeta$ the two conditions that we wish to impose upon the cells of $C_p$ relatively to $K_q$ and $\eta$. Since $T$ is continuous the required result follows for $C_p$.

Consider now the boundary relations mod $L$, where $L$ is a subcomplex of $K$. Let $\Gamma_p$ be a (singular) cycle mod $L$. By (a) there is a subdivision $\Gamma_p'$ of $\Gamma_p$ homotopically deformable into a subcycle $\Gamma_p^*$ of $K$, its points on $L$ remaining on $L$, and $\Gamma_p'\sim\Gamma_p$ mod $L$ on $\Gamma_p$ itself (*Topology*, p. 87) and hence a fortiori on $K$. By (c), if $C_{p+1}\rightarrow\Gamma_p'$ mod $L$, there is a subdivision $C_{p+1}'$ of $C_{p+1}$ with the same boundary $\Gamma_p'$, deformable into a subchain $C_{p+1}^*$ by a homotopy leaving $\Gamma_p'$ invariant, so that $C_{p+1}^*\rightarrow\Gamma_p'$ mod $L$. Therefore if the initial cycle$\sim 0$ mod $L$ in the topological sense, the reduced cycle$\sim 0$ mod $L$ in the combinatorial sense. From this follows immediately as in *Topology* (p. 88), that the topological and combinatorial homology groups of the same types are simply isomorphic and hence have the same numerical invariants. Therefore the combinatorial homology characters are topological invariants.

8. *Remarks*. I. Once the notion of singular cell has become familiar one will naturally abandon the explicit (too explicit) $(e, T, E)$ notation in favor of the simpler $E$ of *Topology*.

II. The following circumstance may arise in connection with our definition of singular cell. Taking for the sake of simplicity $p=2$, let $e_2=ABC$ be an (oriented) isosceles triangle with $AB=AC$ and let $AD$ be the altitude issued from $A$. Let $U$ be the symmetry about $AD$ and $T$ a transformation $=1$ on $ADB$, $=U$ on $ADC$. If we set $T'=T\cdot U$, $e_2'=ACB$, we have

$$(e_2, T, E_2) = (e_2', TU, E_2) = (-e_2, TU, E_2) = (-e_2, T, E_2)$$

and therefore

$$2(e_2, T, E_2) = 0.$$

Owing to this, E. Čech, who pointed out this circumstance to us, suggested that in the present and in the similar instance for any $p$, the singular cell be also considered as degenerate. The more extended meaning to be thus attached to degenerate cells, while justifiable, is not however essential.

# XI

# XI

## ON GENERALIZED MANIFOLDS

The object of the present paper is to extend to a larger class of spaces certain results recently obtained for topological manifolds.† The extension consists in replacing the requirement that every point possess a combinatorial cell for neighborhood by certain weaker conditions on the chains through the point. Roughly speaking they amount to demanding that locally any $p$-chain be deformable (in a certain very general sense) into one which does not meet any assigned $q$-space ($= q$ dimensional space), where $p + q < n$, the dimension of the manifold. This extension is made in Part III of the present paper. In Part I we take up again, partly as a preparation to the second Part, the homology theory of metric spaces from the standpoint initiated in our Colloquium Lectures *Topology,* Ch. VII. The notation and terminology are as in our book.‡

### § 1. The Approximating Complexes of a Metric Space.

1. The homology properties of a compact metric space are intimately related to the homology properties of certain subchains of an infinite complex, the fundamental complex of the space (*Topology,* Ch. VII), or to certain sequences of chains of approximating complexes (Alexandroff). We shall first show how these may be selected in a certain convenient way for the sequel.

Let for the present $\mathfrak{R}$ be a compact metric $n$-space and let $U$, $V$, $W$, denote generically its open sets, and $F(U)$, $F(V)$, $F(W)$, their boundaries.

We shall repeatedly consider various aggregates of subsets, $\Sigma = \{A^\alpha\}$, of $\mathfrak{R}$. The mesh of $\Sigma$ is max diam $A^\alpha$. If the set of $A$'s covers $\mathfrak{R}$ we call $\Sigma$ a *covering,* an *$\epsilon$-covering* if its mesh $< \epsilon$. Of particular importance are the finite coverings by open sets ($=$ f. c. o. s.).

Each set $A^\alpha$ of the aggregate $\Sigma$ may be considered as an abstract point,

---

† S. Lefschetz and W. W. Flexner, *Proceedings of the National Academy,* Vol. 16 (1930), pp. 530-533; W. W. Flexner, *Annals of Mathematics,* Ser. 2, Vol. 32 (1931), pp. 393-406, 539-548.

‡ A very extensive paper by Čech on the same general topic was presented simultaneously with the present one to the *Annals of Mathematics* where his paper is now appearing. While there are many contacts between the two, they differ essentially in method and scope. Čech deals indeed with a much more general type of space, but the restriction to locally compact metric spaces which we have imposed here, has enabled us to proceed much more quickly to the point.

and we may then introduce for each intersection $A^{\alpha_0}\cdots A^{\alpha_p}\neq 0$ an abstract $p$-simplex $\sigma_p = A^{\alpha_0}\cdots A^{\alpha_p}$. It will be convenient to designate the intersection also by $\sigma_p$ : $\sigma_p = 0$ signifies then that the sets $A^{\alpha_0},\cdots,A^{\alpha_p}$ do not intersect.

The aggregate $\{\sigma\}$ has the property that with each $\sigma$ every face of $\sigma$ also belongs to the set. Hence $\{\sigma\}$ is a closed simplicial (abstract) complex $\Phi$, the *skeleton* of $\Sigma$. If another aggregate $\Sigma' = \{A'^{\alpha}\}$ has for skeleton $\Phi'$ a complex whose structure is that of a subcomplex of $\Phi$, we shall briefly say that its skeleton is a subcomplex of $\Phi$. The *dimension* of $\Phi$ is the highest integer $\nu$ such that there is at least one aggregate of $\nu+1$ intersecting $A$'s. $\nu$ is also called the *order* of $\Sigma$. Clearly of course $\Phi$ is finite when and only when $\Sigma$ is finite.

Suppose in particular that $\Sigma = \{U^{\alpha}\}$ is an $\epsilon$-f. c. o. s. It is called *irreducible* (Alexandroff) when there is no $\epsilon$-f. c. o. s. whose skeleton is a proper subcomplex of $\Phi$. If $\Sigma$ is reducible there is an $\epsilon$-f. c. o. s. $\Sigma^1$ whose skeleton is a proper subcomplex $\Phi^1$ of $\Phi$. If $\Sigma^1$ is in turn reducible there is an $\epsilon$-f. c. o. s. $\Sigma^2$ whose skeleton is a proper subcomplex $\Phi^2$ of $\Phi$, etc. Since $\Phi$ has only a finite number of subcomplexes the process must stop after a finite number of steps. Therefore there exists an irreducible $\epsilon$-f. c. o. s. whatever $\epsilon$. If the order of the initial covering is the least possible for an $\epsilon$-f. c. o. s. it will also be the order of the ultimate irreducible covering.

We recall that as $\epsilon\to 0$ the least order $\nu$ tends to an upper limit $n$ or else $\to\infty$. In the first case dim $\mathcal{R} = n$, in the second case dim $\mathcal{R}=\infty$.

Let $\Sigma = \{U^{\alpha}\}$ be a f. c. o. s. whose skeleton $\Phi$ is the same as for $\{\bar{U}^{\alpha}\}$. Then there exists a constant $\eta$, the *characteristic constant* of $\Sigma$, such that: (a) if a set $A$ on $\mathcal{R}$ whose diameter $<\eta$ meets a certain number of $U$'s, these $U$'s have a non-vacuous intersection; (b) any point $x$ of $\mathcal{R}$ is on at least one $U$ such that $d(x, \mathcal{R}-U)>\eta$. As a consequence of (b) if diam $A<\eta$ then some $U\supset A$.

2. Taking $n = \dim \mathcal{R}$ finite, let $\epsilon$ be so small that the least order of an $\epsilon$-f. c. o. s. is $n$, and let $\Sigma$ be an irreducible $\epsilon$-f. c. o. s. There exists another $\epsilon$-f. c. o. s. of order $n$, $\Sigma' = \{V^{\alpha}\}$ consisting of as many sets as $\Sigma$ and such that for every $\alpha$ we have $V^{\alpha}\Subset U^{\alpha}$.† Clearly $\Sigma'$ is an $\epsilon$-f. c. o. s. whose skeleton is $\Phi$ or a subcomplex of $\Phi$, and since $\Sigma$ is irreducible it can only be $\Phi$. Therefore *the order of $\Sigma$ is $n$*. In other words an irreducible f. c. o. s. whose mesh is sufficiently small is of order $n$. Observe incidentally that $\{\bar{V}^{\alpha}\}$ has the same skeleton as $\{V^{\alpha}\}$.

† Menger, *Dimensionstheorie*, p. 160. We shall use his "strong inclusion" symbol $\Subset$ ($A\Subset B$ means that $\bar{A}\subset B$).

Consider now a sequence $\{\Sigma^i\}$, where $\Sigma^i = \{U^{i\alpha}\}$ is an irreducible $\epsilon_i$-f. c. o. s. such that: (a) $\epsilon_1 = \epsilon$; (b) if $\eta_i$ is the characteristic constant of $\Sigma^i$ we have $\epsilon_{i+1} < \frac{1}{2}\eta_i$ and $< \frac{1}{2}\epsilon_i$; (c) $\{\bar{U}^{i\alpha}\}$ has the same skeleton as $\Sigma^i$. As a consequence $\Sigma^i$ is of order $n$ and for every $U^{i+1,\beta}$ there is a $U^{i\alpha} \supset U^{i+1,\beta}$. Let $\Phi^i$ be the skeleton of $\Sigma^i$; choose for each $U^{i+1,\beta}$ a definite $U^{i\alpha} \supset U^{i+1,\beta}$ and define a transformation $t_i$ of the vertices of $\Phi^{i+1}$ into vertices of $\Phi^i$ whereby the vertex $U^{i+1,\beta}$ goes into the vertex $U^{i\alpha}$. Let $\sigma_p = U^{i+1,\beta_0} \cdots U^{i+1,\beta_p}$ be a simplex of $\Phi^{i+1}$. As a consequence, if $U^{i\alpha_h} = t_i U^{i+1,\beta_h}$ then $U^{i\alpha_h} \supset U^{i+1,\beta_h}$ and hence $\sigma'_q = U^{i\alpha_0} \cdots U^{i\alpha_p}$ is a simplex of $\Phi^i$. (It may happen that several of the vertices $U^{i\alpha_h}$ coincide, in which case $q < p$). Thus if certain vertices $U^{i+1}$ belong to a $\sigma_p$ of $\Phi^{i+1}$ the transformed vertices $t_i U^{i+1}$ are vertices of a $\sigma_q$ $(q \leqq p)$ of $\Phi^i$. Consequently $t_i$ may be extended to a simplicial transformation $\tau_i$ of $\Phi^{i+1}$ into $\Phi^i$ or into a subcomplex of $\Phi^i$. We call $\tau_i$ a *projection* of $\Phi^{i+1}$ onto $\Phi^i$, and more generally $\tau_i\tau_{i+1} \cdots \tau_{i+j-1}\Phi^{i+j}$ a projection of $\Phi^{i+j}$ onto $\Phi^i$. The latter is also a simplicial transformation of $\Phi^{i+j}$ into $\Phi^i$ or into a subcomplex of $\Phi^i$.

3. I say that in fact $\tau_i\Phi^{i+1} = \Phi^i$, that is every simplex of $\Phi^i$ is the transform of a simplex of $\Phi^{i+1}$, or, in other words, $\Phi^i$ is completely covered by $\tau_i\Phi^{i+1}$. For let us suppose that $\tau_i\Phi^{i+1} = \Psi$, a proper subcomplex of $\Phi^i$. There exists then a simplex $\sigma_p = U^{i\alpha_0} \cdots U^{i\alpha_p} \subset \Phi^i - \Psi$. Denote generically by $V^\alpha$ the sum of all the sets $U^{i+1,\beta}$ which make up $\tau_i^{-1}U^{i\alpha}$; clearly $V^\alpha \subset U^{i\alpha}$. Since every $U^{i+1}$ corresponds to one (and only one) $V$, $\Sigma = \{V^\alpha\}$ is an $\epsilon_i$-f. c. o. s. and it has a subcomplex $\Phi'$ of $\Phi$ as its skeleton. I say that $\sigma_p$ is not a cell of $\Phi'$. For otherwise we would have $V^{\alpha_0} \cdots V^{\alpha_p} \neq 0$, and hence there would exist a $U^{i+1,\beta_0} \cdots U^{i+1,\beta_p} \neq 0$, where $U^{i+1,\beta_h}$ is a constituent of $V^{\alpha_h}$. Since $\tau_i U^{i+1,\beta_h} = U^{i\alpha_h}$, we would then have in $\sigma'_p = U^{i+1,\beta_0} \cdots U^{i+1,\beta_p}$ a simplex of $\Phi^{i+1}$ such that $\tau_i\sigma'_p = \sigma_p$ and hence $\sigma_p \subset \Psi = \tau_i \cdot \Phi^{i+1}$, contrary to assumption.

Under the circumstances then $\sigma_p \not\subset \Phi'$. It follows that $\Phi'$ is a proper subcomplex of $\Phi^i$ and also the skeleton of an $\epsilon_i$-f. c. o. s. But this is ruled out since $\Sigma^i$ is irreducible. Hence $\sigma_p$ cannot exist, and $\Psi = \tau_i\Phi^{i+1} = \Phi^i$.

*Definitions.* A sequence $\{B^i\}$ of elements, (sets, complexes, etc.) such that $B^i \subset \Phi^i$ and $\tau_i B^{i+1} = B^i$ is called a *projection-sequence* (of sets, of complexes, etc.).

Given any (non-singular) chain $C_p$ we shall designate by $| C_p |$ the complex made up of the cells of the chain. A sequence $\{C_p{}^i\}$ will be called a projection-sequence of chains or cycles whenever

$$\tau_i C_p{}^{i+1} = C_p{}^i.$$

4. Let $U^{i\delta} \supset x$. There exists a set $U^{i-1,\gamma} \supseteq U^{i\delta}$ such that $\tau_{i-1}U^{i\delta} = U^{i-1,\gamma}$; a set $U^{i-2,\beta}$ similarly related to $U^{i-1,\gamma}$, etc., clear up to a certain set $U^{1a}$. Let $k_i$ be for each $i$ the class of all sets $U^{1a}$ thus obtained. The classes $k_i$ are all finite, $\neq 0$ and $k_i \supset k_{i+1}$. Therefore from a certain $i$ on $k_i = k_{i+1} = \cdot\cdot\cdot$. Consequently there exists an infinite sequence $\{U^{ia_i}\}$ such that $U^{ia_i} \supseteq U^{i+1,a_{i+1}}$, $\tau_i U^{i+1,a_{i+1}} = U^{ia_i}$, $\Pi U^{ia_i} = x$. Let $U^{ia_0}, \cdot\cdot\cdot, U^{ia_p}$ be all the sets of $\Sigma^i$ occurring in any such sequence corresponding to the same point $x$ and let $V^i = U^{ia_0} \cdot\cdot\cdot U^{ia_p} \neq 0$, so that $\sigma_p{}^i = U^{ia_0} \cdot\cdot\cdot U^{ia_p}$ is a simplex of $\Phi^i$. Since every $U^{ia}$ here occurring is the $\tau_i$ transform of a similar $U^{i+1,\beta}$ we have $\tau_i\sigma^{i+1} = \sigma^i$, hence $\{\sigma^i\}$ is a projection-sequence of simplexes. Moreover clearly $V^i \supseteq V^{i+1}$, $\Pi V^i = x$.

Conversely if $\{\sigma^{*i}\}$ is a projection-sequence of simplexes, and if $V^{*i}$ is the intersection of the sets $U^i$ associated with $\sigma^{*i}$, then $V^{*i} \supseteq V^{*i+1} \supset x$, hence $\Pi V^{*i} = x$. Clearly also the sets $U^i$ associated with $\sigma^{*i}$ are among those associated with $\sigma^i$, hence $\sigma^{*i}$ is $\sigma^i$ or a face of $\sigma^i$. We call $\{\sigma^{*i}\}$ and $\{\sigma^i\}$ respectively *projection-sequence* and *maximal projection-sequence* for the point $x$.

5. Owing to the choice of $\{\Sigma^i\}$ we may use $\{\Phi^i\}$ to map the space $\mathfrak{R}$ topologically on an Euclidean $S_r$, $r \geqq 2n + 1$.† Choosing $r \geqq 2n + 2$ we may even carry out the mapping so as to be able to construct the joining cell of any simplex of $\Phi^{i+1}$ with its transform under $\tau_i$ (deformation cell corresponding to $\tau_i$), and from there, as the sum of all these cells, the $(n+1)$-complex $K$ or *fundamental complex* of $\mathfrak{R}$, (*Topology*, p. 327) which will be an infinite complex on $S_r$. The part of $K$ obtained on removing $\Phi^i$, $\Phi^{i-1}, \cdot\cdot\cdot$ and the cells joining them will be denoted by $N^i$ and the finite complex $K - N^i$ by $K^i$.

In practice we shall find it more convenient to have a representation of $\mathfrak{R}$ and $K$ on the Hilbert parallelatope

$$\mathcal{H}: \quad 0 \leqq x_i \leqq 1/i, \qquad (i = 1, 2, \cdot\cdot\cdot + \infty).$$

This image is to be constructed as follows. As proved by Urysohn $\mathfrak{R}$ has a topological image $\mathfrak{R}'$ on $\mathcal{H}$. Consider now the following homeomorphism of $\mathcal{H}$:

$$T: \quad x'_i = \tfrac{1}{2}(x_i + 1/i)$$

which transforms it into the subset

$$\mathcal{H}': \quad 1/2i \leqq x_i \leqq 1/i.$$

Then $T\mathfrak{R}' = \mathfrak{R}''$ is a topological image of $\mathfrak{R}$ which possesses no point for which any $x_i$ is zero. We identify henceforth $\mathfrak{R}$ with $\mathfrak{R}''$.

---

† See p. 586.

Let us denote by $S^i$ the subset of $\mathcal{H}$ consisting of all points for which $x_k = 0$ when $k > (2n + n_i)$, where $n_i$ increases so rapidly that we may carry out the construction of $K$, given in *Topology*, p. 325, in such manner that $\Phi^i \subset S^i - S^{i-1}$. As a consequence $K \cdot \mathcal{R} = 0$. Now, with closures referring to $\mathcal{H}$, the only limit-points of $\bar{K}$ not on $K$ are on $\mathcal{R}$, hence $\mathcal{R} = \bar{K} - K$. It is in order to fulfill this condition that the complex $K$ has been constructed in the above special manner.

6. It is convenient to join each point of $\Phi^{i+1}$ to its transform by $\tau_i$ by a segment in $\mathcal{H}$. The sum of these segments coincides with $K$. An infinite arc consisting of a sequence of projecting segments for $\tau_1, \tau_2, \cdots$ plus their coterminal end-points, will be called a *projecting line.* The projecting lines all start at $\Phi^1$, which we designate henceforth by $\Phi$, and continue indefinitely throughout $K$.

If $\{B^i\}$ is a projection sequence of sets or complexes, the set $\mathcal{B}$ obtained by adding to the sequence the projecting segments of the points of the $B$'s is called a *projection-set.* If the $B$'s are complexes, the projecting segments of a definite $p$-cell of $B^{i+1}$ make up a $(p+1)$-cell; these are the joining cells of $B^{i+1}$ and $\tau_i B^{i+1}$ (No. 5). The sum of the closures of all these cells is a *projection-complex* $\mathcal{K}$. If $B^i$ is a subcomplex of $\Phi^i$ for every $i$, $\mathcal{K}$ is a subcomplex of $K$.

We are primarily interested in the relation between various subcomplexes of $K$ and certain associated sets of $\mathcal{R}$. Properly speaking instead of a subcomplex of $K$ we might well take any subset of $K$, but actually the subcomplexes will suffice for our purpose.

With any subcomplex $L$ of $K$ we may associate the closed subset $F = \bar{L} \cdot \mathcal{R}$, and we observe immediately that this set $F$ depends solely upon the "infinite" part of $L$, i. e. it is unchanged when a finite complex is added to or removed from $L$. In the sense of *Topology,* Ch. VII, $\mathcal{R}$ is associated with the total ideal element of $K$, and $F$ with a certain closed ideal element of the complex.

Suppose that we construct a new fundamental $K'$ for $\mathcal{R}$, that we suppose as before on $\mathcal{H}$, and such that $K' \cdot \mathcal{R} = 0$. Applying to $K$ the deformation theorem of *Topology*, p. 328 † (proved for chains but applicable to complexes), we can reduce $L$ to a subcomplex $L'$ of $K'$ by a deformation that $\to 0$ for any particular cell of $K$ as that cell $\to \mathcal{R}$. Therefore $F = \bar{L}' \cdot \mathcal{R} = \bar{L} \cdot \mathcal{R}$, i. e. the set $F$ is in a large measure independent of the complex $K$.

† In the proof *loc. cit.*, $A^j$ should be mapped on $\tau_{i-1}\lambda_i A^j$. Owing to the condition $\epsilon_i < \frac{1}{2}\eta_{i-1}$ which we have imposed, $C_p{}^{i-1}$ will still be mapped as before on a subchain of $\Phi^i$.

7. We shall now reverse the situation: starting with any particular closed set $F$ we shall associate with it a certain projection-complex $L$, such that $F = \bar{L} \cdot \mathcal{R}$ and $\dim F = p = \dim L - 1$, which is the maximum value possible for $p$.

According to Menger (*Dimensionstheorie*, p. 158) there is a f. c. o. s. of order $\leqq p$ of $F$ (not of $\mathcal{R}$), $\Sigma'^i = \{V^{ia}\}$, such that there is one and only one $V^{ia}$ on any $U^{ia}$ that meets $F$. When $i$ exceeds a certain value the skeleton $\Phi'^i$ of $\Sigma'^i$ is a $p$-complex. Associate with each $V^{ia}$ the vertex $U^{ia}$ of the set of same name. Now when a certain aggregate of sets $V^i$ intersect, the same holds as regards the corresponding sets $U^i$. Hence $\Phi'^i$ will thus become a subcomplex of $\Phi^i$. Now take all the subcomplexes $\Psi'^1$ of $\Phi^1$ which are the projections of a $\Phi'^i$. Since their number is infinite and the number of subcomplexes of $\Phi^1$ is finite, at least one, $\Psi^1$ is the projection of an infinity of complexes $\Phi'^i$. Consider the subcomplexes $\Psi'^2$ of $\Psi^2$ such that $\tau_1 \Psi'^2 = \Psi^1$. There is an infinity of complexes $\Phi'^i$, $i \geqq 2$, projected onto $\Psi^1$ and their projections on $\Phi^2$ are each a $\Psi'^2$. Therefore at least one of the latter, $\Psi^2$, is the projection of an infinity of complexes $\Phi'^i$, etc. By this obvious process we obtain an infinite projection-sequence $\{\Psi^i\}$, where $\Psi^i$ is a subcomplex of $\Phi^i$ which is the projection of a $\Phi'^j$, and $\dim \Psi^i \leqq \dim \Phi'^j \leqq p$. Since $\Phi'^j$ is the skeleton of an $\epsilon_j$-f. c. o. s. of $F$, the latter may be $6\epsilon_j$-deformed into $F$.† Moreover, referring to the representation in $\mathcal{H}$, $\Phi'^j$ can be $\xi_i$-deformed into $\Psi^i$, ($\xi_i \to 0$ with $1/i$). Hence $F$ can be $\zeta_i$-deformed into $\Psi^i$, ($\zeta_i \to 0$ with $1/i$). Therefore $\Psi^i$ is the skeleton of a $\theta_i$-f. c. o. s. of $F(\theta_i \to 0$ with $1/i)$ (Alexandroff, *loc. cit.*, p. 18). As a consequence if we put in the joining cells of the $\Psi$'s, we obtain a fundamental complex $L$ for $F$. We have $\dim L = p + 1$, for it is $\geqq p + 1$ since $\dim F = p$, and $\leqq p + 1$, since $\dim \Psi^i \leqq p$.

Since we have but little information regarding the meshes or the characteristic constants of the coverings of $F$ whose skeleta are the $\Psi$'s, it is not easy to show that the deformation theorem applies to $L$. Therefore for the homology theory another similar $(p + 1)$-complex $L^*$ is more suitable. It is constructed as follows: take the skeleton of the aggregate $\{U^{ia} \cdot F\}$ ($i$ fixed) and remove from it all cells of dimension $> p$. What is left is a subcomplex $\Omega^i$ of $\Phi^i$, and we have immediately, owing to the mode of constructing the $\Phi$'s, $\tau_i \Omega^{i+1} \subset \Omega^i$. The complex $L^*$ consists of all the $\Omega$'s plus their joining cells. It is clearly a $(p + 1)$-subcomplex of $K$, which we shall call the *generalized fundamental complex* of the set $F$. The proof of the deformation theorem is directly applicable to $L^*$ for all cycles or complexes of dimension $\leqq p$. Since

† P. Alexandroff, *Annals of Mathematics*, Vol. 30 (1928-29), p. 13.

$\dim F = p$, $F$ possesses a fundamental $(p+1)$-complex $L'$ to which the deformation theorem is applicable. For example $L'$ may be built up out of a subset of the $\Psi$'s. It follows (see No. 9), that the $q$-cycles, $q > p$, of $F$ are all $\equiv 0$ and hence they need not concern us further.

8. *The chains and cycles of* $K$. The only chains of $K$ with which we shall be concerned are its subchains, no others being considered. Whatever $C_p$ we have: $C_p = C'_p + C_p''$, where $C'_p$ is the part of $C_p$ on $K^i$ and $C_p''$ the rest. It is convenient to write: $C'_p = K^i \cdot C_p$, $C_p'' = N^i \cdot C_p$. The part of $F(C'_p)$ which is on $\Phi^i$ will be designated by $\Phi^i \cdot C_p$ and called the *trace* of $C_p$ on $\Phi^i$.

Let us suppose that we have on $K$ an aggregate of chains $\{C_q{}^i\}$, $q = 0, 1, \cdots, p$; $i = 1, 2, \cdots$, such that: (a) $C_p = \Sigma C_p{}^i$ is a true chain of $K$, i. e. includes no cell of $K$ taken with an infinite coefficient; (b) for every $C_q{}^i$, we have

$$C_q{}^i \to \Sigma \eta^q{}_{ij} C^j{}_{q-1}. \tag{8.1}$$

The aggregate $\{C_q{}^i\}$ is called an *elementary decomposition* of $C_p$. An example is of course the decomposition of $C_p$ into its cells. For later purposes a more general decomposition is introduced here.

Two decompositions $\{C_q{}^i\}$, $\{C'_q{}^i\}$ of two chains $C_p$, $C'_p$ are said to have the same structure if they correspond to one another chain for chain (for every $C_q{}^i$ one and only one chain $C'_q{}^i$ and conversely) and if the corresponding incidence numbers $\eta^q{}_{ij}$ are the same. That is to say if the sets are labelled in such manner that $C_q{}^i$ and $C'_q{}^i$ are the associated chains in the correspondence then they have the same incidence matrices $\| \eta^q{}_{ij} \|$.

Suppose now that we have two decompositions $\{C_q{}^i\}$, $\{C'_q{}^i\}$ of $C_p$, $C'_p$ whose structure is the same, and let there exist for every $C_q{}^i$ a $(q+1)$-chain $\mathfrak{D} C_q{}^i$, called a *deformation-chain,* such that

$$\mathfrak{D}\, C_q{}^i \to C'_q{}^i - C_q{}^i - \Sigma \eta^q{}_{ij} \mathfrak{D}\, C^j{}_{q-1}. \tag{8.2}$$

If we agree to write

$$\mathfrak{D}\, \Sigma\, a_i C_q{}^i = \Sigma\, a_i \mathfrak{D}\, C_q{}^i \tag{8.3}$$

then (8.2) assumes the form

$$\mathfrak{D}\, C_q{}^i \to C'_q{}^i - C_q{}^i - \mathfrak{D} F(C_q{}^i). \tag{8.4}$$

Under the circumstances the passage from $C_p$ to $C_p'$ is called a *deformation* of $C_p$ into $C'_p$, and $\mathfrak{D}\,(C_p)$ is called the *deformation-chain* of $C_p$.

A deformation of a subcomplex $L$ of $K$ into another $L'$ could be defined substantially along similar lines. We would merely replace the chains $C_q{}^i$ by the cells $E_q^i$ of $L$, and in (8.2), (8.3), (8.4), the $C$'s would be cells and the $\mathfrak{D}\, C$'s would continue to be chains but otherwise the rest would be as before. Then $\Sigma \mid \mathfrak{D} E \mid + L + L'$ would be called the *deformation-complex* $\mathfrak{D}\, L$ of $L$.

All this is entirely in line with the treatment of deformations in *Topology*, p. 78, except that there we had only cells and obtained (8.2) from direct geometric considerations, essentially by considering the deformation as a "singular" translation, whereas (8.2) serves directly to define the deformation. This departure is justified on the ground that (8.2) is the central property of a deformation as regards the applications to any homology theory.

For purposes of reference, if we agree to neglect everywhere chains on $L$, or else if we only consider integral chains mod $m$ or both we have associated deformations and deformation-chains mod $L$, mod $m$, mod $(L, m)$, as the case may be.

If in a given deformation $\mathfrak{D}$ every deformation chain is of diameter $< \epsilon$ we have a so-called *$\epsilon$-deformation.*

By analogy with ordinary deformations we shall say that $\mathfrak{D}$ leaves a chain $C_q^i$ *invariant* or *does not displace the chain,* whenever the chain $\mathfrak{D}\, C_q^i = 0$.

9. *The chains and cycles of the space* $\mathfrak{R}$. Taking substantially the point of view of *Topology*, Ch. VII, § 4, we consider a $(p+1)$-chain $C_{p+1}$ of $K$ as defining a $p$-chain $c_p$ of $\mathfrak{R}$, a cycle mod $\Phi$, $\Gamma_{p+1}$ of $K$ as defining an absolute $p$-cycle $\gamma_p$ of $\mathfrak{R}$. In particular if

$$K \supset C_{p+2} \rightarrow \Gamma_{p+1} \quad \text{mod } \Phi,$$

and if $C_{p+2}$ determines $c_{p+1}$ of $\mathfrak{R}$ we write

$$c_{p+1} \rightarrow \gamma_p, \qquad \gamma_p = F(c_{p+1}),$$

and say "$\gamma_p$ bounds $c_{p+1}$". A special case is where $\Gamma_{p+1}$ is finite, for it is then $\approx 0$ mod $\Phi$, since it can be deformed along the projecting lines onto $\Phi$. We say that $\gamma_p$ is homologous to zero: $\gamma_p \approx 0$, whenever it is a finite or infinite sum of bounding cycles. The extension to cycles mod $A$, $A$ closed, is in the usual manner: $\Gamma_{p+1}$ is then a cycle mod $L$, where $L$ is any subcomplex of $K$ such that $\bar{L} \cdot \mathfrak{R} = A$.

The $p$-th homology group $\mathfrak{H}_p$ (absolute or mod $A$) is the quotient group $\mathfrak{G}_p \div \mathfrak{G}'_p$ of the Abelian group $\mathfrak{G}_p$ of the *p-cycles* (written additively) by the group $\mathfrak{G}'_p$ of the cycles $\approx 0$. The bases and homology characters are defined as usual.

For $p = n$ the bounding relations between the cycles are reduced to the identical linear relations between them. In terms of the $n$-cycles it is possible to define the generalized absolute orientable $n$-circuit (*Topology*, p. 76): it is a compact metric $n$-space $\mathcal{R}$ such that $R_n(\mathcal{R}) = 1$, and $R_n(A) = 0$ for any *proper* closed subset $A$ of $\mathcal{R}$. As a consequence the circuit has a base for the $n$-cycles consisting of a single $\gamma_n$, i. e. every $n$-cycle of $\mathcal{R}$ is of the form $t\gamma_n$. In place of $\gamma_n$ we might as well take $-\gamma_n$ and either one of the pairs $(\mathcal{R}, \gamma_n)$, $(\mathcal{R}, -\gamma_n)$ is called an *oriented* circuit, the passage from one to the other being described as a reversal of orientation. The non-orientable circuit is obtained by taking the cycles mod 2, and similarly for the circuits mod $m$. Analogous notions hold for the circuit mod $A$, $A$ closed, the circuit conditions being $R_n(\mathcal{R}, A) = 1$, $R_n(B, A) = 0$, where $B$ is now any proper closed subset of $\mathcal{R}$ which $\supset A$.

10. We have taken the chains and cycles of $\mathcal{R}$ as represented by actual chains or cycles mod $\Phi$ of $K$. Their characteristic part corresponds however to the infinite portion of the representative $C_{p+1}$ or $\Gamma_{p+1}$. As a matter of fact the difference is not great: we may always suppress, say $\Phi^1, \cdots, \Phi^k$, with all the cells joining them, and consider $\Phi^k$ as the new $\Phi$, thus converting any $\Gamma$ with a finite boundary into a cycle mod $\Phi$. Another way of looking at the matter is as follows: under our conventions for chains the suppression of any finite part of $C_{p+1}$ is not to affect $c_p$. As for a $\gamma_p$ it is then to be represented by a $C_{p+1}$ with finite boundary $C_p$. But if we slide the points of $C_p$ along the projecting lines down onto $\Phi$, and add the deformation-chain, which is finite, to $C_{p+1}$, we have a cycle mod $\Phi$, $\Gamma_{p+1}$, which also represents $\gamma_p$.

The set $\mathcal{R} \cdot |\overline{C_{p+1}}|$, where as before the closure refers to $\mathcal{H}$, is a closed subset of $\mathcal{R}$ associated with $c_p$, that we shall denote by $|c_p|$. This set depends solely on $c_p$, and not on the particular fundamental complex $K$ chosen (No. 6).†

By the points of $c_p$ we shall always mean the points of $|c_p|$. In particular a set $A$ is said to intersect $c_p$ whenever it intersects $|c_p|$, to be $\subset c_p$ or to $\supset c_p$ whenever $A \subset |c_p|$ or $\supset |c_p|$ as the case may be.

Let $A$ be a closed set. By a $p$-cycle mod $A$ we shall mean a $c_p$ such that $F(c_p) \subset A$. The cycle is said to *bound* mod $A$ whenever there exists a $c_{p+1}$ such that $F(c_{p+1}) - c_p \subset A$. Finally it is $\approx 0$ mod $A$ whenever the cycle is a finite or infinite sum of cycles which bound mod $A$.

We may also consider the absolute cycles of $\mathcal{R} - A$. Such a cycle is $\approx 0$ on $\mathcal{R} - A$ whenever it is $\approx 0$ on some closed subset of $\mathcal{R} - A$.

† The $p$-chains such that $\dim |c_p| \leqq p$ form a topological subclass of the class of all $p$-chains. These special chains played an important part in the initial version of the present paper. We found it simpler since then, to eliminate them entirely, and to replace them everywhere merely by the projection-chains which are introduced in No. 13. As the properties needed in Part II were only those of projection-chains, the only important modifications required were in Nos. 11, 18, 19 (June, 1933).

11. A deformation of a $C_{p+1}$ into $C'_{p+1}$ on $K$ may serve to define two kinds of deformations $\mathfrak{D}$ of the associated chains $c_p$, $c'_p$ on $\mathfrak{R}$. The deformation $\mathfrak{D}$ is of the *first kind* whenever the chains of the associated elementary decompositions $\{C_q{}^i\}$, $\{C'_q{}^i\}$, are all finite; it is of the *second kind* when some or all are infinite.

Consider for the present a $\mathfrak{D}$ of the first kind. If the deformation-chain of $C_q{}^i \to 0$ with $1/i$, we consider the two chains $c_p$, $c'_p$ as identical. If $U$ is any open set $\supset c_p$, and if $L$ is any subcomplex of $K$ such that $\overline{L}\cdot\mathfrak{R} = \bar{U}$, then for $i$ above a certain value $\mathfrak{D}\, C_q{}^i \subset L$, and hence $C_{p+1}$ has at most a finite subchain on $K - L$.

As an application if $C_{p+1}$ is deformed over $\mathfrak{H}$, according to the deformation theorem of *Topology*, p. 328, into a new chain $C'_{p+1}$ of $K$, then the chain $c'_p$ defined by $C_{p+1}$ is identical with $c_p$. For the deformation over $\mathfrak{H}$ gives rise to a certain deformation-chain $\mathfrak{D}\, C_{p+1}$ with a suitable elementary decomposition. If we now reduce $\mathfrak{D}\, C_{p+1}$ to $K$ by the deformation theorem, choosing, as we may, the chains of the decompositions which it demands (the analogues of the chains $C_p{}^i$ of the proof *loc. cit.*) exact sums of chains of the decomposition of $C_{p+1}$, the sole effect of the deformation on $C_{p+1}$, $C'_{p+1}$ may be to subdivide them, and this has no influence on $c_p$, $c'_p$. As a consequence we have on $K$ a deformation-chain for a deformation of $C_{p+1}$ into $C'_{p+1}$ which is of the first kind. Hence $c_p \equiv c'_p$.

Suppose in particular that we have a closed set $A$ with $L^*_A$ as its generalized fundamental complex (No. 7) and let $\gamma_p$ be a cycle mod $A$. If $\Gamma_{p+1}$ is the representative chain of $\gamma_p$, $F(\Gamma_{p+1})$ represents the absolute cycle $F(\gamma_p)$ of $A$. This absolute cycle has a representative image $\Gamma'_p$ which is a cycle of $L^*_A$ mod $\Phi$ (No. 7) and by the above

$$K \supset D_{p+1} \to \Gamma'_p - F(\Gamma_{p+1})\,;$$
$$\Gamma'_{p+1} = \Gamma_{p+1} + D_{p+1} \to \Gamma'_p\,; \quad \overline{|\, D_{p+1}\,|}\cdot \mathfrak{R} \subset A.$$

Hence if $\Gamma'_{p+1}$ represents $\gamma'_p$ of $\mathfrak{R}$ we have $\gamma'_p - \gamma_p \subset A$ so that $\gamma'_p$ represents the same cycle mod $A$ as $\gamma_p$. Therefore we may represent a cycle mod $A$ by a chain $C_{p+1}$ whose boundary is on the generalized fundamental complex $L^*_A$ of the set $A$. This result will be useful later.

The only deformations occurring in the sequel are of the second kind, and the elementary decompositions and deformation-chains on $K$ will always be in finite number. This will be understood throughout. They determine elementary decompositions $\{c_q{}^i\}$, $\{c'_q{}^i\}$, and deformation-chains $\mathfrak{D}\, c_q{}^i$ for the deformation of $c_p$ into $c'_p$, and the rest is as in No. 8. In particular

$$(11.1) \qquad \mathfrak{D}\, c_p \to c'_p - c_p - \mathfrak{D} F(c_p)\,;$$

$$(11.2)\qquad \mathfrak{D}\gamma_p \approx \gamma'_p - \gamma_p \approx 0 \text{ on } \mathfrak{R}.$$

12. With notations as in No. 11, let $\gamma_p$ be a cycle mod $A$ whose representative $\Gamma_{p+1}$ has its boundary on $L^*_A + \Phi$. The N S C in order that $\gamma_p \approx 0$ mod $A$, is that for every $i$

$$(12.1)\qquad \Gamma_{p+1} \approx 0 \bmod (N^i + \Phi + L^*).$$

Whether the cycle is $\approx 0$ or not when (12.1) holds for any particular $i$ it holds also for the lower values of $i$. Therefore there is an $h$, called the *index* of $\gamma_p$, such that (12.1) holds for $i \leqq h - 1$ but not for $i \geqq h$. It implies that there exists an infinite cycle $\Gamma'_{p+1} \subset N^{h-1}$ such that

$$(12.2)\qquad \Gamma_{p+1} \approx \Gamma'_{p+1} \bmod \Phi,$$

while no such cycle exists for any $N^i$, $i \geqq h$. In terms of the traces we have at once

$$(12.3)\qquad \Phi^i \cdot \Gamma_{p+1} \approx \Phi^i \cdot \Gamma'_{p+1} \text{ on } \Phi^i, \quad (i \geqq h),$$

$$(12.4)\qquad \Phi^i \cdot \Gamma_{p+1} \approx 0 \text{ on } \Phi^i, \quad (i < h).$$

Conversely suppose that (12.4) holds for $i < h$ but not for any higher $i$. We have then

$$(12.5)\qquad \Phi^i \supset D_{p+1} \rightarrow \Phi^i \cdot \Gamma_{p+1},$$

$$(12.6)\qquad D'_{p+1} = N^i \cdot \Gamma_{p+1} - D_{p+1} \rightarrow 0.$$

Since the cycle $D'_{p+1}$ is finite it is $\approx 0$ mod $\Phi$ on $K$, for it can be projected onto $\Phi$. It follows that (12.2) holds with $\Gamma'_{p+1} = \Gamma_{p+1} - D_{p+1} \subset N^{i-1}$ and hence the index $\geqq h$. On the other hand the index $\leqq h$, since otherwise (12.4) would hold for some $i \geqq h$. Therefore *the index $h$ of $\Gamma_{p+1}$ is the highest value of $i + 1$ for which* (12.4) *holds.*

13. We may consider $\tau_i$ as a deformation of $\Phi^{i+1}$ into $\Phi^i$ over $K$. The cell joining $E_p$ of $\Phi^{i+1}$ with $\tau_i E_p$, suitably oriented, is the deformation-chain of $E_p$ (*Topology*, p. 78), and the deformation-chain of any subchain $C_p^{i+1}$ of $\Phi^{i+1}$ is then obtained as *loc. cit.* by the condition that it is a linear chain-function. If we designate this function by $\mathfrak{D}$ we have

$$(13.1)\qquad \mathfrak{D} C_p^{i+1} \rightarrow C_p^i - C_p^{i+1} - \mathfrak{D} F(C_p^{i+1}),$$

$$(13.2)\qquad C_p^i = \tau_i C_p^{i+1}.$$

If $k^{i+1}$ is any subcomplex of $\Phi^{i+1}$ the sum of the closed deformation-cells of its cells (deformation-chains of the cells) under $\tau_i$ is a complex $\mathfrak{D} k^{i+1}$, the *deformation-complex* of $k^{i+1}$. If we have an infinite sequence of complexes $\{k^{i+1}\}$, where $k^{i+1}$ is a subcomplex of $\Phi^{i+1}$ and $k^i = \tau_i k^{i+1}$ for every $i$, the sum $k = \Sigma \mathfrak{D} k^{i+1}$ is a *projection-complex.*

Let now $\{C_p{}^{i+1}\}$ be an infinite sequence of chains where $C_p{}^{i+1}$ is a subchain of $\Phi^{i+1}$ and $C_p{}^i = \tau_i C_p{}^{i+1}$ for every $i$. We have then an associated chain

$$C_{p+1} = \Sigma \mathfrak{D} C_p{}^i \tag{13.3}$$

defining a chain $c_p$ of $\mathcal{R}$, called a *projection-chain.* If the chains $C_p{}^i$ for $i$ above a certain value $h$ are cycles $\Gamma_p{}^i$, $C_{p+1}$ defines a $\gamma_p$ called a *projection-cycle* of $\mathcal{R}$. The chains $C_p{}^j$, $j \leqq h$, can be replaced by the projections of $\Gamma_p{}^{h+1}$ without modifying $\gamma_p$, so that when we have a $\gamma_p$ we may assume that all the chains $C_p{}^i$ are cycles.

Let $\{C_p{}^i\}$ define as above a projection-chain $c_p$, with $C_{p+1}$ as the associated chain of $K$. Then $| C_p{}^i |$ is not necessarily the projection of the complexes $| C_p{}^{i+j} |$, but their difference is made up of cells of less than $p$ dimensions. It follows that there exists a projection-complex $k$ such, that for each $i$, $k^i - | C_p{}^i |$ consists of cells of dimension $< p$, while $k^i$ is the projection of some $C_p{}^{i+j}$ on $\Phi^i$. The difference $k - C_{p+1}$ will consist of cells of dimension $< p + 1$.

Let $A$, $L^*_A$ be as before and let $C_{p+1}$ be any chain of $K$ with $C_p{}^i$ as its traces. We may introduce as above the finite chains $\mathfrak{D} C_p{}^i$ and also the infinite chain

$$C'_{p+1} = C_{p+1} - \Sigma \mathfrak{D} C_p{}^i.$$

Let $C_{p+1}$ define $c_p$ of $\mathcal{R}$. Whenever $C'_{p+1} \subset L^*_A$ we shall call $C_{p+1}$ a *projection-chain mod* $L^*_A$, and $c_p$ a *projection-chain mod* $A$, a *projection-cycle mod* $A$ when $F(c_p) \subset A$. When $A = 0$ we have $L^*_A = 0$, $C'_{p+1} = 0$ and $c_p$ becomes an ordinary projection-chain.

14. *Certain properties of chain-moduli.* By a *modulus* of $p$-chains of a complex $K$ we understand a system of rational chains of $K$ forming an abelian group with respect to addition. If $\mathfrak{M}, \mathfrak{N}$ are two such moduli, and if $\mathfrak{N} \subset \mathfrak{M}$ then, as usual, $C_p \equiv 0 \bmod \mathfrak{N}$ or $C_p \equiv C'_p \bmod \mathfrak{N}$, mean that $C_p \subset \mathfrak{N}$ or $C_p - C'_p \subset \mathfrak{N}$. If $\mathfrak{M}'$ is a submodulus of $\mathfrak{M}$, by $\mathfrak{M} \equiv \mathfrak{M}' \bmod \mathfrak{N}$, we shall mean that every element of $\mathfrak{M}$ is congruent to an element of $\mathfrak{M}'$ mod $\mathfrak{N}$ and conversely. If we have a modulus $\mathfrak{M}$ on $\Phi^j$ the projections of its chains on $\Phi^i$, $i < j$, constitute a modulus $\mathfrak{M}'$ called the *projection* of $\mathfrak{M}$ on $\Phi^i$. Regarding these moduli of the $\Phi$'s and their projections we shall prove the following important

THEOREM I. *Let there be given for every $h$ two moduli of $p$-chains of $\Phi^h$,*

*$\mathfrak{M}^h$ and $\mathfrak{N}^h \subset \mathfrak{M}^h$, such that the projection of any $\mathfrak{M}^j$, $j \geqq h$, on $\Phi^h$, is congruent to $\mathfrak{M}^h$ mod $\mathfrak{N}^h$. Then corresponding to every $C_p$ of $\mathfrak{M}^h$, there is a projection-sequence $\{C_p{}^i\}$, $C_p{}^i \subset \mathfrak{M}^i$, such that $C_p \equiv C_p{}^h$ mod $\mathfrak{N}^h$.*

If $C_p \subset \mathfrak{N}^h$ we may take a vacuous sequence as the corresponding $\{C_p{}^i\}$. Therefore we may assume that $C_p \not\subset \mathfrak{N}^h$. Under the circumstances $C_p$ is a proper $p$-chain and so is any chain $C'_p \equiv C_p$ mod $\mathfrak{N}^h$.

Consider then all the chains of $\mathfrak{M}^h$, $D_p \equiv C_p$ mod $\mathfrak{N}^h$, which are projections of chains of some $\mathfrak{M}^j$, $j \geqq h$. The number of projections being infinite and the number of subcomplexes $|D_p|$ of $\Phi^h$ finite, at least one of these subcomplexes must carry an infinity of chains $D_p$. Let $\mathcal{K}$ be such a $|D_p|$ with the least number possible, $s$, of $p$-cells and let $E_p{}^1, \cdots, E_p{}^s$ be its $p$-cells, so that

$$D_p = \Sigma\, t_a E_p{}^a.$$

The chain $D_p$ is the projection of an element say of $\mathfrak{M}^j$. Suppose that there is another similar chain

$$D'_p = \Sigma\, t'_a E_p{}^a$$

which is the projection of an element of $\mathfrak{M}^k$, $k \geqq j$. Then $D'_p$ is likewise the projection of an element of $\mathfrak{M}^j$ and

$$D_p'' = \Sigma (t_a - t'_a) E_p{}^a = \Sigma\, t_a'' \cdot E_p{}^a$$

is the projection of an element of $\mathfrak{M}^j$ which is in $\mathfrak{N}^h$. Therefore, if, no matter how high we take $j$, there are in $\mathfrak{M}^j$ two elements whose projections $D_p$, $D'_p$ are different, and both $\equiv C_p$ mod $\mathfrak{N}^h$, there exists always in $\mathfrak{M}^j$ an element whose projection $D_p''$ is a chain of $\mathcal{K}$ and in $\mathfrak{N}^h$.

Conceivably some, but not all the $t''$'s vanish for $j$ high enough. There will be one, however, say $t_1'' \neq 0$ for an infinity of $j$'s, hence for every $j > h$, and $D_p - t_1 D_p''/t_1''$ will be a subchain of $\mathcal{K} - E'$ which is $\equiv C_p$ mod $\mathfrak{N}^h$. We have thus a complex whose number of $p$-cells $< s$, and which carries an infinity of chains such as $D_p$. As this contradicts the assumption regarding $s$, it follows that for $j$ above a certain value $D_p'' \equiv 0$, $D_p \equiv D'_p$. Therefore there is a unique chain of $\mathcal{K}$ which is the projection of chains of $\mathfrak{M}^j$, $j$ above a certain value, and $\equiv C_p$ mod $\mathfrak{N}^h$.

Let us now write $D_p{}^h$ for $D_p$ and consider the chains of $\mathfrak{M}^{h+1}$ whose projection on $\Phi^h$ is $D_p{}^h$. Their number being clearly infinite, we may again choose one, $D_p{}^{h+1}$, consider the least number possible of its $p$-cells and show that if it is not unique $D_p{}^h$ can be replaced by a similar chain with a smaller $s$, etc. We thus obtain a sequence $\{D_p{}^i\}$ $i = h, h+1, \cdots$. The projection sequence

$\{C_p^i\}$ such that $C_p^i = D_p^i$ for $i \geqq h$; $C_p^{h-i} =$ the projection of $D_p^h$ on $\Phi^{h-i}$, has all the properties required by Theorem I.

15. *Remarks.* I. In the proof the fact that the chains are taken with rational coefficients enters in an essential manner when we multiply chains by the number $s_a/r_a$. *Clearly any ring of coefficients forming a field (i. e. with unique division) would be admissible,* for instance the ring of integers mod $p$, $p$ a prime. But we could not have integers mod $m$, $m$ not a prime.

II. Let us call a projection-sequence $\{C_p^i\}$, $C_p^i \subset \mathfrak{M}^i$, *irreducible* whenever for any other similar $\{C'_p{}^i\}$, such that $C'_p{}^i$ is a subchain of $|C_p^i|$, necessarily $C'_p{}^i = tC_p^i$. In that case of course $t$ is independent of $i$. If we examine our construction we see that the sequence $\{C_p^i\}$ of our theorem has been chosen irreducible. For the irreducibility condition is imposed when $i \geqq h$, and follows, by projection, when $i < h$.

16. If we consider again the elements of $\mathfrak{M}^h$ where $h$ is now fixed, I say that *we can construct for* $\mathfrak{M}^h$ *a finite base mod* $\mathfrak{N}^h$, $C_p^{ha}$, $\alpha = 1, 2, \cdots, r$, *whose elements are members of irreducible projection-sequences* $\{C_p^{ia}\}$.

Let $E_p^\beta$ denote this time all the $p$-cells of $\Phi^h$. By the procedure of *Topology,* p. 302 (method of the "first-cell,") and with a suitable numbering of the cells, Theorem II authorizes us to assume that, except for irreducibility, we already have the required base such that in addition

$$C_p^{ha} = E_p^a + \Sigma\, t_{a\gamma} E_p^{r+\gamma}. \tag{16.1}$$

Consider now the subcomplex $\Psi^i$ of $\Phi^i$ consisting of all the cells of $\Phi^i$ projected onto $|C_p^{ha}|$ and apply the theorem to $\{\Psi^i\}$ taking as modulus $\mathfrak{M}^{*i}$ the aggregate of the elements of $\mathfrak{M}^i$ that are subchains of $\Psi^i$. Since the elements of $\mathfrak{M}^{*h}$ all are, mod $\mathfrak{N}^h$, linear combinations of the chains $C_p^{ha}$, and contain only $E_p^a$ among the first $r$ $p$-cells, they are all, mod $\mathfrak{N}^h$, multiples of $C_p^{ha}$, and those in $\mathfrak{M}^h - \mathfrak{N}^h$ must contain $E_p^a$. Now by Th. I taken together with No. 15 Remark II, we can find precisely an irreducible $\{C_p^{ia}\}$ such that $C_p^{ha}$ is of the form (16.1), and in particular congruent mod $\mathfrak{N}^h$ to the chain $C_p^{ha}$ in (16.1). Therefore $\{C_p^{ia}\}$, $\alpha = 1, 2, \cdots, r$, behaves as required.

17. Consider the $(p+1)$-subchains $C_{p+1}$ of $K$, such that $C_{p+1} \cdot \Phi^h \subset \mathfrak{M}^h$ for every $h$. Thus if $\{C_p^i\}$ are the projection-sequences that we have just considered the closures of the joining cells of the chains $C_p^i$ form a chain such as $C_{p+1}$. The finite or infinite linear combinations of these chains which are chains form a modulus $\mathfrak{M}$ and the similar chains corresponding to the moduli $\mathfrak{N}^h$ form a submodulus $\mathfrak{N}$ of $\mathfrak{M}$. We shall say that any particular projection-

chain $C_{p+1}$ of $\mathfrak{M}$ is *irreducible* if it contains no similar subchain (member of $\mathfrak{M}$) which is not a multiple of $C_{p+1}$. The sequences of No. 16 and the corresponding irreducible chains shall be designated by $\{C^{hai}_p\}$, $C^{ha}_{p+1}$, so that $C^{hai}_p = C^{ha}_{p+1} \cdot \Phi^i$.

THEOREM II. *$\mathfrak{M}$ possesses a base mod $\mathfrak{N}$, which is in general infinite, and whose elements are irreducible projection-chains.*

Given any particular sequence $\{C_p{}^i\}$, $C_p{}^i \subset \mathfrak{M}^i$, there exists an $h$, its *index* such that $C_p{}^i \subset \mathfrak{N}^i$ for every $i < h$ but not for $i = h$. Given $C_{p+1} \subset \mathfrak{M}$, we shall call *index* of $C_{p+1}$ the index of the sequence $\{C_{p+1} \cdot \Phi^i\}$. The $(p+1)$-chains whose index $\geqq h$ form a submodulus $\mathfrak{M}_h$ of $\mathfrak{M}$, and we have $\mathfrak{M}_{h+1} \subset \mathfrak{M}_h \subset \mathfrak{M}$. If $C_{p+1} \subset \mathfrak{M}$ we have

$$C_{p+1} \cdot \Phi^h = \Sigma\, t_{ha} C^{hai}_p \mod \mathfrak{N}^h,$$
$$C'_{p+1} = C_{p+1} - \Sigma\, t_{ha} C^{ha}{}_{p+1} \subset \mathfrak{M}_{h+1}.$$

We can treat similarly $C'_{p+1}$, etc. Ultimately we thus obtain a chain

$$D_{p+1} = \sum_{i \geqq h} t_{ia} C^{ia}{}_{p+1}$$

such that the index of $C_{p+1} - D_{p+1}$ exceeds any positive number. Therefore $C_{p+1} - D_{p+1} \subset \mathfrak{N}$. It is also clear that no element of $\{\{C^{ha}{}_{p+1}\}\}$ can be expressed in terms of those of same or higher index. Therefore $\{\{C^{ha}{}_{p+1}\}\}$ is a base whose elements are irreducible projection-chains.

18. THEOREM III. *There exists a base for the p-cycles mod A, A closed, whose elements are irreducible projection-cycles mod A.*

Let $L^*_A$ $(= L^*)$ be the generalized fundamental complex of $A$. Take for $\mathfrak{M}^h$ the set of all $p$-cycles of $\Phi^h$ mod $\Phi^h \cdot L^*$ such that if $\Delta_p$ is any one of them, every $\Phi^i$, $i \geqq h$, contains a chain $\Delta'_p$ whose projection on $\Phi^h \approx \Delta_p$ mod $\Phi^h \cdot L^*$ on $\Phi^h$. If $\Gamma_{p+1}$ is any cycle of $K$ mod $(L^* + \Phi)$ then $\Gamma_{p+1} \cdot \Phi^h \subset \mathfrak{M}^h$ for every $h$. For take $i > h$, and let $\Gamma_{p+1} \cdot \Phi^i$ be projected onto $\Delta^*_p$ of $\Phi^h$. We have

$$\Gamma_{p+1}(\bar{N}^h - N^i) \rightarrow \Gamma_{p+1} \cdot \Phi^h - \Gamma_{p+1} \cdot \Phi^i \mod L^*.$$

Moreover if $D_{p+1}$ is the deformation-chain corresponding to the projection

$$D_{p+1} \rightarrow \Delta^*_p - \Gamma_{p+1} \cdot \Phi^i \mod L^*.$$

Therefore

$$\bar{N}^h - N^i \supset C_{p+1} \rightarrow \Gamma_{p+1} \cdot \Phi^h - \Delta^*_p \mod L^*.$$

The chain $C_{p+1}$ is finite and by sliding it along the projecting lines we can reduce it to a chain on $\Phi^h$ without modifying its boundary mod $L^*$. Therefore

$$\Gamma_{p+1} \cdot \Phi^h - \Delta^*_p \approx 0 \bmod \Phi^h \cdot L^* \text{ on } \Phi^h; \; \Gamma_{p+1} \cdot \Phi^h \subset \mathfrak{M}^h.$$

The modulus $\mathfrak{M}^h$ also contains all the bounding chains mod $\Phi^h \cdot L^*$ of $\Phi^h$. For $E_{p+1}$ of $\Phi^h$ is the projection of some $E'_{p+1}$ of $\Phi^i$; hence $F(E_{p+1}) \subset \mathfrak{M}^h$, and likewise $F(C_{p+1})$ mod $\Phi^h \cdot L^*$ is in $\mathfrak{M}^h$. These bounding cycles form a submodulus $\mathfrak{N}^h$ of $\mathfrak{M}^h$ and the moduli $\mathfrak{M}^h, \mathfrak{N}^h$ are related as in Theorem I. By what we have shown the corresponding moduli $\mathfrak{M}, \mathfrak{N}$ of $(p+1)$-subchains of $K$ are respectively those of the cycles of $K$ mod $(L^* + \Phi)$ and of the bounding cycles of $K$ mod $(L^* + \Phi)$. The required theorem is then a direct consequence of Theorem II.

THEOREM IV. *Any chain $c_p$ is homologous on itself to a projection-cycle mod its own boundary.*

Let $|c_p| = B$, $|F(c_p)| = A$. By Theorem III there is a base $c_p^1, c_p^2, \cdots$, for the $p$-cycles of $B$ mod $A$ whose elements are irreducible projection-chains which are projection-cycles mod $A$. Moreover, referring to No. 17, the index $h(\alpha)$ of $c_p^\alpha$ increases indefinitely with $\alpha$. We have then

$$c_p \approx c'_p = \Sigma\, t_\alpha \cdot c_p^\alpha \bmod A \text{ on } B. \tag{18.1}$$

Let $C^\alpha_{p+1}$ be the projection-chain of $K$ which represents $c_p$. Among the chains $C^\alpha_{p+1}$ only a finite number have a trace $\neq 0$ on any $\Phi^i$, and each of these chains satisfies the condition for projection-chains. Hence any linear combination of them, and in particular the representative of $c'_p$, satisfies the same condition. Therefore $c'_p$ is a projection-chain, and (18.1) proves our theorem.

THEOREM V. *If a projection-cycle mod $A$ is $\approx 0$ mod $A$ it bounds a projection-chain mod $A$.*

Let $\Gamma_{p+1}$ be a projection-cycle mod $(L^* + \Phi)$ representing the projection-cycle mod $A$, $\gamma_p$. Take for $\mathfrak{M}^h$ the modulus of all $(p+1)$-chains on $\Phi^h$ whose boundary mod $L^* \cdot \Phi^h$ is a multiple of $\Gamma_{p+1} \cdot \Phi^h$ and for $\mathfrak{N}^h$ the $(p+1)$-cycles of $\Phi^h$ mod $L^* \cdot \Phi^h$. Let $C_{p+1}$ be the projection of any element of $\mathfrak{M}^i - \mathfrak{N}^i$ on $\Phi^h$ and let $C'_{p+1} \subset \mathfrak{M}^h - \mathfrak{N}^h$. Then

$$tC_{p+1} - C'_{p+1} \to 0 \bmod L^* \cdot \Phi^h, \; t \neq 0.$$

Hence the $(p+1)$-chain at the left is in $\mathfrak{N}^h$. Hence $\mathfrak{M}^h, \mathfrak{N}^h$ are related in the proper way, and by Theorem I there is a projection-chain $C_{p+2}$ such that

$$t_h(C_{p+2} \cdot \Phi^h) \to \Gamma_{p+1} \cdot \Phi^h \bmod L^* \cdot \Phi^h.$$

Since $\{F(C_{p+2} \cdot \Phi^h)\}$ and $\{\Gamma_{p+1} \cdot \Phi^h\}$ are both projection-sequences (up to a chain of $L^* \cdot \Phi^h$) $t_h$ has a value $t$ independent of $h$, and $tC_{p+2} \rightarrow \Gamma_{p+1} \bmod (L^* + \Phi)$. Therefore $tC_{p+2}$ represents a projection $c_{p+1}$ of $\mathcal{R}$ which $\rightarrow \gamma_p \bmod A$.

19. *Connectedness and circuits.* Let again $A, L^*_A$ be a closed set and its generalized fundamental complex, and let $x, y$ be two points of $\mathcal{R} - A$. If $\{\sigma^i\}$ is a projection-sequence for $x$, any vertex $A^i$ of $\sigma^i$ is the projection on $\Phi^i$ of a vertex $A^{i+1}$ of $\sigma^{i+1}$. Therefore $x$ has a projection-sequence $\{A^i\}$ consisting of vertices of the $\Phi$'s, and there is a similar sequence $\{B^i\}$ for $y$.

Now the N S C in order that $\mathcal{R}$ be not disconnected by $A$ is that for every $i$ above a certain value there exist a sequence $U^{i\beta_1}, \cdots, U^{i\beta_r}$ of the sets of the covering $\Sigma^i$, in which any two consecutive sets intersect and $U^{i\beta_1} = A^i$, $U^{i\beta_r} = B^i$. This condition is equivalent to $A^i \approx B^i$ on $\Phi^i$ for $i$ sufficiently high, and hence for every $i$. Since $\{A^i\}$, $\{B^i\}$ are projection-sequences they are the traces of projection-cycles $\Gamma_1$, $\Gamma'_1$ homologous on $K - L^*_A \bmod \Phi$ and representing respectively $x, y$. By means of Theorem IV and V we find immediately that the above condition is equivalent to $x \approx y$ on $\mathcal{R} - A$. *Therefore the N S C in order that $\mathcal{R} - A$ be connected is that* $R_0(\mathcal{R} - A) = 1$.

Another N S C f,or the connectedness of $\mathcal{R} - A$ is that *the open complexes* $(K - L^*_A) \cdot \Phi^i = \Psi^i$ be all connected. For if they are connected we always have $A^i \approx B^i$ on $\Psi^i$ whatever $x, y$ and hence $x \approx y$ on $\mathcal{R} - A$ so that $\mathcal{R} - A$ is connected. Conversely if $\mathcal{R} - A$ is connected we always have $A^i \approx B^i$ on $\Psi^i$. But any two vertices $A^i$, $B^i$ of $\Psi^i$ belong to two projection-sequences $\{A^i\}$, $\{B^i\}$; hence any two vertices of $\Psi^i$ are homologous on $\Psi^i$; therefore $R_0(\Psi^i) = 1$, and $\Psi^i$ is connected.

20. THEOREM VI. *An n-circuit $\mathcal{R} - A$ is connected and n-dimensional at all points.*

If $\mathcal{R} - A$ is disconnected every $\Psi^i$, for $i$ above a certain value, is the sum of two open complexes without common cells. As a consequence, by suppressing a suitable finite part of $K$, we shall have a new $K$ such that $K - L^*_A = K' + K''$, where $K'$ and $K''$ are open complexes without common cells. Let $\gamma_n$ be the fundamental $n$-cycle mod $A$ of the circuit and let $\Gamma_{n+1}$ be its representative cycle mod $(L^*_A + \Phi)$. The chains

$$\Gamma'_{n+1} = K' \cdot \Gamma_{n+1}, \qquad \Gamma''_{n+1} = K'' \cdot \Gamma_{n+1}$$

are similar cycles which represent cycles mod $A$, $\gamma'_n$ and $\gamma''_n$, such that

$$\gamma_n = \gamma'_n + \gamma''_n, \qquad |\gamma'_n| \cdot |\gamma''_n| \subset A.$$

Any point $x$ of $\gamma'_n$ has a neighborhood $U \subset \mathcal{R} - A - |\gamma''_n|$. Hence $B = \mathcal{R} - U$ is a closed set $\supset A$ and also $\gamma''_n$, so that $R_n(B; A) \neq 0$, which contradicts one of the circuit conditions. Therefore $\mathcal{R} - A$ is connected.

Regarding the dimension of $\mathcal{R} - A$, let $x \subset \mathcal{R} - A$, $\dim_x \mathcal{R} = p < n$. We can find an open set $U \supset x$ such that $U \subset \mathcal{R} - A$, $\dim F(U) \leqq p - 1$. It follows that if we suppress a suitable finite part of $K$, the new $K - L^*_A$ shall be disconnected into two subcomplexes $K'$, $K''$ by a projection-complex $L$ which is a fundamental complex for $F(U)$ and whose dimension is therefore $\leqq p$ (No. 7). Since $p < n$ neither $K'$ nor $K''$ will have $(n+1)$-cells with $n$-faces on $L$. Hence $K' \cdot \Gamma_{n+1}$ and $K'' \cdot \Gamma_{n+1}$ are separately cycles mod $(L^*_A + \Phi)$ determining cycles mod $A$, $\gamma'_n$ and $\gamma''_n$ whose intersection $\subset A$, and we have the same contradiction as before.

21. *Extension to locally compact spaces.* Practically all our results may be extended to a locally compact separable space $\mathcal{R}$. It is known that such a space is metric and that it can be mapped topologically on a compact space $\mathcal{R}^*$ with a point $x^*$ removed. That is to say $\mathcal{R}$ can be identified with $\mathcal{R}^* - x^*$. Moreover, topologically speaking $\mathcal{R}^*$ is unique.† If $U^*$ is a neighborhood of $x^*$ on $\mathcal{R}^*$, $U = U^* - x^*$ is an open set of $\mathcal{R}$ and $F(U^*) = F(U)$. There exists then another such set $V^* \Subset U^*$, and if $V = V^* - x^*$, we have also on $\mathcal{R} : V \Subset U$. Since $\mathcal{R}$ is $n$-dimensional we can find an open set $W$ of $\mathcal{R}$ such that $V \Subset W \Subset U$, $\dim F(W) < n$. Therefore if $W^* = W + x^*$, we have $\dim F(W^*) < n$, $x^* \subset W^* \Subset U^*$. In other words given any neighborhood $U^*$ of $x^*$ there is another $W^* \Subset U^*$ whose boundary is of dimension $< n$. This shows that $\dim_{x^*} \mathcal{R}^* \leqq n$. Any point $x \neq x^*$, has relatively to $\mathcal{R}^*$ a neighborhood which is also a neighborhood relatively to $\mathcal{R}$ and hence $\dim_x \mathcal{R}^* = \dim_x \mathcal{R}$, $\dim \mathcal{R}^* = \dim \mathcal{R} = n$.

Let $K^*$ be a fundamental complex of $\mathcal{R}^*$, and let $\{\sigma^i\}$ be a fundamental sequence for $x^*$ and $L^*$ the sum of the sequence and its joining cells. Then $K = K^* - L^*$ is an open complex which we may consider as a fundamental complex for $\mathcal{R}$. We now have two types of cycles to consider for $\mathcal{R}$: (a) the finite $p$-cycles; they correspond to the $(p+1)$-cycles of $K = K^* - L^*$ mod $\Phi$; (b) the infinite $p$-cycles of $\mathcal{R}$; they are represented by the $(p+1)$-cycles of $K^*$ mod $(L^* + \Phi)$. Both types have essentially the same properties as the cycles previously considered. It is also a simple matter to show that they are topological elements of $\mathcal{R}$ itself, independent of the mode of turning $\mathcal{R}$ into a compact space $\mathcal{R}^*$. Let us state in passing that $\mathcal{R}$ will be called an *open n-circuit* whenever it behaves in the same manner regarding the infinite $n$-cycles as previously regarding the finite (ordinary) $n$-cycles.

It is to be observed that the space $\mathcal{R}$, or rather the set $\mathcal{R}$, may actually

† Urysohn and Alexandroff, *Mémoire sur les spaces topologiques compacts*, Amsterdam Academy, Verhandeligen, Deel XIV, No. 1, 1929.

be given in the form $\mathcal{R}^* - A$, where $\mathcal{R}$ is compact, metric, but not necessarily $n$-dimensional, and $A$ is a closed subset of $\mathcal{R}^*$ that may consist of more than one point. $L^*$ is then merely a fundamental complex for $A$, but otherwise the rest is as before. We can pass to the case where $A$ is a single point by applying to $\mathcal{R}^*$ a continuous single-valued transformation, homeomorphic over $\mathcal{R}^* - A$, and reducing $A$ to a single point. Concurrently we replace the subcomplex $L^* \cdot \Phi^i$ by a single point and $L^*$ by a single projection line. It is clear that this does not affect the cycles which we have introduced above nor their homologies.

## PART II. The Generalized Manifold.

22. *Definition.* A generalized $n$-manifold $M_n$ or $M$ is a locally compact separable $n$-space with the following properties whose topological character is obvious:

I. $M$ is the sum of a countable aggregate of disjoined $n$-circuits.

II. The Betti-number $R_n(M, M - x) = 1$ for every $x \epsilon M$.

III. $M$ is locally connected.

IV. Given any closed $q$-set $F$ on $M_n$ and any open set $U$ there is an open set $V \subset U$, such that every chain $c_p$, $p < n - q$, on $V$, whose boundary does not meet $F$, is deformable over $U$, without moving its boundary, into a chain $c'_p$ which does not meet $F$.

If the circuits in I are absolute we call $M$ an *absolute* manifold, otherwise an *open* manifold. If the circuits are all *orientable* $M$ is orientable, otherwise it is *non-orientable*.

*Interpretation of the manifold conditions.* Condition I requires no comment. Regarding II we may consider, with van Kampen, as $p$-cycle of a point $x$ a $c_p$ whose boundary $\not\supset x$, i. e. a cycle mod $M - x$ in the sense of *Topology*, the homologies being of the type $\approx M - x$ : $c_p \approx 0$ mod $M - x$ means that there is a $c_{p+1}$ such that $F(c_{p+1}) - c_p \not\supset x$. The corresponding Betti-number is the number designated by $R_p(M, M - x)$. Since the fundamental $\gamma_n$ of $M$ is itself a cycle mod $M - x$ whatever $x \epsilon M$, II signifies that if $c_n$ is any chain whose boundary $\not\supset x$, a certain $c_n - t\gamma_n \not\supset x$.

The local connectedness in III is the so-called local zero-connectedness of *Topology*, p. 90. It means explicitly that for every $U$ there is a $V \subset U$ such that any two points of $V$ are on a connected subset of $U$.

When we have an absolute $M$ conditions III and IV are equivalent to:

III′. There exists, for every $\epsilon > 0$, a number $\delta(\epsilon) < \epsilon$, such that any two points not farther apart than $\delta$, are on a connected set of diameter $< \epsilon$.

IV′. For every $F$ and $\epsilon$ there exists a number $\eta(\epsilon) < \epsilon$, such that any $c_p$ of diameter $< \eta$ whose boundary does not meet $F$, where $\dim F = q < n - p$, is $\epsilon$-deformable without displacing its boundary, into a chain which does not meet $F$.

For $0 < p < n$, $F = 0$, this becomes a weak type of local $q$-connectedness with the $p$-cell and $(p-1)$-sphere of *Topology* replaced by a $p$-chain and $(p-1)$-cycle.

Conditions II, III, IV are purely local and serve to characterize the homogeneity properties of $M$. Condition I on the contrary refers to the whole manifold and serves also to separate the different types.

23. *The Kronecker-index. Definition.* Taking, merely for convenience, an absolute $M_n$, let $c_p$, $c_{n-p}$ be two chains on $M_n$ which do not intersect one another's boundaries:

$$|c_p| \cdot |F(c_{n-p})| + |F(c_p)| \cdot |c_{n-p}| = 0. \tag{23.1}$$

Their Kronecker-index is to be a number $(c_p \cdot c_{n-p})$ such that:

(a) $(c_p \cdot c_{n-p}) = 0$ when the chains do not meet.

(b) The index when defined is a bilinear function of the two chains.

(c) If the boundaries of $c_p$, $c_{n-p+1}$ do not intersect then

$$(c_p \cdot F(c_{n-p+1})) = (-1)^p (F(c_p) \cdot c_{n-p+1}). \tag{23.2}$$

(d) If $x \epsilon M$ and $\gamma_n$ is the fundamental $n$-cycle of $M$ then

$$(x \cdot \gamma_n) = 1. \tag{23.3}$$

We shall show that there exists a unique index which is a topological invariant of the two chains and which satisfies conditions (a), $\cdots$, (d), and has the following additional properties:

(e) If $c_{p+1}$ and $F(c_{n-p})$ do not meet

$$(F(c_{p+1}) \cdot c_{n-p}) = 0, \tag{23.4}$$

and similarly with $p$ and $n-p$ interchanged.

(f) The chains being as in the definition:

$$(c_p \cdot c_{n-p}) = (-1)^{p(n+1)} \cdot (c_{n-p} \cdot c_p). \tag{23.5}$$

The existence proof as well as properties (e), (f), will be established by induction.

In the theory of the index for combinatorial manifolds (*Topology*, Ch. IV), (a), (b), (d) enter more or less in the definition, while (c) is proved explicitly. It is in fact essentially formula (20) *loc. cit.*, which plays an all important part there and is a direct consequence of the fundamental boundary relation (18) for intersections of chains. On the contrary here the same relation serves directly to define recurrently the index without passing through intersections of dimension $> 0$. This is substantially in accord with the definitions suggested *loc. cit.*, p. 216. See also H. A. Newman's recent paper *Cambridge Philosophical Transactions*, Vol. 27 (1931), pp. 491-501.

24. The Kronecker-index $(c_0 \cdot \gamma_n)$, where $c_0$ is a projection-chain and $\gamma_n$ the fundamental $n$-cycle, is readily treated. In the first place if $C_0$ is a finite subchain of a complex,

$$C_0 = \Sigma\, t_i E_0{}^i, \tag{24.1}$$

we define its Kronecker-index as in *Topology*, p. 169, by

$$(C_0) = \Sigma\, t_i, \tag{24.2}$$

and we recall that $C_0 \approx 0$ implies that $(C_0) = 0$. If $\mathcal{K}$ is an orientable and oriented combinatorial manifold we have $(C_0) = (C_0 \cdot \mathcal{K})$.

Let now $c_0$ be a projection-chain of $M$ (projection-zero-cycle), with $\Gamma_1$ as its representative projection-cycle mod $\Phi$ on $K$. Since $\tau_i \Gamma_1 \cdot \Phi^{i+1} = \Gamma_1 \cdot \Phi^i$, we have $(\Gamma_1 \cdot \Phi^{i+1}) = (\Gamma_1 \cdot \Phi^i)$. This index is therefore independent of $i$ and its value is by definition $(c_0)$.

We shall show below, that $c_0$ can be $\epsilon$-deformed whatever $\epsilon$, into a chain consisting of a finite number of points $x_0, \cdots, x_r$, so that

$$c_0 \approx \Sigma\, s_j x_j. \tag{24.3}$$

Since $M$ is connected, if $x$ is any point of $M$ we have $x \approx x_j$, and hence

$$c_0 \approx x\, \Sigma\, s_0 = sx. \tag{24.4}$$

As we have seen (No. 19) $x$ has a projection-sequence $\{A^i\}$ made up of vertices of the $\Phi$'s. Owing to (23.4) we have

$$\Gamma_1 \cdot \Phi^i \approx sA^i \text{ on } \Phi^i; \tag{24.5}$$

hence $(\Gamma_1 \cdot \Phi^i) = s(A^i) = s = (c_0)$. Since $s$ is clearly a topological function of $c_0$ and does not depend in any sense on the fundamental complex $K$ the same holds for $(c_0)$.

If we now define $(c_0 \cdot \gamma_n)$ by the relation

$$(c_0 \cdot \gamma_n) = s(x \cdot \gamma_n), \tag{24.6}$$

we have by the above and (23.3)

(24.7) $$(c_0 \cdot \gamma_n) = (c_0).$$

This disposes of $(c_0 \cdot \gamma_n)$ and shows in particular that it is a topological invariant of $c_0$.

25. THEOREM VII. (*Deformation theorem*). *Given a projection-chain $c_p$, a closed $q$-set $F$ and any $\epsilon$, there exists an $\epsilon$-deformation of $c_p$ into a chain $c'_p$ with an elementary $\epsilon$-decomposition into projection-chains $\{c'_r{}^i\}$, whose zero-chains are isolated points and whose $r$-chains, $r < n - q$, do not meet $F$.*

Let $c_p$ be represented by $C_{p+1}$ of $K$ and let $E_q^{ha}$ be the cells of $\Phi^h$. We decompose $C_{p+1}$ in a sum of $r$ chains:

(25.1) $$C_{p+1} = C^1{}_{p+1} + \cdots + C^r{}_{p+1},$$

where $r$ is the number of cells of $\Phi^h$, and where $C^a{}_{p+1}$ is a subchain of the complex consisting of all the cells of $K$ on the set of all projecting lines that meet $E^{ha}$. A similar decomposition may then be applied to the chains of $F(C^a{}_{p+1})$, etc., until finally we have an elementary decomposition $\{C^a{}_{q+1}\}$ of $C_{p+1}$ characterized by the property that $C^a{}_{q+1}$ is on the set of cells of $K$ that are on the projecting lines through the points of $E^{ha}$. The chains $C^a{}_{q+1}$ like $C_p$ itself, are projection-chains. There results an elementary decomposition $\{c_q{}^a\}$ of $c_p$ into projection-chains associated with each $\Phi^h$.

Let us now observe that the points of $M$ in whose projection-sequences $\{\sigma^i\}$ the term $\sigma^k$ is $E^{ka}$ or a face of it, are the points of sets $U^{k\beta}$ with a common point. Their sum is an open set $V^{ka}$ whose diameter $< 2\epsilon_k$, where $\epsilon_k = \text{mesh } \Sigma^k$. Let $\eta_k$ be the characteristic-constant of the f. c. o. s. $\{V^{ka}\}$ and take $h$ so high that $\epsilon_h < \frac{1}{2}\delta(\eta_k)$, where $\delta$ is the same as in No. 22, III′. As a consequence any two points $x, y$ on a set $V^{h\gamma}$ will be on a connected subset of some $\bar{V}^{ka} \supset V^{h\gamma}$. By No. 19, $x \approx y$ on $\bar{V}^{ka}$. It follows that if $\{\sigma^i\}$, $\{\sigma'^i\}$ are representative sequences for $x, y$ then for $i > h$, $\sigma^i$ and $\sigma'^i$ can be joined on $\Phi^i$ by a polygonal arc $\lambda$ (sum of vertices and one-cells of $\Phi^i$) whose projections on $\Phi^k$ is on $E^{ka}$. Hence any two vertices of the subcomplex of $\Phi^i$ projected onto $E^{ha}$ can be joined in the above manner by a polygonal arc on $\Phi^i$. For both belong to a pair of simplexes such as $\sigma^i$, $\sigma'^i$.

26. Henceforth $h, k$ are to be kept fixed. Since $M$ is an $n$-circuit it is $n$-dimensional at all points (Theorem VI). Since $q < n$ there exists then on every open set a point not on $F$. Choose such a point $x_a$ on $V^{ha}$, and let $\{A^{ai}\}$ be a projection-sequence of vertices for the point $x_a$.

Consider one of the zero-chains $c_0{}^a$, of the decomposition of $c_p$. It is defined by means of a certain projection-chain $C_1{}^a$ of $K$ so that $\{C_1{}^a \cdot \Phi^i\}$ is a projection-sequence of zero-chains. From the above follows immediately that when $i$ exceeds a certain value we can find a one-chain $D_1^{ai}$ of $\Phi^i$ whose projection is on $E^{ka}$ and such that in addition

$$D_1^{ai} \to (C_1^a \cdot \Phi^i) A^{ai} - C_1^a \cdot \Phi^i = (c_0^a) A^{ai} - C_1^a \cdot \Phi^i. \tag{26.1}$$

By Theorem V we may choose for $D_1^{ai}$ a projection-sequence which determines a projection-chain $D_2^a$ of $K$, and finally a projection-chain $d_1^a$ of $\mathfrak{R}$. As a consequence of (26.1) we have then

$$d_1^a \to (c_0^a) x_a - c_0^a. \tag{26.2}$$

We have thus displaced the zero-chains of the elementary decomposition of $c_p$ into points $\not\subset F$. The displacement and the diameters of the chains of the decomposition may be made as small as we please. From this point on and taking account of Theorem V and condition IV of No. 22 for an $M_n$, the proof of the required deformation theorems proceeds as in *Topology*, p. 93. The only modifications are that singular cells and their singular boundary spheres are replaced by the elementary chains $c_q^i$ and their boundaries.

27. As a first application let $x \subset c_p - F(c_p) \neq 0$, where $c_p$ is otherwise arbitrary. The chain is homologous on itself mod its boundary to a projection-chain $c'_p$ so that $x \not\subset F(c'_p)$. By Theorem VII $c_p$ can be $\epsilon$-deformed whatever $\epsilon$ into a projection-chain $c_p'' \not\supset x$. This implies that $c_p \approx 0$ on $M - x$, and also that for every $\xi > 0$ there is an $\eta(\xi)$ such that if $x$ is a point of $c_p$ farther than $\xi$ from $F(c_p)$, then $c_p \approx c'_p$, where $c'_p$ is at a distance $\geqq \eta$ from $x$.

Referring to No. 22 we have by what precedes,

$$R_p(M - x) = \delta_{np}, \tag{27.1}$$

where $\delta_{np}$ is the Kronecker delta ($=1$ for $p = n$, $=0$ for $p \neq n$).

Let us now observe that if we apply the construction of Nos. 24, 25 to a $\gamma_p$ whose diameter is sufficiently small, we may choose all the chains $c'^i_q$ coincident with a single point $x$. As a consequence if $p > 0$, $c'_p = 0$, and by (11.1) the deformation chain

$$\mathfrak{D}\gamma_p \to -\gamma_p, \tag{27.2}$$

*Therefore for every open set $U$ there is another $V \subset U$ such that every $\gamma_p$, $0 < p < n$, on $V$ is $\approx 0$ on $U$.*

The preceding statement is valid for any manifold. For an absolute manifold owing to compactness we have: *for every $\theta$ there is a $\tau(\theta)$ such that every $\gamma_p$, $p < n$, whose diameter $< \tau$ bounds a chain of diameter $< \theta$.* This is merely another formulation of the weak local $p$-connectedness property mentioned in No. 22.

28. Let $\gamma_p$ be one of the irreducible projection-cycles of the base constructed in No. 18 and whose index $i > h$. When we apply the deformation of the preceding numbers with $F = 0$, we find that the deformed cycle $\gamma'_p = 0$, for its chains correspond element for element to the chains of a degenerate simplicial $p$-cycle. Therefore (27.2) will hold here also and hence $\gamma_p \approx 0$ on $M$. In particular the base alluded to can only contain a finite number of cycles $\not\approx 0$, namely those whose indices do not exceed a certain value. This proves the important

THEOREM VIII. *The Betti-numbers of an absolute $M_n$ are all finite.*

29. *Determination of the Kronecker-index.* We propose to give a recurrent determination of $(c_p \cdot c_{n-p})$ for two chains which do not intersect one another's boundaries. Taking first $p > 0$ we shall reduce the case in question to the same for $p - 1$ and ultimately to a $(c_0 \cdot c_n)$ where $c_0$ consists of a finite number of isolated points. This last index shall be treated directly by reduction to the case considered in No. 23. At the same time we shall show that the index has all the properties expected. We assume then first that this holds already for $p - 1$, extend it to $p$, then take up the case $p = 0$ at the end.

Our first move is to replace $c_p$, $c_{n-p}$ by projection-chains, homologous respectively to $c_p$, $c_{n-p}$ on $|c_p|$, $|c_{n-p}|$ mod their boundaries (Theorem IV). To simplify matters we continue to denote the new chains by $c_p$, $c_{n-p}$. We merely recall that after the reductions the new sets $|c_p|$, $|c_{n-p}|$, $|F(c_p)|$, $|F(c_{n-p})|$ are subsets of the old. As a consequence in what follows, $|c_p|$ for example, may designate indifferently the new or the old set $|c_p|$.

Let now $\xi$ be the least of the two positive numbers $d(c_p, F(c_{n-p}))$, $d(F(c_p), c_{n-p})$. Since every point $x$ of $|c_p|$ is at least as far as $\xi$ from $F(c_{n-p})$, $x$ has a neighborhood $V$ such that $c_{n-p} \approx 0 \bmod M - V$. Since $|c_p|$ is self-compact it can be covered with a finite number of neighborhoods $V^1, \cdots, V^r$, such that $c_{n-p} \approx 0 \bmod M - V^j$. That is to say there exists a projection-chain

$$(29.1) \qquad c^j_{n-p+1} \to c_{n-p} \bmod M - V^j.$$

The sets $V^j$ form a f. c. o. s. for $|c_p|$ and there is an analogue $\zeta$ of the characteristic constant for that covering: every point $x$ of $|c_p|$ will be on some $V^j$ such that $d(x, M - V^j) > \zeta$.

We shall now choose a certain fixed $\epsilon > 0$, and the determination of the index will depend upon that $\epsilon$. We shall endeavor to show that the index remains the same for all $\epsilon$'s sufficiently small, and so we shall not hesitate to take this $\epsilon$ arbitrarily small. In particular we shall require that $\epsilon > \frac{1}{4}\xi, \frac{1}{4}\zeta$ or $\tau(\frac{1}{2}\zeta)$ where $\tau$ is the same function as in No. 27.

By Theorem VII we may $\epsilon$-deform $c_p$ into a projection-chain $c'_p$ with an elementary $\epsilon$-decomposition $\{c'_q{}^i\}$ whose elements (all projection-chains also) of dimension $q < p$ do not meet $c_{n-p}$ and with

$$c'_p = \Sigma\, c'_p{}^i. \tag{29.2}$$

The chains $c'_p{}^i$ are in finite number and their boundaries do not meet $c_{n-p}$. We shall set by definition $(c_p \cdot c_{n-p}) = (c'_p \cdot c_{n-p})$, and hence, if (23b) is to hold,

$$(c_p \cdot c_{n-p}) = (c'_p \cdot c_{n-p}) = \Sigma(c'_p{}^i \cdot c_{n-p}), \tag{29.3}$$

where in the sum we preserve only the useful terms, namely those corresponding to chains $c'_p{}^i$ which meet $c_{n-p}$. The problem is to determine the indices in that sum.

Since the points of $c'_p{}^i$ are not farther than $\epsilon < \frac{1}{4}\xi$ from $c_p$, and since its diameter $< \frac{1}{4}\zeta$, $c'_p{}^i$ is on at least one set $V^j$ and farther than $\frac{1}{2}\zeta$ from the corresponding $M - V^j$. Choose one of the sets $V^j$ of this nature and relabel that $V$ and its $c_{n-p+1}$ entering in (29.1), respectively $V'^i$, $c'^i_{n-p+1}$. The situation being as described we have $F(c'^i_{n-p+1}) = c_{n-p} + c'_{n-p}$, where $c'_{n-p}$ does not meet $c'_p{}^i$. Hence, if the basic laws (23a c) for the index are to hold, we must have

$$(c'_p{}^i \cdot c_{n-p}) = (-1)^p\,(F(c'_p{}^i) \cdot c'_{n-p+1}), \tag{29.4}$$

and hence by (23 b)

$$(c_p \cdot c_{n-p}) = (-1)^p\,\Sigma\,(F(c'_p{}^i) \cdot c'^i_{n-p+1}). \tag{29.5}$$

The right hand side is known under the hypothesis of the induction, hence (29.5) determines a value for $(c_p \cdot c_{n-p})$. It remains to be shown that the index thus obtained behaves as expected.

30. We first observe that the index is a linear function of $c_p$. That is to say if the preceding method has yielded $(c_p \cdot c_{n-p})$ and $(c'_p \cdot c_{n-p})$ then it also yields

$$(tc_p + t'c'_p \cdot c_{n-p}) = t(c_p \cdot c_{n-p}) + t'(c'_p \cdot c_{n-p}). \tag{20.1}$$

We notice also that if $c_p$ and $c_{n-p}$ do not meet and if we take $\epsilon$ less than half their distance apart, the value computed for their index is zero, and hence accords with (23 a) independently of the variable elements entering in its determination.

Let us replace $V'^i$ by any other $V$, say $V''^i$, behaving in the same manner relatively to $c'_p{}^i$ and let $c''^i_{n-p+1}$ be the corresponding $c_{n-p+1}$. To show that substituting $V''^i$ for $V'^i$ has not modified the index we must prove that if we substitute $c''^i_{n-p+1}$ for $c'^i_{n-p+1}$ in (29.5) the index remains the same. Since (23 b) holds for $p-1$ this merely requires that we prove

$$(30.2) \qquad \Sigma\,(F(c'^i_p)\cdot c'^i_{n-p+1} - c''^i_{n-p+1}) = 0.$$

Let $W = V'^i \cdot V''^i$. This open set $\supset c'^i_p$ and $d(c'^i_p, M - W) > \frac{1}{2}\zeta$. Moreover both chains $c'^i_{n-p+1}$, $c''^i_{n-p+1} \to c_{n-p} \bmod M - W$. Hence

$$(30.3) \qquad c'^i_{n-p+1} - c''^i_{n-p+1} \to 0 \bmod M - W.$$

On the other hand since diam $c'^i_p < \tau(\frac{1}{2}\zeta)$ and since, by hypothesis, (24 e) holds for $p-1$ in place of $p$, we have

$$(30.4) \qquad (F(c'^i_p)\cdot c'^i_{n-p+1} - c''^i_{n-p+1}) = 0,$$

from which the required relation (30.2) follows. If $V'^i = V''^i$ but $c'^i_{n-p+1}$ is replaced by $c''^i_{n-p+1}$ the same reasoning holds. Therefore a modification in the chains $c_{n-p+1}$ likewise leaves the index unaltered.

31. Let us now show that (24 e) holds: if

$$(31.1) \qquad M - F(c_{n-p}) \supset c_{p+1} \to c_p$$

then we have

$$(31.2) \qquad (c_p \cdot c_{n-p}) = 0.$$

Take a $U \supset c_{p+1}$ and $\Subset M - F(c_{n-p})$, then apply Theorem V with $\bar{U}$ as the basic space. As a consequence we find that we may assume that $c_{p+1}$ is a projection-chain. By Theorem VII $c_{p+1}$ is $\epsilon$-deformable into $c'_{p+1}$ with an $\epsilon$-decomposition $\{c'^i_q\}$ whose chains of dimension $< p$ do not meet $c_{n-p}$. It is to be observed that the construction of the deformed chains is such that the chains $c'^i_r$ depend solely on those of dimension $< r$. Hence $c_p$ *is* thus deformed into any $c'_p$ serving to calculate its index in accordance with No. 29. We shall have as the new $p$- and $(p+1)$-chains

$$(31.3) \qquad c'_{p+1} = \Sigma\, c'^i_{p+1}, \qquad c'_p = \Sigma\, F(c'^i_{p+1}),$$

and therefore

$$(31.4) \qquad (c_p \cdot c_{n-p}) = \Sigma\,(F(c'^i_{p+1})\cdot c_{n-p}).$$

If $\epsilon$ is small enough the chains $c'^j_{p-1}$ on $F(c'^i_{p+1})$ will meet a single chain $c_{n-p+1}$ that we may call as before $c'^i_{n-p+1}$. By (29.5) and No. 30

$$(31.5) \qquad (F(c'^i_{p+1})\cdot c_{n-p}) = (-1)^p\,(F(F(c'^i_{p+1}))\cdot c'^i_{n-p+1}) \equiv 0,$$

since $F(F) \equiv 0$, and from this follows (31.2).

As an application suppose that we have obtained, always by means of $K$, two different $\epsilon$-deformations $\mathcal{D}'$, $\mathcal{D}''$ of $c_p$ into $c'_p$ and $c''_p$ serving to calculate the index $(c_p \cdot c_{n-p})$. We have

(31. 6) $\mathfrak{D}'c_p \to c'_p - c_p - \mathfrak{D}'F(c_p), \qquad \mathfrak{D}''c_p \to c''_p - c_p - \mathfrak{D}''F(c_p),$

(31. 7) $\mathfrak{D}'c_p - \mathfrak{D}''c_p \to (c'_p - c''_p) - \cdots$

where, under the limitations upon $\epsilon$, the chain omitted does not meet $c_{n-p}$. Hence

(31. 8) $(c'_p - c''_p \cdot c_{n-p}) = 0,$

whatever the procedure chosen to compute the index. Take as the deformation the process which consists merely in replacing $c'_p$ by the decomposition associated with the deformation $\mathfrak{D}'$, and similarly for $c''_p$ and $\mathfrak{D}''$. As a consequence the index (31. 8) becomes merely the difference of the values of the index $(c_p \cdot c_{n-p})$ as computed by means of the two deformations. Therefore these two values are the same. In other words *$(c_p \cdot c_{n-p})$ is independent of the $\epsilon$-deformations used in computing it.*

32. We have already shown that our index possesses properties (23 a) and part of (23 b e). We still have to show that when $p > 0$ properties (23 b c e f) hold.

Since we have established the linearity of the index in $c_p$, (23 b) will be established if we show that the index is also linear in $c_{n-p}$. Consider two projection-chains $c_{n-p}$, $c'_{n-p}$ and let them not meet $F(c_p)$, ($c_p$ a projection-chain), nor let $c_p$ meet their boundaries. Let the $\epsilon$-deformation of $c_p$ into $c'_p$ be so carried out that the $q$-chains $c'^i_q$, $q < p$, of the decomposition of $c'_p$ meet neither $c_{n-p}$ nor $c'_{n-p}$. Then it follows at once from the definition of the index by (29. 5) that

(32. 1) $$(c_p \cdot tc_{n-p} + t'c'_{n-p}) = (c'_p \cdot tc_{n-p} + t'c'_{n-p})$$
$$= t(c'_p \cdot c_{n-p}) + t'(c'_p \cdot c'_{n-p}) = t(c_p \cdot c_{n-p}) + t'(c_p \cdot c'_{n-p}),$$

which proves the required linearity and hence also that property (b) holds completely.

Consider now property (c): if $c_p$, $c_{n-p+1}$ have non-intersecting boundaries then (23. 2) holds. Here we may take in (29. 5) every $c'^i_{n-p+1} = c_{n-p+1}$, which yields

(32. 2) $$(c_p \cdot F(c_{n-p+1})) = (-1)^p \Sigma\, (F(c'^i_p) \cdot c_{n-p+1}).$$

In the summation in (29. 5) only certain chains $c'^i_p$ whose sum is $c'^i_p$ were preserved, namely those which met $c_{n-p}$. As we have just shown if $c'^i_p$ does not meet $c_{n-p}$ the corresponding contribution of its boundary to the sum in (32. 1) is zero; hence the summation may now be extended to all the chains $c'^i_p$. By the linearity of the index for $p-1$, $n-p+1$ we have:

(32. 3) $$\Sigma(F(c'^i_p) \cdot c_{n-p+1}) = (F(\Sigma\, c'^i_p) \cdot c_{n-p+1})$$
$$= (F(c'_p) \cdot c_{n-p+1}) = (F(c_p) \cdot c_{n-p+1}).$$

This relation together with (32. 2) yields (23. 2) and proves that property (c) holds.

From (a) and (c) follows that if $F(c_p)$ and $c_{n-p+1}$ do not meet then

$$(c_p \cdot F(c_{n-p+1})) = 0. \tag{32.4}$$

This is the analogue of (23. 4) with $p$ and $n - p$ interchanged, and together with the result of No. 31 it embodies the proof of property (e).

We postpone the proof of property (f) till later.

33. We shall now consider the case $p = 0$, i. e. the index $(c_0 \cdot c_n)$, where as before $c_0 \subset M - F(c_n)$. As in No. 24, we have here for an $\epsilon$ sufficiently small the analogue of (24. 3):

$$c_0 \approx \Sigma\, s_j x_j \text{ on } M - F(c_n). \tag{33.1}$$

Now for $x_j$ we have by No. 22, condition II,

$$c_n \approx t_j \gamma_n \bmod M - x_j, \tag{33.2}$$

and we shall set

$$(c_0 \cdot c_n) = \Sigma\, s_j t_j (x_j \cdot \gamma_n) = \Sigma\, s_j t_j. \tag{33.3}$$

Properties (a), (b) are at once verified for this index and we only have to prove (e), (f). Here also (f) shall be treated later.

The proof of (e) consists of two parts:

(a) if $M - F(c_n) \supset c_1 \to c_0$ then $(c_0 \cdot c_n) = 0$. As in No. 31 we may assume that diam $c_1 < \epsilon$ assigned. Now for $c_n$ and any $x$ there is a neighborhood $V \supset x$ such that $c_n - \lambda\gamma_n \approx 0$ on $M - V$. Since $M$ is compact it may be covered with a finite number of such sets $V$ : $V^1, \cdots, V^r$ with $\lambda = \lambda_i$ on $V^i$. Let us take $\epsilon < \frac{1}{2}\eta$, where $\eta$ is the characteristic constant of this f. c. o. s., and let the deformations be $< \frac{1}{2}\eta$. We shall take $c_1 \subset V^h$ and farther than $\frac{1}{2}\eta$ from $M - V^h$. Hence if we calculate the index of $c_0 = F(c_1)$ by our method, the corresponding points $x_j$ are all on $V^h$, and the associated constants $t_j$ all equal to $\lambda_h$. Finally since $c_0 \approx 0$, $(c_0) = 0$. Therefore

$$(c_0 \cdot c_n) = (F(c_1) \cdot c_n) = \lambda_h \Sigma\, s_j = \lambda_h (c_0) = 0. \tag{33.4}$$

(b) if $c_n \approx 0$ then $(c_0 \cdot c_n) = 0$. This is evident for $c_n \approx 0$ implies that the representative projection-cycle $\Gamma_{n+1}$ of $c_n$ is $\equiv 0$ and hence $c_n \equiv 0$. Consequently in the homologies $c_n \approx t\gamma_n \bmod M - x$, we always have $t = 0$, so that $(c_0 \cdot c_n) = 0$.

As in No. 31 the first case considered proves here also that $(c_0 \cdot c_n)$ has a value independent of the particular mode of determining it.

34. Let us return to $(c_p \cdot c_{n-p})$. We have obtained its value by an induction on $p$ in which there appear certain intermediary chains $c_{p-i}$, $c_{n-p+i}$, so that we have:

$$
\begin{aligned}
(c_p \cdot c_{n-p}) &= (-1)^p (c_{p-1} \cdot c_{n-p+1}), \\
&\cdots\cdots\cdots \\
(34.1) \qquad (c_{p-i} \cdot c_{n-p+i}) &= (-1)^{p-i}(c_{p-i-1} \cdot c_{n-p+i+1}), \\
&\cdots\cdots\cdots \\
(c_1 \cdot c_{n-1}) &= - \quad (c_0 \cdot c_n);
\end{aligned}
$$

and hence, in the last analysis,

$$(34.2) \qquad (c_p \cdot c_{n-p}) = (-1)^{p(p+1)/2}(c_0 \cdot c_n) = (-1)^{p(p+1)/2} \cdot \lambda \cdot (x \cdot \gamma_n),$$

the value of $\lambda$ being the number given by (33.3). Observe that if $p > 0$, the various chains of dimension $p-1, \cdots$, zero, introduced in this determination are chains of elementary decompositions in which the zero-chains consist of isolated points taken with finite multiplicities. Therefore in particular $c_0$ is of this nature. It follows that the numbers $s_j$, $t_j$ of No. 33, that serve to compute $\lambda$ are all finite and so is $\lambda$. An immediate consequence is the fact that $(c_p \cdot c_{n-p})$ *is independent of the fundamental complex* $K$, *and hence the Kronecker-index is a topological invariant.* For if we have any index whatever with the properties (a), $\cdots$, (e) of No. 23, it will satisfy the relations (34.1) and (34.2). Since $\lambda$ depends solely on certain homologies but not on $K$ our assertion follows.

Now the above has been obtained as a consequence of an induction on $p$. By means of (23 c), explicitly proved for our index, we may carry through a similar induction on $n-p$. This leads to a formula analogous to (34.2)

$$(34.3) \qquad (c_p \cdot c_{n-p}) = (-1)^{[n(n+1)-p(p+1)]/2} \cdot \mu(\gamma_n \cdot x).$$

If we apply the process just stated to $(c_{n-p} \cdot c_p)$ we find that the geometric operations carried out for its determination are the same as those used in determining $(c_p \cdot c_{n-p})$ by our initial procedure (induction on $p$), and that as a consequence the corresponding $\mu$ is $\lambda$, both being equal to a certain expression $\Sigma s_j t_j$ appearing in (33.3). For each $j$ the number $s_j$ is the multiplicity of a certain point as constituent of $c_0$ and $t_j$ the coefficient $t$ in a certain homology (33.2). Therefore

$$(34.4) \qquad (c_{n-p} \cdot c_p) = (-1)^{p(2n-p+1)/2} \lambda \cdot (\gamma_n \cdot x),$$

$$(34.5) \qquad (c_{n-p} \cdot c_p) = (-1)^{p(n+1)} (c_p \cdot c_{n-p})[(\gamma_n \cdot x)/(x \cdot \gamma_n)].$$

In particular for $p = n$, $c_0 = x$, $c_p = \gamma_n$:

(34. 6) $$(x \cdot \gamma_n)^2 = (\gamma_n \cdot x)^2,$$
and hence finally
(34. 7) $$(\gamma_n \cdot x) = \pm (x \cdot \gamma_n).$$

We shall prove later that the proper sign to be chosen here is $+$. Assuming this for the present we have from (23 d), $(\gamma_n \cdot x) = 1$ and hence finally

(34. 8) $$(c_{n-p} \cdot c_p) = (-1)^{p(n+1)} (c_p \cdot c_{n-p})$$

which is (23 f). Thus except for a certain choice of sign we have finally established that the Kronecker-index has all the properties required.

35. *Duality properties of the absolute $M_n$.* In order to obtain the extension of Poincaré's duality relation for the Betti-numbers, all that is now needed is a converse of property (c) of No. 23; if a cycle $\gamma_p \not\sim 0$ there is a $\gamma_{n-p}$ such that $(\gamma_p \cdot \gamma_{n-p}) \neq 0$. A slightly more general result will now be proved.

Consider first the sequence of the images of the skeleta $\Phi^i$ which we still call $\Phi^i$, on an $S_{2n+1}$, whereby one may map topologically a compact metric $n$-space, here our absolute $M_n$, on the space $S_{2n+1}$ † and let us modify the construction as follows: We take an $S_{2n+2}$ referred to coördinates $x_1, \cdots, x_{2n+2}$ and assume that our $S_{2n+1}$ is the one given by $x_1 = 0$, so that $M$ is now mapped onto that space. We then project $\Phi^i$ onto the space $x_1 = 1/i$, and replace $\Phi^i$ by its projection which we henceforth call $\Phi^i$. The joining cells being inserted as before, if their (linear) spaces happen to have intersections of too high dimension, we may slightly displace the vertices of the $\Phi$'s in their $(2n+1)$-spaces so as to remove this untoward circumstance. We now have $M$ and $K$ immersed in a certain $S_{2n+2}$. We may in fact immerse $S_{2n+2}$ in any $S_r$, $r \geqq 2n+2$ and together with it also both $M$ and $K$. We shall choose $r$ such that $r - n$ is even.

Let us surround each $\Phi^i$ by a closed polyhedral neighborhood $\mathcal{K}^i$ in $S_r$, take a subdivision $\mathcal{K}'^i$ of $\mathcal{K}^i$ having a subdivision $\Psi^i$ of $\Phi^i$ as a subcomplex and such that the $\mathcal{K}'^i$-neighborhood of $\Psi^i$ is normal. By reference to *Topology*, p. 91, it will be seen that both conditions may be fulfilled. Moreover we may assume the $\mathcal{K}$'s taken initially mutually exclusive, so that the closed $\mathcal{K}'$-neighborhoods introduced are all mutually exclusive. To simplify matters we designate henceforth these closed neighborhoods themselves by $\mathcal{K}^i$. Besides being polyhedral these neighborhoods have the following property (*loc. cit.*): if $\mathcal{B}^i = F(\mathcal{K}^i)$ then through every point $P$ of $\mathcal{K}^i - \mathcal{B}^i - \Psi^i$ there passes a unique (open) segment resting on $\mathcal{B}^i$ and $\Psi^i$ and varying continuously with $P$. We call these segments the *projecting segments* on $\mathcal{K}^i$.

† See p. 585.

36. Let us assign to each cell $E_q{}^i$ of $\Phi^i$ one of its vertices $A^{iq}$, and let $\Psi'^i$ be the first derived of $\Psi^i$. A unique simplicial transformation $\theta$ of $\Psi'^i$ into $\Phi^i$ is determined by specifying that the vertices of $\Psi'^i$ on $E_q{}^i$ are all to be transformed by $\theta$ into $A^{iq}$. We shall designate by *projection* of $\Psi'^i$ onto $\Phi^{i-1}$, $\Phi^{i-2}, \cdots$, the simplicial transformation $\tau_{i-1}\theta, \tau_{i-2}\tau_{i-1}\theta, \cdots$. It is to be observed that $\theta$ need not be a fixed simplicial transformation, but merely any simplicial transformation of its type.

Let us specify for each $i$ a definite first derived $\mathcal{K}'^i$. It determines a $\Psi^i$, and also a dual $\mathcal{K}^{*i}$ of $\mathcal{K}^i$. If $\Gamma_{n+1}$ represents the fundamental cycle $\gamma_n$ of $M$, then we have a definite $n$-cycle $\Gamma_n{}^i = \Gamma_{n+1} \cdot \Psi^i$ for each $i$. It is the subdivision induced by $\Psi^i$ on the trace $\Gamma_{n+1} \cdot \Phi^i$ of $\Gamma_{n+1}$.

Now referring to *Topology*, Ch. IV, if $C^*{}_q{}^i$ is any subchain of $\mathcal{K}^{*i}$ there is a uniquely defined intersection-chain $C_s{}^i = \Gamma_n{}^i \cdot C^*{}_q{}^i$, $s = q + n - r$, and we have (*Topology*, p. 169, formula 18):

$$(36.1) \qquad \Gamma_n{}^i \cdot C^*{}_q{}^i \to \Gamma_n{}^i \cdot F(C^*{}_q{}^i).$$

The intersection and its boundary are both subchains of $\Psi'^i$ and hence they have a unique projection on any $\Phi^j$, $j \leqq i$. Moreover, since a projection is a simplicial transformation, the boundary of the projection is the projection of the boundary.

37. In the argument to follow, in addition to the customary associated chains $c_q$, $C_{q+1}$, it will be convenient to make a clear distinction between intersections of chains and traces. We shall therefore designate the former as usual by the "dot" product, and the trace of $C_{q+1}$ of $K$ on $\Phi^i$ by $\mathcal{C}_q{}^i$.

LEMMA. *Let $C^\alpha{}_{p+1}$, $C^\alpha{}_{n-p+1}$ $(\alpha = 1, 2, \cdots, s)$ be projection-chains of $K$ representing chains $c_p{}^\alpha$, $c^\alpha{}_{n-p}$ which do not intersect one another's boundaries so that $(c_p{}^\alpha \cdot c^\alpha{}_{n-p})$ is well defined. Suppose that the traces $\mathcal{C}_p{}^{\alpha i}$ have the following property: whatever $h$ there is a $k_\alpha > h$ such that $\mathcal{C}_p{}^{\alpha i}$ is the projection of an intersection-chain $\Gamma_n{}^{k_\alpha} \cdot C^{*\alpha k_\alpha}_{r-n+p}$ where*

$$(37.1) \qquad \sum_\alpha (C^{*\alpha k_\alpha}_{r-n+p} \cdot \mathcal{C}^{\alpha k_\alpha}_{n-p}) = \lambda$$

*is independent of $h$. Then,*

$$(37.2) \qquad \sum_\alpha (c_p{}^\alpha \cdot c^\alpha{}_{n-p}) = \lambda.$$

Let $\Lambda_p$ designate the Lemma as stated. We shall reduce $\Lambda_p$ to $\Lambda_{p-1}$, hence to $\Lambda_0$, then prove $\Lambda_0$.

Dropping $\alpha$ for the present, designate $C^\alpha{}_{p+1}, \cdots$ by $C_{p+1} \cdots$, and let $V^j$, $c^j{}_{n-p+1}$ correspond as in No. 29 to $c_{n-p}$. We first decompose $c_p$ into elements $\{c_q{}^\alpha\}$ which are projection-chains whose diameters $< \epsilon$. We then $\epsilon$-deform

$c_{n-p}$ and the chains $c^j{}_{n-p+1}$ simultaneously so as not to impair their relation to one another and to the $V$'s, and also so that they meet only the elements $c_p{}^a$ of the decomposition of $c_p$. It is readily seen that all these conditions can be fulfilled with $\epsilon$ as small as we please. We choose it $< \frac{1}{2}\zeta$, where $\zeta$ is the same as in No. 29. As a consequence we now have a pair $c_p, c_{n-p}$ whose intersection consists of a finite number of disjoined closed sets $F^a$, one on each $c_p{}^a$. Since diam $F^a < \frac{1}{2}\zeta$, $F^a$ will be covered by a certain set $V$, which we may call $V^a$, such that $d(F^a, M - V^a) > \frac{1}{2}\zeta$. Since the $F$'s do not intersect we can find for each $F^a$ an open set $W^a$ such that $F^a \subset W^a \Subset V^a$, $\overline{W}^a \cdot \overline{W}^b = 0$ for $a \neq b$. Introduce the closed set $G = M - \sum W^a$ and let $L$ be a fundamental projection-complex for $G$ (No. 7) so that $G = \overline{L} \cdot M$. We now remove from $C_{p+1}$ all the $p$-cells on the $\Phi$'s which are on $L$ and also all their joining cells, and call $C'_{p+1}$ the chain left, $C''_{p+1}$ the chain removed and $c'_p$, $c''_p$ the corresponding chains of $M$ whose sum is $c_p$. We have

$$c_p = c'_p + c''_p, \quad F(c_p) \subset M - \sum V^a \subset M - \sum W^a \subset M - c'_p.$$

Hence $| F(c'_p) | \subset | c''_p |$, and by construction the two chains $c'_p$, $c''_p$ have only boundary points in common. Therefore

$$(37.3) \qquad | c'_p | \cdot | c''_p | = F(c'_p).$$

On the other hand if $c'_p{}^a$ designates the part of $c'_p$ on $\overline{W}^a$, we have

$$(37.4) \qquad c'_p = \sum c'_p{}^a, \quad | c'_p{}^a | \cdot | c'_p{}^b | = 0 \text{ for } a \neq b.$$

Therefore also

$$(37.5) \qquad | c'_p{}^a | \cdot | c''_p | = F(c'_p{}^a).$$

As a consequence of (37.4) (second relation), for $i$ sufficiently large $| \mathscr{C}'_p{}^{ai} |$ and $| \mathscr{C}'^{bi} |$, $a \neq b$, will have $\Phi^i$-neighborhoods without common cells, for otherwise we would have $d(| c'_p{}^a | \cdot | c'_p{}^b |) = 0$. We also know that by construction $\mathscr{C}'_p{}^{ai}$ and $\mathscr{C}''_p{}^i$ have no common $p$-cells. Combining with the construction of $C'_{p+1}$, $C''_{p+1}$, we have

$$(37.6) \qquad | F(C'^a_{p+1}) | \cdot | C''_{p+1} | = | F(C'^a_{p+1}) | \subset L.$$

38. Until further notice we shall impose upon the simplicial transformation $\theta$ of No. 36 the following additional restriction: whenever $E_q{}^i$ is a cell of $\mathscr{C}''_p{}^i$ with vertices on $L$, we choose one of these vertices as the $A^{iq}$ for that cell, that is as the vertex of $\Phi^i$ into which $\theta$ is to transform all the vertices of $\Psi'^i$ that are on $E_q{}^i$.

Consider now $C^{*k}_{r-n+p}$ and let $C'^{*k}_{r-n+p}$ be the chain left on removing from it the cells which do not meet $\mathscr{C}'_p{}^k$. Due to the mode of separation of the chains $\mathscr{C}'^{ak}_p$, for $k$ sufficiently high $C'^{*k}_{r-n+p}$ will be a sum of disjoined chains $C'^{*ak}_{r-n+p}$ consisting respectively of the cells which meet the chain $\mathscr{C}'^{ak}_p$. Therefore $C'^{*ak}_{r-n+p}$ meets $\mathscr{C}'^{ak}_p$ but not $\mathscr{C}'^{bk}_p$ for $b \neq a$.

Now observe that as regards the cells of $\Gamma_n{}^k \cdot C^{*k}_{r-n+p}$ that are on the chains $\mathscr{C}'$ the transformation $\theta$ preserves the same properties as in the Lemma. However it now takes the $p$-cells on a $\mathscr{C}''$ and transforms them like cells on an $F(\mathscr{C}')$, i. e. into cells of dimension $< p$ so that the projections of their boundaries do not affect the projections of the chains $F(\Gamma_n{}^k \cdot C'^{*ak}_{r-n+p})$. It follows that as regards the effect on the intersections $\Gamma_n{}^k \cdot C'^{*ak}_{r-n+p}$ its performance is as before and that this chain is now projected into $\mathscr{C}'^{ai}_p$. It follows also that $F(\Gamma_n{}^k \cdot C'^{*ak}_{r-n+p})$ is projected at the same time into $F(\mathscr{C}'^{ai}_p)$. All this holds of course for $k$ large enough, which is all that we need.

It follows from what precedes that we may replace the initial chain $c_p$ by a set of chains $c'^a_p$ whose boundaries behave in a manner similar to that imposed upon $c_p$ by the Lemma.

39. Since $c'^a_p \subset V^a$ we have from our discussion of the index

(39. 1) $$(c_p \cdot c_{n-p}) = \Sigma\, (c'^a_p \cdot F(c^a{}_{n-p+1})).$$

Similarly since $r - n$ is even by *Topology*, p. 169, formula (20),

(39. 2) $$(C^{*k}_{r-n+p} \cdot \mathscr{C}^k_{n-p}) = \Sigma\, (C'^{*ak}_{r-n+p} \cdot \mathscr{C}^{ak}_{d-u})$$
$$= (-1)^p \Sigma\, (F(C'_p{}^{*ak}) \cdot \mathscr{C}^{ak}_{n-p+1}).$$

Comparing these relations and bringing back the index $\alpha$, we find

(39. 3) $$\Sigma\, (F(C'^{a_\alpha k_\alpha}_{r-n+p}) \cdot \mathscr{C}^{a_\alpha k_\alpha}_{n-p+1}) = (-1)^p \cdot \lambda,$$

and the proof of the Lemma is reduced to showing that

(39. 4) $$\Sigma\, (F(c'^{a_\alpha}_p) \cdot c^{a_\alpha}_{n-p+1}) = (-1)^p \cdot \lambda,$$

the relations between corresponding chains being as for $\Lambda_{p-1}$. That is to say we have reduced $\Lambda_p$ to $\Lambda_{p-1}$, and hence to $\Lambda_0$.

40. We take up $\Lambda_0$, and we shall in fact prove the somewhat more stringent result that $\Lambda_0$ holds with all the numbers $k_\alpha$ equal, i. e. with a single chain $c_0{}^a$. We may go as far as No. 39 in the same manner as previously. Referring to No. 37 we have to prove that when the diameters of the sets $c'_0{}^a$ are small enough, if there is a $k$ arbitrarily high such that $\mathscr{C}_0{}^{ai}$ is the projection of $\Gamma_n{}^k \cdot C^{*ak}_{r-n}$, then

(40. 1) $$\Sigma\, t_a(C^{*ak}_{r-n} \cdot \Gamma_n{}^k) = \Sigma\, t_a(c_0{}^a \cdot \gamma_n).$$

This will follow if we can show that

(40. 2) $$(C^{*ak}_{r-n} \cdot \Gamma_n{}^k) = (c_0{}^a \cdot \gamma_n).$$

In the first place we have (No. 24)

(40. 3) $$(c_0{}^a) = (c_0{}^a \cdot \gamma_n) = (\mathscr{C}_0{}^{ai})$$

for $i$ large enough. Also since $\mathscr{C}_0{}^{ai}$ is the projection of $\Gamma_n{}^k \cdot C^{*ak}_{r-n}$, we have

(40. 4) $$(\mathscr{C}_0{}^{ai}) = (\Gamma_n{}^k \cdot C^{*ak}_{r-n}) = (C^{*ak}_{r-n} \cdot \Gamma_n{}^k),$$

from which (40. 2) and hence (40. 1) follow. This proves $\Lambda_0$ and hence also the Lemma.

41. An important application of the Lemma is the proof, still lacking, of formula (34. 7), and hence of property (23 f), for the index. For take first $p = 0$, and $C^a{}_{p+1} = C_1 =$ a chain made up of a single projecting line whose traces $\mathscr{C}_0{}^i$ are vertices $A^i$ of the complexes $\Phi^i$ such that for $i$ above a certain value $A^i$ is the vertex of an $E_n$ of $\Phi^i$. Then taking $k_a = i$, $C^{*a}_{r-n} = E^*{}_{r-n}$ the cell of $\mathscr{K}^i$ dual to $E_n$, $\mathscr{C}^{ak_a}_{r-n} = \Gamma_n{}^i$ and orientations as in *Topology*, Ch. IV, the condition of the Lemma is fulfilled with a single $c_0 = x$ and a single $c_n = \gamma_n$. Therefore

(41. 1) $$(x \cdot V_n) = (E^*{}_{r-n} \cdot E_n) = +1.$$

Choose now $p = n$ and $C^a{}_{p+1} = C_{n+1} = \Gamma_{n+1}$ the projection-chain defining $\gamma_n$, $k_a = i$, $C^{*ak_a}_{r-n+p} = C^*{}_r{}^i$, the sum of the cells of $\mathscr{K}^{*i}$ oriented like the spaces $S_r$, $\mathscr{C}^{ak_a}_{n-p} =$ the same vertex $A^i$ as previously. This time the conditions of the Lemma are again fulfilled and we find

(41. 2) $$(\gamma_n \cdot x) = (C^*{}_r{}^i \cdot A^i) = +1 = (x \cdot \gamma_n),$$

which is the result that we required.

42. From the Lemma to the duality formula for the Betti-numbers is but a step. Let $\gamma^\alpha{}_{n-p}$, $\alpha = 1, 2, \cdots, R_{n-p}$, be a base for the $(n-p)$-cycles consisting of irreducible projection-cycles (Theorem II). Let $\Gamma^\alpha{}_{n-p+1}$, $\mathscr{G}^{ah}_{n-p}$ be the representative cycle mod $\Phi$, and trace on $\Phi^h$, associated with $\gamma^\alpha{}_{n-p}$. Then for $h$ above a certain value the cycles $\mathscr{G}^{ah}_{n-p}$ are independent on $\Phi^h$, hence also independent on $\mathscr{K}^h - \mathscr{B}^h$. For if say

$$\mathscr{K}^h - \mathscr{B}^h \supset C_{n-p+1} \rightarrow \Sigma\, t_a\, \mathscr{G}^{ah}_{n-p},$$

we could slide down $C_{n-p+1}$ along the projecting segments on $\mathscr{K}^h - \mathscr{B}^h - \Phi^h$ onto $\Phi^h$, and obtain a chain on $\Phi^h$

$$C'{}_{n-p+1} \rightarrow \Sigma\, t_a\, \mathscr{G}^{ah}_{n-p}$$

so that the cycles $\mathscr{G}^{ah}_{n-p}$ would not be independent on $\Phi^h$.

As a consequence of the independence of these cycles on $\mathcal{K}^h - \mathcal{B}^h$, $\mathcal{K}^h$ contains a cycle mod $\mathcal{B}^h$, $\mathcal{G}^{\alpha h}_{r-n+p}$ whose cells intersecting $\Phi^h$ consist of cells of the dual $\mathcal{K}^{*h}$, and such that (*Topology*, pp. 140, 174).

$$(42.1) \qquad (\mathcal{G}^{\alpha h}_{r-n+p} \cdot \mathcal{G}^{\beta h}_{n-p}) = \delta_{\alpha\beta}.$$

Consider now the projection $\Gamma_p{}^i$ of all cycles $t\Gamma_n{}^h \cdot \mathcal{G}^{\alpha h}_{r-n+p}$ ($\alpha$ fixed) on a definite $\Phi^i$. As far as the intersections with $\Phi^h$ go the chain $\mathcal{G}^{\alpha h}_{r-n+p}$ is a $C^{*h}{}_{r-n+p}$. With $\Gamma_p{}^i$ we associate the numbers $(t\delta_{\alpha\beta})$ and if $\Gamma'_p{}^i$ corresponds to $t'$ and the numbers $(t'\delta_{\alpha\beta})$, we associate with $s\Gamma_p{}^i + s'\Gamma'_p{}^i$ the numbers $((st + s't')\delta_{\alpha\beta})$. In this manner if $\mathfrak{M}^i$ is the modulus generated by the cycles $\Gamma_p{}^i$, there corresponds to each member of $\mathfrak{M}^i$ a definite set $(t\delta_{\alpha\beta})$. Clearly members corresponding to $t = 0$ give rise to a submodulus $\mathfrak{N}^i$ of $\mathfrak{M}^i$. Also by construction the moduli $\mathfrak{M}^i, \mathfrak{N}^i$ are in the very relationship demanded by Theorem I. Therefore there exists a projection-sequence $\{\Gamma_p{}^{\alpha i}\}$ such that the cycle $\Gamma_p{}^{\alpha i}$ is a member of $\mathfrak{M}^i$ corresponding to $t = 1$. This sequence gives rise to a projection-cycle mod $\Phi$, $\Gamma^\alpha{}_{p+1}$, which defines a normal cycle $\gamma_p{}^\alpha$. Owing to (42.1) and to the mode of defining the moduli $\mathfrak{M}^i$, we have by the Lemma

$$(42.2) \qquad (\gamma_p{}^\alpha \cdot \gamma^\beta{}_{n-p}) = \delta_{\alpha\beta}.$$

Hence (No. 23 property *e*) the cycles $\gamma_p{}^\alpha$ are independent and therefore $R_p \geqq R_{n-p}$. Similarly $R_p \leqq R_{n-p}$ and therefore we have proved Poincaré's duality relation for an absolute $n$-manifold:

$$(42.3) \qquad R_p(M) = R_{n-p}(M).$$

43. *Extension to open manifolds.* Take first an open $M_n$ and let $U$ be an open subset of $M$ whose closure $\bar{U}$ is self-compact. Then if $V \Subset U$, $\bar{V}$ is likewise self-compact. As the manifold conditions hold over $U$ we may apply Theorem VII with the following slight restrictions: $c_p \subset V$, $\epsilon < d(V, M - U)$. From this we conclude, as in No. 28, that there are at most finite numbers: (a) of absolute $p$-cycles of $U$ independent mod $M - V$; (b) of $p$-cycles of $U$ mod $M - U$, independent mod $M - V$. We can then show as in the preceding number that the two numbers are equal.

The sequences of open sets $\{U^i\}$ such that $U^{i+1} \Subset U^i$, $\Pi U^i = 0$, may serve to define the different types of ideal elements as we have done in *Topology*, Ch. VII. In the terminology there used let $\Lambda$ be the total ideal element, and let $\mathcal{L}^1, \mathcal{L}^2$ designate complementary closed and open ideal elements. Let also $L$ be any closed subset of $M$ which $\supset \Lambda$ and let $L^1$ be any closed subset of $L$ with $\mathcal{L}^1$ for ideal element. Then if $L^2 = L - L^1$, $\mathcal{L}^2$ will be the ideal element of $L^2$. By means of properties (a), (b), (c), and by unimportant adaptations of the treatment in *Topology*, Ch. VII, § 3, we prove:

FUNDAMENTAL DUALITY THEOREM. *Let* $\Gamma_p$, $G_{n-p}$ *be associated cycles of the dual types* $M - L^1 \bmod L^2$ *and* $M - L^2 \bmod L^1$ *in any ring of rational coefficients forming a field. There exists two associated dual bases* $\{\Gamma_p{}^{\alpha}\}$, $\{G^{\beta}{}_{n-p}\}$ *made up of true normal cycles whose indices satisfy the relations*

$$(\Gamma_p{}^{\alpha} \cdot G^{\beta}{}_{n-p}) = \delta_{\alpha\beta}. \tag{43.1}$$

*Whatever* $\Gamma_p$, $G_{n-p}$ *we have*

$$\Gamma_p = \Sigma\,(\Gamma_p \cdot G^{\alpha}{}_{n-p}) \cdot \Gamma_p{}^{\alpha}, \tag{43.2}$$

$$G_{n-p} = \Sigma\,(\Gamma_p{}^{\alpha} \cdot G_{n-p}) \cdot G^{\alpha}{}_{n-p}, \tag{43.3}$$

*and the Betti-numbers satisfy the duality relations*

$$R_p(M - L^1, L^2) = R_{n-p}(M - L^2, L^1). \tag{43.4}$$

*In particular:* (a) *when* $L^1 = \Lambda$, $L^2 = 0$ we have

$$R_p(M - \Lambda) = R_{n-p}(M, \Lambda), \tag{43.5}$$

*where the Betti-numbers refer at the left to the finite cycles and at the right to the infinite cycles;* (b) *when* $\Lambda = 0$, *i. e. when* $M$ *is absolute, the bases are finite and the duality relation reduces to that of Poincaré*

$$R_p(M) = R_{n-p}(M). \tag{43.6}$$

*These results hold also when* $M$ *consists of a countable aggregate of circuits.*

The last part of the statement, regarding an $M$ consisting of a countable aggregate of circuits, is an immediate consequence of the following: when $M = \Sigma M^i$, where $M^i$ is a connected $n$-manifold and $M^i \cdot M^j = 0$, then the $p$-th homology group of $M$ is the direct sum of those of the manifolds $M^i$ (the groups are assumed written additively).

# XII

# XII

## CHAIN-DEFORMATIONS IN TOPOLOGY

In topology one has repeated occasion to consider homotopic deformations of chains. They give rise to a basic boundary relation[1] between the extreme positions of a chain $c_p$ in the homotopy and what might be termed the loci of $c_p$ and of its boundary $F(c_p)$. All the consequences of the homotopy that concern algebraic topology (i.e., boundary relations and the associated homologies) may be derived from the fundamental relation. It seems natural therefore to call *chain-deformation* any scheme wherein two $p$-chains $c_p$, $c_p'$ and two other chains that are to take the part of the loci mentioned above, satisfy a boundary relation formally identical with the fundamental relation of homotopy.[2] This notion has already been exploited in a recent paper.[3] We return to it here, first to develop it more fully and then to apply it to the study of the sets that are obtained whenever, in the definition of locally connected sets, singular cells and spheres are replaced by chains. These new sets may be described as locally connected in the sense of homology, and their types correspond substantially to the locally connected types that we have recently investigated.[4] The passage from the first class to the second corresponds also to a substitution of chain-deformation for homotopy.

One of the important results of L2 was the identification of certain locally

[1] Given for the first time in our Colloquium Lectures, *Topology*, New York, 1930, p. 78.

[2] While chain-deformations have most of the properties that their name suggests, they are essentially different from homotopy. This is clearly seen by noting the different effect in the very simple case of the circuits on an orientable surface of genus $p \geqq 2$. Homotopy leads, in this case, to the non-commutative Poincaré group, chain-deformation to the much simpler abelian group with $2p$ free generators.

[3] S. Lefschetz, *On generalized manifolds* (= L1 in the sequel), American Journal of Mathematics, vol. 55 (1933), pp. 475–499 (= pp. 487–511 of the present work).

[4] S. Lefschetz, *On locally connected and related sets* (= L2 in the sequel), Annals of Mathematics, vol. 35 (1934), pp. 118–139. We call attention to the following errata: p. 119, line 23, replace LC by $LC^\infty$; p. 126, suppress line 3 from bottom; in line 4 from bottom, suppress "convex"; in line 5 from bottom, replace "convex sets of $\mathfrak{H}$" by "spheres"; p.127, line 13, replace $K^*$ by $\mathfrak{K}^*$. (L2 = pp. 610–631 of the present work).

Local connectedness in the sense of homology was introduced by P. S. Alexandroff in his paper: *Untersuchungen über Gestalt und Lage abgeschlossener Mengen beliebiger Dimension*, Annals of Mathematics, vol. 30 (1929), pp. 101–187. See also his recent paper: *On local properties of closed sets*, Annals of Mathematics, vol. 36 (1935), pp. 1–35, §3. The same property for euclidean domains plays a central part in R. L. Wilder's recent work. See in particular his last paper: *Generalized closed manifolds in n-space*, Annals of Mathematics, vol. 35 (1934), pp. 876–903.

connected sets with the absolute neighborhood retracts or absolute retracts in the sense of Borsuk. (See in this connection footnote 13.) A similar identification is possible here with generalized retracts in the sense of chain-deformations. Rather than to press this analogy, we preferred to investigate the mutual relations between the two kinds of local connectedness as well as with Borsuk's locally contractible sets, but there remains still much to be done along that line.

One of the most useful notions introduced in L2 was that of the semi-singular complex (singular complex with only part of the expected cells present). It is extended here to aggregates of chains related like the oriented cells of a complex and is found no less useful in the present investigation.

In the endeavor to free our results, as far as possible, from any specific choice of chains, we have presented the theory of chains in axiomatic form at the beginning of the paper. This has the additional advantage of making the paper less dependent upon our previous writings.

Our general notations are those of *Topology*, with the abbreviations of L1 and L2: LC, and NR stand for "locally connected" and "neighborhood retract." In addition we shall write HLC for "LC in the sense of homology", and similarly for HNR. Our two other abbreviations are c.s.v.t. for "continuous single-valued transformation", f.c.o.s. for "finite covering by open sets". The reader will have no difficulty in getting accustomed to these alphabetical notations whose advantages, after all, need not be reserved for the political domain.

## §1. The chains of a topological space

1. There are various ways of extending to a topological space $\mathfrak{R}$ the basic properties of the chains of a geometric complex, their cycles, their boundary relations and the like. Regardless of the procedure adopted certain properties are preserved. As it is with these common properties that we are chiefly concerned, we shall recall them briefly and state them as postulates for the chains of $\mathfrak{R}$:[5]

I. *There exists for every* $p = 0, 1, \cdots$, *a set of topological invariants of* $\mathfrak{R}$, *its p-chains* $c_p$, *and their collection* $\{c_p\}$ *constitutes a free additive abelian group.*

II. *There exists an operation F defined topologically for all the chains and such that* $F\{c_p\}$ *is a homomorphism of* $\{c_p\}$ *into* $\{c_{p-1}\}$, *and of* $\{c_0\}$ *into the identity.*

$F$ is the *boundary-operator*, $Fc_p$ or $F(c_p)$ is the *boundary* of $c_p$, their mutual relation being indicated by a *boundary relation*,

$$c_p \to F(c_p), \tag{1.1}$$

[5] These properties are fully developed in *Topology*, Chapters I, II. For a more systematic exploitation of the abstract viewpoint see A. W. Tucker, *An abstract approach to manifolds*, Annals of Mathematics, vol. 34 (1933), pp. 191–243, where further references, notably to the papers of W. Mayer, will be found. It is to be noted that whereas they consider only abstract complexes and chains, we have always tied them up with definite point-sets. It might be advisable to use different terms, such as abstract complex or chain, geometric complex or chain, for the two concepts.

while to express merely that $c_p$ is a boundary we write a *homology*,

$$c_p \sim 0. \tag{1.2}$$

The chains $c_p$ such that $Fc_p = 0$ are called *p-cycles*, and generically denoted by $\gamma_p$. In particular, every $c_0$ is a $\gamma_0$. From II follows that $\{\gamma_p\}$ is a subgroup of $\{c_p\}$.

III. $FF = 0$.

This means that boundaries are cycles. Moreover if $\beta_p$ is a generic boundary, from II follows again that $\{\beta_p\}$ is a subgroup of $\{\gamma_p\}$. The difference-group (factor-group of the customary terminology) $\{\gamma_p\} - \{\beta_p\}$ is the $p^{\text{th}}$ *homology group* of $\mathfrak{R}$. When it is a free group the number $R_p(\mathfrak{R})$ of its independent generators is called the $p^{\text{th}}$ *Betti-number* of $\mathfrak{R}$.

It is to be kept in mind that in the present paper, *all homologies imply bounding*. This is the reason why we use the symbol $\sim$ and not $\approx$, for example as in *Topology*, Chapter VII, or L1, for $\gamma_p \approx 0$ merely meant that $\gamma_p$ was a finite or infinite sum of bounding cycles, or neglected chains, without being itself strictly in one or the other category. From the group viewpoint, our earlier procedure corresponds to topologizing the groups, replacing $\{\beta_p\}$ by its closure, say $\{\beta_p'\}$, and taking as $p^{\text{th}}$ homology group the difference-group $\{\gamma_p\} - \{\beta_p'\}$.

IV. *There exists a numerical topological invariant linear function of zero-chains, the Kronecker-index* $(c_0)$, *and* $(c_0) = 0$ *when* $c_0 \sim 0$.

In the case of complexes $(c_0)$ is the number of points of $c_0$ each counted with its coefficient in the expression of the chain.

V. *With every* $c_p$ *there is associated a unique closed subset* $|c_p|$ *of* $\mathfrak{R}$ *such that:*

(a) $|0| = 0$; (b) $|c_p + c_p'| \subset |c_p| + |c_p'|$,

(c) $|F(c_p)| \subset |c_p|$; (d) $\dim |c_p| \geqq p$ when $c_p \neq 0$.

If $A$ is any closed subset of $\mathfrak{R}$ we say that $c_p \subset A$ whenever $|c_p| \subset A$. Taken together with V, this enables us to define the boundary relations, homologies and homology-groups mod $A$, or *relative* relations, by contrast with the previous type called *absolute*.

From V(d) follows that *when* $\dim \mathfrak{R} = n$ *is finite every* $c_p = 0$ *for* $p > n$. In particular there are no homology groups, absolute or relative, for dimensions $> n$.

2. There remains one more property, but it is most conveniently expressed in terms of the very useful notion of *quasi-complex*. A quasi-complex $\mathfrak{K}$ is a collection of chains of $\mathfrak{R}$ such that: (a) its $p$-chains form a subgroup of $\{c_p\}$; (b) $F\,\mathfrak{K} \subset \mathfrak{K}$: when $c_p \subset \mathfrak{K}$ likewise $Fc_p \subset \mathfrak{K}$. The quasi-complex is *finite* whenever the dimension of its chains is bounded, and in addition for every $p$ there is a finite base for its $p$-chains whose elements are independent. In that case we frequently reserve the name "chain of $\mathfrak{K}$" for the chains of the bases, and call *subchains* of $\mathfrak{K}$ the other chains of the quasi-complex.

If $c_q^i$ is any base chain of a finite $\mathfrak{K}$, the chains $c_{q-1}^j$ entering in the composition of $F(c_q^i)$, those entering in the composition of $F(c_{q-1}^j)$, etc., are called the *boundary-chains* of $c_q^i$. The sum $|c_q^i| + \Sigma\,|c_{q-1}^j| + \cdots$ has a least upper bound called the *mesh* of $\mathfrak{K}$.

Our last axiom may now be stated:

VI. (*Subdivision axiom.*) *Every chain of* $\mathfrak{R}$ *is a subchain of a finite quasi-complex* $\mathfrak{K}$ *whose mesh is arbitrarily small.*

The complex $\mathfrak{K}$ is called an *elementary decomposition* of the chain, an elementary $\epsilon$-decomposition when its mesh $< \epsilon$.

3. The following are noteworthy examples of systems of chains for which all the axioms hold:

($\alpha$) $\mathfrak{R}$ is a topological space and its chains are the singular chains on $\mathfrak{R}$ in the sense of *Topology*, Chapter II,[6] that is to say, the linear forms in the singular cells on $\mathfrak{R}$ with coefficients members of an additive abelian group $\mathfrak{M}$. All the axioms except the last are readily verified, and the last is verified also provided that we agree to identify a singular $c_p$ with all its subdivisions and call *chain* the class thus obtained. Otherwise we merely have as a theorem that all the members of the class are equivalent regarding boundary relations and homologies (*Topology*, p. 88). It is understood of course that each cell of a subdivision of $c_p$ is to be oriented concordantly with the carrying cell of $c_p$.

A noteworthy case is that where $\mathfrak{R}$ is a simplicial complex $K$. It is then shown that the homology groups derived from the chains made up of the oriented simplexes of $K$, or *combinatorial* homology groups, are isomorphic with those defined above and hence the former are topological invariants.

($\beta$) $\mathfrak{R}$ is a compact metric space and the chains are the *projection-chains* of L1. They are certain specific subchains, taken here with coefficients in a field $\mathfrak{M}$, of a fundamental infinite complex $K$ which consists of the skeleta $\Phi^i$ of f.c.o.s. together with their joining cells (pp. 482–484).[7] The configuration $\bar{K} = \mathfrak{R} + K$ may be identified for convenience with its topological image on the Hilbert parallelotope $\mathfrak{H}$ (see loc. cit. for details). The projection-chains $C_{p+1}$ of $K$ determine the chains $c_p$ of $\mathfrak{R}$, with $|c_p| = \overline{|C_{p+1}|} \cdot \mathfrak{R}$ (p. 489). For $p = 0$ we define the Kronecker-index by $(c_0) = (F(C_1))$ which is equivalent to the definition of p. 501.

We shall call chains of type ($\alpha$) *singular* and chains of type ($\beta$) *regular*. Unless otherwise specified regular chains shall be the type usually considered in the sequel.

The only non-compact metric spaces that interest us are the separable locally compact spaces. For these we may consider the so-called *finite* cycles, or cycles on self-compact subsets of the space, and they are the only kind needed later.

Our definition of regular chains is not intrinsic in that it depends upon a specific fundamental complex $K$. It may be made intrinsic as follows. Suppose that we pass to a new fundamental complex $K'$. By means of the deformation theorem of *Topology*, p. 328 we

[6] See also pp. 479–484.

[7] In the construction of $K$ (loc. cit., beginning of No. 2) the finiteness of dim $\mathfrak{R}$ plays no rôle. All that matters is that the number $n$ is the least order of any $\epsilon$ —f.c.o.s.

first reduce a projection-chain $C_{p+1}$ of $K$ defining $c_p$ of $\mathfrak{R}$ to a chain $C'_{p+1}$ of $K'$, then by Theorems IV, V of L1 (p. 484), we reduce $C'_{p+1}$ to a projection-chain $C''_{p+1}$ of $K'$. The construction, which depends upon a choice of bases as in L1, No. 16, is made unique by adopting a fixed numbering of all the cells of $K'$. It is the class of all projection-chains $C''_{p+1}$ thus obtained which is $c_p$ under the new definition. It is not clear, however, that $| c_p |$ is independent of $K$. Let us assume that this is not necessarily so and let $A$, $A'$ be the sets as determined respectively by $K$ and $K'$. We have immediately $A \subset A'$. If we return from $K'$ back to $K$ we show readily that we reduce $C''_{p+1}$ to a projection-chain on a fundamental complex for $A$ necessarily identical with $C_{p+1}$. Hence $A' \subset A$ and therefore $A' = A$.

It may be observed that projection-chains are imposed by the subdivision axiom. They are in fact the only chains of $K$ which we are able to subdivide into elements of the same nature, notably as regards the subdivision of the elements $C_2$ into elements with the proper $C_1$ boundaries. On the other hand their uniqueness is only proved when $\mathfrak{M}$ is a field, so that the restriction imposed upon $\mathfrak{M}$ is largely due, in the last analysis, to the subdivision axiom.

Besides the two classes ($\alpha$), ($\beta$) others may well be introduced. We mention notably: ($\gamma$) the Vietoris-chains for compact metric spaces;[8] ($\delta$) the chains recently defined by Čech[9] for spaces even more general than topological spaces. In the case ($\gamma$) there is an associated group $\mathfrak{M}$ as general as in ($\alpha$), while in ($\delta$) $\mathfrak{M}$ is again restricted to a field. In fact, when $\mathfrak{M}$ is a field the homology theories yielded by ($\beta$), ($\gamma$), ($\delta$) are the same. Furthermore when $\mathfrak{M}$ is a field, and $\mathfrak{R}$ is $\mathrm{LC}^\infty$ in the sense of L2 the four types coalesce (*Topology*, p. 333).

All but the last of our axiomatic properties are readily verified for the four types mentioned. Regarding the last, subdivision, for the singular type it is a consequence of the definition and for the regular type it is proved as on p. 502. Since they are the only two considered in the present paper this will suffice for our purpose.

## §2. Chain-deformations

4. As a matter of convenience we shall assume henceforth that the basic space $\mathfrak{R}$ is separable, metric, locally compact, and later even, a good part of the time, that it is compact. Until further notice the chains are finite chains in the sense of No. 3 and, unless otherwise stated, merely subjected to our basic axioms.

In the applications, even when dealing with a single chain, one is constrained to deform a whole quasi-complex. For example to have the analogue of $\epsilon$-homotopy, we must break up the chains into the elements of a $\mathfrak{K}$ of mesh $\epsilon$ which is then to undergo the deformation. Owing to this it is best to define at the outset a *chain-deformation* $\vartheta$ of a quasi-complex $\mathfrak{K} = \{c_q\}$, into another $\mathfrak{K}' = \{c'_q\}$. We understand then by such a chain-deformation, (merely "deformation" when there is no ambiguity), a homomorphism of the chain-groups of $\mathfrak{K}$ into those of the same dimension for $\mathfrak{K}'$, which commutes with $F$ and which is associated with a linear operator $\mathfrak{D}$ on the chains of $\mathfrak{K}$ having the property that

[8] L. Vietoris, *Über den höheren Zusammenhang kompakter Räume und eine Klasse von zusammenhangstreuen Abbildungen*, Math. Annalen, vol. 97 (1927), pp. 454–472.

[9] E. Čech, *Théorie générale de l'homologie dans les espaces abstraits*, Fundamenta Mathematicae, vol. 19 (1932), pp. 149–183.

whatever $c_q$ of $\mathfrak{K}$, $\mathfrak{D} c_q$ is a uniquely defined $(q + 1)$-chain of $\mathfrak{R}$, called the *deformation-chain* of $c_q$, such that

$$F \mathfrak{D} c_q = \vartheta c_q - c_q - \mathfrak{D} F c_q. \tag{4.1}$$

Written as relations between operators on $\mathfrak{K}$ we have then

$$\vartheta F = F \vartheta, \tag{4.2}$$

$$F \mathfrak{D} + \mathfrak{D} F = \vartheta - 1. \tag{4.3}$$

$\mathfrak{D}$ may also be considered as inducing for each $q$ a homomorphism of the group of the $q$-chains of $\mathfrak{K}$ into the group of the $(q + 1)$-chains of $\mathfrak{R}$ satisfying (4.3).

The preceding properties, and in particular (4.1), written also

$$\mathfrak{D} c_q \to \vartheta c_q - c_q - \mathfrak{D} F(c_q), \tag{4.4}$$

are those satisfied by chains under a homotopy (*Topology*, p. 78) which we are thus merely taking as postulates for chain-deformation.

5. A cycle $\gamma_p$, a chain $c_p$ + its boundary $F(c_p)$, form special quasi-complexes. For the former $\vartheta$ merely demands that

$$F \mathfrak{D} \gamma_p = \vartheta\gamma_p - \gamma_p. \tag{5.1}$$

Since a boundary is a cycle, $\vartheta\gamma_p - \gamma_p$ is a cycle, and hence $\vartheta\gamma_p$ is one also. Therefore *the chain-deform of a cycle is a cycle.* Regarding $c_p$, $\vartheta$ demands two deformation-chains $\mathfrak{D} c_p$, $\mathfrak{D} Fc_p$ such that

$$F \mathfrak{D} c_p = \vartheta c_p - c_p - \mathfrak{D} Fc_p, \tag{5.2}$$

$$F \mathfrak{D} Fc_p = F\vartheta c_p - Fc_p. \tag{5.3}$$

Of these relations the second follows from the first, since it merely expresses the fact that the boundary of the right-hand side in (5.2) is a cycle. Therefore for $c_p$ also a chain-deformation is specified by the single relation (5.2). It is clear that in the sense just considered a chain-deformation of a quasi-complex $\mathfrak{K}$ induces a chain-deformation of its individual chains.

6. Whenever $\mathfrak{K}$, $\mathfrak{K}'$ and the $\mathfrak{D}$-chains are on a given set $A$ we say that the chain-deformation is *over* $A$. The chains in any given set $B$ form a quasi-complex, and its *chain-deformations* are called *chain-deformations* of $B$. Thus if $\vartheta$ deforms all the chains of $B$ into chains of a set $B$ over $A$, we say that $B$ is chain-deformed into $B'$.

Whenever $\mathfrak{K}$, $\mathfrak{K}'$ are of mesh $< \epsilon$ and all the chains $\mathfrak{D}$ involved are of diameter $< \epsilon$, we say that $\vartheta$ is an $\epsilon$ chain-deformation of $\mathfrak{K}$ or of any subchain of $\mathfrak{K}$. An $\epsilon$ chain-deformation of $c_p$ consists in imposing upon it an elementary $\epsilon$-decomposition reducing it to a $\mathfrak{K}$ of mesh $< \epsilon$, and then $\epsilon$ chain-deforming $\mathfrak{K}$.

7. Let $\vartheta$ be as before and let $\vartheta'$ deform $\mathfrak{K}'$ into $\mathfrak{K}''$ with $\mathfrak{D}'c'_q$ as the deformation-chains. We have now a transformation $\vartheta'' = \vartheta'\vartheta$ of $\mathfrak{K}$ into $\mathfrak{K}''$ and we shall show that it is a chain-deformation. Clearly $\vartheta''$ is a homomorphism commuting with $F$ so (4.3) alone must be verified. We have

$$\mathfrak{D}'F + F\mathfrak{D}' = \vartheta' - 1 \tag{7.1}$$

operating on $\mathfrak{K}'$, but considered as operating on $\mathfrak{K}$ this must be written

$$\mathfrak{D}'F\vartheta + F\mathfrak{D}'\vartheta = \vartheta'\vartheta - \vartheta = \vartheta'' - \vartheta. \tag{7.2}$$

Adding (7.2) and (4.3), we have

$$(\mathfrak{D} + \mathfrak{D}'\vartheta)F + F(\mathfrak{D} + \mathfrak{D}'\vartheta) = \vartheta'' - 1. \tag{7.3}$$

Hence if we introduce the linear chain-operator $\mathfrak{D}'' = \mathfrak{D} + \mathfrak{D}'\vartheta$, we have

$$\mathfrak{D}''F + F\mathfrak{D}'' = \vartheta'' - 1, \tag{7.4}$$

so that (4.3) holds, with $\mathfrak{D}''$ as the $\mathfrak{D}$ operation. From its definition $\mathfrak{D}''$ is linear, hence $\mathfrak{D}''$ is a deformation-chain.

Clearly $\vartheta = 1$ is a chain-deformation corresponding to $\mathfrak{D} = 0$. Regarding the inverse $\vartheta^{-1}$ it can only be defined, if at all, when $\vartheta$ is one-one. In particular, this holds when $\mathfrak{K}$ consists of a single chain plus its boundary. When $\vartheta$ is one-one $c'_q = \vartheta c_q$ determines uniquely $c_q$ and hence $\mathfrak{D}\, c_q$. If we set

$$\mathfrak{D}'c'_q = \mathfrak{D}'\vartheta\, c_q = -\mathfrak{D}\, c_q$$

we have

$$\mathfrak{D}'\vartheta F c_q = \mathfrak{D}'F\vartheta c_q = -\mathfrak{D}F c_q,$$

and therefore

$$F\mathfrak{D}'c'_q = c_q - c'_q - \mathfrak{D}\, F(c'_q).$$

This shows that the passage from $\mathfrak{K}'$ to $\mathfrak{K}$ is a chain-deformation with $-\mathfrak{D}c_q$ as its deformation-chain. Thus in the case in question $\vartheta^{-1}$ is a chain-deformation.

If we call two $q$-chains of $\mathfrak{R}$ *equivalent* whenever one of them can be chain-deformed into the other, the results of the two preceding paragraphs show that this equivalence is *transitive, reflexive,* and *symmetric.*

8. **Homotopy and chain-deformation.** Let a set $A$ undergo a homotopy $T$ into a set $A'$ over $\mathfrak{R}$ and let $B$ be its locus throughout the deformation. If $c_p$ is a chain of $A$ the homotopy will determine a chain $c'_p = Tc_p$. One suspects intuitively that, as is obvious for singular chains, $c'_p$ is a chain-deform of $c_p$ over $B$. We shall in fact prove:

THEOREM I. *When the chains adopted are regular or singular a homotopic deformation of a compact*[10] *set induces a chain-deformation of all its chains.*

[10] We mean here that the set is compact as a space, or, as it is sometimes called, self-compact.

For singular chains the theorem is practically proved in *Topology*, p. 79, so that we may limit our treatment to regular chains. This being the case, we may clearly replace $A$, $B$, $A'$ by $|c_p|$, its locus, and $T \cdot |c_p|$, and hence assume $A$, $A'$ closed and the locus the whole space, that is $B = \mathfrak{R}$. Under the circumstances let $L$ be a fundamental complex for $A$ which is a subcomplex of $K$ (L1, p. 474) and let the whole configuration be assumed immersed in the Hilbert parallelotope $\mathfrak{H}$ (No. 3). Let $P^1, P^2, \cdots$ be the vertices of $K$, which we take to be points of $A$ and consider the product $L \times \lambda = L^*$, where $\lambda$ is a unit-segment. It is an infinite convex complex, which we subdivide in such a manner that if $E$ is any cell of $L$, the cell $E \times \lambda$ of $L^*$ becomes a sum of cells $E \times \lambda'$, where $\lambda'$ is an interval of $\lambda$. That is, every "prismatic" cell of $L^*$ is subdivided by sections "parallel" to its bases. This is carried out in such manner that if the cells of $K$, numbered in some order, are $E^1, E^2, \cdots$, then $E^k \times \lambda$ is subdivided into cells equally spaced (i.e. with intervals $\lambda'$ of equal length) and whose number $> 2^k$. Finally we subdivide in any manner $L^*$ simplicially without introducing new vertices, which may always be done since its cells are all convex. We continue to call $L^*$ the ultimate complex. On $L^*$ we have a representative $C_{p+1}$ of $c_p$, and also a chain $C'_{p+1}$ obtained by translation from $C_{p+1}$ and deformation-chains $\mathfrak{D}\, C_{p+1}$, $\mathfrak{D}FC_{p+1}$ such that

$$\mathfrak{D}\, C_{p+1} \rightarrow C'_{p+1} - C_{p+1} - \mathfrak{D}\, F(C_{p+1}).$$

We shall now transform $L^*$ barycentrically as follows. Let us suppose that the deformation $T$ of $A$ into $A'$ depends upon a parameter $t$ varying from 0 to 1. On the cell $P^k \times \lambda$ there will be a certain number of vertices $P^{k0}, \cdots, P^{kr}$ of $L^*$. Mark now on the path of $P^k$ as $t$ varies, the positions of $P^k$ corresponding to the values $i/r$, $(i = 0, 1, \cdots, r)$, and let them be $P^k = Q^{k0}, \cdots, Q^{kr}$. We map $P^{ki}$ into $Q^{ki}$, thus obtaining a unique image for every vertex of $L^*$, and for every simplex with certain vertices $P^{ki}$ we insert the corresponding simplex with vertices $Q^{ki}$. The various chains on $L^*$ are thus mapped into chains defining chains, of one dimension less on $\mathfrak{R}$, in the same sense as those on $K$ (*Topology*, p. 324). Furthermore by a deformation in $\mathfrak{H}$, (*Topology*, p. 332) together with Theorems IV, V of L1, all leaving $C_{p+1}$ invariant, we may reduce them to projection-chains of $K$. By the very definition of the different chains, $\mathfrak{D}\, C_{p+1}$ and $C'_{p+1}$ go into chains of $K$ which define a $\mathfrak{D}\, c_p$ and $c'_p$ satisfying (4.1), with $|\mathfrak{D}\, c_p| \subset \mathfrak{R}$, $|c'_p| \subset A'$. This proves the theorem.

## §3. Retraction properties

9. Henceforth we restrict the chains definitely to the *regular* or *singular* types, and until further notice, we also assume that the space $\mathfrak{R}$ is *compact, metric.*

We say that a set $A$ is *chain-shrinkable* onto a subset $B$ whenever every one of its chains is deformable onto $B$ over $A$. We also say that $B$ is an HR of $A$. The HR property is the analogue of the retract property, however, with the difference that in retraction of $A$ onto $B$ under a c.s.v.t. the set $B$ remains fixed point for point, while no such condition is imposed under chain-shrinking. The difference is more apparent than real as shown by:

THEOREM II. *If $c_p$ is chain-deformable onto a closed set $A$ on $\mathfrak{R}$, the chain-deformation $\vartheta$ of $c_p$ onto $A$ may be so chosen as to be merely over $\overline{\mathfrak{R} - A}$. More precisely it may be associated with an $\epsilon$-elementary decomposition, $\epsilon$ assigned, $\{c_q^i\}$ of $c_p$, such that the elements $c_q^{2i}$ alone meet $\mathfrak{R} - A$ and that the others are not deformed: $\vartheta c_q^{2i+1} = c_q^{2i+1}$ ($\vartheta = 1$ on $A$).*

The basic step is the derivation of the elementary $\epsilon$-decomposition. When the chains are singular we merely take a subdivision of mesh $\epsilon$ and call $c_q^{2i}$ its elements on $A$ and $c_q^{2i+1}$ the rest. The problem is more difficult when the chains are regular.

Let first $L$ be a fundamental complex of $A$ which is also a subcomplex of $K$ (L1, p. 474) and let there be given a chain-deformation $\vartheta$ of $c_p$ associated with an elementary $\epsilon$-decomposition $\mathfrak{K}' = \{c_q'^i\}$ and the deformation-chains $\mathfrak{D}' c_q^i$. If $C'^i_{q+1}$ is the representative of $c_q'^i$ in $K$ and $c_q^i \not\subset A$, we may write

$$C'^i_{q+1} = C^{2i}_{q+1} + C^{2i+1}_{q+1}, \tag{9.1}$$

where $C^{2i}_{q+1} = L \cdot C'^i_{q+1}$ and $C^{2i+1}_{q+1}$ is the remaining part. The two chains at the right of (9.1) determine chains $c_q^{2i}$, $c_q^{2i+1}$ such that

$$c_q'^i = c_q^{2i} + c_q^{2i+1}, \tag{9.2}$$

where $c_q^{2i+1}$ alone meets $\mathfrak{R} - A$. If $c_q^i \subset A$, we set $c_q^{2i} = c_q^i$, $c_q^{2i+1} = 0$. In any case $c_q^{2i}$, $c_q^{2i+1}$ are both on $|c_q'^i|$, hence their diameters $< \epsilon$, so that $\mathfrak{K} = \{c_q^i\}$ is likewise an $\epsilon$ elementary decomposition of $c_p$. Moreover except for the existence of a suitable $\vartheta$ it behaves as demanded by the theorem. We shall now construct $\vartheta$.

Regarding the operation $\mathfrak{D}$ to be associated with $\vartheta$ we first specify that $\mathfrak{D}\, c_q^{2i} = 0$. Next we treat $\mathfrak{D}' c_q^{2i+1}$ like $c_q'^i$ above and obtain

$$\mathfrak{D}'\, c_q^{2i+1} = d^i_{q+1} + d'^i_{q+1}, \tag{9.3}$$

where $d^i_{q+1} \subset A$, $d'^i_{q+1} \subset \overline{\mathfrak{R} - A}$, and we now set $\mathfrak{D}\, c_q^{2i+1} = d'^i_{q+1}$. We now determine the chain-decomposition $\{c_q''^i\}$ of the new transform chain $\vartheta c_q$ by the boundary relations

$$\mathfrak{D}\, c_q \rightarrow \vartheta\, c_q^i - c_q^i - \mathfrak{D}\, F(c_q^i), \tag{9.4}$$

in which for every combination $i$, $q$, the only unknown term is $c_q''^i = \vartheta c_q^i$, which is thus uniquely determined by (9.4).

Since the chains $F(c^{2i})$ are of $c^{2i}$ type, their $\mathfrak{D}$ in (9.4) are both zero, hence $c_q''^{2i} = c_q^{2i}$. Similarly we have $c_q''^{2i+1} \subset \overline{\mathfrak{R} - A}$. Finally,

$$\mathfrak{D}'\, c_q^i \rightarrow \vartheta'\, c_q^i - c_q^i - \mathfrak{D}'\, F c_q^i \tag{9.5}$$

for every value of $i$. Hence

$$(\mathfrak{D}' - \mathfrak{D})\, c_q^i \rightarrow (\vartheta' - \vartheta)\, c_q^i - (\mathfrak{D}' - \mathfrak{D})\, F c_q^i\,. \tag{9.6}$$

Since the chains $(\mathfrak{D}' - \mathfrak{D})\, c_q^i \subset A$, $\vartheta'\, c_q^i \subset A$, (9.6) shows that $\vartheta\, c_q^i \subset A$ also and hence $\vartheta\, c_p \subset A$. Therefore all the conditions of the theorem are effectively fulfilled.

REMARK. Since $|\, \mathfrak{D}\, c_q^i \,| \subset |\, \mathfrak{D}'\, c_q^i \,|$, if the initial chain-deformation $\vartheta'$ is $\epsilon$, so is the modified one $\vartheta$.

COROLLARY. *If $A$ is open instead of closed, we may choose $\{c_q^i\}$ such that $c_q^{2i} \subset A, c_q^{2i+1} \subset \mathfrak{R} - A$, with $\vartheta\, c_q^{2i} = c_q^{2i}$* ($\vartheta = 1$ on $A$).

For all that is necessary is to apply the theorem to $A$.

APPLICATION. *If $F(c_p) \subset A$ the chain-deformation may be so chosen as to leave $F(c_p)$ unchanged.*

10. The notion of chain-shrinking as here presented suffers from the disadvantage of not being *local.* The "local" properties are, however, frequently the most important, and so we shall consider them now.

Let $A$, $B$ be subsets of $\mathfrak{R}$. We say that $A$ may be *chain-shrunk away from $B$* whenever there is an open set $U \supset B$ such that $A$ may be chain-shrunk onto $A - U$. We have at once

THEOREM III. *If $A$ may be chain-shrunk away from every point of a compact set $B$, it may be shrunk away from $B$.*

By the Borel covering theorem, $B$ has a f.c.o.s. $\{U^i\}$, such that $A$ is shrinkable onto every $A - U^i$. We can find another f.c.o.s. $\{V^i\}$ of $B$ such that $\bar{V}^i \subset U^i$.[11] We shall use both coverings simultaneously.

By assumption $A$ may be chain-shrunk onto $B - U^1$ and as a consequence every chain $c_p$ of $A$ will have been displaced to the exterior of a certain open set $W^1 \supset \bar{V}^1 \cdot B = B^1$.

Suppose that we have succeeded in showing that the displacement may be carried out similarly to outside a certain open set

$$W^{k-1} \supset B^{k-1} = (\bar{V}^1 + \cdots + \bar{V}^{k-1}) \cdot B\,.$$

If $W^{k-1} \supset B$ we are through. In the contrary case, there is at least one more $V$, which we may assume to be $V^k$ such that $\bar{V}^k \cdot B \not\subset W^{k-1}$. In any case however the chains of $A$ not on $W^{k-1}$ will be at a positive distance $\delta$ from $B^{k-1}$. By assumption we can displace all the chains of $A$ onto $A - U^k$. However by Theorem II, this displacement may be replaced by one onto the exterior of the spherical neighborhood $\mathfrak{S}(U^k \cdot B, 1/2\, \delta)$. The chains thus displaced will be at a positive distance from $B^k$, hence outside of some $W^k \supset B^k$. Proceeding thus we shall ultimately have a $W^n \supset B$, with all chains deformable onto $A - W^n$. The theorem is therefore proved.

## §4. HLC spaces and their relations to other spaces

11. The definition of the HLC sets of various types is entirely similar to that of LC sets. Here also, however, we need the *local* characterization.

[11] Menger, *Dimensionstheorie*, p. 160.

We shall say then that $\mathfrak{R}$ is:

$p$-HLC at the point $x$ whenever every open set $U \supset x$ contains another $V \supset x$ also, such that every $p$-cycle, $p > 0$, of $V$ is $\sim 0$ on $U$, and every zero-cycle $\gamma_0$ of $V$ is $\sim$ on $U$ to a point of $V$ taken $(\gamma_0)$ times;

HLC$^p$ at $x$ whenever it is $q$-HLC for every $q \leqq p$;

HLC$^\infty$ at $x$ whenever it is $q$-HLC for every $q$ whatsoever;

weak HLC at $x$ whenever $V$ can be determined as above independently of $q$.

The strong HLC (merely HLC) will be defined for a compact metric $\mathfrak{R}$ in No. 15.

Finally $\mathfrak{R}$ itself is $p$-HLC, $\cdots$ whenever it has the corresponding property at all its points.

The $\epsilon$, $\eta$ formulations in the compact metric case are as usual: in place of "for every $U$ there is a $V$" we must have "for every $\epsilon > 0$ there is an $\eta > 0$". In particular the HLC$^p$ condition may be reduced to "for every $\epsilon$ there is an $\eta(\epsilon, p)$ such that every $q$-cycle $q < p$ (a $\gamma_0$ whose $(\gamma_0) = 0$) of diameter $< \eta$ bounds a chain of diameter $< \epsilon$", while the weak HLC condition will be of the same form with $\eta$ independent of $p$.

12. If $c_p$ and $c_{p-r}^j$ are two chains of a finite quasi-complex $\mathfrak{K}$, we shall say that they are *incident* whenever $c_{p-r}^j$ is a boundary chain of $c_p^i$. The aggregate of the specifications of the incidences of $\mathfrak{K}$ is called the *pattern* of $\mathfrak{K}$.

Suppose that we have given a $\mathfrak{K}$ whose pattern we are to reproduce on $\mathfrak{R}$, and of all the elements expected let there be present all the zero-chains and some, but not necessarily all, of the rest, so that the chains present form a quasi-sub-complex $\mathfrak{K}'$ of $\mathfrak{K}$. We call $\mathfrak{K}'$ a *partial realization* of $\mathfrak{K}$. Let us suppose that out of an expected $c_p^i$ and its boundary chains, there are already present the chains $c_r'^{\,j}$ in $\mathfrak{K}'$. We call max diam $\Sigma \mid c_r'^{\,j} \mid$ the *mesh* of the partial realization $\mathfrak{K}'$ of $\mathfrak{K}$.

If $c_q^i$ is an expected chain of $\mathfrak{K}$ and $F(c_q^i)$ is already present in $\mathfrak{K}'$ a necessary condition is that it be a cycle. Furthermore if $q = 1$ and if its imposed boundary relations are

$$c_1^i \rightarrow \eta_{ij}\, c_0^j \;, \tag{12.1}$$

where, by assumption, the chains $c_0^j$ are already present, from (12.1) follows

$$\eta_{ij}\,(c_0^j) = 0\,, \tag{12.2}$$

and these relations must be satisfied if the data are consistent. They shall naturally be assumed if $\mathfrak{K}'$ is given, or must be verified whenever a $\mathfrak{K}'$ is constructed.

13. THEOREM IV. *N.a.s.c. for a compact metric space $\mathfrak{R}$ to be* HLC$^p$ *are that for every $\epsilon > 0$ there exist an $\eta(\epsilon, p) > 0$ such that every partial realization $\mathfrak{K}'$ on $\mathfrak{R}$ of a $\mathfrak{K}_{p+1}$ whose mesh $< \eta(\epsilon, p)$ may be completed to make up the expected $\mathfrak{K}_{p+1}$ of mesh $\epsilon$.*[12]

[12] Compare with the analogous proposition for LC$^p$-sets, p. 612.

Since $F(c_q)$, $q \leqq p$, is a special $\mathfrak{K}'$ corresponding to a $\mathfrak{K}$ which consists of $c_q$ and $F(c_q)$, the condition of the theorem is clearly sufficient for an $\mathrm{HLC}^p$. We must therefore merely show its necessity, and as it is trivial for $p = -1$, we may take $p > 0$ and use induction on $p$.

Under our assumptions the theorem holds for $p - 1$, and there is a corresponding $\eta(\epsilon, p - 1)$. Moreover since $\mathfrak{R}$ is $p$-HLC we have also the constant $\xi(\epsilon, p)$ of the $p$-HLC property ($\eta(\epsilon, p)$ in the definition).

Let the mesh $\alpha$ of $\mathfrak{K}'$ be $< \eta(\beta, p - 1) < \beta, \beta > 0$ assigned. Under the hypothesis of the induction we may insert all the missing chains of dimension $\leqq p$ and have their diameters $< \beta$. We thus obtain a new partial realization $\mathfrak{K}''$ of $\mathfrak{K}_{p+1}$ in which only the $(p + 1)$-chains may be missing. But if $c_{p+1}^h$ is any expected chain, the sum of the sets of its boundary chains is of diameter $< \alpha + 2\beta < 3\beta$. Therefore, if we choose $\beta < \frac{1}{3}\, \xi(\epsilon, p)$, we may insert the missing chains and choose them of diameter $< \epsilon$. Hence the condition of the theorem is fulfilled whenever the mesh $\alpha < \eta\, (\frac{1}{3}\, \xi(\epsilon, p), p - 1) = \eta(\epsilon, p)$.

14. Theorem V. *Given a compact metric* $\mathrm{HLC}^p$ *space* $\mathfrak{R}$ *there exists for every* $\epsilon > 0$ *a quasi-complex* $\Psi_p$ *into a subchain of which every* $c_q$, $q \leqq p$, *of* $\mathfrak{R}$, *is* $\epsilon$*-deformable over* $\mathfrak{R}$.

Let $\{U^i\}$ be a f.c.o.s. of the $\mathrm{HLC}^p$ space $\mathfrak{R}$ of Theorem IV, and let $\alpha$ be the mesh and $\Phi$ the skeleton of the covering. We shall assume that $\{\bar{U}^i\}$ has the same skeleton $\Phi$, a choice always possible whatever $\alpha$ (see L1, p. 474). This means that any group of $U$'s intersect when and only when their closures do. As to the latter, we know that there exists a constant $\gamma < \alpha$ such that whenever any set of diameter $< \gamma$ meets a group of $U$'s, these $U$'s intersect. From the preceding we conclude then that $\gamma$ has also the same property as regards the $U$'s themselves.

Let us mark on $U^i$ a point $A^i$ which we consider as an oriented zero-cell. If $\Phi_p$ is the sub-complex obtained after removing all simplexes of dimension $> p$ from $\Phi$, we may consider the set $\{A^i\}$ as a partial realization $\Phi_p'$ of a quasi-complex $\Psi_p$ with the same incidence pattern as $\Phi_p$. It may likewise be considered as a partial realization $\Psi'$ of a quasi-complex $\Psi$ with the same pattern as $\Phi$. From the construction of $\Phi$ we conclude that the meshes of the two partial realizations do not exceed the maximum diameter $2\alpha$ of the sum of any group of intersecting $U$'s. It follows in particular that if $\alpha < \frac{1}{2}\eta(\beta, p)$ we may complete $\Psi_p'$ to a $\Psi_p$ of mesh $< \beta$.

It is now a simple matter to show that if $\delta =$ mesh $\mathfrak{K}_p < \gamma$ then $\mathfrak{K}_p$ may be $\epsilon$-chain deformed into a quasi-complex whose elementary chains are subchains of $\Psi_p$. In fact the chain-deformation $\vartheta$ may be so chosen that: (a) every elementary $c_0^i$ of $\mathfrak{K}_p$ goes into $(c_0^i)$ times a vertex of $\mathfrak{K}_p$; (b) if $c_q^j$ is an elementary chain of $\mathfrak{K}$ with zero-chains in its boundary then $\vartheta\, c_q^j$ is the image in the transformation $\Phi_p \rightarrow \Psi_p$ of a chain of $\Phi_p$ which is on the $F(\sigma)$ of the simplex $\sigma$ whose vertices have for images the points $A^i$ corresponding to the boundary zero-chains of $c_q^j$; (c) if $c_q^j$ is not of type (b) then $\vartheta\, c_q^j = 0$.

We satisfy (a) by choosing for $\vartheta\, c_0^i$ a chain $(c_0^i)A^h$ where $A^h$ is such that $U^h$ meets $c_0^i$. By the same token (b) holds for $q = 0$, and (c) does not occur for that value. From this moment on the proof proceeds essentially like that of the deformation theorem of *Topology,* p. 92. The only point that may raise a question is the following. Granting that we have case (b) and that $\vartheta\, F(c_q^j)$, $q < p$, has already been described we must describe $\vartheta\, c_q^j$. Owing to $\delta < \gamma$ the vertices $A^i$ to which $\vartheta\, F(c_q^j)$ belongs are on a set of intersecting $U$'s, and hence they are the images of those of a $\sigma$ of $\Phi$. It follows that $\vartheta\, F(c_q^j)$ is the image of a cycle $\Gamma_{q-1}$ of $F(\sigma)$. Now $\sigma + F(\sigma)$ has a subchain $C_q \to \Gamma_{q-1}$, and the image of $C_q$ (under the chain-transformation $\Phi_p \to \Psi_p$) is a chain $c_q^{\prime\, i} \to F(c_q^i)$. We set $\vartheta\, c_q^i = c_q^{\prime\, i}$ and extend $\vartheta$ to all $q$-chains by the linearity condition. The rest of the proof is as loc. cit.

15. Paraphrasing the treatment of L2, No. 5, we now define a set as *strong* HLC or merely HLC when it is $\mathrm{HLC}^p$ for every $p$, and when in addition the function $\eta(\epsilon, p)$ of Theorem IV has a lower bound $\eta(\epsilon) > 0$, independent of $p$.

The results of our paper regarding the comparison with retracts may also be extended here, provided that absolute retraction is defined relatively to imbedding not in an arbitrary set but in an arbitrary HLC set. We shall not stop to develop this point further. From Theorem V follows readily:

THEOREM VI. *Every chain of an* HLC *is deformable into a subchain of a definite chain-complex.*

For in the present instance we may complete $\Psi_p$ up to the dimension $n$ of $\Phi$, and have a complete chain-image $\Psi$ of $\Phi$. We then apply the proof of Theorem V with the supplementary condition that $\vartheta c_q^i = 0$ for $q > n$ which is consistent since $\Gamma_{q-1} = \Gamma_n$, being now a chain of a $\sigma_n$, will be $= 0$. It follows that every $\mathfrak{K}$ is deformable into a chain-complex whose elements of dimension $p \leqq n$ are *subchains* of $\Phi$, and the others are zero.

THEOREM VII. *If* $\mathfrak{R}$ *is* $\mathrm{HLC}^n$ *and* $n = \dim \mathfrak{R}$ *is finite the space is* HLC.

For $\mathfrak{R}$ possesses no chains of dimension $> n$, hence in the construction connected with Theorem IV, one need never go beyond chains of dimension $n$ and we may take for $\eta(\epsilon)$ the least of the $n + 1$ positive numbers

$$\eta(\epsilon, p),\, p = 0\,,\, 1,\, \cdots\,,\, n\,.$$

16. THEOREM VIII. *The homology groups of an* $\mathrm{HLC}^p$*, absolute or mod a closed* $\mathrm{HLC}^p$ *subset, for dimensions* $\leqq p$*, or of an* HLC*, absolute or mod a closed* HLC *subset, all have the same structure as for a finite complex.*

By this we mean that they have finite bases with a finite number of relations between the elements of the bases, and furthermore for an HLC set that they are zero for dimensions above a certain integer $n$. In particular when the groups are free groups the Betti-numbers are all finite.

The proof is very simple. Take first the absolute groups and regular chains. The chains of $\vartheta\mathfrak{K}$ are the images in one of the correspondences $\Phi_p \to \Psi_p$, $\Phi \to \Psi$

of certain chains of $\Phi_p$ or $\Phi$ and the correspondence preserves boundary relations. Hence the groups $\{\gamma_q\}$ of $\mathfrak{R}$ ($q \leqq p$ for an HLC$^p$) are isomorphic with certain subgroups of the same for $\Phi_p$ or $\Phi$ respectively. They are therefore additive groups generated by a finite number of linear forms with coefficients in the abelian group $\mathfrak{M}$, corresponding say to chains $\gamma_q^i$, $i = 1, 2, \cdots, r$. The homologies between the $\gamma$'s are those on $\mathfrak{R}$, all of the form

$$x_i \gamma_q^i \sim 0, \qquad x_i \subset \mathfrak{M}. \tag{16.1}$$

Since $\mathfrak{M}$ is a field this system may be reduced by a change of base to the form

$$\gamma_q^{s+i} \sim 0, \qquad i = 1, 2, \cdots, r - s \tag{16.2}$$

and the first $s$ elements of the new base form a minimum base for the $q^{\text{th}}$ homology group.

If we deal with singular chains, we may reason substantially as in the proof of the invariance of the combinatorial characters of a complex (*Topology*, p. 88) and show that (16.1) implies that there exists a chain of $\Psi_p$, or $\Psi$ as the case may be,

$$c_{q+1} \to x_i \gamma_q^i . \tag{16.3}$$

Since the $\gamma$'s may after all include all the $q$-subcycles of $\Psi_p$ or $\Psi$, this shows that the $q^{\text{th}}$ homology group is the same as for $\Psi_p$ or for $\Psi$ and hence the same as for $\Phi_p$ or for $\Phi$, as the case may be.

17. Consider now a closed HLC$^p$ subset $G$ of an HLC$^p$ space $\mathfrak{R}$ and let $\{U^i\}$ be the f.c.o.s. of No. 14. It may happen that certain sets $U$ which meet $G$ have an intersection which does not meet $G$. In that case their closure $H$ also fails to meet $G$. Since the number of sets $H$ is finite and they are closed, the distance $\delta$ of their sum from $G$ is $> 0$. Let now, for every $i$, $V^{2i}$ be the set of the points of $U^i$ nearer than $2/3\ \delta$ from $G$, and $V^{2i+1}$ the set of those farther than $1/3\ \delta$ from $G$. One of the two sets may well be vacuous. In any case however the non-vacuous sets $V$ make up a new f.c.o.s. $\{V^i\}$ of $\mathfrak{R}$, which like the initial covering has the same incidence pattern as the covering $\bar{V}^i$ of the closures. Moreover, now if any aggregate of $V$'s intersect $G$, they intersect on $G$ itself. Let us assume then that the initial f.c.o.s. $\{U^i\}$ already possesses this property.

We now modify the construction of $\Psi_p$ in No. 14, by choosing the point $A^i$ on $G$ whenever $U^i$ meets $G$. Then, when the meshes are suitably small, we utilize the HLC$^p$ property of $G$ to place on that set every chain of $\Psi_p$ whose vertices $A^i$ are on $G$. This is always possible since, after all, the f.c.o.s. $\{U^i \cdot G\}$ behaves relatively to $G$ like $\{U^i\}$ relatively to $\mathfrak{R}$ itself. As a consequence, $\vartheta\,\mathfrak{K}$ will transform a chain $c_q^i$ of $\mathfrak{K}$ on $G$ into chains of $\Psi_p$ on $G$. Then, for suitably small meshes throughout, we shall be able to assign deformation-chains $\mathfrak{D}c_q^i$ likewise on $G$. Under the circumstances the chains on $G$ will be merely deformed over $G$ and hence the cycles mod $G$ will be deformed into cycles of $\Psi_p$ mod $G$. From this point on the rest of the proof is as for absolute cycles.

If $\mathfrak{R}$ and $G$ are both HLC the treatment is the same except that $\Psi$ is to take the place of $\Psi_p$ throughout and the conclusion is again the same.

COROLLARIES. I. *If $\mathfrak{R}$ is* HLC$^p$ *and $G$ is* HLC$^q$, *$q < p$, the theorem regarding the homology groups of $\mathfrak{R}$ mod $G$ holds for dimensions $\leqq q$.*

II. *If $n = \dim \mathfrak{R}$ is finite and both $\mathfrak{R}$ and $G$ are* HLC$^n$, *the Betti-numbers absolute, or of $\mathfrak{R}$ mod $G$, if any occur, are all finite.*

APPLICATION. *If $A$ is an* HLC$^p$*-set, $p \leqq n$, in an $n$-sphere $H_n$ the Betti-numbers of $H_n - A$ are all finite; in particular $R_0$ is finite, and hence $H_n - A$ consists of a finite number of regions.*

For dim $A \leqq n$ and the rest follows from the preceding results together with the extension of Alexander's duality relation (*Topology*, p. 339).

18. **HNR-sets.** The closed set $A$ shall be called an HNR of $\mathfrak{R}$ whenever for every positive $\epsilon$ there is a positive $\eta$ such that every chain $c_p$ within a spherical neighborhood $\mathfrak{S}(A, \eta)$ of $A$ is $\epsilon$-deformable onto $A$. These are obvious, but not complete, analogies of the NR property. The most important difference is that we have to use two neighborhoods of $A$, whereas in the NR there is only one.

THEOREM IX. *Let $\mathfrak{R}$ be compact, metric* HLC. *Then any closed subset $A$ of $\mathfrak{R}$ which is also* HLC *is an* HNR *of the space.*

Let $\eta_1(\epsilon)$, $\eta_2(\epsilon)$ be the constants of the HLC definition relative to $\epsilon$ and respectively to $\mathfrak{R}$ and $A$. Let $c_p \subset \mathfrak{S}(A, \eta)$ and let us impose upon it an elementary decomposition of mesh $\xi$ making up a quasi-complex $\mathfrak{K}$. For each $c_0$ of the subdivision we take a point $B$ on $A$ among those nearest to $c_0$. These points constitute a partial realization of mesh $< 2\eta + \xi$ of a $\mathfrak{K}'$ having the same structure as $\mathfrak{K}$. By the HLC property of $A$, if we have a given $\alpha > 0$ and if $2\eta + \xi < \eta_2(\alpha)$ we may complete the partial realization to form $\mathfrak{K}'$ of mesh $< \alpha$. Then $\mathfrak{K}' + \mathfrak{K}$ form a partial realization of a certain $\mathfrak{D}\mathfrak{K}$ of mesh $< \eta + \xi + \alpha$, and if this quantity $< \xi_1(\epsilon)$ we may complete to form $\mathfrak{D}\mathfrak{K}$ of mesh $\epsilon$. As a consequence $\mathfrak{K}$, and hence $c_p$, will then have been $\epsilon$-deformed onto $A$. The two required conditions may be fulfilled by choosing

$$\alpha < \tfrac{1}{3}\eta_1(\epsilon)\,; \qquad \xi, \eta < \tfrac{1}{3}\eta_2\left(\tfrac{1}{3}\eta_1(\epsilon)\right).$$

Therefore the theorem holds, with the $\eta(\epsilon)$ of the HNR definition here chosen as $\frac{1}{3}\eta_2(\frac{1}{3}\eta_1(\epsilon))$.

REMARK. Here also as in Theorem II, the construction may be so modified that the elements of $c_p$, or rather of its elementary decomposition, already $\subset A$ are unmodified. For example, if $F(c_p) \subset A$ then it remains unmodified and $c_p' \sim c_p$.

19. **Relations between the classes LC$^p$, LC and HLC$^p$, HLC.**

THEOREM X. *In the system of singular chains an* LC$^p$ *is an* HLC$^p$ *and an* LC *is an* HLC.

The proof is by means of Theorem I of p. 612 which is the analogue for LC-sets of our present Theorem IV. Assuming then $\mathfrak{R}$ to be LC$^p$ let $\xi(\epsilon, p)$

be the constant called $\eta(\epsilon, p)$ on page 613 (proof of Theorem I), the change of notation being in order to avoid confusing that constant with the $\eta(\epsilon, p)$ of Theorem IV. Let $\mathfrak{K}_p'$ be a partial realization of mesh $< \xi(\frac{1}{3}\epsilon, p)$ of a certain $\mathfrak{K}_p$. We may consider $\mathfrak{K}_p$ as the potential singular image of a certain simplicial complex $K_p$ and the chains of $\mathfrak{K}_p'$ correspond to certain chains of $K$ whose simplexes make up a subcomplex $K'$ of $K_p$. We may also consider $\mathfrak{K}_p'$ as a singular image of certain chains and cycles of $K'$. By subdividing the chains of $\mathfrak{K}_p'$ and inserting suitable vertices on the chains already present we turn it into a singular image of the oriented simplexes which make up the chains in question in $K$. Now let $c_1^i$ be any expected chain of $\mathfrak{K}_p$ and let $C_0^i$ be the chain of $K'$ whose image, already present, is $F(c_1^i)$. We build up a new $K_p$ out of $K'$ by taking the join of some point $P^i$ with $C_0^i$ and adding this new simplicial chain to $K'$. We then proceed similarly by adding new two-chains for any $c_2^i$ whose boundary is now already represented in the complex already obtained, and so on until we have a new $K_p$ which may manifestly replace the former.

Now let $c_q^i$ be an expected chain of $\mathfrak{K}_p$ whose boundary is represented by certain chains in $\mathfrak{K}'$ and which has necessitated the addition of a certain number of vertices $P^j$ in building up the new $K_p$. We represent these vertices by certain points on $F(c_q^i)$, and if there are chains wholly unrepresented in the process we represent them by zero. We have now in the modified $\mathfrak{K}'$ a semi-singular image of $K_p$ in the sense of L2, and hence we may complete it to form a singular image on $\mathfrak{R}$ whose mesh $< \epsilon/3$, and which contains a collection of chains making up a chain-image of $\mathfrak{K}_p$. The chains are now singular with cells of diameters $< \xi(\epsilon/3, p) < \epsilon/3$, or else directly of diameters $< \epsilon/3$, and if $c_q^i$ is new, its singular cells meet the boundary chains of $c_q^i$ already present which are on a set of diameter $< \xi(\epsilon/3, p)$. Therefore mesh $\mathfrak{K} < \epsilon$ and by Theorem IV, $\mathfrak{R}$ is HLC$^p$, with $\eta(\epsilon, p) = \xi(\epsilon/3, p)$.

If $\mathfrak{R}$ is LC the constant $\xi$ has a positive lower bound $\xi(\epsilon)$ independent of $p$, hence $\eta(\epsilon, p) \geqq \xi(\epsilon)$ and $\mathfrak{R}$ is HLC. Our theorem is therefore proved.

20. **Locally contractible spaces.** According to Borsuk to whom this notion is due,[13] a topological space $\mathfrak{R}$ is called locally contractible whenever for every

[13] See K. Borsuk, *Über eine Klasse von lokal zusammenhängenden Räumen*, Fundamenta Mathematicae, vol. 19 (1932), pp. 220–242, notably p. 236.

Borsuk has shown (loc. cit., p. 240) that if dim $\mathfrak{R} = n$ is finite, we have in an obvious sense ANR $\sim$ LC$^\infty$. He then states as open (footnote 39) the question of the equivalence of the two properties when dim $\mathfrak{R}$ is infinite. Now we have shown (page 613) that ANR $\sim$ LC and, as we shall show by an example, LC $\nsim$ LC$^\infty$, hence Borsuk's question must be answered negatively. Thus our criterion for ANR is definitely the only one which holds independently of the dimension. This emphasizes once more the importance of the *semi-singular* complex introduced in L2, and of its analogue, the *partial realization* of chain-complexes of the present paper.

The example alluded to above was communicated to us by Borsuk and is as follows. Consider in the Hilbert parallelotope a sequence of points $\{x_p\}$ on a segment $Ox_1$ from the origin $O$, $\rightarrow O$ monotonely, and let $H_p$ be a euclidean $p$-sphere of center $x_p$ and radius

open set $U$ containing any point $x$ there is another, $V \supset U$, which is homotopic to a point on $U$. Let $\mathfrak{R}$ be compact, metric, locally contractible and consider the class of all spheroids $\{\mathfrak{S}(x, \epsilon)\}$ of $\mathfrak{R}$ with fixed radius $\epsilon$. For every point $x$ there is a spheroid $\mathfrak{S}(x, \xi)$ which is $\epsilon$-homotopic to a point. As the space $\mathfrak{R}$ is compact it has a f.c.o.s. $\{\mathfrak{S}(x_i, \xi)\}$ and there is a constant $\eta$ such that every subset of $\mathfrak{R}$ whose diameter $< \eta$ is on one of the spheroids $\mathfrak{S}(x_i, \xi)$ and hence $\epsilon$-homotopic to a point on $\mathfrak{R}$.

It follows immediately from what precedes that given $\epsilon$ every singular $p$-sphere of $\mathfrak{R}$ whose diameter $< \eta$ bounds a singular $(p + 1)$-cell of diameter $< \epsilon$, where $\eta(\epsilon)$ is independent of $p$. We might call this property weak LC.[14] Owing to Theorem X this implies that $\mathfrak{R}$ is a weak HLC in the system of singular chains. That it is a weak HLC in any system is readily shown as follows. Any cycle $\gamma_p$ such that diam $|\gamma_p| < \eta$ is $\epsilon$-chain deformable, by virtue of Theorem I, into a point, which is a $\gamma_p \sim 0$, when $p > 0$, and which is taken $(\gamma_0)$ times when $p = 0$. Therefore (No. 11) $\mathfrak{R}$ is $p$-HLC for every $p$, and since we have been able to choose $\eta(\epsilon)$ independently of $p$, we have

THEOREM XI. *A locally contractible compact metric space is both a weak* LC *and a weak* HLC.

While local contractibility is not sufficient to insure the LC or HLC properties, this will be the case if we strengthen it in the following manner. Let us call the open set $U$ *self-contractible* whenever it is homotopic to a point on itself, i.e., whenever it is its own set $V$ in the definition of local contractibility. A class of open sets $\{U\}$ will be called *self-contractible* whenever it consists of self-contractible sets and is closed with respect to intersection (if $U'$, $U''$ are in the class so is $U' \cdot U''$). *A compact metric space $\mathfrak{R}$ which possesses for every $\epsilon$ a self-contractible class whose sets are all of diameters $< \epsilon$, is both strong* LC *and strong* HLC. The proof is essentially parallel to the analogue on pp. 618, 619 (No. 17), and need not be repeated here.

21. **Application to locally polyhedral spaces.** We shall say that $\mathfrak{R}$ is locally polyhedral whenever every point $x$ has a neighborhood whose closure is a finite complex $K^x$. We may replace $K^x$ by a subdivision with $x$ as a vertex and then the neighborhood by the star of $x$ in $K^x$. Therefore $\mathfrak{R}$ may be defined as a space of which every point has for neighborhood the star of a vertex in a finite complex. Since every star is homotopic on itself to its own vertex, the space $\mathfrak{R}$ is locally contractible and it is also clearly locally compact.

Suppose now $\mathfrak{R}$ to be also compact and metric. Since it may then be covered

$< \frac{1}{2} d(x_p, x_{p-1} + x_{p+1})$. $\mathfrak{R}$ is the sum of the spheres $+ O +$ the segment with the open diameters of the spheres on the segment removed. It is readily shown that the $p$-LC condition holds for every $\epsilon$ with the corresponding $\eta(\epsilon, p) <$ diam $H_p$. Hence $\eta(\epsilon, p) \to 0$ with increasing $p$ and $\mathfrak{R}$ is $\text{LC}^\infty$ but not LC.

[14] Is the converse true, i.e., is every weak LC locally contractible? Is weak LC $\sim$ LC? For the present these questions must remain unanswered.

with a finite number of the stars, its dimension is finite. Combining this with No. 17 Corollary II and Theorem XI, we have then

THEOREM XII. *A locally polyhedral compact metric space is both* LC and HLC.

COROLLARY. *The homology groups of the space of Theorem* XI, *absolute or mod a closed subset of same type, have the structure of those of a finite complex. In particular its Betti-numbers, if any occur, are also finite.*

IMPORTANT SPECIAL CASE: The space and its subset are *absolute topological manifolds.*

Of course in both theorem and corollary the space is assumed compact.

22. **Extension to locally compact separable spaces.** Substantially all the results of the present section may also be extended to these spaces. We recall that any locally compact separable space is also metric, although the reverse need not hold.[15] As far as our definitions are concerned, the $\mathrm{HLC}^p$ or HLC conditions formulated in No. 11 in terms of $\epsilon$, $\eta$ must now refer to the cycles meeting a specific self-compact subset $A$ of $\mathfrak{R}$, and must hold for every such $A$. However $\eta(A, \epsilon, p)$ or $\eta(A, \epsilon)$ may well vary with $A$. The $U$, $V$ conditions applied to every point of $A$ may be replaced in fact by $\epsilon$, $\eta$ conditions on the chains meeting $A$, but not on all chains. Keeping this in mind we verify at once that Theorem IV holds when the elementary chains of $\mathfrak{K}'$ all meet $A$. In Theorem V the chains considered must now all be chains of $A$ and the deformation is on any preassigned open set $U \supset A$. The subsequent results on the homology groups hold regarding the homologies between chains of $A$ on $U$, that is to say, when bounding is allowed not merely on $A$ but also on $U$. Theorem X may be derived, provided that we extend, as we may readily do, Theorem I of L2 to locally compact separable spaces. It is always understood that the semi-singular complex loc. cit. must have all its (realized) elements meeting $A$. Similarly Theorem XI holds also, it being always understood that the constants $\epsilon$, $\eta$ are related to $A$ as described above. Since Theorem XII is a mere corollary of Theorem XI, it is likewise applicable to locally compact separable spaces.

23. **Concluding remarks.** The theory developed in the preceding pages was not intended to be exhaustive, but only to cover enough ground for some fairly immediate applications. Thus by means of quasi-complexes mod $A$, we could introduce chain-deformations mod $A$ and extend in an obvious manner the results of the present paper. As already observed, another extension would be to non-compact spaces, say to the very general spaces whose homology theory has been developed in recent years by Čech. These extensions however would be strictly mechanical and may safely be left to any experienced reader.

[15] See Alexandroff-Urysohn, *Mémoire sur les espaces topologiques compacts,* Verhandelingen der Koninklijke Akademie van Wetenschappen te Amsterdam, afdeeling natuurkunde, deel XIV, No. 1 (1929), p. 83.

# XIII - XVIII

# XIII

## CLOSED POINT SETS ON A MANIFOLD.*

### I. Introduction.

1. The object of this paper is to investigate certain topological simultaneous invariants of a manifold $M_n$ and a closed subset $G$ on $M_n$. These invariants are integers analogous to the Betti numbers and include them as special cases. They satisfy certain relations (formulas (4), (5), (6), of No. 8) which include practically all duality relations now known in Combinatorial Analysis Situs.

When $G$ is a complex $C_k$ on $M_n$ the invariants can be readily defined in terms of the complexes on $C_k$. For an arbitrary $G$ however they may be defined in several modes depending upon the meaning attributed to the terms "cycle on $G$, complex on $G$ or with boundary on $G$". The desideratum to be fulfilled is to preserve the properties obtained in the case of a $C_k$, and it is fulfilled by the sequences of "fractionary" cycles and complexes by means of which we define the cycles and complexes in question. Similar sequences of ordinary cycles were recently investigated by Vietoris† but appear to be inadequate for the purpose. However, in his paper (footnote 18) he refers to an invariant due to Brouwer which as we have shown (No. 27) coincides with our number $R_\mu(G)$ (generalized Betti number or connection index). Alexandroff‡ has also defined recently the Betti numbers of a closed set on an $S_n$ as limits of those of suitable approximating complexes. It is not known as yet however if this mode of approach leads to invariants satisfying our relations except in a very special case,§ nor would it seem easy to prove. As a matter of fact the problem of the equivalence of the various modes of defining the invariants offers considerable interest.

Our general method consists in treating first $C_k$ by isolating it within a neighborhood whose residual is a manifold with boundary. From a few

* Received December 1, 1927. See the two related preliminary Notes in Proc. Nat. Acad., vol. 13 (1927), pp. 614–622 and 805–807. We shall also mention repeatedly our two recent papers Trans. Amer. Math. Soc., vol. 28 (1926), pp. 1–49; vol. 29 (1927), pp. 429–462; and shall refer them as Trans. I, Trans. II. The parts actually vitally needed for the present paper are Trans. I, Part I, §§ 1, 2, 3, and Trans. II, Section II (mostly the duality theorems). See pp. 199–247 and 248–281 of the present volume.

† Math. Ann., vol. 97 (1927), pp. 454–472.

‡ Comptes Rendus, vol. 184 (1927), pp. 317–319.

§ Alexandroff. Comptes Rendus, vol. 184 (1927), pp. 425–427.

of the simpler results obtained for the latter in Trans. II, we then derive the desired properties of $C_k$ by means of a certain reduction theorem (Nos. 12, 13). In a somewhat similar fashion we pass from $G$ to the general $G$.

One of our most suggestive results consists in two very general theorems on Kronecker-indices (intersections counted with multiplicities that may be positive or negative) for the cycles $M_n - G$ and the complexes with boundary on $G$. They are the direct extension of a noted theorem stated by Poincaré* and first proved independently by Veblen and Weyl,† and both reduce to it when $G$ is a vacuous set. Once this proposition is established the duality formulas (4), (5) can be derived by essentially the same scheme, as indicated by Poincaré for the simplest case, where $M_n$ has no boundary and $G$ is vacuous.

2. **Notations and terminology.** The subscript in the symbol of a configuration shall always denote its dimensionality. It will frequently be omitted whenever there is no danger of confusion. We assume the reader familiar with what is meant by the terms *cell, complex, manifold, cycle* (Trans. I, p. 3; II, p. 437). The generic letters for these entities will be $E$, $C$, $M$, $\Gamma$, with variations when special types are involved. It is important to bear in mind that cells and complexes may be singular unless the contrary is explicitly stated. We recall that a singular $E_k$ or $C_k$ on $A$ are subsets of $A$ which are the continuous one way image (not necessarily one-one way image) of some ordinary $E_k$ or $C_k$.

Concerning boundary relations we shall adopt instead of Poincaré's $\equiv$ the more convenient $\rightarrow$ recently introduced by Alexander.‡ Thus "the boundary of $A$ is $B$" shall be expressed by $A \rightarrow B$ or $B \leftarrow A$. Let $B$ be on $D$ and let $\lambda B$, $\lambda \neq 0$, be the boundary of some $A$ on $D$. We shall express this relationship by $B \approx 0$, mod $D$. The symbol $\approx$ is the homology with division allowed or with zero divisors dropped, of our earlier writings.

Another convenient symbol is $\subset$ and its reverse to denote "is contained in" or "contains". Thus $B \subset A$, $A \supset B$ mean "$B$ is a subset of $A$".

3. **Fractionary complexes and cycles.** We shall now introduce certain generalisations of great help in the sequel. We first extend the term *$\mu$-complex* to cover a symbol $\dfrac{C_\mu}{q}$, where $C_\mu$ is an ordinary complex and $q$ an arbitrary non zero integer. Let us call the usual type: *integral*

---

* Journal de l'Ecole Polytechnique, (2), vol. 1 (1895).

† Veblen, Trans. Amer. Math. Soc., vol. 25 (1923), pp. 540–550 (read before the Amer. Math. Soc. in 1920). Weyl. Revista Matem. (1923).

‡ Trans. Amer. Math. Soc., vol. 28 (1926), pp. 301–329.

complex, the new type *fractionary*—a distinction which we shall rarely need. By definition:

$$\frac{pC_\mu}{pq} = \frac{C_\mu}{q}; \qquad p \cdot \frac{C_\mu}{q} = \frac{pC_\mu}{q}; \qquad \frac{C_\mu}{q} \pm \frac{C'_\mu}{q'} = \frac{q'C_\mu \pm qC'_\mu}{qq'}.$$

A fractionary complex $\frac{C_\mu}{p}$ is called a *cycle,* whenever its numerator, the integral complex $C_\mu$ is a cycle. By "$\frac{C_\mu}{p}$ is on a configuration $A$ or on $\frac{C_\nu}{q}$" we understand that the cells of $C_\mu$ are on $A$ or on those of $C_\nu$, it being unterstood that $C_\mu$ and $C_\nu$ are integral complexes. Henceforth all complexes and cycles are assumed fractionary unless otherwise stated, and we shall continue to apply the notations $C_\mu$, $\Gamma_\mu$, to them.

Regarding boundary relations and homologies we shall henceforth understand by $C_\mu \rightarrow C_{\mu-1}$ that there exists a $p \neq 0$ such that $pC_\mu \rightarrow pC_{\mu-1}$ in the ordinary sense: There is an integral complex $pC_\mu$ whose boundary is $pC_{\mu-1}$. When $C_\mu$ and $C_{\mu-1}$ are on the configuration $A$, we write corresponding to the boundary relation, $C_{\mu-1} \approx 0$ mod. $A$. $\rightarrow$, $\approx$ shall alone be used in the present paper to the exclusion of $\sim$ (ordinary homology). The effect of this limitation plus the introduction of fractionary cycles and complexes will be that in congruences and homologies coefficients may assume rational values and are not confined any more to integers. This is evidently no restriction in any investigation, such as the present one, wholly unaffected by the presence or absence of zero-divisors, and it is amply justified by the greater freedom acquired and by the simplification of many formulas.

It may also be said that fractionary cycles not essential when dealing with complexes alone, could not be dispensed with in the case of the general closed set. See in this connection No. 27.

Regarding the intersection theorems of Trans. I, II, we may apply them directly to the new types of complexes and cycles if we are careful to limit ourselves to the relation $\approx$.

**4. Topological invariants.** Let $A$ be any point set such that we have assigned a meaning to the statement "a complex is on $A$". Consider the boundary relation

$$(1) \qquad \sum \lambda_i C^i_{\mu+1} \rightarrow \sum \nu_j C^j_\mu$$

the complexes being all on $A$. The right hand side is a $\Gamma_\mu \approx 0$ mod. $A$. If $\Gamma^h_\mu$, $h = 1, 2, \cdots, r$, are several cycles of $A$ such that $\sum t_h \Gamma^h_\mu \approx 0$, mod. $A$, they are said to be *dependent* mod. $A$; when no such relation exists they are *independent.* The maximum number of independent $\mu$-cycles of $A$ shall be denoted by $R_\mu(A)$. It is a topological invariant of $A$, provided that the *on* relation has been defined in an invariant manner for $A$, a remark which applies to our other invariants. It is certainly invariant when *on* means *carried by*. This will not always be the case in the sequel. Therefore with special definitions of the *on* relation it will be necessary to prove invariance directly.

Consider now a set $B$ that need not intersect $A$ but is such that, with the *on* relation for the complexes of $A$ as to $B$ is also defined. Let us neglect in (1) all complexes that are on $B$. Then in place of (1) we write

$$\sum \lambda_i' C^i_{\mu+1} \to \sum \nu_j' C^j_\mu, \text{ rel. } B. \tag{2}$$

The left side is then a cycle relatively to $B$ (cycle rel. $B$), whenever the right is identically zero. An ordinary cycle is also a relative cycle but the reverse is not necessarily true. Since the right side of (1) is always an ordinary cycle (Trans. I, p. 5), the right side of (2) is always a relative cycle. The fact that it appears in a relation (2) is expressed by $\Gamma_\mu \approx 0$, mod. $A$, rel. $B$. The definition of dependence and independence is then as previously and the maximum number of independent relative cycles is denoted by $R_\mu(A; B)$. It is a topological simultaneous invariant of $A$ and $B$, a statement to which must be applied the same reservations as for $R_\mu(A)$.

5. Further invariants arise when we consider the relations between the cycles of $B$ and those of $A - B$. The carrying configuration $A$ of all cycles is in general understood in any discussion and no provision is made to indicate it in the notation for invariants.* The latter are respectively maximum numbers of $\mu$ cycles:

On $B$, independent mod. $B$ but $\approx 0$ mod. $A$, or $r_\mu(B)$.

On $A - B$, independent mod. $A$ of the cycles of $B$, or $S_\mu(A - B)$.

On $A$ with no combination mod. $A$ to a cycle on $B +$ a cycle on $A - B$, or $T_\mu(B)$.

Evidently $T_\mu(B) = T_\mu(A - B)$.

In regard to the various types of dependence, bounding or vacuous cycles are always to be taken in consideration. Thus if $\Gamma_\mu \subset A - B$, $\approx 0$ mod. $A$, it is to be considered as $\approx$ a cycle on $B$ mod. $A$ — namely a vacuous cycle on $B$. It follows that $1 + R_\mu(A)$ $\mu$-cycles of $A - B$ are always dependent mod. $A$, upon cycles of $B$ and therefore $S_\mu(A - B) \leqq R_\mu(A)$. Hence *if* $R_\mu(A)$ *is finite the numbers* $S_\mu(A - B)$ *are all finite* and $\leqq R_\mu(A)$.

---

* In our Proceedings Notes, provision was made in the notation for the carrying configuration and for example instead of $r_\mu(B)$, we had $r_\mu(B; A)$ but the less precise notation adopted here has the advantage of being clearer and less cumbersome.

In particular when $A$ is a complex as throughout our paper, $S_\mu(A-B)$ is always finite. Similarly $T_\mu(B) \leqq R_\mu(A)$ and is finite when the latter is finite, in particular for a complex.

6. A different class of invariants is obtained when all coefficients in boundary relations and homologies are reduced mod. $p$ a fixed integer, which we assume prime in order to avoid complications, although this restriction is not essential. If the coefficients are fractions or if the cycles are fractionary the denominators must be assumed not divisible by $p$. This was done for $p=2$ by Veblen,* for $p$ arbitrary by Alexander.† The new invariants thus obtained are denoted by $R_\mu(A;p)$, $R_\mu(A;B,p)$ etc.

The following generalisation suggests itself: Let $\mathfrak{C}$ be any additive property of the complexes $\subset A$. i. e. such that if $C_\mu$ and $C'_\mu$ possess it so does $\lambda C_\mu + \lambda' C'_\mu$, whatever the rational numbers $\lambda, \lambda'$. If we neglect throughout all complexes with the property $\mathfrak{C}$ we obtain what might be called boundary relations, homologies, mod. $A$ rel. $\mathfrak{C}$, cycles rel. $\mathfrak{C}$. The maximum number of independent $\mu$-cycles of this nature is a number that may be designated by $R_\mu(A;\mathfrak{C})$. Thus we get the invariants $R_\mu(A;B)$ when $\mathfrak{C}$ means: $C_\mu \subset B$, the invariants $R_\mu(A;p)$ when $\mathfrak{C}$ means: $C_\mu$ is $p$ times some other complex. It would appear to be highly interesting to find out if the total class of topological invariants thus obtainable, say for a complex, includes any that are essentially different from those of the present paper, i. e. corresponding to the two special cases of $\mathfrak{C}$ just mentioned or to their combinations.

7. **Generalized Euler-Poincaré formula.** Let $A = C_n$, a non singular complex, and $B = C_k$, a subcomplex of $C_n$. Let $\alpha_i$ denote the number of $i$-cells of $C_n - C_k$. Then we have the generalized Euler-Poincaré relation,

$$\sum(-1)^i \alpha_i = \sum(-1)^i R_i(C_n; C_k, p). \tag{3}$$

The ordinary Euler-Poincaré relation corresponds to $C_k$ vacuous, $p=1$, i. e. involves the numbers $R_i(C_n)$.‡ For $C_k$ vacuous, $p=2$, the formula has been derived by Veblen and Alexander,§ and for $p>2$ by Alexander.|| Their proofs apply to (3) directly and need not be repeated here.

8. **Duality relations.** Let $M_n$ be the set of points on a complex $C_n$, and $G$ a closed set $\subset M_n$ which includes all cells of $C_n$ that do not verify

* Annals of Math., (2), vol. 14 (1912), pp. 86–94; see also Veblen-Alexander, ibid pp. 163–178; Veblen, Colloquium Lectures on Analysis Situs, Chs. 1, 2, 3.

† In the Transactions paper already quoted.

‡ It was first proved by Poincaré in the paper already quoted.

§ In the joint paper just quoted. See also Veblen, Coll. Lect. p. 78.

|| In the paper of the Trans. just quoted.

the conditions for an $n$-dimensional manifold with boundary (Trans. II p. 437). $M_n$ may be considered as an $n$-dimensional manifold not necessarily connected with a boundary $F_{n-1} \subset G$ or with a subcomplex of singularities also $\subset G$. Strictly speaking the notation "$M_n$" should not be applied to it, but if one keeps in mind the more general meaning adopted here no harm will follow. Orientability can be defined for $C_n$ as for any manifold, if one applies the usual scheme only to $(n-1)$-cells incident with two $n$-cells. Unless otherwise stated we shall assume $M_n$ orientable. $C_n$ is the *defining* complex of $M_n$, and in order to have no doubt as to what we shall do, we agree that given $G$, any other defining $C_n$ to be considered must have its associated subcomplex of singularities likewise $\subset G$. In order to bring out clearly the degree of generality implied here we may point out that our treatment applies to complexes whose non boundary cells alone satisfy the manifold conditions. For example they may be manifolds in the sense of Veblen: Manifolds without boundary with some $n$-cells of a defining $C_n$ removed. The fact is that we shall actually operate constantly not on $M_n$ but on certain manifolds $M_n - G$ and of the more restricted types of Trans. II, p. 437. The actual behavior of $M_n$ at the boundary is therefore an indifferent matter.

This being understood, the simultaneous invariants of $M_n$ and $G$ satisfy the following two basic duality relations

$$S_\mu(M_n - G) = S_{n-\mu}(M_n - G) \tag{4}$$

$$R_\mu(M_n - G) = R_{n-\mu}(M_n; G). \tag{5}$$

In addition we have the formula

$$R_\mu(M_n;\ G) = r_{\mu-1}(G) + S_\mu(M_n - G) + T_\mu(G). \tag{6}$$

One of our chief objects will be to prove these three relations. From them we derive easily

$$R_\mu(M_n - G) - S_\mu(M_n - G) = r_{n-\mu-1}(G) + T_{n-\mu}(G) \tag{7}$$

which is essentially (4) of our first Proceedings Note, where the left hand side has been designated by $s_\mu$: It is the maximum number of distinct $\mu$-cycles of $M_n - G$ dependent upon cycles of $G$ mod. $M_n$.

When $G$ is vacuous, which requires that $M_n$ be without boundary, both (4) and (5) reduce to Poincare's noted duality relation

$$R_\mu(M_n) = R_{n-\mu}(M_n).$$

On the other hand when $M_n$ is an $n$-sphere $H_n$, (7) embodies the generalisation of Alexander's duality relation between the invariants of a $C_k \subset H_n$ and those of $H_n - C_k$ (formula (9) below). Thus *our group of formulas is, so to speak, the point of convergence of two apparently only distantly related questions of topology.*

By merely inserting the integer $p$ in the proper place in each invariant we have new formulas for the cycles mod. $p$ which are also true and derived exactly like the others. The same is also true when taking the invariants for cycles rel. $H$, a closed set having no points in common with $G$. When $H$ and $G$ have common points the problem becomes more complicated and will not be taken up here.

An important observation due to Veblen and Alexander is that the invariants mod. 2 can also be considered for a non orientable $M_n$. In a sense they embody results of a strictly point-set theoretic nature, with all questions of orientation left out.

9. In our treatment we shall first take up at length the case of a $C_k \subset M_n$, then the general $G$. But before doing so we shall give bibliographical indications.

Important cases of (4), (5) occur in the literature: When $G$ is a vacuous set, both reduce to the noted duality relation of Poincaré.* The similar relation for cycles mod. 2 has been derived by Veblen and Alexander,† for cycles mod. any $p$ by Alexander.‡ When $G = F_{n-1}$ boundary of $M_n$, (4) and (5) reduce to certain duality relations due to the author.§ For $G = C_k$ a subcomplex of the defining $C_n$ of $M_n$ they were first derived in our first Proceedings Note.

When $M_n = H_n$ a hypersphere, $S_\mu(H_n - G) = 0$ for every $\mu$; $T_\mu(G) = 0$ for $\mu \neq n$, $T_n(G) = 1$, since the only cycles not $\approx 0$ carried by $H_n$ are a point and $H_n$ itself. This is likewise true for the invariants mod. $p$. Let $T_\mu = 0$ for $\mu \neq n$, $T_n = 1$. Then (5) together with (6) and the remark that in the present case $r_\mu(G) = R_\mu(G)$, give:

$$R_\mu(H_n - G) = R_{n-\mu-1}(G) + T_{n-\mu}. \tag{8}$$

The same formula for $G = C_k$ and cycles mod. 2

$$R_\mu(H_n - C_k;\, 2) = R_{n-\mu-1}(C_k;\, 2) + T_{n-\mu} \tag{9}$$

was first derived by Alexander.|| Its importance is manifest when it is remembered that by means of it he showed how to derive the generalized

* Palermo Rendiconti, vol. 13 (1899).

† In their joint paper, also Veblen, Coll. Lect., p. 91.

‡ In his paper of the 1926 Transactions.

§ Trans. II, pp. 449, 450, first and second duality theorems. (See pp. 268, 269 of the present volume.)

|| Trans. Amer. Math. Soc., vol. 23 (1922), pp. 333–349.

Jordan-Brouwer theorem in a few lines. A still different method has been recently applied by Alexandroff* to the derivation of (9) for $\mu = 0$ and the most general $G$. The numbers $R_0(H_n - G; p)$ are in fact all equal and their common value $R_0(H_n - G)$ is the number of disconnected domains into which $H_n$ is separated by $G$. This invariant is therefore the simplest and at the same time the most important simultaneous topological invariant of $H_n$ and $G$.

Another important result implied by (9), and the similar formula with $p$ in place of two, is that they give expressions for the topological invariants $R_\mu(G; p)$ of $G$ in terms of those of its residual set $H_n - G$,—a fundamental remark due to Alexander.

## II. The general $C_k$ on $M_n$.

10. From now on we are considering complexes and sets on an $M_n$ which we assume orientable unless otherwise stated. The manifold possesses a certain defining complex $C_n$. On $M_n$ we shall have occasion to consider intersections of complexes but the situation will always be as in Trans. I: The boundaries of intersecting complexes will not meet. Let $C_\mu$, $C_\nu$ be two such intersecting complexes. We shall usually be concerned with the intersection cycle $C_\mu \cdot C_\nu$, or Kronecker index $(C_\mu \cdot C_\nu)$ when $\mu + \nu = n$. In view of what has been shown loc. cit. $C_\mu$ may be replaced by $\bar{C}_\mu$, $(C_\mu^0 + C_\mu'$ of Trans. I) with the same boundary and polyhedral down to cells as close as we please to that boundary. There is a similar $\bar{C}_\nu$ for the other complex and the two approximations may be so chosen that the intersecting cells be all polyhedral and in the most general position relatively to one another on $C_n$. Whenever intersecting complexes occur in the sequel all complexes will be assumed replaced by the approximations in general position relatively to one another.

11. Let then $C_k$ be a fixed complex on $M_n$. As a matter of fact this particular complex might be replaced throughout by the homeomorph of any polyhedral configuration in some Euclidean space, composed of a finite number of faces of dimension up to $k$ and constituting a closed point set. If $M_n$ *has a boundary it must be assumed* $\subset C_k$. The latter is non singular but not necessarily polyhedral, i. e., it has the same degree of generality as, say a Jordan curve. Our first objective is an important reduction theorem for complexes very near $C_k$ which will play an important

* Comptes-Rendus, vol. 184 (1927), pp. 425–427; earlier for $n = 2$, $G$ an arbitrary closed set Brouwer, Math. Ann., vol. 72 (1912), p. 422; Alexandroff from another point of view, Math. Ann., vol. 96 (1926), pp. 527–534. As a matter of fact as he pointed out to me his $R_{n-1}(G; 2)$ is not defined like ours. But (9) being proved for both it follows that they are equal, their common value being $R_0(H_n - G) - 1$. Further references at the end of this paper.

part in the discussion. In its proof we need the following property which is best taken up separately: *Let $H_i$ be an i-sphere (singular cell boundary) on $C_k$. For every $\xi$ sufficiently small there is an $\eta$ (tending to zero with $\xi$) such that when the diameter of $H_i < \eta$, $C_k \subset$ an $E_{i+1}$ (in general singular) bounded by $H_i$ and of diameter $< \xi$.* To show it subdivide $C_k$ into $C'_k$ with cells of diameter $< \frac{1}{2}\xi$. Then there exists an $\eta$ tending to zero with $\xi$, and such that any subset of $C'_k$ of diameter $< \eta$, $\subset$ a set of cells of $C'_k$ having a common vertex (star of cells with that vertex for center). If $H_i$ is that set and $A$ the vertex all that is necessary is to join it to all points of $H_i$ by one-cells rectilinear with respect to $C_k$ and its definition of straightness. Their locus is the required $E_{i+1}$.

12. **Reduction theorem***. *Let $N^\varepsilon$ denote the set of all points of $M_n$ at a distance $< \varepsilon$ from $C_k$. To each $\varepsilon > 0$ corresponds $\varepsilon' > 0$, tending to zero with $\varepsilon$, and such that if $C_\mu \subset N^\varepsilon$, $C_\mu \to \Gamma_{\mu-1} \subset C_k$, then also $C_\mu \approx C'_\mu$ mod. $N^{\varepsilon'}$, $C'_\mu \subset C_k$.*

In other words a cycle rel. $C_k$ which is sufficiently near $C_k$ is $\approx 0$ mod. $N^{\varepsilon'}$ rel. $C_k$. (If the cycles and complexes were all integral we could replace $\approx$ by $\sim$.)

Subdivide $C_\mu$ into cells of diameter $< \varepsilon$, and continue to call the new complex $C_\mu$. This will lead to no confusion since it is $\approx$ the initial one mod. $N^\varepsilon$. Let then $A^1, \cdots, A^\alpha$ be the vertices of $C_\mu$. Associate with each $A^i$ the point $B^i$ nearest to it, on $C_k$, or one of the set of points at the least distance if there is more than one. Let $A^iA^j$ be a one-cell of $C_\mu$. The points $B^i$, $B^j$ are at distance $< 3\varepsilon$ from one another. Hence by No. 11 if $\varepsilon$ is sufficiently small there will be a one-cell $B^iB^j$ on $C_k$ and furthermore its diameter can be made as small as desired. Similarly to any $E_2$ say $A^iA^jA^h$ of $C_\mu$ can be made to correspond an $E_2$; $B^iB^jB^h$ on $C_k$; etc. The only condition is that $\varepsilon <$ a certain number which depends upon several $\xi$'s such as in the preceding number, but not upon $C_\mu$. Furthermore we assume that when any cell of $C_\mu$ is on $C_k$ the corresponding $B$ cell coincides with it.

Apply now the result of No. 11 to $M_n$ itself: We can find one-cells $A^iB^i$ whose diameter can be made as small as we wish by taking $\varepsilon$ sufficiently small. If $A^iA^j$ is an $E_1$ of $C_\mu$, $A^iA^jB^jB^i$ is an $H_1$ whose diameter depends upon $\varepsilon$ and tending to zero with $\varepsilon$. Hence for $\varepsilon$ sufficiently small there is an $E_2$ of which it is the boundary, etc. Proceeding thus we shall have a $C_{\mu+1}$ whose cells are of a diameter $< \varepsilon'$ tending to zero with $\varepsilon$, hence on $N^{\varepsilon'}$. We assume again that where an $E_h$ of $C_\mu$ is on $C_k$, the

*See our Proceedings Notes whose proof of the theorem we have greatly simplified; the present treatment is very close to Alexander's, Trans. Amer. Math. Soc., vol. 16 (1915), pp. 148–154. In a similar order of ideas see Veblen, Coll. Lect., pp. 94, 119.

corresponding $E_{h+1}$ of $C_{\mu+1}$ is a singular cell coincident with $E_h$. The incidence relations between the $\mu+1$ cells of $C_{\mu+1}$ and its $\mu$ cells of type $A^1 \ldots A^\mu B^1 \ldots B^\mu$ are the same as between the $\mu$ and $\mu-1$ cells of $C_\mu$. Let us then orient each $E_{\mu+1}$ of $C_{\mu+1}$ positively relatively to the associated $E_\mu$ of $C_\mu$. Since in $C_\mu$ the boundary cells of the $\mu$-cells not on $C_k$ cancel one another the same will hold for the boundary cells of $C_{\mu+1}$ not on $C_k$ and not cells of $C_\mu$. In other words $C_{\mu+1} \rightarrow C_\mu +$ a complex on $C_k$, which proves the theorem.

13. **Polyhedral neighborhood of $C_k$.*** In order to be able to apply the results of Trans. II we must replace $N^\varepsilon$ by a suitable polyhedral neighborhood. To obtain it we first subdivide $C_n$ into $C_n^1$ with cells of diameter $< \frac{1}{2}\varepsilon$ and take the subcomplex $C_n'$ of $C_n^1$ made up of all cells which carry a point of $C_k$ or have one on their boundary, plus the boundaries of all such cells. The cells of $C_n^1$ abutting on $C_n'$ constitute together with their boundaries and $C_n'$ a new $C_n''$ also a subcomplex of $C_n^1$. The cells of $C_n''$ not on $C_n'$, nor abutting on it constitute a $C_{n-1}$. Let $PQ$ be a segment of $C''-C'$ whose end points are respectively on $C'$ and $C_{n-1}$, and let $R$ be a point dividing it in any fixed ratio $\lambda < 1$. As $PQ$ describes any particular $E_h$ of $C''-C'$, $R$ describes an $E_{h-1}$, polyhedral, convex and flat, and the locus of all these cells is a complex $\Phi_{n-1}$. The set of all closed segments $PR$ plus $C_n'$ is an $n$-complex $N$ *which is the required polyhedral neighborhood of* $C_k$. It is a *closed* neighborhood, the corresponding open neighborhood being $N - \Phi$. Both $N$ and $\Phi$ have all their points within a distance $\varepsilon$ of $C_k$ and as near to it as we please. $\varepsilon$ is the *width* of $N$. Moreover, the points of $\Phi$ are all at a distance $>0$ from $C_k$, and $C_k$ is wholly interior to $N$.

Of equal importance with $N$ is the $n$-complex $K_n = M_n - N + \Phi$, likewise polyhedral and with convex, flat, cells. Its points are also all at a distance $>0$ from $C_k$. We need the following properties:

I. *In the reduction theorem of No.* 12, $N^\varepsilon$, $N^{\varepsilon'}$ *may be replaced by polyhedral neighborhoods $N$ and $N' \supset N$ of widths* $\varepsilon$, $\varepsilon'$.

II. $K_n$ *is an* $M_n$. The definition of manifolds here used is strictly that of Trans. II, p. 437. It implies that the boundary $\Phi$ of $K$ is also a manifold if $K$ itself is one. To prove the proposition we need only to consider the behavior of the cells of $\Phi$ relatively to $K$. Any cell $E_j'$ of $\Phi$ is on a certain $E_{j+1}$ of $C_n'$. Every $E_h'$, $h > j+1$, of $K - \Phi$ incident with $E_j'$ is on an $E_h$ of $C_n'$ incident with $E_{j+1}$ and the incidence relations between corresponding cells of the two sets $\{E_h\}$, $\{E_h'\}$ are the same. Hence since

* The neighborhood here given was considered at length in our first Proceedings Note. It was also obtained independently and by the same construction by Alexandroff, Math. Ann., vol. 98.

$C'_n$ defines a manifold with $E_{j+1}$ for interior cell, the two sets have the structure of an $H_{n-j-1}$. Hence, $E'_{j+1}$ being the subcell of $K$ on $E_{j+1}$, (the part of $E_{j+1}$ on $K$), the set $E'_{j+1}+\{E'_h\}$ has the structure of an $E_{n-j-1}$ plus its boundary, every $E_i$ of the set corresponding to an $E_{i-j-1}$ of a subdivision of ($E_{n-j-1}+$its boundary) into cells. Referring to our definition of an $M_n$ with boundary, this proves that $K$ is such a manifold with $\Phi$ for boundary.

In the same manner it may be shown that $N$ is an $M_n$, but that is immaterial for the sequel.

*Remark.* It will frequently happen that some of our statements referring to a neighborhood $N$ of $C_k$ will hold only when its width $\varepsilon$ is chosen sufficiently small. Such a choice, always possible wherever it will be implied, shall be automatically assumed without anything further being said about it.

## III. Theorems on Kronecker indices.

14. From now on in this section we shall use $\Delta$ for cycles rel. $C_k$ and continue to use $\Gamma$ as the generic letter for ordinary cycles.

In the theorems to be stated presently there appears the Kronecker index of two complexes. We recall that the index $(C_h \cdot C_{n-h})$ of $C_h$, $C_{n-h} \subset M_n$ is the number of their intersections each counted with a certain multiplicity that may be positive or negative (Trans. I, pp. 6, 12). In the proof we shall make repeated application of the following proposition also true for fractionary cycles and complexes (Trans. I, p. 20, Th. IV): *If* $C_{\mu+1}\rightarrow\Gamma_\mu$ *and if* $C_{n-\mu}$ *is a complex whose boundary does not meet* $C_{\mu+1}$, *then* $(\Gamma_\mu \cdot C_{n-\mu}) = 0$. *In particular this holds when* $C_{n-\mu}$ *is a cycle.* To illustrate this basic theorem by a very simple example, let $\gamma$ be the boundary of a spherical region $R$, $\delta$ a spherical arc whose end points are not on $R+\gamma$. Then the theorem merely states the obvious fact that if a point describes $\delta$ in a definite direction it enters $R$ as often as it leaves it.

15. THEOREMS. I. *In order that* $\Gamma_\mu$, *a fixed cycle of* $M_n - C_k, \approx 0$ *mod.* $M_n - C_k$ *it is necessary and sufficient that* $(\Gamma_\mu \cdot \Delta_{n-\mu}) = 0$ whatever $\Delta_{n-\mu}$.

II. *Conversely in order that* $\Delta_{n-\mu}\approx 0$ mod. $M_n$ rel. $C_k$, *it is necessary and sufficient that the index be zero whatever* $\Gamma_\mu$.

When $C_k$ is vacuous I and II reduce to the theorem of Veblen and Weyl mentioned in the Introduction. From the theorem recalled in No. 14 follows immediately that the conditions in I, II are necessary. For when $M_n - C_k \supset C_{\mu+1}\rightarrow \Gamma_\mu$, where $C_{\mu+1}$ does not meet the boundary of $\Delta$ which is on $C_k$, necessarily $(\Gamma \cdot \Delta) = 0$. Next when $\Delta \approx 0$ mod. $M_n$ rel. $C_k$, $C_k \supset C_{n-\mu}$ and $M_n \supset C_{n-\mu+1} \rightarrow \Delta_{n-\mu} + C_{n-\mu}$. Since $C_{n-\mu}$ does not meet $\Gamma_\mu$, $(\Gamma_\mu \cdot C_{n-\mu}) = 0$, hence $(\Gamma_\mu \cdot \Delta_{n-\mu} + C_{n-\mu}) = 0 = (\Gamma \cdot \Delta)$.

Therefore we merely have to prove the sufficiency of the conditions.

16. *Consider first the case where $C_k = F_{n-1}$, boundary of $M_n$*, a manifold in the strict sense of Trans. II p. 437. As p. 439 of that paper we take a copy $\bar{M}_n$ of $M_n$ and match the two along associated boundary points so as to have a manifold without boundary $V_n = M_n - \bar{M}_n$. The image of any particular element of $\bar{M}_n$ on $M_n$ will be denoted by the same letter barred.

We have shown (Trans. II No. 19 Th. I), that every cycle of $V_n$ is homologous to a sum of: (a) A cycle on $M_n$, which may be assumed interior to it (ibid. No. 23 Coroll. I), and therefore is a $\Gamma$ in our notation. (b) A skew-symmetric cycle, i. e., of type $\Delta - \bar{\Delta}$ in our present notation. Hence on $V_n$ every $n - \mu$ cycle $\Gamma'_{n-\mu} \approx \Gamma_{n-\mu} + \Delta_{n-\mu} - \bar{\Delta}_{n-\mu}$, a very simple result that may be proved directly with ease.

17. Let now $\Gamma_\mu$ behave as in I. Then $(\Gamma_\mu \cdot \Gamma_{n-\mu}) = (\Gamma_\mu \cdot \Delta_{n-\mu}) = 0$ by assumption, and $(\Gamma_\mu \cdot \bar{\Delta}_{n-\mu}) = 0$ because the two complexes do not meet, the first being on $M_n - F$, the second on $\bar{M}_n$. It follows $(\Gamma_\mu \cdot \Gamma'_{n-\mu}) = 0$ whatever $\Gamma'$ and by the theorem of Veblen and Weyl, $\Gamma_\mu \approx 0 \bmod. V_n$. Hence (loc. cit. No. 19, Th. IV) $\Gamma_\mu \approx 0 \bmod. M_n$ which completes the proof of I in the present instance.

18. Let now $\Delta_{n-\mu}$ behave in accordance with II. Let $\delta_{n-\mu} = \Delta - \bar{\Delta}$. Then, whatever the skew-symmetric cycle $\delta_\mu$, $(\delta_\mu \cdot \delta_{n-\mu}) = 0$ (Trans. II, No. 24, Th. II). Furthermore since every cycle of $M_n \approx \bmod. M_n$ to an interior cycle, $(\Gamma_\mu \cdot \delta_{n-\mu}) = (\Gamma_\mu \cdot \Delta_{n-\mu}) = 0$ by assumption whatever $\Gamma_\mu$, on $M_n$. By the theorem on the cycles of $V_n$ recalled above, this holds then whatever $\Gamma_\mu$ on $V_n$. Therefore $\delta_{n-\mu} \approx 0 \bmod. V_n$. Hence $V_n \supset C_{n-\mu+1} \to \delta_{n-\mu}$. Since in $\Delta$ all $n - \mu$ cells on $F_{n-1}$ may be neglected (as $\approx 0 \bmod. M_n$, rel. $F_{n-1}$), we may assume that it has none. Furthermore we can manifestly take all complexes polyhedral and $C$ in the form $C' + C''$, $C' \subset M_n$, $C'' \subset \bar{M}_n$. The cells of $\Delta$ not on $F_{n-1}$ can only belong to the boundary of $C'$. Hence $C' \to \Delta$, rel. $F_{n-1}$, and $\Delta \approx 0 \bmod. M_n$, rel. $F_{n-1}$ which completes the proof of II for the case considered.

19. We now take up the case *where $C_k$ is our former general complex on $M_n$* and reduce it to the case just considered. It must be remembered that when $M_n$ has a boundary $F_{n-1}$, always $F_{n-1} \subset C_k$.

Let then $\Gamma_\mu$ be such that $(\Gamma_\mu \cdot \Delta_{n-\mu}) = 0$ for every $\Delta$. Consider two polyhedral neighborhoods $N$, $N^1$ associated after the manner of No. 13 ($N^1 \supset N$, but their widths tend simultaneously to zero), and so chosen that $\Gamma_\mu \subset M_n - N^1$. Let $C^1_{n-\mu}$ be an arbitrary complex of $K$ with its boundary $\Gamma_{n-\mu-1}$ on $\Phi$. From No. 13 property I follows that $N^1$ carries a $C^2_{n-\mu} \to -\Gamma_{n-\mu-1} +$ a cycle on $C_k$. Hence $C^1 + C^2$ is a $\Delta_{n-\mu}$ so that $(\Gamma \cdot C^1 + C^2) = 0$. But since $C^2 \subset N^1$ and $\Gamma \subset M_n - N^1$, $C^2$ and $\Gamma$ do not meet. Hence $(\Gamma \cdot C^2) = 0$, and finally $(\Gamma \cdot C^1) = 0$. But $C^1$ is an arbitrary cycle of $K$, rel. $\Phi$. Hence by the part of the theorem already proved, $\Gamma \approx 0$ mod. $K$, hence mod. $M_n - C_k$ which completes the proof of I.

Regarding II, let $\Delta_{n-\mu}$ be such that $(\Gamma \cdot \Delta) = 0$ whatever $\Gamma$. We have $\Delta_{n-\mu} = C^1_{n-\mu} + C^2_{n-\mu}$; $C_1 \subset K$; $C_2 \subset N$. An arbitrary $\Gamma_\mu$ on $K - \Phi$ does not meet $C^1$, hence $(\Gamma \cdot C^2) = 0 = (\Gamma \cdot \Delta) = (\Gamma \cdot C^1 + C^2) = (\Gamma \cdot C^1)$. Therefore by the part of II already proved $C^1 \approx 0$, mod. $K$, rel. $\Phi$. This means that $C^1 \approx C^3 \subset \Phi$, mod. $K$ hence mod. $M_n$. Therefore $\Delta \approx C^2 + C^3$ mod. $M_n$; $C^3 \subset N$. The boundary of $C^2 + C^3$ is that of $\Delta$, hence it is on $C_k$. Therefore (No. 12, property I of No. 13), $C^2 + C^3 \approx$ a complex on $C_k$ mod. $N^1$, hence mod. $M_n$; therefore finally $\Delta \approx 0$, mod. $M_n$, rel. $C_k$ which completes the proof of II.

## IV. Proof of the relations between the invariants of $C_k$.

20. **Proof of (4).** We have seen that the numbers $S_\mu(M_n - C_k)$ are all finite (No. 5). Let $\Gamma^j_\mu$, $j = 1, 2, \cdots, S_\mu$, be a maximal set of cycles of $M_n - C_k$, independent mod. $M_n$ rel. $C_k$, consider $\|(\Gamma^j_\mu \cdot \Gamma^h_{n-\mu})\|$ and assume $S_\mu > S_{n-\mu}$. There exists then a linear combination of the rows composed wholly of zeros, or a non-trivial $\Gamma_\mu = \sum \lambda_j \Gamma^j_\mu$ such that every $(\Gamma_\mu \cdot \Gamma^h_{n-\mu}) = 0$. Since every $\Gamma_{n-\mu} \approx \sum \nu_h \Gamma^h_{n-\mu}$ mod. $M_n$ rel. $C_k$, $(\Gamma_\mu \cdot \Gamma_{n-\mu}) = \sum \nu_h (\Gamma_\mu \cdot \Gamma^h_{n-\mu}) = 0$, and by Th. II on Kronecker-indices, $\Gamma_\mu = \sum_j \lambda_j \Gamma^j_\mu \approx 0$, mod. $M_n$ rel. $C_k$, contrary to assumption.—Hence $S_\mu \leqq S_{n-\mu}$. Similarly of course $S_{n-\mu} \leqq S_\mu$. Hence the two numbers are equal or

$$S_\mu(M_n - C_k) = S_{n-\mu}(M_n - C_k) \tag{10}$$

which is (4) for $G = C_k$.

21. **Proof of (5).** Let $\Gamma^h_\mu$, $h = 1, 2, \cdots, R_\mu(M_n - C_k)$; $\Delta^j_{n-\mu}$, $j = 1, 2, \cdots$ $\cdots, R_{n-\mu}(M_n; C_k)$ be maximal sets of independent cycles of $M_n - C_k$, and of $M_n$ rel. $C_k$. Consider the associated matrix $\|(\Gamma^h \cdot \Delta^j)\|$ and let $R_\mu(M_n - C_k) > R_{n-\mu}(M_n; C_k)$. As a consequence there exists a linear combination of the rows of the matrix composed wholly of zeros. Hence there is a non trivial linear combination $\Gamma_\mu$ of the cycles $\Gamma^h_\mu$, such that every $(\Gamma \cdot \Delta^j) = 0$. Let $\Delta_{n-\mu}$ be an arbitrary cycle rel. $C_k$. Then $\Delta \approx \sum \lambda_j \Delta^j$, mod. $M_n$ rel. $C_k$. Hence (Th. II on Kronecker indices), $(\Gamma \cdot \Delta) = \sum \lambda_j (\Gamma \cdot \Delta^j) = 0$. Hence (Th. I), $\Gamma \approx 0$ mod. $M_n - C_k$. This implies dependence of the cycles $\Gamma^h$ mod. $M_n - C_k$ contrary to assumption. The inequality assumed is therefore false. That its inverse is likewise false is proved in exactly the same way with the parts of the $\Gamma$'s and $\Delta$'s as well as homologies mod. $M_n - C_k$ and mod. $M_n$ rel. $C_k$ interchanged. Hence

$$R_\mu(M_n - C_k) = R_{n-\mu}(M_n; C_k) \tag{11}$$

which is (5) for $G = C_k$.

22. **Proof of (6).** In our case it reads

$$R_\mu(M_n; C_k) = r_{\mu-1}(C_k) + S_\mu(M_n - C_k) + T_\mu(C_k). \tag{12}$$

We can immediately single out relative cycles in number equal to the right hand side: It consists of:

(a) The set $\Gamma^j_\mu$, $j = 1, 2, \cdots, S_\mu$ of Nr. 20.

(b) A set $D^j_\mu$, $j = 1, 2, \cdots, T_\mu(G)$, composed of cycles independent of the $\Gamma$'s mod. $M_n$ rel. $C_k$.

(c) In addition a set of relative cycles $\Delta^j_\mu$, $j = 1, 2, \cdots, r_{\mu-1}$, obtained thus: There is on $C_k$ a maximal set of cycles $\gamma^j_{\mu-1}$, $j = 1, 2, \cdots, r_{\mu-1}$, independent mod. $C_k$ but each $\approx 0$ mod. $M_n$. There exists then a $\Delta^j_\mu \to \gamma^j_{\mu-1}$, and these $\Delta$'s are the relative cycles in question.

To prove (12) we must show that the relative cycles of the three sets are independent mod. $M_n$ rel. $C_k$, and that any other depends upon them.

Assume then first a relation

$$\sum \lambda_j \Gamma^j + \sum \nu_j D^j + \sum \varrho_j \Delta^j \approx 0, \text{ mod. } M_n \text{ rel. } C_k. \tag{13}$$

Choose again $K$ such that the $\Gamma$'s are all on $K - \Phi$. By taking intersections with $\Phi$ and applying Th. I of Trans. I, p. 12, we derive from (13)

$$\sum \nu_j D^j \cdot \Phi + \sum \varrho_j \Delta^j \cdot \Phi \approx 0, \text{ mod. } \Phi. \tag{14}$$

Now $D^j \cdot \Phi \leftarrow$ the part of $D^j$ on $N$ and $\Delta^j \cdot \Phi - \gamma^j \leftarrow$ the part of $\Delta^i$ on $N$; hence they are $\approx 0$ mod. $N$. From this and (14) follows $\sum \varrho_j \gamma^j \approx 0$ mod. $N$. Hence $N \supset C_\mu \to \sum \varrho_j \gamma_j$. By the reduction theorem (Nos. 12, 13) $C_k$ carries a complex $\approx C_\mu$, hence $\to \sum \varrho_j \gamma^j$ like $C_\mu$. Therefore $\sum \varrho_j \gamma^j \approx 0$ mod. $C_k$ contrary to our assumption as to the $\gamma$'s. It follows that every $\varrho = 0$. We conclude then that (14) reduces to

$$D \cdot \Phi \approx 0, \text{ mod. } \Phi, \qquad D = \sum \nu_j D^j.$$

Hence $\Phi \supset C_\mu \to D \cdot \Phi$. Let $D = D' + D''$, $D' \subset K$, $D'' \subset N$. We have $D' \to D \cdot \Phi$, $D'' \to -D \cdot \Phi$. Also $D = (D' - C_\mu) + (D'' + C_\mu)$. Each parenthesis represents a cycle, the first $\subset K$, hence dependent upon the cycles $\Gamma^j$ mod. $M_n - C_k$, the second $\subset N$, hence $\approx$ a cycle on $C_k$ mod. $M_n$ (Nos. 12, 13). Therefore $D = \sum \nu_j D^j$ turns out to be dependent upon the $\Gamma$'s mod. $M_n$ rel. $C_k$ — contrary to the assumption of mutual independence of the sets $\Gamma^j$, $D^j$. It follows that every $\nu = 0$. But (13) is then reduced to a relation on the $\Gamma$'s alone, which we have assumed that they do not satisfy, hence every $\lambda$ is zero. Thus (13) cannot hold and the relative cycles considered are independent.

Let now $\Delta_\mu$ be any relative cycle whatever, $\rightarrow \gamma_{\mu-1} \subset C_k$. In view of the maximal property of the cycles $\gamma^j$, $\gamma \approx \sum \varrho_j \gamma^j$, mod. $C_k$. Hence $C_k \supset C_\mu \rightarrow \gamma - \sum \varrho_j \gamma^j$. Therefore $\Delta - \sum \varrho_j \Delta^j - C_\mu$ is a cycle and from the definition of the $\Gamma$'s and $D$'s follows that it depends upon them and upon cycles on $C_k$ itself. Hence

$$\Delta \approx \sum \varrho_j \Delta^j + \sum \lambda_j \Gamma^j + \sum \nu_j D^j, \text{ mod. } M_n, \text{ rel. } C_k.$$

Thus the $\Delta$'s, $\Gamma$'s, $D$'s do constitute a maximal set of independent relative cycles from which (12) follows.

## V. The closed set $G$ on $M_n$.

**23. Complexes and cycles on $G$.** There is no a priori reason for not making the definition of the topological invariants rest upon complexes whose cells are on $G$ in the ordinary sense and indeed there is thus obtained a definite class of topological invariants. In general however it is not possible to prove for such complexes a reduction theorem like that of No. 12 which has played a basic part in our treatment of the case of $C_k$. The related invariants will therefore fail to satisty our basic relations (4), (5), (6) except in special cases. It points to a lack of connection between the cycles and complexes on $G$ in the ordinary sense and those in a suitable neighborhood of $G$.

It will be profitable to examine in this regard a simple example due to Alexandroff.* $G$ consists of the following subset of the Euclidean plane $H_2$: $y = \sin \frac{1}{x}$, $0 < x \leqq 1$, plus the segment $-1 \leqq y \leqq +1$, $x = 0$, plus a polygonal line joining the points $(0, 1)$, $(1, \sin 1)$ and not carrying any of the points already assigned to the set. The total set, as is well known, is not locally connected from which readily follows that it carries no linear cycle not deformable into a point. Hence by means of *arcs on $G$* in the strict sense one can only obtain the value $R_1(G) = 0$. On the other hand any neighborhood of width $\varepsilon$ of $G$ carries a $\Gamma_1$ which is $\approx 0$ mod. a neighborhood of width $\varepsilon'$ that *does not* tend to zero with $\varepsilon$. From this and in the light of what we shall say below (No. 28) it follows that the number $R_1(G)$ to be defined presently is equal to *one.* Since our basic formulas will be proved for our topological invariants we can then conclude from (9), $R_0(H_2 - G) = 2$. This shows that $G$, like the Jordan curve, decomposes the plane into *two* connected regions—a conclusion which could not be deduced from any known relation for invariants of $G$ obtained when the *on* relation is taken in its strict sense.

* Math. Ann., vol. 96 (1926), pp. 489-511.

The key to what follows is now plain: We must define the notion "complex on $G$" in such a manner as to make the basic relations (4), (5), (6) hold. This condition, as we shall show, is fulfilled by our definitions.

24. Let us again subdivide $C_n$ into $C_n'$ with cells of diameter $< \varepsilon$, then remove all cells carrying a point of $G$ in their interior or on their boundary. The totality of these cells plus their boundaries is now the closed polyhedral neighborhood to be designated by $N$. The common boundary of $N$ and $M_n - N$ is an $n-1$ complex $\Phi$, and $M_n - N + \Phi = K$ is an $n$-complex. These complexes will play the same part as those of same name considered previously. We could carry the construction farther and obtain manifolds $K, N, \Phi$, but by taking advantage of the results already acquired we can dispense with the extra step.

Consider now a sequence $\{N^i\}$ of our neighborhoods of widths tending to zero with $\frac{1}{i}$, and such that $N^i \supset N^{i+1}$ for every $i$, but otherwise unrestricted. Thus they do not even have to be constructed by means of the same defining $C_n$ of $M_n$. One of them may for example be polyhedral in regard to a given $C_n$, another not, etc.

We now define a *$\mu$-cycle on $G$* as a sequence $\gamma_\mu = \{\Gamma_\mu^i\}$ such that $\Gamma^i \subset N^i$ while for every $p$ there is a $q$ with the property that when $i, j \geqq q$, $\Gamma^i \approx \Gamma^j$ mod. $N^p$.

We shall free the definition from the particular $\{N^i\}$ considered by agreeing that if cycles $\{\Gamma_\mu^i\}$ defined by means of $\{N^i\}$ and $\{\Gamma'^i_\mu\}$ defined by means of $\{N'^i\}$ have a common infinite subsequence of elements $\Gamma$ they define one and the same $\gamma_\mu$. Each sequence $\{\Gamma_\mu^i\}$ may then be considered as a *representation* of $\gamma_\mu$ relatively to $\{N^i\}$. The cycle appears thus as a class of representations such that any two are extremes of a chain of representations with every pair of consecutive terms related as above, the number of terms being of course finite.

*From a representation $\{\Gamma_\mu^i\}$ relatively to $\{N^i\}$ there can always be derived one relatively to $\{N'^i\}$.* All that is necessary is to construct $N'^{j_1} \supset N^{i_1} \supset N'^{j_2} \supset N^{i_2} \cdots$ and take $\Gamma'^h = \Gamma^{i_r}$, $j_{r-1} < h \leqq j_r$; $j_0 = 1$.

As a matter of notation we shall always designate a cycle on $G$ by replacing in the generic term $\Gamma^i$ by $\gamma$. Thus $\{\Gamma_\mu^{ij}\}$ will be designated by $\gamma_\mu^j$, etc.

25. **Homologies and related invariants.** We write down by definition $\gamma_\mu \approx 0$ mod. $G$ whenever for every $p$ there is a $q$ such that for $i \geqq q$, $\Gamma^i \approx 0$ mod. $N^p$. It is a simple matter to verify that if this condition is fulfilled by the given representation $\{\Gamma^i\}$ relatively to $\{N^i\}$ it is likewise for any subsequence $\{\Gamma^{s_i}\}$ as a representation relatively to $\{N^{s_i}\}$ or in fact as a representation relatively to any other $\{N'^i\}$ to which it

may belong. The converse follows also at once by making use of the definition of cycles.

Let $\gamma_\mu^j$, $j = 1, 2, \cdots, r$, be several cycles on $G$, with representations $\{\Gamma_\mu^{ij}\}$ relatively to $\{N^i\}$. Then $\{\sum \lambda_j \Gamma_\mu^{ij}\}$ is also the representation of a cycle relatively to $\{N^i\}$. If $\{N^i\}$ is replaced by $\{N'^i\}$ the representation of the same cycle preserves its form: Namely if $\gamma^j$ is represented by $\{\Gamma_\mu'^{ij}\}$ then the new cycle is represented by $\{\sum \lambda_j \Gamma_\mu'^{ij}\}$. We are therefore justified in denoting it by $\sum \lambda_j \gamma_\mu^j$. When there exists a non trivial sum of this nature $\approx 0$ mod. $G$ the $\gamma$'s are *dependent* mod. $G$, otherwise they are independent. The maximum number of independent $\mu$-cycles of $G$ is as usual denoted by $R_\mu(G)$.

Let $\Gamma_\mu$ be a cycle of $M_n$. We shall define $\gamma_\mu \approx \Gamma_\mu$ mod. $M_n$, as meaning that for every $i$ above a certain limit $\Gamma_\mu^i \approx \Gamma_\mu$ mod. $M_n$. In particular $\gamma_\mu \approx 0$ mod. $M_n$ whenever under the same conditions $\Gamma_\mu^i \approx 0$ mod. $M_n$. These definitions are again manifestly independent of the representation. They make it possible to consider mixed homologies involving cycles of both $M_n$ and $G$. There follows from them a precise meaning for the numbers $r_\mu(G)$, $S_\mu(M_n - G)$, $T_\mu(G)$. We have the following noteworthy properties.

I. *$r$, $S$, $T$ are simultaneous topological invariants of $M_n$ and $G$. $R$ is a topological invariant of $G$ alone.*

The property of $R$ is the only one which is not an immediate consequence of the definitions. To prove it we observe that by reasoning as in No. 12 we can show that for $i$ sufficiently great any $\Gamma_\mu \approx$ mod. some $N^{i-h}$ to a $\Gamma'_\mu$ with its vertices on $G$. More generally if $N^i$ carries a $C_{\mu+1} \to \Gamma_\mu$, $C_{\mu+1}$ is turned into a similar $C'_{\mu+1}$ on an $N^{i-h}$ and with its vertices on $G$, whose boundary is the $\Gamma'$ associated with $\Gamma$. In both cases $i - h$ tends to $\infty$ with $i$. By taking advantage of this we can limit ourselves to $\gamma_\mu$'s whose sequences $\{\Gamma_\mu^i\}$ consist of cycles with vertices on $G$, with the added restriction that when $\gamma_\mu \approx 0$, the complexes bounded by the terms for $i$ above a certain limit also have their vertices on $G$. But if $G \subset M_n$ is homeomorphically transformed into $G' \subset M'_n$, we may establish a one-one correspondence between such specialized sequences associated with suitable neighborhood sequences on $M_n$ and $M'_n$, and of such nature that sums and multiples of cycles go into sums and multiples of the corresponding cycles, and cycles $\approx 0$ mod. $G$ go into cycles $\approx 0$ mod. $G'$ and conversely. This is sufficient to prove that $R_\mu$ is a topological invariant of $G$ alone.

The above reasoning is essentially the same as used by Brouwer and Vietoris in a similar order of ideas. For further details see their papers quoted in the Introduction (Nos. 10,1).

II. *When* $G = C_k$ *the four invariants become the invariants defined in the usual manner, i. e. with the "on" relation as ordinarily understood.* For given $\gamma_\mu$, when $i$ exceeds a certain limit $\Gamma^i \approx \overline{\Gamma}^i$, mod. $N^{i-h}$, $\overline{\Gamma}^i \subset C_k$; given $p$ sufficiently great there is a $q$ such that for $i, j \geqq q$ $\overline{\Gamma}^i \approx \overline{\Gamma}^j$ mod. $N^p$, hence mod. $C_k$ (reduction theorem). It follows that $\gamma_\mu$ identifies a unique $\Gamma_\mu$ on $C_k$; conversely to $\Gamma_\mu$ on $C_k$ corresponds $\gamma_\mu$ with every $\Gamma^i = \Gamma$. That the numbers in question are unchanged under the correspondence plus the change in the *on* relation is shown for example as in the similar case of I.

26. **Relation between the cycles very near $G$ and the cycles on $G$. Reduction theorem.** We have repeatedly pointed out the importance played in the treatment of $C_k$ by the reduction theorem of No. 12 for the cycles very near $C_k$. We shall now show that if $\{N^i\}$ is properly chosen a similar result holds for $G$.

Consider first three neighborhoods $N \supset N' \supset N''$. A set of $\mu$-cycles on $N'$ independent mod. $N$ is likewise on $N'$ independent mod. $N'$. Hence the maximum number $\varrho$ of $\mu$-cycles of $N'$ independent mod. $N \geqq$ the same number for $N'' \subset N'$. Let $R^N$ be its minimum: If $\varrho = R^N_\mu$ already for $N'$, it will assume the same value for any neighborhood $\subset N'$.

Since a set of cycles of $N''$ independent mod. $N$, is necessarily independent mod. $N'$ as well, $R^N \leqq R^{N'}$. Hence $R^N$ does not decrease when we pass from $N$ to any $N' \subset N$. We shall denote its maximum by $R'_\mu$ and return to it in a moment.

Take now $\{N^i\}$ such that $N^{i+1} \supset$ precisely $R^{N^i}$ cycles independent mod. $N^i$. As observed above, since $N^{i+2} \subset N^i$ it will carry the same maximum number $R^{N^i}$ of cycles independent mod. $N^i$ as $N^{i+1}$. Hence every $\Gamma_\mu$ of $N^{i+1} \approx$ mod. $N^i$ to a $\Gamma^{i+2}_\mu$ on $N^{i+2}$. The same reasoning may be continued and we shall thus have a sequence $\gamma_\mu = \{\Gamma^h_\mu\}$; $\Gamma^h = \Gamma^{i+1}$, $h \leqq i+1$; with the property $\Gamma^h \approx \Gamma^{h+1}$, mod. $N^{h-1}$. It defines then a unique cycle on $G$ and by its very construction $\gamma_\mu \approx \Gamma_\mu$, mod. $M_n$. But $\Gamma_\mu$ is after all any cycle of $M_n$ whose points are sufficiently near $G$. Hence

*Reduction Theorem*: *Any* $\Gamma_\mu$ *of* $M_n$ *on a neighborhood of* $G$ *whose width is sufficiently small, is* $\approx$ *a cycle on* $G$ *mod.* $M_n$.

*Remark.* We can extract from any sequence $\{N^i\}$ a subsequence $\{N^{s_i}\}$ having the special property of the sequence considered above. Therefore we can always construct a sequence $\{N^i\}$ having that property and with every $N^i$ polyhedral.

27. *In the preceding proof fractionary cycles played an essential part.* Suppose indeed that $\Gamma_\mu$ were an integral cycle which is not an actual restriction in view of the nature of the theorem sought. Then we can only assert in general that it is $\approx$ a fractionary cycle $\Gamma^{i+2}$. Or remaining in the domain of integral cycles we can only assert that there

is an integer $q_{i+1} \neq 0$ such that $q_{i+1} \Gamma_\mu \approx \Gamma_\mu^{i+2}$, mod. $N^i$. Similarly $q_{i+1} \Gamma_\mu^{i+2} \approx \Gamma_\mu^{i+3}$, mod. $N^{i+1}$, the new cycle being again integral, etc. If the sequence of $q$'s is bounded we can find a $\{\Gamma_\mu^i\} \approx$ a multiple of $\Gamma_\mu$, and with terms all integral cycles, but not otherwise, and the theorem of reduction would fail to hold. These sequences of integral cycles give rise to the cycles introduced by Vietoris.

Another way of looking at the matter is this: A $\Gamma_\mu$ may be such that a multiple $q\Gamma_\mu \approx$ a cycle on a neighborhood of width $\varepsilon$ of $G$; if $q$ remains bounded as $\varepsilon$ tends to zero Vietoris-cycles (sequences of integral cycles) are adequate. If not our more general type must be used.

28. In his paper (footnote 18) Vietoris indicates that Brouwer introduced a number which is $R'_\mu$ of No. 25 (maximum of $R_\mu^N$), his procedure being substantially the same as ours, i. e, without making use of any entities definable as cycles on $G$.* Let us show that its value is $R_\mu(G)$. Let indeed $N^{i+1}$ carry $r$ cycles independent mod. $N^i$: $\Gamma_\mu^{h+1,j}$, $j = 1, 2, \dots, r$. By the above process they give rise to $r$ cycles on $G$, $\gamma_\mu^j$, certainly independent mod. $G$. For if $\sum \lambda_j \gamma^j \approx 0$ mod. $G$ there is a certain $p$ which we may take $> h$, such that for $i > p$, $\sum \lambda_j \Gamma^{ij} \approx 0$, mod. $N^p$, hence mod. $N^h$, since $N^p \subset N^h$. For a similar reason $\Gamma^{ij} \approx \Gamma^{h+1,j}$, mod. $N^h$. Hence $\sum \lambda_j \Gamma^{h+1,j} \approx 0$, mod. $N^h$, contrary to assumption. Therefore the $\gamma$'s are independent and $R_\mu(G) \geqq r$, hence $\geqq R'_\mu$ maximum of $r$.

On the other hand if we possess $r$ independent cycles $\gamma_\mu^j$ on $G$, the $r$ cycles $\Gamma_\mu^{i+1,j}$ are on $N^i$ and independent mod. a certain $N^p \subset N^i$, hence mod. $N^i$ also. Therefore $r \leqq R'_\mu$ maximum of $R^{N_i}$ and consequently $R_\mu(G) \leqq R'_\mu$. This proves that $R'_\mu = R_\mu(G)$ as asserted.

29. **Cycles relatively to $G$.** By definition a $\mu$-cycle rel. $G$ or a complex with boundary on $G$, is a sequence of complexes $\delta_\mu = \{\Delta_\mu^i\}$, with $\Delta^i \to \Gamma_{\mu-1}^i \subset N^i$ and furthermore such that for every $p$ there is a $q$ having the property that for $i, j \geqq q$, $\Delta^i - \Delta^j \subset N^p$. From this follows $\Gamma^i - \Gamma^j \approx 0$ mod. $N^p$, so that $\gamma_{\mu-1} = \{\Gamma_{\mu-1}^i\}$ is a cycle on $G$, obviously $\approx 0$ mod. $M_n$. We express the relation between $\gamma$ and $\delta$ by $\delta_\mu \to \gamma_{\mu-1}$.

In regard to notation we make the same convention as for cycles except that $\delta$ will replace $\gamma$.

As in the case of cycles we agree that when two sequences $\{\Delta_\mu^i\}$, $\{\Delta_\mu'^i\}$ with the above properties relatively to two sequences $\{N^i\}$, $\{N'^i\}$, have

---

* I was not aware of this until the present paper was finished. The relation $R'_\mu = R_\mu(G)$ to be proved presently may then be stated thus: Brouwer's generalized Betti numbers or connection indices are the same as those that we have obtained by means of our sequences of fractionary cycles. It may be observed that, when dealing with cycles mod. 2, Vietoris-cycles are sufficient and do not differ from our type.

an infinite subsequence of common elements, they define one and the same $\delta_\mu$ of which they are distinct representations. Then the corresponding sequences $\{\Gamma^i_{\mu-1}\}$, $\{\Gamma'^i_{\mu-1}\}$ have also an infinite subsequence of common elements and therefore are representations of the same $\gamma_{\mu-1}$. Hence the boundary relation $\delta_\mu \to \gamma_{\mu-1}$ is independent of the representations of $\gamma$ or $\delta$.

As already observed from $\delta_\mu \to \gamma_{\mu-1}$ follows $\gamma_{\mu-1} \approx 0$ mod. $M_n$. Conversely to any $\gamma_{\mu-1} \approx 0$ mod. $M_n$ corresponds a relative cycle $\delta_\mu \to \gamma_{\mu-1}$. For since $\gamma \approx 0$ mod. $M_n$ there is an integer $\alpha$ such that for every $i \geqq \alpha$: (a) there is a $p_i$ tending to $\infty$ with $i$ such that $\Gamma^i \approx \Gamma^{i+1}$ mod. $N^{p_i}$. We specifically denote by $p_i$ the greatest integer having that property. (b) $\Gamma^i \approx 0$ mod. $M_n$. In view of (a) $N^{p_i}$ carries a complex $C^i_\mu \to \Gamma^{i+1}_{\mu-1} - \Gamma^i_{\mu-1}$. Also owing to (b), $M_n$ carries a relative cycle $\Delta^\alpha_\mu \to \Gamma^\alpha_{\mu-1}$. Let now

$$\Delta^i_\mu = \Delta^\alpha_\mu + \sum_{h=\alpha}^{i-1} C^i_\mu; \qquad i > \alpha.$$

This defines a $\delta_\mu$ since it assigns the $\Delta$'s for an infinite sequence of $N$'s. Since $\Delta^i_\mu \to \Gamma^i_{\mu-1}$ $i > \alpha$, $\delta_\mu \to \gamma_{\mu-1}$ and it is therefore the required relative cycle.

30. The following definitions will enable us to consider relative boundary relations and homologies:

(a) We define $\delta_\mu \approx 0$, mod. $M_n$ rel. $G$ as meaning that for every $p$ there is a $q$ such that when $i \geqq q$, $\Delta^i_\mu \approx 0$ mod. $M_n$ rel. $N^p$.

(b) Let $\delta^j_\mu \to \gamma^j_{\mu-1}$, $j = 1, 2, \cdots, \nu$. Then $\{\sum \lambda_j \Delta^{ij}_\mu\}$ is a $\delta_\mu$ to be denoted by $\sum \lambda_j \delta^j_\mu$ and we verify at once that $\sum \lambda_j \delta^j_\mu \to \sum \lambda_j \gamma^j_{\mu-1}$. The $\delta$'s are dependent mod. $M_n$ rel. $G$ whenever a non trivial sum

$$\sum \lambda_j \delta^j_\mu \approx 0 \text{ mod. } M_n \text{ rel. } G.$$

With $p$, $q$, as above it means that for $i > q$,

$$\sum \lambda_j \Delta^{ij} \approx 0, \text{ mod. } M_n \text{ rel. } N^p.$$

When no such condition is fulfilled the relative cycles $\delta^j_\mu$ are independent mod. $M_n$ or mod. $M_n$ rel. $G$ respectively. The maximum number of $\delta$'s independent mod. $M_n$ rel. $G$ is to be denoted according to our conventions by $R_\mu(M_n; G)$.

(c) In regard to relations between the $\delta$'s and ordinary cycles of $M_n$, we identify any $\Gamma_\mu$ of $M_n$, with the relative cycle $\{\Delta^i_\mu\}$ such that every $\Delta^i = \Gamma$. Thus the homology

$$\sum \lambda_j \Gamma^j_\mu + \sum \nu_j \delta^j_\mu \approx 0, \text{ mod. } M_n \text{ rel. } G,$$

means that with $p$, $q$ as above, for $i \geqq q$,

$$\sum \lambda_j \Gamma^j + \sum \nu_j \Delta^{ij} \approx 0, \text{ mod. } M_n, \text{ rel. } N^p.$$

The preceding definitions are freed from the particular $\{N^i\}$ chosen in precisely the same manner as in the analogous cases of the cycles. It follows in particular that $R_\mu(M_n; G)$ is a simultaneous topological invariant of $M_n$ and $G$.

31. **Kronecker indices and intersection theorems.** We denote henceforth by $K^i$, $\Phi^i$, the $K$ and $\Phi$ of No. 24 corresponding to $N^i$. Let $\Gamma_\mu \subset M_n - G$. There exists then a $p$ such that $\Gamma_\mu \subset K^p - \Phi^p = M_n - N^p$. To $p$ there corresponds a $q$ such that $\Delta^i - \Delta^j \subset N^p$ for $i, j \geqq q$. Then $(\Gamma \cdot \Delta^i - \Delta^j) = 0$ since the two complexes have no common point. Therefore $(\Gamma \cdot \Delta^i) = (\Gamma \cdot \Delta^j)$. Thus $(\Gamma \cdot \Delta^i)$ has a value independent of $i$ provided $i$ exceeds a certain limit, and this value is that we define as the Kronecker-index $(\Gamma_\mu \cdot \delta_{n-\mu})$. If we were to take in place of the $\Delta$'s any infinite subsequence of them we would arrive at the same index. From this we conclude immediately that it is independent of the special $\{N^i\}$ and the corresponding representation of $\delta$.

Having shown by this time that all notions introduced are independent of $\{N^i\}$, it will be convenient to take for it a sequence with the special property of No. 26 and with every $N^i$ polyhedral. This is always possible and will imply no loss of generality. See in this connection the Remark at the end of No. 26.

Having defined the index $(\Gamma \cdot \delta)$, we can state the same intersection theorems as for $C_k$ and we proceed to prove them.

I. *In order that $\Gamma_\mu \approx 0$ mod. $M_n - G$ it is necessary and sufficient that $(\Gamma_\mu \cdot \delta_{n-\mu}) = 0$ for every $\delta_{n-\mu}$.*

(a) Let $\Gamma_\mu \approx 0$ mod. $M_n - G$, so that $M_n - G$ carries a $C_{\mu+1} \to \Gamma_\mu$. There exists an $\alpha$ such that $C_{\mu+1} \subset K^{\alpha-1} - \Phi^{\alpha-1}$. Then the boundary of $\Delta^{\alpha+i}$, which is on $N^\alpha$, does not meet $C$. Hence (No. 14), $(\Gamma \cdot \Delta^{\alpha+i}) = 0$ for every $i$, which means by definition $(\Gamma \cdot \delta) = 0$. The condition is thus necessary.

(b) Let now, $\Gamma_\mu$ being fixed, $(\Gamma_\mu \cdot \delta) = 0$ for every $\delta$. There exists an $\alpha$ such that $\Gamma_\mu$ is wholly exterior to $N^\alpha$. Let $\Delta_{n-\mu}$ be an arbitrary cycle rel. $N^\alpha$. It is sufficient to show that from $(\Gamma_\mu \cdot \delta) = 0$, follows $(\Gamma_\mu \cdot \Delta) = 0$, for as a consequence of the theorem as proved for $C_k$ we have then $\Gamma_\mu \approx 0$ mod. $M_n - N^\alpha$, hence also mod. $M_n - G$.

To prove that $(\Gamma_\mu \cdot \Delta) = 0$, we observe that if $\Delta \to \Gamma_{n-\mu-1} \subset N^\alpha$ in view of the reduction theorem of No. 26 we are assured of the existence of a $\gamma_{n-\mu-1}$ such that $\Gamma^i_{n-\mu-1} = \Gamma_{n-\mu-1}$, $i \leqq \alpha$, and that $\Gamma^{i+1}_{n-\mu-1} \approx \Gamma^i_{n-\mu-1}$

mod. $N^{i-1}$ whatever $i$. Since $\Gamma_{n-\mu-1} \approx 0$ mod. $M_n$, likewise $\gamma_{n-\mu-1} \approx 0$ mod. $M_n$. Hence there exists a $\delta_{n-\mu} \rightarrow \gamma$, and as shown in No. 29, $\delta$ can be so chosen that for $i > \alpha$ $\Delta^i = \Delta + C_{n-\mu}$, $C_{n-\mu} \subset N^\alpha$, for the integer $p_i$ of No. 29 is here $i-1$. But since $\Gamma_\mu$ has no point on $N^\alpha$ it will not meet $C$, hence $(\Gamma_\mu \cdot C) = 0$. But the assumption $(\Gamma_\mu \cdot \delta) = 0$ is equivalent to $(\Gamma_\mu \cdot \Delta^i) = 0$ for $i$ sufficiently great. It follows $(\Gamma_\mu \cdot \Delta) = (\Gamma_\mu \cdot \Delta_i - C)$ $= (\Gamma_\mu \cdot \Delta^i) - (\Gamma_\mu \cdot C) = 0$, and this as we have observed completes the proof of I.

II. *In order that* $\delta_{n-\mu} \approx 0$ mod. $M_n$ rel. $G$ *it is necessary and sufficient that* $(\Gamma_\mu \cdot \delta) = 0$ *for every* $\Gamma_\mu \subset M_n - G$.

(a) Let $\delta_{n-\mu} \approx 0$ mod. $M_n$ rel. $G$. Given any particular $\Gamma_\mu$ let $\alpha$ again be such that $\Gamma_\mu$ has no points on $N^\alpha$. For $i$ above a certain limit, $\Delta^i \approx 0$ mod. $M_n$ rel. $N^\alpha$. Hence by the theorem for $C_k$ (here $N^\alpha$), $(\Gamma_\mu \cdot \Delta^i)$ $= 0$, which means $(\Gamma_\mu \cdot \delta) = 0$. The condition is therefore necessary.

(b) Let $(\Gamma \cdot \delta) = 0$ whatever $\Gamma$. Let $p$ be arbitrary and $\Gamma$ fixed on $K^p$. There exists a $q > p$ such that $(\Gamma \cdot \Delta^i) = 0$ for $i \geqq q$. Let $\Gamma_\mu^j$, $j = 1, 2, \cdots, R_\mu(K^p)$ be a maximal set of independent cycles on $K^p$. Then $\Gamma \approx \sum \lambda_j \Gamma^j$, mod. $K^p$. For each $\Gamma^j$ there is a $q_j$ analogous to the above $q$ for $\Gamma$. Let $q'$ be their maximum. Then for $i \geqq q'$, every $(\Gamma^j \cdot \Delta^i) = 0$, hence $(\Gamma \cdot \Delta^i) = \sum \lambda_j (\Gamma^j \cdot \Delta^i) = 0$. Hence by II as proved for an $M_n$ and $C_k$ with $C_k = N^p$, $\Delta^i \approx 0$, mod. $M_n$ rel. $N^p$, which means $\delta \approx 0$ mod. $M_n$ rel. $G$ and completes the proof of II.

32. **Proof of formulas (4), (5), (6).** The proofs of (4) and (5) are exactly as for $C_k$, (Nos. 20, 21) except that $N$, $K$, $\Phi$ corresponding to $\varepsilon$ sufficiently small, are to be replaced by $N^i$, $K^i$, $\Phi^i$, $i$ sufficiently great. There is no need to repeat them here. Slight modifications are necessary in the derivation of (6). We have as in No. 22 three sets independent mod. $M_n$ rel. $G$: $\Gamma_\mu^j$, $D_\mu^j$, $\delta_\mu^j \rightarrow \gamma_{\mu-1}^j$ with the $\gamma$'s a maximal set of $\mu-1$ cycles independent mod. $G$ but $\approx 0$ mod. $M_n$ and we must show their independence.

Assume

$$\sum \lambda_j \Gamma^j + \sum \nu_j D^j + \sum \varrho_j \delta^j \approx 0, \text{ mod. } M_n \text{ rel. } G.$$

This means that for every $p$ there is a $q$ such that for $i \geqq q$

$$\sum \lambda_j \Gamma^j + \sum \nu_j D^j + \sum \varrho_j \Delta^{ij} \approx 0, \text{ mod. } M_n \text{ rel. } N^p.$$

There exists an $\alpha$ such that every $\Gamma^j \subset K^\alpha - \Phi^\alpha$. Choose then $p$ such that $N^p \subset N^\alpha - \Phi^\alpha$. By taking the intersections with $\Phi^\alpha$ we derive from the last homology

$$\sum \nu_j D^j \cdot \Phi^\alpha + \sum \varrho_j \Delta^{ij} \cdot \Phi^\alpha \approx 0, \text{ mod. } \Phi^\alpha.$$

Observing as in No. 22 that $D^j \cdot \Phi^\alpha \approx \Delta^{ij} \cdot \Phi^\alpha - \Gamma^{ij} \approx 0$ mod. $N^\alpha$, since they bound respectively the parts of $D^j$ and $\Delta^{ij}$ on $N^\alpha$, we find $\sum \varrho_j \Gamma^{ij} \approx 0$, mod. $N^\alpha$, $i \geqq \alpha + 1$. This means by definition $\sum \varrho_j \gamma^j \approx 0$, mod. $G$ contrary to assumption. Therefore every $\varrho = 0$. The rest of the proof of the independence of the three sets is as for $C_k$.

The second part of the proof requires that we show that every relative cycle $\delta_\mu$ depends upon those of the given three sets. In view of the maximal property of the cycles $\gamma^j_\mu$, we have $\gamma \approx \sum \varrho_j \gamma^j$ mod. $G$. For every $p$ there exists then a $q$ such that when $i \geqq q$, $\Gamma^i_{\mu-1} - \sum \varrho_j \Gamma^{ij}_{\mu-1} \approx 0$, mod. $N^p$. Therefore $N^p$ carries a $C_\mu \to \Gamma^i - \sum \varrho_j \Gamma^{ij}$. Hence $\Delta^i - \sum \varrho_j \Delta^{ij} - C \to 0$. The cycle of $M_n$ so obtained depends mod. $M_n$ rel. $N^p$ upon the cycles of the two sets $\Gamma^j$, $D^j$. Therefore for every $i > q$

$$\Delta^i \approx \sum \lambda_j \Gamma^j + \sum \nu_j D^j + \sum \varrho_j \Delta^{ij}, \text{ mod. } M_n \text{ rel. } N^p$$

which in effect means that

$$\delta \approx \sum \lambda_j \Gamma^j + \sum \nu_j D^j + \sum \varrho_j \delta^j, \text{ mod. } M_n \text{ rel. } G$$

and as in No. 22 completes the proof of (6).

33. (Added in proof.) I have recently obtained a noteworthy extension of (5) which has the advantage of being more symmetrical and of requiring a single intersection theorem. Given $F$, $G$, closed subsets of $M_n$, the formula reads

$$R_\mu(M_n - F;\, G) = R_{n-\mu}(M_n - G;\, F).$$

The proof entirely analogous to that of (4) or (5) need not be given here.

I am indebted to Alexandroff for the following references to papers in the Göttinger Nachrichten: (a) Alexandroff, Nov. 25, (1927). (b) Pontrjagin, same date. (c) Pontrjagin presented in March 1928. We may also mention a paper (d) Frankl, Wiener Berichte, December 1927, pp. 689–699. All these papers deal with duality relations but only for invariants mod. 2. The argument in (b), (c), (d) is substantially the same as ours. In Pontrjagin's second paper alone is the containing $M_n$ not an $S_n$. See also a closely related paper by Alexandroff, Fundamenta Mat., vol. 11 (1928), pp. 222–228.

# XIV

## ON TRANSFORMATIONS OF CLOSED SETS.*

In two papers published in recent years† we gave a fundamental formula for the number of signed coincidences and fixed points of transformations of manifolds. For a pair $T$, $T'$ of transformation of an orientable $M_n$ without boundary into another, $M'_n$, it is as follows: Let $\Gamma_n$, $\Gamma'_n$ be the cycles representing the two transformations on the product manifold $M \times M'$; $\boldsymbol{f}^p$, $\boldsymbol{g}^p$ the matrices of the coefficients of the transformations induced on the $p$-cycles by $T$ and $T'$, $\alpha^p$, $\beta^p$ the Kronecker index matrices for associated fundamental sets of dimensions $p$ and $n-p$ on $M_n$ and $M'_n$ respectively. Then the number of signed coincidences is the Kronecker index

$$(1) \qquad (\Gamma_n \cdot \Gamma'_n) = \sum (-1)^{n-p} \text{ trace } (\alpha^p)^{-1} \boldsymbol{g}^p \beta^p (\boldsymbol{f}^{n-p})'.$$

To have the number of signed coincidences of two transformations of $M_n$ into itself, we merely consider them as transforming $M_n$ into another copy, $M'_n$, of itself. Then $\beta^p = \alpha^p$ and in place of (1) we have

$$(2) \qquad (\Gamma_n \cdot \Gamma'_n) = \sum (-1)^{n-p} \text{ trace } (\alpha^p)^{-1} \boldsymbol{g}^p \alpha^p (\boldsymbol{f}^{n-p})'.$$

The number of signed fixed points of $T'$ is obtained by taking for $T'$ the identity. Denoting the corresponding cycle by $\Gamma^0_n$, and since $\boldsymbol{g}^p = 1$, we have after replacing $p$ by $n-p$, the number of signed fixed points

$$(3) \qquad (\Gamma_n \cdot \Gamma^0_n) = \sum (-1)^p \text{ trace } \boldsymbol{f}^p.$$

I have recently extended (1) to transformations of the manifold part of any complex, i. e. to transformations of subsets $K - L$, where $L$ is a sub-

* Received December 21, 1929. See Comptes Rendus, vol. 190 (1930), p. 99.

† Trans. Amer. Math. Soc., vol. 28 (1926) and vol. 29 (1927), to be designated by *Tr.* I, *Tr.* II. Another paper related to these questions appeared in these *Annals*, (2), vol. 29 (1928). We shall designate it by *Annals* and its notations and terminology will be adopted throughout with the following mild modification: An oriented complex $C_k$, of the type of *Tr.* I No. 6, will be called *k-chain*, a convenient term due to Alexander. Its boundary will be designated by $F(C_p)$. Finally we shall find it convenient to introduce, as now quite customary, the *closure* $\bar{A}$, of a point-set $A$, or sum of $A$ plus all its limit points.

complex of $K$ and all cells of $K - L$ satisfy the manifold conditions. This can even be done when the manifold conditions are taken in the strictly combinatorial sense introduced by Alexander, with cells and spheres merely defined by homology characters. The extension demands essentially that we substitute for ordinary or *absolute* cycles, the relative cycles of *Annals* and prove the corresponding duality relations loc. cit. However, in this extension the manifold conditions continue to be essential and there is no hint as to what happens when they are not satisfied.

Now Hopf* took an arbitrary complex $K_n$ and proved that (3) is valid for single valued continuous transformations when the fixed points are isolated and on the $n$-cells of $K$, a curious result. If $K'$ is a copy of $K$ and $\Delta_n$, $\Delta_n^0$, the images of $T$ and the identity on $K \times K'$, the index $(\Delta_n \cdot \Delta_n^0)$ is found here to be independent of the mode of orienting the cells of $K$ and again

$$(\Delta_n \cdot \Delta_n^0) = \sum (-1)^p \text{ trace } \boldsymbol{f}^p, \tag{4}$$

as if we had a manifold.

Now Hopf's proof is interesting but appears to apply exclusively to the case that he considers. Indeed Hopf expressed it as his opinion that his result stands by itself and breaks down for the more general type.

We have recently taken up the question from a wholly different angle and shown that Hopf was unduly pessimistic. *It is possible to generalize the situation so as to make our formulas hold for a very general class of continuous transformations of closed sets whose Urysohn-Menger dimension and Betti numbers are finite. That class includes all continuous single valued transformations and in particular the number of signed fixed points is still given by* (3).

It is scarcely necessary to emphasize the importance of the preceding result. Formula (3) would seem thus to aquire a fundamental importance in the whole domain of real analysis.

The complete extension that we have just mentioned will appear in our forthcoming *Colloquium Lectures* (the Providence Colloquium) now going to press. In the present paper we shall derive the extension of (3), i. e. the number of signed fixed points for a closed set.

*What happens when the Urysohn-Menger dimension and the Betti numbers are infinite*? This is still an open question, possibly not amenable to the methods of combinatorial topology.

**1. Intersection of a chain and a closed set. The pseudocycles.** Let $G_n$ be a compact metric space whose Urysohn-Menger dimension is $n$. According to Menger† $G_n$ is the homeomorph of a subset of a Euclidean

* Proc. Nat. Acad., vol. 14 (1928); Math. Zeitschr., vol. 29 (1929).

† Menger: Dimensionstheorie, p. 295. By $S_r$ we understand a Euclidean $r$-space.

$S_{2n+1}$, therefore also of a bounded subset of an $S_r$, where $r$ is any integer $\geqq 2n+1$. We choose $r$ such that $r-n$ is even, but otherwise wholly arbitrary, and identify $G_n$ with its image on $S_r$.

Regarding the Euclidean $S_r$ we are only interested in its finite part. However for various purposes it is convenient to have an $M$ as the basic space. To obtain one we close up $S_r$ at infinity by a point, which turns it into an $r$-sphere $H_r$. Our theory of Kronecker indices and intersections (*Tr.* I, Part I) can then find free play in $H_r$. See pp. 201 – 227.

We need the following result, easily derived from the proof of the theorem *Annals* No. 26. There can be found a sequence $\{N^i\}$ of polyhedral neighborhoods of $G$ whose total intersection is $G$ such that: (a) $\bar{N}^{i+1} \subset N^i$; (b) Any cycle $\Gamma_p$ on $N^i$ is $\approx$ to a cycle on $N^{i+1}$ mod $\bar{N}^{i-1}$. See p. 562.

2. If $C_{r-\mu}$ is any chain whose boundary $F(C_{r-\mu})$ does not meet $G$ then $F \subset S_r - N^i$ for $i$ large enough. On the strength of this, if $\gamma_\mu = \{\Gamma^i_\mu\}$ is a cycle on $G$, we find as in *Annals*, No. 31, that $(\Gamma \cdot C)$ is constant for $i$ large enough. We define that constant value as the Kronecker index $(\gamma_\mu \cdot C_{r-\mu})$. Similarly of course for $(C \cdot \gamma)$, and these indices behave under permutation or change of signs of $C$, $\gamma$, as other indices. See p. 565.

We now propose to characterize a $C_{n-\mu}$ such that $(\gamma \cdot C) = 0$ for every $\gamma$. Let $i$ be such that the $F(C) \subset S - N^i$. Any cycle $\Gamma_\mu \subset N^{i+1}$ can be considered as the $\Gamma^{i+1}_\mu$ of some sequence $\{\Gamma^j_\mu\}$ defining a $\gamma_\mu$. Due to the choice of $\{N^j\}$, all the $\Gamma$'s whose $j > i$ are on $N^i$. Since $F(C) \subset S - N^i$ the numbers $(\Gamma^j \cdot C)$, $j > i$, are all equal. By assumption their value is zero, hence $(\Gamma \cdot C) = 0$. It follows that if $\Phi^{i+1}$ is the boundary of $N^{i+1}$, $C$ is a cycle on $N^{i+1}$ rel $\Phi$ such that $(\Gamma \cdot C) = 0$ whatever the absolute cycle $\Gamma$ of $N^{i+1}$. But $N^{i+1}$ is a manifold with $\Phi$ as its boundary. Hence (*Annals*, No. 15, Th. II) $C \approx 0$ rel $\Phi$ on $N^{i+1}$. Therefore there exists a chain $C'$ on $\Phi$, *hence not meeting* $G$, such that $C \approx C'$. Conversely let $C \approx C'$ which does not meet $G$. Then for $i$ sufficiently high $C \approx 0$, rel $\Phi^{i+1}$, $(\Gamma^i \cdot C) = 0$ and therefore $(\gamma \cdot C) = 0$. Therefore we have proved

THEOREM. *In order that a chain $C_{r-\mu}$ whose boundary does not meet $G$ be $\approx$ to one which does not meet $G$ it is necessary and sufficient that* $(\gamma_\mu \cdot C_{r-\mu}) = 0$ *for every cycle $\gamma_\mu$ of $G$.*

There is a curious analogy with the theorems of *Annals*, Nos. 15, 31. However the situation here is exactly the reverse. Formerly the comparison of cycles and complexes was as regards their behavior *outside* of $G$, the indices being taken on the complementary set of $G$; here on the contrary it is only the points near $G$ that matter.

3. *We now introduce the finiteness of the Betti numbers of $G$.* Under the circumstances we can find a finite maximal independent set $\gamma^h_\mu$, $h = 1, 2, \cdots, R_\mu(G)$, where $\gamma^h_\mu = \{\Gamma^{hj}_\mu\}$. There exists then an $i$ such that

the cycles $\Gamma_\mu^{hj}$ ($j$ fixed) are independent on $N^i$ for every $j$ sufficiently large. It follows then by a simple argument resting in essence on the theorems of *Annals*, No. 15, that there exists a $C_{r-\mu}$ with $F(C) \subset S - N^i$, such that

$$(\gamma^h \cdot C) = t_h$$

where the $t$'s are arbitrary rational numbers. Thus the $t$'s determine on $S_r$ a true family of chains whose boundary does not meet $G$. On that $S_r$, $G$ is $\approx$ to a chain not meeting $G$, when and only when the $t$'s are all zero. There is an analogous family on any $S \supset G$. The totality of all these families, corresponding to assigned $t$'s will be called an $(n-\mu)$-*pseudocycle*, $\delta_{n-\mu}$ of $G$. We define $\delta \approx 0$ by the condition that every $t_h = 0$. An equivalent requirement is that in the associated family $\{C_{r-\mu}\}$ there is a representative not meeting $G$.

If $\delta_{n-\mu}^i$ are several pseudocycles and $C_{r-\mu}^i$ the associated $C$'s we define

$$\delta_{n-\mu} = \sum \lambda_i \, \delta_{n-\mu}^i$$

as corresponding to

$$C_{r-\mu} = \sum \lambda_i \, C_{r-\mu}^i .$$

Linear dependence is then introduced as usual and we have a corresponding Betti-number, $\varrho_{n-\mu}(G)$. The duality relation $\varrho_{n-\mu} = R_\mu$ follows practically by definition. As for the Kronecker indices we set

$$(\gamma \cdot \delta) = (\gamma \cdot C), \qquad (\delta \cdot \gamma) = (C \cdot \gamma).$$

From these conventions plus the fact that $r - n$ is even we find readily:

I. *The Kronecker-indices* $(\gamma \cdot \delta)$, $(\delta \cdot \gamma)$ *behave as regards the distributive law and permutation of terms like ordinary indices of pairs of cycles of the same dimensions on an* $M_n$.

II. *In order that* $\gamma \approx 0$ $(\delta \approx 0)$, *it is necessary and sufficient that* $(\gamma \cdot \delta) = 0$ *whatever* $\gamma$ (*whatever* $\delta$).

The replacement of $r - \mu$ by $n - \mu$ may seem puzzling. Actually it does not greatly matter what value of $r$ is taken provided that $r - n$ is chosen once for all even or odd. The choice here made has the advantage that when $G$ is an $M_n$ without boundary everything goes back to the situation of *Tr.* 1 without any change whatever.

The chains $C$ are cycles rel $S - G$ in the sense of *Annals*. When $G = P$ a point they become the *cycles of the point* in the sense of Van Kampen (Leyden thesis 1929) who made use of them in his proof of the invariance of combinatorial manifolds.

4. **Product* of two closed sets.** We will now consider a second closed set $G'$ whose dimension and Betti-numbers are likewise finite and denote by accents all analogues of the elements associated with $G$ so far. Of course again $G' \subset S_{r'}$ with $r' - n'$ even. Then $G \times G'$ is a closed set $\subset S_r \times S_{r'}$. The analogue of $\{N^i\}$, $\{N'^i\}$ for $G \times G'$ is the sequence $\{N^i \times N'^i\}$ each term of which is a neighborhood of the product set on the product space.

5. Consider now the cycles of $G$, $G'$:

$$\gamma_\mu = \{\Gamma^i_\mu\}, \qquad \gamma'_\nu = \{\Gamma'^i_\nu\}$$

and two chains $C_\sigma$, $C'_\tau$ defining two respective pseudocycles $\delta_{\sigma+n-r}$, $\delta'_{\tau+n'-r'}$. The products $\Gamma^i \times C'$, $i$ sufficiently large, all define one and the same pseudocycle of $G \times G'$. For if $j > i$

$$N^{i-1} \supset C_{\mu+1} \to \Gamma^j_\mu - \Gamma^i_\mu .$$

Hence (Formula 59.2, p. 236):

$$C_{\mu+1} \times C'_\tau \to (\Gamma^j - \Gamma^i) \times C' + (-1)^\mu C_{\mu+1} \times F(C');$$
$$(\Gamma^j - \Gamma^i)\, C' \approx (-1)^{\mu-1} C_{\mu+1} \times F(C'),$$

a chain which does not meet $G \times G'$, since $F(C')$ does not meet $G'$.

The new pseudocycle thus introduced will be designated by $\gamma_\mu \times \delta'_{\tau+n'-r'}$. We have a similar pseudocycle $\delta \times \gamma'$ associated with all products $C \times \Gamma'^i$.

If $D$, $D'$ are two pseudocycles of the respective types $\gamma \times \delta'$, $\delta \times \gamma'$, whose dimensions have $n + n'$ for sum, it is a simple matter to verify that the index $(\Gamma^i \times C' \cdot C \times \Gamma'^i)$, $i$ sufficiently large, is fixed. We designate it naturally by $(\gamma \times \delta' \cdot \delta \times \gamma')$. If we combine this definition with (54.3) of *Tr.* I we find that that formula holds here as if $G$, $G'$ were manifolds of $n$, $n'$ dimensions. See p. 233.

An identical treatment is applicable to all combinations of subscripts $\mu, \nu, \sigma, \tau$. In place of indices we have intersection cycles on $G \times G'$ of the same dimensions if as we were dealing with manifold products. It all comes down to showing that $\{\Gamma^i \times C' \cdot C \times \Gamma'^i\}$ determines a unique cycle on $G \times G'$. We call the intersection $\gamma \times \delta' \cdot \delta \times \gamma'$, and find that the formula of *Tr.* I, No. 55, holds for it. In short as far as the intersection theory goes everything is as if the $G$'s were manifolds provided that in each expression $\gamma \times \delta$, $\gamma \cdot \delta$, $(\gamma \cdot \delta)$, one term stands for a cycle, the other for a pseudocycle. See p. 233.

* By product $A \times B$ of two sets $A = \{a\}$, $B = \{b\}$, is meant the set of all ordered pairs $(a, b)$. For details regarding product complexes and sets, see p. 227.

6. As a preliminary to the extension of the coincidence and fixed point formulas, it is necessary to obtain an analogue of the theorem of *Tr.*I, No.52, for certain pseudocycles of $G \times G'$, For our purpose it is sufficient to consider a chain $D_s$ such that the *projection** of its points on $S'$ is always at finite distance. See p. 230.

We first return to the independent sets $\{\gamma_\mu^h\}$ of No. 3, and select for each $r-\mu$ an associated set $C_{r-\mu}^h$ defining a maximal independent set of pseudocycles, $\{\delta_{n-\mu}^h\}$, of $G$. Since the cycles for $\mu \leqq r$, are in finite number, and independent for each set, there exists an $N^i$ such that the cycles $\Gamma_\mu^{hj}$ with the same $j$ are independent on $N^i$ for some $j$ sufficiently large. We can then take for $H_r$ a simplicial complex of which they and the chains $C_{r-\mu}^h$ are all subchains and $N^i$ a subcomplex. As for the $C$'s we can even omit their cells exterior to $N^i$. A wholly parallel situation takes place as regards $S'_{r'}$ and $G'$ and we have $\Gamma'^{hj}$, $N'^i$, etc. We choose however, the simplicial complex $H'$ with the infinite point $Q'$ of $S'_{r'}$, as vertex and the simplexes with $Q'$ for vertex all exterior to $N'^i$.

The argument of *Tr.* 1, No. 50, enables us at the outset to replace $D_s$ by a subchain of $H \times H'$ having exactly the same properties. This may necessitate replacing $H$ or $H'$ by subdivisions with smaller cells, but there is no objection to that. We assume then that $D$ is a subchain of $H \times H'$ whose projection on $H'$ is finite. See p. 229.

7. Let us subdivide the $t$ subchains of $H_r$ into two categories, the first to consist of a maximal linearly independent set $\{B_t^i\}$ of which no boundary meets $G$, the second of a set of cells $\{E_t^j\}$ no linear combination of which is a $B$. Every $t$-subchain is a linear combination of these two types with rational coefficients. We do likewise for $H'$ except that the $B'$'s are to be *cycles.*

Therefore

$$D_s = \sum x_{ij}\, B_t^i \times B'^j_{s-t} + \sum y_{ij}\, B_t^i \times E'^j_{s-t} + \sum u_{ij}\, E_t^i \times B'^j_{s-t} + \sum v_{ij}\, E_t^i \times E'^j_{s-t}.$$

Our reduction will consist in adding to $D_s$ chains corresponding to pseudocycles $\approx 0$ and which are projected at finite distance on $S'_{r'}$.

If in any term of the sum the first factor does not meet $G$ or the second does not meet $G'$, that term is a chain which corresponds to a pseudocycle $\approx 0$. Since its second factor must not include $Q'$ as vertex, the term may freely be suppressed. We can then assume that every $E'$ in the expression of $D_s$ has a closure $\bar{E}'$ which meets $G'$. By assumption also

* The projection of a point $A \times A'$ on $S$ is $A$ and on $S'$ it is $A'$.

every $\bar{E}$ meets $G$. If we calculate $F(D_s)$ by (49.2) of *Tr.* I we find then that it includes a chain

$$E_t^i \sum v_{ij} F(E'^j_{s-t})$$

which meets $G \times G'$. Since $D_s$ must define a pseudocycle this chain must be $\equiv 0$, i. e. every $v_{ij} = 0$. Therefore the last sum in $D_s$ can be suppressed.

After the preceding reduction $F(D_s)$ will include a set of terms $F(B) \times B'$, $B \times F(E')$, $F(E) \times B'$ whose sum is again $\equiv 0$. There are no terms with $F(B')$ as factor since $B'$ is a cycle.

Suppose that we have cancellation of $B \times F(E')$ with terms of another type. Symbolically there will arise a relation of type

$$\sum y_{ij} B_t^i + \sum F(B) + \sum F(E) \equiv 0.$$

Therefore there is a subchain of $H_r$

$$C_{t+1}^i \to \sum y_{ij} B_t^i.$$

We have then by (49.2), p. 229,

$$C_{t+1}^i \times E'^j_{s-t} \to \sum y_{ij} (B_t^i \times E'^j_{s-t} + (-1)^{t+1} C_{t+1}^i \times F(E'^j_{s-t})).$$

The right hand side is $\approx 0$ on $S_r \times S_{r'}$, hence determines a pseudocycle $\approx 0$. We can subtract it from $D_s$, as a consequence of which the terms in $B_t^i \times E'^j_{s-t}$ will be replaced by terms $C^i \times F(E')$. Since $F(E')$ is a $B'$, these go into the third sum in the expression of $D_s$.

In the $D_s$ as now reduced only the first and third sums will be present. This time we obtain an identical relation for the coefficient of $B'^j_{s-t}$:

$$\sum x_{ij} F(B_t^i) + \sum u_{ij} F(E_t^i) \equiv 0$$

and this is ruled out, since the first sum represents a chain that does not meet $G$, whereas for the second the reverse is true. Therefore every $u = 0$.

Thus $D_s$ has been reduced to its first sum, i. e. to a linear combination of terms $B \times B'$, where $B$ defines a pseudocycle of $G$ and $B'$ is a cycle. The consequence for $\Delta_\sigma$, ($\sigma = s + n - r$) the pseudocycle of $D_s$, is manifestly a homology:

$$\Delta_\sigma \approx \sum \varepsilon_{ij}^t \delta_t^i \times \gamma'^j_{\sigma-t}$$

which expresses the analogue of the theorem of *Tr.* I, No. 52.

If the projection of $D$ on $S$ were bounded the result would be the same except for an interchange of $\gamma$ and $\delta$.

6. **Application to transformations.** Let $T$ be a single valued continuous transformation of $G$ into itself.✡ If we choose a system of coördinates for $S_r$, the transformation will have the form

$$x'_i = f_i(x_1, \cdots, x_r), \qquad i = 1, 2, \cdots, r$$

where the $f$'s are continuous and single valued on $G$. Since $G$ is closed and bounded, we can find single valued functions $F_i(x_1, \cdots, x_r)$, continuous and bounded for all values of the $x$'s and coinciding on $G$ with the respective $f$'s.† From this it follows readily that we can associate with $T$ a new single valued and continuous transformation $T^*$ of $S$ into $S'$ such that $T^*S$ is a bounded chain of $S'$.

We now take advantage of the fact that $R_0(G)$ is finite.—It has for consequence that our problem is reducible to the case where $G$ and $G'$ are connected. This being the case we can establish the existence of a continuous single valued transformation $J$ of $S$ into itself which is the identity on $G$ and such that $JS \subset N^i$. There exists also a similar transformation $J'$ for $G'$, $S'$, $N'^i$.—From this we infer: (a) Since $T^*$ may be replaced by $J'T^*$ we may assume that $T^*S \subset N'^i$. (b) The representative chains of any pseudocycles of $G$ may be replaced by their $J$ transforms and hence assumed on $N^i$ and similarly for $G'$ and $N'^i$. (c) If $S \times S' \supset A$, projected on $N'^i$ and representative of a pseudocycle of $G \times G'$ which is $\approx 0$, then $S \times S' \supset B, C$, chains projected on $N'^i$, such that $B$ does not meet $G \times G'$, while $C \to A + B$.

7. Let now $S'_r$ be a copy of $S_r$ with $G'$ as the image of $G$ on it. As a consequence of what we have just said the image of the transform $T^*S$ on $S \times S'$ in the sense of *Tr.* I, Part II, is a chain $D_r$ with a bounded projection on $S'$. It follows from No. 5 that if $\Delta_n$ is the pseudocycle of $D_r$, we have

$$\Delta_n \approx \sum \varepsilon^t_{ij}\, \delta^i_t \times \gamma'^j_{n-t}.$$

The parallel with (57.1) of *Tr.* I is now complete.

Let $I$ designate the identity on $G$. The associated starred transformation, $I^*$, can be taken as a single valued transformation of $S'$ into $S$ which reduces to the identity within a sphere of very large radius on $S'$ and then so modified that the projection of $D^0_r$ its image, on $S_r$ is bounded. Hence if $\Delta^0_n$ is the pseudocycle that it determines,

$$\Delta^0_n \approx \sum \eta^t_{ij}\, \gamma^i_t \times \delta'^j_{n-t}.$$

The index $(\Delta_n \cdot \Delta^0_n)$ is by definition the contribution to $(D_r \cdot D^0_r)$ of the intersections of $D_r$, $D^0_r$, on a neighborhood $N^i \times N'^i$ of $G \times G'$, whose $i$ is

† Carathéodory, Vorlesungen über reelle Funktionen, 2nd edit. 1927, p. 620, th. 2.

✡ See p. 578.

sufficiently large. Owing to (a) of No. 6, these intersections occur as near as we please to the images of the fixed points of $T^*$ on $G$, i. e. as near as we please to the fixed points of $T$. If $T$ had no fixed points, then $T^*$ would have none on $G$ and $(\Delta_n \cdot \Delta_n^0) = 0$. *Therefore, when* $(\Delta_n \cdot \Delta_n^0) \neq 0$, *$T$ has at least one fixed point on $G$.*

To calculate the index in terms of the effect of $T$ on the cycles of $G$, all that is needed is the extension of (59.1) of *Tr.* I. But that is now an elementary matter since we merely have to compare contributions to indices such as $(D_r \cdot \Gamma_\mu \times C'_{r-\mu})$, on $N^i \times N'^i$ and $(T^* \Gamma_\mu \cdot C'_{r-\mu})$ on $N'^i$, which is done as in *Tr.* I, No. 60 (p. 236).

From this point on and on the strength of (c) of No. 6 the derivation of the fixed point formula (3) is absolutely as in *Tr.* I and it is therefore completely established.

8. **Conclusion.** The right hand side of (3) or (4) shows that $(\Delta_n \cdot \Delta_n^0)$ is a topological invariant of $T$, since the matrices $f^p$ are invariants. Therefore

THEOREM. *Every continuous single valued transformation $T$ of a closed set $G$, whose Betti numbers and dimension are finite, possesses an index given by* (4) *which is a topological invariant. When it is $\neq 0$ $T$ has at least one fixed point.*

9. **Application.** *If the dimension of $G$ is finite and every $R_p(G) = 0$ for $p > 0$, while $R_0(G) = 1$, every continuous single valued transformation of $G$ has at least one fixed point.*

For here $\boldsymbol{f}^p = 0$ when $p > 0$. Moreover $G$ is a continuum since $R_0 = 1$, hence every zero-cycle is a multiple of any one point. Therefore $\boldsymbol{f}^0$ is an integer $\neq 0$ and $(\Delta_n \cdot \Delta_n^0)$, equal to that integer, $\neq 0$.

The question of the existence of a fixed point in the case just considered was raised with us two years ago by Alexandroff, and as we see the reply is affirmative.

This extends and includes of course the noted result due to Brouwer concerning the cell and the projective plane.

10. The extension to coincidences and fixed points in general is quite straightforward. Naturally restrictions such as those of *Tr.* II, No. 30, have to be put on the transformations, but these are inherent in the method and cannot be avoided. See p. 272.

The identification of the result just obtained with that of Hopf when $G$ is a complex, merely demands that we show that in his case the chains $D_r$, $D'_r$ intersect $K \times K'$ in chains whose index is $(\Delta_n \cdot \Delta_n^0)$. This is done however by a transparent comparison of indicatrices of the sort repeatedly given in our papers.

11. **Remarks.** I. One may define Brouwer's degree, $\alpha$, in the special case where $R_n(G) = 1$. If $\Gamma_n$ is an $n$-cycle not $\approx 0$, $\alpha$ is determined

by the condition $T\Gamma - \alpha\Gamma \approx 0$. It is likely that this definition with $\Gamma$ taken as a suitable relative cycle, gives at one stroke the degree of the topological manifolds recently investigated at great length by Wilfrid Wilson* and Hopf.†

II. The part played by the dimension $n$ is merely to locate a homeomorph $G$ in some Euclidean space, through Menger's theorem. We might as well have assumed at the outset that $G$ had that property and then the dimension would not have figured in the problem.

12. (Added in the proofs.) We have just succeeded in removing all restrictions as to the dimensions and finiteness of the Betti numbers of the spaces. The proofs will appear in our *Colloquium Lectures,* Ch. VII.

* Math. Ann., vol. 100, (1928).

† Ibid.

Erratum: The set G of page 576 may require some restriction when T is single-valued, as kindly pointed out to the Author by H. Hopf. It is sufficient, in particular, that G be locally connected, a condition satisfied by any complex. For further details, see the Author's Colloquium Lectures, Chapter VII, nos. 49-50.

# XV

## ON COMPACT SPACES

The chief purpose of the present paper is to give certain applications of the sequences of complexes approximating compact metric spaces introduced by Alexandroff.[1] From an important lemma regarding the complexes we derive anew certain deformation theorems of Alexandroff's, then prove Menger's immersion theorem and his surmise regarding a certain type of universal $n$-space.[2] The paper ends with the study of an interesting class of spaces defined by approximating sequences analogous to, but in general different from the Alexandroff sequences of complexes.

The unavoidable link between dimensionality and combinatorial topology is the Lebesgue-Urysohn-Menger theorem on the order of covering sets. This theorem is all that we borrow from that theory. Aside from that, up to and including the universal space, we need only a few elementary properties of complexes. These are completely established in our Colloquium Lectures *Topology,* whose notations and terminology are used wherever practicable.[3] Indeed had we obtained our basic lemma early enough the present paper would have been incorporated in Ch. VII of the book after § 4.

1. Let $L$ be a compact metric space, $\Sigma = \{F^\alpha\}$ a covering of $L$ by a finite number of closed sets $F^\alpha$. With the covering there is associated a certain complex $\Phi$, the *skeleton* of $\Sigma$, introduced by Alexandroff,[4] as follows: There is a vertex $A^\alpha$ of $\Phi$ for each $F^\alpha$, and a $k$-simplex $\sigma_k = A^{\alpha_0} A^{\alpha_1} \cdots A^{\alpha_k}$ of $\Phi$ for every intersection $F^{\alpha_0} F^{\alpha_1} \cdots F^{\alpha_k} \neq 0$.

---

[1] These Annals vol. 30 (1928–29), pp. 101–187.

[2] See Menger: Dimensionstheorie. As the present manuscript was practically finished there appeared a paper by Nöbiling, Math. Ann., vol. 104 (1930), pp. 71–80, in which there is a different proof of the immersion theorem, and also a proof of the existence of another type of universal $n$-space, likewise surmised by Menger. We show (No. 29) that this space contains one similar to our own, hence Nöbiling's result follows from ours. For $n = 0$ all these results coalesce and were established by Sierpiński; for $n = 1$ our results go back to Menger; he also sketched a proof of the immersion theorem for $n > 1$ but did not carry it through.

[3] We shall have occasion to consider only finite, simplicial complexes, no two cells of any complex $K$ having the same vertices. The successive derived of $K$ are denoted by $K'$, $K''$, $\cdots$ the new vertices of $K^{(i)}$ being the centroids of the carrying cells of $K^{(i-1)}$. The chief purpose of this last condition is to insure that mesh $K^{(i)} \to 0$ with $1/i$. When we state that $K \subset S_r$ we mean that its cells are simplexes of $S_r$.

[4] The results of the present number are found in his paper already quoted. He uses the term "nerve" for $\Phi$ but "skeleton" seems more descriptive.

$\Phi$ is the sum of the $\sigma$'s and it is closed. The dimension $n$ of $\Phi$ is the *order* of $\Sigma$. It is characterized by the property that there are groups of $n+1$ sets $F$, but no more, whose intersection $\neq 0$.

Let $F^{\alpha_0} \cdots F^{\alpha_k} = 0$, and let $P$ be any point of $L$. Since $L$ is compact the average of the distances $d(F^{\alpha_i}, P)$ has a lower bound $\eta > 0$. Since the total number of sets $F$ is finite there is only a finite number of $\eta$'s, —let $\xi (>0)$, be the least of them. $\xi$ has the following important property: If any subset $A$ of $L$ whose diameter $<\xi$ meets a group of $F$'s, their intersection $\neq 0$. For the average distance from any point of $A$ to one of the $F$'s is $<\xi$.

Let $\Sigma^* = \{F^{*\beta}\}$ be an $\varepsilon$-covering ($d(F^{*\beta}) < \varepsilon$) of $L$, $\varepsilon < \frac{1}{2}\xi$, and let $\Phi^*$ be the skeleton of $\Sigma^*$, with $A^{*\beta}$ as its vertices. We associate with each $A^{*\beta}$ one of the vertices $A^{\alpha}$ of $\Phi$ belonging to an $F$ which meets $F^{*\beta}$. To the vertices of any $\sigma$ of $\Phi^*$ correspond those of a $\sigma$ of $\Phi$. From this follows immediately that by barycentric extension to the points of the simplexes, the correspondence defines a continuous single-valued simplicial transformation $T$ of $\Phi^*$ into $\Phi$. In the sequel we shall merely say "simplicial transformation", it being understood that it is barycentric and single-valued.

2. LEMMA. *Given any $\varepsilon$-covering $\sum^*$, $\varepsilon$ sufficiently small, there exists a simplicial transformation $U$ of its skeleton $\Phi^*$ into the derived $\Phi'$ of the skeleton $\Phi$ of $\sum$, such that if $A^*$ is any vertex of $\Phi^*$, either $T \cdot A^*$ coincides with $U \cdot A^*$, or else they are joined by an edge of $\Phi'$.*

To establish the first part it is only necessary to show that there exists a $U$ transforming the vertices in the correct way.

Consider one of the sets $E_k = F^{\alpha_0} \cdots F^{\alpha_k} - \sum F^{\beta_0} \cdots F^{\beta_{k+1}} \neq 0$, where the sum is extended to all the combinations of upper indices taken $k+2$ at a time. $E_k$ represents the part of $F^{\alpha_0} \cdots F^{\alpha_k}$ which belongs to no similar set of more than $k+1$ intersecting $E$'s. The total number of $E$'s is finite and we designate them by $E_k^i$. The following properties are obvious:

I. $E_h^i \cdot E_h^j = 0, \quad i \neq j$.

II. If $\bar{E}_{h-1} \cdot \bar{E}_h \neq 0$, the $F$'s intersecting in $E_h$ include the same for $E_{h-1}$.

3. From I, II plus an elementary induction we conclude that there may be associated with each $E_h^i$ a closed set $G_h^i$ whose points are not farther than $\xi/4$ from $E_h^i$ and such that

III. $E_h^i \subset G_h^i + \sum G_{h+1}^l$, where, if $G_{h+1}^l$ is any term of the sum, $\bar{E}_h^i \cdot \bar{E}_{h+1}^l \neq 0$.

IV. $G_h^i$ meets no $G_h^j$, $i \neq j$, and the only sets $G_{h+1}$ which it meets are those present in the sum in III.

V. $G_h^i \cdot E_k = 0$ for $k \geqq h$ unless $E_k = E_h^i$.

From II, III, IV and from V follow respectively

VI. If $G_h^i \cdot G_k^j \neq 0$, $k > h$, the $F$'s intersecting in $E_k^j$ include the same for $E_h^i$.

VII. The only $F$'s which $G_h^i$ meets are those intersecting in $E_h^i$.

4. We will now prove the lemma for $\varepsilon < \xi/4$ and also $< 1/2$ the distance from any $G$ to a $G$ or to an $F$ which it does not meet. With $\varepsilon$ thus restricted we have:

VIII. If $F^{*\beta} \cdot F^{*\gamma} \neq 0$, $F^{*\beta} \cdot G_h^i \neq 0$, $F^{*\gamma} \cdot G_k^j \neq 0$ then

$h \neq k$, say $h < k$, implies that the $F$'s intersecting in $E_h^i$ are included among the same for $E_k^j$;

$h = k$ implies $i = h$ (if two $F^*$'s intersect they cannot intersect two different $G$'s with the same subscript $h$).

In particular a given $F^*$ can only meet one $G_h$ with given $h$ and if it meets $G_h^i$, $G_k^j$, $h < k$, then the two $G$'s are related as above.

IX. If $F^{*\beta}$ meets $G_h^i$, the only sets $F^\alpha$ which $F^{*\beta}$ can meet are those intersecting in $E_h^i$.

Now for each $E_h^i$ we have a definite group $F^{\alpha_0}, \cdots, F^{\alpha_h}$ intersecting in it. To their intersection there corresponds $\sigma_h = A^{\alpha_0} \cdots A^{\alpha_h}$ of $\Phi$, and hence a unique vertex $B^\beta$ of $\Phi'$, on that simplex.

Let now $A^{*\gamma}$ be any vertex of $\Phi^*$ with its associated set $F^{*\gamma}$. Among the sets $G_h$ which $F^{*\gamma}$ meets there is one with the highest subscript $h$ and it is unique by VIII; let it be $G_h^\delta$. We will define as the transform of $A^{*\gamma}$ the vertex $B^\beta$ of $\Phi'$ corresponding to $E_h^\delta$ in the manner just described. Let $T$ be the transformation of vertices thus defined. In order to be able to extend $T$ barycentrically over $\Phi^*$, and to have a single-valued transformation such as announced, all that is necessary is to show that if $\sigma_k = A^{*\gamma_0} \cdots A^{*\gamma_k}$ is any simplex of $\Phi^*$, its vertices have for transforms the vertices of a simplex of $\Phi'$ (*Topology* p. 85).

Since $\sigma_k$ is a simplex of $\Phi^*$, we have $F^{*\gamma_0} \cdots F^{*\gamma_k} \neq 0$. Let then $G_{h_i}^{\delta_i}$ correspond to $F^{*\gamma_i}$ in the same manner as above. According to VIII if the $h$'s are in increasing order, we have that the $F$'s intersecting in $E_{h_i}^{\delta_i}$ are included among the same for $i+1$. The first set of $F$'s determines a $\sigma_{h_i}$ of $\Phi$, the second a $\sigma_{h_{i+1}}$, and since the second set includes the first, $\sigma_{h_i}$ is a face of $\sigma_{h_{i+1}}$, or else coincides with it. As a consequence the simplexes $\sigma_{h_i}$ constitute a mutually incident set and therefore the corresponding $B$'s are vertices of a simplex of $\Phi'$ (*Topology*, p. 117). But these $B$'s are precisely the vertices $T \cdot A^{*\gamma_i}$, so that the latter are actually the vertices of a simplex of $\Phi'$. The barycentric extension of $T$ to the whole of $\Phi^*$ is the required transformation. We shall also call it $T$.

5. We pass to the last part of the lemma. Let $A = U \cdot A^*$. If $F$, $F^*$ are the sets of $\Sigma$ and $\Sigma^*$ associated especially with $A$ and $A^*$, we know that $F^*$ meets $F$. Therefore, by IX, $F$ is one of the sets of $\Sigma$ intersecting in $E_h^\delta$. This implies that $B = T \cdot A^*$ is a vertex of $\Phi'$ on

a simplex of $\Phi$ which has $A$ for vertex. Owing to the structure of $\Phi'$, $B$ is then on the closed star of $A$ as to $\Phi'$ and either $B = A$ or else $AB$ is an edge of $\Phi'$. This completes the proof of the lemma.

6. We will now consider a sequence of coverings $\{\Sigma^i\} = \{\{F^{i\alpha}\}\}$, where the mesh $\varepsilon_i$ of $\Sigma^i \to 0$ with $1/i$ monotonely and so rapidely that any two $\Sigma^i$, $\Sigma^{i+1}$ are related like the two coverings of the lemma, with $T^i$, $U^i$ as the two associated transformations. We shall further restrict the $\varepsilon$'s in a moment.

Consider the sets $E_h^{i\alpha}$ such as in No. 3, corresponding to $\Sigma^i$. The set $E_h^{i\alpha}$ is the intersection of a certain number of sets $F^{i\beta}$, and these in turn are met by certain sets $F^{i\gamma}$ (among which we include the sets $F^{i\beta}$ themselves). Let $V^{i\alpha}$ be the open *core* of $\Sigma F^{i\gamma}$:

$$V^{i\alpha} = L - \overline{(L - \Sigma F^{i\gamma})}.$$

Evidently $V^{i\alpha} \supset \bar{E}^{i\alpha}$ and hence $d(E^{i\alpha}, L - V^{i\alpha}) > 0$. For a given $i$ these distances are in finite number; if $\eta_i$ is the least of them, we shall take $\varepsilon_{i+1} < \frac{1}{2}\eta_i$. This is the other restriction on the $\varepsilon$'s mentioned above.

It is to be observed that from any $\{\Sigma^i\}$ whatever, whose $\varepsilon$'s merely $\to 0$, we can extract a subsequence whose $\varepsilon$'s are restricted in the present manner. For all practical purposes the subsequence can take everywhere the place of the initial sequence.

We will now define a new space whose points are those of $L$ and whose determining neighborhoods are the $V$'s. A neighborhood of a point $P$ is a $V^{i\alpha}$ whose $E^{i\alpha} \supset P$. There is one and only one for each $i$. Since the $F^i$'s whose sum is $\bar{V}^{i\alpha}$ are not farther than $\varepsilon_i$ from $E^{i\alpha}$, $d(V^{i\alpha}) \leq 4\varepsilon_i$. From this follows readily that the space just defined is identical with $L$.

7. Let $V^{i\alpha}$, $V^{i+1,\beta}$ be two neighborhoods of $P$. Since $E^{i+1,\beta}$ intersects $E^{i\alpha}$ and since $\varepsilon_{i+1} < \frac{1}{2}\eta_i$, the $F^{i+1}$'s whose sum is $\bar{V}^{i+1,\beta}$ are on $V^{i\alpha}$ and hence $\bar{V}^{i+1,\beta} \subset V^{i\alpha}$. We have then $P = \Pi V^{i\alpha}$.

Let $\sigma^{i\alpha}$ be the simplex of $\Phi^i$ corresponding to $E^{i\alpha}$, that is whose vertices correspond to the sets $F^i \supset E^{i\alpha}$ and let $N(\bar{\sigma}^{i\alpha})$ be the $\Phi^i$-neighborhood of $\bar{\sigma}^{i\alpha}$, (= the sum of the simplexes of $\Phi^i$ having a vertex in common with $\sigma^{i\alpha}$). The simplexes of $\bar{N}(\bar{\sigma}^{i\alpha})$ correspond to the sets $F^{i\gamma}$ and their intersections. Since the sets $F^{i+1}$ whose sum is $V^{i+1,\beta}$ meet only the sets $F^i$ whose sum is $\bar{V}^{i\alpha}$, $T^i \cdot \bar{N}(\bar{\sigma}^{i+1,\beta}) \subset \bar{N}(\sigma^{i\alpha})$.

8. Given any point $P$ of $L$ there is a unique set $E^i$, say $E^{i\alpha}$, carrying $P$ and $P = \Pi E^{i\alpha}$. Conversely if we have a sequence $\{E^{i\alpha}\}$ such that $\Pi E^{i\alpha} \neq 0$, since $d(E^i) \to 0$ with $1/i$, $\Pi E^{i\alpha} = P$, a single point of $L$. If $\sigma^{i\alpha}$ corresponds to $E^{i\alpha}$ in $\Phi^i$, each $\bar{N}$ of $\{\bar{N}(\bar{\sigma}^{i\alpha})\}$ is projected onto its predecessor by the proper $T$. A sequence of such $\bar{N}$'s associated with a sequence of $E$'s whose intersection $\neq 0$, is called a *projection sequence*

(= p. s.). The set of all p. s. and the points of $L$ are in $(1-1)$ correspondence.

Let $P \subset V^{i\alpha}$. The sets $F^i$ through $P$ are all included among the sets $F^{i\gamma}$, and hence, if $\{\bar{N}(\bar{\sigma}^{i\beta})\}$ is the p. s. for $P$, $\sigma^{i\beta} \subset \bar{N}(\bar{\sigma}^{i\alpha})$. Conversely if the latter condition is satisfied $P \subset V^{i\alpha}$. Thus the points of $V^{i\alpha}$ are in $(1-1)$ correspondence with the set of all p. s. $\{\bar{N}(\bar{\sigma}^{i\beta})\} = Q$ such that $\sigma^{i\beta}$ is a simplex of $\bar{N}(\bar{\sigma}^{i\alpha})$. Let $W^{i\alpha}$ designate the set of all p. s. $\{\bar{N}(\bar{\sigma}^{i\beta})\}$ such as just considered. Let us consider the $Q$'s as points of a new space $\mathfrak{L}$, with the $W$'s as the determining neighborhoods, where $W^{i\alpha}$ is a neighborhood of any $Q' = \{\bar{N}(\bar{\sigma}^i)\}$ whose $\sigma^i = \sigma^{i\alpha}$. Between the points of $L$ and $\mathfrak{L}$ there is a $(1-1)$ correspondence wherein $V^{i\alpha}$ and $W^{i\alpha}$ correspond to one another. Therefore $\mathfrak{L}$ is homeomorphic to $L$.[5]

9. So far nothing has been said concerning dim $L$. *We now assume that $L$ is an $n$-space.* As a consequence, we can choose an infinite sequence $\{\Sigma^i\}$ whose coverings are all of order $n+1$ (Lebesgue-Urysohn-Menger). Assume then that such a choice has been made. One of the first and simplest applications of our lemma is the proof of certain mapping theorems brought out by Alexandroff. The first is expressed in terms of *$\varepsilon$-mapping*. We understand by $\varepsilon$-mapping $\mu$ of $L$ onto a metric space $L'$ a continuous single-valued transformation of $L$ into $L'$, such that whatever the point $Q$ of $L'$, $d(\mu^{-1}Q) < \varepsilon$. In speaking of $\varepsilon$-deformation we shall understand that $L$ and the $\Phi$'s are on the Hilbert parallelotope $\mathfrak{H}$, the $\Phi$'s being constructed as in *Topology* p. 326. Then

THEOREM. *Necessary and sufficient conditions in order that $L$ be an $n$-space, are that $n$ be the least integer such that for every $\varepsilon > 0$,*

(a) *$L$ is $\varepsilon$-mappable on an $n$-complex whathever $\varepsilon$, or else,*

(b) *$L$ is $\varepsilon$-deformable onto an $n$-complex over $\mathfrak{H}$.*

10. We begin with a preliminary observation. Let $K$, $K^*$ be two complexes, $K'$, $K^{*\prime}$, their first derived, and suppose that $K = T \cdot K^*$, $T$ simplicial. Then $K' = T' \cdot K^{*\prime}$, where $T'$ is simplicial and transforms the cells of $K^*$ like $T$ itself. Take indeed the $T'$ transforming the vertices of $K^*$ like $T$, and the vertex of $K^{*\prime}$, $K'^*$ on $\sigma^*$ of $K^*$ into that of $K'$ on $\sigma = T \cdot \sigma^*$, then extend $T'$ barycentrically over $K^{*\prime}$. The transformation thus obtained behaves manifestly as required.

We need also the following property: Referring to the lemma and in the notations used in its proof, $U \cdot A^*$ is on the star of $T \cdot A^*$ relatively to $\Phi$. Hence if $\sigma$ is any simplex of $\Phi^*$, $U \cdot \bar{\sigma}$ is on the $\Phi$-neighborhood

[5] A space analogous to $\mathfrak{L}$ has been considered by Alexandroff, loc. cit., but in his treatment the elements of the p. s. are simplexes, which requires that $\{\Sigma^i\}$ be a so-called *subdivision* sequence. The p. s. here introduced do not demand that and hence enable us to proceed more speedily.

of $T \cdot \bar{\sigma}$, and therefore the closed $\Phi'$-neighborhood of $U \cdot \bar{\sigma}$ is on the closed $\Phi$-neighborhood of $T \cdot \bar{\sigma}$.

11. When $L$ is an $n$-space we can assume that the $\Phi$'s are $n$-complexes. Let us write for convenience $\Phi^{ik}$ for $(\Phi^i)^{(k)}$, the $k$th derived of $\Phi^i$. By repeated application of the lemma together with the second property in No. 10, we find that $\Phi^{i+k+1}$ is simplically transformable into $\Phi^{ik}$ in such a manner that if $\{\bar{N}(\bar{\sigma}^j)\}$ is any p. s., $\bar{N}(\bar{\sigma}^{i+1})$, $\bar{N}(\bar{\sigma}^{i+2})$, $\cdots$ are transformed into mutually inclusive subcomplexes of $\Phi^{i1}$, $\Phi^{i2}$, $\cdots$. Owing to the structure of the $N'$s the diameter of the transform of $\bar{N}(\bar{\sigma}^{i+j}) < 3$ mesh $\Phi^{ij}$, and hence $\rightarrow 0$ with $1/j$. Therefore the subcomplexes converge on a certain point $Q$ of $\Phi^i$. Let $\tau_i$ be the transformation of $L$ into $\Phi^i$ carrying into $Q$ the point $P$ of $L$ corresponding to the p. s. Through $\tau_i$ two points, $P$, $P'$ corresponding to $\{\bar{N}(\bar{\sigma}^j)\}$, $\{\bar{N}(\bar{\sigma}'^j)\}$ go into points on a closed cell of $\Phi^{ik}$ provided that $\sigma^j \subset \bar{N}(\bar{\sigma}'^j)$ for $j \leqq i+k+1$. Since mesh $\Phi^{ik} \rightarrow 0$ with $1/k$, $\tau_i$ is continuous. Two points $P$, $P'$ of $L$ cannot have the same transform unless they belong to the same neighborhood $V^{i\alpha}$ of No. 6. Hence if $\xi_i$ is the maximum diameter of $V^{i\alpha}$ for $i$ fixed, $\tau_i$ is a $\xi_i$-mapping of $L$ onto the $n$-complex $\Phi^i$. Since $\xi_i \rightarrow 0$ with $1/i$, we see that the condition of part (a) of the theorem is necessary.

The sufficiency of (a), is proved as follows (Alexandroff loc. cit.): By what precedes since $L$ cannot be mapped on a complex of dimension $<n$, $\dim L \geqq n$. Let $\mu$ be an $\varepsilon$-mapping of $L$ onto $K_n$. There exists an $\eta > 0$ such that if $Q$, $Q'$ are points on $K_n$ not farther than $\eta$ apart, $\mu^{-1}(Q+Q')$ is of diameter $< 2\varepsilon + \eta$. Replace $K_n$ by a simplicial subdivision, still to be called $K_n$, and so chosen that its first derived $K_n'$ is of mesh $< \eta/2$. Then the system $\{s^\alpha\}$, of the closed stars of the vertices of $K_n$ relatively to $K_n'$, is an $\eta$-covering of $K_n$ whose order is $n$. Now the system $\{\mu^{-1} s^\alpha\}$ is a $3\varepsilon$-covering of $K_n$ whose order is likewise $n$. Since such a covering exists for every $\varepsilon$, $\dim L \leqq n$, hence $\dim L = n$. This proves part (a) of the theorem.

Regarding part (b) it is readily seen that if $P$ is any point of $L$, $d(P, \mu P) \rightarrow 0$ with $1/i$, and hence, for $i$ large enough, $\mu$ is an $\varepsilon$-deformation. The converse of (b) follows from the fact that when $L$ is $\varepsilon$-deformable onto $K_n$ it is $\eta$-mappable onto it, where $\eta \rightarrow 0$ with $\varepsilon$.

12. Remark. The transformation $\tau_{ij} = U_i \cdots U_{i+j-2} U_{i+j-1}$ represents a simplicial transformation of $\Phi^{i+j}$ into $\Phi^{ij}$ such that $\tau_i = \tau_{ij} \tau_{i+j}$. If $P$, $Q$ are points of $L$ or $\Phi^j$, the points $\tau_i P$, $\tau_{ij} Q$ of $\Phi^i$ are called their *projections* on $\Phi^i$.

13. We shall need the following elementary properties:

(a) Let $K_n$ be an $n$-complex and $A^1, \cdots, A^q$ its vertices. Given $q$ arbitrary points $B^1, \cdots, B^q$ of $S_{2n+1}$ we can construct a $K_n^*$ on $S_{2n+1}$ having the same structure as $K_n$ and such that if $A^{*i}$ is the vertex of $K_n^*$ corresponding to $A^i$, every $A^{*i}$ is in a prescribed neighborhood of $B^i$.

*Proof.* Choose $q$ arbitrary points $A^{*i}$ in $S_{2n+1}$ and take all simplexes $A^{*i} \cdots A^{*j}$ such that $A^i \cdots A^j$ is a simplex of $K_{2n}$. Their totality will fail to add up to a complex having the structure of $K_n$ when and only when the $A^*$'s satisfy certain algebraic equations in finite number. Hence in any vicinity of $q$ points of $S_{2n+1}$ there are groups of $q$ points not restricted like the special groups.

(b) If $K_n \subset S_r$ has for mesh $\eta$, its derived $K_n'$ has its mesh $\leqq \eta\left(1-\frac{1}{n+1}\right)$.

*Proof.* It is sufficient to take for $K_n$ a $\bar{\sigma}_n$. Now the diameter of a $\sigma$ is equal to the length of its longest edge. For in the first place, given a segment on a convex polyhedral region of $S_n$ there is one at least as long, parallel to it and with an end point at a vertex. Hence the longest segment will be issued from a vertex. But in a triangle $ABC$ the length of any segment issued from $A$ is included between $AB$ and $AC$. Hence in a $\sigma$ with $A$ for vertex the longest segment issued from $A$ will be on the boundary of $\sigma$, hence on a $\sigma$ of lower dimension. Therefore by induction it is proved that $d(\sigma)$ = the longest edge of $\sigma$.

Now if $G$ is the centroid of $\sigma$ and $AB$ the segment through $G$ joining the vertex $A$ to the opposite face, $G$ divides $AB$ in the ratio 1 to $n$. Therefore $AG$, $BG < d(\sigma) \cdot \left(1-\frac{1}{n+1}\right)$. This implies property (b) for $\bar{\sigma}$ and hence for $K$.

14. THE IMMERSION THEOREM. *Every compact metric n-space can be mapped topologically on an* $S_{2n+1}$ (Menger).

The proof rests on the construction of a sequence $\{\Psi^i\}$ where $\Psi^i$ is a complex on a fixed $S_{2n+1}$ whose structure is that of $\Phi^i$, and with the following properties: Let $\sigma^i$, $N(\sigma^i)$, $T^i$, etc., apply now to the $\Psi$'s as before to the $\Phi$'s. There exists on $S_{2n+1}$ a neighborhood $\mathfrak{N}(\sigma^i)$ of $\bar{N}(\bar{\sigma}^i)$ whose points are not farther than $\eta_i$ from $\bar{N}(\bar{\sigma}^i)$ and such that:

I. $\eta_i \leqq \varepsilon_i \leqq \left(1-\frac{1}{n+1}\right)^{i-1} \varepsilon_1$, where $\varepsilon_i$ is the mesh of $\Psi^i$;

II. if $\bar{N}(\bar{\sigma}^i) \cdot \bar{N}(\bar{\sigma}'^i) = 0$, likewise $\mathfrak{N}(\sigma^i) \cdot \mathfrak{N}(\sigma'^i) = 0$;

III. if $\sigma^i = T^i \cdot \sigma^{i+1}$, $\mathfrak{N}(\sigma^i) \supset \mathfrak{N}(\sigma^{i+1})$.

From the construction of the $\Phi$'s in terms of the coverings follows that no two cells of $\Phi^i$ have the same vertices. We can therefore construct $\Psi^1$ as required. Suppose now that we have already constructed every $\Psi$ up to the index $i$. Let $\Psi^{i\prime}$ designate the derived of $\Psi^i$ and let $T'^i$, $U'^i$ denote the simplicial transformation of $\Phi^{i+1}$ into $\Psi^i$ and $\Psi^{i\prime}$ whereby each vertex $A^{i+1}$ of $\Phi^{i+1}$ goes into the images of $T^i A^{i+1}$ and $U^i A^{i+1}$ on $\Psi^i$ and $\Psi^{i\prime}$.

By No. 13a, we can construct $\Psi^{i+1}$ with the structure of $\Phi^{i+1}$ and such that every $B^{i+1}$ is not farther than $\zeta_i$ from the corresponding $U'^i A^{i+1}$. Then if $\sigma'^{i+1}$ is any cell of $\Phi^{i+1}$, for $\zeta_i$ small enough, its image $\sigma^{i+1}$ will be as near as we please to $U'^i \sigma'^{i+1}$, and hence, referring to No. 10, $\bar{N}(\bar{\sigma}^{i+1})$ will be as near as we please to $\bar{N}(T'^i \cdot \bar{\sigma}'^{i+1})$ where the two $N$'s are respectively $\Psi^i$- and $\Psi^{i+1}$-neighborhoods.

It follows that for suitably small $\zeta_i$, $\bar{N}(\bar{\sigma}^{i+1}) \subset \mathfrak{N}(T'^i \cdot \sigma'^{i+1})$, and $\varepsilon_{i+1}$ will differ as little as we please from $\left(1 - \dfrac{1}{n+1}\right)\varepsilon_i$.

Therefore we can fulfill condition I as regards $\varepsilon_{i+1}$, then choose $\eta_{i+1} < \varepsilon_{i+1}$ and at all events so small that $\mathfrak{N}(\sigma^{i+1}) \subset \mathfrak{N}(T'^i \cdot \sigma'^{i+1})$, thus complying with III, and also so small that II holds. The required construction of the $\Psi$'s is thus completed.

We now apply the term p. s. to the image of a p. s. made up of subcomplexes of the $\Psi$'s. If $\{\bar{N}(\bar{\sigma}^i)\}$ is a p. s., according to I, $d(\bar{N}(\bar{\sigma}^i))$ and hence also $d(\mathfrak{N}(\sigma^i)) \to 0$ with $(1/i)$. Therefore, by III, $\mathfrak{N}(\sigma^i)$ converges on a point $P^*$ of $S_{2n+1}$. We designate by $L^*$ the set of all points $P^*$.

Consider $L^*$ as a space whose defining neighborhoods are $L^* \cdot \mathfrak{N}(\sigma^i)$, the latter being a neighborhood for every $P^*$ associated with it in the preceding manner ($\mathfrak{N}$ is member of a sequence $\to P^*$). The association of $P^*$ defined by $\{\mathfrak{N}(\sigma^{i\alpha})\}$ with $Q = \{\bar{N}(\bar{\sigma}^{i\alpha})\}$ is a $(1-1)$ correspondence between $L^*$ and the space $L$ of No. 8, in which as a consequence of II, $W^{i\alpha}$ and $L^* \cdot \mathfrak{N}(\sigma^{i\alpha})$ correspond to one another point for point. Therefore $L$ and $L^*$ are homeomorphic, and so are $L$ and $L^*$. Since $L^* \subset S_{2n+1}$, *this proves Menger's immersion theorem.*

15. We shall now transfer to the relations between the $\Psi$'s and $L^*$ the terminology of the Remark No. 12, In particular the $\tau$'s are now projections of $L^*$ and $\Psi^j$ onto $\Psi^i$. Now it is apparent that the preceding construction of $L^*$ could be modified as follows: Replace $\Psi^1$ by $(\Psi^1)^{(j+1)}$ and $\Psi^2$ by $\Psi^j$. Then the neighborhoods $\mathfrak{N}(\sigma)$ being chosen with their points never farther from $\sigma$ than, say, three times its diameter, we would obtain an $L^*$ such that no point is farther than three times the mesh of $(\Psi^1)^{(j)}$ from its projection on $\Psi^1$. Since that mesh $\to 0$ with $1/j$, and since $\Psi'$ can be replaced by any $\Psi^i$ we have:

THEOREM. *There exists a topological image $L^*$ of $L$ on $S_{2n+1}$ such that no point is farther than an assigned $\varepsilon > 0$ from its projection on $\Psi^i$.*

16. **Universal $n$-space.** The term "universal space" is now in general use in the following sense: Given a class $\{\mathfrak{N}\}$ of metric sets or spaces, if there exists a member $\mathfrak{N}_0$ of the class such that every $\mathfrak{N}$ can be topologically mapped on $\mathfrak{N}_0$, the latter is said to be a *universal* set or space

for the whole class. The class to be considered here is that of all compact metric $n$-spaces and we propose to construct a *universal n-space* for that class.

17. Let $G$ be a point-set coincident with a subcomplex of $K_n$. The sum $N$ of all cells of $K$ with a vertex on $G$ is an open subcomplex of $K$, the *K-neighborhood* of $G$. We have already considered such an $N$ in No. 7, when $G$ consists of a single closed cell. If $G$ has the property that every cell of $K$ with all its vertices on $G$ is a cell of $G$, $N$ is said to be a *normal* neighborhood. An equivalent property is that every cell of $N-G$ has some vertex $\not\subset G$. When $N$ is normal through every point $P$ of $N-G$ there passes a unique segment $QR$, on the same simplex $\sigma$ of $K$ as the point, with $Q$ on $G$ and $R$ on the boundary of $N$. $Q$ is on the face of $\sigma$ on $G$, $R$ on the opposite face; both points vary continuously when $P$ so varies on $N-G$. The segment $QR$ is the *projecting segment of $P$ relatively to $N$*, $Q$ and $R$ the *projections* of $P$ on $G$ and on the boundary of $N$. As regards the boundary of $N$ it is the sum of the faces of the simplexes of $N$ without vertices on $G$.

In the sequel the neighborhoods to be considered are $K'$-neighborhoods of a subcomplex of $K$ and these are always normal (*Topology* p. 91). The preceding situation will then hold automatically.

18. We introduce an infinite sequence $\{K^i_{2n+1}\}$ defined as follows:— (a) $K^0_{2n+1}$ is a $\bar{\sigma}_{2n+1}$; (b) if $K^i_n$ is the sum of the closed $n$-cells of $K^i_{2n+1}$ then $K^{i+1}_{2n+1}$ is the closure of the $(K^i_{2n+1})''$-neighborhood of $K^i_n$. Setting

$$\Delta = \Pi K^i_{2n+1} = \overline{\Sigma K^i_n},$$

we shall show that *$\Delta$ is a universal compact n-space.* This space, obviously compact, differs only from the one announced by Menger in that in his construction our simplexes are replaced by parallelotopes. As a matter of fact, with insignificant modifications, our discussion would hold if $\{K^i_{2n+1}\}$ were any sequence of convex complexes, each a subdivision of its predecessor, their mesh $\to 0$ with $1/i$. However the sequence here adopted is decidedly the most convenient.

19. We will establish the universal space property of $\Delta$ independently of the considerations of this number. Granting that result it is a simple matter to show that $\Delta$ is an $n$-space. For if $L$, $L'$ are compact metric spaces and $L' \subset L$, we have $\dim L' \leqq \dim L$. This is indeed an immediate consequence of the Urysohn-Menger definition plus an elementary recurrence. Since $\Delta$ contains compact metric $n$-spaces, $\dim \Delta \geqq n$. On the other hand, if $\varepsilon_i$ is the mesh of $K^i_{2n+1}$, $K^{i+1}_{2n+1}$ can be $\varepsilon_i$-deformed into $K^i_n$ by sliding its points along the projecting segments of the $(K^i_{2n+1})'$-neighborhood

of $K_n^i$. Since $\Lambda \subset K_{2n+1}^{i+1}$, $\Lambda$ is thus $\varepsilon_i$-deformable onto a $K_n$, and as $\varepsilon_i \to 0$ with $1/i$, $\dim \Lambda \leqq n$, and hence $= n$.

20. Any simplex of $K_{2n+1}^i$ not on $K_n^{i-1}$ is of type $\sigma_p \sigma_q$ where $\sigma_p \subset K_n^{i-1}$, and $\bar{\sigma}_q$ does not meet $K_n^{i-1}$. The sum of the simplexes $\sigma_q$ will be designated by $\mathfrak{B}_{2n}^i$ and the sum of the (open) simplexes $\sigma_p \sigma_q$ by $\mathfrak{K}_{2n+1}^i$. The first will also be called the *boundary* of $K_{2n+1}^i$.

21. Let now $L$ be the same $n$-space as before. We propose to construct a sequence of homeomorphs $\{L^i\}$ of $L$, $L^i \subset \mathfrak{K}_{2n+1}^i$, and show that $\lim L^i = L^*$, a space homeomorphic to $L$. Since $L^* \subset \Lambda$, this will prove that $\Lambda$ is a universal $n$-space. At the same time as $L^i$ we shall consider topological images on $\mathfrak{K}_{2n+1}^i$ of the sequence set $\{\Psi^j\}$ associated with $\{\Sigma^j\}$, or of points and sets on $L$. These will all be designated by an additional superscript $i$, as $\Psi^{ij}$, $P^i$, $\cdots$.

In the construction we need a special sequence of neighborhood-pairs on $L$. If $\{V^j\}$ is an enumerable determining set of neighborhoods for $L$, the pairs in question $(V^j, V^k)$ are those such that $\bar{V}^j \cdot \bar{V}^k = 0$. They constitute likewise an enumerable set which, ranged in some order, shall be designated by $\{(V^j, W^j)\}$. Given any two distinct points $P$, $Q$ of $L$, there exists at least one pair $(V^j, W^j)$ such that $P \subset V^j$, $Q \subset W^j$. This is the essential property of the sequence for our purpose.

22. To obtain $L^1$ we first construct $\Psi^{11} = \Psi^1$ as in No. 14, except that its vertices are taken on $\mathfrak{K}^1$. Since the latter is a neighborhood of the sum of the $n$-faces of $K_{2n+1}^0 = \bar{\sigma}_{2n+1}$, if the vertices are taken near enough to those of the simplex, $\Psi^{11} \subset \mathfrak{K}^1$. We then construct the other $\Psi$'s as in No. 14, taking care that all are on $\mathfrak{K}^1$. As a consequence there results the topological image $L^1$, of $L$, on $K_{2n+1}^1$. Suppose that for each index $\leqq i$, and large enough $j$ in each case, we have constructed $L^i$, $\Psi^{ij}$ and the nested neighborhoods $\mathfrak{N}(\sigma^{ij})$ of the construction in No. 14, all on $\mathfrak{K}^i$. We shall show how to obtain the same for $i+1$.

23. Let $A^i$, $B^i$ designate the closures of the $(K_{2n+1}^i)''$-neighborhoods of $K_n^{i-1}$ and $\mathfrak{B}_{2n}^i$. I say that $A^i \cdot B^i = 0$. It is only necessary to show that their cells on any simplex $\sigma_p \sigma_q$ of $\mathfrak{K}^1$, $\sigma_p \subset K_n^{i-1}$, $\sigma_q \subset \mathfrak{B}^i$, are distinct. This amounts to proving that two cells of $(\overline{\sigma_p \sigma_q})''$ having respectively a vertex on $\sigma_p$ and $\sigma_q$ can have no common vertices. But the cells of $(\overline{\sigma_p \sigma_q})'$ so related have only one vertex in common—the centroid of $\sigma_p \sigma_q$. Since the centroid is not a vertex of any cell of $(\overline{\sigma_p \sigma_q})''$ with a vertex on $\sigma_p$ or $\sigma_q$ our assertion follows.

If we set $C^i = K_{2n+1}^i - A^i - B^i$, $C^i$ will be an interior subset of $K_{2n+1}^i$, i.e. on $\mathfrak{K}^i$. By sliding the elements, $L^i$, $\Psi^{ij}$, $\cdots$, along the projecting lines of $\mathfrak{K}^i$ we can bring them onto $C^i$. We shall assume this done and continue to designate them as before. We observe that *the deformation of every point is over a closed cell of* $K^i_{2n+1}$.

24. Consider the pair of neighborhoods $V^i$, $W^i$ and their images $V^{ii}$, $W^{ii}$. Now, on the one hand the closures of the latter do not meet, on the other hand $\Psi^{ij}$ approximates indefinitely to $L^i$ with increasing $j$, in such manner that the distance from any point of $L^i$ to its projection on $\Psi^{ij}$ tends uniformly to zero with $1/j$—a direct consequence of the construction of the $\Psi$'s and $L^i$. Therefore, for $j$ above a certain value, the distances of the projections of $V^{ii}$ and $W^{ii}$ on $\Psi^{ij}$ will differ from $d(V^{ii}, W^{ii})$ by an arbitrary small quantity. Since mesh $\Psi^{ij} \to 0$ with $1/j$, for $j$ sufficiently large, it will be possible to decompose $\Psi^{ij}$ into a sum of three (closed) complexes $\theta$, $\theta'$, $\theta''$ such that: (a) $\theta'$ and $\theta''$ have no common cells; (b) every cell of $\theta$ is of the form $\sigma' \sigma''$ where $\sigma'$ is a cell of $\theta'$ and $\sigma''$ a cell of $\theta''$; (c) the projection of $V^{ii}$ on $\Psi^{ij}$ is on $\theta'$ and that of $W^{ii}$ is on $\theta''$.

25. We recall that if $K$ is any finite complex there exists an $\eta > 0$ such that every complex (singular or not) of mesh $< \eta$ on $K$ is barycentrically deformable over $K$ into a subcomplex of $K$ in such a manner that no point leaves the closure of its carrying cell throughout the deformation. (Alexander-Veblen, see *Topology* p. 86.)

There exists in particular an $\eta$ such as just considered for $\mathfrak{B}^i_{2n} + (K^{i-1}_n)''$ (the latter term represents the subcomplex of $K^i_{2n+1}$ in coincidence with $K^{i-1}_n$). For reasons of continuity we can find an $\eta' > 0$ such that if we slide onto $K^{i-1}_n$ or $\mathfrak{B}^i_{2n}$ along the projecting segments of $K^i_{2n+1}$, a subset of $C^i$ whose diameter $< \eta'$, the resulting set has its diameter $< \eta$. We choose $j$ such that mesh $\Psi^{ij} < \eta'$, which is clearly possible since the two previous conditions imposed upon $\Psi^{ij}$ merely demand that $j$ be high enough.

26. Let us then first slide $\theta'$ and $\theta''$ along the projecting segments until they come respectively onto $K^{i-1}_n$ and $\mathfrak{B}^i_{2n}$. There result two singular complexes whose mesh $< \eta$. We can then deform them in the manner indicated in No. 25 so as to reduce them to subcomplexes $\theta'^*$, $\theta''^*$ of $(K^{i-1}_n)''$ and $\mathfrak{B}^i_{2n}$. Under these circumstances the vertices of a generic simplex of $\theta$ are deformed onto vertices of $\theta'^*$ and $\theta''^*$ which belong to a closed cell of $K^i_{2n+1}$. It follows that, associated with the barycentric deformations of $\theta'$, $\theta''$, there is one of $\theta$ into a subcomplex $\theta^*$ of $K^i_{2n+1}$ of such a nature that $\Psi^{ij} = \theta + \theta' + \theta''$ has been barycentrically deformed into the subcomplex $\theta^* + \theta'^* + \theta''^*$ of $K^i_n$, in such manner that no point leaves the closure of its cell on $K^i_{2n+1}$.

If $P$ is any point of $\Psi^{ij}$ the corresponding point $Q$ on the proper $\theta^*$ is on the closure of the cell that carries $P$, and hence the segment $PQ$ is on that closed cell. If $Q$ is on $\theta'^*$ or $\theta''^*$ there is an intervall $QQ'$

of the segment $QP$ on $A^i$ or $B^i$ respectively. Since $P$ remains on $C^i$ we can find on each $PQ$ a $Q'$, $Q' \not\subset C^i$, whose locus $\Omega$ is a homeomorph of $\Psi^{ij}$ and such that

$$Q' \subset A^i - K_n^i \text{ when } P \subset \theta',$$

$$Q' \subset B^i - \mathfrak{B}_{2n}^i \text{ when } P \subset \theta'',$$

$$Q' \subset \mathfrak{K}_{2n+1}^{i+1} \text{ when } P \subset \theta.$$

The third condition can always be taken care of since $Q'$ can be chosen as near as we please to $K_n^i$ and hence on the $(K_{2n+1}^i)''$-neighborhood of the latter. Under the circumstances $\Omega \subset \mathfrak{K}_{2n+1}^{i+1}$.

27. By the "projection" of a point of $\Psi^{ih}$ or of $L^i$ on $\Omega$ we shall now understand the image of its projection on $\Psi^{ij}$ under the homeomorphism between $\Omega$ and $\Psi^{ij}$.

Since $\mathfrak{K}^{i+1}$ is a region of $S_{2n+1}$ and $\Omega$ is a closed set on the region, there exists a positive $\zeta$ such that if a $\sigma_k$, $k \leq n$, is not farther than $\zeta$ from $\Omega$ and its diameter also $< \zeta$, then $\sigma_k \subset \mathfrak{K}^{i+1}$. Now by (a) of No. 13, we can construct a complex $\Psi^{i+1,h}$, $h > j$, (which implies that it has the structure of $\Psi^h$), whose vertices are as near as we please to the projections of the corresponding vertices of $\Psi^{ih}$ on $\Omega$. But the mesh of the projection of $\Psi^{ih}$ on $\Psi^{ij} \to 0$ with $1/h$, and hence this holds likewise for the mesh of the projection on $\Omega$. It follows that for $h$ large enough: (a) $\Psi^{i+1,h} \subset \mathfrak{K}^{i+1}$; (b) if $P'$, $P''$ are the projections of any point $P$ of $L^i$ on $\Omega$ and on $\Psi^{i+1,h}$ (i. e. $P''$ is the image of the projection of $P$ on $\Psi^{ih}$), then $d(P', P'') < \varepsilon$ arbitrarily assigned positive number. We now construct the complexes $\Psi^{i+1,k}$, $k > h$, and $L^{i+1}$ as before and we can again so construct them that no point of $L^{i+1}$ or of any $\Psi^{i+1}$ is farther than $\varepsilon$ from its projection on a given $\Psi^{i+1}$, in particular from its projection on $\Psi^{i+1,h}$. But for a proper choice of $\varepsilon$ the various projections of the points of $V^{ii}, W^{ii}$ will remain as near as we please to those of $\theta', \theta''$. Therefore we can so dispose of the situation that the projections of $V^i, W^i$ on $\Psi^{i+1,h}$ be respectively on $A^i$ and $B^i$ without points on $C^i$. Since we can also manage to bring every point of $L^{i+1}$ arbitrarily near to its projection on $\Psi^{i+1,h}$, we can so construct $L^{i+1}$ that $V^{i,i+1} \subset A^i$, $W^{i,i+1} \subset B^i$. Finally as the deformation from $\Psi^{ij}$ to $\Omega$ is such that no point leaves the closure of its carrying cell on $K_{2n+1}^i$, and so the points of $L^{i+1}$ are arbitrarily close to their projections on $\Omega$, we can also assume that we have:

I. If $P$ is any point of $L$, then $P^{i+1}$ is on the $(K_{2n+1}^i)''$-neighborhood $N^i$ of the closure of the cell of $K_{2n+1}^i$ that carries $P^i$.

In addition, according to the above:

II. If $P \subset V^i$, $Q \subset W^i$ are two points of $L$ then $P^{i+1} \subset A^i$, $Q^{i+1} \subset B^i$.

Since no cell of $(K^{i+1}_{2n+1})''$ can have vertices on both $A^i$ and $B^i$, we have from II:

III. The situation being as in II, $N^{i+1}$ does not meet its analogue $N'^{i+1}$ for $Q$.

Now there exists an $\eta > 0$ such that when $d(P^i, Q^i) < \eta$, $P^i$ and $Q^i$ are on the same star of a cell of $K^i_{2n+1}$ and therefore $P^{i+1}$, $Q^{i+1}$ will be on the $(K^i_{2n+1})''$-neighborhood of the closed star. Since $L^i$ is a homeomorph of $L$ we have:

IV. There exists a $\xi_i \to 0$ with $1/i$ such that if $d(P, Q) < \xi_i$ $P^{i+1}$ and $Q^{i+1}$ are on neighborhoods $N^i$, $N'^i$ which intersect.

We have thus constructed $L^{i+1}$ from $L^i$ so as to satisfy the preceding properties in addition to the condition $L^{i+1} \subset \mathfrak{K}^{i+1}$. In particular we have thus an inductive construction for $\{L^i\}$.

28. Let $\mathfrak{N}^i$ denote the $K^i_{2n+1}$-neighborhood of the closed cell of $K^i_{2n+1}$ that carries $P^i$. From I and the fact that $K^{i+1}_{2n+1}$ is a subcomplex of $(K^i_{2n+1})''$, we conclude that $\mathfrak{N}^i \supset \mathfrak{N}^{i+1}$. Furthermore $d(\mathfrak{N}^i) < 3$ mesh $K^i$, and hence $\to 0$ with $1/i$. Therefore the $\mathfrak{N}$'s converge on a single point $P^* \subset \varDelta$, since $P^* \subset K^i_{2n+1}$ whatever $i$. The set $L^*$ of all points $P^*$ is then a subset of $\varDelta$. I say that it is homeomorphic to $L$.

We have already that every point $P$ of $L$ has a single image $P^*$ on $L^*$. Let $Q$ be a point of $L$ other than $P$. There exists then a pair $V^i$, $W^i$ such that $V^i \supset P$, $W^i \supset Q$. It follows from III that $P^{i+1}$ and $Q^{i+1}$ are on certain non-intersecting neighborhoods of cells of $(K^{i+1}_{2n+1})''$. But these neighborhoods include the neighborhoods $\mathfrak{N}^{i+2}$ and its analogue $\mathfrak{N}'^{i+2}$ for $Q$. Therefore $\mathfrak{N}^{i+2} \cdot \mathfrak{N}'^{i+2} = 0$ and as the first carries $P^*$ and the second $Q^*$, $P^* \neq Q^*$. Therefore the correspondence between $L$ and $L^*$ is $(1 - 1)$.

Finally let $d(P, Q) < \xi_i$. From IV and $N^i \subset \mathfrak{N}^i$, $N'^i \subset \mathfrak{N}'^i$, follows $\mathfrak{N}^i \cdot \mathfrak{N}'^i \neq 0$. Since $P^* \subset \mathfrak{N}^i$, $Q^* \subset \mathfrak{N}'^i$, $d(P^*, Q^*) < 6$ mesh $K^i_{2n+1}$, and hence $\to 0$ with $1/i$, i.e. with $\xi_i$. Therefore the correspondence from $L$ to $L^*$ is continuous and as $L$ is compact it is a homeomorphism. Thus $L$ has been mapped topologically on a subset $L^*$ of $\varDelta$. *Therefore $\varDelta$ is a universal n-space.*

29. We have already alluded (footnote 2) to a universal space whose existence has been established by Nöbiling. That space, which we shall call $\varDelta^*$, is the set of all points of $S_{2n+1}$ having at most $n$ rational coördinates. It is a simple matter to derive Nöbiling's result from ours, as we shall now show.

In the first place it is clear that, if in the construction of $\varDelta$ in No. 18, we replace $(K^i_{2n+1})''$ by $(K^i_{2n+1})^{(h_i)}$, $h_i \geqq 2$, the ulterior treatment goes through as before, and therefore the resulting space, say $\varDelta'$, is still a universal $n$-space. We will take advantage of this new degree of freedom in a moment.

Suppose now that we have a certain number $s$ of positive transcendental numbers $\alpha_1, \cdots, \alpha_s$, between which there exists no relation $f(\alpha_1, \cdots, \alpha_s) = 0$, where $f$ is a polynomial with integer coefficients. The set of all positive numbers $\alpha_{s+1}$ satisfying an equation $f_1(\alpha_1, \cdots, \alpha_{s+1}) = 0$, where $f_1$ is of the same type as $f$, is enumerable. Hence we can always increase the number of $\alpha$'s having the property of the initial set by one and therefore find $(2n+1)^2$ such numbers. It follows that we can find a $\sigma_{2n+1}$ such that the coördinates of its vertices are transcendental numbers that satisfy no polynomial equation with integral coefficients.

Under the circumstances if we take any derived $(\bar{\sigma}_{2n+1})^{(h)}$ its $n$-simplexes will have no point with more than $n$ rational coördinates. It follows that $K_n^i \subset \Lambda^*$ whatever $i$, where $K_n^i$ is as in No. 18 but for the modified construction corresponding to $\Lambda'$.

Now the set of points having more than $n$ rational coördinates is the sum of all $S_n$ of $S_{2n+1}$ of the form $x_{k_j}$ = a rational number, $j = 1, 2, \cdots, n+1$. These spaces constitute an enumerable aggregate $\{S_n^j\}$, and we have (Nöbiling): $\Lambda^* = S_{2n+1} - \Sigma S_n^j$. According to the above $K_n^{i-1}$ does not meet any space $S_n^j$, so that the distance $\delta_i$ from $K_n^{i-1}$ to $S_n^i$ is positive. We will choose $h_{i-1}$ such that mesh $(\bar{\sigma})^{(h_{i-1})} \leqq \frac{1}{2}\delta_i$. Since $K_{2n+1}^i$ is the closure of the $(K_{2n+1}^{i-1})^{(h_{i-1})}$-neighborhood of $K_n^{i-1}$ it will not meet $S_n^i$, and hence $S_n^i$ will not meet $\Lambda' = \Pi K_{2n+1}^i$. This implies that $\Lambda' \subset S'_{2n+1} - \Sigma S_n^i = \Lambda^*$, and since $\Lambda'$ is a universal $n$-space this is likewise true for the space $\Lambda^*$ of Nöbiling.

**30. Separable spaces.** There is in existence an incomplete proof of the following proposition: *Every separable metric n-space can be mapped topologically on a compact metric n-space*[6]. Granting this result, it follows from what we have proved that a separable metric $n$-space can be mapped topologically on the universal space $\Lambda$, and, in particular, can be topologically imbedded in an $S_{2n+1}$.

It is possible to obtain the preceding result for an important class of separable metric spaces, the locally compact metric spaces, independently of the theorem of Hurewicz. These spaces can in fact be approximated by sequences $\{\Phi^i\}$ analogous to those of Alexandroff except that the $\Phi$'s are infinite. With relatively unimportant modifications in the treatment we can prove in particular the imbedding theorem. In constructing the initial complex $\Psi^1$ of No. 14, we choose its vertices $A^1, A^2, \cdots \to \infty$ with

[6] Hurewicz, Amsterdam Proc., vol. 30 (1927), pp. 425. See the observation in Menger's book at the end of p. 284.

the order, but the rest of the construction is substantially as before, distance conditions being replaced wherever need be, by suitable neighborhood restrictions.

31. **A certain class of metric spaces.** Let $\{\Omega^i\}$ be a sequence of compact metric spaces, such that for every $i$ there exists a continuous single-valued transformation $\tau_i$ of $\Omega^{i+1}$ into the *whole* of $\Omega^i$, i. e. such that $\Omega^i = \tau_i \Omega^{i+1}$. If $P$ is any point of $\Omega^{i+1}$, the points $\tau_i P$, $\tau_{i-1}\tau_i P, \cdots$ are called the *projections* of $P$ on $\Omega^i$, $\Omega^{i-1}$, etc. By a *projection sequence* (= p. s.) we now understand a sequence $\{P^i\}$, where $P^i$ is a point of $\Omega^i$ and $\tau_i P^{i+1} = P^i$ for every $i$.

We will now consider the set of all p. s. as points of a new space $L$. The result of No. 11 implies that every compact metric space is of this type. We shall show that the converse also holds: every space such as $L$ is compact and metric. What we have then is a construction of compact metric spaces in terms of (generally simpler) spaces of same type. By means of it we may hope to control to some slight extent the properties of $L$.

We may think of the consecutive points of a p. s. as joined by arcs which vary continuously with their extremities. The points of $L$ correspond then to the curves of a certain continuous family, in some respects similar to a family of dynamical trajectories.

32. Through immersing the $\Omega$'s in the Hilbert parallelotope (*Topology* p. 323) we can choose for their totality a metric such that the distances between the points of any $\Omega^i$ are less than a certain constant $A$ independent of $i$. We will now define $d(Q, Q')$, $Q = \{P^i\}$, $Q' = \{P'^i\}$ by the expression

$$d(Q, Q') = \sum_{i=1}^{+\infty} \frac{d(P^i, P'^i)}{i!}.$$

The verification of the two basic distance axioms (*Topology* p. 5) is immediate, hence $L$ is a metric space.

We will now show that $L$ is compact. For that purpose it is sufficient to prove that it possesses the following two properties[7]: (a) it is *totally bounded,* i. e. it can be $\varepsilon$-covered by a finite number of closed sets whatever $\varepsilon$; (b) it is *complete,* i. e. every Cauchy-sequence of points on $L$ has an actual limit.

Now given $\varepsilon$ we can choose $i$ such that

$$A \sum_{j=i+1}^{+\infty} \frac{1}{j!} < \frac{1}{2}\varepsilon.$$

[7] Hausdorff, Grundzüge der Mengenlehre, 1st ed., p. 314.

Since the $\Omega$'s are compact, we can cover every $\Omega^h$, $h \leqq i$, with a finite number of closed sets $F^{\alpha h}$ whose diameter $< \dfrac{\varepsilon h!}{2i}$. Consider now the set $G$ of all points $Q$ of $L$ such that $P^h \subset F^{\alpha_h h}$, where the $\alpha$'s are given. We will have $G \neq 0$ for certain sets of $\alpha$'s and then $G$ is closed. Furthermore every point of $L$ belongs to a $G$. Therefore the $G$'s $\neq 0$ constitute a finite covering of $L$. But if $Q$, $Q'$ belong to the same $G$, we find immediately $d(Q, Q') < \varepsilon$. Therefore $L$ is totally bounded.

Now let $\{Q^\alpha\} = \{\{P^{\alpha i}\}\}$ be a Cauchy-sequence. From the expression of the distance-function follows that $\{P^{\alpha i}\}$ is a Cauchy-sequence for $\Omega^i$. Since $\Omega^i$ is compact the sequence has a limit $P^i$. Owing to the continuity of $\tau_i$, $P^i = \tau_i P^{i+1}$, so that $Q = \{P^i\}$ is a point of $L$. Now in order that $d(Q, Q^\alpha) < \varepsilon$ it is sufficient that $d(P^i, P^{i\alpha})$ be less than a certain $\eta$ for $i$ less than a certain $h$. Since $P^{\alpha i} \to P^i$ we can choose an $\alpha_0$ such that these conditions be fulfilled for every $\alpha > \alpha_0$ and $i < h$. Hence $d(Q, Q^\alpha) < \varepsilon$ for $\alpha > \alpha_0$, and $Q^\alpha \to Q$. Thus $L$ is complete, and hence it is compact.

33. A basic property of $L$ is that *it is $\varepsilon_i$-mappable on $\Omega^i$, where $\varepsilon_i \to 0$ with $1/i$.* Let indeed $T^i$ be the mapping of $L$ on $\Omega^i$ whereby if $Q = \{P^i\}$, $T^i \cdot Q = P^i$. Then if $Q' = \{P'^j\}$ is also mapped on $P^i$ by $T^i$, we have $P^j = P'^j$ for $j \leqq i$, hence $d(Q, Q') \leqq A \sum_{j=i+1}^{+\infty} \dfrac{1}{j!} = \varepsilon_i \to 0$ with $1/i$.

An immediate consequence is that *if dim $\Omega^i \leqq n$, whatever $i$, likewise dim $L \leqq n$.* For $\Omega^i$ is $\varepsilon$-mappable on an $n$-complex whatever $\varepsilon$, and hence this holds likewise for $L$.

34. It appears probable that when dim $\Omega^i = n$ for $i$ above a certain value likewise dim $L = n$, but we have not been able to establish the fact. We shall then endeavor to restrict the $\Omega$'s in such manner as to have dim $L \geqq n$, and hence dim $L = n$. This will be done by choosing for the $\Omega$'s appropriate complexes.

Let indeed every $\Omega^i$ be an $n$-complex and suppose that there exist for every $i$ an $n$-cycle $\Gamma_n^i$ of $\Omega^i$ such that we have homologies:

$$\alpha_i \tau_i \Gamma_n^{i+1} \sim \beta_i \Gamma_n^i \text{ on } \Omega^i. \tag{1}$$

Under the circumstances *$L$ is an $n$-space.* For each $\Omega^i$ has a subdivision which is the skeleton of some covering $\Sigma^i$ of $L$ whose mesh is $3\varepsilon_i$, where $\varepsilon_i$ is as in No. 33, (Alexandroff, loc. cit. p. 18). Replacing if need be, every $\Omega^i$ by the subdivision in question, still called $\Omega^i$, and $\{\Omega^i\}$ by a subsequence, we can treat the new $\{\Omega^i\}$ as the sequence of *Topology* Ch. VII § 4. In our treatment indeed we assumed that we had a subdivision sequence. Now the only place where that mattered at all was in the proof of the deformation theorem, p. 328. It is however easily verified that this proof goes through provided $\varepsilon_i \to 0$ sufficiently fast, as we may well assume.

Under the circumstances we conclude from (1) to the existence of a Vietoris sequence $\{\Delta_n^i\}$, where $\Delta_n^i$ is a rational cycle of $\Omega^i$ and we have an $\varepsilon_i$-homology

$$\Delta_n^{i+1} \approx \Delta_n^i \text{ on } L,$$

in the sense of Vietoris (see *Topology* p. 330). It follows that $\{\Delta_n^i\} = \gamma_n$ is an $n$-cycle on $L$. Referring to the discussion loc. cit., it will be seen that $\gamma_n \not\approx 0$, on $L$, since $n$ is the common dimension of all the $\Omega$'s. Therefore the Betti-number $R_n(L) \geqq 1$, and consequently dim $L \geqq n$, (*Topology* p. 335). Hence finally dim $L = n$ as asserted.

It is thus seen that we have obtained the inequality dim $L \geqq n$ as a consequence of the fact that some homology character of $L$ for the dimension $n$ is $\neq 0$. As it happens it was the Betti-number, but any other character would do equally well.

35. A very simple method to construct effectively an $n$-space $L$ of the preceding type is as follows: Let $K_{n-1}$ be an absolute $(n-1)$-circuit. We will take $\Omega^i = K_{n-1}^i \times \gamma^i$ where $K_{n-1}^i$ is a copy of $K_{n-1}$ and $\gamma^i$ a circumference. Let $u_i$ be an angular parameter for $\gamma^i$, $A^{i+1} \times B^{i+1}$ any point of $\Omega^{i+1}$ where $A^{i+1}$ is the image of a point $A$ of $K$ and $B^{i+1}$ corresponds to the value $u_{i+1}$ of the parameter. We define $\tau_i$ by $\tau_i A^{i+1} \times B^{i+1} = A^i \times B^i$, where $A^i$ is the image of $A$ and $B^i$ corresponds to $u_i = k_i u_{i+1}$, $k_i$ an arbitrary integer. The number $k_i$ is the Brouwer degree of $\tau_i$. If $\Gamma_{n-1}$ is the fundamental $(n-1)$-cycle on $K_{n-1}$ and $\Gamma_{n-1}^i$ its image on $K^i$, the fundamental $n$-cycle on $\Omega^i$ is $\Gamma_{n-1}^i \times \gamma^i$ and we have

$$\tau_i \Gamma_{n-1}^{i+1} \times \gamma^{i+1} \sim k_i \Gamma_{n-1}^i \times \gamma^i \text{ on } \Omega^i, \tag{2}$$

which is (1) for the present system. Therefore $L$ is an $n$-space.

36. The space just constructed depends upon the arbitrary sequence of integers $\{k_i\}$. For the infinite $K_{n+1}$ of *Topology* Ch. VII, § 4, associated with $\{\Omega^i\}$, we have finite chains

$$C_{n+1}^{i+1} \to \Gamma_n^{i+1} - k_i \Gamma_n^i \tag{3}$$

where in the notation loc. cit., $C^i \subset N^i - N^{i+1}$. Furthermore there are no other boundary relations except (3) between the $\Gamma$'s. It follows that

$$C_{n+1} = \sum h_i C_{n+1}^i \to -k_1 \Gamma_n^1, \quad \frac{1}{h_i} = k_1 k_2 \cdots k_i.$$

Thus $C_{n+1}$ is a rational infinite chain with finite boundary. This chain defines an $n$-cycle $\Gamma_n$ on $L$, and from what preceeds we conclude that every other $n$-cycle on $L$ depends on $\Gamma_n$. Therefore the absolute Betti-number $R_n(L)=1$. It is not difficult to show that the same holds for the numbers $R_n(L;m)$ whatever $m$. However, unless all the numbers $k_i$ for $i$ above a certain value, are $\pm 1$, there is no integral $n$-cycle and the Betti-number for the integral cycles (Vietoris-cycles proper) is zero.

By taking products $L\times K$, where $L$ is of the type just constructed and $K$ a complex, we can obtain spaces with non-zero Betti-numbers for dimensions $<$ the dimension of the spaces themselves, and also having a numerical value $>1$.

37. Another interesting modification is in a different direction: Let $\Omega^i$ have a subcomplex $\omega^i$, such that $\{\omega^i\}$ constitutes a sequence analogous to $\{\Omega^i\}$ attached to the same sequence of transformations $\{\tau_i\}$. Then, $\Gamma_n^i$ designating now a cycle of $\Omega^i$ mod $\omega^i$, let (1) hold mod $\omega^i$. Under the circumstances $\{\omega^i\}$ defines a closed subset $l$ of $L$ and we have $R_n(L;l)=1$, hence again $\dim L=n$. The extension of the construction of No. 35 and the considerations of No. 36 hold here also with boundary relations, etc., taken mod $l$, instead of absolute.

Suppose that we merely know that the $\Omega$'s are all $n$-complexes and the $\tau$'s simplicial, it being always understood that $\tau_i\,\Omega^{i+1}=\Omega^i$. Then I say that we have a space of the type just considered. Let indeed $\sigma_n^1$ be any $n$-simplex of $\Omega^1$. Then there exists a simplex $\sigma^2$ of $\Omega^2$ of which it is the projection. But the projection of a $\sigma_p$ is a simplex of dimension $\leqq p$. Hence $\sigma^2$ is of dimension $\geqq n$ and since it is a simplex of an $n$-complex its dimension can only be $n$, so that its proper designation is $\sigma_n^2$. Similarly there exists a $\sigma_n^3$ of $\Omega^3$ whose projection on $\Omega^2$ is $\sigma_n^2$, etc. We have thus a sequence of simplexes $\{\sigma_n^i\}$, each the projection of its successor. If we set $\omega^i=\Omega^i-\sigma_n^i$, the sequence $\{\omega^i\}$ has the same properties as above except that now we have as the basic boundary relation

$$\sigma_n^i\to 0 \quad \bmod(\omega^i,2) \quad \text{on} \quad \Omega^i.$$

Therefore here $R_n(L;l,2)=1$ and $L$ is again an $n$-space.

# XVI

## ON SEPARABLE SPACES.[1]

We owe to E. H. Moore the introduction of a very general type of double sequence $\{P^{i\alpha}\}$ which he called a *development*. With such sequences there may be associated certain abstract spaces investigated at length by Chittenden and Pitcher with noteworthy results particularly in connection with the problem of metrization. They dealt at considerable length with so-called *regular* developments of a space $\mathfrak{R}$. In a regular development the $P$'s are neighborhoods of $\mathfrak{R}$ and $\alpha$ has a finite range for every $i$. Our first object in the present paper is to investigate a type of development called *normal* whose sets are subjected to more stringent conditions of convergency than with the regular type. It turns out, however, that every separable metric space possesses normal developments. As a consequence they seem to be just what is needed for the treatment of many questions on separable metric spaces.

Making use in part of normal developments, we have put on a solid basis the theory of the order for separable spaces, and in particular, completely extended to them the fundamental order theorem (Lebesgue order theorem). It was then a simple matter to prove that every separable metric space can be mapped topologically on a compact metric space of the same dimension. As a consequence certain mapping theorems that we have obtained previously for compact spaces[2] hold for separable spaces. By means of

---

[1] Received January 13, 1932. The results of the present paper were communicated in a Note to the Proceedings of the National Academy of Sciences, vol. 18, February 1932.

[2] See our paper *On Compact Spaces*, pp. 579–596 of the present work, also Nöbiling, Math. Ann., vol. 104 (1930), pp. 71–80, and a recent paper by Pontrjagin-Tolstova, Math. Ann., vol. 105 (1931), pp. 734–746. Regarding our paper, the sets $G$ of No. 3 are better introduced as follows: Define first the sets

$$(1) \qquad [F^{\alpha_0} \cdot \ldots \cdot F^{\alpha_k}] = F^{\alpha_0} \cdot \ldots \cdot F^{\alpha_k} - \Sigma F^{\beta}, \qquad \beta \text{ not an } \alpha_i,$$

then introduce closed sets $G$ such that

$$(2) \qquad G^{\alpha_0 \cdots \alpha_k} \supset [F^{\alpha_0} \cdot \ldots \cdot F^{\alpha_k}] - \Sigma G^{\alpha_0 \cdots \alpha_k \beta_1 \cdots \beta_h}, \qquad \beta_j \text{ not an } \alpha_i,$$

while $F^{\beta} \cdot G = 0$. We now select the $G$'s so as to satisfy IV, VI, VII loc. cit., with $E$ in VII replaced by [ ]. For $k = n$ the sets [ ] are closed and do not intersect, hence we can take for the $G$'s closures of non-intersecting neighborhoods of the [ ]'s. Let everything go through down to $k+1$. On the one hand no two $\overline{[\;]}$ with $k+1$ terms meet; on the other hand $[F^{\alpha_0} \cdot \ldots \cdot F^{\alpha_k}]$ meets only sets $G^{\alpha_0 \cdots \alpha_k \beta_1 \cdots \beta_h}$. Hence we can find suitable $G$'s

our results it is a simple matter to complete a noteworthy intersection theorem due to Menger and Hurewicz.[3]

## § 1. Generalities and Two Lemmas.

1. We shall denote our fundamental separable metric space by $\mathfrak{R}$, its points by $x, y, \cdots$, its distance-function by $d(x, y)$. If $A, B$ are subsets of $\mathfrak{R}$, by $A \Subset B$ is meant that the closure, $\bar{A}$, of $A$ is a subset of $B$. If $\sigma = \{A^\alpha\}$, $\sigma' = \{A'^\beta\}$ are two aggregates of subsets of $\mathfrak{R}$, we write $\sigma \subset \sigma'$ or $\sigma \Subset \sigma'$, whenever respectively every $A$ or every $\bar{A}$ is a subset of an $A'$. We shall also denote by $\bar{\sigma}$ the aggregate $\{\bar{A}^\alpha\}$ consisting of the closures of the sets of $\sigma$.

An open subset of $\mathfrak{R}$ will be generically designated by the letters $U$, $V$, $W$ with indices. An aggregate $\{U^i\}$, such that every open set of $\mathfrak{R}$ is a sum of sets $U^i$, is called a *base* of $\mathfrak{R}$. Separability is defined by the possession of an enumerable base.

A *covering* of $\mathfrak{R}$ is an aggregate of subsets $\{A\}$, whose sum is $\mathfrak{R}$; its *mesh* is the maximum diameter of the $A$'s. The covering is *finite* or *infinite* according as the number of $A$'s is finite or infinite. Of basic importance for the sequel are the finite coverings by open sets (= f. c. o. s.).

2. We recall that according to Urysohn $\mathfrak{R}$ can be mapped topologically on the Hilbert parallelotope $\mathfrak{H}$:

$$|y_i| < \frac{1}{i}, \qquad i = 1, 2, \cdots, +\infty,$$

which is itself compact and metric. The mapping is obtained as follows: Given any two closed subsets $A, B$ of $\mathfrak{R}$, whose intersection $A \cdot B = 0$, there exists a continuous single-valued function $f(x)$ of the point $x$ on $\mathfrak{R}$ such that

$$f(x) = 0 \text{ on } A, \quad f(x) = 1 \text{ on } B, \quad 0 < f(x) < 1 \text{ on } \mathfrak{R} - A - B.$$

We will call $f(x)$ a *characteristic function* of the pair $A, B$. Let $\{V^n\}$ be an enumerable base, and consider all the pairs $(V^p, V^q)$ such that $V^q \Subset V^p$. We have $\bar{V}^q \cdot \mathfrak{R} - V^p = 0$, and since the two intersecting sets are closed they possess a characteristic function. The totality of these functions is enumerable; range them in a certain order and call them $f_i(x)$, $i = 1, 2, \cdots, +\infty$. The subset $R$ of $\mathfrak{H}$ represented by

by taking closures of neighborhoods of the sets at the right in (2). The proof of the Lemma I is then carried out as before except that on p. 523 $T$ and $U$ should be interchanged.

[3] For accurate references to the literature the reader is asked to consult the recent article by Tietze-Vietoris, Encykl. der Math. Wiss. III A B 13, which appeared prior to the papers quoted in the preceding footnote. The notation and terminology of our paper are those now generally current in Topology.

(1) $$y_i = \frac{f_i(x)}{i}, \qquad i = 1, 2, \cdots, +\infty,$$

is the required topological image of $\mathfrak{R}$.

3. LEMMA I. *Given an enumerable aggregate of non-intersecting pairs of closed sets* $\{(A^i, B^i)\}$, *the metric can be so chosen that every* $d(A^i, B^i) > 0$.

Let $g_i(x)$ be the characteristic function for the $i$th pair, and let the $f$'s be as in (1). The equations

(2) $$y_{2i} = \frac{g_i(x)}{2i}, \qquad y_{2i+1} = \frac{f_i(x)}{2i-1}, \qquad x \subset \mathfrak{R},$$

represent a subset $R$ of $\mathfrak{H}$. By paraphrasing Urysohn's reasoning in the theorem just recalled, it is shown that $R$ is homeomorphic to $\mathfrak{R}$. Let us adopt for $\mathfrak{R}$ the metric of $R$, i. e. take as the distance of any two points of $\mathfrak{R}$ that of their images on $R$. If $x \subset A^i$, $x' \subset B^i$, we have $d(x, x') \geqq \left| \frac{g_i(x) - g_i(x')}{2i} \right| = \frac{1}{2i}$, which proves the Lemma.

It is hardly necessary to observe that when $\mathfrak{R}$ is compact the Lemma holds with the initial metric.

LEMMA II. *Given a f. c. o. s.* $\Sigma = \{U^\alpha\}$, *the metric can be so chosen that there exists an* $\eta$, *such that every subset of* $\mathfrak{R}$ *whose diameter* $< \eta$, *is a subset of some* $U$. *When* $\mathfrak{R}$ *is compact this holds without modifying the metric.*

We make use of the following property due to Menger.[4] *For every* $U^\alpha$ *there is a* $V^\alpha \Subset U^\alpha$ *such that* $\sigma = \{\overline{V^\alpha}\}$ *is also a covering.* If we set $A^\alpha = \mathfrak{R} - U^\alpha$, a closed set, we have $\overline{V}^\alpha \cdot A^\alpha = 0$. By Lemma I we can choose a metric such that $d(\overline{V}^\alpha, A^\alpha) > 0$ for every $\alpha$. These distances being in finite number they will have a lower bound $\eta > 0$. Let $B$ be any set of diameter $< \eta$. Since $\sigma$ is a covering $B$ meets at least one $V$, say $V^\alpha$. Since $d(V^\alpha, A^\alpha) \geqq \eta$, we will have $B \cdot A^\alpha = 0$, hence $B \subset U^\alpha$, which proves Lemma II for the separable case. When $\mathfrak{R}$ is compact we always have $d(\overline{V}^\alpha, A^\alpha) > 0$ without modifying the metric, and this completes the proof.

LEMMA III. *Given a finite covering by closed sets* $\sigma = \{A^\alpha\}$, *the metric can be so chosen that there exists an* $\eta > 0$ *such that if* $A^{\alpha_0} \cdots A^{\alpha_p} = 0$ *no set of diameter* $< \eta$ *meets all the sets* $A^{\alpha_i}$. *When* $\mathfrak{R}$ *is compact this holds without modifying the metric.*

An equivalent property is: *The largest distance* $\delta(x)$ *from a point* $x$ *to one of the sets* $A^{\alpha_i}$ *has a positive lower bound* $\eta$ *as* $x$ *ranges over* $\mathfrak{R}$.

Consider first a finite number of f. c. o. s. $\Sigma^i = \{U^{i\alpha}\}$, $i = 1, 2, \cdots, k$. I say that the metric can be so chosen that Lemma II holds simultaneously for all the $\Sigma$'s. For the sets $U^{1\alpha} \cdot U^{2\beta} \cdots U^{k\gamma}$ constitute a f. c. o. s. Apply

[4] *Dimensionstheorie*, p. 160.

Lemma II to it and let $\eta$ be as in the lemma. Then if diam $B<\eta$, there is a $U^{1\alpha}\cdot U^{2\beta}\ldots U^{k\gamma}\supset B$, and hence a $U^{i\delta}\supset B$ for every $i$.

Let now $U^{\alpha}=\Re-A^{\alpha}$, an open set. We have

$$A^{\alpha_0}\ldots A^{\alpha_p}=\Pi(\Re-U^{\alpha_i})=\Re-(U^{\alpha_0}+\cdots+U^{\alpha_p}).$$

Hence if the sets $A^{\alpha_i}$ do not intersect, their complements $U^{\alpha_i}$ constitute a f. c. o. s. The number of the latter being finite, let the metric be so chosen that Lemma II applies simultaneously to all, with $\eta$ as in the lemma. If diam $B<\eta$, there is then a $U^{\alpha_i}\supset B$ and hence $B\cdot A^{\alpha_i}=0$, which proves Lemma III. For the same reasons as previously, in the compact case no modification of metric is called for.

## § 2. Normal Developments.

5. By a *decreasing sequence of open sets* we shall understand a sequence of open sets $\{U^n\}$ such that $U^{n+1}\Subset U^n$ for every $n$. For such a sequence $\Pi U^n=\Pi\bar{U}^n$, and the intersection $\neq 0$ when $\Re$ is compact. Whether $\Re$ is compact or not, let the intersection exist and denote it by $A$. This set is necessarily closed since it is the intersection of an enumerable group of closed sets. If every open set $V\supset A$ contains all but a finite number of the sets $U$, we say that the sequence *converges* to $A$ and write $\{U^n\}\to A$. When $A=0$ we say that the sequence converges to zero.

It is now a simple matter to define a *normal development* $\Delta$. It consists of a sequence $\{\Sigma^i\}=\{\{U^{i\alpha}\}\}$ of f. c. o. s. with the following properties:

(a) $\{\{U^{i\alpha}\}\}$ is a base;

(b) every $\bar{U}^{i+1,\alpha}$ is on a $U^{i\beta}$;

(c) if $U^{i\alpha}\Subset U^{j\beta}$, then for $i$ above a certain value, every $\bar{U}^{h\gamma}$ meeting $\bar{U}^{i\alpha}$ is likewise on $U^{j\beta}$;

(d) every decreasing sequence $\{U^{n_i\alpha_i}\}$ converges to zero or to a point.

Evidently (a) implies separability. It is important to observe that it may be replaced by:

(a′) *there is a decreasing sequence converging to any given point $x$ of* $\Re$.

For $\Sigma^1$ being a covering there is a $U^{1\alpha_1}\supset x$. Owing to (a) then and to the regularity of $\Re$, there is a $U^{n_2\alpha_2}$ such that $x\subset U^{n_2\alpha_2}\Subset U^{1\alpha_1}$, then a $U^{n_3\alpha_3}$ such that $x\subset U^{n_3\alpha_3}\Subset U^{n_2\alpha_2}$, etc. Therefore $\{U^{n_i\alpha_i}\}$ is a decreasing sequence all of whose sets $\supset x$, hence by (d) it converges to $x$.

Conversely suppose that (a′), (b), (c), (d) hold for $\Delta$. Whatever the open set $V\supset x$, there is a $U^{n_i\alpha_i}\supset x$ and $\subset V$, hence $V$ is a sum of $U$'s and (a) holds.

Another noteworthy property is that both (a) and (d) are consequences of

(e) *the mesh of* $\Sigma^i\to 0$ with $1/i$.

For let it be $\varepsilon_i$ and let $\Pi U^{n_i\alpha_i} \supset x+y$, $x \neq y$. For a certain $i$ we will have diam $U^{n_i\alpha_i} < d(x, y)$, and hence $U^{n_i\alpha_i} \not\supset x+y$, i. e. the $U$'s cannot converge to more than one point. As regards (a), let $V$ be open and $\supset x$. Since $\mathfrak{R}-V$ is closed, $d(x, \mathfrak{R}-V) = \varepsilon > 0$. For some $i$ mesh $\Sigma^i < \varepsilon$. Since $\Sigma^i$ is a covering there will be an $U^{i\alpha} \supset x$ and hence $\subset V$ since diam $U^{i\alpha} < \varepsilon$. Therefore $V$ is a sum of sets $U^{i\alpha}$, which shows that (a) follows likewise from (e).

6. **Existence theorem.** *Every separable metric space possesses a normal development.* Before we take up the formal proof we will make an observation as regards Lemma I. Suppose that we have a first finite aggregate of non-intersecting pairs of closed sets $(A^i, B^i)$, $i = 1, 2, \cdots, p_1$, then increase it by another $(A^{p_1+i}, B^{p_1+i})$, $i = 1, 2, \cdots, p_2$, etc. We can construct the characteristic functions $g_i(x)$, for the first $p_1$ sets, then the new functions $g_{p_1+i}(x)$ arising for the second set, etc. Consider now the set $R$ represented by (2) as before and let $R^i$ denote the locus obtained when in (2) all functions $g(x)$ beyond $g_{p_i}$ are replaced by zero. It has been shown in the course of the proof of Lemma I that the $R$'s are all homeomorphs of $\mathfrak{R}$, and $R^i$ is the $R$ of the lemma corresponding to the first $p_i$ pairs. Now let $d_i(x, x')$ denote the distance-function for $\mathfrak{R}$ corresponding to the metric of $R^i$: distance equal to that of the images of the points $x, x'$ on $R^i$. By writing down the expressions of the distances, we verify immediately that $d_{i+1}(x, x') \geqq d_i(x, x')$. Hence *for fixed* $x, x'$, $d_i(x, x')$ *is a non-decreasing function of* $i$. This is the property which we need in the immediate sequel.

7. To prove our theorem we must construct the successive $\Sigma$'s of $\Delta$. We take for $\Sigma^1$ a wholly arbitrary f. c. o. s. and choose a metric, i. e. a distance-function $d_1(x, y)$, such that Lemma II becomes applicable to $\Sigma^1$, the $\eta$ of the lemma being denoted by $\eta_1$. We then choose for $\Sigma^2$ a f. c. o. s. whose mesh $\varepsilon_2 < \eta_1$, so that we will have $\Sigma^2 \Subset \Sigma^1$. $\Sigma^2$ may be determined thus: $d_1(x, y)$ is the actual distance-function of a certain topological image $R$ of $\mathfrak{R}$ on $\mathfrak{H}$. Since $\mathfrak{H}$ is compact it possesses an $\varepsilon_2$-f. c. o. s., and the sets of the latter intersect $R$ in sets which are open, of diameter $< \varepsilon_2$, and hence their images on $\mathfrak{R}$ are the sets of a suitable $\Sigma^2$.

We now impose a second modification upon our metric, that is choose a new distance function $d_2(x, y)$, as outlined in No. 6, such that: (a) Lemma II becomes applicable to $\Sigma^2$, thus giving rise to an $\eta_2$ for it; (b) whenever $U^{2\alpha} \Subset U^{1\beta}$ and hence $\bar{U}^{2\alpha} \cdot A^{1\beta} = 0$, $A^{i\beta} = \mathfrak{R} - U^{i\beta}$, we have also $d_2(\bar{U}^{2\alpha}, A^{1\beta}) > 0$. The distances mentioned in (b) being in finite number they will have a lower bound $\xi_2 > 0$. The passage from $d_1$ to $d_2$ is by augmenting the functions $g_i$ after the manner of No. 6, so that we will have $d_2(x, y) \geqq d_1(x, y)$ whatever $x, y \subset \mathfrak{R}$. We now choose for $\Sigma^3$ any f. c. o. s. whose mesh $\varepsilon_3 < \xi_2$ or $\eta_2$. Suppose that we have thus reached $\Sigma^i$. We choose $d_i(x, y)$ as outlined in No. 6, such that: (a) Lemma II holds for $\Sigma^i$ with a corresponding $\eta_i$; (b) whenever $U^{i\alpha} \Subset U^{j\beta}$, $j < i$, and hence $\bar{U}^{i\alpha} \cdot A^{j\beta} = 0$, also $d_i(\bar{U}^{i\alpha}, A^{j\beta}) > 0$. There will be again a positive lower bound $\xi_i$ for the latter distances, and we choose $\Sigma^{i+1}$ of mesh $\varepsilon_{i+1} < \xi_i$ or $\eta_i$. We do this for every $i$ with the added restriction that $\varepsilon_i \to 0$ monotonely with increasing $i$. Since $\varepsilon_{i+1} < \eta_i$ we have $\Sigma^{i+1} \Subset \Sigma^i$ for every $i$. From $\varepsilon_{i+k} \leqq \varepsilon_{i+1} < \xi_i$ follows that every $\bar{U}^{i+k,\beta}$ which meets $\bar{U}^{i\alpha}$ is on any $U^{j\gamma} \Supset U^{i\alpha}$, $j < i$, and this for every $i$. Therefore conditions (b), (c) for a normal development hold throughout $\Delta = \{\Sigma^i\}$.

The choice of our metric is such that $d_{i+1}(x, y) \geqq d_i(x, y) \geqq d_1(x, y)$. Therefore the mesh of $\Sigma^i$, as measured by the fixed distance-function $d_1(x, y)$, $\to 0$ monotonely like $\varepsilon_i$. From this follows that conditions (a), (d) for a normal $\Delta$ hold also (No. 5). Therefore $\Delta$ is normal.

REMARK. Referring to the lemmas we find that when $\mathfrak{R}$ is compact there is no need to modify the metric. Therefore

THEOREM. *When $\mathfrak{R}$ is compact any development $\Delta$ consisting of a sequence of f. c. o. s. is normal provided that mesh $\Sigma^i$ tends rapidly enough to zero with $1/i$.*

8. **Criterion for compactness.** *$\mathfrak{R}$ is compact when and only when every decreasing sequence $\{U^{n_i\alpha_i}\}$ converges to a point.*

When $\mathfrak{R}$ is compact $\Pi U^{n_i\alpha_i} = \Pi \bar{U}^{n_i\alpha_i} \neq 0$, and hence it consists of a point.

Conversely suppose that the intersection is always a point. Let $A$ be an infinite subset of $\mathfrak{R}$, and let $K^1$ be the aggregate of sets $U^{1\alpha}$ each of which carries an infinite number of points of $A$. Since $A$ is infinite, $K^1 \neq 0$. In the class $K^1$ there is a subclass $K^2$ consisting of sets $U^{1\alpha}$ such that $U^{1\alpha} \Supset U^{2\beta}$, where $U^{2\beta}$ carries an infinite number of points of $A$. Similarly $K^2$ has a subclass $K^3$ consisting of all sets $U^{1\alpha} \Supset U^{2\beta} \Supset U^{3\beta}$, where $U^{3\beta}$ carries an infinite number of points of $A$, etc. Whatever $i$, there is a $U^{i\alpha_i}$ with an infinite number of points of $A$. By condition (b) there is a $U^{i-1,\alpha_{i-1}} \Supset U^{i\alpha_i}$ and thus a sequence $U^{1\alpha_1} \Supset U^{2\alpha_2} \cdots \Supset U^{i\alpha_i}$. Therefore $K^i \neq 0$, and hence $k_i$, the number of sets $U^{1\alpha}$ in $K^i$, $> 0$. Also $K^i \subset K^{i-1}$. Hence $k_1 \geqq k_2 \geqq \cdots > 0$. Therefore the sequence $k_1$, $k_2, \cdots$ has a least positive integer say $k_p$. The sets of the class $K^p$, will be sets $U^{1\alpha}$, such that there corresponds to each a decreasing infinite sequence $\{U^{i\alpha_i}\}$ on each of whose sets there is an infinite number of points of $A$. We have by assumption $\Pi U^{i\alpha_i} = x$, a point, and owing to condition (c), every neighborhood of $x$ contains a set $U^{i\alpha_i}$, hence an infinite number of points of $A$. The point $x$ is then a condensation-point of the arbitrary infinite subset $A$ of $\mathfrak{R}$, and $\mathfrak{R}$ is therefore compact.

9. With any normal development $\Delta$ there is associated a mapping of $\mathfrak{R}$ on a certain compact space which is basic in the next section. Consider the following sets of non-intersecting pairs of closed sets: $(\bar{U}^{i\alpha} \bar{U}^{j\beta})$; $(\bar{U}^{i\alpha_0} \cdot \ldots \cdot \bar{U}^{i\alpha_p}, \bar{U}^{i\beta})$; $(\bar{U}^{i\alpha}, A^{j\beta})$, where $A^{j\beta} = \mathfrak{R} - U^{j\beta}$. They include the pairs of Urysohn's mapping described in No. 2. If $f_i(x)$, $i = 1, 2, \cdots + \infty$, are characteristic functions corresponding to all these pairs, (1) will again represent a topological image $R$ of $\mathfrak{R}$ on $\mathfrak{H}$. If $U^{i\alpha}$, $A^{j\beta}$ continue to designate the images of the sets thus called previously, we will have now that whenever two closed sets in a pair such as above do not meet their distance on $R$ is $> 0$.

10. Let $R^*$ designate the closure of $R$ on $\mathfrak{H}$. Since the latter is compact, $R^*$ is self-compact. We have thus mapped $\mathfrak{R}$ topologically on a set $R^*$ which is compact as a space. This is the mapping that we have been seeking.

Now the closure $A^{*i\alpha}$ of $A^{i\alpha}$ on $\mathfrak{H}$ is closed also relatively to $R^*$. Hence $U^{*i\alpha} = R^* - A^{*i\alpha}$ is relatively open. The points of $R$ on $U^{*i\alpha}$ are all its points at a distance $> 0$ from $A^{*i\alpha}$, hence also from $A^{i\alpha}$, and their sum is $U^{i\alpha}$, so that $U^{i\alpha} \subset U^{*i\alpha}$. Moreover if $x$ is a condensation-point of $R$ on $U^{*i\alpha}$, the points of $R$ very near to $x$ are at a distance $> 0$ from $A^{i\alpha}$, and hence again $\subset U^{i\alpha}$, so that $U^{*i\alpha} = U^{i\alpha} +$ condensation-points of $U^{i\alpha}$ on $\mathfrak{H}$. *Therefore the closures of $U^{i\alpha}$ and $U^{*i\alpha}$ relatively to $\mathfrak{H}$ coincide.*

Let the closure bars refer respectively to closures on $R$ or $R^*$ as regards their subsets. Since $R^*$ is closed the closures relatively to it and to $\mathfrak{H}$ are the same. It follows that $\bar{U}^{*i\alpha}$ is the closure of $U^{i\alpha}$ on $\mathfrak{H}$.

From the closure relations between the $U$'s and $U^*$'s, we deduce:

$$\left.\begin{array}{ll}(a) & \bar{U}^{*i\alpha} \cdot \bar{U}^{*j\beta} \neq 0, \\ (b) & U^{*i\alpha} \Subset U^{*j\beta}, \\ (c) & \bar{U}^{*i\alpha_0} \cdot \ldots \cdot \bar{U}^{*i\alpha_p} \neq 0,\end{array}\right\} \text{ are equivalent to the same non-starred relations;}$$

(d) mesh $\Sigma^{*i} \to 0$ with $1/i$, $\Sigma^{*i} = \{U^{*i\alpha}\}$.

The last property is evident. From our basic closure relations between the $U$'s follows that

$$d(\bar{U}^{i\alpha}, \bar{U}^{j\beta}) = d(\bar{U}^{*i\alpha}, \bar{U}^{*j\beta}), \tag{3}$$

$$d(\bar{U}^{i\alpha}, A^{j\beta}) = d(\bar{U}^{*i\alpha}, A^{*j\beta}). \tag{4}$$

Now to impose the existence of one or the other intersection in (a) is equivalent to demanding that the distances (3) vanish, which proves (a). Similarly (b) is a consequence of the fact that to impose one of the inclusions in it is equivalent to demanding that the distances in (4) be

$>0$. As regards (c) since the closures as to $R^*$ are also closures as to $\mathfrak{H}$, we have

$$\bar{U}^{i\alpha_0}\cdot\ldots\cdot\bar{U}^{i\alpha_p}=\bar{U}^{*i\alpha_0}\cdot\ldots\cdot\bar{U}^{*i\alpha_p}\cdot R.$$

Therefore when the sets $\bar{U}^{i\alpha}$ intersect so do the sets $\bar{U}^{*i\alpha}$. Conversely if the intersection of the non-starred sets $=0$, there is a $q<p$ such that $\bar{U}^{i\alpha_0}\cdot\ldots\cdot\bar{U}^{i\alpha_q}\neq 0$, $\bar{U}^{i\alpha_0}\cdot\ldots\cdot\bar{U}^{i\alpha_{q+1}}=0$. Therefore by our construction of $\mathfrak{R}$ (see No. 9)

$$d(\bar{U}^{i\alpha_0}\cdot\ldots\cdot\bar{U}^{i\alpha_q},\ \bar{U}^{i\alpha_{q+1}})>0,$$

and since the distances are the same as for the corresponding starred sets, necessarily

$$\bar{U}^{*i\alpha_0}\cdot\ldots\cdot\bar{U}^{*i\alpha_{q+1}}=0\cdots=\bar{U}^{*i\alpha_0}\cdot\ldots\cdot\bar{U}^{*i\alpha_p}.$$

From (a), (b), (d) we conclude that conditions (a′), (b), (c), (d) of No. 5 for a normal development hold as regards $\Delta^*=\{\Sigma^{*i}\}$.

The set $R$ is clearly dense on $R^*$. Conversely if $R^*$ is self-compact, with $\Delta^*$ as a normal development, and $R$ is a subset of $R^*$ which is dense on the latter, the sets $U^{i\alpha}=U^{*i\alpha}\cdot R$ define on $R$ a normal development $\Delta$ related to $\Delta^*$ in the same manner as above. We have therefore proved:

THEOREM. *To every normal development $\Delta$ of $\mathfrak{R}$ there corresponds a dense topological mapping of $\mathfrak{R}$ on a compact metric space $\mathfrak{R}^*$, on which there exists a normal development $\Delta^*$ such that the i-th coverings of each have the same intersection properties (expressed by condition* (c)*). Conversely given $\Delta^*$ on $\mathfrak{R}^*$, there is defined on any dense subset $R$ of $\mathfrak{R}^*$ a normal development $\Delta$ related to $\Delta^*$ in the preceding manner.*

## § 3. THE ORDER OF A SPACE.

11. The *order* of an aggregate of sets $\sigma=\{A^\alpha\}$, is the largest integer $\varrho(\sigma)$ such that there are $1+\varrho(\sigma)$ sets $A$ with a non-vacuous intersection.

Let $\Sigma$ be a fixed f. c. o. s., $\sigma$ a variable one $\subset\Sigma$. The number $\varrho(\sigma)$ being a positive integer has a least value $\omega(\Sigma)$ which is actually reached for some $\sigma$. As $\Sigma$ itself varies $\omega(\Sigma)$ has a highest value finite or infinite, $\omega$, called the *order* of $\mathfrak{R}$. When $\omega$ is finite it is reached for some $\Sigma$. In any case $\omega$ is a topological invariant of $\mathfrak{R}$.

Instead of a variable f. c. o. s. $\sigma$, we might have coverings by other types of sets. Any topological class $\mathfrak{K}$ of such sets gives rise to a topological invariant $\omega(\mathfrak{R},\mathfrak{K})$ of $\mathfrak{R}$ analogous to $\omega$. The following two classes are noteworthy:

(a) $\mathfrak{K}^1$, the class of all closures of open sets.

(b) $\mathfrak{K}^2$, the class of all closed sets.

The second class is the one selected by all previous writers. Let $\omega_1$, $\omega_2$ be the $\omega$'s corresponding to the two new classes. I say that $\omega_1 = \omega_2 = \omega$.

12. Since $\mathfrak{R}^1 \subset \mathfrak{R}^2$ it follows immediately from the definition of the $\omega$'s that $\omega_2 \leqq \omega_1$. Take now two f. c. o. s., $\sigma$, $\Sigma$, with $\sigma \subset \Sigma$. As recalled in the proof of Lemma II, according to Menger there can be found, for every $\alpha$, a $V^\alpha \Subset U^\alpha$ such that $\sigma' = \{V^\alpha\}$ is likewise a f. c. o. s. Let $\bar{\sigma}' = \{\bar{V}^\alpha\}$. We manifestly have $\varrho(\bar{\sigma}') \leqq \varrho(\sigma)$, hence the lower bound of $\varrho(\bar{\sigma}') \leqq \omega(\Sigma)$. As the sets of $\bar{\sigma}'$ belong to $\mathfrak{R}^1$, passing to the upper bounds of the two orders we find $\omega_1 \leqq \omega$.

13. We will now show that $\omega_2 \geqq \omega$. Let $\sigma = \{A^\alpha\}$ be a finite covering by closed sets $\subset \Sigma = \{U^\beta\}$. By Lemma II the metric of $\mathfrak{R}$ can be so chosen that the distances $d(A^\alpha, \mathfrak{R} - U^\beta)$ corresponding to all pairs $A^\alpha$, $U^\beta$ such that $A^\alpha \subset U^\beta$, have a lower bound $\xi_1 > 0$. By Lemma III it can be so chosen that there exists a $\xi_2 > 0$ such that, if $A^{\alpha_0} \cdots\cdots A^{\alpha_p} \neq 0$, at least one distance $d(x, A^{\alpha_i}) > 0$ whatever the point $x$ of $\mathfrak{R}$. The choice of the two metrics involves a mapping of $\mathfrak{R}$ on $\mathfrak{H}$ by a system (2) with suitable functions $g_i(x)$. The two sets of functions are finite. If we unite them into a single set and take the representation (2) corresponding to that set, we find that both properties above mentioned are fulfilled simultaneously by the metric associated with the new representation, and we have as above a $\xi_1$ and a $\xi_2$; let $\xi$ be the least of the two.

Let now $V^\alpha$ be the set of all points of $\mathfrak{R}$ whose distance from $A^\alpha < \xi$. It is an open set, and if $A^\alpha \subset U^\beta$ we have $V^\alpha \Subset U^\beta$. Hence

$$\sigma \subset \sigma' = \{\bar{V^\alpha}\} \Subset \{U^\beta\}$$

and $\sigma'$ is a f. c. o. s. Moreover $V^{\alpha_0} \cdots\cdots V^{\alpha_p} \neq 0$ when and only when $A^{\alpha_0} \cdots\cdots A^{\alpha_p} \neq 0$. That the second implies the first is clear. Conversely let the first hold, and the second not. Then a point of $V^{\alpha_0} \cdots\cdots V^{\alpha_p}$ is nearer than $\xi \leqq \xi_2$ to the sets $A^{\alpha_i}$, which contradicts the basic property of $\xi_2$. Thus when one of the two intersections exists so does the others. It follows that

$$\varrho(\sigma) = \varrho(\sigma').$$

From this we conclude, as in No. 12, that

$$\omega(\Sigma) \leqq \varrho(\sigma),$$

hence passing to the maxima, $\omega_2 \geqq \omega$. Since $\omega_2 \leqq \omega_1 \leqq \omega$, these three numbers are equal.

14. **Order of a normal development.** If $\Delta = \{\Sigma^i\}$ little can be said in general about $\varrho(\Sigma^i)$. Suppose however that $\Delta$ has the following property: for every $i$,

$$\varrho(\Sigma^{i+1}) = \varrho(\bar{\Sigma}^{i+1}) = \omega(\Sigma^i),$$

its least possible value consistent with $\Sigma^{i+1} \Subset \Sigma^i$. Owing to this inclusion $\varrho(\Sigma^{i+1}) \geqq \varrho(\Sigma^i)$, and hence $\varrho(\Sigma^i)$ is a non-decreasing function. Its limit, $\Omega$, finite or infinite, is called the order of $\Delta$, and a development of the type considered is said to be *minimal*. Since

$$\varrho(\Sigma^{i+1}) = \omega(\Sigma^i),$$

necessarily $\omega \geqq \Omega$.

THEOREM. *A space $\mathfrak{R}$ of finite order $\omega$ possesses minimal normal developments whose order is also $\omega$.*

We obtain $\Delta$ such as required by a suitable modification of the construction in § 2. With the same notations as loc. cit., we take, as we may by No. 13, $\Sigma^1$ such that

$$\varrho(\Sigma^1) = \varrho(\bar{\Sigma}^1) = \omega,$$

its maximum value. Suppose that we have obtained $\Sigma^1, \cdots, \Sigma^{i-1}$, such that

$$\varrho(\Sigma^{i-1}) = \varrho(\bar{\Sigma}^{i-1}) = \omega.$$

We now take an $\varepsilon_i$-f. c. o. s., $\sigma^i$, hence $\Subset \Sigma^{i-1}$, and choose $\Sigma^i \subset \sigma^i$, such that

$$\varrho(\Sigma^i) = \varrho(\bar{\Sigma}^i) = \omega(\sigma^i).$$

We have then

$$\omega(\sigma^i) \geqq \omega(\Sigma^{i-1}) = \omega, \quad \text{and} \quad \omega(\sigma^i) \leqq \omega.$$

Hence

$$\varrho(\Sigma^i) = \omega(\sigma^i) = \omega.$$

For the usual reason likewise

$$\omega(\Sigma^i) \geqq \omega(\sigma^i) = \omega,$$

and also $\leqq \omega$, hence $\omega(\Sigma^i) = \omega$. By this construction then

$$\varrho(\Sigma^i) = \varrho(\bar{\Sigma}^i) = \omega(\Sigma^i) = \omega$$

for every $i > 1$, and hence $\Delta$ is normal and minimal of order $\omega$.

15. **Order of a compact space.** When $\mathfrak{R}$ is compact an order may be defined in somewhat different manner. Take all $\varepsilon$-f. c. o. s., and let $r(\varepsilon)$ be their lowest order. Since an $\varepsilon'$-covering is also an $\varepsilon$-covering when $\varepsilon' < \varepsilon$, we have then $r(\varepsilon') \geqq r(\varepsilon)$. Hence $r(\varepsilon)$ is a non-decreasing function of $\varepsilon$ and it has a maximum $\omega_0$, finite or infinite, called the *metric* order. I say again that $\omega_0 = \omega$, *or the metric and topological orders are equal.*

Let first $\omega_0$ be finite and construct $\Delta$ as follows: $\Sigma^1$ is of mesh $\varepsilon_1$ such that already $r(\varepsilon_1) = \omega_0$; $\Sigma^i$ is taken of mesh $\varepsilon_i < \varepsilon_{i-1}$ or $\eta_{i-1}$, the $\eta$ of Lemma II corresponding to $\Sigma^{i-1}$; we impose $\varrho(\Sigma^i) = \omega_0$, its minimum value since $\varepsilon_i < \varepsilon_1$, and also $\varepsilon_i \to 0$ with $1/i$. By Lemma II the sets of $\Sigma^i$ can be slightly shrunk without its ceasing to be a covering, say $\Sigma'^i$. Again

$$\varrho(\Sigma'^i) = \omega_0,$$

and moreover

$$\varrho(\Sigma'^i) \leqq \varrho(\overline{\Sigma}'^i) \leqq \varrho(\Sigma^i) = \omega_0,$$

hence

$$\varrho(\overline{\Sigma}'^i) = \omega_0.$$

Writing now $\Sigma^i$ for $\Sigma'^i$, we see that the development $\Delta$ will be minimal of order $\omega_0$. Now when $\omega$ is finite it is precisely such a $\Delta$ that we have constructed in No. 15, therefore when $\omega$, $\omega_0$ are finite $\omega_0 = \omega$. When $\omega = \infty$ we have, whatever the integer $n$, a f. c. o. s., $\Sigma$ whose $\omega(\Sigma) > n$. If $\eta$ is the number of Lemma II for $\Sigma$, any $\varepsilon$-f. c. o. s. $\sigma \subset \Sigma$, whatever $\varepsilon < \eta$, hence its order

$$\varrho(\sigma) \geqq \omega(\Sigma) > n,$$

and therefore $r(\varepsilon) > n$ so that limit $r(\varepsilon) = \infty = \omega_0 = \omega$.

If in place of open sets we choose $\varepsilon$-coverings by closed sets we obtain a new metric order $\omega_0'$. This is the order as defined by Urysohn. It is not difficult to show that it is again equal to $\omega$. For let $\sigma = \{A^\alpha\}$ be a finite $\varepsilon$-covering by closed sets. By Lemma III we can shlightly dilate the $A$'s without changing $\varrho(\sigma)$.—$A^\alpha$ is thus replaced by a certain $\overline{V}^\alpha$ and $\sigma' = \{V^\alpha\}$ is an $\varepsilon'$-f. c. o. s., where $\varepsilon' \to 0$ with $\varepsilon$, whose order is also $\varrho(\sigma)$. Hence $r(\varepsilon') \leqq \varrho(\sigma)$ and therefore $\omega_0 \leqq \omega_0'$.—On the other hand if we have an $\varepsilon$-f. c. o. s. $\sigma$, Lemma II authorizes us to contract its sets into a certain covering $\sigma' = \{V^\alpha\} \Subset \sigma$, and $\varrho(\overline{\sigma}') \leqq \varrho(\sigma)$. Since $\overline{\sigma}'$ is a special finite $\varepsilon$-covering by closed sets, we find this time $\omega_0 \geqq \omega_0'$, hence $\omega_0' = \omega_0 = \omega$.

16. **The Lebesgue order theorem.** *The order of $\mathfrak{R}$ is equal to its dimension.* This fundamental theorem was first stated for Euclidean spaces by Lebesgue, then proved for them by Brouwer, later also by Lebesgue. The complete theorem was proved for compact spaces by Urysohn,[5] while for separable spaces Menger[6] showed that $\omega \leqq n = \dim \mathfrak{R}$. By a judicious use of normal developments Urysohn's proof for the compact case can be made to yield the proof of the theorem for separable spaces.

We have to show that when $\mathfrak{R}$ is separable $\omega \leqq n$, and at the same time $\geqq n$. We indicate briefly the mild departures from Urysohn's proof.

---

[5] Fundamenta Matematicae, vol. 8 (1925), p. 292.

[6] *Dimensionstheorie*, Ch. V.

Let $\Sigma$ be a given f. c. o. s. To prove $\omega \leqq n$ we have to find a f. c. o. s. $\sigma \subset \Sigma$ whose sets have boundaries that are at most $(n-1)$-dimensional. According to No. 7 we can find a f. c. o. s. $\sigma' = \{V^\beta\} \Subset \Sigma = \{U^\alpha\}$. Given $V^\beta$ there exists a $U^\alpha \Supset V^\beta$. Since the two closed sets $\bar{V}^\beta$, $\mathfrak{R} - U^\alpha$ do not meet there exists an open set $W^\beta$ such that $V^\beta \Subset W^\beta \Subset U^\alpha$, $\dim F(W^\beta) < n$.[7] Therefore $\sigma = \{W^\beta\}$ is a f. c. o. s. behaving as required. From this point on the proof continues substantially as with Urysohn.

We now pass to $n \leqq \omega$. By Urysohn's method and a simple induction we can establish this result: if $\Delta$ is a normal development having the property that $\varrho(\Sigma^i) = \varrho(\bar{\Sigma}^i) = k$, a fixed integer, then $n \leqq k$. Since there are minimal developments of order $\omega$, it follows that for them $k = \omega$, hence $n \leqq \omega$. This is then in outline the proof of the order theorem as it may now be derived by the aid of normal developments.

17. **The Imbedding theorem for a separable metric space:** *$\mathfrak{R}$ can be mapped topologically on a compact metric space of the same dimension.* We have seen in no. 10 that with any normal developmnt $\Delta$ of $\mathfrak{R}$ there is associated a topological mapping of $\mathfrak{R}$ on a compact metric space $\mathfrak{R}^*$ having the following property: In the notations loc. cit. there is a normal development $\Delta^*$ of $\mathfrak{R}^*$ such that the orders of $\Sigma^i$ and $\Sigma^{*i}$ are the same (no. 10 property c).

Suppose in particular that $\Delta$ is minimal of order $\omega = n$. Then

$$\varrho(\Sigma^i) = \varrho(\Sigma^{*i}) = n.$$

Since mesh $\Sigma^{*i} \to 0$ (No. 10d), order $\mathfrak{R}^* = \dim \mathfrak{R}^* \leqq n$. But as $\mathfrak{R}^*$ contains an $n$-dimensional set, $\dim \mathfrak{R}^* \geqq n$, hence $\dim \mathfrak{R}^* = n$, which proves our theorem.

18. Remark. Combined proofs of the order and imbedding theorems can be obtained along the following lines: (a) $\omega \leqq n$ is to be established as by Menger; (b) $\omega = n$ is to be proved as by Urysohn for the compact case only; (c) it is then proved as above that $\mathfrak{R}$ can be mapped topologically on a compact metric space $\mathfrak{R}^*$ whose order $\leqq \omega$. By (b) we then have $\dim \mathfrak{R}^* \leqq \omega$, and since $n \leqq \dim \mathfrak{R}^*$, this yields $n \leqq \omega$, and hence $n = \omega$, which is the Lebesgue order theorem. We then have $\dim \mathfrak{R}^* \leqq n$ and hence $= n$, which together with (c) yields the imbedding theorem.

19. Corollary. *The various immersion theorems obtained in our paper already quoted (footnote 2) are valid for any separable metric space. In particular the two universal n-spaces there considered are also universal n-spaces for separable metric n-spaces.*

---

[7] Menger, loc. cit., p. 115.

**20. Menger's generalization of the Lebesgue order theorem.** Let $\Sigma$ be a f. c. o. s., $\sigma = \{A^{\alpha}\}$ a finite covering by closed sets $\subset \Sigma$. Instead of the order $\varrho(\sigma)$ we may consider with Menger the maximum dimension $\varrho_k(\sigma)$ of any intersection $A^{\alpha_0} \cdots\cdots A^{\alpha_k}$. By the same considerations as before we obtain an invariant $d_k$ analogous to the order, and we then have, with $\dim \mathfrak{R} = n$:

THEOREM (Menger). $d_k = n-k$.

For $k = n+1$ this yields $d_{n+1} = -1$, which is merely the expression of the Lebesgue order theorem. As in the proof of the latter we have to show that $d_k \leqq n-k$ and also $\geqq n-k$. The first part was proved by Menger, the second also by him for a compact space, (loc. cit. p. 156 and following). Thus all that is now lacking is the inequality $d_k \geqq n-k$ for any separable metric space.

21. By repetition of the argument used in proving the theorem in No. 14 we show that there exists a development $\Delta$ such that $\varrho_k(\bar{\Sigma}^i) = d_k$ for every $i$. To complete the proof there remains to be shown as in the similar case in No. 16, that if there exists a $\Delta$ such that $\varrho_k(\bar{\Sigma}^i) \leqq \delta$ for every $i$, then $n \leqq \delta + k$; since $\delta$ can take the value $d_k$, the required inequality will follow. We have already disposed of $\delta = 0$ (No. 16), hence we can use induction on $\delta$.

Suppose then that $\varrho_k(\bar{\Sigma}^i) \leqq \delta$ for every $k$. Now all the sets $\bar{U}^{i\alpha_0} \cdots\cdots \bar{U}^{i\alpha_k} \neq 0$ constitute an enumerable aggregate of closed sets. Let us designate them, ranged in some order by $\{G^{\beta}\}$, $\beta = 1, 2, \cdots + \infty$. Since $\dim G^{\beta} \leqq \delta$, by a well known theorem[8] $\dim \Sigma G^{\beta} \leqq \delta$. It follows that any point $x$ of $\mathfrak{R}$ possesses a neighborhood $V$ whose boundary intersects the sum, and hence every $G$, in a set whose dimension $\leqq \delta - 1$[9]. But the intersections $U^{i\alpha} \cdot F(V)$ determine on $F(V)$ a normal development such that the corresponding $G$'s are the sets $G^{\beta} \cdot F(V)$, and these sets are $\delta - 1$ dimensional at most. Under the hypothesis of the induction then $\dim F(V) \leqq \delta - 1 + k$, and hence $\dim \mathfrak{R} = n \leqq \delta + k$. This completes the proof of Menger's theorem.

---

[8] Hurewicz, Math. Ann., vol. 96 (1927), p. 760. See also Menger, loc. cit. p. 96.

[9] Menger, loc. cit. p. 82.

(ADDED IN THE PROOF.) Professor Čech recently called our attention to a paper by Hurewicz, Monatshefte für Math. und Physik, 37 (1930), pp. 199–208, which had escaped us, and where he proves the imbedding theorem by a procedure based like ours on a modification of metric. While we have many points of contact with Hurewicz we diverge in scope, and more particularly in the explicit consideration of normal developments and in the treatment of the order.

# XVII

## ON LOCALLY CONNECTED AND RELATED SETS

In our investigations on the fixed point problem[1] we came across a noteworthy type of sets belonging to a class later investigated at length by K. Borsuk[2] under the name of *retracts*. We considered in various places in our book certain generalized locally connected sets and indicated in outline their bearing on the fixed point problem. Our chief object in the present paper is to study more completely the mutual relations of those two classes of sets, and incidentally to exhibit more clearly their relation to the fixed point problem.

Let $A$ be a compact metric space. We call a complex $K$ on $A$ consisting partly of abstract, partly of singular cells a *semi-singular* complex $K$ on $A$ (a more precise definition will be found in No. 3). We shall show that all absolute neighborhood retracts $A$ are intrinsically characterized by the following property (a sort of uniform local connectedness): for every $\epsilon$ there is a $\delta$ such that every finite semi-singular complex $K$ on $A$ of mesh $< \delta$ can be completed to form a singular complex of mesh $< \epsilon$ on $A$. The uniformity property is implicit in the fact that $K$ is arbitrary on $A$. The simplest case of this important property is that in which we know only the vertices of a singular $K$, i.e., the only non-abstract cells are the zero-cells (these are always assumed already present), the abstract cells of positive dimension all being of diameter $< \delta$. Then it is possible to insert all the other cells so as to keep their diameters $< \epsilon$. Moreover this must be possible regardless of the dimension of $K$. The $\epsilon$, $\delta$ relationship has then the uniformity property also with respect to this dimension. A closely related characterization may also be given for the absolute retracts. From this it follows readily that the homology groups of $A$ are of the same type as those of a finite simplicial $p$-complex ($p$ finite) and in particular that its Betti-numbers $R_q(A)$ are all finite, the numbers $R_q(A)$ corresponding to $q > p$ being zero. This enables us to complete a result of *Topology* regarding the fixed point problem.

Our treatment rests upon the following noteworthy result: given a closed subset $A$ of a Hilbert parallelotope $\mathfrak{H}$, we can construct an infinite complex $K$ and a deformation $\Pi$ of $\mathfrak{H}$ into $K$ such that $\Pi = 1$ on $A$ and $\Pi(\mathfrak{H} - A) = K$.

1. We first recall the basic definitions and introduce a certain number of convenient abridged notations.

---

[1] See our Colloquium Lectures on Topology, New York, 1930, p. 347. The general notations and terminology of this treatise will be adopted in the present paper.

[2] Notably in two papers in Fundamenta Matematicae, vol. 17, (1931), vols. 18 and 19 (1932) (= Borsuk I, II, III in the sequel).

*Retracts.* Let $B$ be a self-compact metric set and let $A$ be a closed subset of $B$. We say that $A$ is a *retract* of $B$ if there exists a continuous single-valued transformation (= c. s. v. t.) $T$ of $B$ into $A$ which reduces to the identity on $A(T.x = x \text{ for } x \subset A)$. The set $A$ is an *absolute retract* (= AR) if it is a retract for every $B \supset A$. The set $A$ is a *neighborhood-retract* (= NR) of $B$ if $A$ is a retract of some relatively open subset $U$ $(\supset A)$ of $B$. Finally, $A$ is an *absolute neighborhood retract* (= ANR) if it is a neighborhood retract for every $B \supset A$. In particular, (Borsuk, III p. 223) *an* ANR (*an* AR) *is characterized by the property that its image on the Hilbert parallelotope* $\mathfrak{H}$[3] *is a neighborhood retract* (*a retract*) *for* $\mathfrak{H}$.

*Local Connectedness.* A self-compact metric set $A$ is *locally p-connected* if, for every positive $\epsilon$, there is a positive $\eta$ such that any singular $p$-sphere $H_p$ on $A$ with diameter $< \eta$ bounds a singular cell $E_{p+1}$ of diameter $< \epsilon$. To be precise, a singular sphere $H_p$ is the image of a non-singular sphere $H'_p$ under a certain c. s. v. t. $T$. The pair $T$, $H'_p$ is not unique,[4] and the condition imposed is that among all the possible choices there is one in which $H'_p$ is the point-set boundary of a certain non-singular cell $E'_{p+1}$ and in which $T$ is a transformation defined over $\bar{E}'_{p+1}$, where $TE'_{p+1} = E_{p+1}$, $TH'_p = H_p$, diam $E_{p+1} < \epsilon$.

A condition equivalent to the preceding one, but formulated with much less precision, is that $H_p$ is $\epsilon$-homotopic to a point on $A$.

We shall say that $A$ is an LC$^p$-set if it is locally $q$-connected for every $q$ such that $0 \leqq q \leqq p$. A set $A$ which is locally connected for every $p$ will be called an LC$^\infty$-set. LC$^\infty$-sets are the same as the locally connected sets of *Topology*.

2. Let a simplex $\sigma_q$ have for its singular image (continuous single-valued transform) the singular simplex $\sigma'_q$ on $A$. The corresponding images of the various faces (vertices, one-cells, etc.) of $\sigma'_q$ are the faces of $\sigma'_q$. With reference to the Note just quoted, a face $\sigma'_r$ of $\sigma'_q$ has a certain autonomous existence independently of $\sigma_q$ in the following sense: if $T$ is the transformation sending $\sigma_q$ into $\sigma'_q$, $\sigma'_r$ need not be considered as derived from any face of $\sigma_q$ by means of $T$, but it may equally well be considered as derived from some other simplex $\sigma^*_r$ by a transformation $T^*$, where $T^* = TU^{-1}$ and where $U$ is a barycentric transformation of $\sigma_r$ into $\sigma^*_r$. Thus $\sigma'_r$ is a face of $\sigma'_q$ provided that among all the transformations $T^*$ corresponding to it there exists one like $T$, that is, a transformation of $\sigma_q$ into $\sigma'_q$ which carries a face $\sigma_r$ of $\sigma_q$ into $\sigma'_r$ so that on $\sigma_r$ we have $T = T^*U$, where $U$ is a barycentric transformation of $\sigma_r$ into $\sigma^*_r$. These remarks are necessary for a proper understanding of the sequel.

3. We recall, with Vietoris, that by an abstract $\epsilon$ $p$-simplex $\sigma_p$ on $A$ is meant a set of $p + 1$ points of $A$ whose sum is of diameter $< \epsilon$. Similarly an abstract complex $K$ on $A$ is an abstract simplicial complex made up of abstract $\sigma$'s on $A$.

---

[3] This is the subset of Hilbert space defined by $0 \leqq x_n \leqq 1/n$, $n = 1, 2, \cdots + \infty$.

[4] See pp. 479–484.

The *mesh* of $K$ is then, as usual, the maximum diameter of its $\sigma$'s (i.e., of the sets of their vertices).

Let $K'$ be a subcomplex of $K$ which includes all its vertices, and let $K^*$ be a geometric realization of $K$ (a polyhedral simplicial complex whose structure is that of $K$) with $K'^*$ as the realization of $K'$. A singular image $\mathfrak{K}'$ of $K'^*$ on $A$ shall be called a *semi-singular* image of $K$. In relation to $K$ the image $\mathfrak{K}'$ partakes of both the singular and the abstract complexes since some of the cells of $K$ have singular images on $A$, others are represented merely by their vertices, and others by some singular faces and by their vertices. If we had a proper singular image $\mathfrak{K}$ of $K$ and removed from it the images of the cells of $K - K'$, the residue would be $\mathfrak{K}'$.

Any cell $\sigma$ of $K$ is then represented on $A$ in $\mathfrak{K}'$ by certain singular faces, some vertices, and possibly its own singular image. The sum of all these sets, or of as many as are present, is a certain point-set $\bar{\sigma}'$ of $A$, and the largest diam $\bar{\sigma}'$ is called the mesh of $\mathfrak{K}'$.

4. THEOREM I. *$A$ is an $LC^p$ when and only when for every positive $\epsilon$ there exists a positive $\eta$ such that any semi-singular image $K_p'$ of a $K_p$ on $A$ with mesh $< \eta$ can be completed to form a singular image of $K_p$ on $A$ with mesh $< \epsilon$.*

*Important special case: the semi-singular image is an abstract complex with the same structure as $K_p$.*

The condition of the theorem is sufficient. For in the statement $K_p$ may *a fortiori* be replaced by any subcomplex; consequently $p$ may be replaced by any $q < p$. Consider, for any such $q$, a singular $H_{q-1}$ on $A$ with diameter $< \eta$. It may be regarded as the point-set image of an $F(\sigma_q)$ the images of whose faces make up a semi-singular image of $\bar{\sigma}_q$. By hypothesis the image may be completed by a singular $\sigma_q$ of diameter $< \epsilon$ to form the singular image of $\bar{\sigma}_q$. Hence $A$ is locally $q$-connected for every $q \leqq p$, that is, it is an $LC^p$.

We proceed to show that the condition of the theorem is necessary. This is trivial for $p = 0$ and hence we may assume $p > 0$. This being the case, let $\eta_q(\epsilon)$ be the number $\eta$ entering into the definition of local $q$-connectedness. We shall choose successive constants $\zeta_0, \cdots, \zeta_q (q \leqq p)$, where $\zeta_i < \zeta_{i+1}$, $\zeta_q = \epsilon$, and the other $\zeta$'s are constrained to satisfy certain inequalities to be considered presently.

In the first place we require $\zeta_0 < \eta_0(\zeta_1)$ and in any case we impose the condition $\eta(\epsilon, p) \leqq \zeta_0$. Let $x_0x_1$ be an abstract one-cell of $K_p'$, that is to say, let $x_0$ and $x_1$ be the images of the end-points of a one-cell of $K_p$. By hypothesis $d(x_0, x_1) \leqq \zeta_0 < \eta_0(\zeta_1)$; hence we may join the two points on $A$ by a singular one-cell $\sigma_1'$ with diameter $< \zeta_1$. That is, we may replace the abstract one-cells of $K_p'$ by singular one-cells of diameters $< \zeta_1$. Consequently in the new complex, which we continue to call $K_p'$, all the one-cells will be singular and none will be abstract. Now let $x_0x_1x_2$ be an abstract two-cell of $K_p'$. Its one-faces $\sigma_1'^i (i = 0, 1, 2)$ are all singular and their sum is of diameter $< 2\zeta_1$. A necessary condition that they bound a singular 2-cell of diameter $< \zeta_2$ is that

(1) $$\operatorname{diam}\,(\sigma_1'^0 + \sigma_1'^1 + \sigma_1'^2) < \eta_1(\zeta_2).$$

This will certainly hold if

(2) $$2\zeta_1 < \eta_1(\zeta_2).$$

If (1) holds for all similar triples of singular one-cells, and this will certainly be true if (2) holds, we shall be able to insert all the requisite two-cells and have them be of diameters $< \zeta_2$. As to those already present, they have diameters $\leqq \eta(\epsilon, p) < \zeta_0 < \zeta_2$. Therefore all the singular two-cells will be of diameters $< \zeta_2$. The same reasoning may evidently be repeated: if the $m + 1$ vertices of the simplexes $\sigma_{m-1}'^{i}$ are the vertices of a potential $\sigma_m'$, we must have

(3) $$\operatorname{diam} \Sigma\, \sigma_{m-1}'^{i} < \eta_{m-1}(\zeta_m),$$

and these inequalities will certainly hold relative to each $\sigma_m'$ if we have

(4) $$2\zeta_{m-1} < \eta_{m-1}(\zeta_m),$$

although (4) may be stronger than necessary. Given $\epsilon$ and $p$ we can find $\eta$ and the $\zeta$'s so as to satisfy all the inequalities (4) for $m = 1, 2, \cdots, p$, and the $\zeta_0$ thus obtained is a suitable $\eta(\epsilon, p)$. When the inequalities (4) hold, the same is true with regard to the weaker inequalities (3), and the successive faces may be inserted one by one: first all the one-faces, then all the two-faces, etc., and finally $\sigma_q'$ itself. This proves the necessity of the condition of the theorem.

5. Now let $\eta(\epsilon)$ denote the greatest lower bound of the $\eta(\epsilon, p)$ of Theorem I for a given $\epsilon$ and all values of $p$. If $\eta(\epsilon) > 0$, i.e., if in Theorem I we can always choose a fixed $\eta(\epsilon, p)$ independent of $p$, we shall call $A$ an LC-set. This implies that $A$ is an $LC^p$-set for every $p$, and moreover that it is such a set in a certain uniform manner relative to $p$.

Suppose that $A$ is an LC-set and let $\alpha$ be its diameter. The largest value that $\epsilon$ may take is $\alpha$, but as $\epsilon \to \alpha$ it may well happen that $\eta$ remains less than a certain $\beta < \alpha$. If, however, $\eta \to \alpha$ at the same time as $\epsilon$, we shall call $A$ an $\overline{LC}$-set. The basic theorem to be established in the sequel is

THEOREM II. *The classes LC and ANR, likewise the classes $\overline{LC}$ and AR are respectively identical* ($LC = ANR$; $\overline{LC} = AR$).

This proposition embodies an intrinsic characterization of absolute retracts and absolute neighborhood retracts. The proof will rest upon a lengthy discussion which we shall take up presently.

6. By a classical result of Urysohn's, a self-compact metric set $A$ has a topological image $A'$ on the Hilbert parallelotope $\mathfrak{H}$. It will be convenient to denote by $x, y, \cdots$ the points of $\mathfrak{H}$ whose coordinates are $(x_1, x_2, \cdots), (y_1, y_2, \cdots), \cdots,$ and by $\mathfrak{H}_n$ the $n$-face of $\mathfrak{H}$ on which all coordinates $x_{n+i} = 0$. In particular, $\mathfrak{H}_0 = (0, 0, \cdots)$ is the origin.

By the *projection* of $x$ onto the face $\mathfrak{H}_n$ we shall mean the point $(x_1, \cdots, x_n,$

$0, 0, \cdots)$. The projection $A^n$ of any set $A$ onto $\mathfrak{H}_n$ is the set of the projections of all its points. We observe that $A^n$ is closed if $A$ is closed.

7. Now let $A$ be the set of the theorem. The discussion to follow is of no interest when $A = \mathfrak{H}$, so that we may assume that $A$ is a proper subset of $\mathfrak{H}$. As a consequence, from a certain $\nu$ on (kept fixed henceforth), $A^\nu$ will be a proper subset of $\mathfrak{H}_\nu$.

We subdivide $\mathfrak{H}_\nu$ simplicially into a complex $\mathfrak{H}'_\nu$ of mesh $\epsilon_\nu = \epsilon/2^\nu > 0$, then take the sum of the closed cells of the subdivision that meet $A^\nu$, and erect on each of these cells a rectangular prism in the space of $\mathfrak{H}_{\nu+1}$, thus obtaining a certain closed convex complex $\mathfrak{H}'_{\nu+1}$ on $\mathfrak{H}_{\nu+1}$ which $\supset A^{\nu+1}$. The distance $d_{\nu+1} = d(A^{\nu+1}, \mathfrak{H}_{\nu+1} - \mathfrak{H}'_{\nu+1})$ is such that $0 < d_{\nu+1} < \epsilon_\nu$. We subdivide $\mathfrak{H}'_{\nu+1}$ simplicially into a complex of mesh $\epsilon_{\nu+1} < \frac{1}{2}d_{\nu+1} < \frac{1}{2}\epsilon_\nu$ which we still call $\mathfrak{H}'_{\nu+1}$, take the sum of the closed cells of $\mathfrak{H}'_{\nu+1}$ which meet $A^{\nu+1}$, erect rectangular prisms on them in the space of $\mathfrak{H}_{\nu+2}$, etc., and proceed indefinitely in this way with the following mild modification. Suppose that $\mathfrak{H}'_n$ has a cell of the form $\sigma\sigma'$, where $\sigma'$ is a cell of $\mathfrak{H}'_{n+1}$ but $\sigma$ is not. Through the subdivision of $\mathfrak{H}'_{n+1}$ into a complex of mesh $\epsilon_{n+1} < \frac{1}{2}d_{n+1}$ it may turn out that $\sigma'$ is subdivided into simplexes $\sigma'^i$. We then replace $\sigma\sigma'$ by the sum of the simplexes $\sigma\sigma'^i$, thus preserving simpliciality throughout.

Ultimately we shall have a sequence of complexes $\mathfrak{H}'_\nu, \mathfrak{H}'_{\nu+1}, \cdots$ where $\mathfrak{H}'_n$ is a simplicial closed complex on $\mathfrak{H}_n$ which $\supset A^n$, which is coincident with $\mathfrak{H}_n$ for $n = \nu$, whose mesh $\epsilon_n < \epsilon/2^n$, and which is such that the projections of its points on $\mathfrak{H}'_{n-1}$ are not farther than $\frac{1}{2}\epsilon_{n-1}$ from $A^{n-1}$. Moreover, by construction, $\mathfrak{H}'_n \cdot \mathfrak{H}'_{n+1}$ coincides with a closed subcomplex of both $\mathfrak{H}'_n$ and $\mathfrak{H}'_{n+1}$. Therefore $J_n = \mathfrak{H}'_n - \mathfrak{H}'_n \cdot \mathfrak{H}'_{n+1}$ is an open subcomplex of $\mathfrak{H}'_n$ whose mesh $<\epsilon_n$ and the projections of whose points on $\mathfrak{H}_{n-1}$ are not farther than $\epsilon_{n-1}$ from $A^{n-1}$. Setting $J = J_\nu + J_{\nu+1} + \cdots$, we have $J_p \cdot J_q = 0$ for $p \neq q$, and $J$ is a closed infinite simplicial complex—closed in the sense that if $\sigma$ is any cell of $J$ then every face of $\sigma$ is likewise a cell of $J$.

8. Let $\rho_n$ denote the square root of the remainder after $n$ terms of the convergent series $\dfrac{1}{1^2} + \dfrac{1}{2^2} + \cdots$, and let $y^n$ denote the projection of any point $x$ on $\mathfrak{H}_n$. Taking $x \subset J_{n+1}$ we have $x = y^{n+1} \subset \mathfrak{H}'_{n+1}$ and $y^n \subset \mathfrak{H}'_n \cdot \mathfrak{H}'_{n+1}$. Let $z^n$ be a point of $A^n$ such that $d(y^n, A^n) = d(y^n, z^n)$. That a point such as $z^n$ exists follows from the fact that $A^n$ is closed. Let $z$ be a point of $A$ with projection $z^n$ on $A^n$. We have $d(x, A) \leq d(x, z)$ and hence

$$(5) \qquad d(x, A) = d(y^{n+1}, A) \leq d(y^{n+1}, y^n) + d(y^n, z^n) + d(z^n, z)$$

$$\leq \frac{1}{n+1} + \epsilon_n + \rho_n \leq \frac{1}{n+1} + \frac{\epsilon}{2^n} + \rho_n,$$

a quantity which $\to 0$ with $1/n$. Therefore $J_n \to A$ with increasing $n$, while $J_n \cdot A = 0$, and hence $J \cdot A = 0$.

9. Consider a point $x$ of $\mathfrak{H} - A$ and its projection $y^n$. Since all the points of $\mathfrak{H}'_{n+1}$ can be projected onto $\mathfrak{H}'_n$, whenever $y^{n+1} \subset \mathfrak{H}'_{n+1}$ then likewise $y^n \subset \mathfrak{H}'_n$. Hence there exists a largest $n$ such that $y^{n+1} \subset \mathfrak{H}'_{n\;1}$, for, if the contrary were true, (5) would hold for every $n$, and hence it would strictly be the case that $d(x, A) = 0$ and $x \subset A$. Since $y^{n+2} \not\subset \mathfrak{H}'_{n+2}$ we likewise have $y^{n+1} \not\subset \mathfrak{H}'_{n+2}$ and hence $y^{n+1} \subset J_{n+1}$. Thus for each $x$ there exists a unique $n$, and hence a unique $y^n$, such that $y^n \subset J_n$. This $y^n$ can be on no other $J$ since no two $J$'s intersect. Moreover, owing to (5), $d(x, y^n) \to 0$ as $x \to A$.

We now define $P$ to be the transformation: $x \to y = y^n$. The point $y = Px$ is uniquely defined for every $x \not\subset A$ so that the transformation $P$ is single-valued over $\mathfrak{H} - A$. It is evident that $P$ reduces to the identity on $J$. Furthermore $P$ is defined as the identity for any point $x \subset A$. Thus $P$ has the following properties:

(a) It is single-valued on $\mathfrak{H}$.
(b) $P = 1$ on $A + J$; $P(\mathfrak{H} - A) = J$.
(c) $d(x, Px) \to 0$ as $x \to A$.

10. Unfortunately $P$ is not continuous. To prove this, let $U^n$ be the open region of $\mathfrak{H}$ whose projection onto $\mathfrak{H}_n$ has $\mathfrak{H}'_n$ for closure. Since $\epsilon_n < \frac{1}{2}d_n$, we have at once that $U^n \Subset U^{n-1}$, so that the boundaries $F(U^n)$ do not intersect one another. If $x_0 \subset F(U^n)$, we have $y_0 = Px_0 = y_0^n$. However, in any vicinity of $x_0$ there are points $x$ such that $y = Px = y^n$, and points such that $y = y^{n-1}$. For two points of these two distinct types the corresponding $y$'s are at a distance apart which is as near as we please to $\delta = d(y_0^{n-1}, y_0^n)$, and $\delta$ may reach the value $1/n$ without, however, exceeding it. This is a measure of the discontinuity in $y$ when $x$ crosses $F(U^n)$ and we notice that $\delta \to 0$ with $1/n$.

We shall restore continuity in the transformation by approximating $P$ in the vicinity of $F(U^n)$ by a new continuous transformation $\Pi$ so chosen as to preserve the three properties (*a*), (*b*), (*c*). To do this we first subdivide $\mathfrak{H}'_{n+1}$ regularly and take the sum $N$ of all the cells of the subdivision with a vertex on $J_n$. $N$ is a polyhedral neighborhood of $J_n$ on $\mathfrak{H}'_{n-1}$ with the following property: if $\sigma = \sigma'\sigma''$ is any simplex of $N$, where $\sigma' \subset J_n$ and $\sigma''$ has no vertex on $J_n$, then through every point of $N$ there passes a unique segment $\xi'\xi''$ with $\xi'$ on $\sigma'$ and $\xi''$ on $\sigma''$.[5] Now if $y^{n-1}$ coincides with the point $\xi$ of $\sigma$ we shall choose as the point $y = \Pi x$ the point $y'^n = (x_1, \cdots, x_{n-1}, x'_n, 0, \cdots)$ of $\mathfrak{H}'_n$, where $x'_n = \dfrac{\xi\xi''}{\xi'\xi''} \cdot x_n$ (the terms of the fraction are the lengths of the segments indicated). It follows that on $\sigma$ the point $y$ varies continuously with $x$ and coincides with $Px$ on $\sigma'$ and $\sigma''$. This gives new points $\Pi x$ on $\mathfrak{H}_n$ when $y_{n-1} \subset \sigma$, and their totality constitutes a convex cell whose dimension $= \dim \sigma + 1$. The closure of this cell and of the similar cells for all $\sigma$'s of $N$ are added to $J_n$, the sum, $J'_n$, is subdivided simpli-

[5] This is the "normal neighborhood" property of *Topology*, p. 91.

cially into cells of diameters $< \epsilon_n$, and this is done in such manner that $J'_n \cdot J'_{n+1}$ coincides with subcomplexes of both $J'_n$ and $J'_{n+1}$. Again setting

$$K_n = J'_n - J'_n \cdot J'_{n+1}, \qquad K = K_\nu + K_{\nu+1} + \cdots,$$

we have, as before, that $K$ is a simplicial closed infinite complex with $K \cdot A = 0$.

11. Since $\epsilon_{n-1} < \frac{1}{2}d_{n-1}$, no cell of $J_{n-1}$ has vertices on both $J_n$ and $J_{n-2}$; hence $\bar{N} \cdot J \subset J_{n-1} + J_n$ and the modifications entailed in $P$ in correlation with $N$ concern only the points $x$ whose $Px \subset J_{n-1}$. Since $J_p \cdot J_q = 0$ for $p \neq q$, the modifications are consistent throughout $\mathfrak{H} - A$. For any point $x$ for which no modification is assigned we specify $\Pi x = Px$. Finally for $x \subset A$ we specify $\Pi x = x$. To sum up:

I. $\Pi$ is continuous and single-valued on $\mathfrak{H}$;
II. $\Pi = 1$ on $A$;
III. $\Pi(\mathfrak{H} - A) = K$; $\Pi\,\mathfrak{H} = \bar{K} = K + A$.

The only question that may be raised concerns the continuity of $\Pi$; it is proved as follows:

(a) let $x_0$ be neither on $A$ nor such that $Px_0$ is on an $N$; then, in a suitable vicinity of $x_0$, $\Pi$ coincides with $P$. Since $P$ is continuous, $\Pi$ is also continuous;

(b) let $x_0 \subset N$; then in a small neighborhood of $x_0$, $\Pi x$ is the point $y'^n$ of No. 11, $\Pi x_0 = y_0'^n$, and $\Pi x \to \Pi x_0$ as $x \to x_0$; hence $\Pi$ is continuous at $x_0$;

(c) let $x_0 \subset F(U^n)$, so that $\Pi x_0 = Px_0 = y_0^n = y_0'^n$; a point $x$ in a small vicinity of $x_0$ may be on $U^n$ or on $\mathfrak{H}'_n - \bar{U}^n$. In the first case $\Pi x = y^n$, in the second, $\Pi x = y'^n$, and both have the same limit $y_0^n$ as $x \to x_0$;

(d) let $x_0 \subset A$; if $Px \subset J_n$ then $d(Px, \Pi x) = x'_n \leqq 1/n$; now as $x \to x_0$, $n \to \infty$, $d(Px, \Pi x) \to 0$, and, since $d(x, Px) \to 0$, it follows that $d(x, \Pi x) \to 0$ also. This means that as $x \to x_0$, $\Pi x \to \Pi x_0 = x_0$. Therefore $\Pi$ is continuous at $x_0$ and consequently throughout $\mathfrak{H}$.

12. If we combine what we have already proved, we have the comprehensive and noteworthy

THEOREM III. *Given a closed subset $A$ of the Hilbert parallelotope $\mathfrak{H}$, there exists a closed infinite simplicial complex $K$ on $\mathfrak{H} - A$, and a continuous single-valued transformation $\Pi$ over $\mathfrak{H}$ such that:* (a) *$K$ has at most a finite number of cells exterior to any neighborhood $U$ of $A$ and their maximum dimension $\to 0$ as $U \to A$, while the diameters of the cells on $U \to 0$ at the same time;* (b) $\Pi(\mathfrak{H} - A) = K$; (c) $\Pi = 1$ *on $A$ (all points of $A$ are fixed points of $\Pi$).*

When $A = \mathfrak{H}$ the proposition is still true with $K = 0$; in this case $\Pi = 1$. If dim $A = n$ is finite the set has a topological image $A'$ on a parallelotope $R$ of a bounded $S_n$.[6] The decomposition of mutually inclusive polyhedral neighborhoods of $A'$ tending to $A'$ into simplicial complexes of meshes $\to 0$ gives rise to a $K_n$ analogous to $K$ for which $\Pi = 1$ over the whole of $R$, so that the analogous theorem becomes trivial.

[6] See, for example, pp. 569-578.

13. Returning to the situation of No. 11 and earlier, let us now introduce the hypothesis that $A$ is an LC-set. We first define a single-valued transformation $T$ of the vertices of $K$ into points of $A$ as follows: if $x$ is any vertex of $K$ then we choose for $x' = Tx$ any point of $A$ such that $d(x, x') = d(x, A)$. Such a point always exists since $A$ is closed. If $x$ is a vertex of $K_{n+1}$ we have (No. 8)

$$d(x, A) = d(y^{n+1}, A) \leqq d(y^{n+1}, y^n) + d(y^n, z^n) + d(z^n, z)$$

$$\leqq \frac{1}{n+1} + 2\epsilon_n + \rho_n.$$

Since mesh $K_{n+1} < \epsilon/2^{n+1}$, $T$ transforms the $q + 1$ vertices of any $\sigma_q$ of $K_{n+1}$ into $q + 1$ points of $A$ whose sum is of diameter $< 2/(n + 1) + 5\epsilon/2^n + 2\rho_n = \lambda_n$, where $\lambda_n \to 0$ monotonely with $1/n$.

From the construction of $K_n$ it follows that if any $\sigma$ of $K_n$ has vertices that do not belong to $K_n$, they must be vertices of $K_{n+1}$. The sum of these special $\sigma$'s is an open subcomplex $K_n''$ of $\bar{K}_n$, and the $\sigma$'s whose vertices are all vertices of $K_n$ make up a closed subcomplex $K_n'$ of $K_n$.

Now consider the function $\eta(\epsilon)$ of No. 5, take a sequence $\{\xi_n\}$ which approaches 0 monotonely, and set $\eta_m = \eta(\xi_m)$, so that $\eta_m$ likewise $\to 0$. Let $r$ be the smallest integer such that $\lambda_{r+1} < \eta_1$. For every $m$ there exists an $n$ such that $\lambda_n \leqq \eta_m$. Allowing for repetitions among the $\xi$'s and $\eta$'s, but not among the $\lambda$'s, we may say that for every $m$ we shall have $\lambda_{r+m} < \eta_m$, where still $\eta_m \to 0$ with $1/m$.

14. Now let $x^1, x^2, \cdots$ be the vertices of $K'_{r+m}$. It follows from Theorem I that there exists on $A$ a singular image $\mathfrak{K}'_{r+m}$ of $K'_{r+m}$ with $Tx^i$ as the image of the vertex $x_i$ and with mesh $\mathfrak{K}'_{r+m} < \xi_m$. We now define as transform by $T$ of $K'_{r+m}$ on $A$ the c. s. v. t. whereby $K'_{r+m}$ goes into $\mathfrak{K}'_{r+m}$.

Now suppose that $\sigma = \sigma'\sigma''$ is a cell of $K''_{r+m}$, with $\sigma'$ a cell of $K'_{r+m}$ and $\sigma''$ a cell of $K'_{r+m+1}$. Then $\sigma'$ and $\sigma''$ have for images $T\sigma'$ and $T\sigma''$ which are singular cells of diameters $< \xi_m$. On the other hand, the vertices of $\sigma$ have for their transforms a set of diameter $< \lambda_{r+m}$. Therefore diam $(T\sigma' + T\sigma'') < 2\xi_m + \lambda_{r+m} \to 0$ with $1/m$.

According to Theorem I, for $m$ above a certain value we shall be able to join $\mathfrak{K}'_{r+m}$ and $\mathfrak{K}'_{r+m+1}$ by a singular image $\mathfrak{K}''_{r+m}$ of $K''_{r+m}$ such that the transforms of the cells $\sigma'$ and $\sigma''$ will be as before, and moreover that mesh $\mathfrak{K}''_{r+m} \to 0$ with $1/m$. We now extend $T$ as previously to the cells of $K''_{r+m}$, so that for $m$ sufficiently high it is defined over $K_{r+m}$, and we set $\mathfrak{K}_{r+m} = \mathfrak{K}'_{r+m} + \mathfrak{K}''_{r+m} = TK_{r+m}$.

15. We now have a transformation $T$ defined over a certain set

$$K^{(p)} = K_p + K_{p+1} + \cdots$$

provided that $p$ is sufficiently large, and it is a c. s. v. t. over $K^{(p)}$. Observe that when $x \subset K_n$, $n > p$,

$$d(x, Tx) < \lambda_n + \text{mesh } \mathfrak{K}_n + \text{mesh } K_n \to 0$$

with $1/n$. Therefore, if we extend again $T$ to $A$ by the condition that $T = 1$ on $A$, $T$ is a continuous single-valued transformation of $K^{(p)} + A$ into $A$.

Consider now the transformation $\Pi$ of No. 11. Since it is continuous and $= 1$ on $A$, any relative neighborhood of $A$ with respect to $\Pi\mathfrak{H}$ will be the transform of a neighborhood of $A$ on $\mathfrak{H}$. It follows that there exists an open set $U \supset A$ such that $\Pi U \subset K^{(p)}$. Consequently $T\Pi \cdot U \subset A$, $T\Pi = T = \Pi = 1$ on $A$, $T\Pi \cdot U = A$. *Therefore the* LC-*set $A$ is an* NR *as to $\mathfrak{H}$ and hence it is an* ANR.

16. Conversely, suppose that $A$ is an ANR, and hence an NR as to $\mathfrak{H}$, and let $U$ be an open set $\supset A$ such that there exists a c.s.v.t. $T$ of $U$ into $A$ which reduces to the identity on $A$. We shall show that $A$ is an LC-set.

We first observe that there is an open set $U'$ such that $U \supset \bar{U}' \supset A$, so that $U'$ may replace $U$ above. In other words, we may assume that in place of $U$ we have $\bar{U}$ throughout. Now, *given any $\epsilon$, there exists a $\xi > 0$ such that if $C$ is any subset of $\bar{U}$ whose diameter $< \xi$, then diam $TC < \epsilon$.* Since $A$ is closed it possesses a finite $\epsilon$-covering by relatively open sets $u^i$. Since $T$ is continuous and $= 1$ on $A$ the sets $U^i = \bar{U}T^{-1}u^i$ are relatively open for $\bar{U}$ and $u^i = TU^i$. Moreover, since $T\bar{U} = A$, $\{U^i\}$ is a finite covering of $\bar{U}$ by relatively open sets. The characteristic constant[7] $\xi$ of that covering has the required property. For if diam C $< \xi$, some $U^i \supset C$, $U^i \supset TC$, hence diam $TC <$ diam $u_i < \epsilon$.

17. Now that we have $\epsilon$ and $\xi$, let us consider a finite covering of $A$ by spherical regions $\{\mathfrak{S}^i\}$ with diameters $< \xi$ and also $< d(A, \mathfrak{H} - U)$ whose centers are on $A$, and let $\eta$ be the characteristic constant of the covering. Every subset $C$ of $U$ which meets $A$ and whose diameter $< \eta$ will be on some $\mathfrak{S}^i$ and we shall have diam $TC < \epsilon$.

It is now a simple matter to show that $\epsilon$ and $\eta$ are related as required by the condition for an $\mathrm{LC}^p$-set no matter what may be the value of $p$, and hence that $A$ is an LC-set. For, let $K_p$ be an abstract simplicial complex and let $\mathfrak{K}'$ be a semi-singular image of $K_p$ on $A$ with mesh $< \eta$. The problem is to complete it to a singular image $\mathfrak{K}$ of $K_p$ with mesh $< \epsilon$.

We first take $\bar{U}$ itself as our set $A$ and construct a singular image $\mathfrak{K}^*$ of $K_p$ on $\bar{U}$ by the process of No. 4 when modified so that each new singular cell introduced will lie on an $\mathfrak{S}^i$.

Given any set $C$ on $\mathfrak{H}$, the intersection of all spheres containing $C$ is a set which we shall denote by $(C)$. Clearly if $C' \subset C$ then $(C') \subset (C)$.

Now consider any potential cell $\sigma_q$ of $\mathfrak{K}^*$, and suppose that among its faces it already possesses the realized singular cells $\sigma', \cdots, \sigma$ of $\mathfrak{K}'$ among which are

[7] Let $A$ be a closed subset of a compact metric space. With any finite covering of $A$ by open sets $\{U^i\}$, there are associated two positive constants $\alpha$ and $\beta$ defined as follows: any set $B$ which meets $A$ and whose diameter $< \alpha$ is on some $U^i$; if $B \subset A$, diam $B < \beta$ and $B$ meets a certain number of sets $U^i$, these sets intersect on $A$. We call the smaller of the two numbers $\alpha, \beta$ the *characteristic constant* of $\{U^i\}$.

included all the vertices of $\sigma_q$. If $\zeta = \Sigma\sigma^i$, $(\zeta)$ is on some $\mathfrak{S}^i$. Hence to show that $\sigma_q$ can always be chosen so as to be on some $\mathfrak{S}^i$ it is sufficient to prove that it can be chosen on $(\zeta)$. Now the process of No. 4 consists in introducing first the singular one-cells, then all the singular two-cells, etc. It is therefore sufficient to show that if the condition $\sigma_q \subset (\zeta)$ has already been fulfilled for the dimensions $< q$, it can be fulfilled for $q$.

Consider again $\sigma_q$ and let $\sigma'^i$ be those of its singular faces already present of all dimensions up to and including $q - 1$, where if $\zeta^i$ is the analogue of $\zeta$ for $\sigma'^i$ we have $\sigma'^i \subset (\zeta^i)$. Since $\zeta^i \subset \zeta$, we have $(\zeta^i) \subset (\zeta)$, and hence $\Sigma\sigma'^i \subset (\zeta)$. Thus the singular boundary sphere $\eta_{q-1}$ of $\sigma_q$ is already introduced and it lies on $(\zeta)$. The segments from a fixed point $x$ of $(\zeta)$ to the points of $\eta_{q-1}$ generate a suitable $\sigma_q \subset (\zeta)$. This shows that we can construct $\sigma_q$ as required.

The rest is now obvious. Take for $\mathfrak{K}$ the singular complex on $A$: $\mathfrak{K} = T\mathfrak{K}^*$. Since $\mathfrak{K}' \subset A$, $T\mathfrak{K}' = \mathfrak{K}'$, hence $\mathfrak{K}$ is a singular image of the type required. Moreover, since each cell of $\mathfrak{K}^*$ is on an $\mathfrak{S}^i$, mesh $\mathfrak{K}^* < \eta$, and hence mesh $\mathfrak{K}$ $< \epsilon$. Therefore $\epsilon$ and $\eta$ are related as required, and $A$ is an LC-set.

18. We have thus completely identified LC-sets and ANR-sets. We shall now identify $\overline{\text{LC}}$-sets and AR-sets. Referring to No. 15 we find that in order that $A$ be an AR it is sufficient that the transformation $T$ be defined not merely on $K^{(p)}$ for $p$ above a certain value, but on every $K^{(p)}$, that is, on $K$ itself. This in turn will certainly occur provided that $\epsilon$ and $\eta(\epsilon)$ tend simultaneously to diam A, that is, provided that $A$ is an $\overline{\text{LC}}$-set.

Conversely let $A$ be an AR. Since an AR is also an ANR, $A$ is certainly an LC-set. On the other hand the extension from the semi-singular $\mathfrak{K}'$ of No. 17 to the singular $\mathfrak{K}$ can be carried out here whatever $\mathfrak{K}$ may be. Therefore $A$ is an $\overline{\text{LC}}$-set provided that there exists a $\mathfrak{K}'$ of mesh $\alpha =$ diam $A$. Since $A$ is closed it contains two points $x$, $y$ whose distance is $\alpha$, and since $A$ is a retract of $\mathfrak{H}$ there is a c. s. v. t. $T$ of $\mathfrak{H}$ into $A$ with $T = 1$ on $A$. The image under $T$ of the closed segment $xy$ is a closed singular $\sigma_1 \subset A$ whose diameter $= \alpha$, and $\sigma_1$ plus any point is precisely a suitable $\mathfrak{K}'$ of mesh $\alpha$ which is the semi-singular image of a closed two-cell. This shows that AR $= \overline{\text{LC}}$ and completes the proof of Theorem II.

19. **Homology characters of LC-sets.** Let $A$ and $U$ be as before and let us observe that there exists a constant $\alpha > 0$ such that if $d(x, A) < \alpha$, then the segment $\overline{x, Tx} \subset U$. For otherwise we could find a sequence of points $x_i$ converging to a point $x_0$ such that $d(x_i, Tx_i) \to 0$ with $1/i$, and that for every $i$ the segment $\overline{x_i, Tx_i}$ contains points of $\mathfrak{H} - U$. Therefore we would have $d(x_0, \mathfrak{H} - U) = 0$. But $x_0$ is the limit of the points $Tx_i$ of $A$, and hence $x_0 \subset A$ since $A$ is self-compact. We would then have $d(A, \mathfrak{H} - U) = 0$, which is ruled out by the fact that $U$ is a neighborhood of the closed set $A$ on the compact space $\mathfrak{H}$. We shall assume not only that $\alpha$ has the property just described, but that in addition it is $< d(A, \mathfrak{H} - U)$.

Let $\Sigma = \{U^i\}$ be a finite $\epsilon$-covering of $A$ by open sets of $\mathfrak{H}$, where $\epsilon < \alpha/8$ and where the covering is such that its skeleton $\Phi$ in the sense of Alexandroff is the same as for the aggregate $\{\bar{U}^i\}$.[8] We denote by $\xi$ the characteristic constant of $\Sigma$, and assume $\Phi$ to be constructed as in *Topology*, p. 326, with the vertex corresponding to $U^i$ not farther than $\alpha/8$ from the set. Consequently mesh $\Phi < \frac{1}{2}\alpha$; since each cell of $\Phi$ has a vertex not farther than $\alpha/8$ from $A_1$, every point $x$ of $\Phi$ is nearer than $\alpha$ to $A$ and the segment $\overline{x, Tx} \subset U$.

Since an LC-set is also an LC$^\infty$-set, or a locally-connected set in the sense of *Topology*, its homology groups defined by sequences of abstract chains are simply isomorphic with those defined by means of the singular chains on the set (*Topology*, p. 333, Remark I).

20. Let $C_p$ be a singular chain on $A$ with mesh $< \alpha/8$ and $\xi$, and let $\tau$ be a transformation of the vertices $x^i$ of $C_p$ into vertices of $\Phi$ such that $\tau x^i$ is the vertex of $\Phi$ corresponding to one of the sets $U^i \supset x^i$. It follows from the construction of $\Phi$ that, to the vertices of a cell $\sigma$ of $C_p$, there correspond the vertices of a simplex $\tau\sigma$ of $\Phi$. From this we conclude, as in *Topology*, p. 86, that $C_p$ may be $\beta$-deformed into a subchain $C'_p$ of $\Phi$, where $\beta < \alpha$, and that $\tau$ may be extended to a c. s. v. t. of $C_p$ into $C'_p$, still to be called $\tau$. Consequently if $x$ is any point of $C_p$, the segment $\overline{x, \tau x} \subset U$ and the totality of these segments gives rise to a deformation $\mathfrak{D}$ of $C_p$ into $C'_p$ in the sense of *Topology*, p. 78, with a deformation-chain $\mathfrak{D}C_p \subset U$.

Since $\tau$ is a linear, boundary-preserving operation on chains, it transforms cycles into cycles. That is, if $\Gamma^i_p$ are cycles of $A$, the chains $\tau\Gamma^i_p$ are cycles of $\Phi$.

Suppose that we have

$$\Phi \supset C_{p+1} \rightarrow \Sigma\, s_i \cdot \tau\Gamma^i_p.$$

We then have successively:

$$U \supset \Sigma\, s_i \cdot \mathfrak{D}\Gamma^i_p \rightarrow \Sigma\, s_i\,(\tau\Gamma^i_p - \Gamma^i_p);$$

$$U \supset D_{p+1} = C_{p+1} - \Sigma\, s_i \cdot \mathfrak{D}\Gamma^i_p;$$

$$A \supset TD_{p+1} \rightarrow \Sigma\, s_i \cdot \Gamma^i_p \sim 0 \text{ on } A.$$

Consequently any formal homology between the $\tau\Gamma$'s on $\Phi$ is also satisfied by the $\Gamma$'s on $A$. Therefore the absolute homology groups of $A$ are subgroups of those of the finite complex $\Phi$, and hence they have the same general structure: the Betti-numbers and torsion coefficients are all finite, and those above a certain $p$ are zero.[9] In particular, for an $\overline{\text{LC}}$-set we may take for $\Sigma$ the single set $\mathfrak{H}$, so

[8] To have a suitable $\Sigma$ we may, for example, take a first covering by spherical regions of radii $\frac{1}{2}\,\epsilon$, then replace these by concentric regions whose radii are slightly smaller.

[9] Added in proof, Jan. 5, 1934: I learn from a preprint of a paper by Borsuk, due to appear in Fundamenta Matematicae, vol. XXI, that he has proved the slightly more general result, that the homology groups are those of a finite complex. This may readily be obtained by a slight extension of our treatment, the complex being a suitable $\Phi$.

that $\Phi$ is a single point and the homology characters are those of a point, that is, the set is *cell-like* in the sense of A. W. Tucker. Combining our results we have

THEOREM IV. *For every* LC-*set* (= ANR) *there exists an integer* $p$ *such that all the absolute homology characters for the dimensions* $\leqq p$ *are finite and the others are zero. An* $\overline{\text{LC}}$-*set* (= AR) *is cell-like.*

The least integer $p$ for which the preceding results hold is clearly a topological invariant of the LC-set $A$. We shall call it the *characteristic integer* of the set.

21. **Applications.** Taking account of Alexander's generalized duality relations (*Topology*, p. 339) we have, as a consequence of Theorem IV,

THEOREM V. *If* $A$ *is an* LC-*set in a sphere* $H_n$, *the Betti-numbers of* $H_n - A$ *are all finite and* $p \leqq n$. *If* $A$ *is an* $\overline{\text{LC}}$-*set,* $H_n - A$ *has the Betti-numbers of a point.*

As a special case, $R_0(H_n - A)$ is finite, that is, $H_n - A$ consists of a finite number of regions, and if $A$ is an $\overline{\text{LC}}$-set, $H_n - A$ is connected (Borsuk III, p. 230).

Finally, if we recall that our basic fixed point formula is strictly applicable to an ANR-set (*Topology*, p. 347) we have

THEOREM IV. *Let* $T$ *be a c.s.v.t. of an* LC-*set* $A$ *into itself or into part of itself, and let* $p$ *be the characteristic integer of* $A$. *If* $\varphi^q$ *is the matrix of the transformation induced by* $T$ *on a base for the q-cycles of* $A$, $\varphi^q = 0$ *for* $q > p$, *and* $\varphi^q$, *is finite for* $q \leqq p$. *The signed number of fixed points* (*Topology*, *p. 359*) *is given by a finite sum:*

$$\Theta = \Sigma (-1)^q \operatorname{trace} \varphi^q.$$

*In particular,*

$$\Theta = \operatorname{trace} \varphi^0 \neq 0 \text{ for an } \overline{\text{LC}}\text{-set}.$$

*There is at least one fixed point when* $\Theta \neq 0$. *In particular, every c.s.v.t. of an* $\overline{\text{LC}}$-*set has at least one fixed point.*

APPLICATION. Every acyclic continuous curve (= Peano continuum) is an $\overline{\text{LC}}$-set (Borsuk II, p. 211). Hence *every c.s.v.t. of a continuous curve into itself or into part of itself has at least one fixed point.* (See Borsuk II, p. 205.)

# XVIII

## ON THE FIXED POINT FORMULA

The object of the present note is to give a new and final proof of the fixed point formula for LC spaces (= absolute neighborhood retracts). For the sake of simplicity we consider only continuous single valued transformations (= c.s.v.t.). A comparison with our earlier proof[1] will show that practically the whole burden may now be thrown on the abstract complex, the extension to abstract spaces being reduced to the barest minimum.

1. Our starting point is the fixed element formula for abstract complexes.[2] Let $K$ be a finite complex and let $E_p^i$ be its elements. We shall only consider rational chains and their transformations. A transformation $T$ of the type envisaged has the form

$$(1.1) \qquad TE_p^i = x_{p,j}^i E_p^j; \qquad x_{p,j}^i \text{ rational.}$$

In order that $T$ possess some fixed element, i.e. that some $E_p^i$ be a member of the chain $TE_p^i$, it is necessary and sufficient that some number $x_{p,i}^i \neq 0$ ($i$ unsummed). A sufficient condition is then that

$$(1.2) \qquad \theta = \sum (-1)^p \text{ trace } x_p \neq 0, \qquad x_p = \| x_{p,j}^i \| .$$

The transformations that really matter are those which permute with $F$, the boundary operator: $TF = FT$. This property means that $T$ transforms a chain and its boundary into a chain and its boundary, and in particular a cycle into a cycle. It may be shown then that if $\{\gamma_p^i\}$ is a base for the $p$-cycles and

$$T\gamma_p^i = y_{p,j}^i \gamma_p^j, \qquad \| y_{p,j}^i \| = y_p,$$

we have trace $x_p$ = trace $y_p$ and hence

$$(1.3) \qquad \theta = \sum (-1)^p \text{ trace } y_p .$$

The precise result which we need may now be explicitly formulated as

[1] See our Colloqium Lectures, *Topology*, New York (1930), Chapter VII, pp. 347, 359.

[2] Details and references regarding the author's proof for manifolds, Hopf's extension to geometric complexes and the author's proof of the same will be found in *Topology*, Chapter VI. The proof for abstract complexes is due to A. W. Tucker: Annals of Math., vol. 34 (1933) p. 238. He also introduced the transformations permutable with $F$. See in regard to the abstract complex: S. Lefschetz, Bull. American Math. Soc., vol. 43 (1937).

THEOREM 1. *If a chain-transformation $T$ permutable with $F$ has no fixed elements its invariant $\theta$ computed from* (1.3) *is zero. Therefore a sufficient condition for the presence of fixed elements is $\theta \neq 0$.*

It is this theorem which we propose to extend to LC spaces.

2. We recall briefly certain needed definitions, notably of the singular elements and of LC spaces.[3] We are dealing throughout with a basic compact metric space $R$. The *singular* elements of $R$ may be introduced in two distinct manners. Under one definition the singular cells are made the basic elements and the chains, cycles, complexes defined in terms of the cells.[4] Under the other definition, which we shall adopt here, each element is considered as the c.s.v.t. of a simplicial polyhedral antecedent determined up to a simplicial homeomorphism.[5] It may be shown that any element of the first type has a subdivision of the second type so that as regards homology groups the difference between the two types is not very great.

A *semi-singular* complex $K$ is a singular image of a closed subcomplex $L$ of a given simplicial complex $K^*$ which includes all the vertices of $K^*$. The *mesh* of a singular complex is the maximum diameter of its cells. The mesh of a semi-singular complex $K$ associated as above with $K^*$ is the maximum of the diameters of the realized cells and of the sum of the realized faces of the non-realized cells.

The space $R$ is said to be LC (= locally connected) if for each $\epsilon > 0$ there exists an $\eta(\epsilon) > 0$ such that any semi-singular complex $K$ of mesh $<\eta$ can be completed to a full singular complex of mesh $<\epsilon$.

Suppose $R$ is an LC space and $\epsilon > 0$ is given. Choose $\delta = \eta(\epsilon)/4$, $\mu = \eta(\delta)$ and $\nu = \min(\mu/3, \delta)$. Then choose a finite set of points $\{A^i\}$ in $R$ so that every point is within a distance $\nu$ of this set. Let $\Phi'$ be the complex having the $A$'s for its vertices, it being agreed that any set of vertices form a simplex if their diameter is $<\mu$. We may regard $\Phi'$ as a semi-singular complex on $R$, in this case the closed subcomplex is the set of vertices. As its mesh is $<\mu$, it may be completed to a singular complex $\Phi$ of mesh $<\delta$.

Let $\tau$ be a point transformation defined throughout $R$ and sending each point into one of the points $A^i$ nearest to it. Let $K$ be a singular complex on $R$ of mesh $<\nu$. Consider the transformation $\tau$ applied only to the vertices of $K$. For any simplex of $K$ the diameter of the set into which its vertices are mapped is $<3\nu \leqq \mu$. Hence this is a simplicial map of $K$ into $\Phi'$. Construct the abstract complex $\bar{K}$ composed of $K$ and $\Phi'$ together with the prismatic cells joining each simplex of $K$ to its image in $\Phi'$. $\bar{K}$ may be simplicially subdivided

---

[3] See pp. 610–621.

[4] See pp. 479–484.

[5] See notably Alexandroff-Hopf: *Topologie*, p. 332. They call this second type of singular element "continuous".

without touching $K$ or $\Phi'$ and without introducing further vertices.[6] In $R$ we have the partial map of $\bar{K}$ composed of $K$ and $\Phi$. The mesh of this semi-singular complex is $<3\nu + \delta \leqq \eta(\epsilon)$; hence the missing cells of $\bar{K}$ may be inserted so as to have a diameter $<\epsilon$. Since every singular complex has a subdivision of mesh $<\nu$, we see that *every singular complex has a subdivision which is $\epsilon$-deformable into a chain of $\Phi$.*

We shall base the homology theory of $R$ on singular chains and cycles.[7] An immediate consequence of the above result is that every cycle of $R$ is homologous to a cycle of $\Phi$. Therefore the *homology groups of $R$ are isomorphic with subgroups of the homology groups of a finite simplicial complex.* This implies that all the homology groups for dimensions above a certain $n$ vanish.

It follows that we may find a base for the $p$-cycles of $\Phi$ consisting of two sets: $\{\Gamma_p^i\}$, $i = 1, 2, \cdots, r$, a maximal independent set for $R$ and $\{\Delta_p^i\}$, $i = 1, 2, \cdots, s$ which bound in $R$. In terms of Betti-numbers: $r = R_p(R)$, $r + s = R_p(\Phi')$. These sets need only to be considered for $p \leqq n$. We have then explicitly certain chains $C_{p+1}^i$ such that $F(C_{p+1}^i) = \Delta_p^i$.

3. We are now ready to take up the proof of the fixed point formula for the LC space $R$. Let $T$ be a c.s.v.t. of $R$ into itself. Its effect upon the cycles of $R$ is described by homologies

$$T\Gamma_p^i \sim y_{p,j}^i \Gamma_p^j, \qquad y_p = \| y_{p,j}^i \|, \tag{3.1}$$

and we may therefore introduce the number

$$\theta = \sum (-1)^p \text{ trace } y_p . \tag{3.2}$$

$\theta$ is clearly independent of the choice of bases and hence is a function of the "homology-class" of $T$. We have then the following extension of Theorem 1:

THEOREM 2. *If $T$ is a c.s.v.t. of the LC space $R$ into itself without fixed points the invariant $\theta(T) = 0$. Therefore a sufficient condition for the presence of fixed points is $\theta \neq 0$.*

4. Suppose $T$ is without fixed points. We may then choose $\epsilon$ so small that every point is displaced a distance $>4\epsilon$ by $T$, and $\nu$ so small that any set of diameter $<\nu$ is at a distance $>\epsilon$ from its transform. This is to be in addition to the other inequalities imposed in No. 2.

Let now $\Psi$ be the finite singular complex consisting of $\Phi$ and of the cells of

---

[6] Let $P^1, \cdots, P^r$ be the vertices of $K$, let $Q^i = \tau P^i$, and let $\zeta = P^h, \cdots, P^k$ be any simplex of $K$. The sets of vertices $P^k, \cdots, P^i Q^i \cdots Q^k$ are vertices of certain simplexes whose dimension is $\dim \zeta + 1$. Their realized faces are those in $K$ or $\Phi$. The LC condition enables us to insert them for all cells of $K$ at once and the realized cells corresponding to $\zeta$ have for sum the deformation cell $D\zeta$ of $\zeta$ in the sense of *Topology* p. 78. To have $D\zeta$ oriented we must take the alternate sum of the cells.

[7] For LC spaces these cycles and their bounding relations yield the same homology groups as Vietoris cycles. See *Topology*, p. 333.

the chains $C^i_{p+1}$ $(p = 0, 1, \cdots, n)$. We may $\epsilon$-deform a $\lambda$-subdivision of $T\Psi$ into $\Phi$ (No. 2). Let $\overline{T}$ be the induced chain-transformation on $\Phi$. That is to say, any cell $\zeta^i_p$ of $\Phi$ becomes first $T\zeta^i_p$ then, after subdivision and $\epsilon$-deformation, goes into a certain new chain $\overline{T}\zeta^i_p$ of $\Phi$. We have then

$$\overline{T}\zeta^i_p = x^i_{p,j}\zeta^j_p. \tag{4.1}$$

Since both $T$ and the $\epsilon$-deformation are permutable with $F$ the same holds for $\overline{T}$. Therefore $\overline{T}\Gamma^i_p$ is a cycle of $\Phi$ and

$$F(\overline{T}C^i_{p+1}) = \overline{T}\Delta^i_p. \tag{4.2}$$

Since a deformation transforms a cycle into a homologous cycle we have

$$\overline{T}\Gamma^i_p \sim T\Gamma^i_p \text{ on } R.$$

It follows that the cycle of $\Phi$

$$\overline{T}\Gamma^i_p - y^i_{p,j}\Gamma^j_p \sim 0 \text{ on } R$$

and hence it is $\sim$ to a linear combination of $\Delta$'s on $\Phi$, or

$$\overline{T}\Gamma^i_p \sim y^i_{p,j}\Gamma^j_p + z^i_{p,j}\Delta^j_p \text{ on } \Phi. \tag{4.3}$$

From (4.2) follows also

$$\overline{T}\Delta^i_p \sim 0 \text{ on } \Phi. \tag{4.4}$$

Hence

$$\theta(\overline{T}) = \sum (-1)^p \text{ trace } y_p = \theta(T).$$

Now in view of the choice of $\epsilon$, $\lambda$, $\mu$, $\nu$, every chain at the right in (4.1) is at a distance $> \epsilon > 0$ from $\zeta^i_p$. Therefore $\overline{T}$ has no fixed element, so that by Theorem 1, $\theta(\overline{T}) = 0 = \theta(T)$. This proves Theorem 2.

# BIBLIOGRAPHY (TO 1955)

# BIBLIOGRAPHY (TO 1955)


1912

[1] *Two theorems on conics.* Ann. of Math. (2), 14: 47–50.

[2] *On the $V_3^3$ with five nodes of the second species in $S_4$.* Bull. Amer. Math. Soc., 18: 384–386.

[3] *Double curves of surfaces projected from space of four dimensions.* Bull. Amer. Math. Soc., 19: 70–74.

1913

[4] *On the existence of loci with given singularities.* Trans. Amer. Math. Soc., 14: 23–41. Doctoral dissertation, Clark University, 1911.

[5] *On some topological properties of plane curves and a theorem of Möbius.* Amer. J. Math., 35: 189–200.

1914

[6] *Geometry on ruled surfaces.* Amer. J. Math., 36: 392–394.

[7] *On cubic surfaces and their nodes.* Kansas Univ. Science Bull., 9: 69–78.

1915

[8] *The equation of Picard-Fuchs for an algebraic surface with arbitrary singularities.* Bull. Amer. Math. Soc., 21: 227–232.

[9] *Note on the n-dimensional cycles of an algebraic n-dimensional variety.* Rend. Circ. Mat. Palermo, 40: 38–43.

1916

[10] *The arithmetic genus of an algebraic manifold immersed in another.* Ann. of Math. (2), 17: 197–212.

[11] *Direct proof of De Moivre's formula.* Amer. Math. Monthly, 23: 366–368.

[12] *On the residues of double integrals belonging to an algebraic surface.* Quart. J. Pure and Appl. Math., 47: 333–343.

1917

[13] *Note on a problem in the theory of algebraic manifolds.* Kansas Univ. Science Bull., 10: 3–9.

[14] *Sur certains cycles à deux dimensions des surfaces algébriques.* Accad. dei Lincei. Rend. (5), 26, 1° sem.: 228–234.

[15] *Sur les intégrales multiples des variétés algébriques.* Acad. des Sci. Paris, C.R., 164: 850–853.

[16] *Sur les intégrales doubles des variétés algébriques.* Annali di Mat. (3), 26: 227–260.

1919

[17] *Sur l'analyse situs des variétés algébriques.* Acad. des Sci. Paris, C.R., 168: 672–674.

[18] *Sur les variétés abéliennes.* Acad. des Sci. Paris, C.R., 168: 758–761.

[19] *On the real folds of Abelian varieties.* Proc. Nat. Acad. Sci. U.S.A., 5: 103–106.

[20] *Real hypersurfaces contained in Abelian varieties.* Proc. Nat. Acad. Sci. U.S.A., 5: 296–298.

1920

[21] *Algebraic surfaces, their cycles and integrals.* Ann. of Math. (2), 21: 225–228. *A correction.* Ibid., 23: 333.

1921

[22] *Quelques remarques sur la multiplication complexe.* p. 300–307 of Comptes Rendus du Congrès International des Mathématiciens, Strasbourg, Sept. 1920. Toulouse, É. Privat, 1921.

[23] *Sur le théorème d'existence des fonctions abéliennes.* Accad. dei Lincei. Rend. (5), 30, 1° sem.: 48–50.

☼[24] *On certain numerical invariants of algebraic varieties with application to Abelian varieties.* Trans. Amer. Math. Soc., 22: 327–482. Awarded the Bôcher Memorial Prize by the American Mathematical Society in 1924. A translation, with minor modifications, of the memoir awarded the Prix Bordin by the Académie des Sciences, Paris, for the year 1919; for announcement, see Acad. des Sci. Paris, C.R., 169: 1200–1202, and Bull. Sci. Math., 44: 5–7.

1923

[25] *Continuous transformations of manifolds.* Proc. Nat. Acad. Sci. U.S.A., 9: 90–93.

[26] *Progrès récents dans la théorie des fonctions abéliennes.* Bull. Sci. Math., 47: 120–128.

[27] *Sur les intégrales de seconde espèce des variétés algébriques.* Acad. des Sci. Paris, C.R., 176: 941–943.

[28] *Report on curves traced on algebraic surfaces.* Bull. Amer. Math. Soc., 29: 242–258.

1924

☼[29] L'Analysis Situs et la Géométrie Algébrique. Paris, Gauthier-Villars. vi, 154 pp. (Collection de Monographies publiée sous la Direction de M. Émile Borel.) Based in part on a series of lectures given in Rome in 1921 under the auspices of the Institute of International Education, and also on research conducted under the auspices of the American Association for the Advancement of Science. Nouveau tirage, 1950.

[30] *Sur les intégrales multiples des variétés algébriques.* J. Math. Pures Appl. (9), 3: 319–343.

1925

[31] *Intersections of complexes on manifolds.* Proc. Nat. Acad. Sci. U.S.A., 11: 287–289.

[32] *Continuous transformations of manifolds.* Proc. Nat. Acad. Sci. U.S.A., 11: 290–292.

1926

☼ [33] *Intersections and transformations of complexes and manifolds.* Trans. Amer. Math. Soc., 28: 1–49.

[34] *Transformations of manifolds with a boundary.* Proc. Nat. Acad. Sci. U.S.A., 12: 737-739.

1927

[35] *Un théorème sur les fonctions abéliennes.* p. 186--190 of *In Memoriam N. I. Lobatschevskii.* Kazan', Glavnauka.

☼ [36] *Manifolds with a boundary and their transformations.* Trans. Amer. Math. Soc., 29: 429–462, 848.

[37] *Correspondences between algebraic curves.* Ann. of Math. (2), 28: 342–354.

[38] *The residual set of a complex on a manifold and related questions.* Proc. Nat. Acad. Sci. U.S.A., 13: 614–622, 805–807.

[39] *On the functional independence of ratios of theta functions.* Proc. Nat. Acad. Sci. U.S.A., 13: 657–659.

1928

[40] *Transcendental theory; Singular correspondences between algebraic curves; Hyperelliptic surfaces and Abelian varieties.* Chap. 15–17, p. 310–395, vol. 1, of Selected Topics in Algebraic Geometry; Report of the Committee on Rational Transformations of the National Research Council. Washington (NRC Bulletin no. 63).

[41] *A theorem on correspondences on algebraic curves.* Amer. J, Math., 50: 159–166.

☼ [42] *Closed point sets on a manifold.* Ann. of Math. (2), 29: 232–254.

1929

[43] Géométrie sur les Surfaces et les Variétés Algébriques. Paris, Gauthier-Villars. 66 pp. (Mémorial des Sciences Mathématiques, Fasc. 40.)

[44] *Duality relations in topology.* Proc. Nat. Acad. Sci. U.S.A., 15: 367–369.

1930

[45] Topology. New York, American Mathematical Society. ix, 410 pp. (Colloquium Publications, vol. 12.) See **[100]**.

[46] *Les transformations continues des ensembles fermés et leurs points fixes.* Acad. des Sci. Paris. C.R., 190: 99–100.

[47] *On the duality theorems for the Betti numbers of topological manifolds* (with W. W. Flexner). Proc. Nat. Acad. Sci. U.S.A., 16: 530–533.

☼ [48] *On transformations of closed sets.* Ann. of Math. (2), 31: 271–280.

1931

[49] *On compact spaces.* Ann. of Math. (2), 32: 521–538.

1932

[50] *On certain properties of separable spaces.* Proc. Nat. Acad. Sci. U.S.A., 18: 202–203.

☼[51] *On separable spaces.* Ann. of Math. (2), 33: 525–537.

☼[52] *Invariance absolue et invariance relative en géométrie algébrique.* Rec. Math. (Mat. Sbornik), 39, no. 3: 97–102.

1933

☼[53] *On singular chains and cycles.* Bull. Amer. Math. Soc., 39: 124–129.

[54] *On analytical complexes* (with J. H. C. Whitehead). Trans. Amer. Math. Soc., 35: 510–517.

☼[55] *On generalized manifolds.* Amer. J. Math., 55: 469–504.

1934

[56] Elementary One- and Two-Dimensional Topology (a course given by Prof. Lefschetz, Spring 1934; notes by H. Wallman). Princeton University. 95 pp., mimeographed.

☼[57] *On locally connected and related sets.* Ann. of Math. (2), 35: 118–129.

1935

[58] Topology (lectures 1934–35; notes by N. Steenrod and H. Wallman). Princeton University. 203 pp., mimeographed.

[59] *Algebraicheskaia͡ geometriia͡: metody, problemy, tendentsii.* pp. 337–349, vol. 1, of Trudy Vtorogo Vsesoiu͡znogo Matematicheskogo S''ezda, Leningrad, 24–30 June 1934. Leningrad-Moscow. An invited address at the Second All-Union Mathematical Congress.

☼[60] *Chain-deformations in topology.* Duke Math. J., 1: 1–18.

[61] *Application of chain-deformations to critical points and extremals.* Proc. Nat. Acad. Sci. U.S.A., 21: 220–222.

[62] *A theorem on extremals. I, II.* Proc. Nat. Acad. Sci. U.S.A., 21: 272–274, 362–364.

[63] *On critical sets.* Duke Math. J., 1: 392–412.

1936

[64] *On locally-connected and related sets* (second paper). Duke Math. J., 2: 435–442.

☼[65] *Locally connected sets and their applications.* Rec. Math. (Mat. Sbornik) n.s., 1: 715–717. A contribution to the First International Topological Conference, September 1935.

[66] *Sur les transformations des complexes en sphères.* Fund. Math., 27: 94–115.

[67] *Matematicheskaia͡ dei͡atel'nost' v Prinstone.* Uspekhi Mat. Nauk vyp. 1, pp. 271–273.

1937

[68] Lectures on Algebraic Geometry 1936–37; notes by M. Richardson and E. D. Tagg. Princeton University. 69 pp., planographed.

[69] *Algebraicheskaia͡ geometriia͡.* Uspekhi Mat. Nauk vyp. 3, pp. 63–77.

[70] *The role of algebra in topology.* Bull. Amer. Math. Soc., 43: 345–359. Address of retiring president of the American Mathematical Society.

☼[71] *On the fixed point formula.* Ann. of Math. (2), 38: 819–822.

1938

[72] Lectures on Algebraic Geometry (Part II) 1937–1938. Princeton University Press. 73 pp., planographed.

[73] *On chains of topological spaces.* Ann. of Math. (2), 39: 383–396.

[74] *On locally connected sets and retracts.* Proc. Nat. Acad. Sci. U.S.A., 24: 392–393.

[75] *Sur les transformations des complexes en sphères (note complémentaire).* Fund. Math., 31: 4–14.

☼[76] *Singular and continuous complexes, chains and cycles.* Rec. Math. (Mat. Sbornik) n.s., 3: 271–285.

1939

[77] *On the mapping of abstract spaces on polytopes.* Proc. Nat. Acad. Sci. U.S.A., 25: 49–50.

1941

[78] *Abstract complexes.* pp. 1–28 of Lectures in Topology; the University of Michigan Conference of 1940. Ann Arbor, University of Michigan Press; London, Oxford University Press.

1942

[79] Algebraic Topology. New York, American Mathematical Society. vi, 389 pp. (Colloquium Publications, vol. 27.)

[80] Topics in Topology. Princeton University Press. 137 pp. (Annals of Mathematics Studies, no. 10). A second printing, 1951. London, Oxford University Press.

[81] *Émile Picard (1856–1941): Obituary.* Amer. Phil. Soc. Yearbook 1942, pp. 363–365.

1943

[82] Introduction to Non-linear Mechanics, by N. Kryloff and N. Bogoliuboff; a free translation by S. Lefschetz of excerpts from two Russian monographs. Princeton University Press. 105 pp. (Annals of Mathematics Studies, no. 11). London, Oxford University Press.

[83] *Existence of periodic solutions for certain differential equations.* Proc. Nat. Acad. Sci. U.S.A., 29: 29–32.

1946

[84] Lectures on Differential Equations. Princeton University Press. viii, 210 pp. (Annals of Mathematics Studies, no. 14.) London, Oxford University Press.

1949

[85] Introduction to Topology. Princeton University Press. viii, 218 pp. (Princeton Mathematical Series, no. 11.) London, Oxford University Press. A work originating from a short course delivered in 1944 before the Institute of Mathematics of the National University of Mexico.

[86] Theory of Oscillations, by A. A. Andronow and C. E. Chaikin; English language edition edited under the direction of S. Lefschetz. Princeton University Press. ix, 358 pp.

[87] *Scientific research in the U.S.S.R.: Mathematics.* Amer. Acad. Polit. and Soc. Sci. Annals, 263: 139–140.

1950

[88] Contributions to the Theory of Nonlinear Oscillations, edited by S. Lefschetz. Princeton University Press. ix, 350 pp. (Annals of Mathematics Studies, no. 20.) London, Oxford University Press.

[89] *The structure of mathematics.* American Scientist, 38: 105–111.

1951

[90] *Numerical calculations in nonlinear mechanics.* pp. 10–12 of Problems for the Numerical Analysis of the Future. Washington, Govt. Printing Office. (National Bureau of Standards, Applied Math. Series, no. 15.)

1952

[91] Contributions to the Theory of Nonlinear Oscillations, vol. 2, edited by S. Lefschetz. Princeton University Press. 116 pp. (Annals of Mathematics Studies, no. 29.) London, Oxford University Press.

[92] *Notes on differential equations.* pp. 61–73 of Contributions to the Theory of Nonlinear Oscillations, vol. 2.

1953

[93] Algebraic Geometry. Princeton University Press. ix, 233 pp. (Princeton Mathematical Series, no. 18.)

[94] *Algunos trabajos recientes sobre ecuaciones diferenciales.* pp. 122–123, vol. 1 of Memoria de Congreso Cientifico Mexicano, U.N.A.M., Mexico.

[95] *Las grades corrientes en las matemáticas del siglo XX.* pp. 206–211, vol. 1 of Memoria de Congreso Cientifico Mexicano, U.N.A.M., Mexico.

1954

[96] *Russian contributions to differential equations.* pp. 68–74 of Proceedings of the Symposium on Nonlinear Circuit Analysis, New York, 1953. New York, Polytechnic Institute of Brooklyn.

☼ [97] *Complete families of periodic solutions of differential equations.* Comment. Math. Helv. 28: 341–345.

[98] *On Liénard's differential equation.* pp. 149–153 of Wave Motion and Vibration Theory. New York, McGraw-Hill. (Amer. Math. Soc. Proceedings of Symposia in Applied Math., vol. 5.)

# BIBLIOGRAPHY (1956-1970)

# BIBLIOGRAPHY (1956-1970)


1956

[99] *On a theorem of Bendixson.* Bol. Soc. Mat. Mexicana, 1: 13–27.

[100] Topology, 2nd ed. New York, Chelsea Publishing Company. xi, 413 pp. (Cf. [45] of the present Bibliography.)

1957

[101] *On coincidences of transformations.* Bol. Soc. Mat. Mexicana, 2: 16–25.

[102] *The ambiguous case in planar differential systems.* Bol. Soc. Mat. Mexicana, 2: 63–74.

[103] *Withold Hurewicz: in memoriam.* Bull. Amer. Math. Soc. 63: 77–82.

[104] *Sobre la modernizacion de la geometría.* Revista Matemática, 1: 1–11.

[105] Differential Equations: Geometric Theory. New York, Interscience. 364 pp. (Cf. [116] of the present Bibliography.)

1958

[106] *On the critical points of a class of differential equations.* Pp. 19–28, vol. 4 of, Contributions to the Theory of Non-Linear Oscillations. Princeton Univ. Press. Princeton.

[107] *Liapunov and stability in dynamical systems.* Bol. Soc. Mat. Mexicana (2), 3: 25–39.

[108] The Stability Theory of Liapunov. University Institute for Fluid Dynamics and Applied Mathematics, College Park, Md. (Lecture Series No. 37.)

1960

[109] *Controls: an application to the direct method of Liapunov.* Bol. Soc. Mat. Mexicana, 5: 139–43.

[110] *Algunas consideraciones sobre las matemáticas modernas.* Revista de la Unión Matemática Argentina, 20: 7–16.

[111] *Resultados nuevos sobre casos críticos en ecuaciones diferenciales.* Revista de la Unión Matemática Argentina, 20: 122–24.

1961

[112] *The critical case in differential equations.* Bol. Soc. Mat. Mexicana, 6: 5–18.

[113] Geometricheskaia Teoriia Differentsial'nykh Uravnenii. Moskva, Izd-vo Inostrannoi Lit-ry. 387 pp. (Trans. of [105].)

[114] Stability by Liapunov's Direct Method; written with J. P. Lasalle. New York, Academic Press. 133 pp.

[115] *Recent Soviet contributions to ordinary differential equations and non-linear mechanics;* written with J. P. Lasalle. J. Math. Analysis and Appl., 2: 467–99.

1962

[116] Differential Equations: Geometric Theory, 2nd (rev.) ed. New York, Interscience. 390 pp.

[117] Recent Soviet Contributions to Mathematics; edited with J. P. Lasalle. New York, Macmillan. 324 pp.

1963

[118] *On indirect automatic controls.* Trudy Mezhdunarodnogo Simpoziuma po Nelineinym Kolebaniyam. Kiev, Izdat. Akad. Nauk Ukrain. SSR, 23–24.

[119] *Some mathematical considerations on nonlinear automatic controls.* Pp. 1–28 of, Contributions to Differential Equations, vol. 1. New York, Interscience.

[120] Elementos de Topologia. México, Universidad Nacional Autónoma de México. 159 pp.

[121] Proceedings of International Symposium on Nonlinear Differential Equations and Nonlinear Mechanics, Colorado Springs, 1961; edited with J. P. Lasalle. New York, Academic Press. 505 pp.

1964

[122] Stability of Nonlinear Automatic Control Systems. New York, Academic Press. 150 pp.

1965

[123] *Liapunov stability and controls.* SIAM J. Control, Ser. A, 3: 1–6.

[124] *Planar graphs and related topics.* Proc. Nat. Acad. Sci. U.S.A., 54: 1763–65.

[125] *Recent advances in the stability of nonlinear controls.* SIAM Review, 7: 1–12.

[126] *Some applications of topology to networks.* Pp. 1–6 of, Proc. Third Annual Allerton Conference on Circuit and System Theory. Urbana, Univ. of Illinois.

1966

[127] Stability in Dynamics. Princeton, Princeton University School of Engineering and Applied Science. 52 pp. (William Pierson Field Engineering Lectures, March 3, 4, 10, 11, 1966.)

1967

[128] Stability of Nonlinear Automatic Control Systems [In Russian]. Moscow, Izdat. "Mir." 183 pp. (A translation of [122].)

1968

[129] *On a theorem of Bendixson.* J. Diff. Equations, 4: 66–101.
[130] *A page of mathematical autobiography.* Bull. Amer. Math. Soc., 74: 854–79.

1969

[131] *The Lurie problem on nonlinear controls.* Pp. 9–19, vol. I, of, Lectures in Differential Equations, edited by A. K. Aziz. New York, Van Nostrand–Reinhold.
[132] *The early development of algebraic geometry.* Amer. Math. Monthly, 76: 451–60.
[133] *Luther Pfahler Eisenhart, 1876–1965: A biographical memoir.* Nat. Acad. Sci. U.S.A., Biographical Memoirs, 40: 69–90.